The World of Large Scale Systems

OTHER IEEE PRESS BOOKS

Electronic Switching: Digital Central Office Systems of the World, *Edited by Amos E. Joel, Jr.*
A Guide for Writing Better Technical Papers, *Edited by C. Harkins and D. L. Plung*
Low-Noise Microwave Transistors and Amplifiers, *Edited by H. Fukui*
Digital MOS Integrated Circuits, *Edited by M. I. Elmasry*
Geometric Theory of Diffraction, *Edited by R. C. Hansen*
Modern Active Filter Design, *Edited by R. Schaumann, M. A. Soderstrand, and K. R. Laker*
Adjustable Speed AC Drive Systems, *Edited by B. K. Bose*
Optical Fiber Technology, II, *Edited by C. K. Kao*
Protective Relaying for Power Systems, *Edited by S. H. Horowitz*
Analog MOS Integrated Circuits, *Edited by P. R. Gray, D. A. Hodges, and R. W. Brodersen*
Interference Analysis of Communication Systems, *Edited by P. Stavroulakis*
Integrated Injection Logic, *Edited by J. E. Smith*
Sensory Aids for the Hearing Impaired, *Edited by H. Levitt, J. M. Pickett, and R. A. Houde*
Data Conversion Integrated Circuits, *Edited by Daniel J. Dooley*
Semiconductor Injection Lasers, *Edited by J. K. Butler*
Satellite Communications, *Edited by H. L. Van Trees*
Frequency-Response Methods in Control Systems, *Edited by A.G.J. MacFarlane*
Programs for Digital Signal Processing, *Edited by the Digital Signal Processing Committee*
Automatic Speech & Speaker Recognition, *Edited by N. R. Dixon and T. B. Martin*
Speech Analysis, *Edited by R. W. Schafer and J. D. Markel*
The Engineer in Transition to Management, *I. Gray*
Multidimensional Systems: Theory & Applications, *Edited by N. K. Bose*
Analog Integrated Circuits, *Edited by A. B. Grebene*
Integrated-Circuit Operational Amplifiers, *Edited by R. G. Meyer*
Modern Spectrum Analysis, *Edited by D. G. Childers*
Digital Image Processing for Remote Sensing, *Edited by R. Bernstein*
Reflector Antennas, *Edited by A. W. Love*
Phase-Locked Loops & Their Application, *Edited by W. C. Lindsey and M. K. Simon*
Digital Signal Computers and Processors, *Edited by A. C. Salazar*
Systems Engineering: Methodology and Applications, *Edited by A. P. Sage*
Modern Crystal and Mechanical Filters, *Edited by D. F. Sheahan and R. A. Johnson*
Electrical Noise: Fundamentals and Sources, *Edited by M. S. Gupta*
Computer Methods in Image Analysis, *Edited by J. K. Aggarwal, R. O. Duda, and A. Rosenfeld*
Microprocessors: Fundamentals and Applications, *Edited by W. C. Lin*
Machine Recognition of Patterns, *Edited by A. K. Agrawala*
Turning Points in American Electrical History, *Edited by J. E. Brittain*
Charge-Coupled Devices: Technology and Applications, *Edited by R. Melen and D. Buss*
Spread Spectrum Techniques, *Edited by R. C. Dixon*
Electronic Switching: Central Office Systems of the World, *Edited by A. E. Joel, Jr.*
Electromagnetic Horn Antennas, *Edited by A. W. Love*
Waveform Quantization and Coding, *Edited by N. S. Jayant*
Communication Satellite Systems: An Overview of the Technology, *Edited by R. G. Gould and Y. F. Lum*
Literature Survey of Communication Satellite Systems and Technology, *Edited by J. H. W. Unger*
Solar Cells, *Edited by C. E. Backus*
Computer Networking, *Edited by R. P. Blanc and I. W. Cotton*
Communications Channels: Characterization and Behavior, *Edited by B. Goldberg*
Large-Scale Networks: Theory and Design, *Edited by F. T. Boesch*
Optical Fiber Technology, *Edited by D. Gloge*
Selected Papers in Digital Signal Processing, II, *Edited by the Digital Signal Processing Committee, IEEE*
A Guide for Better Technical Presentations, *Edited by R. M. Woelfle*
Career Management: A Guide to Combating Obsolescence, *Edited by H. G. Kaufman*
Energy and Man: Technical and Social Aspects of Energy, *Edited by M. G. Morgan*
Magnetic Bubble Technology: Integrated-Circuit Magnetics for Digital Storage and Processing, *Edited by H. Chang*
Frequency Synthesis: Techniques and Applications, *Edited by J. Gorski-Popiel*
Literature in Digital Processing: Author and Permuted Title Index (Revised and Expanded Edition), *Edited by H. D. Helms, J. F. Kaiser, and L. R. Rabiner*
Data Communications via Fading Channels, *Edited by K. Brayer*
Nonlinear Networks: Theory and Analysis, *Edited by A. N. Willson, Jr.*
Computer Communications, *Edited by P. E. Green, Jr. and R. W. Lucky*

The World of Large Scale Systems

Edited by

James D. Palmer

Vice President and General Manager
Mechanical Technology, Inc.

Richard Saeks

Paul Whitfield Horn Professor
Texas Tech University

A volume in the IEEE PRESS Selected Reprint Series, prepared under the sponsorship of the IEEE Systems, Man, and Cybernetics Society.

IEEE
PRESS

The Institute of Electrical and Electronics Engineers, Inc., New York.

PRINTED IN THE UNITED STATES OF AMERICA

Sole Worldwide Distributor (Exclusive of the IEEE)

JOHN WILEY & SONS, INC.
605 Third Ave.
New York, NY 10158

Wiley Order Numbers: Clothbound: 471-87298-9
Paperbound: 471-87299-7

IEEE Order Numbers: Clothbound: PC01560
Paperbound: PP01578

Library of Congress Cataloging in Publication Data
Main entry under title:

The World of large scale systems.

(IEEE Press selected reprint series)
Includes indexes.
1. Large scale systems. I. Palmer, James D.
II. Saeks, R.
QA402.W67 621.3 82-6169
ISBN 0-87942-161-4 AACR2
ISBN 0-87942-162-2 (pbk.)

Contents

Preface

THE PAST decade has witnessed the initiation of major research and development in large scale systems. Unlike most research and development that is initiated in response to the requirements of a single industry, the recent activity in the large scale systems area is a response to the needs of at least five distinct industries—aerospace, computer, power, semiconductor, and communications. Hence, large scale systems research has developed simultaneously along several distinct lines of inquiry, each motivated by a different application. The multiplicity of parallel, but uncoordinated, research activities in large scale systems constitutes a major new thrust in systems research and development. There has been, however, only minimal communication among the various industries, which have gone off in separate directions, often duplicating, but rarely benefiting from each other.

Recognizing the potential benefits of integrating the various large scale systems research and development activities within electrical and electronics engineering, the IEEE created a Committee on Large Scale Systems (COLSS) to "unify and coordinate" IEEE activities in the large scale systems area. This volume of selected reprints, *The World of Large Scale Systems,* represents COLSS's effort to spotlight the variety of activity in the large scale systems area. The book is therefore divided into five application areas, each of which is represented by about six reprints which illustrate the research and development undertaken in those areas. No attempt has been made at completeness. Rather, we have attempted to give the reader a sampling of the wide variety of tools, techniques, ideas, and philosophies that underlie large scale systems engineering.

J. D. Palmer
R. Saeks
June 1981

Introduction

ALTHOUGH much has been said about the exponential strides made by technology during the past century, little note has been taken of the concomitant expansion in the complexity of technological systems. The communications receiver, made up of a dozen or so vacuum tubes and passive Radio Launch Control components, has given way to avionics systems containing hundreds of Integrated Circuits (IC's), each of which exceeds an entire communications receiver in complexity. Individual power companies serving a single city or county have been interconnected into transcontinental power grids. The simple machine governor has been superseded by city-wide traffic control problems. The complexities of a single telephone exchange have given way to international dialing and computer communication networks. Indeed, even the socioeconomic interface with technology, which could once be dealt with on an intuitive level, has given way to complex economic/energy system considerations in the civilian world and the command/control/communications system of the military world.

In each of the above examples, system complexity and the complexity of the individual system components have increased simultaneously. From the point of view of the circuit theorist, the complexity of a single chip rivals that of a national power grid. As such, one might heuristically argue that system complexity increases with the square of technological progress. Moreover, the computational requirements for system analysis and design increase in a polynomial progression with system complexity. As such, the modern systems engineer must cope with a problem whose size increases several orders of magnitude for each order of magnitude advance in technology.

Needless to say, the once tantalizing solution of applying classical systems techniques to large scale systems with the aid of still larger and faster computers has never materialized. Moreover, the modern systems engineer faces new "problems of scale" which have not previously been encountered. Although well defined for classical systems, the underlying physical and economic motivation for the study of decentralized control, automated system maintenance, etc., is the complexity of the modern system. As such, techniques for studying these problems in a classical system theoretic context do not exist. To deal with these phenomena of scale, systems engineers have developed entirely new approaches to system analysis and design over the past decade, specifically formulated for application to the modern large scale system and its associated problems. These include both theoretical techniques for decentralized control or stability analysis, and numerical techniques for sparse matrix inversion, system identification and modeling, etc.

Not surprisingly, the electric power industry was among the first to recognize the impact of scale in systems engineering. Indeed, the power system is not only large in complexity, but also geographically and in energy processing capability. Initially, the power system was modeled as a linear circuit operating at a single frequency, with sophisticated sparse matrix techniques being employed to deal with the large matrices encountered. The power industry was the first to significantly employ sparse matrix methods and, to this day, most sparse matrix methods used by systems engineers in all fields of endeavor are derived from techniques originally developed for application to power system analysis and design.

Although most power system studies still employ a linear system model, nonlinear models are now commonly used for generators, and nonlinear techniques are used for system stability studies and the analysis of transient effects. Moreover, decentralized or partially decentralized control systems abound in today's national power grids.

Although a power grid is intuitively large, an integrated circuit is usually characterized by its degree of smallness. Nonetheless, from a circuit theoretic point of view, an IC is every bit as large as a power grid. Indeed, a typical large scale integration (LSI) chip has as many nodes, branches, and nonlinearities as the largest of power grids. While in application these digital devices may be simulated by straightforward Boolean models, it is necessary to know the factors that will determine the device speed and reliability (such as rise times, parasitic effects, operating temperatures, etc.) at the design stage. As such, a full nonlinear analog simulation of these digital devices is required by the IC design engineer. This, in turn, has led to the development of a family of sophisticated computer codes for the analysis and design of LSI circuits. Although such codes were initially based on classical circuit analysis algorithms, nonlinear techniques were soon developed to model the switching transistors, and the sparse matrices of the power engineer were adapted to the problem. New circuit analysis algorithms which are especially amenable to the application of sparse matrix techniques were eventually developed. Today such large scale circuit simulation codes are used on a daily basis in the world's IC houses, while researchers among the circuits and systems community have turned their attention to questions of system stability and the development of new nonlinear transistor models.

The main thrusts of large scale system science as practiced by the control engineer are in the areas of stability and decentralized control. Although the stability criteria of classical control theory (Lyapunov, Hurwitz, Popov, Nyquist, etc.) are theoretically applicable to systems of arbitrary size, they are, in fact, of little use in any system of more than a dozen or so degrees of freedom. As such, stability theory in a large scale system context has taken on a "structural character," wherein one attempts to show that an entire system is stable if its components are stable and its connection equations satisfy an appropriate inequality. Although this process results in rather loose sufficient conditions, the technique is applicable to systems of considerable size and has therefore proven useful in practice.

Like stability theory, optimal control, as classically formulated, is theoretically applicable to systems of arbitrary size and, in this case—with the aid of sparse matrix techniques—the required Ricatti equations can be solved for systems of surprisingly large dimension. Physically, however, large control systems are usually large geographically and dimensionally, thus rendering the implementation of a classical controller, with its required communication between all state variables, prohibitively expensive. As such, in a large scale system, one looks for a decentralized controller wherein only the states within geographically centralized subsystems need communicate with one another. Examples that require such decentralized controllers include regional and national power grids, traffic control systems, and computer communications networks.

Although the study of socioeconomic systems has often suffered from the "yellow taxi-cab syndrome," modern technological systems often have a socioeconomic interface which cannot be neglected. This includes economic interface with an energy system, psychological interface with a traffic system, and sociological interface with a military command, control, and communications system. As such, socioeconomic considerations have become a significant aspect of large scale systems engineering. Since little "hard information" is available about these systems, much of the engineering effort in the area has been directed towards modeling. This includes direct system identification from input-output data sequences (possibly corrupted by noise) and the "so-called" causal modeling, wherein one attempts to deduce a graphical model interrelating various socioeconomic phenomena.

In each of the large scale systems engineering activities described above, the computer is the fundamental tool which makes it possible to deal with systems of "real world" size. The computer itself however, is an equally large system. Moreover, the computer network in which hundreds or thousands of computers are interconnected is today's reality; thus, a further face of the large systems coin has evolved with the analysis and design of computer networks. Here, one studies routing problems, information flow, information retrieval, etc. As with the case of the IC, one is dealing with an interconnected system and the study draws much from graph theoretic techniques. On a smaller scale, the communications network for a distributed control system often takes the form of a computer network. As such, the efficiency and information transmission capabilities of such a computer network may determine the degree of decentralization required for a control system.

If one were to draw a graph with each of the above described large scale systems research activities taken as a node and overlap between the activities defining the branches, the classical Kurotowski graph would result. If, however, the branches were defined by the amount of communications between engineers and researchers associated with the various activities, the result might easily be a set of isolated nodes. The present volume has been prepared in the hope that it will help to initiate the process of adding to the branches of this graph.

Part I

Circuits
(Simulation, Sensitivity, Testing)

The Sparse Tableau Approach to Network Analysis and Design

GARY D. HACHTEL, MEMBER, IEEE, ROBERT K. BRAYTON, AND FRED G. GUSTAVSON

Abstract—The tableau approach to automated network design optimization via implicit, variable order, variable time-step integration, and adjoint sensitivity computation is described. In this approach, the only matrix operation required is that of repeatedly solving linear algebraic equations of fixed sparsity structure. Required partial derivatives and numerical integration is done at the branch level leading to a simple input language, complete generality and maximum sparsity of the characteristic coefficient matrix. The bulk of computation and program complexity is thus located in the sparse matrix routines; described herein are the routines OPTORD and 1-2-3 GNSO. These routines account for variability type of the matrix elements in producing a machine code for solution of $Ax=b$ in nested iterations for which a weighted sum of total operations count and round-off error incurred in the optimization is minimized.

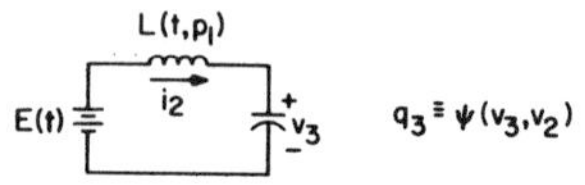

Fig. 1. Nonlinear network.

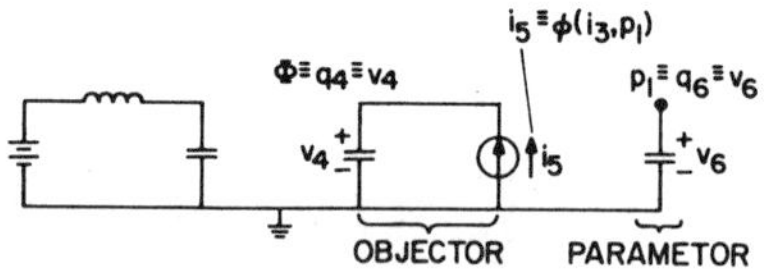

Fig. 2. Augmented network.

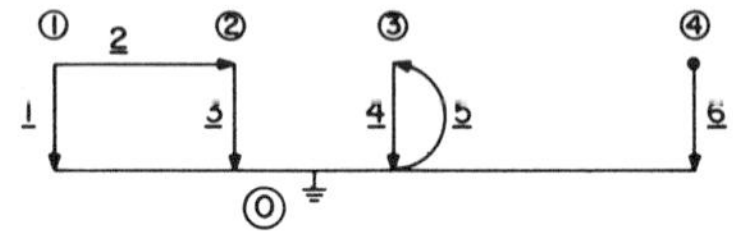

Fig. 3. Node-branch labeling for the directed graph.

I. Introduction

COMPUTER-AIDED design of electrical networks has recently been undergoing a rapid evolution due to the development of highly efficient methods for solving systems of linear equations where the matrix of coefficients is sparse [1]–[4]. These new techniques, when fully absorbed in a computer-aided design program, drastically affect even the initial formulation of the problem.

In this paper we describe a research investigation aimed at designing and implementing a computer network design program (NDP) for automated network optimization which fully incorporates the sparse matrix methods. The formulation is unorthodox yet simple and is applicable to any system described mathematically by a set of algebraic and differential equations. A general nonlinear electrical network is such a system, and it follows that the basic procedure for solving for the currents and voltages in a network can be viewed in terms of a simple mathematical procedure: Gaussian elimination. This has both a simplifying and unifying effect and makes possible a simple yet truly general-purpose computer program. By combining the concept of variability type with sparse matrix techniques, this program can achieve practically optimum efficiency by choosing the order of elimination so as to minimize the total operations count required for simulation and/or automated optimization.

In this section we give an overview of our method for orientation purposes. Basic to our approach is the concept of a tableau which includes all network information in a nonreduced form.

For illustration consider the automatic design of the network of Fig. 1. The nonlinear elements were chosen to demonstrate that there are no restrictions on element type. In general, we consider design objectives of the class

$$\underset{p_m \leq p \leq p_M}{\text{minimize}}\ \Phi \equiv \int_0^T \phi(w, p, t)\, dt \qquad (1)$$

where Φ represents a generalized scalar objective functional and ϕ an objective function of the unknown vector w and the vector of designable parameters p. In Fig. 1, $\phi=\phi(i_3, p_1)$. It is convenient in automated network design to add these quantities directly to the tableau of unknowns. This is done (Fig. 2) by augmenting the original network with a unit capacitance driven by a dependent current source of value ϕ (the integrand in (1)) and another unit capacitance connected only to the reference node. The reason for this last augmentation has to do with adjoint sensitivity computation [5] and will be made clear in Section II. The first augmentation (cf., Fig. 2) is called herein an *objector* and the second a *parametor*. The directed graph of the augmented network is shown in Fig. 3 with integer node and branch labels. Fig. 4 shows the tableau of algebraic and differential equations which describe the network, where V, i, v stand

Manuscript received May 27, 1970; revised August 10, 1970.
The authors are with the IBM T. J. Watson Research Center, Yorktown Heights, N. Y. 10598.

Reprinted from *IEEE Trans. Circuit Theory*, vol. CT-18, pp. 101–113, Jan. 1971.

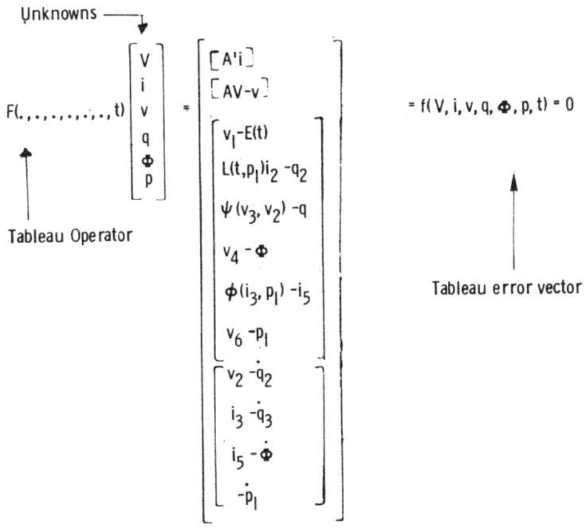

Fig. 4. Tableau operator, unknown vector, and tableau vector.

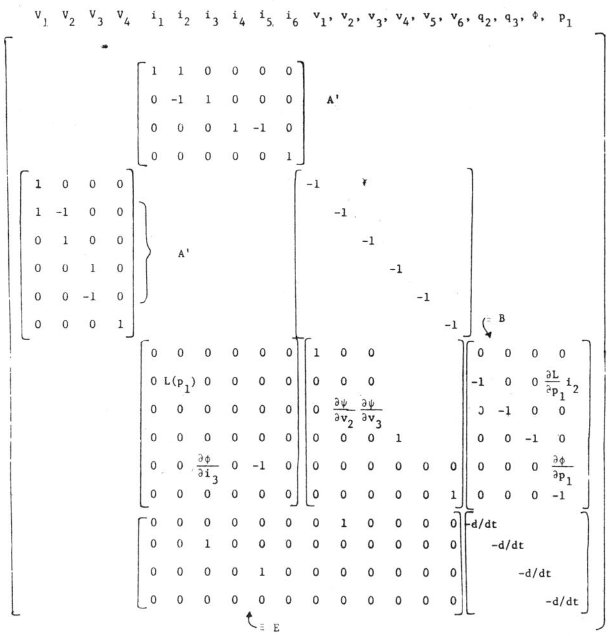

Fig. 5. Tableau matrix.

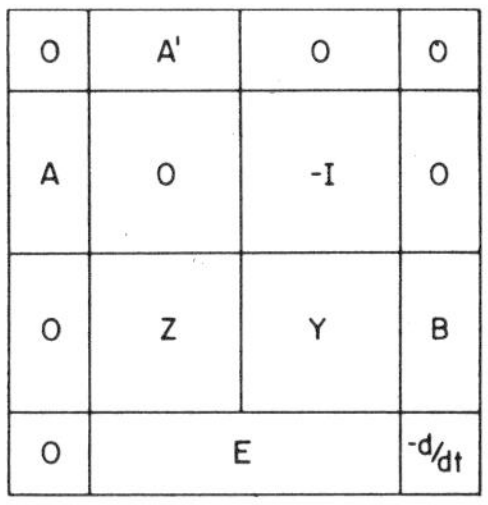

Fig. 6. Form of general tableau matrix.

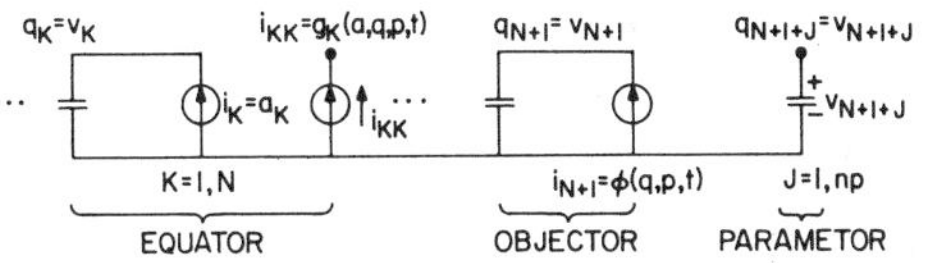

Fig. 7. Network analog of general ODE optimization problem.

for the vectors[1] of node voltages, branch currents, and branch voltages, respectively. Energy storage is denoted by the generic term q which stands, in network applications, either for magnetic flux or electronic charge. However, since the objector and parameter charges are of special interest, the labels Φ and p are retained for these elements.

The tableau operator $F(.,.,.,.,.,.,.,t)$ (Fig. 4) is an algebraic–differential operator which operates to the right on the unknown vector $w \equiv \text{col}(V, i, v, q, \phi, p)$ to form the tableau error vector $f(w, t)$. The network equations may thus be stated in the form

$$f(w(t), t) \equiv f(V, i, v, q, \Phi, p, t) = 0, \qquad t\in[0, T]. \quad (2)$$

The tableau matrix $F_l(w, t)$ is the linearization of the tableau operator F around the operating point w, i.e.,

$$F_l(w, t) \equiv \frac{\partial f(w, t)}{\partial w}. \quad (3)$$

The tableau matrix for the network of Fig. 1 is illustrated in Fig. 5. Figs. 4 and 5 show that (for this case) the first ten rows represent the Kirchoff's current ($A'i=0$) and voltage ($AV-v=0$) laws. The matrix A is the branch versus node incidence matrix [6], which has the advantage of being directly obtainable from the network description input data cards (which also establish the labeling of Fig. 3.) Rows 11–16 of f (Fig. 4) state the branch constitutive relations (one per branch) of the network, while rows 17–20 give the dynamic relations describing the generalized energy storage elements (e.g., $i_3-\dot{q}_3=0$ (row 18)).

For general networks, the tableau matrix F_l has the form of Fig. 6, wherein for network applications submatrices A, B, and E have all elements ± 1 or 0. Note F_l is very sparse, averaging in general about three nonzero elements per row.

Note that the optimization problem (1) for any system of ordinary differential equations (ODE) of the form $g(\dot{q}, q, p, t)=0$ can be put in the form of Fig. 6 by the following substitution:

$$f(a, q, \Phi, p, t) \equiv \begin{bmatrix} g(a, q, p, t) \\ a \quad -\dot{q} \\ \phi \quad -\dot{\Phi} \\ -\dot{p} \end{bmatrix} = 0, \qquad t\in[0, T]. \quad (4)$$

The network analog of the general ODE optimization problem is shown in Fig. 7, wherein each of the N components g_K, $K=1, N$, of the system $g(a, q, p, t)=0$ is represented by an *equator*, i.e., a unit capacitance driven by a current source, $i_K=a_K$, and a floating current source of value $i_{KK}=g_K(a, q, p, t)$. The objector and parameters (one for each of the np designable parameters) are as previously defined for the network problem.

The tableau approach to time-domain simulation is to discretize, at the branch level, the derivative operator d/dt. The nth discrete time point is denoted by t_n and $w_n \equiv w(t_n)$. Thus the last $nq+1+np$ rows (nq charges, 1 objector, np

[1] A lower case letter is a column vector unless otherwise specified, "$'$" denotes the transpose operation. Thus v' is a row vector.

parametors) have d/dt replaced by the kth-order backward differentiation formula [7]. For example, for charges

$$\dot{q}(t_n) \equiv \frac{dq_n}{dt} \equiv -\left(\sum_{j=0}^{k} \alpha_j q_{n-j}\right)/h \tag{5}$$

where $h \equiv t_n - t_{n-1}$. This transforms the tableau operator F, the tableau error vector f, and the tableau matrix F_l, into their discrete time forms F_d, f_d, and F_{dl}. Here f_d is obtained from f by replacing $\dot{q}$, $\dot{\Phi}$, and $\dot{p}$ by sums of the form (5), and F_{dl} is obtained from F_l (Figs. 5 and 6) by replacing d/dt by $-\alpha_0/h$.

Thus the time-domain simulation problem (2) is transformed into that of solving the difference equations

$$\begin{aligned} w_0 &\equiv w(t_0) \equiv w(0) \\ F_d(\cdot, p, t_n)w_n &= f_d(w_n, p, t) \equiv 0 \end{aligned} \tag{6}$$

at each time point t_n in the interval $[0, T]$. (Actually, f_d depends on $w_{n-1}, \cdots, w_{n-k}$ because of (5), but this dependence is suppressed here for convenience of notation.) A Newton iteration is used to solve the nonlinear algebraic difference equations so that the overall computational task of time-domain simulation is the following double iteration.

$$\begin{array}{ll}
t_0 = 0 & \\
n = 0 & \\
w_0 = \text{initial condition} & \\
\rightarrow n = n + 1 & \text{(time iteration loop start)} \\
t_n = t_{n-1} + h_n & \\
K = 0 & \\
w_n^{(0)} = w_{n-1} & \text{(or } w_n^{(0)} \text{ is obtained by prediction, cf., Section II).} \\
\rightarrow K = K + 1 & \text{(Newton iteration loop start)} \\
F_{dl}(w_n^{(K-1)})\Delta w_n^{(K-1)} = -f_d(w_n^{(K-1)}) & \\
w_n^{(K)} = w_n^{(K-1)} + \Delta w_n^{(K-1)} & \\
\text{converged? no} \rightarrow \text{Newton iteration;} \ \text{yes} \downarrow & \\
t_n \le T? \ \text{yes} \rightarrow \text{time iteration;} \ \text{no} \downarrow &
\end{array} \tag{7}$$

With only trivial modifications (7) applies to dc and ac cases also. In Section II we show how the routine for inverting F_{dl} may be used also to obtain the performance gradient $\partial\Phi/\partial p$ by the efficient adjoint approach [5]. (With this approach the cost of computing the np-dimensional vector $\partial\Phi/\partial p$ is roughly equal to that involved in (7)). Thus the tableau approach achieves dc, ac, and transient optimization with one matrix operation, namely, the inversion of F_{dl} in (7). Note (7) is essentially a two-step algorithm. The first step is to evaluate f_d and F_{dl}. A virtue of the tableau approach is that f_d and F_{dl} may be read directly from a simple input language using one input card per branch. The second computationally critical step is the inversion of the sparse matrix F_{dl}. This operation is handled by package programs for sparse Gaussian elimination. The tableau format allows these packages to be general-purpose programs not special to the network application. In Section III we describe prediction and control of truncation error in a variable-order variable-time-step application of the differentiation formulas (5) to algebraic–differential systems. This approach is similar to that developed by Gear [8] for strictly differential systems except that we store backward differences instead of the Nordsieck vector. Except for systems with only a small number of differential equations, the use of backward differences is more efficient computationally. This can be established by a simple operations count. Section IV discusses the sparse matrix algorithms, and in Section V we show how the tableau f_d, F_{dl} for the network optimization may be straightforwardly set up from a SCEPTRE-like input language [9].

Sparse matrix packages OPTORD and 1-2-3 GNSO are used to invert F_{dl} (actually the L/U factorization method is employed rather than direct inversion). Section IV gives the specific algorithms employed. The output of OPTORD is a pivoting order, i.e., the order in which the rows and columns of F_{dl} are eliminated in inverting F_{dl}. The ordering selected gives an optimized compromise between 1) minimum roundoff error in the solution Δw, and 2) minimum total operations count in performing the double iteration (7). 1-2-3 GNSO produces the required machine code SOLVE in three partitions: C-SOLVE, to be executed at the beginning of each optimization step, i.e., before entering the time loop of (7); T-SOLVE, to be executed in the time loop of (7); and X-SOLVE, to be executed in the Newton loop of (7). All elimination of variables is done by the Gaussian elimination routines. For example, the program does not require the determination of state variables, trees, nodal admittance matrices, capacitor loops, or any of the various specialized network concepts which tend to complicate a basically simple procedure. The sparse matrix packages OPTORD and 1-2-3 GNSO allow us to dispense with these procedures. In this sense the tableau matrix–Gaussian elimination approach tends to have a simplifying effect on network analysis not only computationally but to some extent theoretically as well. In Section VI we demonstrate that nodal analysis [6] and state-variable analysis [10] may be regarded as special methods of performing Gaussian elimination with different pivoting strategies. From this point of view there are many methods of analysis, and the most appropriate one for purposes of computation is the one which gives the best combination of accuracy and efficiency.

Comparisons of unconstrained OPTORD analysis with OPTORD constrained to eliminate state variables or nodal variables last are among the numerical results given in Section VII. Also, results showing how storage requirements grow with problem size are given for a sequence of examples.

II. The Tableau and Adjoint Sensitivity Computation

The main computation in minimizing $\Phi(p)$ is to compute, given p, $\Phi(p)$ and $\partial\Phi/\partial p$. With these quantities, nonlinear programming algorithms (e.g., Davidon–Fletcher–Powell [11]) can be utilized in minimizing $\Phi(p)$. In this section we present a theorem which demonstrates the role of the tableau matrix F_{dl} in the adjoint-method computation of $\partial\Phi/\partial p$. It is convenient for this purpose to group the algebraic network equations and variables so that

$$a \equiv \mathrm{col}\,(V, i, v)$$

and

$$0 = f(a, q, \Phi, p, t) \equiv \begin{bmatrix} g(a, q, p, t) & & \\ Ea & -\dot{q} & \\ \phi(a, q, p, t) & & -\dot{\Phi} \\ & & -\dot{p} \end{bmatrix}, \quad 0 \le t \le T. \quad (8)$$

Thus

$$\Phi = \Phi(T) \equiv \int_0^T \phi(a, q, p, t)\, dt \quad (9)$$

and the format of the tableau matrix F_{dl} is as shown in Fig. 8.

In these terms we may now state, without proof, the following theorem.

Theorem 1

Assume that the system (8) starts at equilibrium, i.e., $\dot{q}(0)=0$, and that T is independent of p, i.e., $\partial T/\partial p = 0$. Then

$$\frac{\partial\Phi}{\partial p} = \hat{p}'(0) + \overline{p'} \quad (10)$$

where $\hat{p}'(t)$, $0 \le t \le T$, satisfies the adjoint differential equations[2]

$$[\hat{a}', \hat{q}', \hat{\Phi}, \hat{p}'] \begin{bmatrix} \frac{\partial g(t)}{\partial a} & \frac{\partial g(t)}{\partial q} & 0 & \frac{\partial g(t)}{\partial p} \\ E & d/dt & 0 & 0 \\ \frac{\partial \phi(t)}{\partial a} & \frac{\partial \phi(t)}{\partial q} & -d/dt & \frac{\partial \phi(t)}{\partial p} \\ 0 & 0 & 0 & -d/dt \end{bmatrix} = [0, 0, 0, 0]. \quad (11)$$

with initial conditions

$$\hat{a}'(T) = \hat{q}'(T) = \hat{p}'(T) = 0, \qquad \hat{\Phi}(T) = -1 \quad (12)$$

[2] $\partial g(t)/\partial a$, etc., denotes $\partial/\partial a\{g(a(t), q(t), p, t)\}$, where $a(t)$, $q(t)$ is the solution of (8).

VARIABLES

$$\begin{array}{cccc} [\; a & q & \Phi & p \;] \end{array} \qquad \text{EQUATION}$$

$$F_{d\ell} = \begin{bmatrix} \frac{\partial g}{\partial a} & \frac{\partial g}{\partial q} & 0 & \frac{\partial g}{\partial p} \\ E & \frac{\alpha_0}{h} & 0 & 0 \\ \frac{\partial \phi}{\partial a} & \frac{\partial \phi}{\partial q} & \frac{\alpha_0}{h} & \frac{\partial \phi}{\partial p} \\ 0 & 0 & 0 & \frac{\alpha_0}{h} \end{bmatrix} \begin{bmatrix} g(a, q, p, t) = 0 \\ Ea - \dot{q} = 0 \\ \phi(a, q, p, t) - \dot{\Phi} = 0 \\ -\dot{p} = 0 \end{bmatrix}$$

Fig. 8. Algebraic–differential format of tableau.

and $\overline{p'}$ satisfies the adjoint algebraic equations

$$[\overline{a'}, \overline{q'}, \overline{\Phi}, \overline{p'}] \begin{bmatrix} \frac{\partial f_a(0)}{\partial a} & \frac{\partial f_a(0)}{\partial q} & 0 & \frac{\partial f_a(0)}{\partial p} \\ E & 0 & 0 & 0 \\ \frac{\partial \phi(0)}{\partial a} & \frac{\partial \phi(0)}{\partial q} & -1 & \frac{\partial \phi(0)}{\partial p} \\ 0 & 0 & 0 & -1 \end{bmatrix} = [0, \hat{q}'(0), 0, 0]. \quad (13)$$

Note that except for the d/dt entries, the coefficient matrices in (11) and (13) are identical to the tableau operator matrices $F_l(t)$ and $F_l(0)$, respectively. Thus the columns of F_l, operating to the left on the adjoint unknown vector $[\hat{a}', \hat{q}', \hat{\Phi}, \hat{p}']$ give the adjoint sensitivity equations just as the rows of F, (cf., Fig. 4) operating to the right on col $[a', q', \Phi, p']$, give the original differential equations (8). Thus to solve (11) and (13) as well as (8) involves Gaussian elimination applied only to F_{dl} (of Fig. 8), modified as shown in the d/dt entries.

This theorem establishes the actual flow of the computations leading to $\partial\Phi/\partial p$. First (7) is solved yielding $a(t_n)$, $q(t_n)$, $\Phi(t_n)$ at the time points t_n in the interval $[0, T]$. Those components of $a(t_n)$ and $q(t_n)$ which cause F_{dl} to vary are stored on disk. These are returned to core during solution of (11). Since the times t_m involved in solving (11) do not generally coincide with the t_n involved in (7), kth-order interpolation formulas, similar to (5), are used to obtain $a(t_m)$ and $q(t_m)$. The final value of $\hat{q}'(t)$, i.e., $\hat{q}'(0)$, is then passed into the right-hand side of (13) which is solved to give $\overline{p'}$. Finally, $\hat{p}'(0)$ and $\overline{p'}$ are added to give $\partial\Phi/\partial p$.

At this point both $\Phi(p)$ and $\partial\Phi/\partial p$ are available, and an optimization step may be taken. It is important to emphasize that in the iterations involved in solving (7) and (11) only one matrix structure, that of F_{dl}, is subject to sparse Gaussian elimination, and no other matrix operations, e.g., addition, transposition, and multiplication, are necessary.

III. Discretization of the Time Derivative Operator

Two basic assumptions have guided the development of the tableau approach to network simulation: for most networks the tableau matrix 1) is sparse and 2) has widely separate eigenvalues (i.e., it describes a stiff system of ODE). These assumptions dictate that for efficiency, implicit quasi-A-stable integration formulas be used [7]. The backward differentiation formulas (5) of order $k(1 \leq k \leq 6)$ satisfy these requirements. Using the Nordsieck vector [12], Gear [8] has demonstrated their advantage in providing a variable-order variable-time-step approach to minimizing the number of time steps subject to specified allowable integration errors. In the present context of simulation and optimization such an approach is especially necessary because the computation of $\partial\Phi/\partial p$ (cf., discussion of *Theorem 1*) requires storage of all arguments of the nonlinearities at every time step.

Gear's basic strategy, with minor modifications, of step size and order selection based on prediction and control of truncation error have been incorporated into the tableau approach. However, the Nordsieck vector has been replaced with stored backward differences to reduce operations counts when integrating a large number of equations. This replacement gives rise to new difference formulas for prediction and truncation error, and these will be presented along with the description of the discretization of the d/dt operator. Since this operator can act on true charges (q), the objector (Φ), or parameters (p), we state the development in terms of a generic vector x. Thus restated (5) reads

$$-h\dot{x}_{n+1} = \sum_{i=0}^{k} \alpha_i x_{n+1-i} \tag{14}$$

where $x_{n+1-i} \equiv x(t_{n+1-i})$, $h_i \equiv t_{i+1} - t_i$, and $h \equiv h_n$. Note the h_i depend on i in a variable-step method. The coefficients α_i are determined by the requirement that (14) annihilate polynomials of degree $\leq k$. In particular, this means that (14) must be satisfied for each of the $k+1$ substitutions $x_i = (t_{n+1} - t_{n+1-i})^j$, $j = 0, 1, \cdots, k$, thus yielding $k+1$ equations in the $k+1$ unknowns $\alpha_0, \alpha_1, \cdots, \alpha_k$. These can be solved in closed form, yielding an efficient method for computing the α_i. Since in practice one solves for the vector Δx_i instead of x_i, higher precision is obtained by expressing (14) in terms of the Δx_i. To this end, let

$$x_{n+1-i} \equiv x_{n-i} + \Delta x_{n-i}, \qquad i = 0, 1, \cdots, k$$

and

$$\beta_i \equiv \sum_{j=0}^{i} \alpha_j. \tag{15}$$

Then (14) becomes

$$\sum_{i=0}^{k-1} \beta_i \Delta x_{n-i} + h\dot{x}_{n+1} = 0. \tag{16}$$

Thus in (7) the effect of d/dt operating on q in forming f_d is given by

$$\frac{dq_{n+1}}{dt} = -\frac{\beta_0}{h}(q_{n+1} - q_n) - \frac{1}{h}\sum_{i=1}^{k-1} \beta_i \Delta q_{n-i}. \tag{17}$$

Correspondingly, the d/dt elements in F_l are replaced by $-\beta_0/h$ to obtain F_{dl} (note $\beta_0 = \alpha_0$).

In (7), one has the option of starting the Newton iteration with $w_{n+1}^{(0)} = w_n$, or of predicting from past history an alternative starting value. Experience shows that this prediction step, if efficiently executed, is usually worthwhile. The kth-order prediction formula

$$x_{n+1}^P \equiv -\sum_{i=1}^{k+1} \gamma_i x_{n+1-i} \tag{18}$$

has the same order of accuracy as (17). Again, the γ_i are chosen to annihilate polynomials of degree $\leq k$. The coefficients γ_i are also used in computing the truncation error

$$h((\dot{x}_{n+1})_{\text{true}} - (\dot{x}_{n+1})_{\text{approximate}}) \equiv E_k + O(h^{k+2}) \tag{19}$$

where $(\dot{x}_{n+1})_{\text{approximate}}$ denotes the value given by (14). E_k is obtained by expanding the terms x_{n+1-i}, $i = 0, 1, \cdots, k$, in (14) in a Taylor series around t_{n+1}. Using the fact that (18) and (14) annihilate polynomials of degree $\leq k$, one can show that

$$E_k = H_k \frac{h^{k+1}}{k+1} \frac{d^{(k+1)}x_{n+1}}{dt^{k+1}}$$

$$= -\delta_0\left(x_{n+1} + \sum_{i=1}^{k+1} \gamma_i x_{n+1-i}\right) \tag{20}$$

where

$$H_k \equiv \frac{1}{k!}\sum_{i=0}^{k} \alpha_i \left(\frac{t_{n+1-i} - t_{n+1}}{h}\right)^{k+1}$$

$$\equiv \frac{1}{k!}\prod_{i=1}^{k} \frac{(t_{n+1} - t_{n+1-i})}{(t_{n+1} - t_n)}$$

and where the extra coefficient δ_0 is determined by requiring the right-hand equation of (20) to be satisfied identically for polynomials of degree $k+1$. As in Gear's work on automatic step size and order control [7], the new time step h and order k are chosen so that h is maximum, consistent with

$$\frac{E_k}{h} \leq \frac{E_T}{T} \tag{21}$$

where T is the total integration time and E_T a specified total absolute error. In practice, to reduce roundoff error, the prediction and truncation error formulas (18) and (20) are rewritten in terms of the Δx_i of (16), just as (17) was obtained from (14).

The *Lemma* presented demonstrates that it is unnecessary to predict any components of the w vector (cf., (7)) other than those which cause the tableau matrix F_{dl} to vary with w. In the *Lemma* u_{n+1} represents the subset of w_{n+1}, on which depend the nonlinear components of the tableau error vector $f(w, t)$. (For the example of Fig. 4, $u = \text{col}(v_3, v_2, i_3) = \text{col}(w_{12}, w_{11}, w_6)$.)

Lemma

Let w_{n+1}^{P1} be the result of predicting (cf., (18)) all components of w_{n+1}, and w_{n+1}^{P2} be the result of predicting only the subset u_{n+1} of w_{n+1}. Then $w_{n+1}^{P1(1)}$ (obtained from the first Newton iteration step of (7) with $w_{n+1}^{(0)} = w_{n+1}^{P1}$) is equal to $w_{n+1}^{P2(1)}$ (obtained from (7) with $w_{n+1}^{(0)} = w_{n+1}^{P2}$).

Thus in contrast to Gear's method [7], it is not necessary to predict all quantities operated upon by d/dt. Since u is generally a very small subset of w (3 components of 20 in Fig. (4), the computational advantage of predicting u instead of predicting either 1) all the variables, or 2) all the nonlinearities plus all the charges [13] is substantial. The proof of the *Lemma*, which will be omitted here, is based on the fact that the tableau matrix is identical for the two modes of prediction, i.e.,

$$F_{dl}(w_{n+1}^{P1}) \equiv F_{dl}(w_{n+1}^{P2}).$$

In summary, three difference operators are used, one each for $\dot{x}_{n+1}$ (as in (17)), $x_{n+1}^{(0)}$ (as in (18)), and $H_k(h^{k+1}/(k+1)) \cdot (d^{(k+1)}x_{n+1}/dt^{k+1})$ (as in (20)). Each of these formulas use stored backward differences Δx_{n+1-i}. The coefficients α_i, γ_i, δ_0 are computed at each time step. The time step and order (k, $1 \leq k \leq 6$) are determined as recommended by Gear using the inequality (21). Prediction is only done for the variables u_{n+1} which cause F_{dl} to vary.

IV. Sparse Matrix Methods

There are basically two sparse matrix programs used. The first program OPTORD deals with finding an optimal pivot order, i.e., an order of elimination of rows and columns so that "fill-in" and "operations count" are minimized at the same time numerical stability is retained. Qualitatively, in terms of reducing fill-in, the program OPTORD uses a strategy due to Markowitz [4]. Similar techniques can be found in Tinney–Walker [3] and Dantzig *et al.* [4]. OPTORD is novel, however, in its introduction of variability type to identify and avoid redundant operations in cyclic computations. The use of a numerical test in OPTORD is also not usually incorporated in these methods.

The second program 1-2-3 GNSO takes the permutation generated by OPTORD and, taking into account the variability types of the matrix elements, writes five machine code programs (called C-SOLVE, T-SOLVE, X-SOLVE, B-SOLVE, and A-SOLVE) which execute only the nontrivial arithmetic operations necessary in the Crout factorization algorithm for Gaussian elimination [15]. This program is based on a similar program GNSO [1], which generated only a single FORTRAN program, and did not recognize variability type.

We discuss in turn OPTORD and 1-2-3 GNSO, both of which recognize three basic properties of the nz nonzero elements of the ns × ns matrix S, namely, 1) location I, J; 2) numerical value $S[I, J]$; and 3) variability type $VT[I, J]$. The six variability types are listed in Table I.

TABLE I

Type 1	elements which are $+1$ } called topological type
Type 2	elements which are -1 } called topological type
Type 3	elements which never change, called c type
Type 4	elements which change with design parameters, called p type
Type 5	elements which change with time, called t type
Type 6	elements which change with the unknown, called x type

These types are represented in the example of the matrix F_l of Fig. 6. In this example, the 1, 5 and 5, 11 elements of F_l are the topological type ($VT[1, 5]=1$, $VT[5, 11]=2$), the 12, 5 element is p type ($VT[12, 5]=4$), the 17, 17 element is t type (because d/dt is replaced in F_{dl} by $-\alpha_0/h$, which changes with time in a variable-step method, $VT[17, 17]=5$), and element 15, 7 is of x type ($VT[15, 7]=6$). Note that in a time-domain optimization, topological-type elements are invariant, p-type elements change every optimization step, t-type elements every time step, and x-type elements every Newton iteration (cf. (7)).

Ideally, an arithmetic operation is done only when and if it must be done. Thus multiplication of a t-type element by a p-type element should properly be executed in the time loop of (7) but not in the inner Newton loop. To this end there is associated with each multiplication of the term $S[k, L]*S[K, L]$, a variability type

$$VTM \equiv \max\left[VT[k, L], VT[K, L]\right]$$

which identifies when the multiplication must be done.

In order that OPTORD may reduce the total operations count in a given design problem, a set of weights wt$[i]$, $i=1, \cdots, 6$, are specified, one for each of the six variability types. Thus the weight wt$[i]$ can be an estimate of relatively how often i-type operations must be executed during an optimization run.

The permutation vectors rp and cp are used to specify the pivoting order. Thus the ith pivot is in row $rp[i]\equiv I$ and column $cp[i]\equiv J$. The task of OPTORD is to select I and J so as to minimize the ith weighted pivot cost

$$PC[I, J] \equiv \mu M[I, J] + \nu R[I, J] \tag{22}$$

where $\mu, \nu \geq 0$ are specified input parameters, M is a weighted multiplication count, and R is a roundoff error factor. We define

$$M[I, J] \equiv \sum_{j=1}^{m_{IJ}} wt_j \tag{23}$$

with wt_j standing for the weight associated with the highest variability type involved in the jth of the m_{IJ} multiplications required in eliminating row I and column J. Also,

$$R[I, J] \equiv \frac{1}{r_I} \frac{\sum_j |S[I, j]|}{|S[I, J]|} \tag{24}$$

where j runs over the r_I nonzero column indices in row I.

The ith of the n formally identical steps in the process of selecting rp and cp is described by the following algorithm.

OPTORD Algorithm

At the end of the $(i-1)$ step, $i-1$ eliminations have been carried out. Thus in the ith step the $(n+1-i)\times(n+1-i)$ matrix $S^{(i)}$ is to be considered (note $S^{(1)}\equiv S$) in which the nonzero locations, the values, and the variability types of its elements are known.

1) For each row K determine the subset $l[j], j=1, \cdots, \hat{r}_K$ of the r_K nonzero columns of the Kth row of $S^{(i)}$ for which

$$TOL \cdot R[K, l[j]] \leq 1 \tag{25}$$

where $0 \leq TOL \leq 1$ is a threshold factor. The remaining $r_K - \hat{r}_K$ elements in row K have numerical values which are considered too small to be candidate pivots.

2) Determine the $\hat{r}_K$ values of $M[K, l[j]]$ (cf., (23)), $j=1, \hat{r}_K$, associated with the elimination of row K and column $l[j]$ (element $K, l[j]$ is the candidate pivot). In Gaussian

elimination a multiple of the pivot row K is added to each row k in $S^{(i)}$, which has a nonzero element in column $l[j]$, in such a way that the k, $l[j]$ element becomes zero. Thus for each such row k we compute

$$S^{(i)}[k,L] - S^{(i)}[k,l[j]]*(S^{(i)}[K,L]]/S^{(i)}[K,l[j]]) \quad (26)$$

where L runs over the union of column indices in rows k and K. Associate with each nontrivial multiplication (represented by $*$ in (26)) its variability type

$$VTM[k,L] \equiv \begin{cases} \max\{VT[k,l[j]], VT[K,l[j]], VT[K,L]\} \\ 0, \quad \text{if } S^{(i)}[K,L]=0. \end{cases} \quad (27)$$

If there are c_j nonzero elements in the pivot column $l[j]$ of $S^{(i)}$, then $(c_j-1)(r_K-1)$ such multiplications would be required to use element K, $l[j]$ as the next pivot. The weighted multiplication count for each of the $\hat{r}_K$ pivots K, $l[j]$, $j=1,2,\cdots,\hat{r}_K$, is therefore given by[3]

$$M[K,l[j]] = wt[VT[K,l[j]]] + \sum_k\left(\sum_L wt[VTM[k,L]]\right) \quad (28)$$

where k runs over the nonzero row indices in column $l[j]$, and L runs over the union of column indices in rows k and K. The first term accounts for the computation of $(S^{(i)}[K,l[j]])^{-1}$ and for $k=K$; the sum accounts for the multiplications $(S^{(i)}[K,l[j]])^{-1}*S^{(i)}[K,L]$.

3) Compute $PC[K,l[j]]$, $j=1,\cdots,\hat{r}_K$, using (24) and find the minimum pivot cost column $JC[K]$ for each row K, $K=1,2,\cdots,n+1-i$, i.e.,

$$PC[K,JC[K]] = \min_j\{PC[K,l[j]]\}. \quad (29)$$

4) Find the minimum operations-count row I, i.e.,

$$M[I,JC[I]] = \min_K\{M[K,JC[K]]\}. \quad (30)$$

5) Execute the numerical Gaussian elimination of row $I\equiv rp(i)$ and column $J=JC(I)\equiv cp(i)$, creating the $(n-i)\times(n-i)$ matrix $S^{(i+1)}$.

6) Update the variability types of each nonzero element in $S^{(i+1)}$ (cf., (26)), according to

$$VT[k,L] = \max\{VTM[K,L], VT[k,L]\}.$$

Note that if the operation count weight is set to zero (i.e., $\mu=0$ in (22)), step 3 selects the largest pivot element in row K. Thus partial pivoting [17] is effected in each row, and the row is chosen (cf., (30)) to minimize the weighted multiplication count. The decision to have OPTORD minimize only the number of multiplications required at each step of Gaussian elimination was made on the basis of some comparison with other methods. Roughly, the result of this comparison was that it was important to base the next pivot choice on the result of executing the previous pivot. However, how one then chose the best pivot was not so critical and many methods seemed to give the same benefits. These conclusions are in agreement with those reported in [15].

[3] We define $wt[0]=0$ since VTM can be zero (cf., (27)).

The choice of minimizing multiplications was in keeping with minimizing the length of the SOLVE code and maximizing execution speed.

1-2-3 GNSO

1-2-3 GNSO creates an executable nonlooping program for solving both $b=Sx\equiv LUx\equiv Ly$ and $\hat{b}'=\hat{x}'S\equiv\hat{x}'LU\equiv\hat{y}'U$. 1-2-3 GNSO differs from GNSO [1] mainly in its use of variability type in creating five distinct lists of machine instructions: C-SOLVE; T-SOLVE; X-SOLVE; B-SOLVE; and A-SOLVE (GNSO creates a single list of FORTRAN instructions). C-SOLVE, T-SOLVE, and X-SOLVE compute the factorization of S into the product of upper and lower triangular matrices L and U. C-SOLVE, T-SOLVE, and X-SOLVE need be executed only once per optimization step, time step, and Newton step, respectively (cf. (7)). B-SOLVE carries out the back substitutions $Ly=b$, $Ux=y$ required in simulation (cf., (7)). A-SOLVE carries out the adjoint back substitution $\hat{b}'=\hat{y}'U$, $\hat{y}'=\hat{x}'L$ required in the sensitivity computation (11) or (13).

The logic of 1-2-3 GNSO is based on the Crout method of factorizing S into LU. For purposes of presentation here it will be assumed that S has been rearranged so that the pivots $S[rp[K], cp[K]]$ appear on the diagonal of S. Thus the Crout formulas are for $K=1,2,\cdots,n$,

$$L_{IK} = \left(S_{IK} - \sum_{J=1}^{K-1} L_{IJ}U_{JK}\right), \qquad I=K,\cdots,n$$

$$U_{KI} = \left(S_{KI} - \sum_{J=1}^{K-1} L_{KJ}U_{JI}\right) \div L_{KK}, \qquad I=K+1,\cdots,n. \quad (31)$$

(Note $U_{KK}\equiv 1$.) The index ordering indicates how the Kth column of L and Kth row of U are to be computed sequently and assures that all right-hand-side terms of (31) are known.

As in OPTORD the variability type of each new L_{IK} or U_{KI} term is equated to the highest type of any term in (31). Similarly, each product in (31) is given the highest type of its two factors. These variability types are used to decide to which lists, C-SOLVE, T-SOLVE, or X-SOLVE, the instructions created to compute each L_{IK} (or U_{KI}) are to be added. For a multiplication, this is done by placing the required machine instructions (LOAD, MULTIPLY, ADD) at the end of the appropriate list. For example, instructions for x-type multiplications are added to X-SOLVE. If a multiplication involves a topological type factor (± 1), the LOAD and MULTIPLY instructions are suppressed. If a number of topological type terms are to be added, the result must be topological (± 1 or 0) (zero if the number of such terms is even).[4] In this

[4] That the sum of two or more terms of topological type in a Crout formula is also of topological type (± 1 or 0) is due to the fact that the tableau with all nontopological type elements replaced by zero is a totally unimodular matrix [18]. This is because in the tableau we have made the restriction that not more than one topological element may appear in a branch-constitutive-relation row. Therefore, since the incidence matrix A is totally unimodular and any totally unimodular matrix augmented by rows or columns of the identity matrix is totally unimodular, it follows that the topological part of the tableau matrix is also totally unimodular. The result then follows from the fact that the elements of L and U can be expressed in terms of subdeterminants of the tableau matrix.

case the only instruction generated is a LOAD of ± 1 (or no instruction if the result is 0).

If a given formula (31) involves multiplications of various types, we have

$$L_{IK} \text{ (or } U_{KI}) \equiv TC(IC) + TT(IT) + TX \tag{32}$$

where $TC(IC)$ is a temporary storage location for the sum of terms of ± 1 p type or c type and $TT(IT)$ a temporary location for the sum of t-type terms. TX represents the sum of the remaining terms, but no temporary storage location is required. The instructions required for computing $TC(IC)$, $TT(IT)$, and TX are placed at the end of the C-SOLVE, T-SOLVE, and X-SOLVE lists, respectively. The detailed procedure for $TT[IT]$ (the procedure is similar for TC and TX) is given in the following list. We add the following instructions to the T-SOLVE list.

1) LOAD, ADD S_{IK} (if S_{IK} is t type and nonzero)
2) LOAD, MULTIPLY, ADD (for each t-type product in (31))
3) ADD $TC(IT)$ (if present in (32))
4) STORE L_{IK} (if TX absent)
5) STORE $TT(IT)$ (if TX present).

A development similar to that based on (31) and (32) describes the creation of the instruction lists for the back solve programs B-SOLVE (for solving $Sx=b$) and A-SOLVE (for solving $\hat{x}'S=\hat{b}'$). Like X-SOLVE, B-SOLVE (or A-SOLVE) is executed within the Newton iteration since the right-hand-side vector b and the auxiliary vector y change whenever the unknown variable x changes.

Thus by creating the five lists of instructions to be executed in the appropriate loops of the computation, 1-2-3 GNSO achieves near-ultimate efficiency: multiplications are performed only if they are absolutely necessary.

V. An Input Language for the Tableau Approach

An automated network design program (NDP) has been written, incorporating the approach to automated network optimization previously described. NDP interfaces the sparse matrix, implicit integration, and adjoint sensitivity routines through a simple SCEPTRE-like [9] input language. As described in the following paragraphs, the input language processor passes a complete description of the tableau matrix F_l, which is identified as the S matrix of OPTORD and 1-2-3 GNSO. 1-2-3 GNSO then produces a machine code program, SOLVE (cf., Section IV). To conserve storage, the SOLVE code is designed to work only on the nonzero nontopological entries of F_l.

Before describing the input language in detail, the threaded list structure used to compact F_l is described. A vector s is used to store the nonzero nontopological values in the (row-wise) order in which they appear in F_l. A vector SP, which points to s, and a column index vector CI are associated with the nonzero elements of F_l. Finally, a row pointer vector RP is used to identify which column indices belong to a given row. For the example of Figs. 1–5, the array s and the components of SP, CI, and RP, belonging to rows 11–17 of F_l (cf., Fig. 5), are

$$s' \equiv \left(1, -1, L, i_2, \frac{\partial L}{\partial p_1}, \frac{\partial \psi}{\partial v_2}, \frac{\partial \psi}{\partial v_3}, \frac{\partial \phi}{\partial i_3}, \frac{\partial \phi}{\partial p_1}, d/dt\right)$$

$$\begin{array}{rl} SP(21\text{–}36) = & (1,\ 3,\ 2,\ 4,\ 5,\ 6,\ 2,\ 1,\ 2,\ 7,\ 2,\ 8,\ 1,\ 2,\ 1,\ 9) \\ & \ \uparrow\ \ \uparrow \qquad\quad \uparrow \qquad\qquad \uparrow \qquad \uparrow \qquad\quad \uparrow \quad \uparrow \\ CI(21\text{–}36) = & (11,\ 6,\ 17,\ 20,\ 12,\ 13,\ 18,\ 14;\ 19,\ 7,\ 9,\ 20,\ 16,\ 20,\ 12,\ 17) \\ & \ \uparrow\ \ \uparrow \qquad\quad \uparrow \qquad\qquad \uparrow \qquad \uparrow \qquad\quad \uparrow \quad \uparrow \\ RP(11\text{–}17) = & (21,\ 22,\ \ 25,\ \ 28,\ \ 30,\ \ 33,\ \ 35,\ ). \end{array} \tag{33}$$

The row pointers point to the first nonzero column index in a given row. Thus RP and CI combine to define a unique I, J location in F_l for each component of the pointer SP. This pointer in turn identifies the component of s to be associated with $(F_l)_{IJ}$.

1-2-3 GNSO requires knowledge of SP, CI, and RP, as well as a similarly compacted variability-type array VT. Nevertheless, the SOLVE code generated refers only to the compacted s array (which thus replaces the S elements in (31)). Thus, although in the present example, location, value, and variability type must be known for 41 of the 400 elements in F_l, only 10 quantities need be in core during the simulation–optimization phase of NDP. Note also (cf. Fig. 4) that in (33) only 4 quantities, v_3, i_3, v_2, and i_2, need be stored on disk for the purpose of recreating s (i.e., F_l) for the backward time adjoint integration (11).

In the NDP language the input description of the time-domain network optimization problem of Figs. 1–5 is shown in Table II. The first data card defines the reference node (to be node 0 of Fig. 3). The next nb cards, one for each branch in the network, are free format with successive data fields giving the branch label, the start node, the finish node, the element type, and the name and argument list of a user defined function subprogram. The fifth card gives the name of the subprogram which defines the objective function to appear as integrand in (1) and signals NDP to add the objector and parametor to the network. The sixth and seventh cards state that there is one designable parameter and that in the constrained optimization problem (1) the parameter labeled 1 has a lower bound of 0.01, a starting value of 1, and an upper bound of 100. The eighth card identifies the unknown variables required to recreate the matrix F_l in (11). According to the *Lemma* of Section III, those are also the variables which are predicted in the forward time integration (7). Note NDP translates bracketed function arguments into unknown vector components. In the example, $/T2/ \to i_2 \to x(6)$, $/A3/ \to v_3 \to x(13)$, and $/P1/ \to p_1 \to x(20)$. Other examples of notation are $N1$ (voltage of node 1) and $Q3$ (charge on the capacitor in branch 3).

TABLE II

```
NETWORK DESCRIPTION, NREF = 0
1, 1-0, 3 = E(t)
2, 1-2, 12 = QL(/T2/P1/t)
3, 2-0, 11 = PSI (/A2/A3/)
OPTIMIZE PHI (/T3/P1/)
DESIGN PARAMETERS, NP = 1
P1 (0.01, 1, 100)
NONLINEARITIES /A3/A2/T3/T2/
```

TABLE III

```
f(11) = x(11) - FTIME(t)
f(12) = QL(x(6), s(3), x(20), s(4), t) - x(17)
f(13) = PSI(x(12), s(5), x(13), s(6))
f(15) = PHI(x(7), s(7), x(20), s(8)) - x(9)
```

TABLE IV

```
FUNCTION QL(X1, S1, X2, S2, T)
S1 = FL(X2, S3, T)
QL = X1 * S1                 (≡i2*L)
S2 = X1 * S3                 (≡i2*∂L/∂p1)
RETURN
END
```

Given the card input of Table II, NDP creates a subroutine VARQ and input to two permanent subroutines FTOP and QTVARQ. These subroutines compute the tableau error vector f and the compacted version s of the tableau matrix F_l. The input to FTOP is the (suitably compacted) incidence matrix A. Note there is a $+1$ (-1) in the column of A corresponding to the start (finish) node. Thus A is essentially read directly from the input cards. The result of FTOP is the evaluation of the first $nn+nb$ components of f.

The generated subprogram VARQ for the present example comprises the list of FORTRAN instructions shown in Table III. There is a one-to-one correspondence between these FORTRAN instructions and input cards 2–5 of Table II, except that the argument lists in the user-defined function subprograms, QL, PSI, and PHI, differ from those on the input cards. This difference is for user convenience. The convention is that given the unknown vector x the subprograms must return not only the function value (e.g., QL) needed for computing f, but also the partial derivatives (needed for evaluating F_l) of the function value with respect to members of the argument list which are components of the unknown vector x, e.g., $s(3) \equiv \partial QL/\partial x(6) \equiv L$, $s(4) \equiv \partial QL/\partial x(20) \equiv i_2(\partial L/\partial p_1)$. Thus unknown vector components in the argument list are always followed by a component of the compacted tableau matrix array s. For user convenience these extra arguments are left out of the input cards.

The user-specified FORTRAN function subprogram QL is described as an example in Table IV. In this function subprogram, another FORTRAN function subprogram FL is called to evaluate the inductance $L \equiv FL$ (dummy argument $S1$) as a function of the time T and the design parameter p_1 (dummy argument $X2$). The value of L is stored in $S(3)$ (cf., Table III). The magnetic flux for this linear time-varying inductor is then computed and stored in QL. Then the derivative $\partial QL/\partial p_1$ (dummy argument $S2$) is computed in terms of $\partial L/\partial p_1$ (dummy argument $S3$) and stored in $S(4)$.

The third executable program QTVARQ creates the $Ea\text{-}\dot{q}$ (cf. (8)) rows of the tableau vector f and the d/dt elements of the tableau matrix. QTVARQ computes $Ea\text{-}\dot{q}$ by replacing $\dot{q}$ by a summation as described by (17).

Note that FTOP, VARQ, and QTVARQ operate directly on the input data of Table II with a minimum of intervening data processing. There are no instructions in NDP which refer to excess or link capacitors or tree inductors. Thus the tableau approach imposes no topological restrictions whatsoever on the type or connectivity of the network elements.

Note also that functions like those in Table IV are entirely user defined. Any function of any group of variables may thus be defined by the user so that there is no restriction on the type of elements treatable by the tableau approach.

VI. Relation Between Pivoting Order and Network Methods of Analysis

When the time discretization procedure (cf., Section III) is carried out before any reduction is made on the network equations, the only constraint on pivoting order is that the numerical value of the pivot element should not be too small. From the point of view of pivot order selection, we examine two of the principal methods of network analysis: nodal analysis and state-variable analysis. For ease of illustration consider the tableau of Fig. 9 which describes a network of only capacitances and other admittance-type elements but no forcing terms. Using the pivoting order (4, 6), (2, 4), (3, 5), (5, 3), (6, 2) we obtain the nodal equations

$$\left\{A_2'YA_2 + A_1'CA_1\frac{d}{dt}\right\}V = 0 \tag{34}$$

where the matrix operating on node voltages V is known as the nodal admittance matrix.

The state equations are more difficult to describe in terms of a pivoting order since it requires finer groupings to be made on the variables. Thus a proper tree must be obtained,[5] the tree variables separated from the link variables, and all variables except the capacitor voltages in the proper tree and the inductor currents in its links eliminated. For a

[5] To illustrate that in the general case the state-variable method corresponds to a particular set of pivots and this set gives rise to a proper tree, consider the following constraints on the pivot order.

1) Select the first nb (number of branches) pivots from $AV-Iv=0$ rows of F_{dl} (Fig. 6). If possible, let the kth pivot fall in the first nonzero column in the kth row of A. In this case, which will occur nn (number of independent nodes) times, the kth branch is in the tree. If after $k-1$ pivots, the kth row of A has been annihilated, then pivot on $I_{kk}=(F_{dl})_{nn+k,nn+nb+k}$. In this case the kth branch is a link.

2) Select the next nn pivots to fall in locations in the $A'i=0$ rows which are the transpose of the nn pivot locations in A.

These constraints cause a tree to be selected whose identity depends on the order of branches on the input cards. A proper tree [10] can be obtained by rearranging the tableau so that voltage sources come first in the list of branches followed by capacitors, conductors, resistors, inductors, and current sources.

simple example, it is now assumed that the capacitor branches of Fig. 9 form a tree; hence this is a proper tree. Thus for the tableau of Fig. 9, the pivoting order (2, 1),

V	i_1	i_2	v_1	v_2	q
	A_1^T	A_2^T			
A_1			-1		
A_2				-1	
			C		-1
		-1		Y	
	1				$-\frac{d}{dt}$

Fig. 9. Simplified tableau.

(3, 5),(1, 2), (4, 6), (5, 3) yields the equations

$$C\frac{dv_1}{dt} = -(A_1')^{-1}A_2'YA_2A_1^{-1}v_1 \tag{35}$$

which are the state equations in this case.

In both nodal and state-variable analysis the equations are reduced to their standard form and then the d/dt operator is replaced by a difference operator by some procedure analogous to that described in Section III. The difference equations thus obtained are identical to the difference equations obtained by first replacing the d/dt operator and then pivoting in the same sequence as was done to obtain the state-variable or nodal-variable equations.

Both cases (34) and (35) represent a reduced form of the original tableau matrix F_t. Thus the classical network methods may be regarded as special cases of the tableau approach. From this point of view the particular methods are identified by the constraints they impose on the pivoting order in Gaussian elimination. Since these constraints on the pivoting order are not in general consistent with objectives of minimizing operations count, it follows that these classical network methods, being primarily motivated by analytical simplicity, are intrinsically slower than the sparse tableau approach.[6] The sparse tableau approach also has an advantage in terms of the possibility for higher precision answers. To illustrate this, consider the following system of linear equations

$$\begin{pmatrix} A & B \\ C & D \end{pmatrix}\begin{pmatrix} x \\ y \end{pmatrix} = \begin{pmatrix} b_1 \\ b_2 \end{pmatrix}. \tag{36}$$

Let this represent the unreduced tableau and the system

$$(D - CA^{-1}B)y = b_2 - CA^{-1}b_1 \tag{37}$$

represent the reduced equations, e.g., these would correspond to state-variable or nodal analysis. Since the unreduced tableau is taken directly from the input data, the matrices A, B, C, D and vectors b_1, b_2 can be assumed to be precise. The matrix $S \equiv D - CA^{-1}B$ and the the vector $\tilde{b}_2 \equiv b_2 - CA^{-1}b_1$ are the result of some arithmetic operations on the input data and therefore are not precise. Let E be the error in S and ε be the error in $\tilde{b}_2$. Thus in the machine the reduced problem is represented by

$$(S + E)y = \tilde{b}_2 + \varepsilon. \tag{38}$$

To solve (36) or (38) precisely (to the accuracy of the machine) one can use iterative refinement [17]. For (36) this means the iteration

$$\begin{pmatrix} A & B \\ C & D \end{pmatrix}\begin{pmatrix} \Delta x^{(k)} \\ \Delta y^{(k)} \end{pmatrix} = \begin{pmatrix} b_1 & -Ax^{(k)} & -By^{(k)} \\ b_2 & -Cx^{(k)} & -Dy^{(k)} \end{pmatrix}$$
$$x^{(k+1)} = x^{(k)} + \Delta x^{(k)}$$
$$y^{(k+1)} = y^{(k)} + \Delta y^{(k)}.$$

For (38) this iteration is

$$(S + E)\Delta y^{(k)} = \tilde{b}_2 + \varepsilon - (S + E)y^{(k)}$$
$$y^{(k+1)} = y^{(k)} + \Delta y^{(k)}.$$

According to Wilkinson [17], in the iterative refinement of the problem $Mz = b$, if the matrix M being inverted can be factored into LU, where $LU = M(I + F)$, with

$$\|F\|_\infty \le \tfrac{1}{2} \tag{39}$$

and if the error vector $b - Mz^{(k)}$ is computed with accumulation, then $\|z^{(k)} - z^*\|_\infty < 2^{-t}$ for k large enough. (Here z^* is the true solution and t is the number of digits in the machine.) Thus to within the accuracy of the machine and subject to (39), equations (36) and (38) can be solved exactly. In the case of equation (36) we then have an accurate answer. However, in the case of (38) we only have an accurate answer to equation (38) but not necessarily equation (36) which was our original purpose. Thus the initial errors E and ε made in manipulating the input data to form the reduced equations cannot be recovered.

VII. Numerical Results for a Practical Example

The current switch–emitter follower logic chain of Fig. 10 is used to demonstrate optimal design and large-scale application capabilities of the tableau approach and to compare its computational efficiency with competitive methods. Each five-terminal stored model in Fig. 10(c) represents one logic stage (Fig. 10(b)) in the chain. Each of these logic circuit models use three transistors, which in turn are represented by the high-speed transistor model of Fig. 10(a).

The automated design problem selected has eight logic stages and is characterized by the parameter set $p' \equiv (R_1, R_2, R_3)$ (for the lower and upper parameter bounds we use $p_m' \equiv (0.05, 0.05, 0.05)$ and $p_M' \equiv (0.15, 0.15, 0.15)$). The design objective is to realize a specified output voltage waveform $V_{out}(t)$ for a given input voltage waveform $V_{in}(t)$ (Fig. 11). The specified output is characterized by delay, rise time, and up and down levels. Thus the objective instantaneous function $\phi = \frac{1}{2}(V_{realized} - V_{specified})^2$ vanishes when the shaded area of Fig. 11 vanishes. For an eight-circuit logic chain the result of six Fletcher–Powell optimization steps (bottom of Fig. 11) is that $\Phi(p)$ decreases from 0.7 to 0.0003.

[6] It can be seen that the symbolic elimination of topological equations is formally equivalent to the 1-2-3 GNSO process of converting multiplications by ± 1 into additions or subtractions.

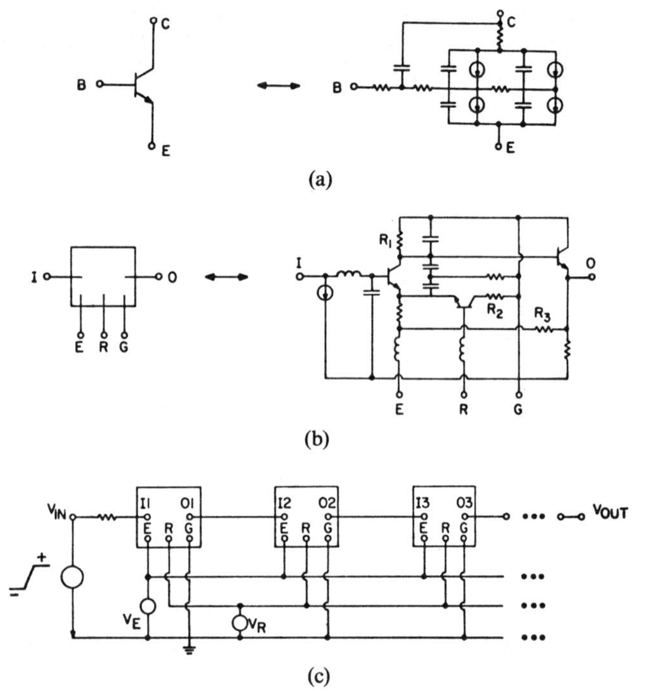

Fig. 10. Current switch–emitter follower circuit logic chain. (a) Two-dimensional transistor model. (b) Current switch–emitter follower logic stage. (c) Logic chain.

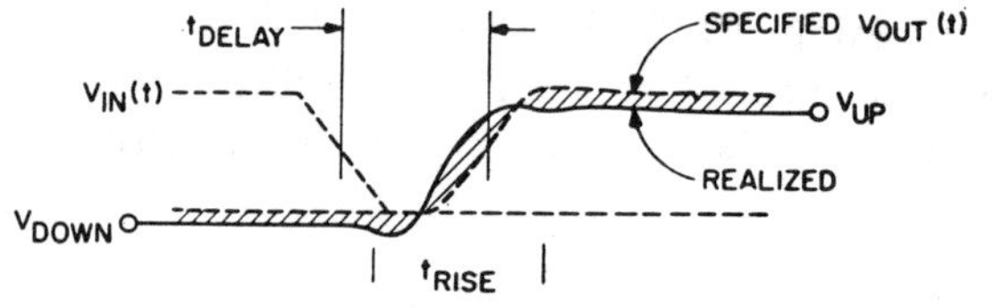

	Φ(p)	R1	R2	R3
INITIAL	.7	75Ω	100Ω	150Ω
FINAL	.3 x 10⁻³	114Ω	100Ω	111Ω

Fig. 11. Three-parameter optimization of eight-stage logic chain.

TABLE V

Problem complexity (8 circuit optimization)	
nodes	180
branches	460
nonlinear elements	112
differential equations	192
total equations	1300
Optimization (Fletcher–Powell with quadratic search)	
designable parameters	3
gradient computations ($\partial\Phi/\partial p$)	6
function evaluations ($\Phi(p)$)	36
time steps	30 000
newton iterations	85 000

Various aspects of problem complexity are given in Table 4. Note that for this example, p-type elements (variability type=4) of the tableau matrix F_l varied 36 times and t- and x-type elements varied 30 000 and 85 000 times, respectively.

To show how code length varies with problem size, we give in Table V the results of applying OPTORD/1-2-3 GNSO to logic chains of 1, 8, and 22 circuits.

For the data given in Table VI we made no attempt to try to minimize the length of the code but chose partial pivoting ($\mu=0$, $\nu=1$ in (22)) to insure accurate answers. The data indicate that for the one-circuit case, F_l has 462 nonzero elements, of which 92 are stored in the threaded list (33). The L/U factorization adds another 450 fill-ins so that there are 912 nonzero elements in L/U, of which 360 must be stored during execution of the generated SOLVE code. The sum of the lengths of the data, and of the C-, T-, X-, B-, and A-SOLVE codes is 27 316 bytes. The central processing unit IBM Model 91 times required for function evaluation (time-domain simulation of 24 ODE) and gradient computation are 5 s and 3 s, respectively. Note that the various data in Table VI (in particular note the total code length) increase approximately linearly with the number of circuits analyzed and, therefore, with the order of F_l. (Such linear growth may be typical for network tableau matrices.) However, the growth of simulation time ($\Phi(p)$ column of Table VI) is supralinear. This is because both the number of time steps and the number of Newton iterations per time step generally increase with problem size.

The results of Table VII illustrate the effects of varying the ordering strategy to minimize the length of x-type code (X-SOLVE and B-SOLVE). The data are for the one-circuit case of Fig. 10(b) (this time without objectors and parametors so the original tableau has only 446 nonzero elements) and serve to quantify the comparison of the tableau approach using unconstrained OPTORD with OPTORD constrained to correspond to competitive network analysis methods given in Section VI. In each case the ordering parameters TOL, μ, ν, and wt[i], $i=1, 2, \cdots, 6$, (cf., (22)–(26)) were chosen by trial and error to minimize code length irrespective of accuracy considerations and are tabulated in Table VIII.

For the nodal-analysis method, the last nn (number of independent nodes) pivots were constrained to come from the upper-left-hand partition of F_l (originally zero in Figs. 6 and 9). For the state-variable method, the last ns (number of state variables is $ns=21$) pivots were constrained to be in the state-variable columns and in the corresponding d/dt rows. Within these constraints, the full OPTORD algorithm was used to select the pivots.

One basis for comparing the performance of the generated codes is the sum of the X-SOLVE and B-SOLVE code lengths[7] since these are executed in the innermost loop of the program (85 000 times in the example of Table V). In Table VII the data (compare cases 1 and 2 to 5 and 6) demonstrate our general experience that the pivot order constraint imposed by the nodal-analysis method did not significantly increase SOLVE code lengths over unconstrained OPTORD. The pivot order constraint (cases 3 and 4) imposed by the state-variable analysis method caused a more substantial increase in code length. In all cases observed to date, unconstrained OPTORD has given the best performance, with nodal analysis a fairly close second. (An explanation of this phenomenon will be given in a subsequent paper.)

[7] Since the generated codes are relatively homogeneous and have no loops, the code length is roughly proportional to execution time.

TABLE VI

PARTIAL PIVOTING ($\mu=0$, $\upsilon=1$) OPTORD/1-2-3 GNSO FOR LOGIC CHAIN OF 1 CIRCUIT, 8 CIRCUITS, 22 CIRCUITS (PARAMETORS AND OBJECTOR INCLUDED)

Nodes/Branches	F_t Nonzero	F_t Stored	L/U Fill	L/U Stored	Total Code in Bytes	Model 91 Time (s) $\Phi(p)$	Model 91 Time (s) $\partial\Phi/\partial p$	Number of ODE
16/61	462	92	450	360	27 316	5	3	24
180/460	3472	680	3735	3430	276 404	120	30	192
492/1260	8632	1736	9332	8617	713 504	900	500	928

TABLE VII

COMPARISON OF SPARSE TABLEAU APPROACH (NDP) WITH CLASSICAL NETWORK METHODS

Case	Analysis Mode	OPTORD/1-2-3 GNSO Output: L/U Fill	L/U Stored	Code Length (bytes)[a] T-SOL	X-SOL	B-SOL
1	Nodal -OPTORD	215	348	2200 / 1564	1868 / 1656	7864 / 6754
2	Nodal -OPTORD	194	390	2812 / 1790	2388 / 1932	7668 / 6786
3	$S-V$ -OPTORD	472	426	4776 / 3968	2828 / 2630	9584 / 8288
4	$S-V$ -OPTORD	500	580	4044 / 3150	14 492 / 14 102	10 620 / 9 414
5	NDP -OPTORD	191	344	2576 / 1652	2048 / 1686	7632 / 6554
6	NDP -OPTORD	204	329	2820 / 1956	1572 / 1400	7792 / 6616

[a] Upper number in bracketed pairs represents output of current 1-2-3 GNSO compiler. Lower number represents estimates based on hypothetical optimum compiler.

TABLE VIII

ORDERING PARAMETERS FOR CASES OF TABLE VII

Case	TOL	μ	υ	$wt(1)$	$wt(2)$	$wt(3)$	$wt(4)$	$wt(5)$	$wt(6)$
1	0	0.5	1	100	100	500	400	450	1500
2	0	1	0	100	100	150	200	150	300
3	0	0.5	1	100	100	500	400	450	1500
4	0	1	1	100	100	500	600	450	900
5	0	1	0.5	100	100	90	101	90	150
6	0	1	0	100	100	102	101	102	150

(All columns from TOL through $wt(6)$ are OPTORD/1-2-3 GNSO Input.)

Note in Table VII a correlation between generated code lengths and the number of stored quantities in L/U. This suggests that minimizing the fill in of non ± 1 quantities may be the best criterion for optimizing code lengths.

When circuits become much larger, as in the example of Table V, accuracy requirements rather than code length can become the dominating factor in selecting ordering parameters. The nodal and state-variable column constraints on the pivoting order prevent the implementation of a true partial pivoting strategy ($\mu=0$, $v=1$) and thus contribute to a reduction in the accuracy of the answers. This consideration may be the cause of problems observed in NODAL OPTORD code for large values of the time step h and with S-V OPTORD for small values of h. Another accuracy consideration is the fact that the number of Newton iterations per time step increases with problem size for large networks. The evidence suggests that the Newton iteration is not only solving the nonlinear difference equations in (7), but also serving to refine the solutions of the linear equations. This consideration has thus far prevented successful runs on very large networks with parameters set for other than partial pivoting. This aspect of the accuracy problem may also prevent extensive exploitation of possible sparsity of the right-hand-side vector. One might think that for linear time invariant rows of the tableau vector f (e.g., the first 11 rows in Fig. (4), the right-hand side would stay zero if started at zero. This is not the case. Instead it has been found necessary to evaluate the full right-hand side at every Newton iteration to obtain rapid convergence.

The preceding results and discussion, combined with the error discussion of Section VI, suggest that nodal or state-variable constraints on the pivoting order offer no appreciable storage or execution speed advantages and create the possibility of accuracy problems.

VIII. CONCLUSIONS

The advantages claimed for the tableau approach are 1) generality, 2) simplicity, 3) speed, and 4) problem size.

Generality is claimed because any system of differential and algebraic equations can be handled. In particular, for electrical networks no topological restrictions are necessary and there are no restrictions on the type of elements allowed. This allows, for example, the inclusion of the objective function in the tableau by introducing the objector element.

Simplicity is achieved with the tableau approach by basing the main computational task of the program on Gaussian elimination. All other computations have a simple direct relation to the input data for the program. Although the programs OPTORD and 1-2-3 GNSO are not simple, they are general in nature and relate to the more general problem of solving $f(x, \alpha)=0$ by Newton's method for different values α where $\partial f/\partial x$ is sparse and has elements with different variability types. It therefore seems wise to develop such programs to a high degree and base more special programs (such as network design programs) around these developments.

The claim of speed is based on the fact that operations count in the basic Newton loop has been minimized so that only necessary operations are executed. This has been done by using variability type and near-optimal ordering algorithms in OPTORD. In addition, the number of time steps required for integration has been made small by using implicit methods and Gear's error control algorithms.

Large problems can be handled under this program because storage is saved by not storing 0's and 1's. In addition, OPTORD saves space by causing little fill in. Since the SOLVE code contains no loops, it can be stored in auxiliary storage and executed as it is brought into core thus requiring no storage. (However, for maximum speed, it should remain in core.)

As was shown in Section VII, the S-V OPTORD and NODAL-OPTORD methods offer no advantages over the unconstrained OPTORD methods. In fact, possibly serious accuracy problems may occur. Note that the NODAL-OPTORD column constraints cause differencing errors in the back substitution which may be serious when arguments of nonlinearities are branch voltages. Thus if such a branch voltage is small in comparison with its terminal node voltages, it will be computed to fewer significant digits.

Note that some speed and storage advantages may accrue to the nodal or state-variable methods if the elimination leading to the appropriate final variable set is done symbolically at the input processor level. Since variables which enter linearly and are neither energy storage nor involved in output can be eliminated, the amount of B-SOLVE code can be reduced. However, the excess B-SOLVE code (which corresponds roughly to solving $Ax=(b_1-By)$ in (36)) could, with some programming effort, be removed from the output of 1-2-3 GNSO. Thus this excess code is a deficiency of the implementation but not of the sparse tableau approach. Note that a priori symbolic reduction introduces topological matrix–vector operations and complicates the forming of the Jacobian matrix and right-hand side vector. (These additional operations correspond roughly to the forming of $D-AC^{-1}B$ and $b_2-CA^{-1}b_1$ in equation (37).) In contrast, the analogous operations for OPTORD analysis are executed by the highly efficient SOLVE code.

In NDP we have chosen to emphasize speed at some cost to problem size by keeping the SOLVE code in core. On a large machine such as 360/91 with 2×10^6 bytes of memory, this turns out to be a good choice. On this machine the capacity limitation of NDP seems to be about 2500 branches. A transient analysis of a network with 1200 branches took about 20 min. An entire optimization involving three parameters of a 460 branch network and requiring 36 function evaluations and 6 gradient evaluations took 1 h. Extrapolating linearly, we calculate that optimization for a 2500 branch network would require 7 h. However, this is low since larger networks usually require more time steps per transient integration and more Newton iterations per time step. Further, for a large circuit more than three parameters will normally be used which surely will increase the number of function and gradient evaluations required. Thus all indications are that 7 h is a very rough lower bound for optimization of a 2500 branch network. It would not be surprising if as many as 20–60 h were required. On the other hand, no serious attempt has been made yet to optimize the various parameter settings in NDP to cut down the number of time steps, Newton iterations, etc., so that this figure may be cut down.

If transient analyses for even larger networks are desired, the 1-2-3 GNSO SOLVE code could be buffered into core. This would save a factor of 2–3 in storage, but cause an unknown degradation in execution speed. An alternative approach would be to not generate the SOLVE code but to use a scheme similar to that suggested by Chang [9]. This would save a factor of 3 in storage, and perhaps a factor of 2 in speed would be lost. (However, with this method, variability type cannot be exploited.) Thus a 7500 branch network transient analysis would be possible and a rough lower bound on execution time would be about 4 h on the 360/91.

Acknowledgment

The authors would like to thank T. E. Grapes, who was responsible for programming the compiler which converts GNSO to 1-2-3 GNSO, and D. A. Calahan, for extensive discussions and the interchange of ideas.

References

[1] F. G. Gustavson, W. Liniger, and R. A. Willoughby, "Symbolic generation of an optimal Crout algorithm for sparse systems of linear equations," *J. Ass. Comput. Mach.*, vol. 17, pp. 87–109, January 1970.

[2] R. K. Brayton, F. G. Gustavson, and R. A. Willoughby, "Some results on sparse matrices," to be published in *Math. Comput.*

[3] W. F. Tinney and J. W. Walker, "Direct solutions of sparse network equations by optimally ordered triangular factorization," *Proc. IEEE*, vol. 55, pp. 1801–1809, November 1967.

[4] *Proc. Sparse Matrix Symp.*, IBM Rep. RA-1, March 1969.

[5] G. D. Hachtel and R. A. Rohrer, "Techniques for the optimal design and synthesis of switching circuits," *Proc. IEEE*, vol. 55, pp. 1864–1877, November 1967.

[6] F. H. Branin, Jr., "Computer methods of network analysis," *Proc. IEEE*, vol. 55, pp. 1787–1801, November 1967.

[7] C. W. Gear, "The automatic integration of stiff O.D.E.'s," *Proc. 1968 IFIPS Congr.*, pp. A81–A85.

[8] ——, "The control of parameters in the automatic integration of ordinary differential equations," *J. Ass. Comput. Mach.*; Department of Computer Science, University of Illinois, Urbana, Internal Rep., p. 14, May 1968.

[9] S. Sedore *et al.*, "SCEPTRE—an automated digital computer program for determining the response of electronic systems to transient nuclear radiation," vol. 11, IBM Space Guidance Center, Oswego, N. Y., Tech. Rep. AFWL-TR-66-126.

[10] E. S. Kuh and R. A. Rohrer, "The state-variable approach to network analysis, *Proc. IEEE*, vol. 53, pp. 672–686, July 1965.

[11] R. Fletcher and M. J. D. Powell, "A rapidly convergent descent method for minimization," *Comput. J.*, vol. 6, pp. 163–168, 1965.

[12] A. Nordsieck, "On numerical integration of ordinary differential equations," *Math. Comput.* vol. 16, pp. 22–49, 1962.

[13] D. A. Calahan, "Optimization of switching circuits," presented at the 2nd Biennial Cornell Conf. Engineering Applications of Electronic Phenomena, 1969.

[14] H. M. Markowitz, "The elimination form of the inverse and its application to linear programming," *Management Sci.*, vol. 3, pp. 255–269, April 1957.

[15] G. B. Dantzig, R. P. Harvey, R. D. McKnight, and S. S. Smith, "Sparse matrix techniques in two mathematical programming codes," *Proc. Sparse Matrix Symp.*, IBM Rep. RA-1, pp. 85–99, March 1969.

[16] P. D. Crout, "A short method for evaluating determinants and solving systems of linear equations with real or complex coefficients," *AIEE Trans.* (Supplement), vol. 60, pp. 1235–1240, December 1941.

[17] J. H. Wilkinson, *The Algebraic Eigenvalue Problem.* New York: Oxford University Press, 1965.

[18] ——, *Rounding Errors in Algebraic Processes.* Englewood Cliffs, N. J.: Prentice-Hall, 1963, pp. 121–126.

[19] S. Seshu, "Topological considerations in the design of driving point functions," *Proc.* AMS, pp. 1068–1073, October 1965.

[20] A. Chang, "Applications of sparse matrix methods in electric power system analysis," *Proc. Sparse Matrix Symp.*, IBM Rep. RA-1, pp. 113–121, March 1969.

A Sparse Matrix Method for Analysis of Piecewise-Linear Resistive Networks

TOSHIO FUJISAWA, ERNEST S. KUH, AND TATSUO OHTSUKI

***Abstract*—Nonlinear resistive networks, which can be characterized by the equation $f(x)=y$, where $f(\cdot)$ is a continuous piecewise-linear mapping of R^n into itself, are discussed. x is a point in R^n and represents a set of chosen network variables and y is an arbitrary point in R^n and represents the input to the network. New theorems on the existence of solutions together with a convergent method for obtaining at least one of the solutions are given. Also dealt with is an efficient computational algorithm which is especially suited for analysis of large piecewise-linear networks. The effectiveness of the method in terms of the amount of computation and data handling and storage is demonstrated.**

I. Introduction

THE PROBLEM of analysis and design of large-scale nonlinear resistive networks is becoming of widespread interest. One encounters such a problem not only in electric or electronic circuits, but also in hydraulic networks, structural analysis, numerical integration, and economic modeling [1]–[5]. Because of the size of equations usually involved, it is crucial to devise an efficient computational method for finding the solutions. Recently, a number of theoretical results have been found concerning the question of the existence and uniqueness of solutions [6]–[11]. It turns out that these results not only provide a basic understanding on inherent properties of networks which possess a unique solution, but also are the paramount importance in obtaining convergent computational algorithms.

Manuscript received February 8, 1972; revised July 14, 1972. This work was supported by the National Science Foundation under Grant GK-10656X1, by the Office of Naval Research under Contract N000-14-A-0200-1035, and by the Joint Services Electronics Program under Contract F44620-71-C0087.

T. Fujisawa is with the School of Engineering Sciences, Osaka University, Toyonaka, Osaka 560, Japan.

E. S. Kuh is with the Department of Electrical Engineering and Computer Sciences and the Electronics Research Laboratory, University of California, Berkeley, Calif. 94720.

T. Ohtsuki is with the Central Research Laboratory, Nippon Electric Co., Shimonumabe, Kawasaki 211, Japan.

In this paper we shall confine our study to piecewise-linear resistive networks. The paper can be divided into two parts. In the first part (Sections II and III) we extend the theory of piecewise-linear resistive networks as developed by Fujisawa and Kuh in [10]. New theorems on the existence of solutions (not necessarily unique) are presented and a convergent method of finding at least one of the solutions is given (the proof of the theorems is given in Appendix I). With these results, the applicability of the piecewise-linear method is enlarged to include not only all networks with unique solution, but also a broad class of networks which possess multiple solutions.

The second part of this paper (Sections IV and V) is concerned with an efficient computational algorithm which is especially suited for analysis of large piecewise-linear networks. The problem is essentially solving linear equations $A\beta=\alpha$ for successive A's. However,

Reprinted from *IEEE Trans. Circuit Theory*, vol. CT-19, pp. 571–584, Nov. 1972.

there exists a special relation between two successive A's which is imposed by a key property of continuous piecewise-linear functions developed in Section III.

The solution method which is based on the idea developed by Bennett [12] is presented in Section IV. The method depends on the conventional Gaussian elimination but takes advantage of the key property of continuous piecewise-linear functions; consequently, it is more efficient than either of the two existing methods, namely, 1) using Gaussian elimination at each step and 2) finding the inverse of the modified matrix at each step. The proof of the new formulas together with a quick review of the Gaussian elimination method is given in Appendix II. In Section V, we demonstrate the effectiveness of the method when the matrix is sparse. The data structure for the computer program is discussed in some length. An example which illustrates the advantage of the method is given.

II. Existence of Solutions

Piecewise-linear resistive networks are assumed to be characterized by equations of the form

$$f(x) = y \tag{1}$$

where f is a continuous piecewise-linear function which maps R^n into itself. x is a point in R^n and represents a set of chosen network variables and y is an arbitrary point in R^n and represents the input to the network.

For a continuous piecewise-linear function f, the whole space R^n is divided into a *finite* number of polyhedral regions by a finite number of boundary hyperplanes. A typical boundary hyperplane can be characterized by the following equation:

$$r^T x = \text{constant} \tag{2}$$

where r is the normal vector of the hyperplane. In each region, say region m (denoted by R_m), the piecewise-linear function f is represented by linear equations:

$$A^{(m)}x + w^{(m)} = y, \qquad m = 1, 2, \cdots, l \tag{3}$$

where $A^{(m)}$ is a constant $n \times n$ matrix (called Jacobian matrix for convenience) and $w^{(m)}$ is a constant n-vector, both defined in region R_m. It is assumed that there are altogether l-regions in the x-space.

It is crucial to note that continuity of the function imposes a constraint on the Jacobian matrices of any two neighboring regions. Suppose that two neighboring regions, say R_1 and R_2, have a common boundary hyperplane H as shown in Fig. 1. For f to be continuous on H it is necessary and sufficient that on the boundary where

$$r^T \Delta x = 0 \tag{4}$$

$$A^{(2)}\Delta x = A^{(1)}\Delta x. \tag{5}$$

Furthermore, it is easy to see that the above constraint is satisfied if and only if

$$A^{(2)} - A^{(1)} = cr^T \tag{6}$$

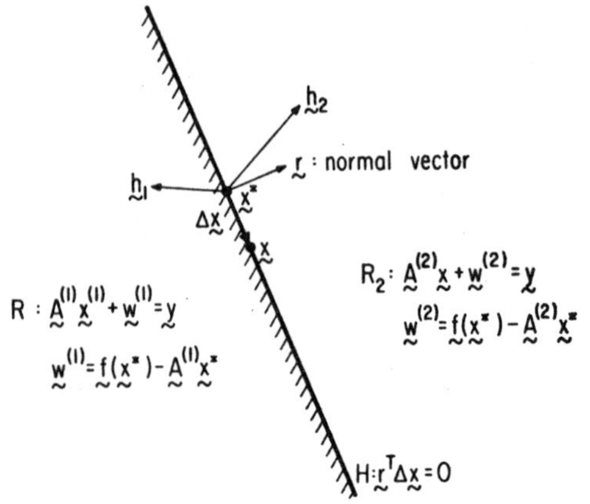

Fig. 1. Boundary hyperplane between two neighboring regions. Tilde under letters corresponds to boldface in text.

where c is an arbitrary constant n-vector. The above relation indicates that the difference of the Jacobian matrices of two neighboring regions is a dyad of a specific form. This turns out to be a key property of continuous piecewise-linear functions, which plays a major role in the theory and computation of piecewise-linear networks.

The key property in (6), together with the well-known determinant formula

$$\det(1 + PQ) = \det(1 + QP) \tag{7}$$

gives the following useful relation between the determinants of Jacobian matrices of two neighboring regions:

$$\det A^{(2)} = \det(1 + cr^T A^{(1)^{-1}}) \det A^{(1)}$$
$$= K \det A^{(1)} \tag{8}$$

where

$$K = 1 + r^T A^{(1)^{-1}} c \tag{9}$$

is a scalar constant.

In the following we shall first review briefly some important concepts of global homeomorphism of continuous piecewise-linear mappings as developed in [10]. Special attention will be given to properties of the function at a boundary hyperplane. Two new theorems on existence of solutions will be presented.

A continuous function f is said to be a homeomorphism of R^n onto itself if and only if (1) has a unique solution for all y. The necessary and sufficient conditions for f to be a homeomorphism as stated by Holzmann and Liu [7] are 1) f is a local homeomorphism for all x and 2) $\lim \|f(x)\| = \infty$ as $\|x\| \to \infty$. The local homeomorphism at x is defined as follows. There exists a neighborhood of x which is mapped homeomorphically onto a neighborhood of $f(x)$. For continuous piecewise-linear functions it is not difficult to see that condition 2) is automatically satisfied if, in all regions,

$$\det A^{(m)} \neq 0, \qquad m = 1, 2, \cdots, l. \tag{10}$$

With this assumption, we only need to study condition 1). In this respect, Fujisawa and Kuh gave the following theorem [10].

Theorem 1

Let f be a continuous piecewise-linear mapping of R^n into itself. A necessary and sufficient condition for f to be a homeomorphism of R^n onto itself is that, for any unit vector α and for any $x \in R^n$, there exists one and only one nonzero vector $\beta(\alpha, x)$ such that

$$f(x + \nu\beta) = f(x) + \nu\alpha \tag{11}$$

for all sufficiently small positive ν.

Let us discuss the implication of this theorem. Obviously, if x is an inner point of a region, say R_m, then $\det A^{(m)} \neq 0$ guarantees local homeomorphism and the condition in (11) of the theorem is satisfied with $\beta = A^{(m)-1}\alpha$.

Next, take the case in which x lies on one and only one boundary hyperplane (called simple boundary). With reference to Fig. 1, consider a point x^* on the simple boundary hyperplane H of the two neighboring regions R_1 and R_2. For a given unit vector α, according to Theorem 1, homeomorphism is equivalent to the existence of a unique nonzero vector β which may lie in R_1, R_2, or on the boundary. As shown in Fig. 1, we define

$$h_1 \triangleq A^{(1)^{-1}}\alpha \tag{12a}$$

in R_1 and

$$h_2 \triangleq A^{(2)^{-1}}\alpha \tag{12b}$$

in R_2. By means of the key property in (6), we can derive a relation between the two vectors:

$$\begin{aligned} h_2 &= A^{(2)^{-1}}\alpha = (A^{(1)} + cr^T)^{-1}\alpha \\ &= [A^{(1)^{-1}} - A^{(1)^{-1}}c(r^T A^{(1)^{-1}}c + 1)^{-1} r^T A^{(1)^{-1}}]\alpha \\ &= h_1 - \frac{1}{K} A^{(1)^{-1}} c r^T h_1. \end{aligned} \tag{13}$$

In the above, we have used the well-known Householder's formula

$$(F+GLM)^{-1} = F^{-1} - F^{-1}G(MF^{-1}G + L^{-1})^{-1}MF^{-1} \tag{14}$$

and the identity in (9). From (13), we obtain

$$r^T h_2 = \frac{1}{K} r^T h_1 \tag{15}$$

where from (8), we recall that K gives the ratio of the two determinants of the neighboring Jacobian matrices. In the following, we shall assume that $K > 0$, i.e., the determinants of the Jacobian matrices of two neighboring regions have the same sign. For a given α, we compute h_1 according to (12a); clearly, there exist three possibilities, namely, 1) $r^T h_1 < 0$, 2) $r^T h_1 > 0$, and 3) $r^T h_1 = 0$.

Under case 1), $r^T h_1 < 0$; from (15), $r^T h_2 < 0$, which implies that there exists no such h_2 in R_2. The unique vector β is simply h_1 in R_1. Under case 2), $r^T h_1 > 0$; from (15), $r^T h_2 > 0$; thus the unique β is simply h_2 in R_2. Under case 3), both h_1 and h_2 lie on the boundary; (15) requires that h_1 and h_2 must have the same direction and the continuity of the function requires that $h_1 = h_2$. Thus the unique β is the vector $h_1 = h_2$ on the boundary.

It is easy to see that if $K < 0$, then Theorem 1 cannot be satisfied. Therefore, we have demonstrated that if a point x lies on a simple boundary, the condition in (11) of Theorem 1 is satisfied if and only if $K > 0$, that is, the determinants of Jacobian matrices of the two neighboring regions have the same sign.

The discussion above might suggest that a necessary and sufficient condition for f to be a homeomorphism of R^n onto itself is the following: $\det A^{(1)}, \det A^{(2)}, \cdots, \det A^{(l)}$ are all positive or all negative. However, a counterexample given in [10] indicates that this is not the case. The crux lies in situations that x is on more than one boundary. Under such a situation, x may not satisfy the condition of Theorem 1 even if all determinants of Jacobian matrices to which x belongs have the same sign. If the function f is continuously differentiable, let the Jacobian matrix be J; then $\det J(x) \neq 0$ implies local homeomorphism at x. However, as far as we know, there exists no corresponding statement in the continuous piecewise-linear case. In [10], Fujisawa and Kuh gave the following sufficient condition for homeomorphism, which, incidentally, is of considerable interest from a computation point of view, as will be indicated in Section IV.

Theorem 2

Let f be a continuous piecewise-linear mapping of R^n into itself and let $A_k^{(m)}$ denote the leading minor of order k of $A^{(m)}$, that is, the matrix composed of the first k-rows and k-columns of $A^{(m)}$. The mapping f is a homeomorphism of R^n onto itself, if, for each $k = 1, 2, \cdots, n$, the determinants of the $k \times k$ matrices

$$A_k^{(1)}, A_k^{(2)}, \cdots, A_k^{(l)}$$

do not vanish and have the same sign.

Clearly, the above condition is satisfied by matrices which are positive or negative definite and matrices of class P[13] (matrices with all positive principal minors). The application of interchange of rows and columns can be made to further extend the applicability of the above theorem. It should be emphasized nevertheless that the condition in the theorem is only sufficient but not necessary.

On the other hand, if we backtrack the determinant sign conditions discussed earlier, we can state the following two theorems concerning not homeomorphism, but existence of solutions. The proof of the theorems is given in Appendix I and their application is given in Section III.

Theorem 3

Let f be a continuous piecewise-linear mapping of R^n into itself. Then there exists at least one solution x to an arbitrary y if $\det A^{(m)}$, $m = 1, 2, \cdots, l$ are all positive or all negative.

In Section III, the above condition will be referred to frequently; therefore, for convenience, the condition is henceforth referred to as "Condition $\boldsymbol{A}$."

Theorem 4

Under the same assumptions as in Theorem 3 (i.e., Condition $\boldsymbol{A}$), for any unit vector $\boldsymbol{\alpha}$ and any point $\boldsymbol{x}$, there exists at least one nonzero vector $\boldsymbol{\beta}$ such that $\boldsymbol{f}(\boldsymbol{x}+\nu\boldsymbol{\beta})=\boldsymbol{f}(\boldsymbol{x})+\nu\boldsymbol{\alpha}$ for all sufficiently small positive ν.

III. Solution Method

In 1965 Katzenelson considered piecewise-linear resistive networks of a special class in which all resistors are uncoupled and are of strictly monotonically increasing type and he developed a convergent computational algorithm [14]. The piecewise-linear approach was further extended to include more general cases by Chua [15] and Ohtsuki and Yoshida [16]. Recently, the Katzenelson algorithm has been used in computing the solution of resistive networks of much broader class [8], [10]. In particular, Fujisawa and Kuh [10] have shown that the algorithm can be applied to (1) and it always converges to a solution as long as the equation has a unique solution for an arbitrary input $\boldsymbol{y}$. In other words, homeomorphism implies the convergence of the algorithm. Then one might ask the following question. Does the algorithm always converge to a solution under the assumption that there exists at least one solution to any input? The answer is no, as a later example will illustrate. Instead, we can state the following theorem.

Theorem 5

The Katzenelson algorithm converges within l-steps to a solution if Condition $\boldsymbol{A}$ is satisfied, i.e.,

$$\det \boldsymbol{A}^{(m)}, \qquad m = 1, \cdots, l$$

are all positive or all negative.

In the following paragraphs a proof of the above theorem is given together with a review of the essence of the Katzenelson algorithm.

The problem is to find a solution of (1) for a given input $\boldsymbol{y}^*$. To begin, choose an arbitrary point $\boldsymbol{x}^{(1)}$, which is an inner point of a region, say R_1. In R_1 the equation which characterizes the network is

$$\boldsymbol{A}^{(1)}\boldsymbol{x} + \boldsymbol{w}^{(1)} = \boldsymbol{y}. \tag{16}$$

Substituting $\boldsymbol{x}^{(1)}$ into the above, we obtain

$$\boldsymbol{A}^{(1)}\boldsymbol{x}^{(1)} + \boldsymbol{w}^{(1)} = \boldsymbol{y}^{(1)} \tag{17}$$

which usually differs from the given $\boldsymbol{y}^*$. Denote the line segment joining $\boldsymbol{y}^{(1)}$ and $\boldsymbol{y}^*$ in the y-space by L_y. The problem is then reduced to one of determining a continuous curve that is really a polygonal curve, starting with $\boldsymbol{x}^{(1)}$ in the x-space such that $\boldsymbol{f}$ gives a one-to-one correspondence between the set of points of the curve in the x-space and the set L_y in the y-space. Then, the other end point of the curve is a solution.

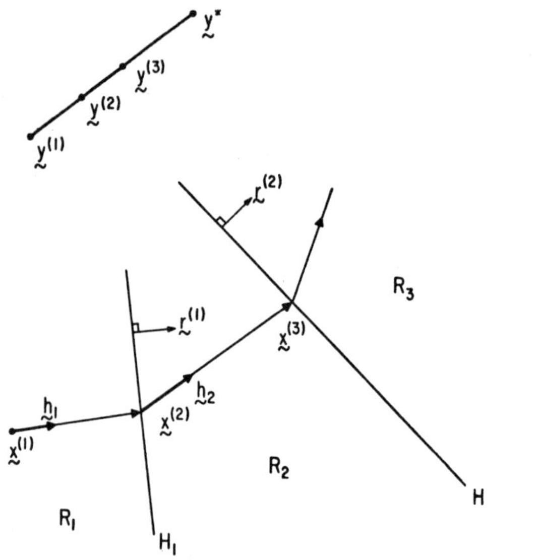

Fig. 2. Construction of the solution curve. Tilde under letters corresponds to boldface in text.

The portion of the solution curve which lies in R_1 is determined by

$$\boldsymbol{x}^{(1)}(\lambda) = \boldsymbol{x}^{(1)} + \lambda \boldsymbol{A}^{(1)-1}(\boldsymbol{y}^* - \boldsymbol{y}^{(1)}) \tag{18}$$

where $\lambda \geqq 0$ is a parameter. If $\boldsymbol{x}^{(1)}(1)$ happens to be in R_1, then $\boldsymbol{x}^{(1)}(1)$ is the desired solution. The line segment joining $\boldsymbol{x}^{(1)}$ and $\boldsymbol{x}^{(1)}(1)$ is the solution curve and the algorithm terminates here. If otherwise, the value of λ has to be determined so that $\boldsymbol{x}^{(1)}(\lambda)$ lies on the boundary of R_1. Denote such value of λ by $\lambda^{(1)}$ and define $\boldsymbol{x}^{(2)}=\boldsymbol{x}^{(1)}(\lambda^{(1)})$ and $\boldsymbol{y}^{(2)}=\boldsymbol{f}(\boldsymbol{x}^{(2)})$. The line segment joining $\boldsymbol{x}^{(1)}$ and $\boldsymbol{x}^{(2)}$ is thus the first portion of the desired solution curve. The next step is to extend the solution curve beyond $\boldsymbol{x}^{(2)}$.

There are two cases to be considered separately: 1) $\boldsymbol{x}^{(2)}$ lies on a simple boundary, and 2) $\boldsymbol{x}^{(2)}$ lies on more than one boundary hyperplane, or, stated in another way, $\boldsymbol{x}^{(2)}$ is at a corner.

In case 1), suppose that $\boldsymbol{x}^{(2)}$ lies on the boundary hyperplane H_1, which is common to regions R_1 and R_2, as shown in Fig. 2. Define $\boldsymbol{h}_1 \triangleq \boldsymbol{A}^{(1)-1}(\boldsymbol{y}^*-\boldsymbol{y}^{(1)})$ and $\boldsymbol{h}_2=\boldsymbol{A}^{(2)-1}(\boldsymbol{y}^*-\boldsymbol{y}^{(1)})=c\boldsymbol{A}^{(2)-1}(\boldsymbol{y}^*-\boldsymbol{y}^{(2)})$, where c is a positive constant equal to $\|\boldsymbol{y}^*-\boldsymbol{y}^{(1)}\|/\|\boldsymbol{y}^*-\boldsymbol{y}^{(2)}\|$. Clearly $\boldsymbol{r}^{(1)T}\boldsymbol{h}_1>0$, and hence from (11), $\boldsymbol{r}^{(1)T}\boldsymbol{h}_2>0$ if $K>0$. However, Condition $\boldsymbol{A}$ implies $K>0$ because of (8). Therefore, if Condition $\boldsymbol{A}$ is satisfied, then

$$\boldsymbol{x}^{(2)}(\lambda) = \boldsymbol{x}^{(2)} + \lambda \boldsymbol{A}^{(2)^{-1}}(\boldsymbol{y}^* - \boldsymbol{y}^{(2)}), \qquad \lambda \geq 0 \tag{19}$$

is the portion of the solution curve in R_2.

In case 2), in which $\boldsymbol{x}^{(2)}$ lies on more than one boundary hyperplane, as shown in Fig. 3, we have to make use of Theorem 4 to prove that the solution curve can be extended into a certain region beyond $\boldsymbol{x}^{(2)}$. Now define $\boldsymbol{\alpha} \triangleq \boldsymbol{y}^*-\boldsymbol{y}^{(2)}$, then it follows from Theorem 4 that there exists at least one nonzero vector $\boldsymbol{\beta}$ such that

$$\boldsymbol{f}(\boldsymbol{x}^{(2)} + \nu\boldsymbol{\beta}) = \boldsymbol{y}^{(2)} + \nu\boldsymbol{\alpha} \tag{20}$$

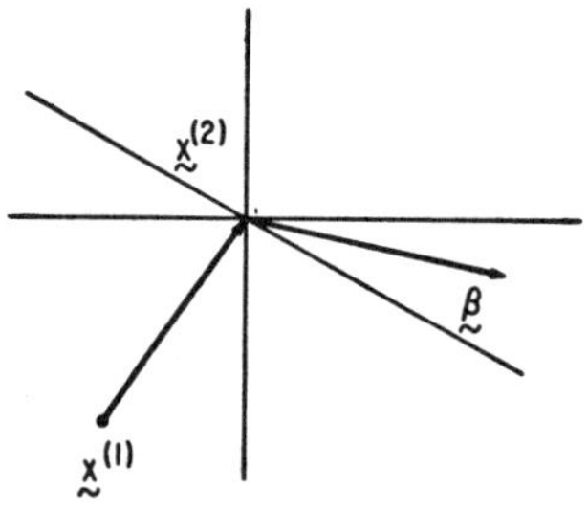

Fig. 3. Solution curve hitting a corner. Tilde under letters corresponds to boldface in text.

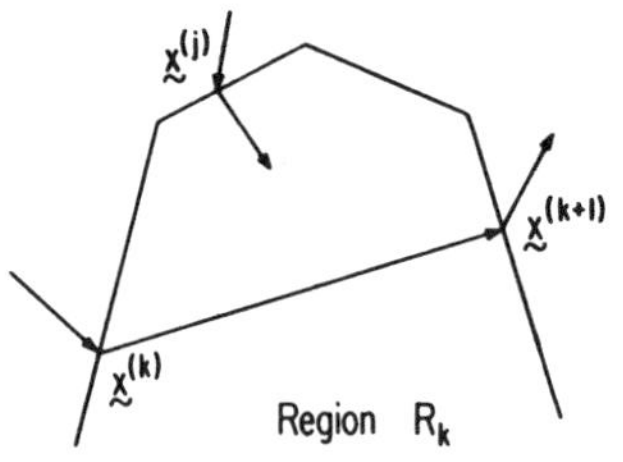

Fig. 4. Solution curve reenters region R_k. Tilde under letters corresponds to boldface in text.

for all sufficiently small positive ν. The underlying assumption is of course that the Condition $\boldsymbol{A}$ is satisfied. One can determine the region in which $\mathbf{x}^{(2)}+\nu\boldsymbol{\beta}$ is contained for all sufficiently small positive ν. The region is denoted by R_2. Then $\boldsymbol{\beta}=\boldsymbol{A}^{(2)-1}(\boldsymbol{y}^*-\boldsymbol{y}^{(2)})$ and, as before, (19) gives an extension of the solution curve.

As mentioned earlier, a point of the intersection of more than one boundary hyperplanes is called a corner. The observations made above then indicate that if the Condition $\boldsymbol{A}$ is satisfied, the solution curve is always extended into another region when it hits a boundary regardless of whether the curve hits a corner or not. Since there exists only a finite number of regions in the $\mathbf{x}$-space, therefore, the proof of Theorem 5 can be completed if it is shown that the solution curve never comes back to a region which the curve once traversed. This is shown as follows. Suppose that the solution curve enters region R_k at $\mathbf{x}^{(k)}$ and leaves it at $\mathbf{x}^{(k+1)}$ and that the curve later enters the same region at $\mathbf{x}^{(j)}$ as is illustrated in Fig. 4. Clearly these three points $\mathbf{x}^{(k)}$, $\mathbf{x}^{(k+1)}$, and $\mathbf{x}^{(j)}$ do not lie on a line. Therefore, the two vectors $\mathbf{x}^{(k+1)}-\mathbf{x}^{(k)}$ and $\mathbf{x}^{(j)}-\mathbf{x}^{(k)}$ are linearly independent, whereas their images under linear mapping $\boldsymbol{A}^{(k)}$ are both constant multipliers of vector $\boldsymbol{y}^*-\boldsymbol{y}^{(1)}$ and hence are linearly dependent. This contradicts the assumption that $\boldsymbol{A}^{(k)}$ is nonsingular. This completes the proof of Theorem 5.

We now present a simple example in which Condition $\boldsymbol{A}$ is not satisfied, yet the equation has at least a solution for any input. It is seen that the Katzenelson algorithm does not work.

Example: Consider a scalar function $f(x)$ of a single variable x as shown in Fig. 5. It is easily seen that there exists at least one solution of the equation $f(x)=y$ for any given y. There are three regions, namely,

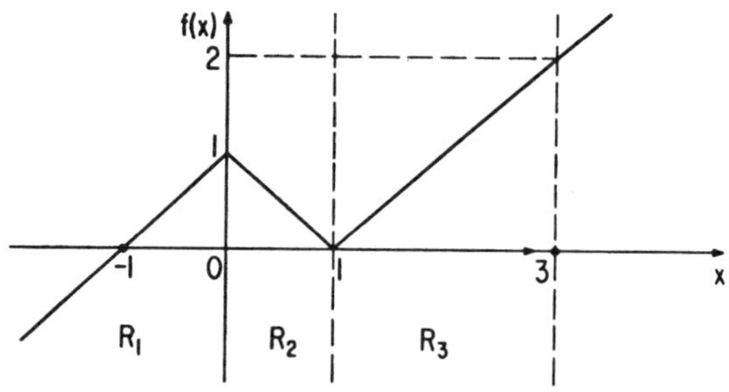

Fig. 5. Continuous piecewise-linear function which has at least one solution for all inputs.

$R_1=(-\infty,\ 0]$, $R_2=[0,\ 1]$, and $R_3=[1,\ \infty)$. In these regions $A^{(1)}=1$, $A^{(2)}=-1$, and $A^{(3)}=1$ are the slopes of the respective segments. Thus the function does not satisfy the Condition $\boldsymbol{A}$. Suppose that the input is $y^*=2$ and that we start at $x^{(1)}=-1$. Then the solution curve starting with $x^{(1)}$ on the real line hits the boundary $x=0$. The curve cannot be extended into region R_2 because the value of $f(x)$ increases for $-1\leqq x\leqq 0$ and decreases for $x\geqq 0$. This indicates that the existence of solution does not generally imply the convergence of the Katzenelson algorithm.

There remain two computational problems to be investigated. The first one is the corner problem. When a solution curve hits a corner, there exists, theoretically, at least one vector $\boldsymbol{\beta}$ as given in (20). The problem from a computational point of view is how to find such a $\boldsymbol{\beta}$. Fujisawa and Kuh gave a systematic perturbation method which forces the solution curve to cross a simple boundary hyperplane one at a time [10]. This method always works but is rather tedious. Probably the best way in practice is to make a small move each time a corner is reached so that the modified solution curve does not hit any corner.

The second computational problem is how to best compute the direction vectors $\boldsymbol{A}^{(1)-1}(\boldsymbol{y}^*-\boldsymbol{y}^{(1)})$, $\boldsymbol{A}^{(2)-1}(\boldsymbol{y}^*-\boldsymbol{y}^{(2)})$, $\cdots$. The straightforward way of computing inverses $\boldsymbol{A}^{(1)^{-1}}$, $\boldsymbol{A}^{(2)^{-1}}$, $\cdots$ each time without using any of the previous information is obviously inefficient. Katzenelson [13] proposed to use the following relation based on the Householder's formula:

$$\boldsymbol{A}^{(2)^{-1}} = \boldsymbol{A}^{(1)^{-1}}\left[1 - \frac{1}{K}\,\boldsymbol{c}\boldsymbol{r}^T\boldsymbol{A}^{(1)^{-1}}\right] \tag{21}$$

where $\boldsymbol{A}^{(2)}$ and $\boldsymbol{A}^{(1)}$ are related to each other as in (6). Once $\boldsymbol{A}^{(1)}$ is inverted, the inverses of the succeeding Jacobians $\boldsymbol{A}^{(2)^{-1}}$, $\boldsymbol{A}^{(3)^{-1}}$, $\cdots$ are obtained successively by the repeated use of the formula (21). We can continue the process until the accumulation of numerical errors becomes significant. This method takes advantage of the key property in (6) of continuous piecewise-linear functions and is definitely more efficient than finding the inverse matrices from scratch at each step. However, it should be pointed out that in large-scale networks, the Jacobian matrix is usually sparse. As is well known, sparse matrix, if it is irreducible, becomes full after inversion [17]. Since sparse matrix technique is crucial in solving large systems of equations, any method which depends on finding the inverse of a large matrix is totally impractical. What is needed is the triangular factorization of a matrix based on Gaussian elimination. Thus a conceivable scheme is to use Gaussian elimination each time we need to solve equations $\boldsymbol{A}^{(m)}\boldsymbol{\beta}=\boldsymbol{\alpha}$. This scheme can be used to exploit the sparsity and thus cut down both the computing time and storage space drastically. However, the conventional Gaussian elimination scheme does not exploit the special relation of successive Jacobian matrices, a key property of continuous piecewise-linear functions.

According to the arguments made above we need to seek a method which not only takes advantage of the sparsity but also uses the key property of continuous piecewise-linear functions. In Section IV such a method is presented. It is demonstrated that the method is more efficient than the one using Householder's formula in the case of full matrix. It is also more efficient than Gaussian elimination in the case of sparse matrix.

Finally it should be remarked that under Condition **A** multiple solutions may exist. By applying the Katzenelson algorithm once, we can find only one of the solutions. The process must be repeated with new initial guesses in order to find the other solutions.

IV. Triangularization of Matrices Modified by Dyads

The problem in this section is to compute the triangular factorization of the matrix

$$\boldsymbol{A}^* = \boldsymbol{A} + \boldsymbol{c}\boldsymbol{r}^T \tag{22}$$

from that of $\boldsymbol{A}$. In (22)

$$\boldsymbol{A} = [a_{ij}] \tag{23}$$

is an $n \times n$ nonsingular matrix;

$$\boldsymbol{c} = [c_1, c_2, \cdots, c_n]^T \tag{24}$$

and

$$\boldsymbol{r} = [r_1, r_2, \cdots, r_n]^T \tag{25}$$

are n-dimensional column vectors. Thus $\boldsymbol{A}^*$ can be expressed as

$$\boldsymbol{A}^* = [a_{ij} + c_i r_j]. \tag{26}$$

The underlying assumption is that $\boldsymbol{A}$ has been factored by means of Gaussian elimination and the following form is obtained:

$$\boldsymbol{U} = \boldsymbol{T}_n \boldsymbol{T}_{n-1} \cdots \boldsymbol{T}_1 \boldsymbol{A} \tag{27}$$

where $\boldsymbol{U} = [u_{ij}]$ is an upper triangular matrix with unit elements on the diagonal and $\boldsymbol{T}_k$ is equal to the identity matrix except for the kth column which is of the form

$$[0, \cdots, 0, t_{kk}, t_{k+1,k}, \cdots, t_{nk}]^T. \tag{28}$$

The objective is to find the triangular factorization of $\boldsymbol{A}^*$, if it exists, and to obtain

$$\boldsymbol{U}^* = \boldsymbol{T}_n^* \boldsymbol{T}_{n-1}^* \cdots \boldsymbol{T}_1^* \boldsymbol{A}^* \tag{29}$$

based on the known factorization of (27). It should be mentioned that in 1965, Bennett considered a more general problem of obtaining the triangular factorization of $\boldsymbol{A}^* = \boldsymbol{A} + \boldsymbol{X}\boldsymbol{G}\boldsymbol{Y}^T$ from that of $\boldsymbol{A}$, where $\boldsymbol{X}$ and $\boldsymbol{Y}$ are $n \times m$ matrices and $m \ll n$ [12]. In his approach, Bennett assumes that $\boldsymbol{A}$ has been decomposed into the matrix triple product $\boldsymbol{LDU}$, where $\boldsymbol{L}$ is a lower triangular matrix with unit elements on the diagonal, $\boldsymbol{U}$ is an upper triangular matrix with unit elements on the diagonal, and $\boldsymbol{D}$ is a diagonal matrix. After $\boldsymbol{A}$ is modified to become $\boldsymbol{A}^*$, the corresponding factorization of $\boldsymbol{A}^*$ into $\boldsymbol{L}^*\boldsymbol{D}^*\boldsymbol{U}^*$ is derived. In our problem, we deal only with the case where $\boldsymbol{A}$ is modified by a dyad $\boldsymbol{c}\boldsymbol{r}^T$ and we prefer to use the Gaussian elimination form rather than $\boldsymbol{LDU}$ factorization. The method is presented in this section and its proof in Appendix II. In Section V, special considerations are given for the sparse matrix case.

Before stating the main theorem which contains all the formulas, we need to introduce three sets of auxiliary variables: two lower triangular matrices $\boldsymbol{C}$ and $\boldsymbol{R}$, and a diagonal matrix $\boldsymbol{D}$. First

$$\boldsymbol{C} = \begin{bmatrix} c_1^{(0)} & & & & & & & \\ c_2^{(0)} & c_2^{(1)} & & & & & & \\ c_3^{(0)} & c_3^{(1)} & c_3^{(2)} & & & & & \\ \vdots & \vdots & \vdots & \ddots & & & & \\ \vdots & \vdots & \vdots & & c_k^{(k-1)} & & & \\ \vdots & \vdots & \vdots & & \vdots & \ddots & & \\ c_n^{(0)} & c_n^{(1)} & c_n^{(2)} & \cdots & c_n^{(k-1)} & \cdots & c_n^{(n-1)} \end{bmatrix} \tag{30}$$

where

$$c_i^{(0)} \triangleq c_i, \qquad i = 1, 2, \cdots, n \tag{31}$$

$$c_i^{(k)} \triangleq c_i^{(k-1)} + t_{ik} c_k^{(k-1)}, \qquad \begin{cases} k = 1, \cdots, n-1 \\ i = k+1, \cdots, n. \end{cases} \tag{32}$$

Note that the first column of $\boldsymbol{C}$ is the given vector $\boldsymbol{c}$. The kth column is given by a vector of $n-k+1$ dimension, whose elements are specified by the elements of the $k-1$th column and t_{ik} of $\boldsymbol{T}_k$. Similarly, for the $\boldsymbol{R}$ matrix, we have

$$\boldsymbol{R} = \begin{bmatrix} r_1^{(0)} & & & & & & & \\ r_2^{(0)} & r_2^{(1)} & & & & & & \\ r_3^{(0)} & r_3^{(1)} & r_3^{(2)} & & & & & \\ \vdots & \vdots & \vdots & \ddots & & & & \\ \vdots & \vdots & \vdots & & r_k^{(k-1)} & & & \\ \vdots & \vdots & \vdots & & \vdots & \ddots & & \\ r_n^{(0)} & r_n^{(1)} & r_n^{(3)} & \cdots & r_n^{(k-1)} & \cdots & r_n^{(n-1)} \end{bmatrix} \tag{33}$$

where

$$r_j^{(0)} \triangleq r_j, \qquad j = 1, 2, \cdots, n \tag{34}$$

$$r_j^{(k)} \triangleq r_j^{(k-1)} - r_k^{(k-1)}u_{kj}, \qquad \begin{cases} k = 1, \cdots, n-1 \\ j = k+1, \cdots, n. \end{cases} \tag{35}$$

The diagonal matrix $\boldsymbol{D}$ is specified by its kth element

$$d_k \triangleq d_{k-1} + c_k^{(k-1)} r_k^{(k-1)} t_{kk}, \qquad k = 1, 2, \cdots, n \tag{36}$$

$$d_0 \triangleq 1.$$

Theorem 6

The modified matrix $\boldsymbol{A}^*$ can be triangularized if the diagonal matrix $\boldsymbol{D}$ is nonsingular, i.e.,

$$d_k \neq 0, \qquad k = 1, 2, \cdots, n \tag{37}$$

and the elements of $\boldsymbol{T}_n^* \cdots \boldsymbol{T}_k^* \cdots \boldsymbol{T}_1^*$ and $\boldsymbol{U}^*$ are given by

$$t_{kk}^* = \frac{d_{k-1}}{d_k} t_{kk} = t_{kk} - \frac{c_k^{(k-1)} r_k^{(k-1)} t_{kk}^2}{d_k}, \qquad k = 1, 2, \cdots, n \tag{38}$$

$$t_{ik}^* = t_{ik} - \frac{r_k^{(k-1)} t_{kk}}{d_k} c_i^{(k)}, \qquad \begin{cases} i = k+1, \cdots, n \\ k = 1, \cdots, n-1 \end{cases} \tag{39}$$

$$u_{kj}^* = u_{kj} + \frac{c_k^{(k-1)} t_{kk}}{d_k} r_j^{(k)}, \qquad \begin{cases} j = k+1, \cdots, n \\ k = 1, \cdots, n-1. \end{cases} \tag{40}$$

A proof is given in Appendix II.

In this section, sparsity of $\boldsymbol{A}$ and $\boldsymbol{A}^*$ is not considered. We treat instead the general full matrix case. We shall first give the complete details of the computation procedure. Next we compare our method with that which is based on the Householder's formula and which is based on the conventional Gaussian elimination.

For efficient use of computer storage, nontrivial entries of $\boldsymbol{U}$ and the $\boldsymbol{T}_k$ are packed in a two-dimensional array as follows:

$$A(\,,\,) = [T \backslash U] = \begin{bmatrix} t_{11} & u_{12} & \cdots & u_{1n} \\ t_{21} & t_{22} & \cdots & u_{2n} \\ \vdots & \vdots & & \vdots \\ t_{n1} & t_{n2} & \cdots & t_{nn} \end{bmatrix} \tag{41}$$

where $A(\,,\,)$ is the name of the array in programming. Two one-dimensional arrays named $C(\,)$ and $R(\,)$ are used to initially store $\mathbf{c}$ and r, respectively. One more variable named $\boldsymbol{D}$ initially stores d_0. As the computational process proceeds, entries of $A(\,,\,)$, $C(\,)$, $R(\,)$, and D are replaced by new ones successively and $A(\,,\,)$ finally gives t_{ij}^* and u_{ij}^* for $\boldsymbol{A}^*$.

Initially, $A(\,,\,)$ contains (41), $C(\,)$ contains elements in the given vector $\mathbf{c}$, i.e., $[c_1^{(0)}, \cdots, c_n^{(0)}]$, $R(\,)$ contains similarly $[r_1^{(0)}, \cdots, r_n^{(0)}]$, and D contains d_0. In the beginning of the kth step, $A(\,,\,)$ contains the following:

$$\begin{bmatrix} t_{11}^* & \cdots & u_{1,k-1}^* & u_{1,k}^* & \cdots & u_{1,n}^* \\ \cdots & & & & & \cdots \\ t_{k-1,1}^* & \cdots & t_{k-1,k-1}^* & u_{k-1,k}^* & \cdots & u_{k-1,n}^* \\ t_{k,1}^* & \cdots & t_{k,k-1}^* & t_{kk} & \cdots & u_{k,n} \\ \vdots & & \vdots & \vdots & & \vdots \\ t_{n,1}^* & \cdots & t_{n,n-1}^* & t_{n,k} & \cdots & t_{nn} \end{bmatrix} \tag{42}$$

$C(\,)$ contains $[c_1^{(0)}, c_2^{(1)}, \cdots, c_k^{(k-1)}, \cdots, c_n^{(k-1)}]$, $R(\,)$ contains $[r_1^{(0)}, r_2^{(1)}, \cdots, r_k^{(k-1)}, \cdots, r_n^{(k-1)}]$, and D contains d_{k-1}. Note that $C(\,)$ and $R(\,)$ contain, respectively, the diagonal entries of matrices $\mathbf{C}$ and $\mathbf{R}$ for columns $1 \cdots k$ and the nonzero entries of the kth column.

At the kth stage, the following computation is done successively.

1) For $i = k+1, \cdots, n$

$$c_i^{(k)} = c_i^{(k-1)} + t_{ik} c_k^{(k-1)} \tag{43}$$

$$r_i^{(k)} = r_i^{(k-1)} - r_k^{(k-1)} u_{ki}. \tag{44}$$

These replace the last $(n-k)$-entries of $C(\,)$ and $R(\,)$, respectively.

2)

$$f_k = t_{kk} c_k^{(k-1)} \tag{45}$$

$$e_k = d_{k-1} \tag{46}$$

$$d_k = d_{k-1} + f_k r_k^{(k-1)} \tag{47}$$

$$g_k = t_{kk}/d_k \tag{48}$$

$$p_k = r_k^{(k-1)} g_k \tag{49}$$

$$q_k = c_k^{(k-1)} g_k. \tag{50}$$

The value of d_k replaces the content of D immediately after it has been computed. Of course, the condition $d_k \neq 0$ has to be checked. The variables f_k, d_k (or e_k), g_k, p_k, and q_k are considered as temporary variables whose values are saved only in the kth step and are used to cut down the arithmetic operation.

3) Finally t_{kk}^*, $t_{k+1,k}^*$, $\cdots$, $t_{n,k}^*$, and $u_{k,k+1}^*$, $\cdots$, $u_{k,n}^*$ are computed and used for updating the content of array $A(\,,\,)$:

$$t_{kk}^* = t_{kk} - p_k f_k. \tag{51a}$$

or

$$t_{kk}^* = g_k e_k. \tag{51b}$$

For $i = k+1, \cdots, n$

$$t_{ik}^* = t_{ik} - p_k c_i^{(k)} \tag{52}$$

$$u_{ki}^* = u_{ki} + q_k r_i^{(k)}. \tag{53}$$

This completes the computation of the kth step. Finally, at the nth step, it is seen that only t_{nn}^* needs to be calculated.

TABLE I
COMPARISON OF HOUSEHOLDER'S FORMULA AND THE NEW METHOD

	Initial inversion or triangularization			Each Modification		
	±	x	÷	±	x	÷
Householder's formula	n^3	n^3	n	$3n^2$	$3n^2$	1
New method	$\frac{1}{3}n^3$	$\frac{1}{3}n^3$	n	$2n^2$	$2n^2$	n

TABLE II
COMPARISON OF GAUSSIAN ELIMINATION AND THE NEW METHOD USING (51B)

	Gaussian		New Method	
n	±	x	±	x
6	55	85	72(66)*	88
7	91	133	98(91)*	117
8	140	196	128(120)*	150

Now comparison is made for the use of Householder's formula and that of our method. Let the number of additions and/or subtractions be N_A, the number of multiplications be N_M, and the number of divisions be N_D. For the new method, these numbers are estimated as follows:

$$N_A \sim 2n^2 \quad N_M \sim 2n^2 \quad \text{and} \quad N_D \sim n. \tag{54}$$

Table I indicates how much we can gain from the use of the new method compared with that of the Householder's formula in the full matrix case.

If one uses the conventional Gaussian elimination each time in the full matrix case, then the necessary arithmetic operation count amounts to the order of $n^3/3$, whereas in our method it is of order n^2. Table II gives exact operation counts for $n=6$, 7, and 8 and indicates that if $n \geq 7$, the new method is superior. The difference becomes very significant as n is large. However, it should be pointed out that this gain in computing time partly is offset by sacrificing numerical accuracy. If the conventional Gaussian elimination is applied to Jacobian matrices each time, a complete control of numerical error can be achieved, for instance, by using partial or complete pivoting [18]. On the other hand, in the use of the new method, we have to keep the same pivoting order as the initial Jacobian matrix used at the first step, which may not be optimal for the succeeding matrices. Furthermore, the roundoff error tends to accumulate as the process goes on. In this respect the observation in the following paragraphs is important.

In Appendix II it is shown that $1/t_{11}^* = (d_1/d_0)(1/t_{11})$, $1/t_{22}^* = (d_2/d_1)(1/t_{22})$, $\cdots$, and $1/t_{nn}^* = (d_n/d_{n-1})(1/t_{nn})$ are the pivoting elements for $\boldsymbol{A}^*$. It is well known [18] that if these pivot elements are small compared with other elements, then the numerical accuracy of triangularization cannot be good. Therefore, a plausible policy is to set a lower bound for these ratios $d_1/d_0, d_2/d_1, \cdots, d_n/d_{n-1}$. Once the bound is violated, we scratch the process and restart the process by applying Gaussian elimination to the Jacobian matrix at this stage where a new solution is obtained by using a new pivoting order.

In Appendix II we point out that the application of the Gaussian elimination is possible if the leading principal minors do not vanish [18]. If otherwise, one has to interchange rows and columns to obtain a triangularization. Under the assumption that the Condition **A** is satisfied, the above interchange, i.e., pivoting, is usually necessary. Furthermore, the change in pivoting orders may be needed from time to time as the process progresses. In this respect Theorem 2 is very interesting for it guarantees the applicability of Gaussian elimination without using interchange of rows and columns. As a matter of fact the leading minor of order k of $\boldsymbol{A}^*$ is the product of d_k and the corresponding minor of $\boldsymbol{A}$. Hence the condition of Theorem 2 in terms of leading principal minors guarantees the condition in (37) of Theorem 6.

V. SPARSE MATRIX METHOD

The method described in Section IV provides a very efficient means of exploiting sparsity of the Jacobian matrices in the analyses of piecewise-linear resistive networks.

Since the sparsity structure of the Jacobian matrices is fixed, in the entire course of computation, $\boldsymbol{A}$ and $\boldsymbol{A}^*$ have the same zero locations or

$$a_{ij} = 0 \Rightarrow c_i r_j = 0. \tag{55}$$

Besides, assuming a fixed pivoting order as well as no numerical cancellation, $[T\backslash U]$ and $[T^*\backslash U^*]$ also have the same sparsity structure, i.e.,

$$t_{ij}(\text{or } u_{ij}) = 0 \Leftrightarrow t_{ij}^*(\text{or } u_{ij}^*) = 0. \tag{56}$$

When a network contains both linear and piecewise-linear resistors as is usually the case in practice, we should further take advantage of this property. By rearranging $\mathbf{x}$ in (1) so that the variables of linear resistors are ordered first, the variable elements of $\boldsymbol{A}$ are pushed to the lower right-hand corner. More specifically,

$$c_i = r_i = 0, \qquad i = 1, 2, \cdots, \mu \tag{57}$$

for some integer $\mu: 1 \leq \mu \leq n$, which implies that

$$c_k^{(k-1)} = r_k^{(k-1)} = 0, \qquad k = 1, 2, \cdots, \mu. \tag{58}$$

Then, from (38)–(42), we have

$$\left.\begin{aligned} t_{kk}^* &= t_{kk} \\ t_{ik}^* &= t_{ik} \\ u_{kj}^* &= u_{kj} \end{aligned}\right\}, \qquad k = 1, 2, \cdots, \mu. \tag{59}$$

Once the first Jacobian matrix is triangularized, the sparsity structure of $[T\backslash U]$ is determined. Thereafter we only have to modify nonzero entries in the lower right-hand corner of $[T\backslash U]$, which is determined by the last $(n-\mu)$ rows and columns.

Before going into the discussion of data structure and computer program, we give a comparison of the arithmetic operation counts of the sparse Gaussian elimination and of the new method. The sparse Gaussian elimination is no different from the usual Gaussian elimination method, except that it skips all trivial calculations such as $0+k=k$ and $0\cdot k=0$ [19]. Let γ_k denote the number of nonzero entries of the kth column of T_k and let ρ_k denote the number of nonzero entries of the kth row of U. Then, the number of multiplications for sparse Gaussian elimination is essentially

$$N_1 \sim \sum_{k=1}^{n-1} (\gamma_k\rho_k + \gamma_k + \rho_k) \tag{60}$$

whereas that of the new method, by the most conservative estimate, is

$$N_2 \sim 2 \sum_{k=1}^{n-1} (\gamma_k + \rho_k). \tag{61}$$

From (60) and (61) we conclude that with rare exceptions the new method is much more efficient than sparse Gaussian scheme. One of the exceptions is a full tridiagonal matrix case, where both algorithms provide a process of order n. For the purpose of testing the sparse matrix version of the new method, a sparse matrix of order 57, which was posed in [19, fig. 1], was processed as an example. The exact number of arithmetic operations involved was counted by computer simulation for ten different pairs of **c** and **r**, which were chosen so that they were consistent with the sparseness structure of **A**. In Table III the results are summarized and compared with the operation count for sparse Gaussian elimination. Thus a considerable amount of saving on machine time may be expected by the use of our method.

In the computer program developed by the authors, $c_i^{(k)}$ and $r_j^{(k)}$ are stored in two one-dimensional arrays $C(\ )$ and $R(\ )$ as in the full matrix case. However, the following three one-dimensional arrays are used to compactly pack nontrivial entries of t_{ij}, u_{ij}, t_{ij}^* and u_{ij}^*.

1) $DT(\ )$: Initially $t_{11}, \cdots, t_{nn}$ are stored and finally replaced by $t_{11}^*, \cdots, t_{nn}^*$.

2) $T(\ )$: Initially off-diagonal nonzero entries t_{ij} are stored columnwise and finally replaced by t_{ij}^*.

TABLE III

Cases	nonzero c_i's	nonzero r_j's	±	×
	Initial Sparse Gaussian Elimination		665	1237
1	c_1	r_1	396	468
2	c_6	r_{46}	163	208
3	c_2	r_1	376	446
4	c_{43}	r_{18}	169	217
5	c_{57}	r_{57}	2	5
6	c_1, c_2	$r_1, r_2, r_8, r_9, r_{11}, r_{12}$ $r_{14}, r_{43}, r_{44}, r_{45}$	396	468
7	$c_{32}, c_{33}, c_{34}, c_{35}, c_{36}, c_{37}$ $c_{38}, c_{39}, c_{40}, c_{41}, c_{42}$	r_{42}	219	277
8	$c_{19}, c_{20}, c_{21}, c_{22}, c_{29}$	$r_{19}, r_{20}, r_{21}, r_{22}, r_{29}$	379	463
9	$c_{32}, c_{33}, c_{34}, c_{35}, c_{42}$	$r_{32}, r_{33}, r_{34}, r_{35}, r_{42}$	356	434
10	$c_{47}, c_{48}, c_{49}, c_{50}, c_{57}$	$r_{47}, r_{48}, r_{49}, r_{50}, r_{57}$	112	145

3) $U(\ )$: Initially off-diagonal nonzero entries u_{ij} are stored row-wise, and finally replaced by u_{ij}^*.

The arrays $T(\ )$ and $U(\ )$ pack all nonzero off-diagonal entries t_{ij}, u_{ij}, and therefore we need to devise a method to locate the address where t_{ij} or u_{ij} is stored. To be more specific, we have to locate the address l, i.e., the lth position in the array $T(\ )$, where t_{ik} is stored. In other words, $t_{ik}=T(l)$. The following four one-dimensional arrays provide a means necessary for address computation. In order to understand our scheme, the example in Table IV is given. Note that the nonzero elements of the matrix $[T\backslash U]$ are marked by crosses in Table IV(a). The four arrays are defined and illustrated as follows.

1) $\xi(\ )$: $\xi(k)$ gives the position in the array $T(\ )$, where the first nonzero entry of $t_{k+1,k}, \cdots, t_{nk}$ is found. In the example, for $k=1$, $\xi(1)=1$. For $k=2$–6, there is only one such nonzero entry for each column; thus $T(\ )$ stores the nonzero entry successively in order. For $k=7$, there are two nonzero entries: we have $\xi(7)=7$ and the next two entries in $T(\ )$ contain t_{87} and t_{97}. Therefore, for $k=8$, $\xi(8)=9$, and $T(9)$ contains t_{98}. If there are no nonzero entries in the kth column, $\xi(k)$ gives a special symbol to indicate the fact.[1]

[1] It should be pointed out that if the original matrix is irreducible, this never happens.

TABLE IV
EXAMPLE TO ILLUSTRATE THE ADDRESS COMPUTATION

$$[T\backslash U] = \begin{bmatrix} x & & & & x & & & & \\ & x & & & & & x & x & \\ & & x & & & & & x & \\ x & & & x & x & & x & x & \\ & & x & & x & & & x & \\ & x & & & & x & x & x & \\ & & & & & & x & & x \\ & & & & & x & x & x & x \\ & & & x & x & & x & x & x \end{bmatrix}$$

(a)

k	DT(k)
1	t_{11}
2	t_{22}
3	t_{33}
4	t_{44}
5	t_{55}
6	t_{66}
7	t_{77}
8	t_{88}
9	t_{99}

(b)

k	ξ(k)
1	1
2	2
3	3
4	4
5	5
6	6
7	7
8	9

ℓ	T(ℓ)	α(ℓ)
1	t_{41}	4
2	t_{62}	6
3	t_{53}	5
4	t_{94}	9
5	t_{95}	9
6	t_{86}	8
7	t_{87}	8
8	t_{97}	9
9	t_{98}	9

(c)

k	η(k)
1	1
2	2
3	4
4	5
5	8
6	9
7	11
8	12

ℓ	U(ℓ)	β(ℓ)
1	u_{15}	5
2	u_{27}	7
3	u_{28}	8
4	u_{38}	8
5	u_{45}	5
6	u_{47}	7
7	u_{48}	8
8	u_{58}	8
9	u_{67}	7
10	u_{68}	8
11	u_{79}	9
12	u_{89}	9

(d)

2) $\alpha(\ )$: $\alpha(l)=i$, if t_{ik} is stored in $T(l)$, that is, $T(l)=t_{ik}$. In Table IV(c), for example, $\alpha(1)=4$, which indicates that $T(1)=t_{41}$. Thus $\alpha(\ )$ gives the designation of the row of the nonzero entry of the lower triangular half of $[T\backslash U]$.

The remaining two arrays $\eta(\ )$ and $\beta(\ )$ are similarly defined, which give the address location of u_{ki}.

3) $\eta(\ )$: $\eta(k)$ gives the position in the array $U(\ )$, where the first nonzero entry of $u_{k,k+1}, \cdots, u_{kn}$ is found.

4) $\beta(\ )$: $\beta(l)=i$, if $U(l)=u_{ki}$.

The order in which nonzero entries of t_{ij}, u_{ij} are stored is consistent with the computation procedure (43)–(53). Therefore, the numerical computation can be performed in the same way as that in the full matrix case. However, trivial operations due to zero t_{ij}, u_{ij}, are automatically skipped in tracing down $T(\ )$ and $U(\ )$. Other trivial computations caused by zero entries of $c_i^{(k)}$, $r_j^{(k)}$ are also skipped simply by testing whether or not $c_i^{(k)}=0$ and $r_j^{(k)}=0$. Unlike sparse Gaussian elimination, address computation in the method developed here is very simple.

The storage requirement for the computer program under consideration can be estimated by the lengths of various one-dimensional arrays. Let m be the number of nonzero elements of t_{ij} and u_{ij} and let n be the order of $\boldsymbol{A}$. Then the storage requirement is estimated to be

$$(m+2n) \text{ floating-point numbers} + (m+2n) \text{ integers}. \quad (62)$$

One of the main advantages of the new method is the surprisingly small storage required to represent the sparseness structure. Note that if we are to use Gaussian elimination for $\boldsymbol{A}^*$, it is necessary to have nontrivial searching procedures for computing addresses in order to write data in or to read data out. If one wishes to avoid the search, some other redundant processing is needed. For instance, a program called GNSO [19] generates a long linear sequence of Fortran statements called SOLVE [19]. Once SOLVE is obtained, it can eliminate all searching procedures and thus speed up the process, provided that the core memory is not overloaded by its length. As is mentioned in Section IV, the pivoting order may have to be changed from time to time. In these cases GNSO has to be called for into operation; thus, overall economy is questionable.

Another important aspect of our method lies in its easy decomposability of data structure, which is needed if data overloads the core memory. The data are stored linearly from the top to the bottom of $T(\ )$ and $U(\ )$ in accordance with the sequence of computation in (43)–(53). Thus, the interchange of data between core and backup memory is very simple. The ordering of t_{ij} and u_{ij} is also consistent with the forward and backward substitutions. It should be noted that the fundamental operation in Gaussian elimination

$$a_{ij}^{(k)} = a_{ij}^{(k-1)} + t_{ik}a_{kj}^{(k-1)}, \qquad i,j > k \tag{63}$$

is inconsistent with this decomposability requirement. Generally t_{ik} and $a_{kj}^{(k-1)}$ are stored near the top, while the $a_{ij}^{(k)}$ are located near the bottom of the array.

VI. Conclusion

In this paper we considered two problems on piecewise-linear resistive networks. First, we developed two new theorems concerning the existence of solutions and presented a convergent method for finding at least one of the solutions. The result can be concisely stated in terms of the determinants of the Jacobian matrices. Essentially, if, in all regions, the determinants are nonzero and have the same sign, there exists at least one solution for an arbitrary input; the Katzenelson algorithm can be used to find a solution. This extends the early work of Fujisawa and Kuh on networks with a unique solution. The question to be asked next is of course for the case in which all Jacobian matrices are nonsingular but have either positive or negative determinants.

The second problem which we presented deals with the development of an efficient computational algorithm for solving *large* piecewise-linear resistive networks. The problem is essentially to solve successively linear equations $\boldsymbol{A}^i\beta=\boldsymbol{\alpha}$, for $i=1, 2, \cdots$, in which two successive $\boldsymbol{A}^i$ differ by a matrix of rank one. The method is based on the triangular factorization of modified matrices developed by Bennett, which makes use of Gaussian elimination and takes advantage of the key property that the $\boldsymbol{A}^i$ possess. When, in addition, the matrices are sparse and have the same sparsity structure, the method is extremely efficient in terms of the required number of arithmetic operations as well as data handling and storage. The method can be used for problems other than networks, for example, linear programming.

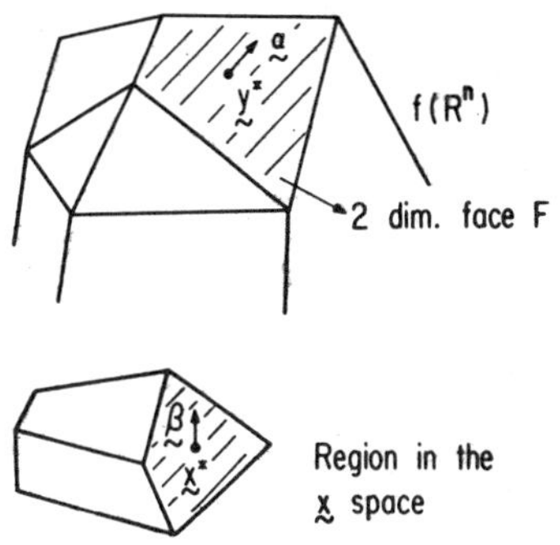

Fig. 6. Three-dimensional illustration of the range space $f(R^n)$. Tilde under letters corresponds to boldface in text.

Appendix I
Proof of Theorems 3 and 4

Proof of Theorem 3

First some general observations need to be made to clarify the nature of continuous piecewise-linear mapping $\boldsymbol{f}$. Each region is a convex polyhedron, either bounded or unbounded. It is to be noted that $\boldsymbol{f}$ is an affine transformation $\boldsymbol{A}^{(m)}\mathbf{x}+\boldsymbol{w}^{(m)}=\boldsymbol{y}$ in each region R_m. It follows that the image of each region under mapping $\boldsymbol{f}$ is also either a bounded closed polyhedron or an unbounded closed polyhedron. Therefore, the range space $\boldsymbol{f}(R^n)$ is the finite union of convex closed polyhedra.

The proof is by contradiction. Assume that $\boldsymbol{f}(R^n)$ does not fill up the whole space R^n. Then there exists an $(n-1)$-dimensional face F, which is a boundary of $\boldsymbol{f}(R^n)$ as is illustrated in Fig. 6. The face F is a finite union of the images of some $(n-1)$-dimensional boundary faces of regions in the $\mathbf{x}$-space. Hence there exists a point $\boldsymbol{y}^*\in F$ which lies on F but not on any other faces and a point $\mathbf{x}^*$ which lies on one and only one boundary hyperplane in the $\mathbf{x}$-space such that $\boldsymbol{y}^*=\boldsymbol{f}(\mathbf{x}^*)$. Let $\boldsymbol{\alpha}$ be the outward normal vector of face F. By assumption, $\boldsymbol{y}^*$ is on F but not on any other faces and hence there exists a positive constant $\epsilon>0$ such that

$$\boldsymbol{y}^* + \nu\boldsymbol{\alpha} \notin \boldsymbol{f}(R^n), \qquad \text{for } 0 < \nu < \epsilon. \tag{I-1}$$ [2]

As $\mathbf{x}^*$ lies on one and only one boundary hyperplane, there exists a neighborhood ∇ of $\mathbf{x}^*$ which is mapped homeomorphically onto a neighborhood of $\boldsymbol{y}^*$. This contradicts (I-1). Thus the proof is complete.

Proof of Theorem 4

From the arguments in Section II it is clear that the statement of the theorem is true if either $\mathbf{x}$ is an interior point or it lies on a simple boundary.

Let us assume that point $\mathbf{x}^*$ lies on mutually distinct boundary hyperplanes $H_1, \cdots, H_k$. The property to prove is a local property at $\mathbf{x}^*$ and hence it can be as-

[2] Though the image of each region is convex, their sum $\boldsymbol{f}(R^n)$ may not be convex. Hence one can not say $\mathbf{x}^*+\nu\boldsymbol{\alpha}\notin\boldsymbol{f}(R^n)$ for all $\nu>0$.

sumed that hyperplanes other than these above do not exist. This is equivalent to the consideration of a modified continuous piecewise-linear function f^* which is obtained by extending the local property of f at x^* to the whole space R^n. It is clear that if f satisfies the condition of Theorem 3, then f^* does too. Therefore, f^* is a mapping of R^n onto itself due to Theorem 3. Therefore, for any unit vector α, there exists at least one nonzero vector β such that

$$f^*(x^* + \nu\beta) = f^*(x^*) + \nu\alpha \qquad \text{(I-2)}$$

for all $\nu > 0$. Since f^* and f coincide with each other in a neighborhood of x^*, (I-2) is valid for f for sufficiently small positive ν. This completes the proof.

Appendix II

Gaussian Elimination and Proof of Theorem 6[3]

The nature of Gaussian elimination can be best seen by examining its major first step. Let

$$A \triangleq A^{(0)} = \left[a_{ij}^{(0)}\right] = \left[\begin{array}{c|c} a_{11}^{(0)} & b^{(0)T} \\ \hline a^{(0)} & G^{(0)} \end{array}\right] \qquad \text{(II-1)}$$

where $a^{(0)}$ and $b^{(0)}$ are $(n-1)$-dimensional column vectors and $G^{(0)}$ is an $(n-1)\times(n-1)$ matrix. Assuming that $a_{11}^{(0)}$ is nonzero, an elementary row operation results in the following:

$$A_1 \triangleq T_1 A = \left[\begin{array}{c|c} \dfrac{1}{a_{11}^{(0)}} & 0 \\ \hline -\dfrac{a^{(0)}}{a_{11}^{(0)}} & 1 \end{array}\right] A = \left[\begin{array}{c|c} 1 & \dfrac{b^{(0)T}}{a_{11}^{(0)}} \\ \hline 0 & A^{(1)} = \left[a_{ij}^{(1)}\right] \end{array}\right] \qquad \text{(II-2)}$$

where

$$A^{(1)} = G^{(0)} - \frac{1}{a_{11}^{(0)}} a^{(0)} b^{(0)T}. \qquad \text{(II-3)}$$

According to the terminology used in Section IV for the T_k and U, the above relations can be written as

$$t_{11} = \frac{1}{a_{11}^{(0)}} \qquad \text{(II-4)}$$

$$t_{i1} = -a_{i1}^{(0)} t_{11}, \qquad i = 2, \cdots, n \qquad \text{(II-5)}$$

$$u_{ij} = t_{11} a_{ij}^{(0)}, \qquad j = 2, \cdots, n \qquad \text{(II-6)}$$

and

$$a_{ij}^{(1)} = a_{ij}^{(0)} + t_{i1} a_{1j}^{(0)}, \qquad i, j = 2, \cdots, n. \qquad \text{(II-7)}$$

For the kth step, the general formulas are

$$A_k = T_k A_{k-1}$$

$$= \begin{bmatrix} 1 & 0 & \cdots & & & & 0 \\ 0 & 1 & \cdots & & & & 0 \\ & & \cdots & & & & \\ & & \cdots & & & & \\ 0 & \cdots & 0 & t_{kk} & 0 & \cdots & 0 \\ & \cdots & & t_{k+1,k} & 1 & \cdots & 0 \\ & \cdots & & & & & \\ 0 & \cdots & 0 & t_{nk} & 0 & \cdots & 1 \end{bmatrix} \left[\begin{array}{ccccc|c} 1 & u_{12} & & & \cdots & u_{1n} \\ 0 & 1 & u_{23} & & \cdots & u_{2n} \\ & & \cdots & & & \\ 0 & \cdots & 0 & 1 & u_{k-1,k} \cdots & u_{k-1,n} \\ \hline 0 & \cdots & & 0 & & \\ & \cdots & & & & A^{(k-1)} = \left[a_{ij}^{(k-1)}\right] \\ & \cdots & & & & \\ 0 & \cdots & & 0 & & \end{array}\right]$$

$$= \left[\begin{array}{ccccc|c} 1 & u_{12} & & & & u_{1n} \\ 0 & 1 & u_{23} & & & u_{2n} \\ & & \cdots & & & \\ & & \cdots & & & \\ 0 & \cdots & 0 & 1 & u_{k,k+1} \cdots & u_{kn} \\ \hline 0 & \cdots & & 0 & & \\ & \cdots & & & & A^{(k)} = \left[a_{ij}^{(k)}\right] \\ & \cdots & & & & \\ 0 & \cdots & & 0 & & \end{array}\right] \qquad \text{(II-8)}$$

[3] It should be pointed out that the notation used here is independent of that of Section II when we considered matrices at different regions. Here superscripts are used to denote iteration steps.

where

$$t_{kk} = \frac{1}{a_{kk}^{(k-1)}}. \quad \text{(II-9)}$$

$$t_{ik} = -a_{ik}^{(k-1)} t_{kk}, \qquad i = k+1, \cdots, n \quad \text{(II-10)}$$

$$u_{kj} = t_{kk} a_{kj}^{(k-1)}, \qquad j = k+1, \cdots, n \quad \text{(II-11)}$$

and

$$a_{ij}^{(k)} = a_{ij}^{(k-1)} + t_{ik} a_{kj}^{(k-1)}, \quad i, j = k+1, \cdots, n. \quad \text{(II-12)}$$

In the above, it is seen that the term $a_{kk}^{(k-1)} = 1/t_{kk}$ is used as the pivoting element at each step, and, by assumption, it is nonzero. Thus, t_{kk} is well defined. It is also of interest to point out the following property of the matrix $\boldsymbol{A}$ in terms of $a_{kk}^{(k-1)} = 1/t_{kk}$: the leading principal minor of order k, i.e., the determinant which is formed by the first k-rows and first k-columns of $\boldsymbol{A}$, is equal to

$$a_{11}^{(0)} a_{22}^{(1)} \cdots a_{kk}^{(k-1)} = \frac{1}{t_{11} t_{22} \cdots t_{kk}}. \quad \text{(II-13)}$$

Thus another way of stating that t_{kk} is well defined for $k = 1, 2, \cdots, n$ is the nonvanishing of all the leading principal minors of $\boldsymbol{A}$.

The problem at hand is to obtain the Gaussian triangularization of $\boldsymbol{A}^*$ where

$$\boldsymbol{A}^* \triangleq \boldsymbol{A}^{*(0)} = \boldsymbol{A}^{(0)} + \mathbf{c}\boldsymbol{r}^T \triangleq \boldsymbol{A}^{(0)} + \frac{1}{d_0} \mathbf{c}^{(0)} \boldsymbol{r}^{(0)^T}. \quad \text{(II-14)}$$

As in (II-1), we may write $\boldsymbol{A}^{*(0)}$ as follows:

$$\boldsymbol{A}^{*(0)} = \left[\begin{array}{c|c} a_{11}^{(0)} + \frac{1}{d_0} c_1^{(0)} r_1^{(0)} & \boldsymbol{b}^{(0)^T} + \frac{1}{d_0} c_1^{(0)} \boldsymbol{r}_1^{(0)^T} \\ \hline \boldsymbol{a}^{(0)} + \frac{1}{d_0} \mathbf{c}_1^{(0)} r_1^{(0)} & \boldsymbol{G}^{(0)} + \frac{1}{d_0} \mathbf{c}_1^{(0)} \boldsymbol{r}_1^{(0)^T} \end{array}\right] \quad \text{(II-15)}$$

where $\mathbf{c}_1^{(0)} \triangleq [c_2^{(0)}, \cdots, c_n^{(0)}]^T$ is the $(n-1)$-vector containing the last $(n-1)$-elements of $\mathbf{c}^{(0)}$ and $\boldsymbol{r}_1^{(0)} \triangleq [r_2^{(0)}, \cdots, r_n^{(0)}]^T$ is the $(n-1)$-vector containing the last $(n-1)$-elements of $\boldsymbol{r}^{(0)}$. Then the major first step for $\boldsymbol{A}^{*(0)}$ is

$$\boldsymbol{A}_1^* = \boldsymbol{T}_1^* \boldsymbol{A}^*$$

$$= \left[\begin{array}{c|c} \dfrac{1}{a_{11}^{(0)} + \frac{1}{d_0} c_1^{(0)} r_1^{(0)}} & \mathbf{0} \\ \hline -\dfrac{\boldsymbol{a}^{(0)} + \frac{1}{d_0} \mathbf{c}_1^{(0)} r_1^{(0)}}{a_{11}^{(0)} + \frac{1}{d_0} c_1^{(0)} r_1^{(0)}} & \mathbf{1} \end{array}\right] \boldsymbol{A}^*$$

$$= \left[\begin{array}{c|c} 1 & \dfrac{\left(\boldsymbol{b}^{(0)} + \frac{1}{d_0} c_1^{(0)} \boldsymbol{r}_1^{(0)}\right)^T}{a_{11}^{(0)} + \frac{1}{d_0} c_1^{(0)} r_1^{(0)}} \\ \hline \mathbf{0} & \boldsymbol{A}^{*(1)} = [a_{ij}^{*(1)}] \end{array}\right] \quad \text{(II-16)}$$

where

$$\boldsymbol{A}^{*(1)} = \boldsymbol{G}^{(0)} + \frac{1}{d_0} \mathbf{c}_1^{(0)} \boldsymbol{r}_1^{(0)T} - \frac{\left(\boldsymbol{a}^{(0)} + \frac{1}{d_0} \mathbf{c}_1^{(0)} r_1^{(0)}\right)\left(\boldsymbol{b}^{(0)} + \frac{1}{d_0} c_1^{(0)} \boldsymbol{r}_1^{(0)}\right)^T}{a_{11}^{(0)} + \frac{1}{d_0} c_1^{(0)} r_1^{(0)}}$$

$$= \boldsymbol{A}^{(1)} + \frac{\left(\mathbf{c}_1^{(0)} - \frac{c_1^{(0)}}{a_{11}^{(0)}} \boldsymbol{a}^{(0)}\right)\left(\boldsymbol{r}_1^{(0)} - \frac{r_1^{(0)}}{a_{11}^{(0)}} \boldsymbol{b}^{(0)}\right)^T}{a_{11}^{(0)} + \frac{1}{d_0} c_1^{(0)} r_1^{(0)}}$$

$$= \boldsymbol{A}^{(1)} + \frac{1}{d_1} \mathbf{c}^{(1)} \boldsymbol{r}^{(1)T}. \quad \text{(II-17)}$$

In (II-17)

$$\mathbf{c}^{(1)} \triangleq \mathbf{c}_1^{(0)} - \frac{c_1^{(0)}}{a_{11}^{(0)}} \boldsymbol{a}^{(0)} \quad \text{(II-18)}$$

and

$$\boldsymbol{r}^{(1)} \triangleq \boldsymbol{r}_1^{(0)} - \frac{r_1^{(0)}}{a_{11}^{(0)}} \boldsymbol{b}^{(0)}. \quad \text{(II-19)}$$

Comparing these with (30)–(35), we recognize that $\mathbf{c}^{(1)}$ is the second column vector of the triangular matrix $\boldsymbol{C}$ and $\boldsymbol{r}^{(1)}$ is the second column vector of $\boldsymbol{R}$ in Section IV. Furthermore using these formulas in Section IV and (II-4)–(II-7), we can easily prove the following:

$$t_{11}^* = \frac{d_0}{d_1} t_{11} = t_{11} - \frac{c_1^{(0)} r_1^{(0)} t_{11}^2}{d_1} \quad \text{(II-20)}$$

$$-\frac{\boldsymbol{a}^{(0)} + \frac{1}{d_0} \mathbf{c}_1^{(0)} r_1^{(0)}}{a_{11}^{(0)} + \frac{1}{d_0} c_1^{(0)} r_1^{(0)}} = -\frac{\boldsymbol{a}^{(0)}}{a_{11}^{(0)}} - \frac{r_1^{(0)} \mathbf{c}^{(1)}}{d_1 a_{11}^{(0)}} \quad \text{(II-21)}$$

$$\frac{b^{(0)} + \frac{1}{d_0} c_1^{(0)} r_1^{(0)}}{a_{11}^{(0)} + \frac{1}{d_0} c_1^{(0)} r_1^{(0)}} = \frac{b^{(0)}}{a_{11}^{(0)}} + \frac{c_1^{(0)} r^{(1)}}{d_1 a_{11}^{(0)}}. \tag{II-22}$$

From (II-21) and (II-22), if we take the ith and jth terms, respectively, we obtain

$$t_{i1}^* = t_{i1} - \frac{r_1^{(0)} t_{11}}{d_1} c^{(1)}, \qquad i = 2, \cdots, n \tag{II-23}$$

and

$$u_{ij}^* = u_{ij} + \frac{c_1^{(0)} t_{11}}{d_1} r_j^{(1)}, \qquad j = 2, \cdots, n. \tag{II-24}$$

Equations (II-20), (II-23), and (II-24) correspond to (38)–(40) in Theorem 6 for $k=1$.

The next step is to start with (II-17) and perform the second Gaussian elimination step. The proof for $k=2$ is identical with the above. Extending this step we can state that (38)–(40) hold for any $k=1, 2, \cdots, n$.

As we mentioned before, the underlying assumption is that the given matrix $\boldsymbol{A}$ can be triangularized, i.e., $a_{kk}^{(k-1)} = 1/t_{kk} \neq 0$ for $k=1, 2, \cdots, n$. Thus t_{kk}, for $k=1, 2, \cdots, n$ must be well defined in order to be able to perform the Gaussian elimination without reordering the rows and columns. In terms of the modified matrix $\boldsymbol{A}^*$, the corresponding condition is that t_{kk}^*, for $k=1, 2, \cdots, n$ are well defined. Since

$$t_{kk}^* = \frac{d_{k-1}}{d_k} t_{kk}$$

the condition is satisfied if and only if $d_k \neq 0$, for $k=1, 2, \cdots, n$. This completes the proof of Theorem 6.

Acknowledgment

The computation for our example was done on IBM 1800. The authors wish to thank J. Sporal for his programming assistance.

References

[1] D. W. Martin and G. Peters, "The application of Newton's method to network analysis by digital computer," *J. Inst. Water Eng.*, vol. 17, pp. 115–129, 1963.

[2] R. K. Livesley, "The analysis of large structural systems," *Comput. J.*, vol. 3, pp. 34–39, 1960.

[3] T. A. Porsching, "Jacobi and Gauss–Seidel method for nonlinear network problems," *SIAM J. Numer. Anal.*, vol. 6, pp. 437–449, 1969.

[4] I. W. Sandberg, "Theorems on the analysis of nonlinear transistor networks," *Bell Syst. Tech. J.*, vol. 49, pp. 95–114, Jan. 1970.

[5] H. Nikaido, *Convex Structures and Economic Theory.* New York: Academic Press, 1968.

[6] R. S. Palais, "Natural operations on differential forms," *Trans. Amer. Math. Soc.*, vol. 92, pp. 125–141, July 1959.

[7] C. A. Holzmann and R. W. Liu, "On the dynamic equations of nonlinear networks with n-coupled elements," in *Proc. 3rd Annu. Allerton Conf. Circuit Syst. Theory*, pp. 536–545, 1965.

[8] E. S. Kuh and I. N. Hajj, "Nonlinear circuit theory: Resistive networks," *Proc. IEEE*, vol. 59, pp. 340–355, Mar. 1971.

[9] T. Fujisawa and E. S. Kuh, "Some results on existence and uniqueness of solutions of nonlinear networks," *IEEE Trans. Circuit Theory*, vol. CT-18, pp. 501–506, Sept. 1971.

[10] ——, "Piecewise linear theory of nonlinear networks," *SIAM J. Appl. Math.*, vol. 22, pp. 307–328, Mar. 1972.

[11] I. W. Sandberg and A. N. Willson, Jr., "Existence of solutions for the equations of transistor–resistor–voltage source networks," *IEEE Trans. Circuit Theory*, vol. CT-18, pp. 619–625, Nov. 1971.

[12] J. M. Bennett, "Triangular factors of modified matrices," *Numer. Math.*, vol. 7, pp. 217–221, 1965.

[13] M. Fielder and V. Ptak, "Some generalizations of positive definiteness and monotonicity," *Numer. Math.*, vol. 9, pp. 163–172, 1966.

[14] J. Katzenelson, "An algorithm for solving nonlinear resistive networks," *Bell Syst. Tech. J.*, vol. 44, pp. 1605–1620, Oct. 1965.

[15] L. O. Chua, "Efficient computer algorithm for piecewise-linear analysis of resistive nonlinear networks," *IEEE Trans. Circuit Theory*, vol. CT-18, pp. 73–85, Jan. 1971.

[16] T. Ohtsuki and N. Yoshida, "DC analysis of nonlinear networks based on generalized piecewise-linear characterization," *IEEE Trans. Circuit Theory*, vol. CT-18, pp. 146–152, Jan. 1971.

[17] R. A. Willoughby, "Some pivot considerations in direct methods for sparse matrices," in *Proc. Mexico Int. Conf. Systems, Networks, Computers*, pp. 387–390, Jan. 1971.

[18] J. H. Wilkinson, *The Algebraic Eigenvalue Problem.* Oxford, England: Clarendon Press, 1965.

[19] F. G. Gustavson, W. Liniger, and R. A. Willoughby, "Symbolic generation of an optimal Crout algorithm for sparse systems of linear equations," *J. Ass. Comput. Mach.*, vol. 17, pp. 78–109, Jan. 1970.

Multiparameter Large-Change Sensitivity Analysis and Systematic Exploration

K. H. LEUNG, MEMBER, IEEE, AND ROBERT SPENCE, MEMBER, IEEE

***Abstract*—The effect on linear circuit response of simultaneous changes in a number (m) of components must often be computed. Two efficient methods for predicting this effect are examined. It is shown that the computational cost of predicting the effect of each set is approximately G_m, where G_m is the cost of a Gaussian elimination of an $m \times m$ matrix. However, if the sets represent combinations of relatively few individual component changes, then systematic exploration is shown to reduce the computational cost *per set* to the order of 2.**

I. Introduction

THE NEED to compute the effect of component parameter change on circuit response arises from the frequent occasions during design and use when component values change either parasitically, manually or automatically (as in circuit optimization). In response to this need, sensitivity algorithms have been devised for calculating the effect of small changes in one or more parameters within a linear circuit [1]–[3]. For *large* changes, algorithms have been proposed and investigated for changes in one [1], [4] or more [1], [5], [6] parameters, but there still exists a need for further assessment and comparative evaluation.

The effect of *simultaneous* large changes in more than one parameter is frequently of interest. The circuit designer may be interactively and dynamically adjusting a number of parameters in order to achieve a desired circuit response. Or, the changes may be automatic, as in a Monte-Carlo statistical circuit analysis [7]. The need to examine the consequence of change in the physical properties (e.g., temperature, doping level) of a circuit also involves the exploration of simultaneous parameter changes [8], as does the calculation of performance contours [9].

This paper examines two algorithms originally employed to predict the effect of single large parameter changes in a linear circuit, and determines their efficiency when extended to deal with simultaneous multiparameter variation. One outcome is an extremely efficient algorithm which is appropriate when the sets of component changes represent *combinations* of individual component changes. Engineering applications of this approach—known as *systematic exploration*—are briefly illustrated.

Manuscript received May 21, 1974; revised November 19, 1974 and March 8, 1975.

The authors are with the Department of Electrical Engineering, Imperial College, Exhibition Road, London SW7 2BT, England.

Reprinted from *IEEE Trans. Circuits & Systems,* vol. CAS-22, pp. 796–804, Oct. 1975.

In what follows, no consideration is given to any advantages that can be secured by the application of sparse matrix techniques [10]. Attention in this paper is focused on the basic operations upon circuit parameters, though in such a way as to make the consideration of sparsity as straightforward as possible.

II. Current Source Substitution Method

A. Single Component Changes

The first of the two large-change sensitivity algorithms to be examined can be explained with reference to the circuit model of Fig. 1. A linear 2-port[1] is excited at port 1 by a unit current source, and generates at port 2 a voltage response equal to the transfer impedance z_{21} (Fig. 1(a)). A single linear 2-terminal component within the circuit is liable to change. The *change* Δy in its admittance is extracted from the 2-port [11], thereby creating a new port (port 3) of the nominal-value circuit N (Fig. 1(b)).

The conceptual basis of the current source substitution method [4] is the replacement of the admittance Δy by a current source whose value is identical with the current I_3 initially flowing in Δy (Fig. 2). According to the Substitution Theorem [12], no disturbance is thereby occasioned to the voltages and currents in the circuit. Since the only change from the nominal circuit N is the connection of the current source I_3 to port 3, and the circuit is linear, the voltage response at port 2 *due to this source acting alone* is the required change ΔV in output voltage.

Expression of I_3 in terms of the unit excitation current source gives the relation between ΔV and Δy as in

$$\Delta V = \frac{z_{23}z_{31}}{z_{33} + \dfrac{1}{\Delta y}}. \tag{1}$$

If we follow the practice of assigning a cost of *one* to multiplication and division, and *zero* to addition and subtraction then, for each new component value, the cost of computing ΔV is two, in addition to the cost of one incurred in the initial calculation of the product $z_{23}z_{31}$. The latter cost quickly becomes insignificant if more than a few values of component change are explored. The same approach and computational costs are associated with large changes in the value of a voltage-controlled current source, the principal difference being that an additional port must then be created across the controlling voltage [4].

Fig. 3 outlines the algorithm and itemizes the associated computational cost. If only one parameter is subject to change, then, at each frequency, pivotal condensation of the circuit's nodal admittance matrix Y_n to obtain the 3×3-port[2] admittance matrix Y_p, followed by matrix inversion to yield the port impedance matrix Z_p, provides the constants appearing in (1), from which ΔV is computed. The alternative approach indicated by a dashed line in Fig. 3 may be more economical if it is not known, beforehand, which parameter is liable to change.

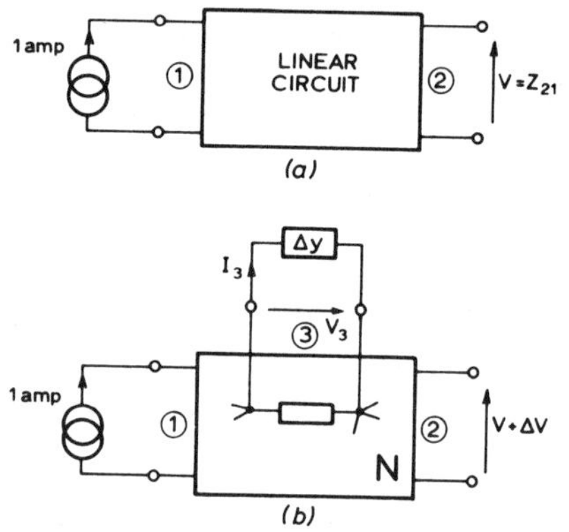

Fig. 1. Linear 2-port, one of whose 2-terminal components is liable to change.

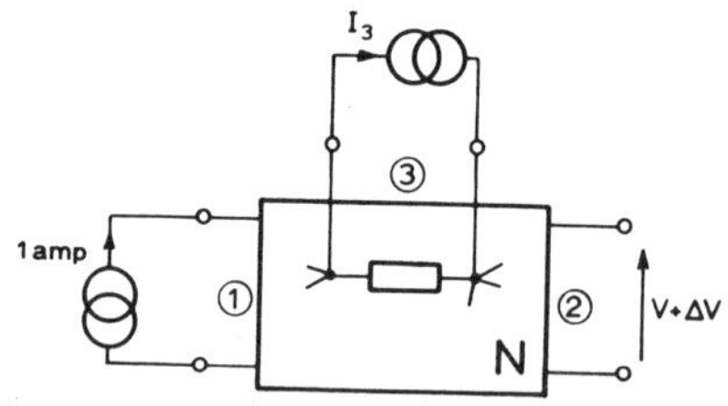

Fig. 2. Simulation of component change by current source.

The computational cost[3] of the substitution current source algorithms of Fig. 3 are compared in Table I with two straightforward methods. One (column A) refers to a repeated nodal analysis for each new parameter value. The other (column B) employs an initial pivotal condensation to obtain the port description of N (Fig. 1(b)) and then, for each parameter change, a Gaussian elimination (Fig. 3, (*Note 1*)): this method, whether used in connection with one or more variable components, is referred to throughout this paper as the reference method. The three rows of the table indicate the costs incurred once, K times and KL times if the effect of L different values of each of K separate components is to be computed: in each case, a single variable parameter is assumed to return to its nominal value before another parameter is varied.[4] Progression to the right over the table shows an increasing "once only" or "K only" cost, and a decreasing "per component change" cost, a situation which is necessary if the effect of many large changes in one or more components is to be explored economically.

[1] The algorithms can most easily be described, and their computational cost identified, if a *port* description is used.

[2] For simplicity, we do not consider the situation in which ports share one or more nodes. Also, in assessing computational cost, we do *not* exploit any symmetry of the matrices involved, despite the fact that we assume that only 2-terminal components are subject to change: the additional cost due to the presence of variable voltage-controlled current sources does not lend itself to simple expression, or change the essential conclusions of this study.

[3] Since many of the computations are either a Gaussian elimination or approximate closely in cost to that of an integer multiple of such eliminations, the cost of a Gaussian elimination is denoted by G_x, where the subscript indicates the dimension of the vectors involved [13].

[4] If a variable parameter is held at a new value, the admittance matrix Y_n must, of course, be reformulated before the effect of a new variable parameter can be computed, or a method described in Section III-B employed.

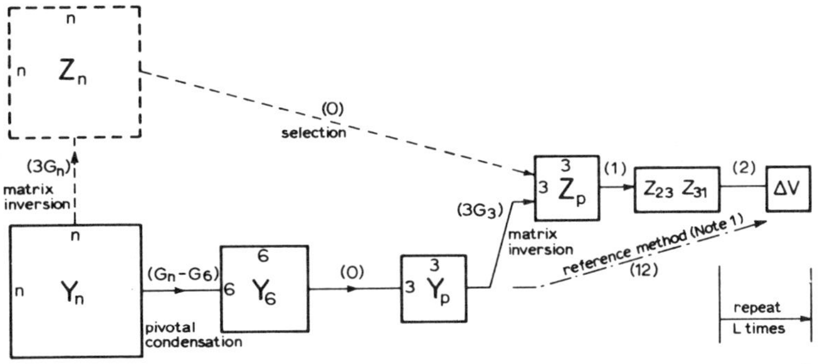

Fig. 3. Approximate cost of computing the effect of changes in one component on the transfer impedance of an $(n + 1)$-terminal circuit. Values in brackets indicate costs. *Note 1*—this path is associated with a straightforward analysis of a 3-port by Gaussian elimination. *Note 2*—arrows indicating repeat calculations associated with L different values of the same component, and with K different components, do not apply to the reference method.

TABLE I

	A Repeated Nodal Analysis of Complete Circuit	B Reference Method (Fig. 3, Note 1)	C Scheme of Fig. 3	D Scheme of Fig. 3 (Dashed Line)
Once-only initial calculation	G_n (to obtain nominal response)	12 (to obtain nominal response)	—	$3G_n$
Preparatory calculation repeated K times	—	$G_n - G_6$	$G_n - G_6 + 3G_3 + 1$	1
Calculation for each component change	G_n	12	2	2

Note: Approximate costs associated with four different methods of computing the effect, on a circuit's transfer impedance, of single component changes.

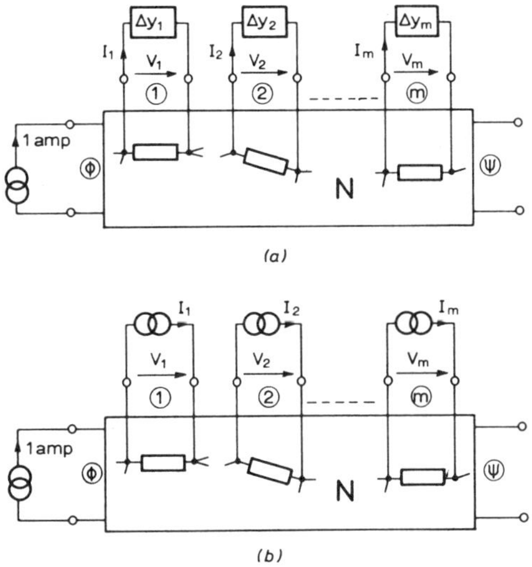

Fig. 4. The use of substitution current sources to simulate simultaneous changes in a number of components.

B. Simultaneous Component Changes

Let m components in a circuit be subject to simultaneous change, and the corresponding admittance changes be denoted by $\Delta y_1,\cdots,\Delta y_m$. Extraction of these admittance changes creates, for the nominal-value circuit N, m new ports numbered 1–m (Fig. 4(a)) in addition to those denoted by ϕ and ψ and associated, respectively, with excitation and response. The effect upon the transfer impedance $z_{\psi\phi}$ is to be computed.

The apparent simplicity of the substitution current source method diminishes rapidly as the number of simultaneously variable parameters increases. Basically, this is because the admittance simulated by a current source is the ratio of its current and voltage: whereas its current is defined, its voltage is a function of all the sources (including those simulating component changes) connected to the circuit. Thus, broadly speaking, the values of the m substitution current sources can be found only by the solution of m equations in m unknowns [5], [14], [15].

Let the $(m + 2)$-port circuit N of Fig. 4(a) be described by a port admittance matrix Y_p, and let the changes $\Delta y_1,\cdots,\Delta y_m$ in the m component admittances be simulated, respectively, by m current sources $I_1,\cdots,I_m$ (Fig. 4(b)). Such a simulation requires that

$$\begin{bmatrix} V_1 \\ \vdots \\ V_m \end{bmatrix} = \begin{bmatrix} -\dfrac{1}{\Delta y_1} & & & 0 \\ & -\dfrac{1}{\Delta y_2} & & \\ & & \ddots & \\ 0 & & & -\dfrac{1}{\Delta y_m} \end{bmatrix} \begin{bmatrix} I_1 \\ \vdots \\ I_m \end{bmatrix} \tag{2}$$

Denote the nominal circuit's port impedance matrix by Z_p, and that partition of Z_p appropriate to the m newly created ports by Z_m. When the circuit is excited by the substitution current sources as well as the original excitation at port ϕ, the voltages generated across the substitution sources are

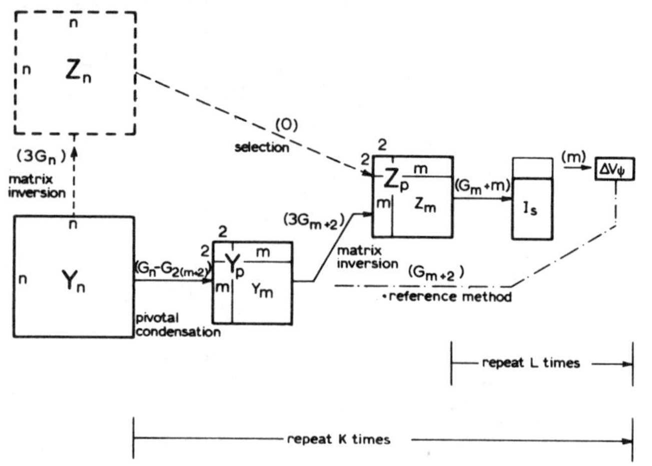

Fig. 5. Use of substitution current source method to compute effect, on circuit's transfer impedance, of simultaneous changes in m components. Arrows indicating repeat calculations associated with L different values of same set of components, and K different groups of components, do not apply to the reference method. Computational costs, some approximate, are indicated in parentheses.

given by

$$\begin{bmatrix} V_1 \\ \vdots \\ V_m \end{bmatrix} = [Z_m] \begin{bmatrix} I_1 \\ \vdots \\ I_m \end{bmatrix} + \begin{bmatrix} z_{1\phi} \\ z_{2\phi} \\ \vdots \\ z_{m\phi} \end{bmatrix} I_\phi. \quad (3)$$

Combination with (2) yields

$$-\begin{bmatrix} z_{1\phi} \\ \vdots \\ z_{m\phi} \end{bmatrix} I_\phi = \left\{ \begin{bmatrix} \frac{1}{\Delta y_1} & & 0 \\ & \ddots & \\ 0 & & \frac{1}{\Delta y_m} \end{bmatrix} + [Z_m] \right\} \begin{bmatrix} I_1 \\ \vdots \\ I_m \end{bmatrix} \triangleq Z' I_S \quad (4)$$

where I_S is the vector of substitution current sources. Thus, once Z_m has been found (by inversion of Y_p) at a cost of about $3G_{m+2}$, the values of the m substitution current sources can be found in two steps: 1) calculation of the reciprocal of each admittance change at a total cost of m, and 2) Gaussian elimination at a cost of G_m. The matrix product $Z_m I_S$ then yields the change in port voltages. If only ΔV_ψ is of interest, the cost is m. This approach is illustrated in Fig. 5, where it is assumed that the starting point is the nominal circuit's nodal admittance matrix Y_n. If there are L sets of values of the *same m* components, the calculation of I_S at a cost of $G_m + m$, and of ΔV_ψ at a cost of m, must be repeated L times. If K *different* groups of components are known beforehand to be of interest—as may well be the case in truly interactive design—then a more economical alternative may be that which is indicated by the dashed lines in Fig. 5.

The computational costs are itemized in the first two columns of Table II: the second column is the more appropriate if the variable parameters are initially unidentified or if K is large. The final column refers, for purposes of comparison, to a straightforward method which is identical

TABLE II

	Current Source Substitution Method		
	Fig. 5	Fig. 5 (Dashed Line)	Reference Method
Once-only calculation	—	$3G_n$	G_{m+2}
Preparatory calculation repeated K times	$G_n - G_{2(m+2)} + 3G_{m+2}$	—	$G_n - G_{2(m+2)}$
Calculation for each set of component changes	$G_m + 2m$	$G_m + 2m$	G_{m+2}

Note: Approximate costs associated with different methods of computing the effect, on a circuit's transfer impedance, of simultaneous changes in m components. The table is equally valid for the matrix modification method (Fig. 8) if the entries $2m$ are omitted from row 3, columns 1 and 2.

in form to that employed as a reference in the previous section (column B of Table I); it involves updating the $(m + 2)$-port's admittance matrix (at no cost) to account for component admittance changes, followed by Gaussian elimination at a cost of G_{m+2} to find the transfer impedance (see Fig. 5). Table II shows that no very clear advantage can be associated with the substitution current source approach.[5]

III. Matrix Modification Method

The basis of this second approach to large-change sensitivity analysis is the fact that, if a matrix Y is modified by the addition of the triple matrix product PHQ, then the inverse of the new matrix is related to that of Y by the expression [1], [16], [17]

$$(Y + PHQ)^{-1} = Y^{-1} - Y^{-1}P(H^{-1} + QY^{-1}P)^{-1}QY^{-1} \quad (5)$$

provided the indicated inverses exist, and the dimensions are compatible. Briefly, the advantage of this relation is the ability, once the original circuit's impedance matrix (Y^{-1}) is found, easily to calculate the impedance description of the new circuit created by the component admittance changes. The individual admittance changes can be characterized by H, and the manner of their association with the nominal circuit by P and Q.

A. Single Component Changes

Refer again to the circuit model of Fig. 1(b), and let the 3-port nominal circuit N be described by admittance and impedance matrixes Y_p and Z_p, respectively. Connection of the admittance Δy to port 3 can be described by the

[5] The extension of the current–source substitution method to simultaneous component changes has been alluded to by Goddard [14], described by Villalaz [15], and explored in detail by Temes *et al.* [5]. Indeed, Temes *et al.* originally derived a form of (4) using the concept of interreciprocity and Tellegen's Theorem: an advantage of so doing is a reduction in the computational effort involved if more than one independent source is present. They also showed how, for small percentage changes in component values, an iterative solution of (4) may be more economic.

addition of a matrix ΔY to Y_p, where

$$\Delta Y = P \, \Delta y Q. \tag{6}$$

P and Q^t are column vectors whose only nonzero elements are $P_3 = Q_3 = 1$, and t denotes transposition. Reference to (5) enables the new port impedance matrix Z_p' to be expressed in terms of its initial value Z_p by

$$Z_p' = Z_p - \gamma^{-1} Z_p P Q Z_p \tag{7}$$

where

$$\gamma = \frac{1}{\Delta y} + Q Z_p P. \tag{8}$$

For the present example, (7) and (8) yield an expression for the new transfer impedance z_{21}'

$$z_{21}' = z_{21} - \frac{z_{23} z_{31}}{z_{33} + \dfrac{1}{\Delta y}} \tag{9}$$

which is identical with the relation, in (1), derived by the current source substitution method. Thus, for computing the effect of changes in a single component, the two theoretical approaches can lead to the same algorithms and computational costs. In this sense, Table I is equally relevant to the matrix modification method.

The Householder relation (5) is, of course, also the basis of the link-at-a-time method of circuit analysis: the approach to large-change sensitivity analysis just described corresponds to the connection of a link [1].

B. Simultaneous Component Changes

Consider a 2-port in which three 2-terminal components are subject to simultaneous change. New ports are created in parallel with these components (Fig. 6(a)) and the relevant admittance changes Δy_1, Δy_2, and Δy_3 simulated by the connection of identical admittances externally to these ports. First, however, we consider the connection of Δy_3 *alone*,[6] so that ports 1 and 2 are left on open circuit.

With reference to (6), this connection is described if P and Q are the vector[7] (00001), and $\Delta y = \Delta y_3$. Equation (7) then becomes

$$Z_p' = Z_p - \gamma^{-1} Z_p \begin{bmatrix} 0 \\ 0 \\ 0 \\ 0 \\ 1 \end{bmatrix} [00001] Z_p \tag{10}$$

where, from (8),

$$\gamma - z_{33} + \frac{1}{\Delta y_3}.$$

[6] A single direct application of the Householder relation (5) to simultaneously account for all component changes is far more costly than the straightforward method used as a reference in Section II-B (see Table II, column 3). We therefore make no further reference to the "once only" application of (5).

[7] The first two elements of P and Q refer to the input and output ports, designated ϕ and ψ, respectively. The remaining elements are numbered 1, 2, 3.

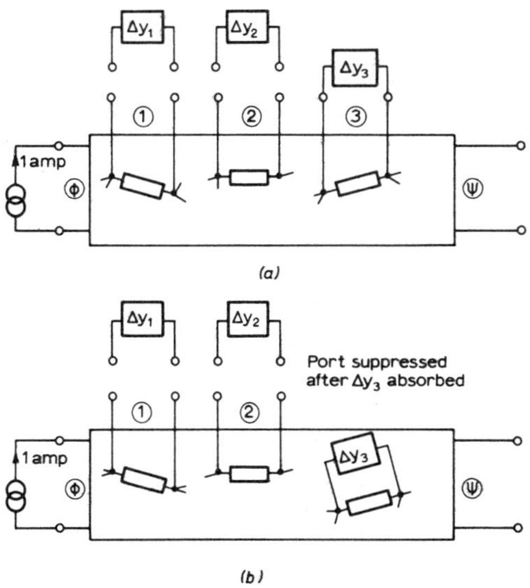

Fig. 6. Relevant to the matrix-modification method of predicting effect of simultaneous component changes.

(Note that z_{33} lies on the fifth row and column of Z_p.) A cost of unity is associated with the calculation of γ. The matrix product $Z_p PQZ_p$ on the right of (10) is

$$\begin{bmatrix} z_{\phi 3} \\ z_{\psi 3} \\ z_{13} \\ z_{23} \\ z_{33} \end{bmatrix} [z_{3\phi} z_{3\psi} z_{31} z_{32} z_{33}]. \tag{11}$$

Examination of (10) shows that if division by γ *precedes* calculation of the product (11), then determination of the new port impedance matrix Z_p' involves a total computational cost of $1 + 5 + 5^2 = 31$. The matrix so obtained describes the result of absorbing the admittance Δy_3 within the original circuit and leaving port 3 on open circuit. In the general case of m variable components, the calculation of Z_p' to account for the connection of one component[8] in the set of m is therefore $1 + (m + 2) + (m + 2)^2$. Starting with the matrix Z_p', the process just described is repeated in turn for the remaining externally connected admittances Δy_1 and Δy_2. The cost for each is the same as for Δy_3. Thus, for m variable components, the cost of computing the new port impedance matrix is easily seen to be $m[(m + 2)^2 + m + 3]$, which exceeds $3G_m$.

Considerable economy can be achieved if it is acknowledged that all but two of the $(m + 2)$ ports are artificially created, and that it may only be the 2×2-port impedance matrix of the *original 2-port* which is of interest. In this case, when modifying the matrix Z_p' to account for the connection of, say, Δy_3, row 3 and column 3 need not be computed, and can be omitted when updating to account for Δy_1 and Δy_2 (Figs. 6(b), 7). If this economy is practiced *successively* as the admittance changes are sequentially

[8] If more than one value of Δy_3 must be explored, the product (11) would be calculated *before* division by γ so that, after a computational cost of $(m + 2)^2$, each value of Δy_3 involves a cost of one (to find γ) plus $(m + 2)^2$.

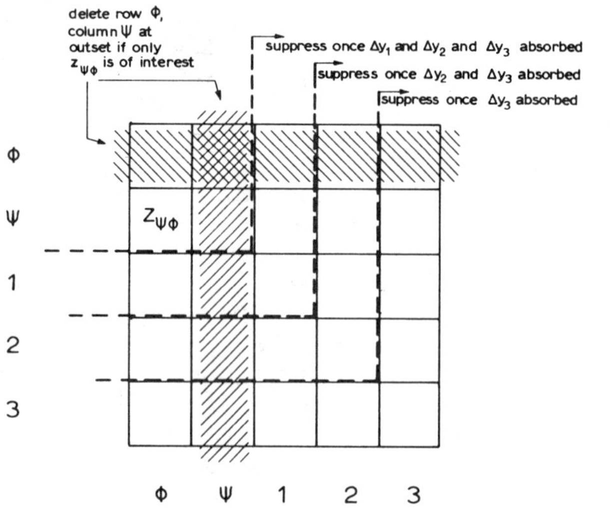

Fig. 7. Identification of rows and columns of port impedance matrix which can be discarded during sequential calculation of effect of component changes (see Fig. 6).

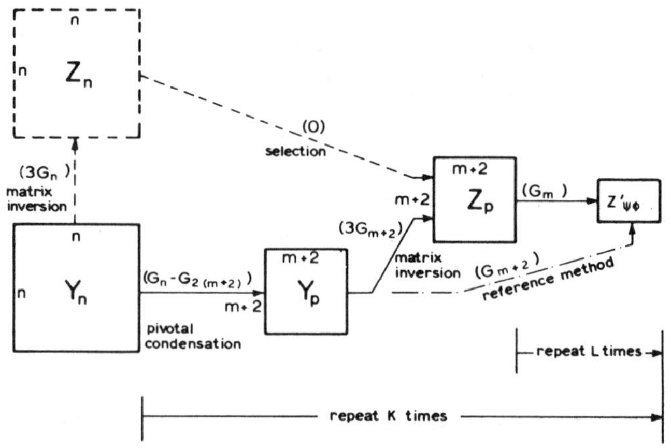

Fig. 8. Matrix-modification method of computing effect on circuit's transfer impedance of simultaneous changes in m components. Arrows indicating repeat calculations associated with L different values of same set of components, and K different groups of components, do not apply to reference method. Computational costs, some approximate, are indicated in parentheses.

absorbed within the circuit, the total cost of computing the new 2×2-port impedance matrix is about G_{m+1}. Moreover, if only a *single* transfer impedance (say $z_{\psi\phi}$) of the original 2-port is of interest, then further economy can be practiced by eliminating row ϕ and column ψ at the outset (Fig. 7); the total cost is then approximately G_m.

These economies can also be achieved if the circuit contains voltage-controlled current sources: each requires the creation of two new ports, one across the controlling voltage and the other across the controlled source. Rows associated with controlled ports, and columns with controlling ports, can be eliminated at the outset. Similarly, during matrix modification to account for change in a current source at port G controlled by a voltage at port H, row H and column G need not be computed, and can be omitted from the new port impedance matrix.

The matrix-modification approach is illustrated in Fig. 8 (full lines), where the computational costs are those associated with the economies described in the previous paragraph. If more than one group of components is known beforehand to be of interest, the alternative algorithm (dashed line) may be more economical. To summarize the costs involved, we make use of the summary (Table II) originally prepared for the current source substitution method, because the only change necessary is the omission of the cost $2m$ in row 3, columns 1 and 2.

C. Comparison of Methods

It appears that there is little to choose between the two methods and the reference method if the effect of *simultaneous* changes in m components is to be computed. Two comments are, however, appropriate. First, one possible advantage of the matrix modification method is that the new circuit description is available after one or more admittance changes have been absorbed: this is not the case with the current source substitution method. The second comment is relevant when each of m potentially variable components is varied *sequentially* rather than simultaneously, as might well happen during interactive, manual exploration by a designer. Following each component change, the new circuit description costs only $m + 3 + (m + 2)^2$ to compute, which is about $3/m$ times the cost of the current source substitution method and the reference method. In other words, the matrix modification method does not exact a penalty if the individual component changes take place one at a time.

An alternative approach to large-change multiparameter sensitivity analysis described by Singhal *et al.* [6] is based on a multilinear expression [18] for transfer impedance in which the numerator and denominator both contain all possible simple multiples of the variable admittances, each multiple being associated with a different coefficient.[9] However, the authors do not identify the computational cost of computing the numerator and denominator coefficients [6, step 4], a cost which appears to be sufficiently high to render the method unattractive. Even if this cost were negligible, the cost of evaluating the multilinear expression is greater than for the methods just reviewed, a conclusion which is supported by Gadenz *et al.* [5].

IV. Systematic Exploration

For multiple sets of simultaneous component changes, a further and often dramatic reduction in computational cost is possible if one condition holds. Namely, if each variable component can only assume one of a limited number of values, and if each set of m component values is some combination of these individual component values. Under these circumstances, *systematic exploration* of the different sets can be achieved at reduced cost.

[9] For three variable parameters, the expression has the form

$$z_{21} = \frac{a_{000} + a_{100}y_1 + a_{010}y_2 + a_{001}y_3 + a_{110}y_1y_2 + a_{101}y_1y_3 + a_{011}y_2y_3 + a_{111}y_1y_2y_3}{b_{000} + b_{100}y_1 + b_{010}y_2 + b_{001}y_3 + b_{110}y_1y_2 + b_{101}y_1y_3 + b_{011}y_2y_3 + b_{111}y_1y_2y_3}.$$

This additional saving is made possible by a fact already discussed in Section III-B: that if a component value is temporarily held constant, there is no further need to consider the port with which it is associated, and the port can be suppressed merely by eliminating the corresponding row and column from the port impedance matrix. In this way, only a smaller impedance matrix need be modified according to (10). More importantly, however, the longer a component can be held constant in value, the lower will be the cost of accounting for changes in the remaining components. Thus there is a computational advantage in rearranging the sequence in which the effect of the sets is computed. Firstly, the sets should be so arranged in sequence that one of the components changes as infrequently as possible. Then, within each group of sets throughout which the aforementioned component is constant in value, a fresh rearrangement into groups is effected so that a new component changes as infrequently as possible. This procedure is repeated as many times as there are components. Calculation of the effect of the component sets in the resulting sequence, while taking advantage of the computational savings associated with port suppression, is referred to as systematic exploration.

With reference to Fig. 6(a) suppose that, for each of C values of Δy_3, B predetermined values of Δy_2 are to be explored, and that for each value of Δy_2, the effect of A predetermined values of Δy_1 on the transfer impedance $z_{\psi\phi}$ is to be calculated. First, of course, since only $z_{\psi\phi}$ is of interest, row ϕ and column ψ of the port impedance matrix are immediately deleted, yielding a 4×4 matrix (the unshaded part of Fig. 7) which we shall denote by Z_4. Systematic exploration of the effect of the sets of component combinations then proceeds as follows. Initially, the first value of Δy_3 is absorbed. This involves the calculation of γ, the deletion of row and column 3 of Z_4, and the calculation of a new 3×3 impedance matrix Z_3 according to (11) and (10). Similarly, for the first value of Δy_2, a new 2×2-port impedance matrix (rows ψ,1; columns ϕ,1) is calculated from Z_3. Finally, the same procedure is applied for *each* value of Δy_1: first, the product $z_{\psi 1} z_{1\phi}$ ((11)) is computed; then, for each Δy_1, the constant γ is found at a cost of one and the ratio of $z_{\psi 1} z_{1\phi}$ and γ ((10)) also at a cost of one to determine the new $z_{\psi\phi}$. After all values of Δy_1 have been considered, the next value of Δy_2 is absorbed, as before, by operating upon the 3×3 matrix Z_3, a matrix which would continue to be stored until Δy_3 took on its second value. Thus the lower bound to which the "per-set" cost can asymptote is the remarkably low figure of *two*; this is the cost of computing, for each Δy_1, the constant γ and the ratio $(z_{\psi 1} z_{1\phi})/\gamma$.

Some appreciation of the savings offered by systematic exploration can be gained from the computational costs shown in Table III, which refers to the 3-component example of Fig. 6; in each case, the starting point is the impedance matrix (Z_p in Fig. 8) describing the original circuit with an additional port across each variable component. In case X, each of the 1000 possible combinations of 10 values each of the 3 variable components is independently evaluated by the matrix modification method. In case Y, the effect of the same combinations is explored systematically. The "per-set" cost shows a dramatic reduction.

TABLE III

	Number of Values of			Computational Cost	
	Δy_3	Δy_2	Δy_1	1000 Combinations	Per Set
X Matrix modification (Fig. 8)	10	10	10	23,000	23
Y Systematic exploration	10	10	10	2,749	2.75
Z Systematic exploration with ordering	4	10	25	2,305	2.31

Note: Computational costs associated with systematic and non-systematic exploration of the effect of simultaneous changes in three components. The cost of Case Z can be expressed as

$$(3^2) + 4\{1 + 3^2\} + 4(2^2) + 40\{1 + 2^2\} + 40(1^2) + 1000\{1 + 1^2\}$$

where normal brackets correspond to the calculation of equation (11) and curly brackets to the calculation of γ and the modification expressed by (10).

In the preceding examples, the *first* admittance considered (Δy_3) was the *most* expensive to absorb and, for ease of exposition, was associated with the *last* row and column of the impedance matrix (Fig. 7). Similarly, the *last* admittance absorbed (Δy_1) involved the *least* recurring "per-component" cost, and was associated with the *first* of the rows and columns pertinent to the variable components. Thus, although it is the *order* in which ports are suppressed that determines the cost of their suppression, it is convenient to (artificially) associate the cost of suppression with the port's row/column position in the impedance matrix.

As a general rule, the component whose value changes *least* frequently as the component sets are explored in sequence should be associated with the *most* expensive port, and vice versa. For the situation, like case Y, in which the number of values which each change can assume is equal, and all possible combinations of those values must be explored, it does not matter, of course, *which* of the components changes least or most frequently, provided the sets are ordered for systematic exploration as described earlier. But the assignment of components to ports would matter, for example, in the situation in which the number of values which a component can assume varies from one component to another. Case Z of Table III, like cases X and Y, concerns the systematic exploration of 1000 combinations of 3 components, but the combinations are those of 4 values of Δy_3, 10 values of Δy_2, and 25 values of Δy_1. The order in which the changes are accounted for (Δy_3, Δy_2, Δy_1; see Fig. 9) is such that the least fluctuating component (with 4 value changes) is associated with the most expensive port, the next least fluctuating component (with 40 value changes) with the next-to-most expensive port, and so on. The cost per set is below that of case Y although it cannot, of course, fall below 2.

The substitution current source method does not appear

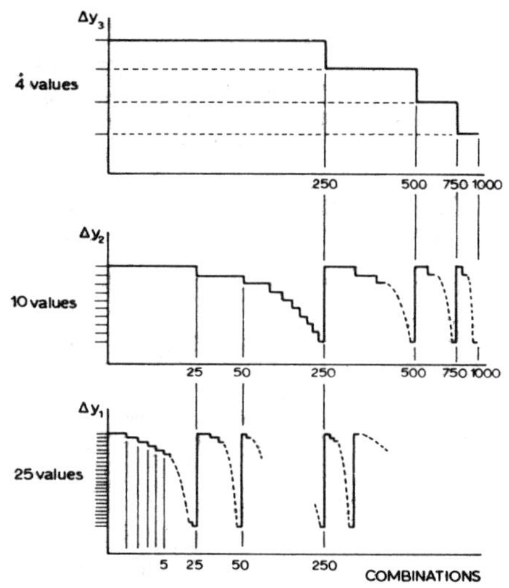

Fig. 9. Ordering of combinations of 4 values of Δy_3, 10 values of Δy_2, and 25 values of Δy_1 for systematic exploration of all possible combinations at lowest cost.

to lend itself to efficient systematic exploration, since a component value held constant does *not* imply that the corresponding substitution current source remains constant in value. The method of Singhal *et al.* [6] can be shown to be less efficient than matrix modification for systematic exploration.

V. Engineering Applications

The algorithm (Fig. 3) for large changes in a single component has already been incorporated within an interactive-graphic design facility [19] and found most effective. The short time delay (approximately, 0.5 s) between the specification of a new value and the display of the new frequency response permits *dynamic* exploration of the effect of component change. Recently, within the same medium and with the same order of time delay, the matrix-modification method has also been implemented to allow one-at-a-time adjustment of up to five parameters.

Two examples can be given of the manner in which circuit design can benefit from the method of systematic exploration. One is the generation of multidimensional performance contours. A two-dimensional performance contour describes how two component values may vary simultaneously—all others retaining their nominal value—without causing the circuit response to move outside a specified tolerance region. Both Butler [9] and Karafin [20] justified the use of such contours in economic circuit design, and Calahan [21] indicated that their display was the best method of communicating large-change sensitivity data to the designer who must decide parameter tolerance tradeoffs. Clearly, it is possible to extend the concept beyond two dimensions, and an efficient way of so doing is the technique of systematic exploration. First, for each frequency of interest; the values of $m-1$ of the m variable components are explored systematically by the matrix modification method. Then, for each set of $m-1$ changes, the corresponding value of the mth is found by the *inverse* of (1). The calculation and display of three-dimensional contours has already been implemented within an interactive graphic medium [19].

Systematic exploration can also be exploited in a new and efficient approach to statistical circuit analysis [22]. To obtain the benefit of systematic exploration, the probability distributions of the component values are first quantized[10] and then appropriately discretized. In this way, a reasonable approximation to component value distribution is obtained within the constraint of the limited values set required by systematic exploration. Then, the details of the individual trials are so ordered as to ensure that, in the calculation of response change, each component change can be held constant for as many trials as possible. Initial results are encouraging [22].

VI. Conclusions

Two approaches to the multiparameter large-change sensitivity analysis of linear circuits have been examined. For single component changes, the "per-component" computational cost can be in the order of 2. For simultaneous changes in m components, the cost per set of component values is at least G_m, and there is little to choose between the methods examined unless changes are also liable to occur one at a time, or the new circuit description is required, in which cases the matrix modification method is preferable. However, if systematic exploration is applicable to the problem in hand, the cost per set of m component changes can again be in the order of 2.

In common with other approaches (e.g., [5], [6]), the matrix-modification method suffers from the limitation that, as the number of variable components approaches and exceeds the number of nodes, the computational saving over straightforward repeated nodal analysis decreases. Research is proceeding on this problem.

Acknowledgment

The comments of Prof. G. Temes, Prof. P. Bryant, and two anonymous reviewers are gratefully acknowledged.

References

[1] J. F. Pinel and M. L. Blostein, "Computer techniques for the frequency analysis of linear electrical networks," *Proc. IEEE*, vol. 55, pp. 1810–1819, Nov. 1967.

[2] S. W. Director and R. A. Rohrer, "Automated network design—The frequency domain case," *IEEE Trans. Circuit Theory*, vol. CT-16, pp. 330–337, Aug. 1969.

[3] S. W. Director, "LU factorization in network sensitivity calculations," *IEEE Trans. Circuit Theory*, vol. CT-18, (Corresp.), pp. 184–185, Jan. 1971.

[4] P. J. Goddard, P. A. Villalaz, and R. Spence, "Method for the efficient computation of the large-change sensitivity of linear nonreciprocal networks," *Electron. Lett.*, vol. 7, pp. 112–113, 1971.

[5] R. N. Gadenz, M. G. Rezai-Fakhr, and G. C. Temes, "A method for the computation of large tolerance effects," *IEEE Trans. Circuit Theory*, vol. CT-20, pp. 704–708, Nov. 1973.

[6] K. Singhal, J. Vlach, and P. R. Bryant, "Efficient computation of large-change multiparameter sensitivity," *Circuit Theory Appl.*, vol. 1, pp. 237–247, 1973.

[7] D. M. Bohling and L. A. O'Neill, "An interactive computer ap-

[10] Since submission of this paper, further support for the use of quantization in statistical circuit analysis has come to our notice [23].

proach to tolerance analysis," *IEEE Trans. Comput.*, vol. C-19, pp. 10–16, Jan. 1970.

[8] K. H. Leung and R. Spence, "Tracking sensitivity: an efficient algorithm for linear nonreciprocal circuits," *Electron. Lett.*, vol. 10, no. 18, pp. 377–378, September 5, 1974.

[9] E. M. Butler, "Realistic design using large-change sensitivities and performance contours," *IEEE Trans. Circuit Theory*, vol. CT-18, pp. 58–66, Jan. 1971.

[10] G. D. Hachtel, R. K. Brayton, and F. G. Gustavson, "The sparse tableau approach to network analysis and design," *IEEE Trans. Circuit Theory*, vol. CT-18, pp. 101–113, Jan. 1971.

[11] C. A. Desoer and E. S. Kuh, *Basic Circuit Theory*. New York: McGraw-Hill, 1969, p. 634.

[12] ——, *Basic Circuit Theory*. New York: McGraw-Hill, 1969, p. 654.

[13] D. A. Calahan, *Computer-Aided Network Design*, Rev. Ed. New York: McGraw-Hill, 1972, p. 13.

[14] P. J. Goddard, "The calculation and reduction of electrical network sensitivity," Ph.D. Thesis, London University, England, Dec. 1971

[15] P. A. Villalaz, "Theory and techniques of network sensitivity applied to electrical device modelling and circuit design," Ph.D. Thesis, London University, England, Sept. 1972.

[16] A. S. Householder, "A survey of some closed methods for inverting matrices," *J. SIAM*, vol. 5, pp. 155–169, 1957.

[17] F. H. Branin, Jr., "The relation between Kron's method and the classical methods of network analysis," *Matrix Tensor Quart.*, vol. 12, no. 3, pp. 69–105, Mar. 1962.

[18] W. J. Troop and E. Peskin, "The transfer function and sensitivity of a network with *n* variable elements," *IEEE Trans. Circuit Theory* (Corresp.), vol. CT-16, pp. 242–244, May 1969.

[19] R. Spence and M. D. Apperley, "On the use of interactive graphics in circuit design," *Proc. 1974 IEEE Int. Symp. Circuits and Systems*, Paper R.14.6.

[20] B. J. Karafin, "The optimum assignment of component tolerances for electrical networks," *Bell Syst. Tech. J.*, vol. 50, no. 4, pp. 1225–1242, Apr. 1971.

[21] D. A. Calahan, *Ibid.* p. 174.

[22] K. H. Leung and R. Spence, "Efficient statistical circuit analysis," *Electron, Lett.*, vol. 10, no. 17, pp. 360–362, Aug. 22, 1974.

[23] T. R. Scott and T. P. Walker, Jr., "Regionalization: a method for generating joint density estimates," *Proc. 1975 IEEE Int. Symp. Circuits and Systems*, Paper S2.4.

Voting Techniques for Fault Diagnosis from Frequency-Domain Test-Data

H. Sriyananda
D.R. Towill
J.H. Williams

Summary—The basic voting technique, which estimates the posterior fault-probability from prior fault-probability plus test data, is developed from first principles and given a pattern-recognition interpretation. This leads to the philosophy that the technique is one of a family suitable for fault diagnosis, if necessary down to component level, using only input-output frequency-domain test-data. A simulation of 39 200 faulty 7-component passive circuits is described, which shows that the two voting techniques considered in detail are superior to previously reported template-matching methods of fault diagnosis by factors of 5:1 or more in situations where nonfaulty components are permitted to have standard deviations of 3% nominal value, which is not untypical of much current practice. The circuit model is similar to modules making up larger systems of several hundred components.

Because voting techniques require relatively little storage of data, they are well suited to quality control schemes based on automatic test equipment. In comparison, data storage for template-matching methods are often prohibitive. Best use of voting techniques is possible when clearly defined fault cases are established which should ideally be considered at the system design stage.

Reader Aids:
Purpose: Advance state-of-the-art
Special Math needed for explanations: Probability theory
Special Math needed for results: Same
Results useful to: Reliability engineers, system test designers, quality assurance managers, system maintenance engineers

1. INTRODUCTION

There is considerable activity in using frequency-domain test-data for fault diagnosis in failed systems. Such tests can lead to reduced numbers of access points, and hence a better engineered and more reliable product [1]. In return, data processing facilities are required, often in the form of an automatic test set. Various methods of obtaining frequency-response test-data have already been reviewed elsewhere [2], and need not be considered here.

The three most popular methods of subsequent fault diagnosis from frequency-domain test-data appear to be the following.

1) Derivation of the coefficients of a transfer-function model describing the system under test [3]. Provided that component values can be related to transfer-function coefficients, a set of nonlinear equations can be solved to estimate the actual component values of the system under test. This is not a trivial problem if only modest computing facilities are provided along with the test equipment. The most likely faulty components are then assumed to be those with the largest changes from nominal values. To obtain realistic estimates of the coefficients, it is imperative that the correct constraints be included in the curve-fit algorithm [3]. A model in transfer-function form must be presumed to exist, and must relate coefficients to component values, or parameter values as appropriate.

2) Complex-plane properties associated with fault conditions isolated via the bilinear transformation [4, 5]. At a specific test frequency, a fault will be identifiable from a unique vector describing the response deviation when plotted in the complex plane. In practice, test frequency selection is often difficult, as is the resolution between vectors for various fault cases [6].

3) Pattern recognition by grading system-response for known fault conditions [7, 8]. The gradings are coded according to the size of the response deviation from nominal, and, for a given fault condition, and the set of test frequencies chosen, one entry into the fault dictionary is constructed. This process is repeated for all conceived fault conditions. The results for the failed system under test are also coded, and compared with the fault dictionary until the matching signature is found, thus indicating the most likely fault.

The purpose of this paper is to introduce a new voting technique for fault diagnosis from frequency-domain test-data; it has two advantages

1) improved fault-diagnosis capability compared with the template-matching pattern-recognition in situations where there is realistic variation in values of nonfaulty components or parameters

2) modest computer requirements which make the technique highly suited for use in automatic test equipment

This paper is essential the condensation, combination, and application of two University of Wales internal publications [9, 10]. In choosing to illustrate the paper by fault diagnosis down to component level, it is not meant to imply that this level is always necessary or even desirable. It does, however, provide a realistic test case for evaluation of diagnostic procedures.

2. NOTATION

SUT	system under test
$H(j\omega)$	frequency-domain description of SUT
$y(\omega)$	deviation in frequency response obtained from measurements on failed SUT
J, K	indexes, always go from 1 to N

Reprinted from *IEEE Trans. Rel.*, vol. R-24, pp. 260–267, Oct. 1975.

$x_J(\omega)$	deviation in frequency response corresponding to fault case J
$\Phi(x_J, y)$	correlation between $y(\omega)$ and $x_J(\omega)$
$\Phi(x_K, x_J)$	correlation between fault cases $x_K(\omega)$ and $x_J(\omega)$
x_{iJ}^*	quantised value of $x_J(\omega)$ at test-datum i
y_{i*}	quantised value of $y(\omega)$ at test-datum i
$[x^*]^T$	recognition matrix ($T \equiv$ transpose)
g_i	fault-diagnosis index indicating worth of test-datum i
a_J	possible fault class
P_J	prior probability of fault case J
P_J^*	posterior probability of fault case J
z	fault-diagnosis vector†
z*	fault-diagnosis vector used in voting scheme†
M	number of test features (frequency-domain test-data)
N	number of fault cases
C, R	circuit parameters
Σ_i	implies sum over i from 1 to M

† Components of z are denoted by z or Z.

3. THEORETICAL DEVELOPMENT OF VOTING SCHEME

Suppose the system nominal response is $H_{NOM}(j\omega)$ and the response of a faulty SUT is $H_{SUT}(j\omega)$, then the information contained in the measurement set y can be defined by

$$y(\omega) \equiv f[H_{SUT}(j\omega) - H_{NOM}(j\omega)] \ . \tag{1}$$

The function sign simply means that $y(\omega)$ may take the form of gain information only, or complex number information at the discretion of the test designer. N distinct fault cases are proposed by the system designer, where N is not less than the number of elements (components or test irreducible subassemblies) in the SUT. For fault case J, the deviation in response, which is known by the test designer beforehand either by direct measurement or simulation, is

$$x_J(\omega) = f[H_J(j\omega) - H_{NOM}(j\omega)] \ . \tag{2}$$

We now correlate $x_J(\omega)$ and $y(\omega)$ to form a fault diagnosis index Z_J:

$$Z_J \equiv \Phi(x_J, y) = \int_{-\infty}^{+\infty} x_J(\omega)\, y(\omega)\, d\omega \ . \tag{3}$$

If the deviations in response corresponding to the fault conditions are uncorrelated, i.e.,

$$\Phi(x_K, x_J) = 0, \text{ for } K \neq J \ , \tag{4}$$

the largest Z_J in z will indicate the source of system failure. For physical systems, (4) rarely if ever holds. Nevertheless, it may be assumed that the element J which corresponds to the largest numerical value of Z_J is more likely to be faulty than any other, given that all elements had the same failure distributions.

4. DISCRETIZATION

In a physical test, measurements are usually made at discrete frequencies. Suppose M experimental data are available (from M/2 spot frequencies if all gain and phase data are used). The discrete form of (3) is then

$$Z_J = \Sigma_i\, x_{iJ}\, y_i \ , \tag{5}$$

with corresponding fault diagnosis vector,

$$\mathbf{z} = [x]^T \mathbf{y} \ . \tag{6}$$

The assumption of (4) reduces to

$$\Sigma_i x_{iJ}\, x_{iK} = 0, \text{ for } K \neq J \ . \tag{7}$$

Equation (7) differs from (4) in one important respect: the spot-test frequencies are chosen by the test designer so that (7) is much more likely to be true than (4). It would generally be sufficient to select the test frequencies so that

$$\Sigma_i (x_{iJ})^2 > \max_K \{|\Phi(x_J, x_K)|\}, \text{ for } J = 1, ..., N, \tag{8}$$

with some reasonable margin in hand. If this is not possible using any combination of test frequencies, additional test access points are needed.

5. QUANTISATION OF DEVIATIONS

For use with automatic test equipment, a fault-diagnosis technique ought to be as simple as possible so that demands for online and post-test computation are realistic. Based on time domain studies already reported [11] the response deviations ought to be quantised into three levels defined by

$$x^*_{iJ} = \begin{cases} 1, \text{ for } x_{iJ} > E_i \\ 0, \text{ for } |x_{iJ}| \leqslant E_i \\ -1, \text{ for } x_{iJ} < -E_i \end{cases} \tag{9}$$

$$y^*_i = \begin{cases} 1, \text{ for } y_i > E_i \\ 0, \text{ for } |y_i| \leqslant E_i \\ -1, \text{ for } y_i < -E_i \end{cases} \tag{10}$$

where * denotes quantisation. E_i is a predetermined deviation level to be set by the test designer, and can be related to gate sizes set for system checkout. Careful choice of E_i can help ensure that (8) is satisfied; indeed, for fault diagnosis in systems with drift in nonfaulty components and only noisy measurements available, an optimum quantisation level may exist which will maximise fault location probability. An example of $[x^*]^T$ is given in Table 1 for a 7-component passive network in which one fault case has been allowed per component.

Equation (7) has been evaluated for this recognition matrix, as shown in Table 2, in which the calculations have been normalised for each column by dividing $\Phi(x_J, x_K)$ by $\Sigma_i(x_{iJ})^2$. By inspection it can be seen that in 38 cases out of 42, $\Sigma_i(x_{iJ})^2 \geqslant |\Phi(x_J, x_K)|$ suggesting that the recognition matrix is based on a reasonable set of test frequencies. The four combinations where $\Sigma_i(x_{iJ})^2 = |\Phi(x_J, x_K)|$ are C_1R_3, C_1C_2, C_1C_3, and R_1R_2. In each case there are a substantial number of distinguishing data points despite the apparently high correlation.

6. ASSESSING THE WORTH OF AN INDIVIDUAL TEST-FEATURE

It is possible to assess the diagnostic worth of an individual test-feature, such as, for example, gain at any spot frequency. A suitable criterion can be intuitively defined as

$$g_i \equiv \log_N [\max_J \{\Pr\{a_J | x^*_{iJ}\}\}] + 1 . \tag{11}$$

The logarithmic form is proposed so as to generate a criterion that would possess additive properties. If $\Pr\{a_J | x^*_{iJ}\} = 1/N$ the proposed criterion is zero which may be thought of as the worst possible result, corresponding to equal probability of failure for all fault cases. On the other hand, if $\Pr\{a_J | x^*_{iJ}\} = 1$, which corresponds to a capability of detecting at least one particular fault with certainty, it can be shown [11] that

$$g_i = \log_N [\max_J \{ |x^*_{iJ} P(a_J)| / (\sum_{J=1}^{N} |x_{iJ}| P_J)\}] + 1; \tag{12}$$

log (0/0) is not defined, but the result for that case is −1 from physical reasoning applied to (12). If all prior probabilities are equal, (12) reduces to the simple form

$$g_i = -\log_N \begin{bmatrix} \text{number of nonzero terms in} \\ \text{row } i \text{ of recognition matrix} \end{bmatrix} + 1 . \tag{13}$$

Previous experimental work has shown that if the test features are ranked according to (13) then fault detection probability correlates well with the cumulative diagnosis index, Σg_i. Thus g_i can be regarded as a useful guide to the worth of test feature i, although the best set of features may not contain the best individual feature [12]. g_i values are shown in Table 1 for the features chosen as reasonable in testing the 7-component circuit.

7. MODIFICATION TO THE FAULT DIAGNOSIS VECTOR

A slight modification to the definition of z is now proposed. If $x^*_{iJ} \neq 0$, and $y_i = 0$, the contribution to z evaluated according to (5) is zero. However, this combination of values contains the information that fault case J is unlikely to exist. This argument cannot be extended to the case where $x^*_{iJ} = 0$ and $y^*_i \neq 0$ due to the errors involved in (7). To allow for inclusion of this information, z is modified as follows. If,

$$z^*_{FJ} = \Sigma_i x_{iJ}{}^* y_i{}^*, \; z^*_{NJ} = \Sigma_i |x_{iJ}{}^*| (1 - |y_i{}^*|)$$

then

$$z^*_J \equiv \begin{cases} (\text{sign } z^*_{FJ})[\,|z^*_{FJ}| - z^*_{NJ}], \text{ for } |z^*_{FJ}| > z^*_{NJ} \\ 0, \text{ otherwise} . \end{cases} \tag{14}$$

The proposed modification leads to a voting technique for fault diagnosis, since z is obtained by casting votes for and against the likelihood of a particular fault existing [9].

TABLE 1

Recognition Matrix for Template Voting

FREQUENCY RAD/SEC	MEASURE-MENT	R_1	R_2	R_3	R_4	C_1	C_2	C_3	g_i (%)
10	A.R.	-1	1						64
20	A.R.	-1	1						64
30	A.R.	-1							100
40	A.R.	-1							100
50	A.R.	-1							100
50	PHASE			-1				-1	64
60	PHASE			-1				-1	64
70	PHASE			-1				-1	64
80	A.R.			-1			-1	-1	44
80	PHASE			-1				-1	64
90	A.R.			-1			-1	-1	44
90	PHASE			-1				-1	64
100	A.R.			-1			-1	-1	44
100	PHASE			-1			-1	-1	44
200	A.R.			-1		1	-1	-1	29
200	PHASE			-1			-1	-1	44
300	A.R.			-1			-1	-1	44
2000	A.R.			-1					100
2000	PHASE				-1		-1	-1	44
3000	PHASE				-1		-1	-1	44
4000	PHASE				-1		-1		64
5000	PHASE				-1		-1		64
6000	PHASE				-1				100
7000	PHASE				-1				100
8000	PHASE				-1				100

(no entry implies zero)

TABLE 2

% Correlation between Fault Cases for Frequency Data Points Shown in Table 1

	R_1	R_2	R_3	R_4	C_1	C_2	C_3
R_1	100	100	0	0	0	0	0
R_2	−40	100	0	0	0	0	0
R_3	0	0	100	0	−100	64	86
R_4	0	0	0	100	0	36	14
C_1	0	0	−8	0	100	−9	−7
C_2	0	0	54	57	−100	100	64
C_3	0	0	92	29	−100	82	100
Σx^2_{ij}	5	2	13	7	1	11	14
max $\lvert\Phi x_J, x_K\rvert$	2	2	12	4	1	9	12

8. FAULT PROBABILITY RANKING

If **P** is the prior relative probability of each fault's occurring (such that $\sum_{J=1}^{N} P_J = 1$), then the posterior relative probability **P*** resulting from applying the voting technique, is

$$P_J^* = \begin{cases} P_J Z_J^*/D_N^*, & \text{if } D_N^* > 0 \\ 0, & \text{if } D_N^* = 0 \end{cases} \tag{15}$$

$$D_N^* \equiv \sum_{J=1}^{N} P_J |z_J^*|$$

Thus if a fault condition is detected, the faults are ranked so as to indicate the relative probability of each fault; and if no fault condition is detected, the fault probabilities are all set to zero.

9. IMPLEMENTING THE BASIC VOTING SCHEME

The setting up of a test schedule using the Basic voting scheme is shown by the following algorithm.

Step

0. a. Model the system under test (SUT).
   b. Obtain reliability data.
1. Tailor test signal to SUT.
2. Response for nominal and faulty systems incorporating reliability data from step # 0b.
3. Determine initial test features.
4. Set tolerance band from checkout.
5. Set up recognition matrix.
6. Check for fault coverage.
7. Calculate prior fault probabilities from reliability data in step #0b.
8. Form diagnosis and cumulative diagnosis index.
9. If features are not satisfactory, alter or add test features and go to step #5. Otherwise proceed.
10. If tolerance band is not satisfactory alter it and go to step #5. Otherwise proceed.
11. Initiate fault diagnosis test schedule using information from step #7.
12. Evaluate on faulty systems.
13. If diagnosis is not acceptable, amend tolerance bands and test features and go to step #5. Otherwise procedure is finished.

Reliability data are used in three ways;

1) in the definition of fault cases,
2) when setting up the recognition matrix, aided by (12),
3) when assessing individual fault likelihood via (15).

For many systems the recognition matrix plus 'prior knowledge leading to feature ranking' is sufficient for setting up a realistic test schedule. In other cases, learning during experimentation with a breadboard model or interactive simulation can be put to good use.

A number of examples of the basic voting technique are fully described in [9]. The method has greatly improved fault diagnosability in systems with measurement noise and parameter drift. Results given later in the paper will indicate the magnitude of the improvement. Experience suggests that for a good test schedule, $M \cong 3N$, so that if all gain and phase data are useful, $M \cong 3N/2$ spot frequencies are required. On the assumption that $3N$ data points are used, the computer requirements for implementing the diagnostic technique are approximately $(200 + 14N + 3N^2)$ words [9], and is thus well within modern minicomputer capacity for a realistically sized system.

10. GENERALISATION OF THE VOTING TECHNIQUE

The basic voting technique can be interpreted as a pattern recognition technique. Using the definitions of [13], **y*** is the feature vector spanning the feature space A_y; a_J are the N possible pattern classes corresponding to a fault case J, thus constituting the fault space. Considering only simple faults, these constitute n mutually exclusive regions. The decision process can be seen to be **P*** which is a nonlinear function of $[x^*]$ and **y***.

The simplest form of pattern recognition as applied to fault diagnosis from frequency-domain test-data is template matching [7, 8, 14] in which the range of possible measurements at each test frequency is graded into a number of divisions, each coded in logical order. Each fault case is then represented by a signature composed of the coded row vector and stored in a fault dictionary. Measurements made on a faulty system are then similarly graded and compared term-by-term with each fault case stored in the fault dictionary until agreement is found. Effective fault diagnosis by this method usually requires a relatively large fault dictionary; but more serious practical objections are that

1) measurement noise and drift in the nonfaulty components quickly reduce the fault diagnosability to an unacceptably low level.

2) if the initial diagnosis is incorrect, there is no second fault suggested, as would be the case for the voting technique when the computer printout includes the complete rankings of fault cases in order of probability of occurrence.

The results shown in Fig. 1 vividly illustrate the rapid fall-off in fault diagnosability for a particular 100 sample test case as the variation of the nonfaulty components increases.

In attempting to improve the efficiency of template-matching to overcome these two deficiences, it was realised that the voting technique is likely to be only one of a family of such techniques which might be profitably used in fault diagnosis [10]. For example, in terms of template-matching, we might vote 'for' the fault case if the measurement and anticipated deviation agree, and vote 'against' the fault case otherwise. Applied to quantisation levels defined by (9) and (10), the fault diagnosis index can be written, for

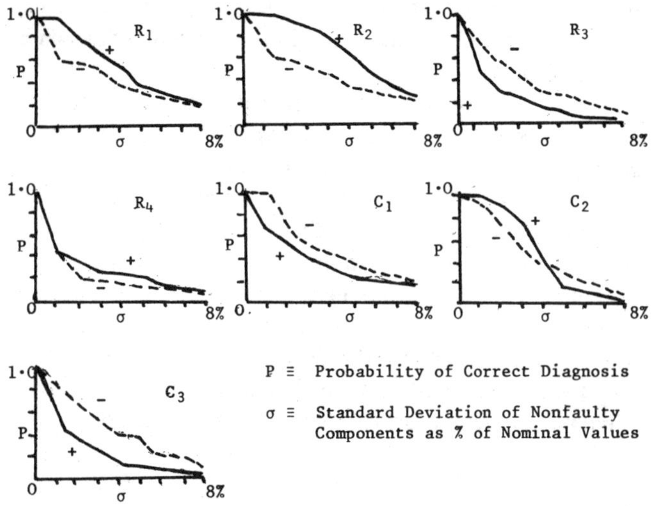

Fig. 1. Fall Off in Diagnosability via Template Matching as Variability of Nonfaulty Components Increases.

$$z^*_{FJ} \equiv \Sigma_i x^*_{iJ} y_i^* + \Sigma_i (1 - |x^*_{iJ}|)(1 - |y_i^*|),$$

$$z^*_{NJ} \equiv M - z^*_{FJ}$$

as,

$$z^*_J \equiv \begin{cases} 2z^*_{FJ} - M, \text{if } (2z^*_{FJ} - M) > 0 \\ 0, \text{ otherwise} \end{cases} \qquad (16)$$

This alternate voting technique is termed template voting.

11. ALTERNATE WAYS OF SELECTING THE RECOGNITION MATRIX

As an alternative to selecting the recognition matrix in the manner of sections 4, 5, 6, it is possible to select test frequencies related to the break points of the Bode plot of the nominal SUT [7]. In [7] it is suggested that extra frequencies must usually be chosen to supplement this basic choice, but a procedure for doing so is not given. A second alternative is to choose test frequencies for which response deviations are a maximum [10]; this will generally lead to $M \cong N$.

In the basic voting scheme, both positive and negative changes in component value are considered as a single fault, and the recognition matrix is set up using the average value of the quantized response deviations. During diagnosis, the sign of the cumulative vote indicates the direction of deviation. With template voting, these two cases are considered as two distinct faults, which results in twice the number of fault cases, because the deviations caused by positive and negative changes are generally not symmetrical. This is illustrated by the recognition matrix shown in Table 3, where only R_1 and R_4 possess the property of symmetry, even if the variations in the quantising levels are ignored.

TABLE 3

Recognition Matrix for Template Voting

FREQUENCY RAD/SEC	MEASURE-MENT	R_1 +	R_1 −	R_2 +	R_2 −	R_3 +	R_3 −	R_4 +	R_4 −	C_1 +	C_1 −	C_2 +	C_2 −	C_3 +	C_3 −
20	A.R.	-1	1	1	-2										
40	A.R.	-1	2		-2										
50	PHASE					-1	1							-1	1
200	A.R.					-2	2			1	-2	-2	2	-2	2
400	A.R.					-2	2	-1	1		-1	-2	2	-2	2
2000	PHASE						2	-2	2			-1	2	-1	2
7000	PHASE							-1	2				2		1

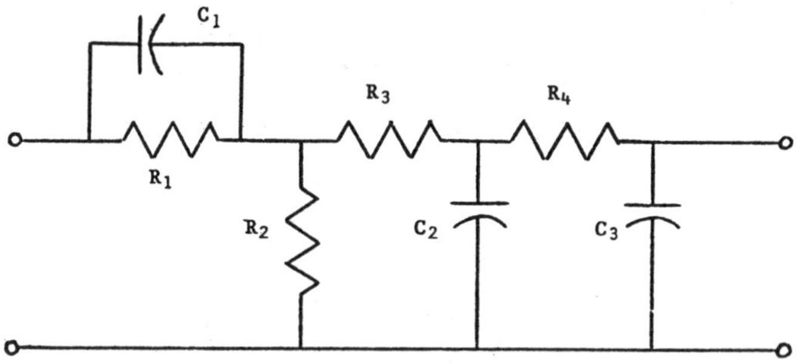

R_1 = 1MΩ, C_1 = 0·01μF
R_2 = 10MΩ, C_2 = 0·001μF
R_3 = 2MΩ, C_3 = 0·001μF
R_4 = 1MΩ,

$$\frac{V_o}{V_i}(s) = \frac{a_o + a_1 s}{b_o + b_1 s + b_2 s^2 + b_3 s^3}$$

$a_o \equiv R_2$
$a_1 \equiv R_1 R_2 C_1$
$b_o \equiv R_1 + R_2$
$b_1 \equiv R_1 C_1 R_2 + R_2 R_4 C_3 + (C_2 + C_3)(R_2 R_3 + R_3 R_1 + R_2 R_1) + R_1 R_4 C_3$
$b_2 \equiv R_1 R_2 R_3 C_1 (C_2 + C_3) + R_1 R_2 R_4 C_3 (C_1 + C_2) + R_3 R_4 C_2 C_3 (R_1 + R_2)$
$b_3 \equiv R_1 R_2 R_3 R_4 C_1 C_2 C_3$

Fig. 2. Circuit used for Simulation Studies and Gain Measurements made on Hardware Model.

12. COMPUTER SIMULATION

The response with which the basic voting technique has so far been received at industrial seminars has prompted questions as to why it works so well relative to alternate methods. Is it the voting logic, choice of recognition matrix, number of test frequencies, the number and/or amplitude of quantizing levels, which is really important; and is fault-case separation into positive and negative sections useful? To answer these questions a large scale simulation was undertaken on the test circuit of Fig. 2.

Experimental results are also shown for a hardware circuit manufactured from typical vendors' components. The results were obtained using a Fourier Response Analyser designed on the correlation principle [15]. The choice of circuit to which

Tables 1, 2, 3 refer, is in line with examples chosen by previous contributors as a suitable system for comparison; it is not intended to imply a restriction on the voting technique, which has already been successfully applied to more complex systems. The selection of fault cases, as distinct from the extensive feature-space scan usually associated with template-matching pattern-recognition, is itself useful in committing system designers to a statement of likely modes of failure. At present there is little guidance in the literature regarding failure modes and their probabilities, yet these data must heavily influence any optimum test schedule. In the absence of such guidance, we feel that test cases can be regarded initially as belonging to one of two sets.

1) There is a fault corresponding to a large change in one component, such as might result in an open or short circuit. If observable at all from input-output measurements only, these faults are usually simple to identify since unique major changes are often observed in the system time or frequency response [16].

2) One component changes in value sufficiently to drive the system just outside specification, but all other components are also permitted to have reasonable production tolerances. This is the more difficult fault detection problem. It has been argued that this is *the* diagnostic test case [17], particularly since component vendors take full advantage of production tolerances. Hence this paper concentrates on this particular case.

Table 4 lists the 7 combinations of voting logic, measurements, number of faults and digitizing states considered. For each combination, 14 possible faults (+ and − changes in 7 component values) were each simulated 100 times giving a total of 1400 simulations per combination, for a total of 9800. Further 4 types of faults were imposed; ±35%, ±50%, ±70% of the nominal values and variable faults where the faulty component value was drawn from a Gaussian distribution with coefficient of variation of 30%.

TABLE 4

SIMULATION PLAN

CODE	VOTING LOGIC	RECOGNITION MATRIX	NO. OF FAULTS	NO. OF DIGITIZING STATES	NO. OF FREQUENCIES	NO. OF MEASUREMENTS
A	BASIC	TABLE 1	7	3	19	25
B	TEMPLATE	TABLE 3	14	5	7	7
C	BASIC	ALL FREQS*	7	3	28	56
D	TEMPLATE	ALL FREQS*	14	5	28	56
E	BASIC	TABLE 3	7	3	7	7
F	TEMPLATE	TABLE 1	14	5	19	25
G	BASIC	ALL FREQS*	14	3	28	56
H	EXACT MATCHING WITH TABLE 1				19	25
I	EXACT MATCHING WITH TABLE 3				7	7

* $10^i \rightarrow 10^{i+1}$ RAD/SEC IN STEPS OF 10^i RAD/SEC, $i = 1,2,3$
GAIN AND PHASE MEASUREMENTS

Two template-matching cases have been included in order to provide a datum against which to measure the improvement obtained by any voting scheme, although the results are strictly comparable only for the ±50% deviations since these are deemed to be the fault cases on which the recognition matrices are based. Throughout the simulation, the coefficient of variation of the nonfaulty components is 3%. The simulation, in flow diagram form, is given below.

COMPUTER SIMULATION FLOW DIAGRAM

Step

0. Calculate system's response with all components on nominal.
1. Impose fault on chosen component*.
2. Randomly perturb non-faulty components**.
3. Calculate system response with simulated component values from steps #1 and #2.
4. Calculate deviations from nominal response using steps #0 and #3.
5. Vote for likely fault using a) Sriyananda scheme or b) template voting.
6. Repeat from step #1 until number of required simulations is complete for chosen component.
7. Impose fault on chosen component and go to step #2. If all component faults are simulated, stop.

13. RESULTS AND DISCUSSION

Simulation results are given in Table 5. As anticipated there is a falling off in diagnosability if the average value of fault is not the same as that chosen when setting up the recognition matrix. Nevertheless, diagnosability using only input-output test data is still feasible. On a purely subjective basis, taking all four test cases into account, the nine combinations for which total results are shown in Table 6 can be grouped into four categories of effectiveness as follows:

	measurements	faults	digitizing states
	Category 1		
B - template voting,	7	14	5
F - template voting,	25	14	5
	Category 2		
A - basic voting,	25	7	3
D - template voting,	56	14	5
C - basic voting,	56	7	3
	Category 3		
E - basic voting,	7	7	3
G - basic voting,	56	14	3

---

*either fixed faults of ±35%, ±50%, ±70% or random faults with coefficient of variation of 30%.

**random perturbation with coefficient of variation of 3%.

	measurements	faults	digitizing states
Category 4			
H - exact template matching,	25	77	3
I - exact template matching,	7	14	5

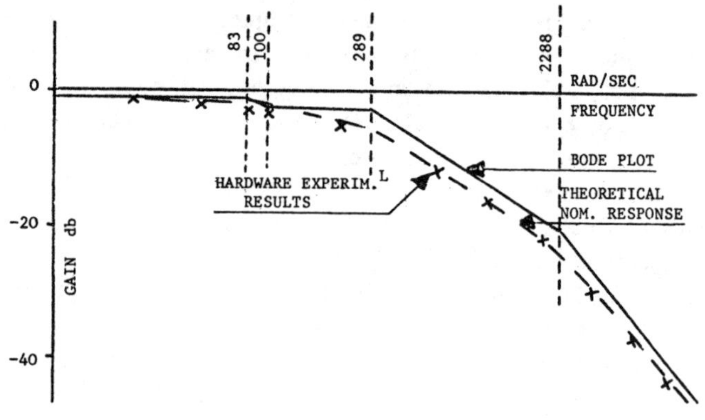

Fig. 3.

TABLE 5

% Successful Diagnosis, Nonfaulty Components, STD. DEV = 3% Nominal Values

±70% CHANGE IN COMPONENT VALUES

	CODE A	B	C	D	E	F	G	H	I
R_1 +	100	45	100	36	100	36	47	2	8
R_1 -	100	100	100	100	100	100	100	0	0
R_2 +	16	98	48	96	16	96	47	1	17
R_2 -	0	100	0	100	0	100	100	0	4
R_3 +	91	56	91	100	57	83	91	0	33
R_3 -	0	73	0	0	0	0	0	0	0
R_4 +	97	100	97	95	100	99	87	0	23
R_4 -	100	88	100	94	100	98	100	5	38
C_1 +	88	100	88	100	75	100	76	0	73
C_1 -	81	2	95	100	0	67	100	0	0
C_2 +	4	41	13	12	36	11	6	0	23
C_2 -	1	13	1	7	12	13	1	0	13
C_3 +	91	97	89	91	100	99	94	1	14
C_3 -	100	100	100	100	100	100	100	0	0

±50% CHANGE IN COMPONENT VALUES

	A	B	C	D	E	F	G	H	I
R_1 +	100	100	100	100	100	100	78	43	95
R_1 -	100	100	100	100	100	100	100	12	66
R_2 +	62	99	74	98	74	99	74	30	66
R_2 -	0	100	0	100	0	100	99	69	88
R_3 +	97	80	97	100	80	96	97	9	67
R_3 -	0	69	0	99	0	98	93	9	34
R_4 +	99	94	96	92	94	100	95	15	29
R_4 -	100	97	100	100	100	100	75	57	25
C_1 +	99	100	99	100	81	100	74	23	80
C_1 -	100	78	100	100	6	100	82	15	62
C_2 +	94	99	92	88	99	89	88	1	76
C_2 -	94	98	84	93	100	100	84	15	97
C_3 +	78	55	77	96	49	88	90	10	45
C_3 -	99	100	99	100	85	100	100	3	83

±35% CHANGE IN COMPONENT VALUES

	A	B	C	D	E	F	G	H	I
R_1 +	92	98	79	91	89	98	71	62	86
R_1 -	100	100	100	87	100	87	87	0	0
R_2 +	31	61	44	77	41	72	34	24	42
R_2 -	1	1	3	1	1	1	15	0	0
R_3 +	86	94	88	37	93	40	93	0	0
R_3 -	54	100	61	79	24	57	13	0	0
R_4 +	88	24	46	32	20	47	46	0	0
R_4 -	100	98	99	44	100	100	18	0	0
C_1 +	100	100	100	100	80	100	27	1	76
C_1 -	100	37	100	74	62	100	40	0	9
C_2 +	69	88	75	48	98	53	58	3	26
C_2 -	69	52	38	31	100	47	45	0	0
C_3 +	27	0	23	15	1	9	22	0	0
C_3 -	3	19	3	0	0	0	5	0	0

RANDOM CHANGES IN COMPONENT VALUES*

	A	B	C	D	E	F	G	H	I
R_1 +	91	90	84	87	92	90	69	44	78
R_1 -	92	87	93	74	87	76	76	2	7
R_2 +	33	88	51	90	38	86	51	18	40
R_2 -	16	40	17	34	16	37	50	8	22
R_3 +	43	66	43	38	88	35	44	1	19
R_3 -	23	52	22	28	47	26	19	1	4
R_4 +	65	45	50	46	33	48	48	4	11
R_4 -	82	54	54	42	68	64	17	8	9
C_1 +	94	98	96	98	71	98	45	6	68
C_1 -	98	38	99	73	46	97	52	2	16
C_2 +	46	70	42	44	87	49	37	0	30
C_2 -	54	33	41	27	81	32	35	0	15
C_3 +	24	18	25	31	18	31	22	1	11
C_3 -	19	21	21	16	16	16	24	0	7

*FAULTY COMPONENT VALUES DRAWN FROM NORMAL DISTRIBUTION WITH MEAN = NOMINAL VALUE, STD. DEV = 30% OF NOMINAL VALUE

The difference between categories 1 and 2 is marginal, but the differences thereafter are great. In some cases any voting technique gives a great improvement in diagnosability compared with template matching; factors of 5:1 are common. In practice, a voting technique is even more superior because the likely faults are ranked in order of probability. For the purpose of this paper, only the highest ranked fault has been considered, which is a pessimistic viewpoint. The simulation confirms that with carefully chosen data points there is little purpose in greatly exceeding $3N$ data points when using basic-voting. Template-voting seems at least as good as basic-voting, or even marginally better, as comparable results are obtained using much fewer data points. This is encouraging, because template-voting may find wider applicability.

14. CONCLUSION

This report confirms the promise of basic-voting for fault diagnosis as a useful tool in fault detection using only input-output frequency-domain test-data. Improvements in diagnosability are possible through the use of an alternate voting logic and/or alternate test frequencies. Compared with template-matching, voting techniques are an order of magnitude better for fault diagnosis of systems where there is important variation in nonfault components. Provided that anticipated fault cases can be reasonably defined, the voting technique can be regarded as fully tested by both analogue and digital simulation. Where such a fault case definition is not feasible, voting is still helpful. Further work on scaling measurement-data would then be appropriate.

TABLE 6

Total Scores and Ranking for Each Simulation

CODE	±50% FAULT SCORE	±50% FAULT RANK	±35% FAULT SCORE	±35% FAULT RANK	±70% FAULT SCORE	±70% FAULT RANK	RANDOM FAULT SCORE	RANDOM FAULT RANK	TOTAL SCORE	TOTAL RANK
A	1122	6	920	1	869	6	780	4	3691	17
B	1269	3	872	2	1013	2	800	1	3954	8
C	1229	4	849	3	922	5	738	5	3738	17
D	1366	2	716	6	1031	1	728	6	3841	15
E	968	7	809	5	796	7	788	2	3361	21
F	1370	1	811	4	1002	3	785	3	3968	11
G	1229	4	574	7	949	4	589	7	3341	22
H	311	9	90	9	9	9	95	9	505	36
I	913	8	239	8	246	8	336	8	1734	32

MAXIMUM POSSIBLE SCORE = 1400

15. ACKNOWLEDGMENTS

This study was partially supported by the Electrical Quality Directorate who financed UWIST Industrial Research Associate J.H. Williams, and by UWIST via a research studentship supporting H. Sriyananda during his Doctoral studies. Thanks are due to Miss G. Eschle for setting out tables and figures and typing the manuscript.

REFERENCES

[1] R.F. Garzia, "Fault Isolation Computer Methods," NASA Contractor Report CR-1758, February 1971, available from NTIS, Springfield, Virginia 22151 USA.

[2] D.R. Towill and P.A. Payne, "Frequency Domain Approach to the Automatic Testing of Control Systems," Proc. Conf. on Automatic Test Systems, Birmingham, U.K. April 1970, IERE, London, Conf. Publication No: 17, pp. 529-548.

[3] P.A. Payne, "An Improved Technique for Transfer Function Synthesis from Frequency Response Data," *IEEE Trans. Automatic Control*, Vol. AC-15, August 1970, pp. 480-483.

[4] Le Phung and Shu-Gar Chan, "On Fault Isolation," *Proc. 1970 Asilomar Conf. on Circuits and Systems,* pp. 41-49, available from Western Periodicals Co., 13 000 Raymer Street, North Hollywood, California USA.

[5] G.O. Martens and J.D. Dyck, "Fault Identification in Electronic Circuits with the Aid of Bilinear Transformations," *IEEE Trans. on Reliability,* Vol. R-21, May 1972, pp. 99-104.

[6] J.H. Williams, "Transform Method of Computerised Fault Location," Conf. on "The Use of Digital Computers in Measurement," University of York, England, 24-27 September 1973. IEE, London, Conference Publication No: 103.

[7] S. Seshu and R. Waxman, "Fault Isolation in Conventional Linear Systems—A Feasibility Study," *IEEE Trans. on Reliability,* Vol. R-15, May 1966, pp. 11-16.

[8] J.H. Maenpaa, C.J. Stehman, and W.J. Stahl, "Fault Isolation in Conventional Linear Systems: A Progress Report," *IEEE Trans. on Reliability,* Vol. R-18, February 1969, pp. 12-14.

[9] H. Sriyananda,* PhD Thesis, UWIST, King Edward VII Avenue, Cardiff, U.K. 1973

[10] J.H. Williams and D.R. Towill,* "Optimisation of Fault Diagnosis Capability Using Frequency Domain Test Data," UWIST, King Edward VII Avenue, Cardiff, U.K. DAG TN 61, September 1973.

[11] H. Sriyananda and D.R. Towill, "Fault Diagnosis Using Time Domain Measurements," *Radio and Electronic Engineer,* Vol. 43, September 1973, pp. 523-533.

[12] G.T. Toussaint, "Note on the Selection of Independent Binary Valued Features for Pattern Recognition," *IEEE Trans. on Information Theory,* Vol. IT-17, September 1971, p. 618.

[13] K.S. Fu, "Sequential Methods in Pattern Recognition and Machine Learning," Academic Press, New York, 1968, Chapter 1.

[14] H. Sriyananda, D.R. Towill, J.H. Williams, "Selection of Test Frequencies for System Fault Diagnosis," AIAA Paper No: 70-864, AIIA Guidance and Control Conference, Key Biscayne, Fl, August 1973, available from TIS, 750 3rd Avenue, New York, N.Y. 10017 USA.

[15] C.S. Elsden and A.J. Ley, "A Digital Transfer Function Based on Pulse Rate Techniques," *Automatica,* Vol. 5, January 1969, pp. 51-60.

[16] P.A. Payne,* PhD Thesis, UWIST, King Edward VII Avenue, Cardiff, U.K. 1972.

[17] E. Booth, "Reliability and Economic Factors in Design of Shipborne Electronic Equipment," Paper presented at IEE Symposium on Electronics, Measurements, and Control in Ships and Shipbuilding, Glasgow, April 1966, availabe from the IEE Library, London, U.K.

*Available from UWIST Library, King Edward VII Avenue, Cardiff, CF1 3NU, U.K.

Fault Diagnosis for Linear Systems Via Multifrequency Measurements

NEERAJ SEN, MEMBER, IEEE, AND RICHARD SAEKS, FELLOW, IEEE

Abstract—**The fault diagnosis problem for a linear system whose transfer function matrix is measured at a discrete set of frequencies is formalized. A measure of solvability for the resultant equations and a measure of testability for the unit under test is developed. These, in turn, are used as the basis of algorithms for choosing test points and test frequencies.**

I. Introduction

CONCEPTUALLY, the fault analysis problem for an analog circuit or system amounts to the measurement of a set of externally accessible parameters of the system from which one desires to determine the internal system parameters or equivalently locate the failed components as illustrated in Fig. 1. Here, the measurements m_i may represent data taken at distinct test points or alternatively, data taken at a fixed test point under different stimuli. Similarly, the r_i represent parameters characterizing the various internal system components. Here, a single parameter may characterize an entire component, say a

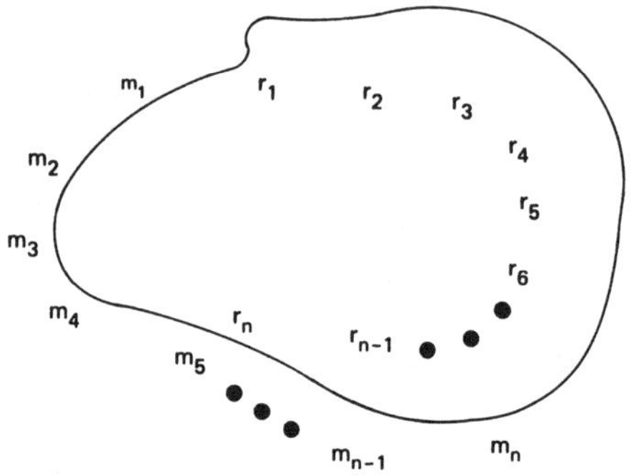

Fig. 1. Conceptual model of fault diagnosis problem.

Manuscript received July 3, 1978; revised April 2, 1979. This research was supported in part by Office of Naval Research Contracts 75-C-0924 and 76-C-1136.

N. Sen was with the Department of Electrical Engineering, Texas Tech University, Lubbock, TX. He is now with the Datapoint Corporation, San Antonio, TX.

R. Saeks is with the Department of Electrical Engineering, Texas Tech University, Lubbock, TX 79409.

Reprinted from *IEEE Trans. Circuits & Systems*, vol. CAS-26, pp. 457–465, July 1979.

resistance, capacitance or inductance. Alternatively, a component may be represented by several parameters: the h parameters of a transistor, the poles and gain of an op-amp, etc. In general, one models a system component by the minimum number of parameters which will allow the failure to be isolated up to a shop replaceable assembly (SRA) with all "allowed" system failures manifesting themselves in the form of some parameter change.

To solve the fault diagnosis problem, one then measures $m = \text{col}\,(m_i)$ and solves a nonlinear algebraic equation

$$m = F(r) \tag{1}$$

for $r = \text{col}\,(r_i)$ to diagnose the fault. The parameters in the resultant r vector which are out of tolerance then indicate the faulty component [6].

The purpose of the present paper is to give an explicit formulation of the *fault diagnosis equations* which arise in the maintenance of linear systems. Here, one measures the system frequency response as observed from a specified set of externally accessible test points at a discrete set of frequencies and it is desired to solve for a vector of internal system parameters r which completely characterize the frequency response matrices of the individual system components; $Z_i(s,r)$, $i = 1,2,\cdots,q$.

In the following section the explicit form for the fault diagnosis equations is derived for a given set of test frequencies. A *measure of solvability* [15] of these equations is then developed in Section III and employed in Section IV in an algorithm for optimally selecting test frequencies. The measure of solvability for the fault analysis equations, given an optimal choice of test frequencies, is then taken as a *measure of testability* [1], [2], [5] for the unit under test (UUT) and is used as the basis of an algorithm for the optical choice of test points [3]–[5]. Finally, a number of examples are presented in Section V.

II. Explicit Form of the Fault Diagnosis Equations

In the case of a linear time-invariant circuit or system, the fault diagnosis equations may be expressed in analytical form [6]. Since the fault diagnosis equations deal with the relationship between the externally measurable system parameters m and the internal component parameters r we adopt a *component connection model* as the starting point for the derivation of the fault diagnosis equations [7], [8]. This is one of several commonly employed large scale system models in which the components and connections in a circuit or system are modeled by distinct equations, thereby permitting one to explicitally deal with the relationship between the individual component parameters and the composite system parameters.

Since the present study is restricted to linear time-invariant systems, we assume that each component is characterized by a transfer function matrix which is dependent on the potentially variable component parameters, $Z_i(s,r)$. For the classical RLC components $Z_i(s,r)$ may take the form R, Ls, or $1/sC$ for the case of a resistor, inductor, or capacitor, respectively. More generally, one may model an op-amp by the transfer function $k/(s-p_1)(s-p_2)$ where the parameter vector r now represents the three potentially variable component parameters; k, p_1, p_2; or a delay by ke^{sT}, etc. Although the symbol Z is used, the components are not assumed to be represented by impedance matrices. Indeed, hybrid models are used in most of our examples. For the purpose of analysis, it is assumed that all faults manifest themselves in the form of changes, possibly catastrophic, in the parameter vector r with the frequency characteristics of the components unchanged. Although not universal, this *fault hypothesis* covers the most commonly encountered situations and subsumes the common industrial practice of assuming that all failures in analog circuits and systems take the form of open and short circuited components [9].

Our system components are thus characterized by a set of simultaneous equations

$$b_i = Z_i(s,r)a_i, \qquad i = 1,2,\cdots,q \tag{2}$$

where a_i and b_i denote the component input and output vectors, respectively. For notational brevity, these component equations may be combined into a single block diagonal matrix equation

$$b = Z(s,r)a \tag{3}$$

where $b = \text{col}\,(b_i)$, $a = \text{col}\,(a_i)$, and $Z(s,r) = \text{diag}\,(Z_i(s,r))$.

Although there are many ways to represent the connection in a circuit or system; say, a block diagram, linear graph or signal flow graph, any such representation is simply a graphical means for displaying a set of connection equations: Kirchhoff laws, adder equations, etc. As such, for our component connection model we adopt a purely algebraic connection model in which the connection equations are displayed explicitly without the intermediary of some kind of graphical connection diagram. This takes the form

$$\begin{aligned} a &= L_{11}b + L_{12}u \\ y &= L_{21}b + L_{22}u \end{aligned} \tag{4}$$

where u and y represent the vectors of accessible inputs and outputs which are available to the test system. In simple systems, the connection matrices L_{ij} are usually obtainable by inspection, whereas, in more complex systems, computer codes have been developed for their derivation [7]. Moreover, they are assured to exist in all but the most pathological systems [8].

It is the pair of simultaneous matrix equations (3) and (4) which are termed the component connection model. By combining (3) and (4) to eliminate the component input and output variables a and b one may derive [6], [7] an expression for the transfer function matrix observable by the test system between the test input and output vectors u and y obtaining

$$S(s,r) = L_{22} + L_{21}(1 - Z(s,r)L_{11})^{-1}Z(s,r)L_{12} \tag{5}$$

where

$$y = S(s,r)u. \tag{6}$$

For a linear time-invariant system the transfer function $S(s,r)$ is a complete description of the measurable data about the unit under test available to the test system. Moreover, being rational it is completely determined by its value at a finite number of frequencies. As such, without loss of generality, we may take our measured data to be of the form

$$\operatorname{col}\left[S(s_1,r),S(s_2,r),\cdots,S(s_k,r)\right]. \tag{7}$$

The *fault diagnosis equations* then take the form

$$\left.\begin{matrix} S(s_1,r) \\ S(s_2,r) \\ \vdots \\ S(s_k,r) \end{matrix}\right| \begin{bmatrix} L_{22}+L_{21}(1-Z(s_1,r)L_{11})^{-1}Z(s_1,r)L_{12} \\ L_{22}+L_{21}(1-Z(s_2,r)L_{11})^{-1}Z(s_2,r)L_{12} \\ \vdots \\ L_{22}+L_{21}(1-Z(s_k,r)L_{11})^{-1}Z(s_k,r)L_{12} \end{bmatrix}. \tag{8}$$

Since $S(s,r)$ is, in general, a matrix, the fault diagnosis equations as derived above take the form of a matrix (col $[S(s_i,r)]$) valued function of a vector valued variable r. Computationally, however, we prefer to work with a vector valued function of a vector valued variable and hence, we transform $S(s,r)$ into a column vector via

$$\operatorname{vec}\left[S(s,r)\right]=\operatorname{col}\left[S^i(s,r)\right] \tag{9}$$

where $S^i(s,r)$ denotes the ith column of the matrix, $S(s,r)$. With the aid of the identity $\operatorname{vec}[XYZ]=[Z^t\otimes X]\operatorname{vec}[Y]$ (8) then transforms into [7], [12]

$$M=\left.\begin{matrix} \operatorname{vec}\left[S(s_1,r)\right] \\ \operatorname{vec}\left[S(s_2,r)\right] \\ \vdots \\ \operatorname{vec}\left[S(s_k,r)\right] \end{matrix}\right| \begin{bmatrix} =\operatorname{vec}\left[L_{22}\right]+\left[L_{12}^t\otimes L_{21}(1-Z(s_1,r)L_{11})^{-1}\right]\operatorname{vec}\left[Z(s_1,r)\right] \\ =\operatorname{vec}\left[L_{22}\right]+\left[L_{12}^t\otimes L_{21}(1-Z(s_2,r)L_{11})^{-1}\right]\operatorname{vec}\left[Z(s_2,r)\right] \\ \vdots \\ =\operatorname{vec}\left[L_{22}\right]+\left[L_{12}^t\otimes L_{21}(1-Z(s_k r)L_{11})^{-1}\right]\operatorname{vec}\left[Z(s_k,r)\right] \end{bmatrix}=F(r) \tag{10}$$

which is the form of the fault diagnosis equations with which we desire to work.

III. Solvability of the Fault Diagnosis Equations

For the fault diagnosis equations derived above to be a viable tool of circuit and system diagnosis two fundamental questions remain to be answered: "What test frequencies should be employed to optimize the solvability of the equations?" and "How solvable are the equations given an optimal choice of test frequencies?" Both of these questions, in turn, hinge on the development of some type of *measure of solvability* [15] for the fault diagnosis equations.

For a set of linear equations

$$m=Fr \tag{11}$$

where r is an n vector, m is a p vector, and F is a p by m matrix one may characterize the solvability of the equations in terms of the number of arbitrary parameters in its solution (if a solution exists). As such, $\delta=n-\operatorname{rank}(F)$ is a natural measure of the solvability for (11). Here, $\delta=0$ implies that the equation has a unique solution, $\delta=1$ implies that the solution is determined up to one arbitrary parameter and so on, with increasing values of δ representing decreasing degrees of solvability.

The fault diagnosis equations are, however, nonlinear even for linear systems—hence we must resort to the *implicit function theorem* to obtain a measure of solvability [15], [16] analogous to the above [13]. Indeed, if r_f is a solution to the fault diagnosis equations, then r_f is determined up to a

$$\delta(r_f)=n-\operatorname{rank}\left[\left[\frac{dF}{dr}(r_f)\right]\right] \tag{12}$$

dimensional manifold (of arbitrary parameters) in a neighborhood of r_f. Here dF/dr is the Jacobian matrix of partial derivatives of F with respect to r. With the aid of the matrix identity $d(M^{-1})/dr=-M^{-1}[dM/dr]M^{-1}$, dF/dr can be computed explicitly from (8) and (10) yielding

$$\frac{dF}{dr}(r_f)=\begin{bmatrix} \left\{\left(\left[1+L_{11}(1-Z(s_1,r_f)L_{11})^{-1}Z(s_1,r_f)\right]L_{12}\right)^t\otimes\left(L_{21}(1-Z(s_1,r_f)L_{11})^{-1}\right\}\left[(d\operatorname{vec}Z(s_1,r_f)\right]/dr \\ \hdashline \left\{\left(\left[1+L_{11}(1-Z(s_2,r_f)L_{11})^{-1}Z(s_2,r_f)\right]L_{12}\right)^t\otimes\left(L_{21}(1-Z(s_2,r_f)L_{11})^{-1}\right\}\left[(d\operatorname{vec}Z(s_2,r_f)\right]/dr \\ \hdashline \vdots \\ \hdashline \left\{\left(\left[1+L_{11}(1-Z(s_k,r_f)L_{11})^{-1}Z(s_k,r_f)\right]L_{12}\right)^t\otimes\left(L_{21}(1-Z(s_k,r_f)L_{11})^{-1}\right\}\left[(d\operatorname{vec}Z(s_k,r_f)\right]/dr \end{bmatrix} \tag{13}$$

where "t" denotes matrix transposition and $\otimes$ denotes the matrix Kronecker (or tensor) product.

The difficulty with the implicit function theorem is that it only yields local information valid in a neighborhood of a solution. Fortunately, however, given the special nature of the Jacobian matrix of (13) coupled with an assumption that the component transfer function matrices $Z_i(s,r)$ are rational in r, it is possible to show that the rank of the Jacobian matrix is "almost constant." This, in turn, allows us to transform the local measure of solvability of (12) into a global measure of solvability. For this purpose we adopt the *algebraic geometric* definition for the term "almost constant;" i.e., we say that a function of r_f is *almost constant* if it is constant except possibly for those values of r_f lying in an *algebraic variety* (the solution space of a finite set of nonzero simultaneous polynomial equations in n variables). More generally, we say that a property holds "almost everywhere" or for *almost all* r_f in n space if it is true for all values of r_f except possibly those lying in an algebraic variety. Since the Lebesque measure of an algebraic variety is zero, this definition for the concept "almost everywhere" is consistent with the more common measure theoretic definition and is more natural in the context of our application [14].

Theorem 1

Let $Z_i(s,r)$; $i=1,2,\cdots,q$; be rational in r. Then $\delta(r_f)$ is almost constant.

Note, the assumption that $Z_i(s,r)$ is rational in r is quite minor being satisfied by all of the examples given in Section II except for the delay (which can be approximated by a function which is rational in r). In practice, the component transfer function matrices will also be rational in s though this is not required for the present theorem since F and dF/dr are formulated in terms of specific test frequencies, $s_1, s_2, \cdots, s_k$. Given our assumption on the $Z_i(s,r)$, together with (13), it then follows that $(dF)/(dr)(r_f)$ is also rational in r_f.

Proof of Theorem 1: We begin by showing that an arbitrary polynomial matrix in r, $P(r)$, has almost constant rank. Since rank $P(r)$ is restricted to the finite set of integers $(0,1,2,\cdots,j$; where j is the minimum of the number of rows and columns in $P(r)$, there exists an r_m which maximizes the rank of $P(r)$

$$\operatorname{rank}[P(r_m)] \geqslant \operatorname{rank}[P(r)]. \tag{14}$$

Now, the rank of a matrix is the dimension of its largest nonsingular square submatrix. As such, $P(r)$ admits a square submatrix $M(r)$, whose dimension is equal to the rank $P(r_m)$ and for which

$$\det M(r_m) \neq 0. \tag{15}$$

Now, det $[M(r)]$ is a polynomial in r which is not identically zero (from (15)) and hence, it is nonzero, almost everywhere. As such,

$$\begin{aligned}\operatorname{rank}[P(r_m)] &\geqslant \operatorname{rank}[P(r)] \geqslant \operatorname{rank}[M(r)]\\ &= \operatorname{rank}[P(r_m)] \quad \text{a.e.}\end{aligned} \tag{16}$$

showing that rank $[P(r)]=$rank $[P(r_m)]$ almost everywhere. As such, rank $[P(r)]$ is almost constant.

Now, to verify that rank $[(dF)/(dr)(r_f)]$ is constant we decompose this matrix as

$$\frac{dF}{dr}(r_f) = \frac{P(r_f)}{d(r_f)} \tag{17}$$

where $P(r_f)$ is a polynomial matrix and $d(r_f)$ is a nonzero common denominator. $P(r_f)$ has almost constant rank while $d(r_f)$ is nonzero almost everywhere and hence can effect the rank of $P(r_f)$ only on an algebraic variety (since the division of a matrix by a nonzero scalar does not effect its rank.) As such, our Jacobian matrix has almost constant rank implying that

$$\delta(r_f) = n - \operatorname{rank}\left[\frac{dF}{dr}(r_f)\right] \tag{18}$$

is also almost constant. The proof of the theorem is therefore complete.

Given the theorem, we may now define a *global measure of solvability* for the fault diagnosis equation δ as the *generic value* of $\delta(r_f)$; i.e., the value $\delta(r_f)$ takes on for almost all r_f. This proves to be a natural measure of solvability since it indicates the ambiguity which will result from an attempt to solve the fault diagnosis equations in a neighborhood of almost any failures. Of course, one requires some sort of equation solving algorithm [10], [11] to locate a neighborhood of an actual failure. The δ parameter, however, represents a bound on the performance of any such algorith. Finally, we note that since δ is independent of r_f, the solution of the fault diagnosis equations, it can be computed at the time the system and its test algorithm are developed by evaluating $\delta(r)$ at a randomly chosen generic point, say r_0. In turn, this parameter may then be employed as an aid in the choice of test frequencies and test points.

IV. Test Frequency Selection

Adopting the measure of solvability δ formulated in the preceeding section, it remains to develop an algorithm for choosing a set of test frequencies; $s_1, s_2, \cdots, s_k$; which maximize the solvability of the fault diagnosis equations (i.e., minimize δ). To this end, let $\delta_{\min}$ denote the minimum value achieved by δ for any set of test frequencies; $s_1, s_2, \cdots, s_k$; $k=1,2,\cdots,$. Since the possible values for δ are restricted to the finite set; $\delta=0,1,\cdots,n$; such a minimum is assured to exist.

The following theorem gives an explicit formula for computing $\delta_{\min}$ while its proof yields an algorithm for choosing a set of test points which achieve $\delta_{\min}$. Since the purpose of this theorem is to formulate an algorithm for choosing test frequencies, the theorem is expressed in terms of

$$\begin{aligned}\operatorname{vec}[S(s,r)] = \operatorname{vec}[L_{22}] + [L_{12}^t\\ \otimes L_{21}(1 - Z(s,r)L_{11})^{-1}]\operatorname{vec}[Z(s,r)]\end{aligned} \tag{19}$$

and

$$\frac{d\text{vec}\left[S(s,r)\right]}{dr}=\left\{\left(\left[1+L_{11}(1-Z(s,r)L_{11})^{-1}Z(s,r)\right]L_{12}\right)^{t}\otimes\left(L_{21}(1-Z(s,r)L_{11})^{-1}\right)\right\}\cdot\left\{d\text{vec}\left[Z(s,r)\right]/dr\right\} \tag{20}$$

viewed as rational functions in s rather than in terms of the function $F(r)$ which is formulated in terms of an *a priori* choice of test frequencies.

Theorem 2

Let $Z_i(s,r)$; $i=1,2,\cdots,q$; be rational in s and r. Then

$$\delta_{\min}=n-\text{col-rank}\left[\frac{d\text{vec}\left[S(s,r)\right]}{dr}\right]$$

where n is the dimension of the parameter vector r and "col-rank" denotes the generic number of linearly independent columns of the rational matrix $[d\text{vec}\,[S(s,r)]/dr]$ over the field of complex numbers. Moreover, $\delta_{\min}$ is achieved by almost any choice of $n-\delta_{\min}$ distinct complex frequencies.

Proof: For the sake of brevity, we will prove the theorem only for the special case where $S(s,r)$ is a scalar transfer function (allowing us to drop the "vec" transformation) though essentially the same proof goes through in the general case modulo some notational complexities [5]. Also, since the rank of the Jacobian matrix is almost constant it suffices to fix the parameter vector r at any generic point, say r_0. This then reduces $[d\text{vec}\,[S(s,r)]/dr]$ to a row vector of rational functions

$$R(s)=\left[\,R_1(s)\quad R_2(s)\quad\cdots\quad R_n(s)\,\right] \tag{21}$$

where

$$R_i(s)=\left[d\text{vec}\left[S(s,r_0)\right]/dr_i\right] \tag{22}$$

and our problem reduces to the verifications of the fact that the number of linearly independent columns of $R(s)$ over the field of complex scalars is equal to the maximum possible rank of the complex matrix

$$\begin{bmatrix} R(s_1)\\ \hline R(s_2)\\ \hline \vdots\\ \hline R(s)\end{bmatrix}=\left[\begin{array}{c|c|c|c} R_1(s_1) & R_2(s_1) & \cdots & R_n(s_1)\\ \hline R_1(s_2) & R_2(s_2) & & R_n(s_2)\\ \hline \vdots & \vdots & & \vdots\\ \hline R_1(s_k) & R_2(s_k) & \cdots & R_n(s_k)\end{array}\right]=\text{col}\left(R(s_i)\right) \tag{23}$$

over all possible choices of the complex frequencies; $s_1,s_2,\cdots,s_k$; $k=1,2,\cdots$. Now, clearly if some column of $R(s)$, say the nth, is dependent on the remaining columns, then

$$R_n(s)=\sum_{j=1}^{n-1}c_jR_j(s) \tag{24}$$

for all s. Then by applying (24) individually for each s_i

$$\text{col}\left(R_n(s_i)\right)=\sum_{i=1}^{n-1}c_j\,\text{col}\left(R_j(s_i)\right) \tag{25}$$

for any possible number or choice of the s_i. The rank of the matrix of (23), therefore, is less than or equal to the number of linearly independent columns of $R(s)$ over the field of complex numbers.

To prove that equality can be achieved with an appropriate choice of $n-\delta_{\min}$ complex test frequencies s_i we invoke our assumption that $S(s,r)$ is a scalar transfer function. Without loss of generality, we may assume that $R_1(s)$ through $R_q(s)$ are the linearly independent entries in $R(s)$ over the field of complex numbers in which case we must show that there exists complex frequencies $s_1,s_2,\cdots,s_k$ ($k=q$ in this case) which make the first q columns of the matrix of (23) linearly independent.

If $q=1$, $R_1(s)$ is not identically zero (since otherwise it would be linearly dependent) and hence for almost all s_1, $R_1(s_1)\neq 0$. As such, the columns in this trivial one by one matrix are linearly independent. With this as a starting point, we will use an inductive argument to show that the theorem holds for all values of q. We, therefore, assume that it has been shown that for $q=p$ there exist complex frequencies; $s_1,s_2,\cdots,s_p$; such that the matrix

$$R_p=\begin{bmatrix} R_1(s_1) & R_2(s_1) & \cdots & R_p(s_1)\\ R_1(s_2) & R_2(s_2) & \cdots & R_p(s_2)\\ \vdots & & & \\ R_1(s_p) & R_2(s_p) & \cdots & R_p(s_p)\end{bmatrix} \tag{26}$$

has linearly independent columns and we desire to show that there exists an s_{p+1} such that the matrix

$$\bar{R}_{p+1}(s)=\begin{bmatrix} R_1(s_1) & R_2(s_1) & \cdots & R_p(s_1) & R_{p+1}(s_1)\\ R_1(s_2) & R_2(s_2) & \cdots & R_p(s_2) & R_{p+1}(s_2)\\ \vdots & & & & \\ R_1(s_p) & R_2(s_p) & \cdots & R_p(s_p) & R_{p+1}(s_p)\\ R_1(s) & R_2(s) & & R_p(s) & R_{p+1}(s)\end{bmatrix} \tag{27}$$

has linearly independent columns for $s=s_{p+1}$. By virtue of our assumption that $S(s,r)$ is a scalar both $\bar{R}_p$ and $\bar{R}_{p+1}(s)$ are square and we may test for linear independence of the columns of $\bar{R}_{p+1}(s)$ by computing its determinant. Expanding (27) in cofactors along its bottom row, we obtain

$$\det\left(\bar{R}_{p+1}(s)\right)=\sum_{j=1}^{p+1}(-1)^{p+j+1}\Delta_{p+1,j}R_j(s). \tag{28}$$

Since $\bar{R}_p$ has linearly independent columns $\Delta_{p+1,p+1}\neq 0$, hence, the coefficients in the summation of (28) are not all zero and thus by the linear independence of the $R_i(s)$ the summation is not identically zero. As such, one can choose almost any s_{p+1} which will make the determinant of $\bar{R}_{p+1}(s_{p+1})$ nonzero thus assuring the $\bar{R}_{p+1}$ has linearly independent columns when its rows are evaluated at the complex frequencies $s_1, s_2, \cdots, s_{p+1}$. The proof of the theorem is thus complete.

Note that the proof of the theorem yields a natural sequential algorithm for choosing test frequencies. Moreover, for the scalar case we have shown that the number of required test frequencies is exactly $n-\delta_{\min}$ (equal to the column rank of the Jacobian matrix). In the general case where $S(s,r)$ is not a scalar, the number of required test frequencies is less than or equal to $n-\delta_{\min}$ [5].

Although the theorem implies that one can randomly choose almost any $n-\delta_{\min}$ test frequencies to maximize the solvability of the fault diagnosis equations, the result does not take cognizence of numerical considerations. Although no theory yet exists for choosing test points with numerical considerations in mind, it has been our experience that the "well posedness" of the fault diagnosis equations is quite sensitive to the choice of test frequencies [5]. In most of our experiments, we have worked with real test frequencies to eliminate the necessity of working in the complex plane. On the other hand, m is most easily measured when values of s_i on the jw axis are employed whereas it has been suggested that test frequencies symetrically spaced around a circle in the complex plane might yield numerically "well posed" equations.

Although the measure of solvability δ for the fault diagnosis equations is dependent on the choice of test frequencies, as well as the properties of the unit under test, $\delta_{\min}$ is determined entirely by the UUT; its components, connections and accessible test points; and is completely independent of the test algorithm employed. As such, $\delta_{\min}$ may be taken as a natural *measure of testability* [1] for the UUT which characterizes the degree to which the fault analysis equations can be solved given an optimal choice of test frequencies and solution algorithm. Moreover, $\delta_{\min}$ may be used as an aid for the optimal selection of test points [3]–[5]. To this end we may choose a set of test points, from several options, so as to minimize $\delta_{\min}$. Alternatively, we may attribute a cost to each input and output test point and then choose the least cost combination of test points which yield a specified $\delta_{\min}$. This latter process reduces to a rather straightforward integer programming problem and is thus readily automated [4], [5]. The technique is illustrated in the examples of the following section.

V. Examples

An initial illustration of the theory consider the *RC* coupled amplifier with inductive load shown in Fig. 2. Here we will take E_i to be the only test input but we will initially allow E_0, i_L, i_C, and V_i to all be taken as test outputs with the measure of testability $\delta_{\min}$ being used to extract a reduced set of test outputs from these options. A

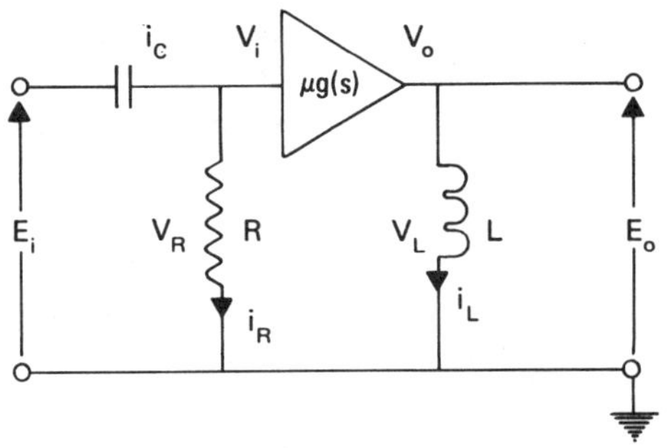

Fig. 2. *RC* coupled amplifier with inductive load.

component connection model for this circuit is given by

$$\begin{bmatrix} V_0 \\ i_L \\ V_C \\ i_R \end{bmatrix} = \begin{bmatrix} \mu g(s) & 0 & 0 & 0 \\ 0 & 1/LS & 0 & 0 \\ 0 & 0 & 1/CS & 0 \\ 0 & 0 & 0 & 1/R \end{bmatrix} \begin{bmatrix} V_i \\ V_L \\ i_C \\ V_R \end{bmatrix} \tag{29}$$

$$\left[\begin{array}{c} V_i \\ V_L \\ i_C \\ V_R \\ \hline E_0 \\ i_0 \\ i_R \\ V_i \end{array}\right] = \left[\begin{array}{cccc|c} 0 & 0 & -1 & 0 & 1 \\ 1 & 0 & 0 & 0 & 0 \\ 0 & 0 & 0 & 1 & 0 \\ 0 & 0 & -1 & 0 & 1 \\ \hline 1 & 0 & 0 & 0 & 1 \\ 0 & 1 & 0 & 0 & 0 \\ 0 & 0 & 0 & 1 & 0 \\ 0 & 0 & -1 & 0 & 1 \end{array}\right] \left[\begin{array}{c} V_0 \\ i_L \\ V_C \\ i_R \\ \hline E_i \end{array}\right]. \tag{30}$$

Taking our vector of potentially variable component parameters to be $r = \mathrm{col}\,(\mu, L, C, R)$ each with unity nominal value, we obtain a nominal transfer function matrix

$$S(s,r) = \begin{bmatrix} \dfrac{s(g(s)+1)+1}{s+1} \\ \dfrac{g(s)}{s+1} \\ \dfrac{s}{s+1} \\ \dfrac{s}{s+1} \end{bmatrix} \tag{31}$$

whereas our Jacobian matrix evaluated at the nominal parameter values is given by

$$\frac{d\mathrm{vec}\left[S(s,r)\right]}{dr} = \begin{bmatrix} \dfrac{sg(s)}{s+1} & 0 & \dfrac{sg(s)}{(s+1)^2} & \dfrac{sg(s)}{(s+1)^2} \\ \dfrac{g(s)}{s+1} & \dfrac{-g(s)}{s+1} & \dfrac{g(s)}{(s+1)^2} & \dfrac{g(s)}{(s+1)^2} \\ 0 & 0 & \dfrac{s}{(s+1)^2} & \dfrac{-s^2}{(s+1)^2} \\ 0 & 0 & \dfrac{s}{(s+1)^2} & \dfrac{s}{(s+1)^2} \end{bmatrix}. \tag{32}$$

Now, an inspection of this matrix will reveal that it has four independent columns over the field of complex numbers and hence if all four possible outputs are used, we will have $\delta_{\min}=0$ implying that the fault diagnosis equations have locally unique solutions. On the other hand, if only two outputs E_0 and i_C are measured, our modified Jacobian matrix will reduce to the first and third rows of the matrix shown in (32) which has column rank 3. As such, if we only use these two test outputs, we obtain $\delta_{\min}=1$ and hence the solution to the fault diagnosis equations will be characterized by a single arbitrary parameter.

In this latter case, with only E_0 and i_C taken as test outputs, Theorem 2 implies that dF/dr will have rank 3 for almost any choice of $3=n-\delta_{\min}$ test frequencies. Choosing $s_1=1$, $s_2=2$, and $s_3=3$, we obtain

$$\frac{dF}{dr}(r_0)=\left[\begin{array}{cccc} g(1)/2 & 0 & g(1)/4 & g(1)/4 \\ 0 & 0 & 1/4 & -1/4 \\ \hline 2g(2)/3 & 0 & 2g(2)/9 & 2g(2)/9 \\ 0 & 0 & 2/9 & -2/9 \\ \hline 3g(3)/4 & 0 & 3g(3)/16 & 3g(3)/16 \\ 0 & 0 & 3/16 & -3/16 \end{array}\right] \tag{33}$$

which has three linearly independent columns as long as $g(1)\neq 0$, $g(2)\neq 0$, and $g(3)\neq 0$. Indeed, in this example, any two of the three frequencies would have sufficed to yield three linearly independent columns. Note, for scalar transfer functions, Theorem 2 implies that $n-\delta_{\min}$ frequencies are actually required but for matrix transfer functions fewer frequencies may suffice.

Of course, for the circuit of Fig. 2, we have a choice of some 15 combinations of the four outputs with which we may choose to work for the diagnosis of the circuit. The resultant $\delta_{\min}$'s for the various combinations of outputs are given in Table I [5].

Finally, with the aid of Table I, one may readily develop a test point selection algorithm for our circuit [4], [5]. For instance, if we desire to find the smallest set of outputs which yield a $\delta_{\min}<1$ an inspection of the table will reveal that E_0 and i_L, i_L and i_C, or E_0 and i_C are the optimal choices. Of course, if one attributes a cost to the various outputs (determined by the convenience of making the required measurements), then we may further distinguish between these three possibilities. For instance, if voltage measurements are deemed to be easier than current measurements, the combination of i_L and i_C may be excluded with the decision between the remaining two options being dependent on whether it is easier to measure the circuit's input current (i_C) or its load current (i_L).

As a second example, consider the one stage transistor amplifier shown in Fig. 3 with the ac equivalent circuit of Fig. 4. Since it is clearly impossible to distinguish between failures in the two parallel bias resistors, R_a and R_b, these two resistors have been combined into the single resistor, R_a' in the component connection model of (34) and (35). Taking all of the component parameters as potentially faulty, r becomes a 12 vector composed of $C_1, r_x, \cdots, R_L$ and as before, we take all parameters to have the nominal value of unity.

TABLE I
MEASURE OF TESTABILITY FOR THE CIRCUIT OF FIG. 2 USING VARIOUS COMBINATIONS OF TEST OUTPUTS

Outputs	$\delta_{\min}$
E_o, i_L, i_C, V_i	0
E_o, i_L, i_C	0
i_L, i_C, V_i	1
i_L, V_i, E_o	1
V_i, E_o, i_C	1
E_o, i_L	1
i_L, i_C	1
i_L, V_i	2
V_i, E_o	2
E_o, i_C	1
i_C, V_i	2
E_o	2
i_L	2
i_C	2
V_i	3

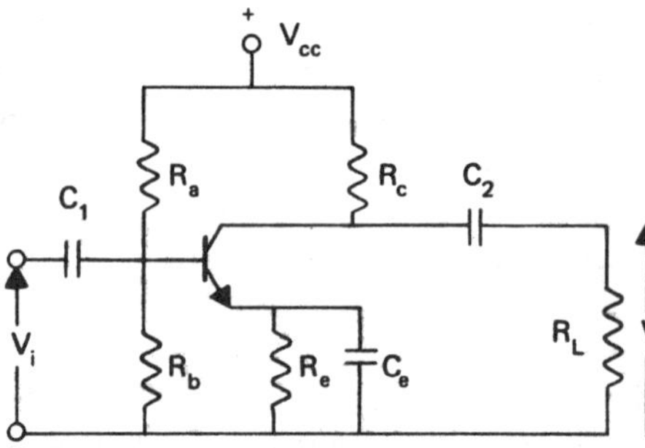

Fig. 3. One stage transistor amplifier.

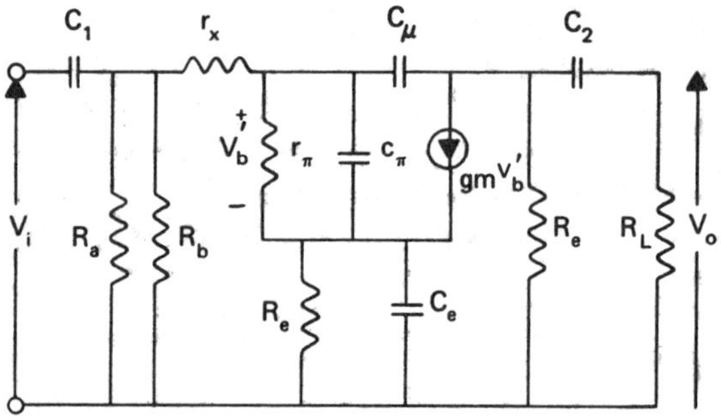

Fig. 4. Amplifier equivalent circuit.

$$\begin{bmatrix} I_{C_1} \\ I_{r_x} \\ I_{r_\pi} \\ I_{C_\mu} \\ I_{C_2} \\ V'_{R_a} \\ V_{R_e} \\ V_{C_\pi} \\ V_{C_e} \\ V_{g_m} \\ V_{R_c} \\ V_{R_L} \\ \hline V_0 \\ I_{C_1} \\ V_{R'_a} \\ I_e \end{bmatrix} = \left[\begin{array}{cccccccccccc|c} 0 & 0 & 0 & 0 & 0 & 1 & 1 & 0 & 1 & 0 & 1 & 1 & 0 \\ 0 & 0 & 0 & 0 & 0 & 0 & 1 & 0 & 1 & 0 & 1 & 1 & 0 \\ 0 & 0 & 0 & 0 & 0 & 0 & 1 & -1 & 1 & -1 & 0 & 0 & 0 \\ 0 & 0 & 0 & 0 & 0 & 0 & 0 & 0 & 0 & 1 & 1 & 1 & 0 \\ 0 & 0 & 0 & 0 & 0 & 0 & 0 & 0 & 0 & 0 & 0 & 1 & 0 \\ -1 & 0 & 0 & 0 & 0 & 0 & 0 & 0 & 0 & 0 & 0 & 0 & 1 \\ -1 & -1 & -1 & 0 & 0 & 0 & 0 & 0 & 0 & 0 & 0 & 0 & 1 \\ 0 & 0 & 1 & 0 & 0 & 0 & 0 & 0 & 0 & 0 & 0 & 0 & 0 \\ -1 & -1 & -1 & 0 & 0 & 0 & 0 & 0 & 0 & 0 & 0 & 0 & 1 \\ 0 & 0 & 1 & -1 & 0 & 0 & 0 & 0 & 0 & 0 & 0 & 0 & 0 \\ -1 & -1 & 0 & -1 & 0 & 0 & 0 & 0 & 0 & 0 & 0 & 0 & 1 \\ -1 & -1 & 0 & -1 & -1 & 0 & 0 & 0 & 0 & 0 & 0 & 0 & 1 \\ \hline -1 & -1 & 0 & -1 & -1 & 0 & 0 & 0 & 0 & 0 & 0 & 0 & 1 \\ 0 & 0 & 0 & 0 & 0 & 1 & 1 & 0 & 1 & 0 & 1 & 1 & 0 \\ -1 & 0 & 0 & 0 & 0 & 0 & 0 & 0 & 0 & 0 & 0 & 0 & 1 \\ 0 & 0 & 0 & 0 & 0 & 0 & 1 & 0 & 1 & 0 & 0 & 0 & 0 \end{array}\right] \begin{bmatrix} V_{C_1} \\ V_{r_x} \\ V_{r_\pi} \\ V_{C_\mu} \\ V_{C_2} \\ I_{R'_a} \\ I_{R_e} \\ I_{C_\pi} \\ I_{C_e} \\ I_{g_m} \\ I_{R_c} \\ I_{R_L} \\ \hline V_i \end{bmatrix}. \tag{34}$$

Once again we let the input voltage be the only test input for the system and we take V_0, I_{C_1}, V_{R_a}, and I_e, to be possible output test points. The resultant $\delta_{\min}$ for each of the 15 possible combinations of these output terminals is tabulated in Table II [5].

$$\begin{bmatrix} V_{C_1} \\ V_{r_x} \\ V_{r_\pi} \\ V_{C_\mu} \\ V_{C_2} \\ I_{R'_a} \\ I_{R_e} \\ I_{C_\pi} \\ I_{C_e} \\ I_{g_m} \\ I_{R_c} \\ I_{R_L} \end{bmatrix} = \begin{bmatrix} 1/C_1s & & & & & & & & & & & \\ & r_x & & & & & & & & & & \\ & & r_\pi & & & & & & & & 0 & \\ & & & 1/C_\mu s & & & & & & & & \\ & & & & 1/C_2s & & & & & & & \\ & & & & & 1/R'_a & & & & & & \\ & & & & & & 1/R_e & & & & & \\ & & & & & & & C_\pi s & & & & \\ & & & & & & & & C_e s & & & \\ & 0 & & & & & & g_m & & 0 & & \\ & & & & & & & & & & 1/R_c & \\ & & & & & & & & & & & 1/R_L \end{bmatrix} \begin{bmatrix} I_{C_1} \\ I_{r_x} \\ I_{r_\pi} \\ I_{C_\mu} \\ I_{C_2} \\ V_{R'_a} \\ V_{R_e} \\ V_{C_\pi} \\ V_{C_e} \\ V_{g_m} \\ V_{R_c} \\ V_{R_L} \end{bmatrix}. \tag{35}$$

From the table it is apparent that no single test output suffices to yield a $\delta_{\min}=0$ (perfect testability) though $\delta_{\min}=0$ can be achieved using two test outputs; V_0 and I_{C_1} or V_0 and I_e.

VI. Conclusions

Our purpose in the preceeding has been to formulate an analytic theory in support of the intuitive art usually associated with the design of a test algorithm. With the aid of the techniques developed above, we believe that it will be possible to develop an automated test program generation (ATPG) algorithm for linear systems [4], [5]. Indeed, such an algorithm could be readily combined with the same computer-aided design (CAD) algorithm used in the system design process [9]. Given the component connection equations such an algorithm could be employed to automatically (or interactively) choose test points and test frequencies and generate the required set of fault diagnosis equations. These could then be stored on tape and supplied to the automatic test equipment (ATE) in which a faulty system would be tested and the fault diagnosis equations solved.

TABLE II
MEASURE OF TESTABILITY FOR THE CIRCUIT OF FIG. 3 USING VARIOUS TEST OUTPUTS

Outputs	δ_{min}
V_o	3
I_{C_1}	2
$V_{R'_a}$	2
I_e	3
V_o, I_{C_1}	0
V_o, $V_{R'_a}$	1
V_o, I_e	0
I_{C_1}, $V_{R'_a}$	2
I_{C_1}, I_e	1
$V_{R'_a}$, I_e	0
V_o, I_{C_1}, $V_{R'_a}$	0
V_o, I_{C_1}, I_e	0
V_o, $V_{R'_a}$, I_e	0
I_{C_1}, $V_{R'_a}$, I_e	0
V_o, I_{C_1}, $V_{R'_a}$, I_e	0

Although we do not propose to discuss the actual solution of the fault diagnosis equations here, it should be pointed out that by assuming that relatively few components have failed, say $p \ll n$, it is possible to develop specialized algorithms for the solution of the fault diagnosis equations which are far more efficient than standard equation solvers in this application [7], [11], [12]. These are typically derived from the *fault simulation* algorithms used in the diagnosis of digital systems and may naturally be classified into "*simulation before test*" and "*simulation after test*" algorithms. Some of the algorithms are discussed in [7] and [9]–[11].

Finally, we note that as formulated above, the measure of testability δ_{min} assumes that any combination component failure is possible. If, however, we assume that at most $p \ll n$ components fail simultaneously, the ambiguity in the solution of the fault diagnosis equations may actually be less than δ_{min}. For instance, in the example of Fig. 3, with only V_0 taken as an output $\delta_{min}=3$, yet the fault diagnosis equations can be solved exactly if we assume that only one parameter is out of tolerance [10]. The point, here, is that even though the solution of the fault diagnosis equations in n space has three arbitrary parameters when the solution is restricted to the one dimensional manifold of parameter vectors in which all but one coordinant are nominal it is unique.

REFERENCES

[1] W. J. Dejka, "Measure of testability in device and system design," in *Proc. 20th Midwest Symp. Circuits Syst.* (Lubbock, TX), pp. 39–52, Aug., 1977.
[2] ——, "A review of measurements of testability for analog systems," *Proc. 1977 AUTOTESTCON* (Hyannis, MA), pp. 279–284, Nov. 1977.
[3] N. Sen and R. Saeks, "A measure of testability and its application to test point selection—Theory," *Proc. 20th Midwest Symp. Circuits Syst.* (Lubbock, TX), pp. 576–583, Aug. 1977.
[4] ——, "A Measure of testability and its application to test point selection—Computation," *Proc. 1977 AUTOTESTCON* (Hyannis, MA), pp. 212–219, Nov. 1977.
[5] N. Sen, M.S. thesis, Texas Tech Univ., Lubbock, TX, 1977.
[6] M. N. Ransom and R. Saeks, "A functional approach to fault analysis in linear systems," in *Rational Fault Analysis*. New York: Marcel Dekker, 1977, pp. 124–134.
[7] R. Saeks and R. A. DeCarlo, *Interconnected Dynamical Systems*. New York: Marcel Dekker, to be published.
[8] S. P. Singh and R.-W. Liu, "Existence of state equation representation of linear large-scale dynamical systems," *IEEE Trans. Circuits and Syst.*, vol. CAS-20, pp. 239–246, 1973.
[9] *Proc. of the Workshop on Automatic Test Technology*, NSIA, San Diego, April, 1978.
[10] H. M. S. Chen and R. Saeks, "A search algorithm for the solution of the fault diagnosis equations," Texas Tech Univ., (unpublished notes), 1978.
[11] H. M. S. Chen, M.S. thesis, Texas Tech Univ., 1977.
[12] M. N. Ransom and R. Saeks, "The connection function—Theory and application," *Int. J. Circuit Theory and its Applications*, vol. 3, pp. 5–21, 1975.
[13] W. Fleming, *Functions of Several Variables*. Reading: Addison-Wesley, 1965.
[14] M. Spivak, *Calculus on Manifolds*. Benjamin: Amsterdam, 1966.
[15] R. S. Berkowitz, "Conditions for network element value solvability", *IRE Trans. Circuit Theory*, vol. CT-9, pp. 24–29, 1962.
[16] S. O. Bedrosian and R. S. Berkowitz, "Solution procedure for single-element-kind networks," *IRE Inter. Conv. Record*, vol. 10, Part 2, p. 16, 1962.

Calculation of Parameter Values from Node Voltage Measurements

TIMOTHY N. TRICK, FELLOW, IEEE, WATARU MAYEDA, MEMBER, IEEE, AND ADEL A. SAKLA

***Abstract*—In this paper a theorem is given for the necessary and sufficient test conditions required to determine the value of a circuit component from node voltage measurements at a single test frequency. Several techniques are then presented for the calculation of circuit component values from voltage measurements, including a new technique based on the adjoint circuit concept. In addition, some results are presented on the measurement of component values using different test frequencies.**

I. Introduction

PRESENTLY, there is considerable interest in the automatic testing and fault analysis of electrical circuits because of the complexity of modern circuits and the availability of low-cost computer systems. This paper addresses itself only to the analog circuit fault-analysis problem. In fault analysis, we want to determine which components are causing the circuit to malfunction.

One of the first theoretical studies of the analog circuit fault-analysis problem was initiated by Berkowitz [1]. Berkowitz mathematically defined the concept of network-element-value solvability and studied the measurement conditions required to solve the problem. Unfortunately, his work 1) only gave some necessary conditions for network solvability, 2) lacked algorithms for network element evaluation, and 3) did not include active elements. Since his work, there have been other theoretical studies which suggested the use of the following approaches: the bilinear transformation, first-order sensitivity, and optimization [2], [3]. Most of these studies allow for the possibility of only one faulty component, and ignore the fact that possibly more than one solution exists which corresponds to the measurements. However, there is a more recent paper [12] which investigates measurement conditions necessary and sufficient for the exact determination of component values using both single and multiple test frequencies. Unfortunately, the method is not very practical for the given test conditions, since it requires the measurement of both branch currents and voltages.

Finally, one of the more popular approaches that has been suggested is the fault dictionary [4], [5]. This is a brute-force approach in which the circuit is simulated with a number of different fault possibilities and the results are stored in a dictionary. The faculty circuit measurements are then correlated with the simulated results in the dictionary, and the faulty components are hopefully identified. The problem with this approach is that it is impossible to simulate all possible failures which may occur. Usually only single element failures are considered, and furthermore, it is assumed that the component has suffered a catastrophic failure (open or short-circuit failure). Thus this method is only reliable when the actual types of failures have been correctly anticipated by the test programmer.

We believe that a more reliable system is one in which the components' values can be computed from the test measurements. This brings us back to the solvability problem. In this paper, we present the *necessary* and *sufficient* test conditions required to determine the value of a component using only voltage measurements and a *single* sinusoidal test frequency. Also, we present several methods for the computation of an element value from these test conditions. A new algorithm is presented which makes use of the adjoint-circuit concept for the efficient computation of element values from node voltage measurements alone. Finally, some results are presented for the computation of element values using multiple-sinusoidal test frequencies.

Our discussion will be confined to linear passive networks, but in most cases the results can easily be extended to active and nonlinear circuits. Furthermore, we assume that only voltage measurements can be made, that is, no branches can be cut to measure current. Finally, we assume that no parallel elements of the same kind exist between two nodes. In practice, it is difficult to isolate such a fault to one of the components without current measurements. Usually both elements are replaced.

II. Conditions for Solvability

Suppose that we wish to measure an impedance Z_0 at a given frequency ω_0. Assume that Z_0 is connected between node c and node r in a linear time-invariant network N which contains passive admittance branches and no controlled sources. Furthermore, suppose that some of the nodes in N are accessible and others are not. An accessible node is one at which one can not only measure the

Manuscript received December 1, 1978; revised March 19, 1979. This work was supported in part by the Joint Services Electronics Program of the U.S. Army, U.S. Navy, and the U.S. Air Force, under Contract DAAG-29-78-C-0016 and by the U.S. Navy Naval Air Engineering Center under Contract N68335-78-C-0524.

The authors are with the Coordinated Science Laboratory and the Department of Electrical Engineering, University of Illinois, Urbana, IL 61801.

Reprinted from *IEEE Trans. Circuits & Systems,* vol. CAS-26, pp. 466–474, July 1979.

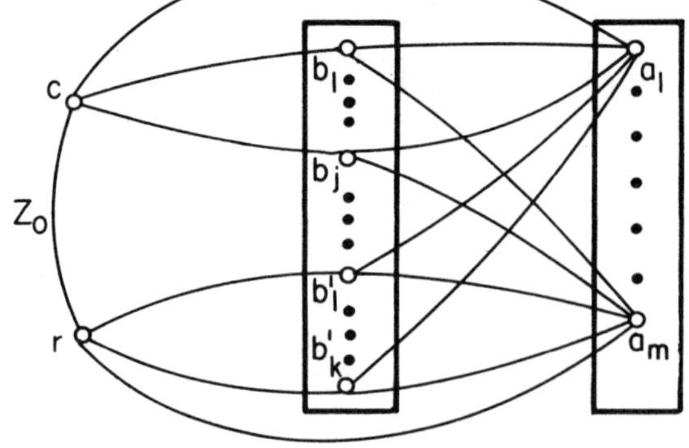

Fig. 1. Measurement of Z_0.

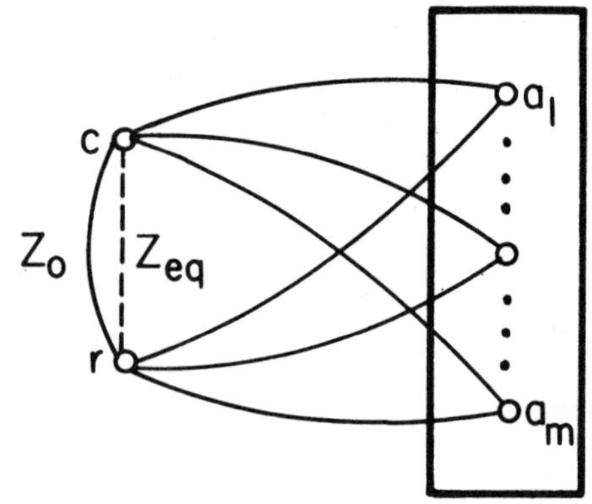

Fig. 2. Elimination of inaccessible nodes.

voltage, but also apply some type of excitation, such as, a current source, a voltage source, or grounding the node.

Theorem 1

The measurement of the impedance Z_0 at its terminals c and r is possible if and only if all paths in the circuit from node c to node r (excluding the path through Z_0) contain at least one accessible node. Mayeda [6] first proved this theorem using topological concepts. In this paper, a slightly different proof is presented.

Consider the circuit in Fig. 1 with the unknown element Z_0 connected between nodes c and r. The nodes a_1–a_m are the accessible nodes, and the b and b' nodes are the unaccessible nodes. If we eliminate the b and b' nodes from the circuit equations, we obtain the equivalent circuit in Fig. 2. Now we will be able to measure the impedance Z_0 provided that in this equivalent circuit there is no impedance in parallel with Z_0, that is, the admittance of the dashed line in Fig. 2 is zero. We will prove that this admittance is zero if and only if there is *no* path from c to r through only b and b' type nodes.

Fig. 1 has been drawn so that branches from c connect only node b_1 through b_j, and branches from node r connect nodes b_1' through b_k', and there is no path from the b nodes to the b' nodes which consists only of inaccessible nodes. In this case with r as the reference node the admittance equations for the circuit have the form

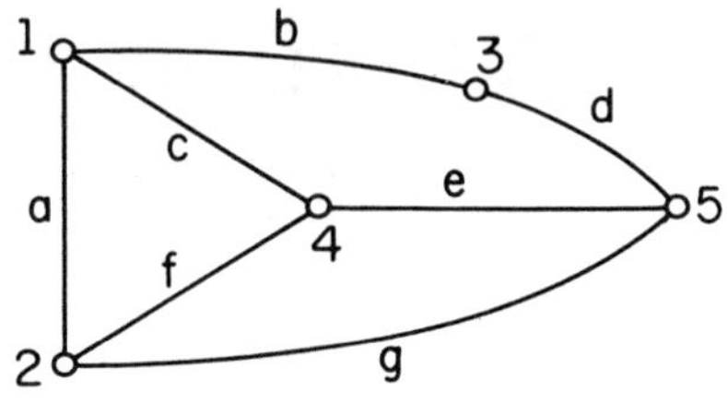

Fig. 3. Graph of a circuit example.

$$\begin{bmatrix} I_c \\ I_a \\ I_b \\ I_{b'} \end{bmatrix} = \begin{bmatrix} Y_{cc} & Y_{ca} & Y_{cb} & 0 \\ Y_{ac} & Y_{aa} & Y_{ab} & Y_{ab'} \\ Y_{bc} & Y_{ba} & Y_{bb} & 0 \\ 0 & Y_{b'a} & 0 & Y_{b'b'} \end{bmatrix} \begin{bmatrix} V_c \\ V_a \\ V_b \\ V_{b'} \end{bmatrix}. \tag{1}$$

Now let us connect identical voltage sources between the accessible nodes and the reference node so that $V_c = V_{a_i}$ for $i = 1, \cdots, m$. We will now show that the current I_c supplied by the voltage source connected to node c is equal to V_c/Z_0, if and only if there is no path between b and b' which consists only of inaccessible nodes. To prove this we simply need to show that $V_{b_i} = V_c$ for $i = 1, \cdots, j$, if and only if there is no path between b and b' which consists only of the inaccessible nodes.

Assuming that no such paths exist, from (1) we write

$$\begin{bmatrix} Y_{bc} Y_{ba} Y_{bb} \end{bmatrix} \begin{bmatrix} V_c \\ V_a \\ V_b \end{bmatrix} = 0.$$

Since the b nodes connect only to node c or to other b- or a-type nodes, it follows that the sum of the admittances in each row of the above array is zero. Thus if $V_c = V_{a_i}$ for $i = 1, \cdots, m$, the above equation has a solution, $V_{b_i} = V_c$ for $i = 1, \cdots, j$. In a practical circuit in which Y_{bb} is nonsingular this solution is unique. It follows that the only current path between nodes c and r is through the element Z_0 under these test conditions. Thus we can measure Z_0. One can prove the only if part of the theorem by showing that if a branch is connected between node c and a b' node, or between a b node and node r, or between a b node and a b' node then a second current path exists, and one can only measure the parallel combination of Z_{eq} and Z_0.

To illustrate the theorem, consider the network graph in Fig. 3. In order to measure the impedance of the branch "a" connected between nodes 1 and 2, either nodes 3 and 4, or nodes 4 and 5 must be accessible in addition to 1 and 2. This theorem will appear intuitively obvious in the next section where we present some typical measurement circuits for the determination of an impedance.

III. Isolation of an Element

The determination of the value of an element is not difficult if both the current through the element and the voltage across the element can be measured. In order to determine the current through an element, one terminal of

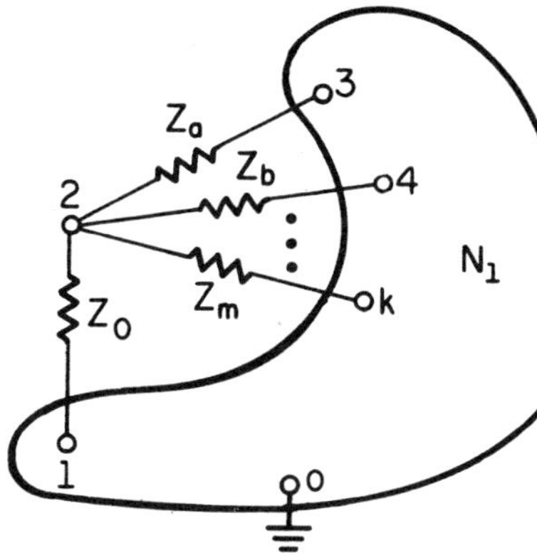

Fig. 4. Arbitrary network containing Z_0.

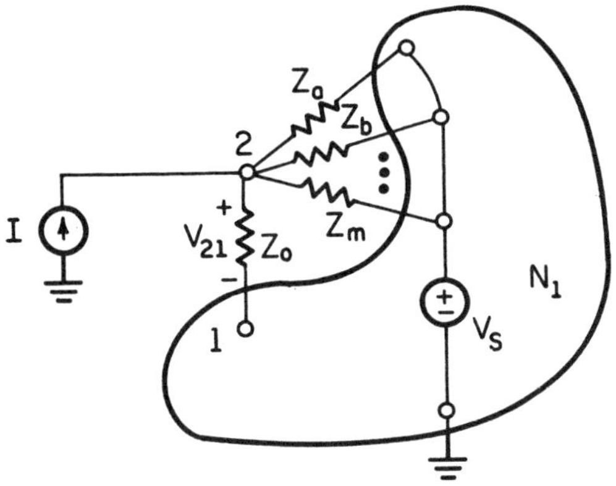

Fig. 5. Equipotential method for element isolation.

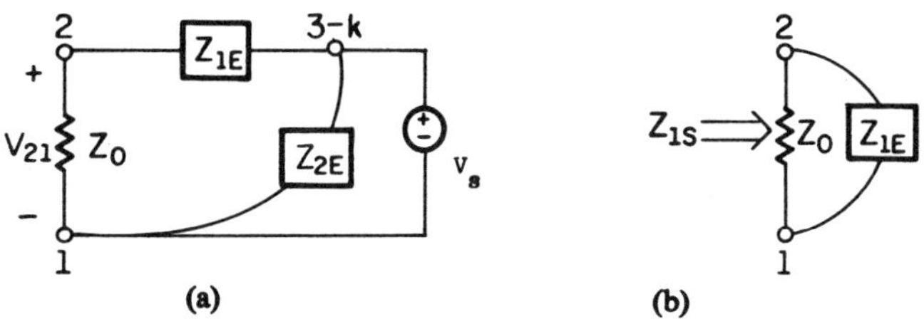

Fig. 6. Determination of Z_0 without iterative adjustments of the voltage source.

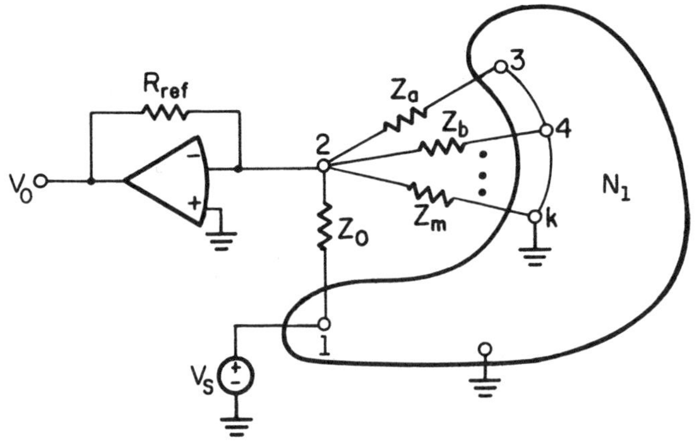

Fig. 7. Isolation with a high-gain amplifier.

the element must be isolated. This can be done by breaking the connection at one end of the element, but this approach is not considered practical in most test situations. However, in this section we give three other methods for effectively isolating an element which do not require that conductive paths be broken.

Consider the circuit in Fig. 4. The unknown element is connected between the nodes numbered 1 and 2. The impedances Z_a–Z_m lie on paths which ultimately return to node 1. Let us inject a current I into node 2. The element Z_0 can be isolated by shorting nodes 3 to k, connecting a voltage source to node 3, and adjusting the value of the voltage source until $V_2 = V_3$, as shown in Fig. 5. Then

$$Z_0 = \frac{V_{21}}{I}.$$

The disadvantage of this approach is that it requires the shorting of nodes in the circuit, and furthermore, the voltage source must be adjusted iteratively until $V_3 = V_2$.

A second approach which does not require an iterative adjustment consists of shorting nodes 3–k, and applying a voltage source between nodes 3 and 1, as shown in Fig. 6(a). For this circuit

$$\frac{V_{21}}{V_s} = \frac{Z_0}{Z_{1E} + Z_0}. \tag{2}$$

Note that Z_{1E} is the equivalent impedance of the parallel combination of impedances Z_a–Z_m. Next, nodes 3–k are shorted to node 1, as shown in Fig. 6(b), and the impedance

$$Z_{1s} = \frac{Z_{1E} Z_0}{Z_{1E} + Z_0} \tag{3}$$

is measured. From (2) and (3) we obtain

$$Z_0 = \frac{Z_{1s}}{1 - \dfrac{V_{21}}{V_s}}. \tag{4}$$

Thus the unknown impedance Z_0 can be determined from a two-step measurement procedure. The accuracy of the method deteriorates when the voltage ratio V_{21}/V_s is nearly unity, or equivalently, $|Z_{1E}| \ll |Z_0|$.

Finally, a third approach to isolating an element consists of connecting the element Z_0 to the input terminal of a high gain amplifier and grounding nodes 3–k, as shown in Fig. 7. Ideally, the amplifier would have infinite gain and zero offset so that node 2 is a virtual ground. Thus assuming negligible current through the impedances $Z_a, Z_b, \cdots, Z_m$, we obtain

$$Z_0 = -R_{\text{ref}} \frac{V_s}{V_0}. \tag{5}$$

This method requires only one measurement. The Hewlett-Packard 3060A Board Test System uses a sophisticated version of this node-isolation technique to make in-circuit component measurements.

It is interesting to note that each of the above measurement schemes requires that there be an accessible node on every path from node 1 to node 2, and Theorem 1 tells us that we cannot measure Z_0 with a single test frequency with any fewer nodes. In the next section, a new algorithm is given for the simultaneous calculation of Z_0, $Z_a, \cdots, Z_m$, without the need for shorting nodes in the circuit. The method is based on the adjoint-circuit concept. Furthermore, the method allows one to easily find the necessary and sufficient test conditions for element-value solvability.

IV. The Adjoint-Circuit Approach

In this section, a new algorithm is proposed for the calculation of the element values in a linear circuit from node voltage measurements. The method requires the measurement of the node voltages of the circuit under several different test conditions. In addition, the simulation of a second circuit, the adjoint circuit, is required under various short-circuit constraints. The adjoint-circuit simulations need to be made only once for any given circuit, and the results can be stored for future use. The element values of the test circuit can be easily computed from the node voltage measurements on the test circuit and the responses obtained from the adjoint circuit simulations.

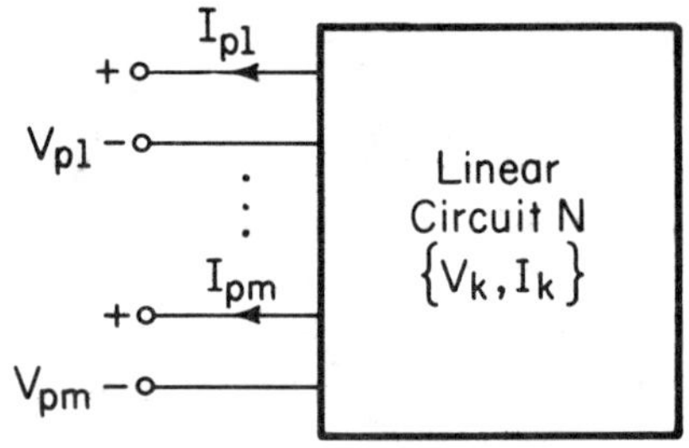

Fig. 8. Linear circuit with *m* input, output, or test ports.

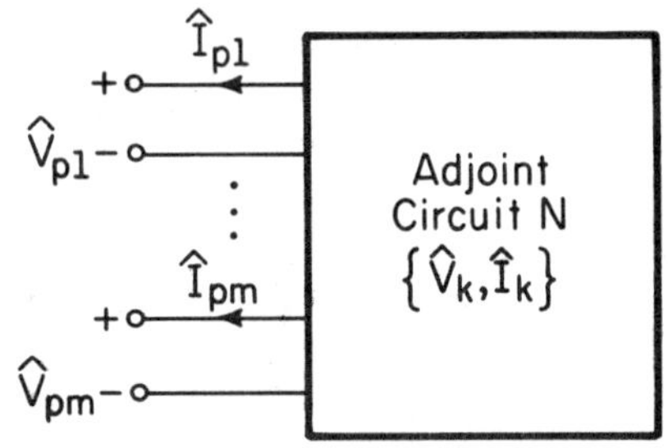

Fig. 9. Adjoint circuit of the circuit under test.

To derive the algorithm consider the linear circuit N in Fig. 8 which has m ports available. One may connect voltage or current sources to these ports or simply make open-circuit voltage measurements at these ports. We will assume that it is not practical to make short-circuit current measurements, because if this is the case, then by placing such a port in series with each element, the branch currents are measurable and the problem becomes trivial. Also, we assume that the circuit N has b internal branches whose respective branch voltages and currents in the frequency domain are denoted by $\{V_k, I_k\}$. Finally, we introduce a second circuit $\hat{N}$, shown in Fig. 9, which has the same topology as N, but not necessarily the same branch or port constraints. It is assumed that these circuits obey Kirchhoff's laws, so that by means of Tellegen's conservation of power theorem we can write [7]

$$\sum_{k=1}^{b} (\hat{V}_k I_k - V_k \hat{I}_k) = -\sum_{j=1}^{m} (\hat{V}_{pj} I_{pj} - V_{pj} \hat{I}_{pj}). \quad (6)$$

Now suppose that one or more faults occur in the circuit N so that $V_k \rightarrow V_k + \Delta V_k$, $I_k \rightarrow I_k + \Delta I_k$, $V_{pj} \rightarrow V_{pj} + \Delta V_{pj}$, and $I_{pj} \rightarrow I_{pj} + \Delta I_{pj}$. Equation (6) becomes

$$\sum_{k=1}^{b} \left[\hat{V}_k (I_k + \Delta I_k) - (V_k + \Delta V_k) \hat{I}_k \right] = -\sum_{j=1}^{m} \left[\hat{V}_{pj} (I_{pj} + \Delta I_{pj}) - (V_{pj} + \Delta V_{pj}) \hat{I}_{pj} \right]. \quad (7)$$

Subtracting (6) from (7) we obtain

$$\sum_{k=1}^{b} (\hat{V}_k \Delta I_k - \Delta V_k \hat{I}_k) = -\sum_{j=1}^{m} (\hat{V}_{pj} \Delta I_{pj} - \Delta V_{pj} \hat{I}_{pj}). \quad (8)$$

Thus by means of Tellegen's theorem, one can obtain a relationship between the voltage and current changes in the original circuit, due to parameter changes in that circuit, and the voltages and currents of the adjoint circuit [8], [9]. Equation (8) is valid for both frequency- and time-domain measurements.

Next we need to introduce the branch constraints into (8). Since we have assumed that only voltage measurements are available, we will assume that we can express every branch current as a function of its branch voltage. In our discussion, we will assume that each branch in N is linear, and has an admittance whose nominal value is Y_k. Although controlled sources can easily be handled [10], [11], and nonlinear elements can be included by using linear approximations at the dc operating point. Now

$$I_k + \Delta I_k = (Y_k + \Delta Y_k)(V_k + \Delta V_k) \quad (9)$$

and with no faults

$$I_k = Y_k V_k.$$

Therefore,

$$\Delta I_k = \Delta Y_k (V_k + \Delta V_k) + Y_k \Delta V_k$$

where V_k denotes the branch voltages when the circuit has *no* faulty components. Furthermore, let us choose the kth branch constraint of the circuit $\hat{N}$, such that it is the adjoint circuit of N [7], that is,

$$\hat{I}_k = Y_k \hat{V}_k \quad (10)$$

then we obtain

$$\sum_{k=1}^{b} (V_k + \Delta V_k) \hat{V}_k \Delta Y_k = -\sum_{j=1}^{m} (\hat{V}_{pj} \Delta I_{pj} - \Delta V_{pj} \hat{I}_{pj}). \quad (11)$$

In (11) we assume that we can measure the branch voltages $V_k + \Delta V_k$, and the admittance changes ΔY_k are the unknowns. In order to solve for the ΔY_k's we need to generate a set of simultaneous linear-algebraic equations with a rank equal to the number of unknowns. This can

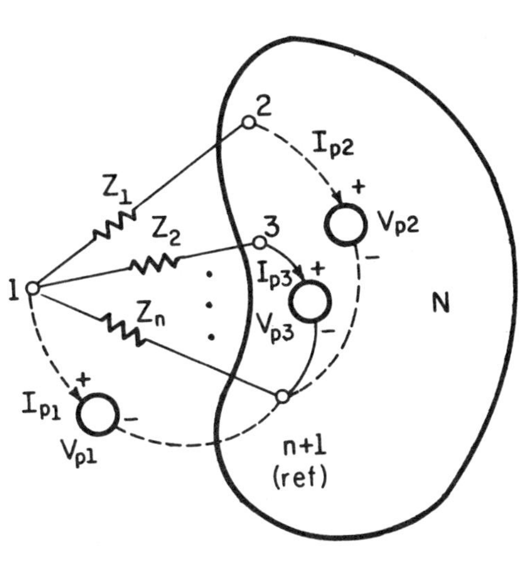

Fig. 10. Network with n unknown impedances connected to a common node.

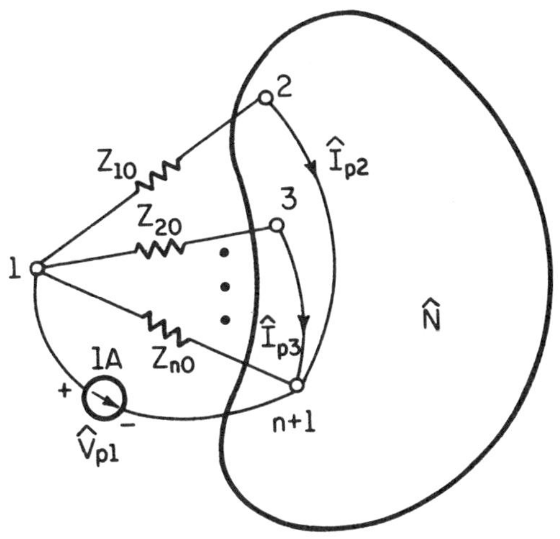

Fig. 11. Selection of adjoint circuit port constraints to isolate the components Z_1–Z_n.

be done by an appropriate choice of test conditions for the circuit N and its adjoint $\hat{N}$. Some examples follow.

A. Impedances Connected to a Common Node

Consider the situation in which we have n impedances connected to a node 1 as shown in Fig. 10. Suppose that we wish to measure the deviation of all n of these impedances from their nominal values. In Fig. 10 we will assume that node $n+1$ is the reference node and we have introduced n test ports at which measurements can be made, and excitations applied to the circuit. In order to isolate the impedances Z_1–Z_n and simplify the calculations, the ports 2–n are shorted, and a $1A$ source is applied at port 1 *in the adjoint circuit* as shown in Fig. 11. With these constraints (11) becomes

$$\hat{V}_{p1}\cdot \sum_{k=1}^{n} (V_k+\Delta V_k)\cdot \Delta Y_k = \Delta V_{p1}\cdot 1A + \sum_{k=2}^{n} \Delta V_{pk}\cdot \hat{I}_{pk} \tag{12}$$

or we can divide by $\hat{V}_{p1}$ and eliminate the adjoint-circuit responses.

$$\sum_{k=1}^{n} (V_k+\Delta V_k)\cdot \Delta Y_k = -(Y_{10}+Y_{20}+\cdots Y_{n0})\Delta V_{p1} + \sum_{k=1}^{n-1} Y_{k0}\cdot \Delta V_{pk+1} \tag{13}$$

or

$$\boxed{\sum_{k=1}^{n} (V_k+\Delta V_k)\cdot \Delta Y_k = -\sum_{k=1}^{n} Y_{k0}\cdot \Delta V_k}. \tag{14}$$

Recall that V_k is the branch voltage in the circuit without faults. Y_{k0} is the nominal value of the admittance, and ΔV_k is the change in the branch voltage in the faulty circuit. *Equation (14) can also be derived by taking the difference between the Kirchhoff's current law equations at node 1 for both the faulted and the unfaulted circuits.*

Note that we have not as yet specified the port constraints in the circuit under test. In order to solve for the unknown admittance deviations from their nominal values, we need to measure the branch voltages $(V_k+\Delta V_k)^{(j)}$ for n independent test conditions. Thus the superscript j denotes a particular test. We write

$$\begin{bmatrix} (V_1+\Delta V_1)^{(1)} & \cdots & (V_n+\Delta V_n)^{(1)} \\ \vdots & & \\ (V_1+\Delta V_1)^{(n)} & \cdots & (V_n+\Delta V_n)^{(n)} \end{bmatrix} \begin{bmatrix} \Delta Y_1 \\ \vdots \\ \Delta Y_n \end{bmatrix} = \begin{bmatrix} -\sum_{k=1}^{n} Y_{k0}\Delta V_k^{(1)} \\ \vdots \\ -\sum_{k=1}^{n} Y_{k0}\Delta V_k^{(n)} \end{bmatrix}. \tag{15}$$

For now we assume that each of these tests is made at the same frequency. In order to obtain a unique solution to (15), the square branch voltage array must have rank n, that is, the rows of this square array must be independent. The necessary and sufficient test conditions required to determine the admittances Y_1–Y_n from branch voltage measurements can be found from the short-circuit admittance equations for the circuit.

$$YV_p = I_p \tag{16}$$

where the port voltages for a particular input excitation $I_p^{(j)}$ is

$$V_p^{(j)} = Y^{-1} I_p^{(j)}. \tag{17}$$

Thus in order to generate n independent port voltage vectors $V_p^{(j)}$, $j=1,2,\cdots,n$, and hence n independent branch voltage vectors $V^{(j)}$, we need to excite the circuit with n independent current source vectors. One possible set of test conditions is given below.

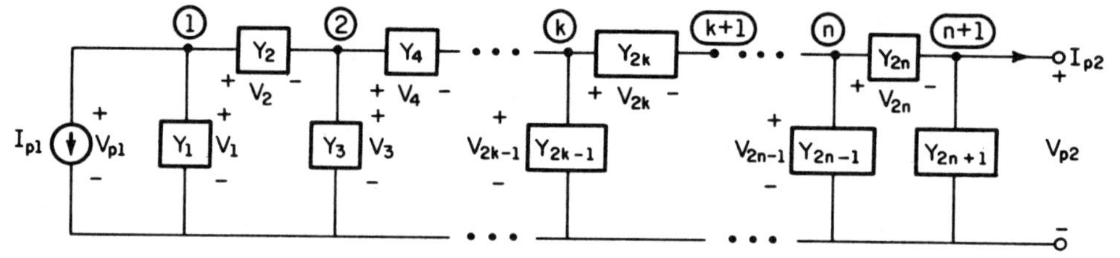

Fig. 12. Ladder network.

Test Conditions: Set $I_{pj}=1A$ and $I_{pi}=0$, $i=1,2,\cdots,n$, but $i\neq j$, and measure the branch voltages $(V_k+\Delta V_k)^{(j)}$. Do this for $j=1,2,\cdots,n$. Substitute this information into (15) and solve for ΔY_k, $k=1,2,\cdots,n$.

Thus the use of Tellegen's theorem and the concept of the adjoint circuit leads to a simple linear relationship between the unknown parameter changes and the branch voltage measurements. The number of linear algebraic equations which must be solved is equal to the number of unknown impedances connected to the node. Note also that the necessary and sufficient conditions for the determination of the impedances Z_1–Z_n from only their branch voltages is that the nodes 1–$n+1$ be accessible. Finally, the above results apply not only to a collection of impedances connected to a common node, but also to any cutset of impedances in the circuit. Below, a simple algorithm is given for the calculation of the impedances in a ladder network from node voltage measurements.

B. Ladder Network Example

Consider the ladder network in Fig. 12. The following theorem states the minimum test conditions that are both necessary and sufficient to determine all the admittances from only node voltage measurements at a single test frequency.

Theorem 2

All the parameters in the ladder circuit can be uniquely determined from voltage measurements at a single frequency under the following two test conditions.

1) $I_{p1}\neq 0$, $I_{p2}=0$ and measure the branch voltages $(V_i+\Delta V_i)^{(1)}$, $i=1,2,\cdots,2n+1$.

2) $I_{p1}=0$, $I_{p2}\neq 0$ and measure the branch voltages $(V_i+\Delta V_i)^{(2)}$, $i=1,2,\cdots,2n+1$, and provided that a) the circuit does not have transmission zeros at the measurement frequency, b) the series arm impedance $Z_{2k}\neq 0$ at the measurement frequency, and c) the equivalent impedance at any node in the circuit is finite.

Condition a) is necessary so that we have nonzero branch voltages to measure at the given test frequency, and b) is necessary in order to avoid having Y_{2k-1} and Y_{2k+1} in parallel at the test frequency. The admittances of parallel elements usually cannot be determined from only voltage measurements at a single test frequency. Finally, in practice accuracy requirements may dictate that we perform more than the above number of minimum tests. Next a proof is given of this theorem.

In order to simplify both the proof and the computational complexity of the method, we will isolate the elements at a node by shorting the other nodes in the *adjoint* circuit and applying a $1A$ source at the nonshorted node as shown in Fig. 13. We do this successively for all $n+1$ nodes in the adjoint circuit. The first adjoint circuit isolates the admittances Y and Y_2 so that (15) becomes

$$\begin{bmatrix}(V_1+\Delta V_1)^{(1)} & (V_2+\Delta V_2)^{(1)}\\(V_1+\Delta V_1)^{(2)} & (V_2+\Delta V_2)^{(2)}\end{bmatrix}\begin{bmatrix}\Delta Y_1\\ \Delta Y_2\end{bmatrix}=\begin{bmatrix}-\Delta V_1^{(1)}Y_{10}-\Delta V_2^{(1)}Y_{20}\\ -\Delta V_1^{(2)}Y_{10}-\Delta V_2^{(2)}Y_{20}\end{bmatrix} \tag{18}$$

so that we can compute ΔY_1 and ΔY_2 from the measurement of ΔV_1 and ΔV_2 for the two test conditions. For the kth adjoint circuit (15) becomes

$$\begin{bmatrix}-(V_{2k-2}+\Delta V_{2k-2})^{(1)} & (V_{2k-1}+\Delta V_{2k-1})^{(1)} & (V_{2k}+\Delta V_{2k})^{(1)}\\ -(V_{2k-2}+\Delta V_{2k-2})^{(2)} & (V_{2k-1}+\Delta V_{2k-1})^{(2)} & (V_{2k}+\Delta V_{2k})^{(2)}\end{bmatrix}\begin{bmatrix}\Delta Y_{2k-2}\\ \Delta Y_{2k-1}\\ \Delta Y_{2k}\end{bmatrix}$$
$$=\begin{bmatrix}+\Delta V_{2k-2}^{(1)}Y_{2k-2}-\Delta V_{2k-1}^{(1)}Y_{2k-1}-\Delta V_{2k}^{(1)}Y_{2k}\\ \Delta V_{2k-2}^{(2)}Y_{2k-2}-\Delta V_{2k-1}^{(2)}Y_{2k-1}-\Delta V_{2k}^{(2)}Y_{2k}\end{bmatrix}. \tag{19}$$

In (19) there are only two unknowns, ΔY_{2k-1} and ΔY_{2k}. The ΔY_{2k-2} is determined in the $k-1$ solution. A necessary and sufficient condition to obtain a unique solution for the ΔY's is that

$$\frac{(V_{2k}+\Delta V_{2k})^{(1)}}{(V_{2k-1}+\Delta V_{2k-1})^{(1)}}\neq\frac{(V_{2k}+\Delta V_{2k})^{(2)}}{(V_{2k-1}+\Delta V_{2k-1})^{(2)}}. \tag{20}$$

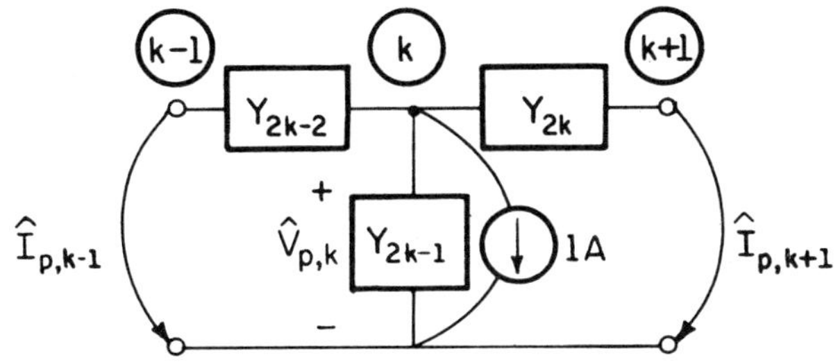

Fig. 13. Isolation of elements by shorting nodes in the adjoint circuit.

The requirement that there be no transmission zeros at the measurement frequency insures that the above voltages are not zero. Also, (20) is equivalent to the requirement that

$$\frac{Z_{2k}}{Z_{2k}+Z'_{2k+1}} \neq \frac{-Z_{2k}}{Z'_{2k-1}} \tag{21}$$

where Z'_{2k-1} is the equivalent impedance at node k with $Y_{2k}=0$, and Z'_{2k+1} is the equivalent impedance at node $k+1$ with $Y_{2k}=0$. We can write (20) as

$$Z_{2k}\left[Z_{2k}+Z'_{2k+1}+Z'_{2k-1}\right]\neq 0. \tag{22}$$

Note that the equivalent impedance at the kth node is

$$Z_{eq,k}=\frac{Z'_{2k-1}\left[Z_{2k}+Z'_{2k+1}\right]}{Z_{2k}+Z'_{2k+1}+Z'_{2k-1}}. \tag{23}$$

Thus condition (22) requires that $Z_{2k}\neq 0$, and that the equivalent impedance at each node be finite at the measurement frequency. The requirement that the voltages not be zero at the measurement frequency also means that there are no transmission zeros at the measurement frequency, that is, Z_{2k} must be finite and $Z_{eq,k}\neq 0$ at this frequency.

In conclusion, Theorem 2 gives two simple test conditions by which all the impedances in a ladder network can be computed provided that the test frequency is not at any of the transmission-zero frequencies, nor is any of the series arm impedances zero at the test frequency. In the next section we examine the possibility of using multiple test frequencies to determine the components in a circuit.

V. Multiple Test Frequencies

Recall that in order to uniquely determine the admittance changes in a faulty circuit, we need to have n test conditions such that the rank of the square matrix in (15) is n, that is, equal to the number of unknowns. In addition to exciting the circuit at different test points we can also use different test frequencies, however, the admittances are frequency dependent. Thus (15) must be written in terms of individual component changes, that is

$$\begin{bmatrix} (j\omega_1)^{\alpha_1}(V_1+\Delta V_1)^{(1)} & \cdots & (j\omega_1)^{\alpha_n}(V_n+\Delta V_n)^{(1)} \\ \vdots & & \\ (j\omega_n)^{\alpha_1}(V_1+\Delta V_1)^{(n)} & \cdots & (j\omega_n)^{\alpha_n}(V_n+\Delta V_n)^{(n)} \end{bmatrix} \cdot \begin{bmatrix} \Delta P_1 \\ \vdots \\ \Delta P_n \end{bmatrix} = \begin{bmatrix} -\sum_{k=1}^{n} Y_{ko}(j\omega_1)\cdot \Delta V_k^{(1)} \\ \vdots \\ -\sum_{k=1}^{n} Y_{ko}(j\omega_n)\cdot \Delta V_k^{(n)} \end{bmatrix} \tag{24}$$

where $\Delta P_j = \Delta G_j$, ΔC_j, or $\Delta \Gamma_j$, and the exponent $\alpha_j = 0$, $+1$, or -1 depending on if the jth branch is a resistor, capacitor, or inductor, respectively. The voltage $(V_j+\Delta V_j)^{(i)}$ is measured at the frequency $j\omega_i$.

If we assume a single unit input function and that the branch voltage is a rational function of the complex frequency variable s, then we can write (24) as

$$\operatorname{diag}\left[\frac{1}{D(s_1)}\cdots\frac{1}{D(s_n)}\right]\begin{bmatrix} (s_1)^{\alpha_1}N_1(s_1) & \cdots & (s_1)^{\alpha_n}N_n(s_1) \\ \vdots & & \\ (s_n)^{\alpha_1}N_1(s_n) & \cdots & (s_n)^{\alpha_n}N_n(s_n) \end{bmatrix} \cdot \begin{bmatrix} \Delta P_1 \\ \vdots \\ \Delta P_n \end{bmatrix} = \begin{bmatrix} -\sum_{k=1}^{n} Y_{ko}(s_1)\cdot \Delta V_k^{(1)} \\ \vdots \\ -\sum_{k=1}^{n} Y_{ko}(s_n)\cdot \Delta V_k^{(n)} \end{bmatrix} \tag{25}$$

where the polynomial $D(s)$ yields the poles of the transfer functions for the branch voltages with respect to the input port, and $N_i(s)$ yields the zeros. A necessary and sufficient condition for the solution of (25) is that the difference between the highest and lowest degree terms of at least one of the polynomials in (25) be $n-1$, and furthermore, the rank of the coefficient matrix representing these polynomials must be n. If these conditions are satisfied, then one can always find a set of n frequencies such that the rank of the square matrix in (25) is n.

Note that by using different measurement frequencies we can determine the value of parallel elements, provided they are different types. For example, consider the parallel RLC circuit combination in Fig. 14. For this circuit (25)

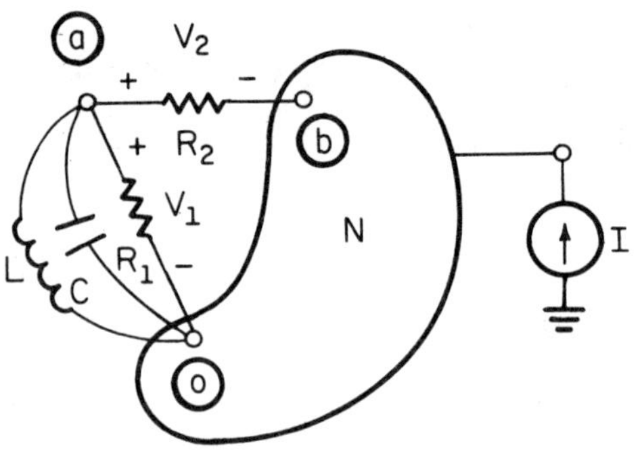

Fig. 14. Calculation of *RLC* parallel elements.

has the form

$$\operatorname{diag}\left[\frac{1}{D(s_1)}\cdots\frac{1}{D(s_4)}\right]\begin{bmatrix}\frac{1}{s_1}N_1(s_1) & s_1N_1(s_1) & N_1(s_1) & N_2(s_1)\\ \vdots & \vdots & \vdots & \vdots\\ \frac{1}{s_4}N_1(s_4) & s_4N_1(s_4) & N_1(s_4) & N_2(s_4)\end{bmatrix}\begin{bmatrix}\Delta\Gamma\\ \Delta C\\ \Delta G_1\\ \Delta G_2\end{bmatrix}$$

$$=\begin{bmatrix}-\frac{1}{s_1L_0}\Delta V_1(s_1)-s_1C_0\Delta V_1(s_1)-G_{10}\Delta V_1(s_1)-G_{20}\Delta V_2(s_1)\\ \vdots\qquad\vdots\qquad\vdots\qquad\vdots\\ -\frac{1}{s_4L_0}\Delta V_1(s_4)-s_4C_0\Delta V_1(s_4)-G_{10}\Delta V_1(s_4)-G_{20}\Delta V_2(s_4)\end{bmatrix}. \tag{26}$$

The square matrix in the above array may not have rank four with a single input test port. However, a rank-four array can be obtained by using more than one test port. For example, one could excite the circuit at port *a* and measure the two branch voltages at three different frequencies. Then one could excite port *b* and measure the two branch voltages at one frequency.

V. Summary

In summary, a method has been presented for the calculation of large circuit component variations. The method requires that the branch voltages of the unknown elements be available. Given these measurements the values of the components can easily be computed by solving a set of linear algebraic equations. Furthermore, the formulation of the method allows one to determine necessary and sufficient test conditions required to determine the component values. Further investigation is required into the accuracy of the method in the presence of measurement errors. Also, in the absence of sufficient test measurements, it is impossible to solve for the component values via linear equations. However, the method could be used to isolate the faulty component to some subset of the components in the circuit. Further study is needed in this area. Finally, the use of this algorithm in the case when there are inaccessible nodes is currently under investigation. In this case the rank of the system of equations is less than the number of parameters. However, by assuming that only some of the components are faulty, one can generate a number of fault possibilities which would produce the voltage deviations measured in the test circuit.

References

[1] R. S. Berkowitz, "Conditions for network-element-value solvability," *IRE Trans. Circuit Theory*, vol. CT-9, pp. 24–29, Mar. 1962.

[2] G. O. Martens and J. D. S. Dyck, "Fault identification in electronic circuits with the aid of bilinear transformations," *IEEE Trans. Rel.*, vol. R-21, pp. 99–104, May 1972.

[3] R. Saeks and S. R. Liberty, Eds., *Rational Fault Analysis*. New York: Marcel Dekker, 1977.

[4] S. Seshu and R. Waxman, "Fault isolation in conventional linear systems-A feasibility study," *IEEE Trans. Rel.*, vol. R-15, pp. 11–16, May 1965.

[5] J. D. Bastian, "Fault isolation of analog circuits," *Proc. Twelfth Asilomar Conf. Circuits, Syst., and Comp.*, Western Periodicals, North Hollywood, Nov. 1978.

[6] W. Mayeda and G. Peponides, "Determination of component values in passive networks under limited measurements," in *Proc. Twelfth Asilomar Conf. Circuits, Syst., and Comp.*, Western Periodicals, North Hollywood, Nov. 1978.

[7] T. N. Trick, *Introduction to Circuit Analysis*. New York: Wiley, Ch. 15, 1977.

[8] S. W. Director and R. A. Rohrer, "Automated network design—The frequency domain case," *IEEE Trans. Circuit Theory*, vol. CT-16, pp. 330–337, Aug. 1969.

[9] R. N. Gadenz, M. G. Rezai-Fakhr, and G. C. Temes, "A method for the computation of large tolerance effects," *IEEE Trans. Circuit Theory*, vol. CT-20, pp. 704–708, Nov. 1973.

[10] T. N. Trick and A. A. Sakla, "A new algorithm for the fault analysis and tuning of analog circuits," in *Proc. IEEE Int. Symp. Circuits and Syst.*, pp. 156–160, May 1978.

[11] T. N. Trick and C. J. Alajajian, "Fault analysis of analog circuits," in *Proc. 20th Midwest Symp. Circuits and Syst.*, Western Periodicals, North Hollywood, Part 1, pp. 211–215, Aug. 1977.

[12] R. Saeks, S. P. Singh, and R. Liu, "Fault isolation via component simulation," *IEEE Trans. Circuit Theory*, vol. CT-19, pp. 634–640, Nov. 1972.

The Evolution of Digital Electronics Towards VLSI

ROBERT W. KEYES

Abstract—Technological trends are extrapolated to the end of this century. Problems of utilizing high levels of integration are noted, and the capabilities of technology are viewed in the perspective of the problems to provide a forecast of the levels of integration that will be found in large computing systems. A physical model and some more speculative system assumptions are used to estimate the performance of the systems. The physical characteristics forecast for the system are summarized.

I. Introduction

PROGRESS in microelectronics has been extremely rapid. All aspects of electronic data processing have benefited from the great improvements that have taken place in the cost, reliability, and performance of electronic components. It is possible to extrapolate the history of integrated-circuit technology into the future and foresee a long continuation of this rapid progress [1]-[5]. However, the extrapolation implies certain changes in the intensive parameters of a system, such as increasing impurity concentrations and electric fields in semiconductor components and increasing density of power dissipation in large systems containing many components. One is thus forced to ask how far extrapolation can be carried before the changing intensive parameters lead to fundamental alterations in the phenomena which are basic to the operation of semiconductor devices. In other words, one must consider the extrapolation of semiconductor technology to ever-smaller dimensions in the light of known semiconductor phenomena and the laws of physics.

The purpose of this paper is to extrapolate the basic properties of increasingly powerful integrated-circuit chips through the remainder of this century. Models will then be used to derive performance, as measured by time per operation, from the physical description of the chip. Such performance projections will also require assumptions about the packaging of the chips. Certain parameters of the model have been derived by fitting to existing systems. Somewhat more speculative system assumptions will be introduced to provide a forecast of the evolution of large computers through the time period considered. Most of the quantitative description of the models is relegated to the Appendixes.

The relation of computer performance to technology requires a physical model of a complete computing system. The physical implementation of a system, however, offers a great many opportunities for choice, so that a wide variety of physical designs are actually encountered. In addition, physical implementations are extremely complex in detail. Thus any model that aims to be generally applicable must be greatly simplified and can only hope to represent a limited number of important features in a semiquantitative way. In most cases neither theory nor experience suffices to provide accurate and unique relationships between the properties of chips and the characteristics of system components built from the chips. Such relationships must be regarded as deterministic in modeling, and are described by simple mathematical formulas fitted to the qualitative and semiquantitative information that is available.

The application of projections of the capability of technology to the highest performance systems will be considered. It will be assumed that the development of large computers will occur by evolution, i.e., that no drastic revolutions will change the basic character of such machines. Thus the subject is "what might happen" rather than "what will necessarily happen." The scenario could be drastically changed by, for example, the development of techniques for using many millions of slow components to process information at the same rate as hundreds of thousands of fast components, or the development of a superconducting computer operating at cryogenic temperatures.

II. Extrapolation to VLSI

Extrapolations are broadly based on the assumptions that present levels of development effort will continue and produce results comparable to those achieved in the past. They also assume that such results will be economically desirable, that they will enable a user to exercise data-processing functions at reduced cost. Extrapolation through the earlier part of the period covered has been substantially guided by available projections of others [1]-[5]. It is influenced by judgements as to where the physical limits will be found in the latter part of the period. Fig. 1 shows an extrapolation of the basic capabilities of semiconductor technology as measured by chip size and minimum dimension produced by the lithographic process. The dates used in plotting the extrapolations of this section are intended to mean the dates of delivery of systems incorporating the technologies in question. There is no fundamental physical barrier to the achievement of the progress shown [6].

Fig. 2 shows how a chip may be regarded as composed of a large number of squares, each with side equal to the minimum lithographic dimension shown in Fig. 1. The character of each small square can be controlled independently of the other squares. Thus the number of such small squares on the chip is the number of independently controllable or resolvable ele-

Manuscript received August 1, 1978; revised November 10, 1978.
The author is with the IBM Thomas J. Watson Research Center, Yorktown Heights, NY 10598.

Reprinted from *IEEE J. Solid-State Circuits,* vol. SC-14, pp. 193–201, Apr. 1979.

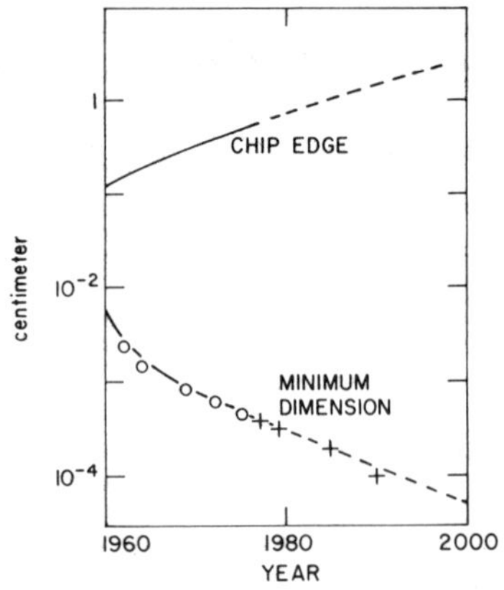

Fig. 1. History and projection of the minimum lithographic dimension and the maximum technologically feasible chip size. The points used to extrapolate minimum dimension are taken from [1] (O) and [3] (+).

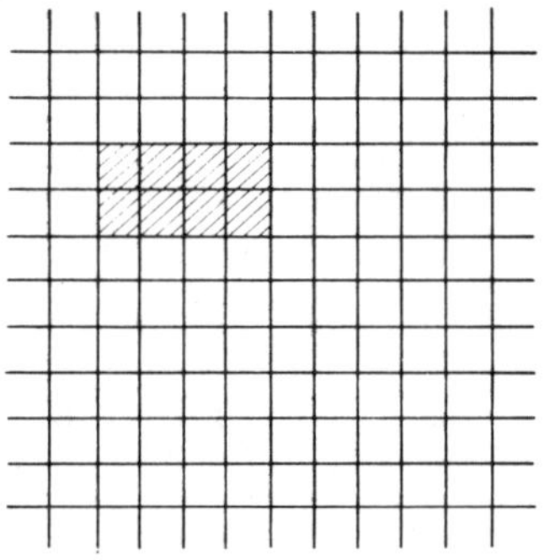

Fig. 2. Illustrating the concept of a resolvable element, a square with side equal to the minimum lithographic dimension. For example, the shaded area may be used to make an ohmic contact of sufficiently low resistance.

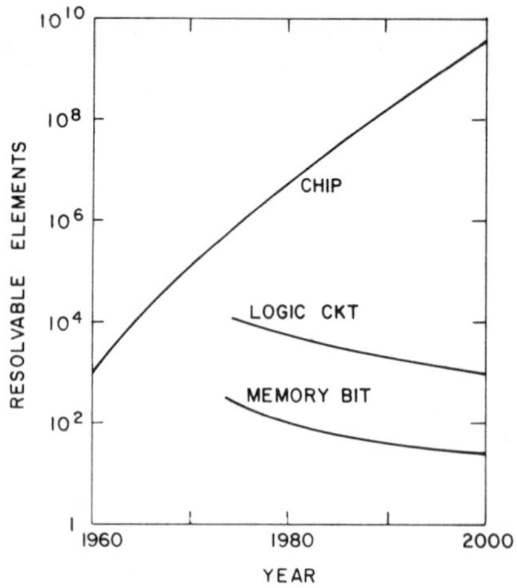

Fig. 3. The number of resolvable elements on a chip of maximum dimension, calculated from Fig. 1, and estimates of the number of such elements needed to construct a logic circuit and a memory bit plus the other elements, such as wires, decoders, and drivers that accompany them on the chip.

ELECTROMIGRATION
HOT ELECTRONS AND BREAKDOWN
WIRING COMPLEXITY
OHMIC RESISTANCE
INTERCONNECTIONS
POWER DISSIPATION

Fig. 4. The phenomena that will limit the advance of integrated digital electronics.

ments. Quite a few such resolvable elements are needed to fabricate, for example, the source, drain, and gate electrodes of a transistor and the insulating regions between circuits.

The number of resolvable elements per chip as derived from the extrapolated technological capabilities of Fig. 1 is shown in Fig. 3. Fig. 3 also shows an extrapolation of the number of elements needed to construct a logic gate or a memory bit together with its associated wiring and off-chip communication devices. Ingenuity has reduced the complexity of circuits in the past, at least as measured by resolvable elements in the surface plane of the chip, and the continuation of the exercise of this kind of ingenuity is anticipated [7], [8]. Increasing chip size, decreasing minimum dimension (miniaturization), and circuit inventiveness all contribute to the progress of integration [1].

Achievement of the extrapolated capabilities shown in Figs. 1 and 3 will require continuous process-development programs and substantial investments in tooling. Thus the projected advances will not take place unless they can be economically utilized to construct functional devices. Some of the phenomena that apparently are likely to limit the functioning of miniaturized, highly integrated semiconductor devices are listed in Fig. 4. It is clear that the degree to which these factors can be analyzed quantitatively varies widely. Consider them in order.

The first, electromigration, the motion of atoms induced by current in metallic wires, we believe will be fully contained by the development of better alloys.

Circuits based on lithography with minimum dimensions down to 1 μm promise to have superior performance that can be advantageously utilized. Beyond this point, high electric fields in devices will produce hot electrons and dielectric breakdown phenomena that will slow progress [9]-[12]. For example, exploiting the ultimate capacity of lithography may necessitate the development of devices in a material other than silicon. Nevertheless, economics favors progress towards the reduction of dimensions, since process costs tend to depend on the area processed rather than the number of devices. It is assumed, as shown in Fig. 1, that the hot-electron phenomenon will not prevent the development of devices with 0.5-μm dimensions in the year 2000.

On the other hand, several factors constitute more immediate limits to progress to larger substrates. In logic the complexity of the design of chips containing many circuits impedes progress. The resistance of interconnecting wires increases as dimensions are reduced. The increasing length of interconnections on a chip (measured in units of the average circuit-to-circuit distance) with increasing number of circuits exacerbates this problem [13], [14]. A guide to the dependence of the length on circuit count is shown in Fig. 5. The total line length on the chip, in the same units, is also shown in Fig. 5. Increasing the number of circuits on a chip increases the number of connections that must be made between the chip and its substrate, and the provision of the

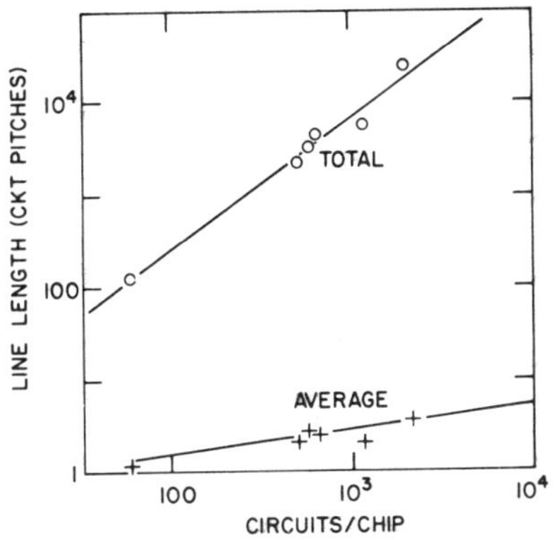

Fig. 5. The average interconnection length and the total wire length on a logic chip as a function of the level of integration, after the work of Donath [13] and Heller *et al.* [14]. The lengths are measured in circuit "pitches." The circuits are assumed to be placed on a square lattice, and the "pitch" is the circuit-to-circuit distance, which is the square root of the area per circuit.

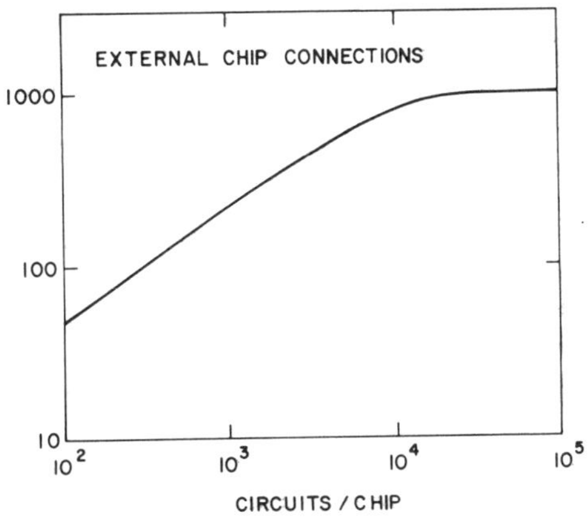

Fig. 6. The number of signal-carrying connections that must be made to a logic chip as a function of the level of integration [15]. The number of connections is proportional to the two-thirds power of the number of circuits at low levels of integration, but is not extrapolated beyond 1000 connections.

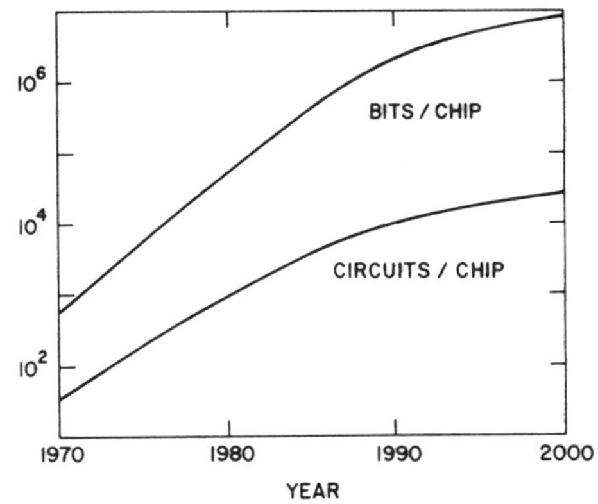

Fig. 7. An extrapolation of the levels of integration of main memory and logic chips, tempering the capabilities presented in Fig. 1 with the difficulties listed in Fig. 4.

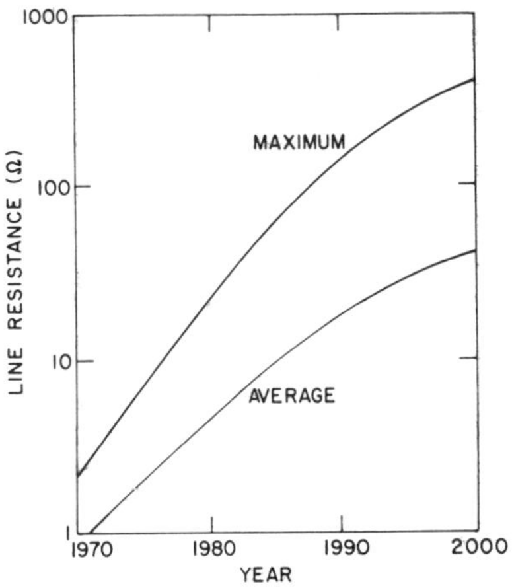

Fig. 8. The resistance of aluminum interconnecting lines, on logic chips, as calculated from the dimensions of Fig. 1, the average line lengths of Fig. 5, and maximum line lengths taken to be half of the chip edge.

hundred or so connections now required is already difficult. An estimate of the dependence of the number of connections required in random logic on the number of circuits on a chip is shown in Fig. 6 [15]. Further, it is not clear that system speed is increased by striving for the highest possible number of circuits on a single substrate. The technical difficulties of utilizing very large random logic chips will prevent levels of integration from achieving the high values that could be attained by straightforward development of material science and lithographic processing.

Fig. 6 is an extrapolation of the results of Landman and Russo [15] at low circuit counts, but it is assumed that the number of connections to the chip will not exceed 1000. If a chip contains a reasonably complete functional entity or subsystem, the number of external connections may be even smaller.

Memory is more directly responsive to the cost per function, rather than to performance as measured by speed of operation, than is logic. Since the achievement of high logic speeds is one of the forces driving miniaturization of logic devices, memory devices will not demand such aggressive exploitation of advanced lithographic capability. However, advances will eventually be slowed by the hot-electron effects that will inevitably become more serious as device dimensions fall towards 1 μm or less.

Thus our guess at the rate of growth of bits/chip and logic circuits per chip, Fig. 7, shows a slower growth for logic than for memory and a declining rate of growth after 1990 as the limits are more firmly felt. These projections do not strain the ability of technology to provide larger chip sizes, and it is assumed that chip size grows at a rate that will accommodate the component growth shown in Fig. 7. For comparison with Fig. 1, the logic-chip edge is 0.25 cm and the memory-chip edge is 1.6 cm in the year 2000. The projections imply that miniaturization of off-chip connections will be achieved; the chip area per connection decreases to a value less than 10^{-4} cm^2.

One of the consequences of the assumption about the size of logic chips presented in Fig. 7 is shown in Fig. 8: The increase in interconnection resistance on the chip. The resistance of the average interconnection increases both because of the decrease in the dimensions of a wire and because of the increase in length shown in Fig. 5. The resistance of the maximum length wire, taken to have a length one-half the chip

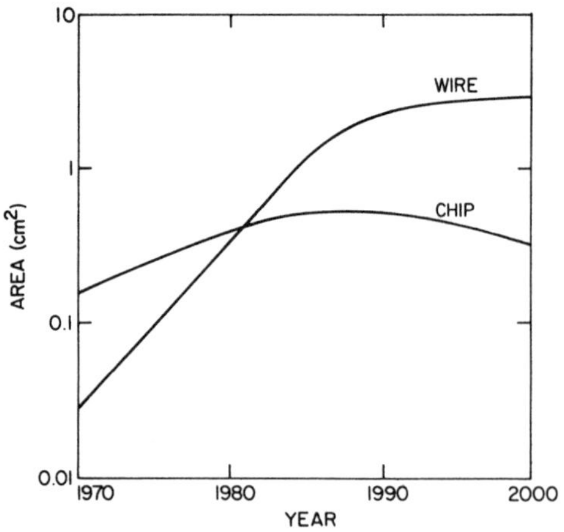

Fig. 9. The area of logic chip wiring, as calculated from the dimensions of Fig. 1 and the total line lengths of Fig. 5, compared with the area of a chip.

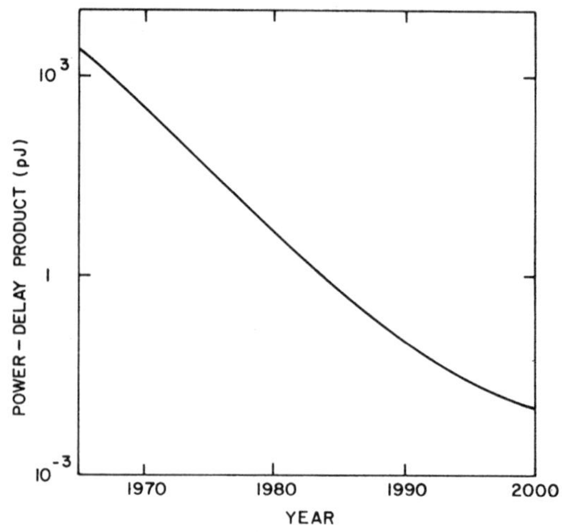

Fig. 10. The power-delay product of logic, including early vacuum tube as well as transistorized computers. The extrapolation is drawn to approach a limit of 0.01 pJ.

edge, becomes quite large, and is intolerable with present methods. Wiring designs and circuit layouts must find ways to avoid these longest wires.

Another consequence of the assumed level of integration of logic chips and the growth of line length shown in Fig. 5 is presented in Fig. 9. Here the area required for the wiring, assuming that the wire width and the spacing between wires are equal to the minimum dimension and that 25 percent of the wiring channels can be utilized, is plotted and compared with the area of the chip. The more rapid growth of area needed for wiring implies that a substantial increase in the number of wiring layers will be required.

The progress in miniaturization can be expected to yield improvements in performance; that is, in speed of logic circuits. Part of the improved performance is a direct result of shorter distances that signals must travel between circuits. Miniaturization also plays a large role in the steady decrease of the energy utilized per switching operation since the introduction of transistors into computers. The trend of this parameter, the product of power per circuit and logic delay (*pt*) is shown in Fig. 10 [16]. The energy dissipated has decreased by about four orders of magnitude since 1957. The sources of this decrease may be very roughly assigned as follows. One order of magnitude to decrease in voltage; we have suggested that the voltage is approaching a lower limit and cannot be decreased much further [6]. Two orders of magnitude to miniaturization. One order of magnitude to improved designs, such as replacing p-n junction area with semiconductor/insulator interface. Historically, the power-delay product of integrated logic chips has decreased nearly in proportion to the area per circuit. Thus it can be expected that circuits will be operated with less power. Nevertheless, because of the projected rapid increase in the number of circuits per chip, it is desirable to increase the density of power dissipation. Difficulties in removing heat directly degrade circuit speed because of the power/speed tradeoff.

There is room for optimism concerning the power-dissipation limit, at least in the near future. The decrease in the *pt* product of high-speed logical operations that has characterized the last 30 years will continue through several more orders of magnitude, as forecast in Fig. 10. The value shown for *pt* in the year 2000 is only about 10^6 times the thermal energy of an electron or molecule at ordinary temperatures.

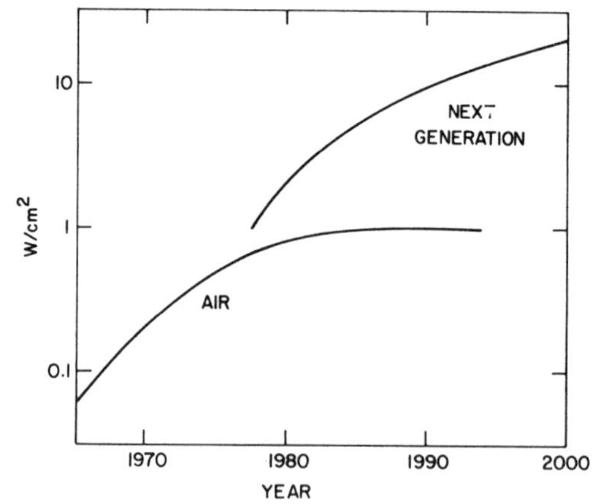

Fig. 11. A projection of cooling capabilities for logic chip packages. Forced-air cooling is probably limited to about 1 W/cm^2. New cooling technologies, in which heat is transferred to a liquid without the intermediary of air, are emerging [17], [18], and it has been assumed that they will extend cooling capability to 20 W/cm^2.

Advances in cooling technology also seem imminent. Circuits are conventionally cooled by transfer of heat from a semiconductor or semiconductor module to air. However, air cooling is limited to a power density of about 1 W/cm^2, as shown in Fig. 11. Emerging technologies, in which heat is transferred directly to a liquid without the intermediary of air [17], [18], break through the 1-W/cm^2 limit and are probably limited to something like the maximum rate of heat transfer to boiling Freon, about 20 W/cm^2, as also illustrated in Fig. 11.

As mentioned, there are many choices to be made in the physical design of a system, and many factors other than performance to be considered. Optimization is not done on a day-to-day basis, so that the technological parameters change in a series of steps rather than as a continuous curve. Thus the projections shown in Figs. 6-11 are trends and can only be expected to represent the state of technology semiquantitatively at any given time.

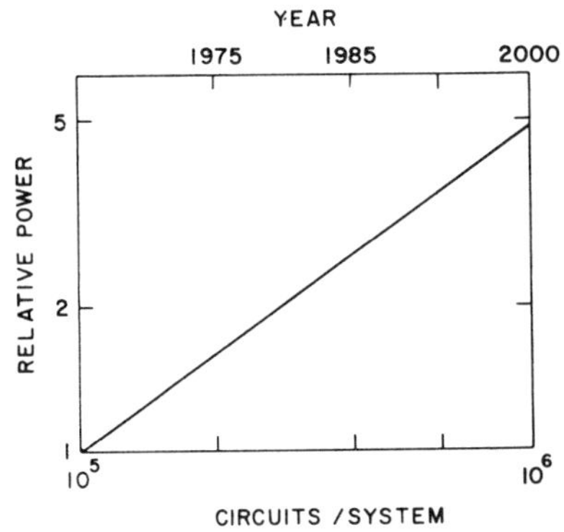

Fig. 12. An assumption that research in the "computer sciences" will make it possible to use increasing amounts of hardware to process information faster. The information processing power of a computer is assumed to increase as the 0.7 power of the number of circuits. As used in the present model, this means that 60 packaged delay times (see Appendix A) are used to process an instruction in a system of 2×10^5 circuits, and that the number decreases to 20 for a system of 10^6 circuits. The projected rate of system learning, only a very rough guess, is also indicated.

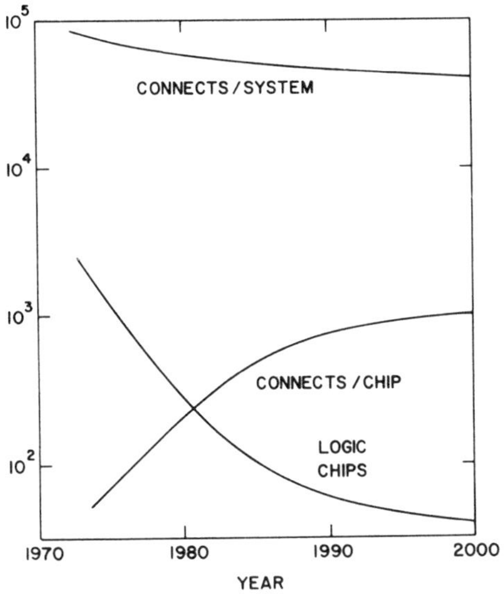

Fig. 13. The provision of chip-to-package connections is regarded as a costly element of computer system manufacture. Thus, one of the objectives of the projection of the level of integration of logic chips shown in Fig. 7 was to decrease the total number of off-chip connections used in the light of the growth of system size shown in Fig. 12. The results, calculated from Figs. 6, 7, and 12, are shown.

III. Modeling System Performance

It is anticipated that system learning will occur, and that it will be possible to utilize ever-increasing numbers of circuits in a single system to speed the instruction stream. The relative power of a system is assumed to grow slightly slower than proportionately to the system circuit count, as shown in Fig. 12, and the circuit count is allowed to grow from 2×10^5 in 1975 to 10^6 in the year 2000. The meaning of this in the model of performance is that the time per instruction is assumed to decrease from 60 packaged logic delays to 20 packaged logic delays during the period.

The implications of this system assumption for the chip-connection problem are shown in Fig. 13. The number of logic chips decreases because of the rapid increase in the level

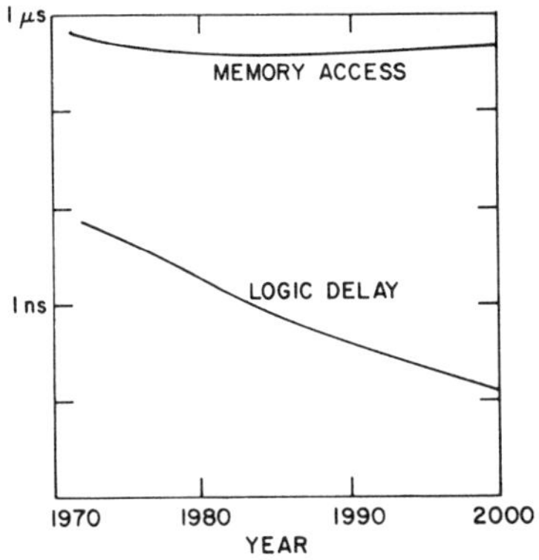

Fig. 14. Packaged logic delays and main-memory access times calculated from the models presented in Appendixes A and B, and the extrapolations of technology.

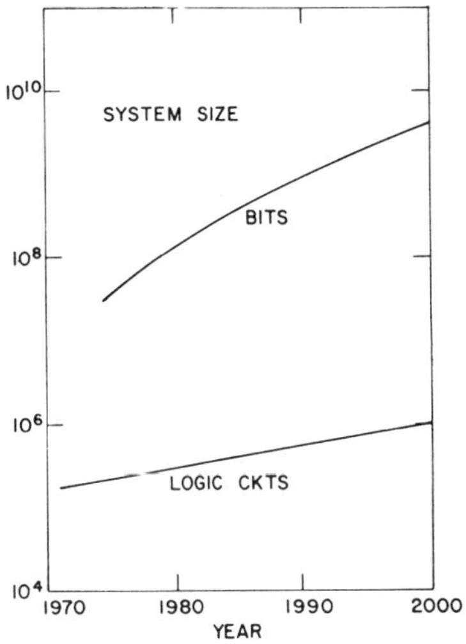

Fig. 15. The rapid growth of memory utilization resulting from the assumption that the number of bits grows in proportion to MIP's.

of integration. The number of chip connections also decreases somewhat. The prevention of a substantial growth in total number of chip connections is regarded as an objective of design in these projections, and implies that a much lower level of integration of the logic chips would be unacceptable.

The forecasts of integration level, *pt* product, and cooling capability (Figs. 7, 10, and 11) can be used in a physical theory to predict the progress of packaged logic delay. The limiting delay is regarded as the sum of two parts, the circuit delay, which is obtained by dividing the *pt* product by power, and the chip-to-chip transit time [19]. The latter, however, increases with power; since the power density is limited to values shown in Fig. 11, the more power expended per chip, the further apart the chips must be placed. The power per circuit can be chosen to minimize the total delay, as described in more detail in Appendix A.

The resulting logic delay is shown in Fig. 14. The calculation predicts that it will decrease through more than another order of magnitude through the remaining part of the century.

For comparison, the main memory access time, estimated in Appendix B, is also shown in Fig. 14. This does not decrease much through the time period in question because of the rapid growth in memory size. It has been assumed that the main

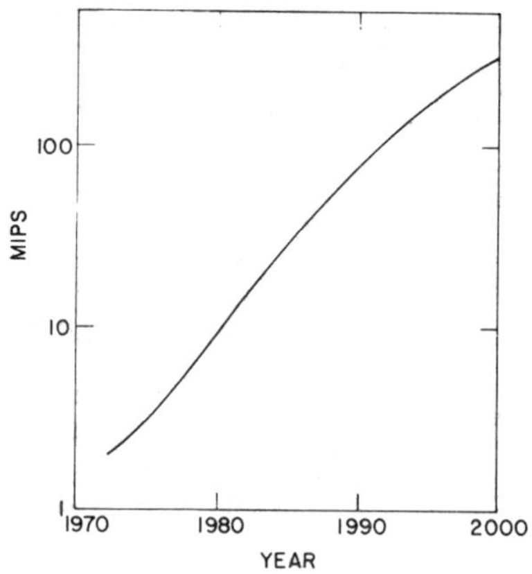

Fig. 16. The projected growth of processing power as measured by millions of instructions performed per second.

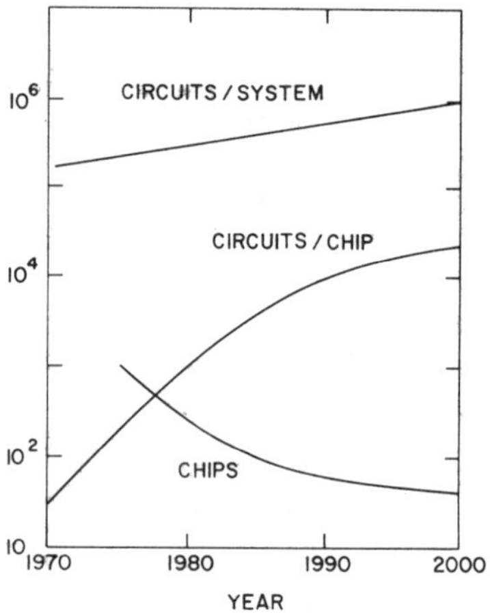

Fig. 17. A summary of the projected evolution of numbers of logic components.

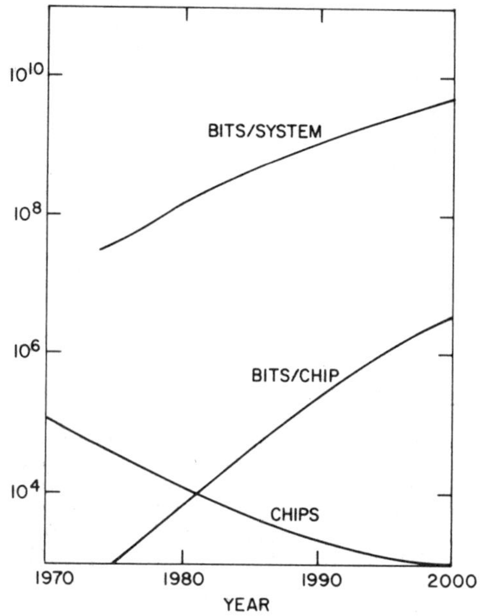

Fig. 18. A summary of the projected future numbers of memory components.

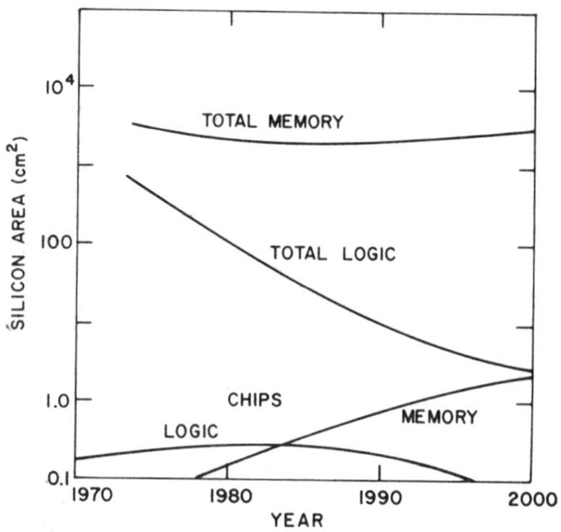

Fig. 19. The projected utilization of silicon area.

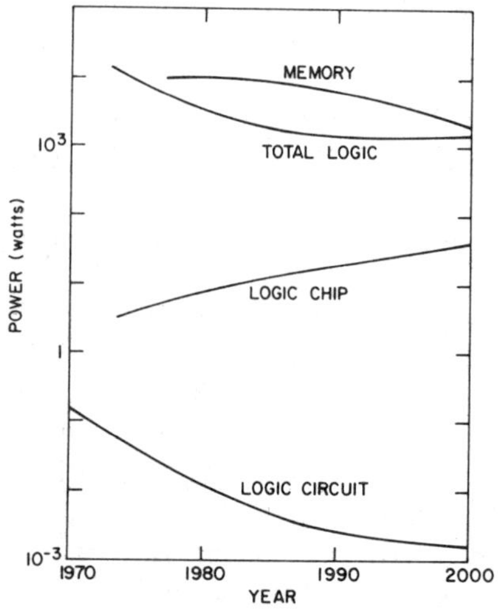

Fig. 20. The power demands predicted by the model.

memory will continue its historical trend to grow in proportion to computer power, resulting in the number of bits per system shown in Fig. 15.

The increasing gap between logic speed and memory access will have to be taken up by increasing attention to buffers and cache memory, and, perhaps, by innovations in memory hierarchy. The model used to determine the effect of memory access time and the interposition of a cache memory on system performance is described in Appendix B. Assuming that this can be successfully done, the growth of large system power, as measured by millions of instructions per second, shown in Fig. 16, can be forecast.

IV. Physical Characteristics of the VLSI System

A few additional results follow from the calculations presented.

The system parameters—that is, total component count, components per chip, and number of chips—are summarized in Figs. 17 and 18 for logic and memory.

The assumptions about lithography and the level of integration lead to values for the chip area. The projections relating to area are summarized in Fig. 19. The memory chip grows steadily in area through the period. The rapid growth of total memory compensates the advantages gained from lithography and ingenuity to keep the total silicon area in the memory nearly constant. On the other hand, the limitation of the size of logic chips to 25 000 circuits implies a decreasing chip size, and the slow growth of circuit count in the system leads to a decrease in the silicon area required for logic. Thus over 100 times more silicon area will be devoted to memory than to logic by the end of the period.

Power utilization by the electronic components is summarized in Fig. 20. Power consumption is increasingly dominated by memory during the period.

V. Conclusions

Utilization of the advanced capabilities for the fabrication of integrated circuits that will become available for random logic chips will require solution of several difficult problems: Design and construction of the increasingly complex chip wiring, provision of many more external electrical connections to increasingly small chips, and devising circuits that can tolerate large interconnection resistances. Assuming that some progress can be made in these areas, the evolutionary development of large machines should lead to single processors capable of handling 300 million instructions per second by the end of the century. Such large machines will be physically dominated by memory; for example, the silicon area in the memory will be three orders of magnitude greater than the area in logic.

Appendix A
Model of Packaged Delay

The removal of the heat generated in the semiconductor devices is a performance-limiting aspect of packaging. It is possible to quantify this limit as follows. The characteristic heat-transfer parameter of a planar packaging technology is the maximum rate per unit area at which heat can be removed from the package. A widely used measure of the performance of semiconductor circuits is their power-delay product. Assume that device-design changes within a given technology can be used to trade speed with power while maintaining a fixed power-delay product, i.e., if p is the power per circuit and t_c is the delay per circuit, then the semiconductor technology only requires that

$$pt_c = U. \tag{A1}$$

The energy dissipated per switching operation, the pt product, is an important parameter of switching circuits. The progress of the pt product in simple circuits, is shown as a function of time in Fig. 10. In complex logic circuits, it depends on circuit fan-out and wire loading. Energy must be supplied to change the capacitance of several other devices and of the wiring that interconnects devices in actual logic circuits. A term proportional to the average wire length is added to the pt product of Fig. 10 to take account of the effect of the wiring. The term proportional to length has the form $C'V^2l$, where C' is the capacitance per unit length, V is a voltage, and l is the total wire length driven by the circuit, taken to be three times the average wire length given by Fig. 5. $C'V^2$ is about 2 pJ/cm in advanced modern chips; a value 1 pJ/cm is used here to allow room for improvement in wire geometry

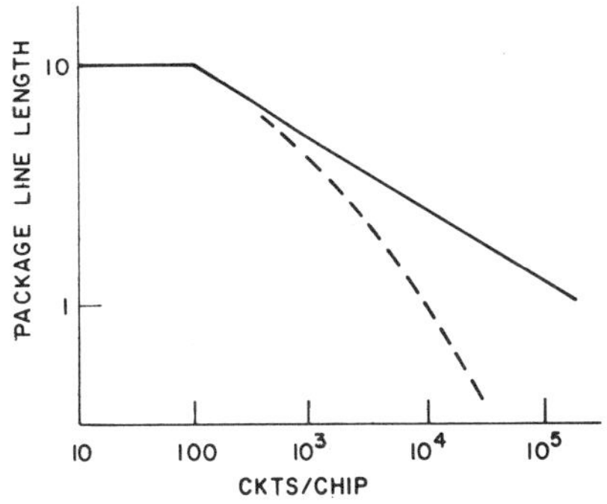

Fig. 21. The weighting factor m for off-chip delays in the minimization of average packaged delays. The solid line is an interpolation between values $m = 10$ for low levels of integration and $m = 1$ for integration approaching the total systems on a single chip. The dotted line shows the additional reduction in m that results from the reduced incidence of off-chip connections as the level of integration increases. The latter factor is based on the assumption that off-chip delays are only encountered with probability (1/12) if the number of circuits per chip exceeds 50 000.

and reduction of voltage. Summing up, the energy per operation used in the model is

$$U = (pt) + 3 \times (1 \text{ pJ/cm}) \times l_{av}. \tag{A2}$$

Here (pt) must be understood to mean the pt product given by Fig. 10.

An analytic representation of l_{av} is used in the calculations. From Fig. 5

$$l_{av} = 0.3N^g \tag{A3}$$

where N is the number of circuits on a chip and g has the value 0.325.

Delays encountered by signals in passing from chip to chip also constitute an appreciable part of the average logic delay. The contribution of the off-chip delays to the average depends on the frequency with which signals leave a chip and the length of the off-chip wiring on the package. Little accurate information that sheds light on either of these quantities is available.

Almost every logic stage sends its output off-chip in systems with low levels of integration, and it appears that the off-chip delays amount to the time needed for a signal to traverse about ten chip pitches. "Chip pitch" is the center-to-center distance between chips on a planar package. This component of delay might be expected to decrease with increasing levels of integration. More lengthy sequences of operations can be completed on a chip if it contains many logic gates, and a signal leaving a chip would seem to have a greater probability of having a receptor at a shorter distance if the level of integration is high. These results are obvious in the limit in which a system can be constructed from only one or a very few chips. Thus the weighting of off-chip delays used here is essentially an interpolation between the simple views of the very low and very high levels of integration just described. The form of the assumed relationships is plotted in Fig. 21. Here the average number of chip pitches included in the logic delay per stage is plotted as a function of the number of logic circuits per chip. These numbers are the values of the quantity m in the theory of packaged delay given below.

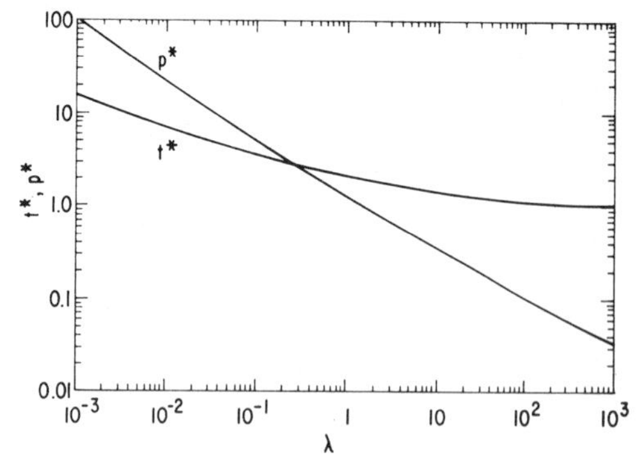

Fig. 22. The values of p^* and t^* which provide the solution of (A6)–(A9).

Let semiconductor chips, each containing N logic gates with the property described by (1) be packaged on a planar substrate from which heat can be removed at a maximum density Q per unit area, with each chip occupying an area L^2. Assume, further, that an amount of power P_D is dissipated in the high-current circuits that drive the transmission lines that connect one chip to another. Then heat removal requires

$$P_D + Np \leqslant QL^2. \tag{A4}$$

The time taken for a signal to propagate from one chip to another is mL/c_1, where c_1 is the velocity of electromagnetic waves on the interconnections and m is the length of interconnections measured in units of the distance between chips L. It is also required that L be larger than the chip edge a. The total packaged delay t_p is the sum of the circuit delay t_c and the chip-to-chip propagation time. All of these effects are taken into account by expressing t_p as a function of the power as

$$t_p = (U/p) + (m/c_1)\,((Np + P_D)/Q)^{1/2} + a) \tag{A5}$$

where m is the factor that weights the chip-to-chip delays shown in Fig. 21. There is a value of p that minimizes t_p, given by the equation

$$(1 + Np/P_D)^{1/2} = mNp^2/2c_1 UQ^{1/2}P_D^{1/2}. \tag{A6}$$

Equation (4) can be solved with the aid of Fig. 22, in which $p^* \equiv Np/P_D$ is plotted as a function of a parameter λ, defined by

$$\lambda \equiv (m/2c_1 NU)\,(P_D^3/Q)^{1/2}. \tag{A7}$$

The propagation time τ_p then is given by

$$\tau_p = (ma/v) + (mP_D^{1/2}/c_1 Q^{1/2})\,\tau^*. \tag{A8}$$

The quantity τ^* here is also plotted in Fig. 22 and is related to λ and p^* by

$$\tau^* = (\tfrac{1}{2}\lambda p^*) + (1 + p^*)^{1/2}. \tag{A9}$$

The results given by (A6)–(A9) are used to determine the packaged delays.

Appendix B
Memory Access and Cache

As is the case with most of the quantitative relationships that are needed for a model of system performance, no simple, unique relation between memory access time and memory size, power dissipation, and chip characteristics can be found. Qualitatively, it is seen that memory access time decreases with an increase in chip power, with miniaturization, and with decreasing memory-system size. Memory access time does not directly enter into the model of system performance. Thus, a relatively simple view of memory access time was adopted. The expression used for memory access time is

$$T_M = K_M C_M^{\alpha} p^{-1} A^{1/2}. \tag{B1}$$

Here K_M is a constant, C_M is the number of chips in the memory, p is the power per chip, and A is the chip area, α is a parameter, estimated to have the value 0.2.

Since the memory-chip size grows steadily in the model, memory access times do not decrease significantly according to (B1). It is assumed that the increasing gap between processing time and memory access time is taken up by a fast buffer or cache memory, built essentially with the technology used for the logic chips.

Assume that a small buffer memory or cache with fast access time holds the information that is currently needed by the processor. The cache is small enough that occasionally the desired information is not found there, and it must be obtained by transfer from the main memory. In such a case, a main memory access time must be added to the cycle time of the processor. The size of the cache is determined in practice by economic considerations. The cache is very expensive, and the minimum amount that will optimize system performance per unit cost is used. Performance is measured by the reciprocal of t_I, the instruction time, and costs are measured roughly by the number of chips in the model used here. Thus the quantity to be optimized is

$$F = t_I(hn_L + n_C) \tag{B2}$$

where n_C is the number of cache chips, n_L is the number of logic chips, and h is a factor that accounts for the added cost per logic chip attributable to the more complex physical design of logic. The instruction time consists of the time calculated without references to the main memory plus the product of the "miss ratio" r and the main memory access time T_M

$$t_I = k_L t_p + k_M r T_M. \tag{B3}$$

Here k_L is the number of logic delays per instruction, t_p is the logic delay calculated in Appendix A, and k_M is the number of memory references per instruction. The miss ratio r is approximately related to the size of the cache by [20], [21]

$$r = K_C/n_C N \tag{B4}$$

where N is the number of bits on a cache chip. When (B3) and (B4) are substituted into (B2) and the resulting expression for F is minimized with respect to n_C, it is found that

$$n_C = (hk_M K_C n_L T_M/Nk_L t_p)^{1/2}. \tag{B5}$$

The following values were used in the application of this model: $h = 5$ and $K_C = 2000$; n_L is obtained by dividing the circuits per system by the circuits per chip; N is taken to be two times the number of logic circuits per chip; t_M is obtained from (B1); and k_L and k_M are both assumed to vary inversely as the "relative power" factor given in Fig. 12 and have the ratio $k_M/k_L = 1/12$.

References

[1] G. E. Moore, "Progress in digital integrated electronics," in *Tech. Digest 1975 Int. Electron Devices Meet.* (IEEE, New York, 1975), pp. 11-13.

[2] R. N. Noyce, "Large scale integration: What is yet to come?", *Science*, vol. 195, pp. 1102-1107, Mar. 18, 1977.

[3] G. Marr, "Perspectives on MOS directions," in *Fall COMPCON 77 Digest* (IEEE Catalog No. 77CH1258-3C), pp. 242-244.

[4] "Semiconductors face the '80s," *IEEE Spectrum*, vol. 14, pp. 42-48, Oct. 1977.

[5] E. Bloch and D. Galage, "Component progress: Its effect on high-speed computer architecture and machine organization," *Computer*, vol. 11, pp. 64-76, Apr. 1978.

[6] R. W. Keyes, "Physical limits in semiconductor electronics," *Science*, vol. 195, pp. 1230-1235, Mar. 18, 1977.

[7] J. M. Oliphant, "Technological impact on semiconductor memories," in *1976 Fall COMPCON Digest* (IEEE, New York, 1976), pp. 174-177.

[8] P. R. Schroeder, "Prospects for the 64K RAM," in *1977 Fall COMPCON Digest* (IEEE, New York, 1977), pp. 114-115.

[9] T. H. Ning, C. M. Osburn, and H. N. Yu, "Effect of electron trapping on IGFT characteristics," *J. Electron. Mat.*, vol. 6, pp. 65-76, Feb. 1977.

[10] B. Hoeneisen and C. A. Mead, "Fundamental limitations in microelectroncs–I. MOS technology," *Solid-State Electron.*, vol. 15, pp. 819-829, Aug. 1972.

—, "Fundamental limitations in microelectronics–II. Bipolar technology," *ibid.*, vol. 15, pp. 981-987, Sept. 1972.

[11] S. A. Abbas and R. C. Dockerty, "Hot carrier instability in IGFET's," *Appl. Phys. Lett.*, vol. 27, pp. 147-148, Aug. 1, 1975.

[12] D. R. Collins, "Excess current generation due to reverse bias p-n junction stress," *Appl. Phys. Lett.*, vol. 13, pp. 264-266, Oct. 15, 1968.

[13] W. E. Donath, "Placement and average interconnection lengths of computer logic," IBM Res. Rep. RC4610 (submitted to *IEEE J. Solid-State Circuits*).

[14] W. R. Heller, W. F. Mikhail, and W. E. Donath, "Prediction of wiring space requirements for LSI," in *Proc. 14th Design Automation Conf.*, (New Orleans, LA), June 20-22, 1977, pp. 32-42.

[15] B. S. Landman and R. L. Russo, "On a pin versus block relationship for partitions of logic graphs," *IEEE Trans. Comput.*, vol. C-20, pp. 1469-1479, 1971.

[16] K.-U. Stein, "Technologischer Stand in der Mikroelektronik," *Electrotech. Maschinenbau*, vol. 93, pp. 240-248, 1976.

[17] E. A. Wilson, "True liquid cooling of computers," in *1977 National Computer Conf. Proc.* (Montvale, NJ.: AFIPS Press, 1977), pp. 341-348.

[18] R. M. Russell, "The CRAY-1 computer system." *Commun. Ass. Comput. Mach.*, vol. 21, pp. 63-72, Jan. 1978.

[19] R. W. Keyes, "A figure of merit for IC packaging," *IEEE J. Solid-State Circuits*, vol. SC-13, pp. 265-266, Apr. 1978.

[20] S. I. Rege, "Cost, performance, and size tradeoffs for different levels in a memory hierarchy," *Computer*, vol. 9, pp. 43-51, Apr. 1976.

[21] T. Makino and N. Ohno, "Characteristics of not found probability in memory hierarchy system," *NEC Res. Develop.*, vol. 20, pp. 51-58, Oct. 1976.

Part II

Control
(Stability, Decentralized Control)

THE APPLICATION OF LYAPUNOV'S SECOND METHOD TO INTERCONNECTED SYSTEMS*

F. N. BAILEY†

Summary. As many engineering systems are made up of an interconnection of simple subsystems, it is natural to attempt to utilize this interconnection structure in developing system analysis techniques. In this paper interconnection information is used in conjunction with properties of the individual subsystems to obtain sufficient conditions for asymptotic stability-in-the-large. This method may be applied to a broad class of linear, nonlinear and time-varying interconnected systems as indicated in the following three steps.

First, the Lyapunov functions and comparison equations are found for the individual subsystems. Second, the comparison equations are interconnected, following the interconnections in the original system, into a system of comparison equations. This system is linear, with constant coefficients and of order equal to the number of subsystems in the original interconnected system. Finally, the stability of the null solution of this auxiliary system is examined. If it is asymptotically stable, then the null solution of the interconnected system is asympotically stable-in-the-large. An example illustrates the use of interconnection information in determining the sufficient conditions for asymptotic stability-in-the-large for a ninth-order interconnected system.

1. Introduction. It has long been realized that in the analysis of complex, high dimensional systems, a straightforward application of general mathematical tools would often become bogged down in the welter of detail. For this reason the full promise of many potentially valuable mathematical tools has never been realized. In attempts to overcome or to circumvent these difficulties, it has frequently been found that the more efficient procedures are those which depend strongly on special properties or structural features present in the particular system under study. However, such approaches have generally ignored a basic structural feature of complex systems, namely, the interconnection structure. Many complex, high order systems are actually composite systems—interconnections of a large number of relatively simple subsystems into a complex whole—with special structural features which might be used to advantage in analysis.

The value of interconnection structure has been recognized by Kron [1] who has developed procedures for determining the behavior of certain types of high order composite systems through an analysis of each of the individual, low order subsystems and their interconnections. This substitution of many low order (and hopefully easier) problems for one high order problem can be of value in cases where difficulties in analysis (computer time, etc.) depend on problem order to a power higher than unity. Although Kron's work has never become popular, the basic concept of studying a complex system through an analysis of its components (subsystems) and their interconnections is quite appealing since the implied piece-by-piece analysis through the subsystems might avoid the formidable difficulties that are often encountered in a straightforward attack.

This paper describes an application of this "composite system approach" to stability problems or, more specifically, to the problem of determining criteria for asymptotic stability-in-the-large (sometimes termed global asymptotic stability) of interconnected systems containing linear, nonlinear, and time-varying elements. The innovation lies in the use of Lyapunov's second method *in conjunction with interconnection information* to simplify the stability analysis of composite systems. Lyapunov functions for the individual subsystems are found and then "interconnected", following the existing interconnections between the subsystems, to obtain a vector Lyapunov function applicable to the composite system. By using this approach the manifold difficulties commonly encountered in a direct application of Lyapunov's second method to the original composite system can be largely circumvented. In §7 the suggested procedure is applied in determining stability criteria for a ninth-order, nonlinear, time-varying composite system.

2. Notation and definitions. The vector notation used is similar to that employed by Hahn [2] or Cesari [3]. Let E^n denote the n-dimensional Euclidean space of n vectors, $x = \text{col}\,[x_1, x_2, \cdots, x_n]$, where the x_i are real numbers or real valued functions on the interval $T = [0, \infty)$ of the real line. The transpose of x is denoted by x' and the inner product is defined as

$$(x, y) = x'y = \sum_{i=1}^{n} x_i y_i .$$

The norm of a vector in E^n is the Euclidean norm $\| x \| = (x, x)^{1/2}$ and if P is an $m \times n$ matrix of real elements, then

$$\| P \| = \min \{\alpha \mid \quad \alpha \| x \| \geqq \| Px \| \quad \text{for all} \quad x \in E^n\}.$$

A useful metric on E^n is $d(x, y) = \| x - y \|$ and when limits and continuity are mentioned the implied topology is taken with respect to this metric. For any subset A of E^n the distance from x to A is

$$d(x, A) = \inf_{y \in A} d(x, y),$$

and for any $\epsilon > 0$,

$$S_\epsilon(A) = \{x \mid d(x, A) < \epsilon\}.$$

If $f(x)$ is defined on $R \subset E^n$ and $B \subset R$, then $f(B) = \{f(x) \mid x \in B\}$. For any clearly defined $t_0 \in T$, let T_0 denote the set $[t_0, \infty)$.

The differential equation $\dot{x} = f(x, t)$ is normally an nth order vector differential equation with $x(t)$ and $f(x, t)$ denoting n-vectors defined on T and $E^n \times T$, respectively. It is generally assumed that all differential equations satisfy conditions sufficient to guarantee the existence, uniqueness, and continuity of all solutions in t, x_0, and t_0 (continuity from the inside is implied at the boundary of any closed region).

A continuous, real valued function $v(x)$ on E^n with continuous first partial derivatives is said to be positive definite [positive semidefinite] if $v(0) = 0$ and $v(x) > 0[v(x) \geqq 0]$ for all $x \neq 0$. A continuous, real valued function $v(x, t)$ on $E^n \times T$ with continuous first partial derivatives is said to be positive definite [positive semidefinite] if $v(0, T) = 0$ and there is a positive definite [positive semidefinite] function $w(x)$ such that $v(x, t) \geqq w(x)$ on $E^n \times T$. A continuous, real valued function $v(x)$ or $v(x, t)$ with continuous first partial derivatives is said to be negative definite [negative semidefinite] if $-v(x)$ or $-v(x, t)$ is positive definite [positive semidefinite]. If $v(x)$ is a definite or semidefinite scalar function on E^n then ∇v is the vector $\text{col}\left[\frac{\partial v}{\partial x_1}, \frac{\partial v}{\partial x_2}, \cdots, \frac{\partial v}{\partial x_n}\right]$.

3. Stability and Lyapunov's second method. Following common practice, all further discussion of stability will refer to the stability of the null solution, $x = 0$, of the vector ordinary differential equation (equation of perturbed motion [4]),

$$\dot{x} = f(x, t), \qquad x(t_0) = x_0, \tag{3.1}$$

where x is an n-vector and $f(0, T) = 0$. Most questions concerning stability of equilibrium points or stability of a given motion can be formulated in the manner of (3.1) by a simple coordinate transformation [2].

The following definitions are standard [2].

DEFINITION 3.1. The null solution, $x = 0$, of (3.1) is *stable*[1] if for every $\epsilon > 0$ there is a $\delta > 0$ such that $x(T_0 ; S_\delta(0), t_0) \subset S_\epsilon(0)$.

DEFINITION 3.2. The null solution, $x = 0$, of (3.1) is *asymptotically stable* if it is stable and there is some $\gamma > 0$ such that

$$\lim_{t \to \infty} \| x(t; x_0, t_0)\| = 0$$

for every $x_0 \in S_\gamma(0)$.

DEFINITION 3.3. The null solution, $x = 0$, of (3.1) is *asymptotically stable-*

* Received by the editors February 12, 1965, and in revised form June 18, 1965.

† Center for Control Sciences, University of Minnesota, Minneapolis, Minnesota.
This research was supported in part by the National Science Foundation under Grant No. GP-540.

[1] In Definitions 3.1, 3.2, and 3.3 the initial time t_0 is not clearly defined. However it is easily shown [2] that if any of these properties hold for one initial time then they must also hold for all future (greater) initial times.

Reprinted with permission from *SIAM J. Contr.*, vol. 3, pp. 443–462, Ser. A., 1965.

in-the-large (abbreviated ASL) if it is stable and for every $x_0 \in E^n$,

$$\lim_{t\to\infty} \| x(t; x_0, t_0)\| = 0.$$

Lyapunov's second (or direct) method, abbreviated LSM, is simply a method of determining certain qualitative features, such as stability, of the solutions of (3.1) without actually solving the equations. In recent years LSM has received considerable attention in both the control engineering and mathematics literature and a wide variety of results are available [2], [5], [6], [7]. In addition, there is a generalization of LSM following Conti [8], [9] who views the positive definite function v (here called a Lyapunov function[2]) as the dependent variable in a first-order auxiliary equation (Brauer [9] uses the more descriptive term *comparison equation*). For example, let $v(x, t)$ be a positive definite function on $E^n \times T$ and $\dot{v}$ be the total derivative of v with respect to (3.1). If there is a function $\omega(v, t)$, with $\omega(0, T) = 0$, such that along the solutions of (3.1),

$$\dot{v} \leqq \omega(v, t), \tag{3.2}$$

then under quite weak conditions it can be shown that the null solution of (3.1) has the same stability and boundedness properties as the null solution of the first-order (scalar) auxiliary differential equation

$$\dot{r} = \omega(r, t). \tag{3.3}$$

This reduces the problem of determining the stability of an nth order system to that of determining the stability of a first-order system—a very significant simplification.

While the auxiliary equation is an important generalization of the earlier formulations of LSM [the earlier versions of LSM are the special cases where $\omega(v, t)$ is required to be a constant] the application of this new formulation requires the choice of two functions $v(x, t)$ and $\omega(v, t)$. For high order systems this choice will generally be very difficult and a major limitation on the practical value of LSM. (As in the earlier theory, the important theorems generally give only sufficient conditions for stability so that negative results yield little useful information.) The remaining sections of this paper describe a procedure for simplifying the choice of the two functions $v(x, t)$ and $\omega(v, t)$ by applying the composite system approach suggested above. Through an effective use of interconnection information this approach circumvents some of the current obstacles to the application of LSM and comparison equations in high order systems.

4. Composite systems, transfer systems, and models. As mentioned in the introduction, a composite system is a complex system composed of interconnections of simpler subsystems. The basic building blocks of the composite systems considered here will be called transfer systems.

DEFINITION 4.1. A *transfer system* is an input-output device whose terminal variables may be characterized by relations of the form

$$\dot{x} = f(x, t, u(t)), \tag{4.1}$$

$$y = h(x(t), t), \tag{4.2}$$

where $x(t)$ is an n-dimensional state vector, $u(t)$ is a p-dimensional input vector, and $y(t)$ is a q-dimensional output vector.

The terminal relations (4.1) and (4.2) characterizing the transfer system are called the *transfer system model*. A composite system can now be defined as an interconnection of transfer systems.

DEFINITION 4.2. Consider a set of m transfer systems, S_i, $i = 1, \cdots, m$. A *composite system* is an interconnection of these transfer systems so that for the ith transfer system the (vector) input u_i is given as

$$u_i = \sum_{j=1}^{m} B_{ij} y_j + G_i u, \qquad i = 1, \cdots, m,$$

where y_j is the (vector) output of the jth transfer system, u is an external (vector) input to the composite system and B_{ij}, G_i are constant matrices. (Note that only time-invariant linear interconnections are allowed.)

The partitioned matrix

$$B = \begin{bmatrix} B_{11} & B_{12} & \cdots & B_{1m} \\ B_{21} & B_{22} & \cdots & B_{2m} \\ \cdot & \cdot & \cdot & \cdot \\ B_{m1} & B_{m2} & \cdots & B_{mm} \end{bmatrix}, \tag{4.3}$$

where the submatrices B_{ij}, $i, j = 1, \cdots, m$, are the same as those used in Definition 4.2, will be termed the composite system interconnection matrix or simply the interconnection matrix since it indicates the interconnection structure of the composite system.

[2] This terminology is not uniform. Many authors call a positive definite function a Lyapunov function only if its derivative meets certain requirements. In this paper the terms *Lyapunov function* and *positive definite function* are interchangeable.

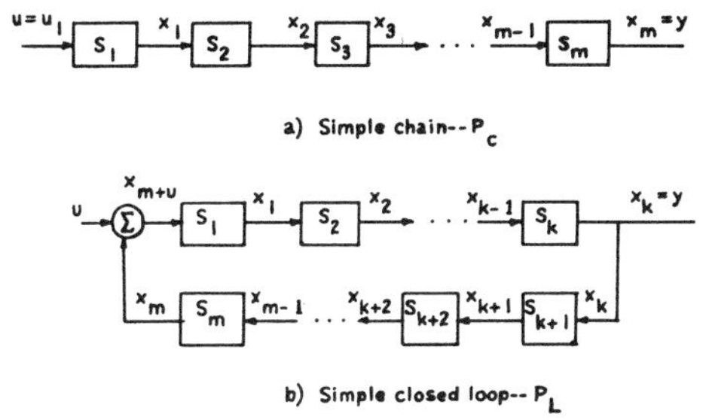

FIG. 4.1. *Two simple composite systems*

It should be pointed out that this sort of interconnection implies the usual system theory assumption [10] that *the individual transfer system models are not affected by the various types of interconnections; that is, there is no "loading" effect of one system on another.* While this in itself greatly simplifies the mechanisms through which instability of the composite system can occur it is a reasonable assumption for a large number of interconnected physical systems. The external input u is included to emphasize the fact that the composite system itself might be a transfer system in a larger composite.

In this paper the individual transfer system models may be linear, nonlinear and/or time-varying but will be assumed to have the special form

$$\dot{x}_i = f_i(x_i, t) + D_i u_i, \tag{4.4}$$

$$y_i = H_i x_i, \tag{4.5}$$

for $i = 1, \cdots, m$, where D_i and H_i are matrices (the subscript i denotes the ith transfer system in a given composite system). When these transfer systems are interconnected,

$$D_i u_i = \sum_j D_i B_{ij} H_j x_j + D_i G_i u,$$

and the composite system model takes on the form

$$\begin{aligned} \dot{x}_1 &= f_1(x_1, t) + C_{11}x_1 + C_{12}x_2 + C_{13}x_3 + \cdots + C_{1m}x_m + K_1 u, \\ \dot{x}_2 &= f_2(x_2, t) + C_{21}x_1 + C_{22}x_2 + C_{23}x_3 + \cdots + C_{2m}x_m + K_2 u, \\ &\qquad \cdot \quad \cdot \quad \cdot \\ \dot{x}_m &= f_m(x_m, t) + C_{m1}x_1 + C_{m2}x_2 + C_{m3}x_3 + \cdots + C_{mm}x_m + K_m u, \\ y &= h(x_1, x_2, \cdots, x_m, t), \end{aligned} \tag{4.6}$$

where

$$C_{ij} = D_i B_{ij} H_j, \qquad K_i = D_i G_i,$$

and y is the composite system output. If the individual transfer system models are n_ith order, the composite system model will be nth order where $n = \sum_{i=1}^{m} n_i$. By defining a composite system state vector

$$x = \text{col}\,[x_{11}, \cdots, x_{1n_1}, x_{21}, \cdots, x_{2n_2}, \cdots, x_{m1}, \cdots, x_{mn_m}],$$

the composite system model can be further refined to the form

$$\dot{x} = f(x, t) + Cx + Ku, \tag{4.7}$$

$$y = h(x, t), \tag{4.8}$$

where x is the composite system state vector, f is a column vector of the f_i's, C is the partitioned matrix of elements C_{ij}, and K is the partitioned (column) matrix of elements K_i. Since $C_{ij} = D_i B_{ij} H_j$, the matrix C also indicates the interconnection structure of the composite system.

As examples, consider the two simple composite systems shown in Fig. 4.1. Here it is assumed, for simplicity, that $H = I$ (the output of each transfer system is its state vector x). For the simple chain P_c, the matrix C has the form

$$C_c = \begin{bmatrix} 0 & 0 & 0 & \cdots & 0 & 0 \\ C_{21} & 0 & 0 & \cdots & 0 & 0 \\ 0 & C_{32} & 0 & \cdots & 0 & 0 \\ 0 & 0 & C_{43} & \cdots & 0 & 0 \\ & \cdot & & \cdot & & \cdot \\ 0 & 0 & 0 & \cdots & C_{m,m-1} & 0 \end{bmatrix},$$

and for the simple closed loop P_L, the matrix C has the form

$$C_L = \begin{bmatrix} 0 & 0 & 0 & \cdots & 0 & C_{1m} \\ C_{21} & 0 & 0 & \cdots & 0 & 0 \\ 0 & C_{32} & 0 & \cdots & 0 & 0 \\ 0 & 0 & C_{43} & \cdots & 0 & 0 \\ & \cdot & & \cdot & & \cdot \\ 0 & 0 & 0 & \cdots & C_{m,m-1} & 0 \end{bmatrix}.$$

More general interconnections will result in more general patterns of nonzero elements (submatrices) in the matrix C.

5. Properties of transfer systems. In keeping with the composite system approach outlined in the introduction, the basic interconnection structure has been developed in §4. The next step is a study of significant properties of the subsystems, the transfer systems. In the following section interconnection structure and transfer system properties will be combined in the development of stability criteria for the composite system.

At this point it is necessary to restrict attention to a special, but very important, class of transfer systems. The members of this class are related by the property of exponential stability-in-the-large.

DEFINITION 5.1. The null solution of (3.1) is said to be *exponentially stable-in-the-large* (ESL) if there are two positive constants α and β such that

$$\| x(t; x_0, t_0)\| \leqq \beta \| x_0 \| e^{-\alpha(t-t_0)},$$

for all $(x_0, t_0) \in E^n \times T$.

DEFINITION 5.2. A vector function $f(x, t)$ on $E^n \times T$ is said to be in *class B* (denoted $f \in B$) if, for all $(x, t) \in E^n \times T$, $f(x, t)$ is continuous in x and t and has continuous first partial derivatives with respect to $x_1, x_2, \cdots, x_n$ such that $| \partial f_i/\partial x_j | < L$, where L is a constant and $i, j = 1, \cdots, n$.

The following theorem brings out an important feature of equations having a null solution which is ESL.

THEOREM 5.1. *Assume that $f(x, t) \in B$ in* (3.1). *Then* (3.1) *is ESL if, and only if, there is a positive definite function $v(x, t)$ such that*

(a) $c_1\| x \|^2 \leqq v(x, t) \leqq c_2\| x \|^2$,

(b) $\dot v(x, t) \leqq -c_3\| x \|^2$,

(c) $\| \nabla v \| \leqq c_4\| x \|$.

Proof. See [11, p. 59].

DEFINITION 5.3. A system modeled by (4.4) and (4.5) will be said to be in *class E* if the unforced model [(4.4) with $u(t) = 0$] is ESL and $f(x, t) \in B$.

The practical importance of class E is indicated in the following theorem.

THEOREM 5.2. *The following ordinary differential equations are ESL:*

(a) *the linear constant coefficient equation $\dot x = Ax$, where A is stable (i.e., all of its eigenvalues have negative real parts)*;

(b) *the linear time-varying equation $\dot x = A(t)x$, where $A(t)$ is continuous and bounded and $\dot x = A(t)x + u(t)$ has the property that bounded u implies bounded x*;

(c) *the equation $\dot x = f(x, t)$, where $f(x, t) \in B$, $f(0, T) = 0$ and $x'f(x, t) \leqq -c_3\| x \|^2 < 0$ for all $(x, t) \in E^n \times T$ with $x \neq 0$.*

Proof. (a) is a well known result [2]. A proof of (b) is given in [12, p. 518]. Part (c) is easily verified using the Lyapunov function,

$$v(x) = \tfrac{1}{2}\| x \|^2,$$

so that

$$\dot v(x, t) = x'f(x, t) \leqq -c_3\| x \|^2.$$

The desired result then follows immediately from Theorem 5.1.

The list given in Theorem 5.2 is far from exhaustive. A more detailed description of class E and a consideration of other interesting classes of transfer systems will be the subject of future study.

One additional property of systems in class E should be noted here. It can be shown [13] that for such systems it is possible to define a gain η as a ratio of a bound on the size (norm) of the state vector x to a bound on the size (norm) of the input u. Roughly speaking (neglecting transients),

$$\eta = \frac{\sup_{t \geqq t_0} \| x \|}{\sup_{t \geqq t_0} \| u \|}.$$

An upper bound on η can be estimated with the aid of a simple extension of some of the concepts of LSM. A complete discussion of this gain and its estimation may be found in [13]. However, the following theorem (also proved in [13]) is included here because it leads to an interesting interpretation of results obtained in §6.

THEOREM 5.3. *A transfer system in class E has a gain,*

$$\eta \leqq \left(\frac{c_4}{c_3}\right)\left(\frac{c_2}{c_1}\right)^{1/2} \| D \|,$$

where c_1, c_2, c_3, and c_4 are the constants noted in Theorem 5.1.

Note that the accuracy of the gain estimate depends on the constants c_1, c_2, c_3, c_4 and thus on the particular Lyapunov function used in making the estimate.

6. Stability of composite systems. The concepts developed in earlier sections can now be applied to the central problem, stability of the composite system. However, before attacking the most complex situation with general interconnections the salient features of the composite system approach will be illustrated by first considering the two simple systems shown in Fig. 4.1. While both of these systems are shown with an input, only the stability of unforced composite systems ($u = 0$) will be considered.

The following lemmas will be of major importance in this section.

LEMMA 6.1. *Let $x(t; x_0, t_0)$ be a solution of the differential inequality*[3]

$$\dot x \leqq Ax, \tag{6.1}$$

with $x(t_0; x_0, t_0) = x_0$, and let $y(t; y_0, t_0)$ be a solution of the differential equation

$$\dot y = Ay. \tag{6.2}$$

If all the elements a_{ij}, $i, j = 1, \cdots, n$, of A are nonnegative, and $x_0 = y_0$, then $x(t; x_0, t_0) \leqq y(t; y_0, t_0)$ for all $t \in T_0$.

Proof. See [14].

LEMMA 6.2. *Let B be a matrix with negative diagonal elements and nonnegative off-diagonal elements. If $x(t; x_0, t_0)$ and $y(t; y_0, t_0)$ are solutions of*

$$\dot x \leqq Bx, \tag{6.3}$$

$$\dot y = By, \tag{6.4}$$

and $x_0 = y_0$, then $x(t; x_0, t_0) \leqq y(t; y_0, t_0)$ for all $t \in T_0$.

Proof.[4] Let $-d$ be the smallest of the diagonal elements of B. Application of the transformations $v = e^{+dt}x$ and $w = e^{+dt}y$ changes (6.3) and (6.4) to

$$\dot v \leqq (B + dI)v,$$

$$\dot w = (B + dI)w,$$

and $B + dI$ has all nonnegative elements. Then by Lemma 6.1, since $v_0 = w_0$, it follows that $v(t; v_0, t_0) \leqq w(t; w_0, t_0)$ and thus $x(t; x_0, t_0) \leqq y(t; y_0, t_0)$ for all $t \in T_0$.

LEMMA 6.3. *The null solution of the system of linear differential equations*

$$\dot x_1 = -a_1x_1 + b_1x_n,$$
$$\dot x_2 = -a_2x_2 + b_2x_1,$$
$$\cdots$$
$$\dot x_n = -a_nx_n + b_nx_{n-1},$$

with a_i and b_i real, $a_i > 0$ and $b_i \geqq 0$, $i = 1, \cdots, n$, is ASL if and only if

$$\prod_{i=1}^{n} \frac{b_i}{a_i} < 1.$$

A proof of this lemma is given in the Appendix.

LEMMA 6.4. *If $a > 0$ and $b \geqq 0$, then for all $z \in T$,*

$$-az^2 + bz \leqq -\frac{a}{2}z^2 + \frac{b^2}{2a}.$$

Proof.

$$-az^2 + bz \leqq -\frac{a}{2}z^2 + \frac{b^2}{2a}$$

if and only if

$$-\frac{a}{2}z^2 + \frac{b^2}{2a} - \frac{a}{2}z^2 + bz - \frac{b^2}{2a} \leqq -\frac{a}{2}z^2 + \frac{b^2}{2a}$$

if and only if

$$\left(-\frac{a}{2}z^2 + \frac{b^2}{2a}\right) - \frac{1}{2a}(az - b)^2 \leqq -\frac{a}{2}z^2 + \frac{b^2}{2a}.$$

[3] Throughout this section, the notation $x \leqq y$, where x and y are n-vectors, means that $x_i \leqq y_i$ for $i = 1, \cdots, n$.

[4] The author is indebted to J. K. Hale for this proof.

Now consider the simple chain P_c of Fig. 4.1(a). It is intuitively obvious that such an interconnection of individually stable systems will be stable because of the "weak" interconnections involved (no loading assumed, see §4). The following theorem gives a simple proof of this fact.

THEOREM 6.1. *Consider the simple chain P_c with $u = u_1 = 0$ and $u_i = x_{i-1}$ for $i = 2, \cdots, m$. If the individual transfer systems S_i, $i = 1, \cdots, m$, are in class E, then the null solution of the composite system is ASL.*

Proof. Since the individual transfer systems are in class E, the ith transfer system is modeled by (4.4) and (4.5). Moreover, $f_i(x_i, t) \in B$, and when $u_i = 0$, $\dot{x}_i = f_i(x_i, t)$ is ESL. Thus there is a positive definite function $v(x, t)$ satisfying (a), (b), and (c) of Theorem 5.1. Now the total derivative of this $v(x, t)$ with respect to (4.4) is

$$\dot{v}_i = \dot{v}_{i1}(x_i, t) + \nabla v_i' D_i u_i \leqq -c_{i3}\| x_i \|^2 + c_{i4}\| x_i \| \cdot \| D_i \| \cdot \| u_i(t) \|,$$

where $\dot{v}_{i1}$ is the total derivative of $v_i(x, t)$ with respect to $\dot{x}_i = f_i(x_i, t)$. By Lemma 6.4 then,

$$\dot{v}_i \leqq -\frac{c_{i3}}{2} \| x_i \|^2 + \frac{c_{i4}^2 \| D_i \|^2}{2c_{i3}} \| u_i \|^2,$$

or

$$\dot{v}_i \leqq -\alpha_i v_i + \gamma_i \| u_i \|^2,$$

where

$$\alpha_i = \frac{c_{i3}}{2c_{i2}} > 0, \qquad \gamma_i = \frac{c_{i4}^2 \| D_i \|^2}{2c_{i3}} > 0.$$

When the interconnections are made, $u_1 = 0$ and

$$\| u_i \|^2 = \| x_{i-1} \|^2 \leqq \frac{1}{c_{i-1,1}} v_{i-1} \quad \text{for} \quad i = 2, 3, \cdots, m.$$

With $\beta_i = \gamma_i / c_{i-1,1}$, $i = 2, \cdots, m$, the resulting system of differential inequalities becomes

$$\begin{aligned} \dot{v}_1 &\leqq -\alpha_1 v_1, \\ \dot{v}_2 &\leqq -\alpha_2 v_2 + \beta_2 v_1, \\ &\cdots \\ \dot{v}_m &\leqq -\alpha_m v_m + \beta_m v_{m-1}, \end{aligned} \tag{6.5}$$

where $v_i \geqq 0$. Since $\beta_1 = 0$, the trivial solution of the system (6.5) with inequalities replaced by equalities (a system of auxiliary equations) is ASL (Lemma 6.3). Then by Lemma 6.2, the system (6.5) is also ASL, implying that each v_i goes to 0 as t goes to ∞. Since $\| x_i \|^2 \leqq v_i / c_{i1}$, the equilibrium solution $x = 0$ of the composite system is therefore ASL.

A more impressive result is obtained for the simple closed loop P_L. In this case the stability of the individual systems is not sufficient. The additional requirement is that the loop gain estimated from the chosen Lyapunov functions be less than unity—again an intuitively plausible requirement.

THEOREM 6.2. *Consider the simple closed loop P_L with $u = 0$ (no external input), $u_1 = x_m$, and $u_i = x_{i-1}$ for $i = 2, \cdots, m$. If the individual transfer systems S_i are in class E with gain estimates η_i, $i = 1, \cdots, m$, then the null solution of the composite system will be ASL if $\prod_{i=1}^m \eta_i < 1$.*

Proof. As in the proof of Theorem 6.1, there is again a system of differential inequalities

$$\begin{aligned} \dot{v}_1 &\leqq -\alpha_1 v_1 + \beta_1 v_m, \\ \dot{v}_2 &\leqq -\alpha_2 v_2 + \beta_2 v_1, \\ &\cdots \\ \dot{v}_m &\leqq -\alpha_m v_m + \beta_m v_{m-1}, \end{aligned} \tag{6.6}$$

where

$$\alpha_i = \frac{c_{i3}}{2c_{i2}} > 0, \qquad i = 1, \cdots, m,$$

$$\beta_1 = \frac{c_{14}^2 \| D_1 \|^2}{2c_{13}\, c_{m1}} > 0,$$

$$\beta_i = \frac{c_{i4}^2 \| D_i \|^2}{2c_{i3}\, c_{i-1,1}} > 0, \qquad i = 2, \cdots, m,$$

where the $v_i \geqq 0$, for $i = 1, \cdots, m$. The trivial solution of the composite system will be ASL if the system of auxiliary equations [(6.6) with inequalities replaced by equalities] is ASL. But, by Lemma 6.3, this will occur if

$$\prod_{i=1}^m \frac{\beta_i}{\alpha_i} < 1.$$

Now

$$\frac{\beta_i}{\alpha_i} = \frac{c_{i2}\, c_{i4}^2 \| D_i \|^2}{c_{i3}^2\, c_{i-1,1}} = \frac{c_{i1}}{c_{i-1,1}} \eta_i^2,$$

(where $c_{i-1,1} = c_{m1}$ when $i = 1$) and so the requirement that $\prod_{i=1}^m \eta_i^2 < 1$, or $\prod_{i=1}^m \beta_i/\alpha_i < 1$, thereby gives ASL.

The main theorem now gives sufficient conditions for ASL of the null solution of a composite system with *arbitrary interconnections*, that is, a system where C has an arbitrary number of nonzero elements. It will, however, be assumed that $B_{ii} = C_{ii} = 0$ since a nonzero B_{ii} represents an internal connection within the transfer system that can be included in the transfer system model itself.

THEOREM 6.3. *Let P be a general composite system made up of m transfer systems, S_i, of order n_i, $i = 1, \cdots, m$, and modeled by (4.7) with $C_{ii} = 0$. Assume that each transfer system S_i is in class E and has a Lyapunov function $v_i(x_i, t)$ satisfying the bounds listed in Theorem 5.1 with coefficients c_{i1}, c_{i2}, c_{i3}, c_{i4} (all coefficients are positive). Consider the mth order linear system of auxiliary equations,*

$$\dot{r} = Ar, \tag{6.7}$$

where A is an $m \times m$ matrix of elements,

$$a_{ij} = \begin{cases} \dfrac{c_{i3}}{2c_{i2}} & \text{for } i = j, \\[2ex] \dfrac{c_{i4}^2 \sum_{j=1}^m \| C_{ij} \|^2}{c_{i3}\, c_{j1}} & \text{for } i \neq j, \end{cases}$$

with C_{ij} an element (submatrix) of C. The null solution, $x = 0$, of the unforced composite system model [(4.7) with $u = 0$] will be ASL if the null solution, $r = 0$, of the mth order linear auxiliary system (6.7) is ASL.

Proof. The ith transfer system can be modeled by (4.4) and (4.5) and due to the interconnections,

$$u_i = \sum_{j=1}^m B_{ij} y_j. \tag{6.8}$$

Since this ith transfer system is in class E, there is a Lyapunov function $v_i(x_i, t)$ such that, with respect to (4.4),

$$\dot{v}_i \leqq -c_{i3}\| x_i \|^2 + c_{i4}\| x_i \| \cdot \| D_i u_i \|. \tag{6.9}$$

Now a substitution of (4.5) and (6.8) into (6.9) and application of Lemma 6.4 gives

$$\dot{v}_i \leqq -\frac{1}{2} c_{i3} \| x_i \|^2 + \frac{c_{i4}^2}{2c_{i3}} \sum_{j=1}^m \| D_i B_{ij} H_j \|^2 \sum_{\substack{j=1 \\ j \neq i}}^m \| x_j \|^2.$$

A reintroduction of the inequalities

$$c_{i1}\| x_i \|^2 \leqq v_i(x_i, t) \leqq c_{i2}\| x_i \|^2, \qquad i = 1, \cdots, m,$$

and a use of the relation $C_{ij} = D_i B_{ij} H_j$ yields

$$\dot{v}_i \leqq -\frac{c_{i3}}{2c_{i2}} v_i + \frac{c_{i4}^2}{2c_{i3}} \sum_{j=1}^m \| C_{ij} \|^2 \sum_{\substack{j=1 \\ j \neq i}}^m \frac{v_j}{c_{j1}}.$$

The resulting system of inequalities is

$$\begin{aligned} \dot{v}_1 &\leqq -\frac{c_{13}}{2c_{12}} v_1 + \left(\frac{c_{14}^2}{2c_{13}} \sum_{j=1}^m \| C_{1j} \|^2 \right) \left(\sum_{j=1}^m \frac{v_j}{c_{j1}} \right), \\ &\cdots \\ \dot{v}_m &\leqq -\frac{c_{m3}}{2c_{m2}} v_m + \left(\frac{c_{m4}^2}{2c_{m3}} \sum_{j=1}^m \| C_{mj} \|^2 \right) \left(\sum_{j=1}^{m-1} \frac{v_j}{c_{j1}} \right), \end{aligned}$$

or

$$\dot{v} \leqq Av, \tag{6.10}$$

where $v = \operatorname{col}[v_1, v_2, \cdots, v_m]$ and $\leqq$ means that the inequality holds componentwise. Now by Lemma 6.2, $v(t) \leqq r(t)$ if $v(t_0) = r(t_0)$. Thus

$$\| x_i(t; x_0, t_0) \|^2 \leqq \frac{1}{c_{i1}} v(t) \leqq \frac{1}{c_{i1}} r(t), \qquad i = 1, \cdots, m,$$

and so asymptotic stability-in-the-large of the solution $r = 0$ of (6.7) implies asymptotic stability-in-the-large of the solution $x = 0$ of the composite system.

This result does not have the intuitive appeal found in the results of Theorems 6.1 and 6.2. That is, there is nothing like a loop gain to suggest that the resulting stability criterion is reasonable. However, this is not surprising. The same difficulty is encountered even in a linear time-invariant composite system with arbitrary interconnections.

Theorems 6.1 and 6.2 are now corollaries to Theorem 6.3 when the matrix C has the special form C_c or C_L.

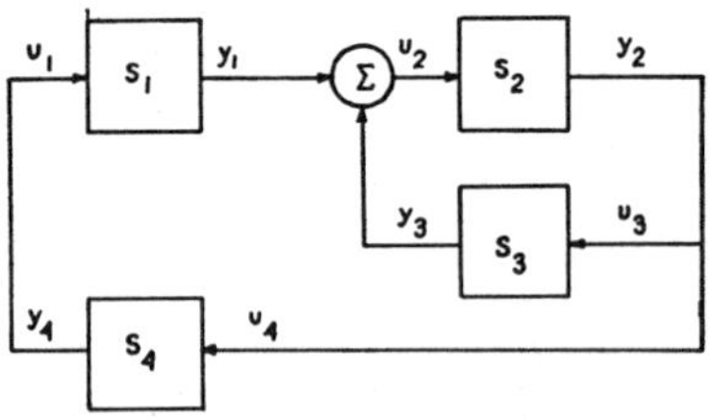

FIG. 7.1. *Composite system for example*

7. Example—a ninth-order composite system. Consider the ninth-order composite system shown in Fig. 7.1. The individual transfer systems are assumed to have the following models:

S_1 : linear, constant coefficient, third order,

$$\dot{x}_1 = A_1 x_1 + D_1 u_1(t), \quad \text{where} \quad A_1 = T\begin{bmatrix} -2 & 0 & 0 \\ 0 & -3 & 0 \\ 0 & 0 & -5 \end{bmatrix} T^{-1},$$

$$y_1 = H_1 x_1,$$

and T is any nonsingular 3×3 matrix;

S_2 : linear, variable coefficient, second order,

$$\dot{x}_2 = A_2(t)x_2 + D_2 u_2(t), \quad \text{where} \quad A_2(t) = \begin{bmatrix} 0 & a(t) \\ -1 & -2a(t) \end{bmatrix},$$

$$y_2 = H_2 x_2,$$

and $a(t)$ is a continuous real valued function such that $a^{-1}(t)$ exists for all $t \in T$ and, in addition,

$$0.5 \leqq a^{-1}(t) \leqq 1 \quad \text{and} \quad 0 \leqq \frac{da^{-1}(t)}{dt} \leqq 1;$$

S_3 : nonlinear, first order,

$$\dot{x}_3 = f_3(x_3) + D_3 u_3(t), \text{ where } f_3(x_3) = -x_3 - \tfrac{1}{4}\sin 2x_3,$$

$$y_3 = H_3 x_3;$$

S_4 : nonlinear, third order,

$$\dot{x}_4 = f_4(x_4) + D_4 u_4(t), \text{ where } f_4(0) = 0,$$

$$y_4 = H_4 x_4.$$

The problem here is to determine a value of the positive constant k that will insure that the composite system is ASL if

$$x_4' f_4(x_4) \leqq -k\| x_4 \|^2.$$

The interconnections suggested in Fig. 7.1 indicate that

$$u_1 = y_4,$$

$$u_2 = y_1 + B_{23} y_3,$$

$$u_3 = y_2,$$

$$u_4 = y_2.$$

These interconnections are described by the matrix C whose partitioned elements are $C_{ij} = D_i B_{ij} H_j$. In this case, C has the partitioned form

$$C = \begin{bmatrix} 0 & 0 & 0 & C_{14} \\ C_{21} & 0 & C_{23} & 0 \\ 0 & C_{32} & 0 & 0 \\ 0 & C_{42} & 0 & 0 \end{bmatrix},$$

and the composite system model is

$$\dot{x} = f(x, t) + Cx, \tag{7.1}$$

where $x = \text{col}\,[x_{11}, x_{12}, x_{13}, x_{21}, x_{22}, x_{31}, x_{41}, x_{42}, x_{43}]$ is the composite system state vector. When written out in detail, the composite system model has the form

$$\begin{aligned}
\dot{x}_{11} &= a_{11}x_{11} + a_{12}x_{12} + a_{13}x_{13} + c_{14}^{11}x_{41} + c_{14}^{12}x_{42} + c_{14}^{13}x_{43}, \\
\dot{x}_{12} &= a_{21}x_{11} + a_{22}x_{12} + a_{23}x_{13} + c_{14}^{21}x_{41} + c_{14}^{22}x_{42} + c_{14}^{23}x_{43}, \\
\dot{x}_{13} &= a_{31}x_{11} + a_{32}x_{12} + a_{33}x_{13} + c_{14}^{31}x_{41} + c_{14}^{32}x_{42} + c_{14}^{33}x_{43}, \\
\dot{x}_{21} &= a(t)x_{22} + c_{21}^{11}x_{11} + c_{21}^{12}x_{12} + c_{21}^{13}x_{13} + c_{23}^{11}x_{31}, \\
\dot{x}_{22} &= -x_{21} - 2a(t)x_{22} + c_{21}^{21}x_{11} + c_{21}^{22}x_{12} + c_{21}^{23}x_{13} + c_{23}^{21}x_{31}, \\
\dot{x}_{31} &= f_3(x_{31}) + c_{32}^{11}x_{21} + c_{32}^{12}x_{22}, \\
\dot{x}_{41} &= f_{41}(x_{41}, x_{42}, x_{43}) + c_{42}^{11}x_{21} + c_{42}^{12}x_{22}, \\
\dot{x}_{42} &= f_{42}(x_{41}, x_{42}, x_{43}) + c_{42}^{21}x_{21} + c_{42}^{22}x_{22}, \\
\dot{x}_{43} &= f_{43}(x_{41}, x_{42}, x_{43}) + c_{42}^{31}x_{21} + c_{42}^{32}x_{22},
\end{aligned} \tag{7.2}$$

where c_{ij}^{mn} is the (m, n)th element of C_{ij}, the (i, j)th submatrix in the partition of C.

A straightforward approach to this problem would involve choosing a Lyapunov function involving the 9 state variables, evaluating its derivative with respect to (7.2), and then studying this derivative to determine conditions under which it is negative definite. Clearly, this approach would involve the solution of some very difficult problems. On the other hand, the techniques developed in §6 provide a method for obtaining a solution to this problem rather quickly.

The first step is to find Lyapunov functions of the type described in Theorem 5.1 for each of the transfer systems. This will, of course, be possible if and only if these transfer systems are in class E.

S_1 : Since this is a constant-coefficient linear system, standard techniques [2] can be used to obtain the Lyapunov function $v_1(x_1) = x_1'Px_1$ with $\dot{v}_1(x_1) = -x_1'Qx_1$, where

$$P = (T^{-1})'\begin{bmatrix} 1 & 0 & 0 \\ 0 & 2 & 0 \\ 0 & 0 & 3 \end{bmatrix} T^{-1} \quad \text{and} \quad Q = (T^{-1})'\begin{bmatrix} 4 & 0 & 0 \\ 0 & 12 & 0 \\ 0 & 0 & 30 \end{bmatrix} T^{-1}.$$

For this Lyapunov function the inequalities of Theorem 5.1 are satisfied with $c_1 = 1$, $c_2 = 3$, $c_3 = 4$, and $c_4 = 6$.

S_2 : In this case, choose $v_2(x_2, t) = x_2'P(t)x_2$ with $\dot{v}_2(x_2, t) = -x_2'Q(t)x_2$, where

$$P(t) = \begin{bmatrix} 2 + a^{-1}(t) & 1 \\ 1 & 1 \end{bmatrix} \quad \text{and} \quad Q(t) = \begin{bmatrix} 2 & 0 \\ 0 & 4a(t) \end{bmatrix}.$$

In this case the inequalities of Theorem 5.1 are satisfied with $c_1 = 0.5$, $c_2 = 3.5$, $c_3 = 2$, and $c_4 = 7$.

S_3 : Since this model is first order, $x_3 = x_{31}$. Choose

$$v_3(x_3) = \tfrac{1}{2}x_{31}^2,$$

and then $\dot{v}_3(x_3) = x_{31}f_3(x_{31})$. The inequalities are now satisfied with $c_1 = c_2 = \tfrac{1}{2}$, $c_3 = 0.56$, and $c_4 = 1$.

S_4 : In this case, choose

$$v_4(x_4) = \tfrac{1}{2}x_4'x_4,$$

and note that

$$x_4'f(x_4) \leqq -k\| x_4 \|^2.$$

Then $c_1 = c_2 = \tfrac{1}{2}$, $c_3 = k$, and $c_4 = 1$.

With Lyapunov functions chosen for each of the transfer systems, it is possible to apply Theorem 6.3. The system of auxiliary equations is, in this case, the fourth order system

$$\begin{aligned}
\dot{r}_1 &= -\alpha_1 r_1 + \beta_1 r_4, \\
\dot{r}_2 &= -\alpha_2 r_2 + \beta_2 r_1 + \gamma_2 r_3, \\
\dot{r}_3 &= -\alpha_3 r_3 + \beta_3 r_2, \\
\dot{r}_4 &= -\alpha_4 r_4 + \beta_4 r_2.
\end{aligned} \tag{7.3}$$

In (7.3)

$$\alpha_1 = \frac{2}{3}, \qquad \alpha_2 = \frac{1}{3.5}, \qquad \alpha_3 = 0.56, \qquad \alpha_4 = k,$$

and

$$\beta_1 = 18\,\| C_{14} \|,$$

$$\beta_2 = \frac{49}{2}(\| C_{21} \| + \| C_{23} \|),$$

$$\beta_3 = \frac{2}{0.56}\| C_{32} \|,$$

$$\beta_4 = \frac{2}{k}\| C_{42} \|,$$

$$\gamma_2 = 49(\|C_{21}\| + \| C_{23} \|),$$

where C_{ij} are submatrices of C. The numerical values in the above constants are obtained from the bounds on the Lyapunov functions chosen for the individual transfer systems. (See Theorem 6.3.) According to Theorem 6.3, the null solution of the composite system (Fig. 7.1) will be ASL if the null solution of the system of auxiliary equations is ASL. The latter will occur if and only if the roots of the characteristic equation,

$$[-\alpha_4(k) - \lambda][-\alpha_1 - \lambda][(-\alpha_1 - \lambda)(-\alpha_3 - \lambda) - \gamma_2\beta_3] + \beta_4(k)[(-\alpha_3 - \lambda)\beta_1\beta_2] = 0,$$

all have negative real parts. [Here the dependence of α_4 and β_4 on k is indicated as $\alpha_4(k)$ and $\beta_4(k)$.] Thus the original problem has been reduced to a much simpler problem that can be solved by root locus or numerical techniques.

8. Conclusions. In the study of complex physical systems frequently it is found that the difficulties encountered are strongly order dependent. Stability study through LSM is a typical example of a technique that is theoretically applicable to problems of arbitrary order but actually encounters formidable practical limitations that grow rapidly with the order of the system being treated. Thus the application of LSM to low order (third, fourth and sometimes fifth) problems has received a considerable amount of attention while high order problems are still inaccessible.

This paper describes an attempt to circumvent these order dependent limitations of LSM by noting that many high order systems are actually, or effectively, an interconnection of lower order systems. The recognition of this interconnection structure has led to the definition of a composite system as an interconnection of simpler subsystems, termed transfer systems, and a study of the properties of these transfer systems with the aid of LSM and the associated concept of the auxiliary equation. For a composite system made up of transfer systems belonging to a reasonably broad class, the auxiliary equations for the individual transfer systems can be interconnected, following the interconnections existing in the given composite system, to obtain a system of auxiliary equations whose solutions have the same stability properties as those of the given composite system. Moreover, in the construction of this system of auxiliary equations it has been necessary to apply LSM only to the lower order transfer systems. The difficult problem of constructing a single Lyapunov function for the high order composite system has actually been avoided.

As might be suspected, this simplification of the original problem is not obtained without some sacrifice. The main disadvantage of this approach, a fault common to most attempts to obtain general sufficient conditions for stability, is that it may sometimes be overly restrictive (overly sufficient) due to a failure to make the best use of available information about the fine structure of the system under analysis. This fine structure is "washed out" at points where matrix norms or absolute values are used. Hopefully, this is compensated by the fact that the introduction of transfer systems and interconnection information into stability analysis has increased the class of problems to which LSM has practical application.

Appendix. The following proof of Lemma 6.3 uses the root locus concept described in [10].

Proof of Lemma 6.3. The characteristic equation for this system is

$$\prod_{i=1}^{n} (\lambda + a_i) - \prod_{i=1}^{n} b_i = 0.$$

Now replace b_1 with μb_1 and consider the root locus obtained for $\mu \geqq 0$. If $\mu = 0$ there are n negative real roots $\lambda = -a_i$, $i = 1, \cdots, n$. As μ increases the most positive root moves to the right along the real axis reaching the origin when

$$\text{(A-1)} \qquad \prod_{i=1}^{n} a_i - \mu \prod_{i=1}^{n} b_i = 0.$$

It is now only necessary to ascertain that no root has crossed the imaginary axis for a smaller value of μ. However, this cannot occur since at all points λ *on the root locus* it is easily shown that

$$\mu = \prod_{i=1}^{n} \frac{\lambda + a_i}{b_i}.$$

Thus the point where the locus crosses the imaginary axis with the smallest value of μ is at the origin and the root moving along the real axis must be the first to cross. Now if $\mu = 1$ the necessary and sufficient condition (A-1) becomes

$$\prod_{i=1}^{n} \frac{b_i}{a_i} < 1,$$

completing the proof.

Acknowledgment. The author would like to thank K. B. Irani, A. W. Naylor, and B. F. Barton of the Cooley Electronics Laboratory, Department of Electrical Engineering, The University of Michigan, for their encouragement and support throughout the course of this research.

REFERENCES

[1] G. Kron, *Diakoptics*, Macdonald, London, 1963.
[2] W. Hahn, *Theory and Application of Lyapunov's Direct Method*, Prentice-Hall, Englewood Cliffs, New Jersey, 1963.
[3] L. Cesari, *Asymptotic Behavior and Stability Problems in Ordinary Differential Equations*, Academic Press, New York, 1963.
[4] A. M. Lyapunov, *Problème général de la stabilité du mouvement*, Annals of Mathematics Studies, vol. 17, Princeton University Press, Princeton, 1947.
[5] R. E. Kalman and J. E. Bertram, *Control system analysis and synthesis via the second method of Lyapunov, I. Continuous time systems*, Trans. ASME Ser. D. J. Basic Engrg., 82D (1960), pp. 371–393.
[6] A. I. Lur'e, *Some Nonlinear Problems in the Theory of Automatic Control*, Her Majesty's Stationery Office, London, 1957.
[7] T. Yoshizawa, *Lyapunov's functions and boundedness of solutions*, Funkcial. Ekvac., 2 (1959), pp. 95–142.
[8] R. Conti, *Sulla prolingabilità della soluzioni di un sistema di equazioni differentiali ordinarie*, Boll. Un. Mat. Ital., 11 (1956), pp. 510–514.
[9] F. Brauer, *Global behavior of solutions of ordinary differential equations*, J. Math. Anal. Appl., 2 (1961), pp. 145–158.
[10] J. J. D'Azzo and C. H. Houpis, *Feedback Control System Analysis and Synthesis*, McGraw-Hill, New York, 1960.
[11] N. N. Krasovskii, *Stability of Motion*, Stanford University Press, Stanford, California, 1963.
[12] J. L. Massera and J. J. Schaffer, *Linear differential equations and functional analysis, I*. Ann. of Math., 67 (1958), pp. 517–573.
[13] F. N. Bailey, *Stability of interconnected systems*, Tech. Rpt. 152, Cooley Electronics Laboratory, The University of Michigan, Ann Arbor, 1964.
[14] E. F. Beckenbach and R. Bellman, *Inequalities*, Springer-Verlag, Berlin, 1961.

THE METHOD OF VECTOR LYAPUNOV FUNCTIONS IN ANALYSIS OF COMPOSITE SYSTEMS WITH DISTRIBUTED PARAMETERS (SURVEY)*

V. M. Matrosov

UDC 62-50:517.917

Some recent developments in the method of Lyapunov functions, stimulated by problems in qualitative study of dynamical and controlled systems with distributed parameters, are surveyed. A nonlinear control system with distributed parameters, consisting of several interconnected subsystems described by evolutionary differential equations or partial integro-differential equations etc., is discussed. Vector Lyapunov functions are used systematically for investigating the dynamic properties of the system, and some general theorems are stated.

Introduction

In A. M. Lyapunov's classical work [1] an energy approach was used for investigating stability in systems with distributed parameters. While creating the theory of stability of motion for systems described by ordinary differential equations, the great mathematician outlined a much more general second method for investigating stability, closely linked with the ideas of Poincaré's qualitative theory of differential equations [2]. In its basic form, this second method is the method of Lyapunov functions (or in N. G. Chetaev's terminology [3], Lyapunov's direct method); it was primarily developed by Lyapunov himself and came to play a dominant role later.

As a result of work by K. P. Persidskii [4], E. A. Barbashin [5], N. N. Krasovskii [6], V. V. Rumyantsev [7], V. I. Zubov [8], D. M. Volkov [9], A. A. Movchan [10], T. K. Sirazetdinov [11], P. K. C. Wang [12] and other authors [13], the method of Lyapunov functions has been extended to infinite systems of equations [4], functional-differential systems [6], integro-differential equations [5, 11], evolutionary partial differential equations [8-13], and certain composite systems with distributed parameters [7]. In suitable conditions, the nonlinear equations and systems mentioned may be written as a single ordinary differential equation in the appropriate functional space, and investigated by the methods of nonlinear functional analysis. Starting with work by K. P. Persidskii [4], M. G. Krein [14], and J. L. Massera [15], the method of Lyapunov functions was extended to differential equations in a Banach space, while E. A. Barbashin [5], V. I. Zubov [8], A. A. Movchan [10], J. Auslender, P. Silbert [6], N. P. Bhatia [17] and other authors [18-28] extended the method of dynamic, generalized dynamic, and general systems in metric or topological spaces, or to abstract processes in sets with two "distances" specified in them. Various new types of stability were thereby brought to light and investigated.

The method of Lyapunov functions has also been successfully employed for studying asymptotic behavior of processes in systems with distributed parameters (relative boundedness in T. Yoshizawa's detailed classification [29], dissipativeness, convergence, positive invariance of a set) [16, 18, 21, 22, 25, 27, 28, 30-33], and also for proving theorems on existence, uniqueness, correctness, nonlocal continuability of solutions of the equations of motion [32, 34-42] (of processes [10, 25]), on the influence of small perturbations in them, and on the convergence of approximations [38-40]. Here, as a rule, a functional of Lyapunov's type satisfies a differential inequality of Chaplygin's type [43-45].

As a result, M. A. Krasnosel'skii [46] and his associate [34, 36, 38], K. Kordunyanu [47], H. A. Antosiewicz [39], and V. Lakshmikantham [30] were led to establish an extremely general principle, called the principle

*Presented at the First Volga District Conference on Automatic Control, Kuibyshev, April, 1971.

Kazan'. Translated from Avtomatika i Telemekhanika, No. 1, pp. 5-22. Original article submitted November 18, 1971.

Reprinted with permission from *Automat. Remote Contr.,* vol. 34, pp. 1–15, Part I, Jan. 1973.

of comparison with a Lyapunov type functional. From this may be easily derived most of the classical and modern basic theorems in the theory of nonlinear differential equations with continuous right side.

The modern method of Lyapunov functions is thus an extremely general and effective tool for investigating nonlinear systems with distributed parameters. It is widely used in stability problems relating to elastic systems [10-13, 25, 48], magnetohydrodynamic flows [11], rigid and deformable bodies, notably satellites, with liquid-filled cavities [7], chemical and nuclear reactors [13], and many other objects [13, 48]. Lyapunov-Bellman functionals are used when solving problems of optical control synthesis, e.g., for elastic rocket vehicles [11]. Effective methods for constructing Lyapunov functionals have been described for a number of concrete systems with distributed parameters [6, 7, 9-13, 41-42, 48]. Nevertheless, just as in Lyapunov's lifetime, when applying the method to concrete problems a fundamental difficulty remains, due to the lack of a general algorithm for constructing the Lyapunov functional for any given nonlinear system. In view of this, the latest trend in the method, towards weakening the conditions on the Lyapunov functionals, is of considerable interest. In the case of ordinary differential equations, various aspects of this problem have been investigated by numerous mathematicians at different times (see [3-8, 15-22, 28-30, 44-47]).

Another trend involves the idea of using several scalar Lyapunov functions (functionals) [49], or in Bellman's terminology [50], a vector Lyapunov function, satisfying a differential inequality (see e.g., [51-53]). The conditions imposed on each component functional are much weaker than those in the corresponding theorems with a single Lyapunov functional. This leads to a more flexible apparatus and sometimes enlarges the class in which the required functionals exist, while it can greatly simplify the central problem of constructing the functionals.

Vector Lyapunov functions have been used by V. Lakshmikantham [54], W. Hahn [20], D. Bushaw [24], S. K. Persidskii [55], and other authors [22, 45, 56-62], for studying the stability and boundedness of the solutions of differential equations [45, 54-61] and processes in systems with distributed parameters [20, 22, 24, 62], and in addition, were used by P. Hartman [63] for investigating the existence of solutions (processes).

This work led in turn to development of the principle of comparison with a vector Lyapunov function. The principle has been established in a general form in recent years, for nonlinear differential equations in a Banach space with an unbounded discontinuous operator [64].

This principle is particularly effective in the qualitative investigation of composite dynamic and controlled systems, the multi-connectness and other properties of which make it difficult to construct a single Lyapunov functional for them. The reason is that the structural features of the system can be taken into account, and the decomposition principle [65] applies when constructing the vector Lyapunov function [5, 66, 67].

The following typical features of composite systems with distributed parameters will be borne in mind in our dynamic analysis: 1) composite structure, 2) multiconnectness, 3) the presence of infinite-dimensional subsystems, 4) the varied nature of the subsystems, and 5) the fact that the subsystems and internal connections of the system have nonlinear and nonstationary characteristics (see A. I. Kukhtenko [68]).

1. Vector Lyapunov Functions

Given a composite system consisting of m interconnected subsystems with input vectors $u_1, \ldots, u_{m'}$, phase vectors $x_1, \ldots, x_m$, and output vectors $z_1, \ldots, z_{m''}$. Let the phase space E_μ of each subsystem be a Banach space, in which we distinguish a set H_μ, usually dense in E_μ (the domain of definition of all the operators) [69]. Since our aim is to analyze the dynamic properties of the system, it will be assumed that the input vectors (controls, perturbations, interconnections) and output vectors are given nonlinear functions (operators) with values in certain sets U_μ and Z_μ,

$$u_\mu(t, z_1, \ldots, z_{m''}) : \mathrm{I}_0 \times Z_1 \times \ldots \times Z_{m''} \to U_\mu,\quad z_\mu(t, x_\mu) : \mathrm{I}_0 \times \mathrm{H}_\mu \to Z_\mu. \tag{1}$$

Here $I_0 = [0, +\infty)$.

Every process (motion) in the μ-th subsystem is described by the solution of a Cauchy problem for a differential equation in (real) Banach space E_μ:

$$\frac{dx_\mu}{dt} = F_\mu(t, x_\mu, u_\mu),\qquad (x_\mu(t_0) = x_{\mu 0} \in \mathrm{H}_{\mu 0} \subset \mathrm{E}_\mu), \tag{2}$$

where $t \in I_{t_0} = [t_0, +\infty)$ is time, and $F_\mu : I_0 \times H_\mu \times U_\mu \to E_\mu$ is a given nonlinear operator.

For example, let the first subsystem be finite-dimensional:

$$x_1 = \{x_1^1, \ldots, x_1^{n_1}\} \in \mathrm{H}_1 = \mathscr{R}^{n_1} = \mathrm{E}_1, \quad U_1 \subset \mathscr{R}^{k_1}, \quad Z \subset \mathscr{R}^{s_1},$$

where $\mathscr{R}^n$ is n-dimensional Euclidean space. With $\mu = 1$, (2) is a system of n_1 ordinary differential equations with right sides dependent on u_1.

Let the second subsystem be described by a vector integro-differential equation in a bounded closed domain

$$\frac{\partial y_2(t, q)}{\partial t} = Y_2(y_2(t, q), t, q, u_2) + \int_{\bar{Q}_2} K[y_2(t, p), t, q, p, u_2]\, dp,$$

$$p, q \in \bar{Q}_2, \; y_2 \in \mathscr{R}^{n_2}, \; y_2(t_0, q) = y_{02}(q) \quad (q \in \bar{Q}_2),$$

where Y_2 and K are continuous vector functions, and y_2 is the vector of "parameters distributed in the domain $\bar{Q}_2$." In this case it is convenient to take $x_2(t) = \{y_2(t, q) : q \in \bar{Q}_2\}$, $\mathrm{E}_2 = \mathscr{C}_{R^{n_2}}(\bar{Q}_2)$ and to investigate the solutions in the classical sense. If the functions Y_2 and K are discontinuous, we can understand the partial derivative $\partial y_2(t, q)/\partial t$ in the distribution sense and take $E_2 = L_p(\bar{Q}_2)$ (the space of n_2-dimensional vector functions with modulus to the p-th power summable in $\bar{Q}_2$). Under suitable conditions, a second equation of the type (2) is obtained.

The subsequent subsystems may be described by equations with many-valued right sides (systems with incomplete information) [5, 21], by equations with after-effect (functional differential equations) [6, 28], or by various classes of evolutionary partial differential equations [34, 41, 46], etc.

Let the m-th subsystem be described by a nonlinear system of parabolic equations with strongly elliptic operator

$$\frac{\partial y_m^i(t, q)}{\partial t} = -\sum_{\substack{j=1 \\ s,\nu=1}}^{n_m \; n} A_{ij}^{s,\nu}(t, q)\frac{\partial^2 y_m^j(t, q)}{\partial q^s \partial q^\nu} + Y_m^i\left(t, q, y_m(t, q), \frac{\partial y_m(t, q)}{\partial q}, u_m\right),$$

$$q \in \bar{Q}_m \subset \mathscr{R}^n, \; y_m = \{y_m^1, \ldots, y_m^{n_m}\} \in \mathscr{R}^{n_m} \quad (i = 1, \ldots, n_m)$$

under the initial conditions $y_m(t_0, q) = y_{0m}(q)$ $(q \in \bar{Q}_m)$, and boundary conditions $y_m(t, q) = 0$ for $t \geq t_0$, $q \in \bar{Q}_m \setminus Q_m$ and with real functions $A_{ij}^{s\nu}$, Y_m^i continuous or superpositionally measurable. Here we can take $x_m(t) = \{y_m(t, q) : q \in \bar{Q}_m\}$ and $E_m = \mathscr{C}_{\mathscr{R}^{n_m}}(\bar{Q}_m)$, $H_m = \overset{\circ}{\mathscr{C}}_{\mathscr{R}^{n_m}}(\bar{Q}_m)$ or $E_m = L_2(\bar{Q}_m)$, $H_m = \overset{\bullet}{W}{}_2^{(2)}(\bar{Q}_m)$ ($W^{(2)}(\bar{Q}_m)$ is a Sobolev space), depending on whether classical or distributed-type solutions of the problem are to be investigated.

On substituting the functions (1) in (2), a system of differential equations in Banach spaces $E_1, \ldots, E_m$ is obtained:

$$\frac{dx_\mu}{dt} = X_\mu(t, x_1, \ldots, x_m) \; (x_\mu(t_0) = x_{0\mu} \in \mathrm{H}_{0\mu}, \; \mu = 1, \ldots, m) \tag{3}$$

with nonlinear operators $X_\mu : I_0 \times H_1 \times \ldots \times H_m \to E_\mu$. Denote $x = \{x_1, \ldots, x_m\}$, $X(t, x) = \{X_1(t, x_1, \ldots, x_m), \ldots, X_m(t, x_1, \ldots, x_m)\}$ and introduce the direct product of B spaces $E_1 \times E_2 \times \ldots \times E_m = E$ with naturally induced linear operations and the norm $\|x\| = \|x_1\| + \ldots + \|x_m\|$. Since m is finite, E is also a Banach space, in which the system (3) may be written as a single equation:

$$\frac{dx}{dt} = X(t, x) \qquad (x(t_0) = x_0 \in \mathrm{H}_0 \subset \mathrm{E}), \tag{4}$$

where $X : I_0 \times H \to E$, $H = H_1 \times \ldots \times H_m$ (usually a set dense in E), and $H_0 = H_{01} \times \ldots \times H_{0m}$. Generally speaking, X is a discontinuous unbounded operator, defined in $\Omega = I_0 \times H \subset \Gamma$. The initial data t_0, x_0 are taken from the sets $T_0 \subseteq I_0$, $H_0 \subseteq E$, $\Gamma = I_0 \times E$.

The solution x(t) of (4) in the interval $T = [t_0, \tau)$ is understood in Caratheodory's sense, i.e., as an abstract function, absolutely continuous in T and satisfying (4) almost everywhere in T (the process x(t) in the system (3)).

Real functionals $v^1 : \Gamma \to \mathcal{R}^1, \ldots, v^k : \Gamma \to \mathcal{R}^1$ and the vector function $v = \{v^1, \ldots, v^k\} : \Gamma \to \mathcal{R}^k$ will be considered. Introduce into $\mathcal{R}^k$ the concept of norm and partial ordering between vectors $y, y' \in \mathcal{R}^k$: $\|y\| = \max_{s=1,\ldots,k} |y^s|$, $y \leqslant y' : y^s \leqslant y'^s$ $(s = 1, \ldots, k)$.

The commonly employed functionals, such as even the norm, become nondifferentiable in a Banach space. Henceforth, therefore, so-called semi-differentiable functions [35-38, 64] will be used.

We call v a comparison vector function for Eq. (4) if it satisfies the following conditions:

I. A function $f : A \to \mathcal{R}^k$ exists such that, given any solution x(t) of (4), defined in an interval T,

$$\bar{D}_+ v(t, x(t)) \mathrel{\dot{\leqslant}} f(t, v(t, x(t))) \qquad \text{for} \quad t \in T.$$

Here $\bar{D}_+ v^s(t, x(t))$ is the upper right Dini derivative number with respect to t; the dot above a relationship indicates that it is satisfied for almost all t. Condition I is satisfied, for instance, if the upper right derivative

$$v'(t, x) = \overline{\lim_{\Delta t \to 0+}} \frac{1}{\Delta t} [v(t + \Delta t, x + \Delta t X(t, x)) - v(t, x)], \qquad \{t, x\} \in \Omega$$

of the function v (or the upper right semiderivative in the sense of [35-38, 62, 64]) satisfies, by virtue of (4), the inequality $v'(t, x) \leq f(t, v(t, x))$ for $t \in I_0$, $x \in H$, and the function v is locally Lipschitz in x.

II. The majorizing function f has an open domain of definition $A \subseteq I_0 \times \mathcal{R}^k$, the section A(t) of which, with t arbitrary, contains the set of values of the function v, is bounded in any bounded closed domain $\bar{B} \subset A$, is superpositionally measurable, and satisfies Wazewski's condition [51] in A:

$$f^s(t, y) \leqslant f^s(t, y') \quad \text{for} \quad y^s = y'^s, \; y^\nu \leqslant y'^\nu \quad (\nu \neq s, \; \nu, s = 1, \ldots, k).$$

For example $A = I_0 \times \mathcal{R}^k$, $f^s(t, y)$ is bounded, satisfies Caratheodory's condition and is nondecreasing with respect to any nondiagonal component y^ν $(\nu \neq s)$ of the vector $y \in \mathcal{R}^k$.

III. Given any solution x(t) of (4) in an interval $\bar{T}_1 = [t_0, \tau_1] \subset T$, the vector function v(t, x(t)) is absolutely upper semicontinuous in this interval in Alekseev's sense [32].

For example, v(t, x) satisfies a one-sided Lipschitz condition with respect to $\{t, x\}$ in any bounded closed domain $\bar{\Psi} \subset \Gamma$:

$$v(t', x') - v(t, x) \leqslant \bar{l} \cdot [t' - t + \|x' - x\|] \quad (t \leqslant t', \; \bar{l} \in \mathcal{R}^k_+ = \{y \in \mathcal{R}^k : 0 < y^s \; (s = 1, \ldots, k)\}).$$

Using the (in general discontinuous) majorizing function f, we form the comparison system

$$\frac{dy}{dt} = f(t, y), \qquad y \in \mathcal{R}^k. \tag{5}$$

Its solutions can be understood in various ways, e.g., in Caratheodory's sense or in Filippov's sense, etc. In our problem it is more convenient to consider so-called generalized solutions of the second kind (G2 solutions) [70]: absolutely continuous functions y(t), satisfying almost everywhere inequalities of the type

$$\underline{\lim_{y \to y(t)}} f(t, y) \mathrel{\dot{\leqslant}} \frac{dy(t)}{dt} \mathrel{\dot{\leqslant}} \overline{\lim_{y \to y(t)}} f(t, y) \equiv \bar{f}_y(t, y(t)).$$

It can be shown that, under condition II, G2 solutions y(t) of the system (5) pass through every point $\{t_0, y(t_0)\} \in \bar{B} \subset A$, and can be continued rightwards to the boundary of $\bar{B}$. Among these solutions there exists an upper G2 solution $\bar{y}(t)$ $(\bar{y}(t_0) = y(t_0), \; y(t) \leq \bar{y}(t))$. In addition, a theorem was proved in [70] regarding Chaplygin-type differential inequalities for comparison with the upper G2 solutions. Hence, as in [64], the following lemma is obtained, generalizing a theorem of [71].

Lemma. Let v be a comparison vector function for Eq. (4), $v(t_0, x_0) \leq y_0$ $(\{t_0, x_0\} \in T_0 \times H_0, \; \{t_0, y_0\} \in \bar{B} \subset A)$, and let x(t) be a solution of (4) $(x(t_0) = x_0)$. Then there exists an upper solution $\bar{y}(t)$ of the comparison system (5) with $\overline{y}(t_0) = y_0$, continuable rightwards up to the boundary of $\overline{B}$, and $v(t, x(t)) \leq \overline{y}(t)$ in the interval $[t_0, \tau)$, in which the two solutions $x(t)$ and $\overline{y}(t)$ are defined.

Consider a functional

$$\bar{v}(t, x) = \max_{s=1,\ldots,l} v^s(t, x) \qquad (1 \leqslant l \leqslant k).$$

Introduce an interval $T = (0, \tau) \subseteq I_0$ $(T_0 \subseteq T)$, sets $M \subset \Gamma$, $M_0 \subseteq T_0 \times H_0$ such that their cross-sections $M(\tau_1)$ (by an arbitrary hyperplane $t = \tau_1 \in T$), and $M_0(t_0) \subseteq H_0$ (by the hyperplane $t = t_0 \in T_0$) are nonempty, and nonnegative functionals $\rho_M: T \times E \to \mathscr{R}^1$, $\rho_{M_0}^0: T_0 \times H_0 \to \mathscr{R}^1$, $\rho_M(t, x) = d_t(x, M(t))$, $d_{t_0}^0(x_0, M_0(t_0)) = \rho_{M_0}^0(t_0, x_0)$, satisfying the conditions:

(a) $d_t(x, M(t)) = 0$ for $\{t, x\} \in M$,

(b) $d_{t_0}^0(x_0, M_0(t_0)) = 0$ for $\{t_0, x_0\} \in M_0$.

These functionals are used respectively for measuring deviations of the phase vector $x(t)$ from the set $M(t)$ and of the initial point $x_0 \in H_0$ from the initial set $M_0(t_0)$ when investigating the stability, correctness and other properties in the system, using a similar approach to that adopted by A. A. Movchan [10], I. E. Gilbert and R. J. Knops [25], and T. K. Sirazetdinov [11] et al., when investigating the stability of an individual process ($M(t) = x(t)$, $M_0(t_0) = x(t_0)$). In particular, we can take [64]

$$d_t(x, M(t)) = d(x, M(t)) \equiv \inf_{x' \in M(t)} \|x - x'\| \quad \text{for} \quad \{t, x\} \in T \times E,$$

$$d_{t_0}^0(x_0, M_0(t_0)) = d^0(x_0, M_0(t_0)) = \inf_{x_0' \in M_0(t_0)} d^0(x_0, x_0') \quad \text{for} \quad \{t_0, x_0\} \in T_0 \times H_0,$$

where $d(x, x')$ is the metric of space E, and $d^0(x_0, x_0')$ is a proper metric, introduced into the set H_0, which converts it into the metric space of initial data.

Henceforth the logical symbols $\forall$, $\exists$, $\gg$, $\Rightarrow$, $\Leftrightarrow$, &, $\bigvee$, will be used, denoting "for any," "there exist (s)," "such that," "implies," "equivalence," "and," and "or."

The comparison function v is called a vector Lyapunov function for Eq. (4) with respect to ρ_M, $\rho_{M_0}^0$ in T, if it has some of the following types of properties:

1°. $\bar{v}$ is positive-definite with respect to ρ_M in T:

$$\exists \alpha > 0 \gg \forall r \in (0, \alpha) \exists \beta_\alpha(r) > 0 \gg (\forall \{t, x\} \in T \times E: r \leqslant d_t(x, M(t)) < \alpha) \bar{v}(t, x) \geqslant \beta_\alpha(r).$$

2°. $\bar{v}$ is positive-definite with respect to ρ_M in T in the large: 1° & $(\alpha = +\infty)$.

3°. For any $t_0 \in T_0$

$\overline{\lim}\, v(t_0, x_0) \leq 0$ (uniformly with respect to $x_0 \in H_0$).

$d_{t_0}^0(x_0, M_0(t_0)) \to 0$,

$(v(t_0, x_0) \leq 0$ for $\{t_0, x_0\} \in M_0)$.

4°. An infinitesimal limit superior with respect to $\rho_{M_0}^0$ in T_0: 3° with uniformity with respect to $t_0 \in T_0$.

Here and henceforth, the colon appearing in a definition replaces the words, "i.e.," or "which means that."

2. Stability

We shall start with definitions, of, and theorems concerning, stability of a set in T, under the assumption of the existence and continuability into $T_{t_0} = [t_0, \tau)$ of the solutions x of Eq. (4) with $\{t_0, x(t_0)\} \in T_0 \times H_0$ (the class of such solutions, passing through the point $\{t_0, x_0\}$, will be denoted by $\mathscr{X}(t_0, x_0)$) and of upper solutions $\overline{y}$ of the comparison system (5) with $\overline{y}(t_0) \in V_0 = \{v(t_0, x_0) \in \mathscr{R}^k: \{t_0, x_0\} \in T_0 \times H_0\}$.

Let the interval of the $y = 0$ axis, $t \in T$ be contained in A along with its η neighborhood $(0 < \eta \leq +\infty)$,

$$M_0(t_0)^\delta = \{x_0 \in H_0 : d_{t_0}^0(x_0, M_0(t_0)) < \delta\},$$

$$V_0^\gamma = \{y_0 \in V_0 : y_0^s < \gamma,\ s = 1, \ldots, k\} \qquad (\gamma, \delta > 0).$$

2.1. ρ_M, $\rho^0_{M_0}$ stability in T: $\forall\varepsilon>0,\ t_0\in T_0\ \exists\delta(\varepsilon,t_0)>0 \Rightarrow \forall x(t_0)\in M_0(t_0)^{\delta(\varepsilon,t_0)}\ (\forall x\in\mathscr{X}(t_0,x(t_0))\ t\in T_{t_0})\, d_t(x(t), M(t))<\varepsilon$.

2.2. ρ_M, $\rho^0_{M_0}$ stability in T, uniform with respect to t_0: 2.1 with $\delta(\varepsilon)$ independent of $t_0 \in T_0$.

2.3. Asymptotic ρ_M, $\rho^0_{M_0}$ stability in T: 2.1 & $\forall t_0\in T_0 \exists\pi(t_0)\in(0,+\infty] \Rightarrow \forall x(t_0)\in M_0(t_0)^{\pi(t_0)}\forall x \in\mathscr{X}(t_0, x(t_0))$

$$\lim_{t\to\tau-} d_t(x,(t),M(t))=0 \qquad (t\in T).$$

2.4. Asymptotic ρ_M, $\rho^0_{M_0}$ stability in T, uniform with respect to $\{t_0, x_0\}$: 2.2 & $\exists\pi\in(0,+\infty]\Rightarrow\forall\mu$, $\delta\in(0,\pi)$, $\exists\theta(\mu,\delta)>0$, $\xi(\mu,\delta)\subset(0,1)\Rightarrow\forall t_0\in T_0 t_0+\theta<\xi\cdot\tau\&(\forall x(t_0)\in M_0(t_0)^\delta$, $x\in\mathscr{X}(t_0, x(t_0))$, $t\in T_{t_0+\theta} d_t(x(t), M(t))<\mu$.

2.5. Asymptotic ρ_M, $\rho^0_{M_0}$ stability in the large in E in T: 2.3 with $\pi(t_0)=+\infty\ \forall t_0\in T_0$.

Necessary conditions for ρ_M, $\rho^0_{M_0}$ stability (or such stability, uniform with respect to t_0) are: (c) $d_{t_0}(x_0, M(t_0)) = 0$ in M_0, (d) the functional $t_0 \in T_0$ must be continuous for $t_0 \in T_0$ (or (e) uniformly continuous with respect to $t_0 \in T_0$) with respect to $d_{t_0}(x_0, M(t_0))$ in $d^0_{t_0}(x_0, M_0(t_0))$ in $M_0(t_0)$. Obviously, 2° & 3° ⇒ (c) & (d), 2° & 4° ⇒ (c) & (e), 2.4 & $(\tau<+\infty)\Rightarrow \sup_{T_0} t_0<\tau$. Sometimes a further condition is introduced: (f) given any solution $x(t)$, the function $d_t(x(t), M(t))$ is continuous in T_{t_0}.

Definitions will now be introduced for upper semi-stability of an interval of the $y = 0$ axis, $t \in T$ by virtue of the comparison system (5) with respect to $y^1,\dots,y^l$ under the condition $y_0 \in V_0$:

2.1⁺ upper semi-stability: $\forall\beta>0,\quad t_0\in T_0\exists\gamma(\beta,t_0)>0\Rightarrow\forall y_0\in V_0^{\gamma(\beta,t_0)}\ (\forall t\in T_{t_0})\ \bar{y}^s(t,t_0,y_0)<\beta$ $s = 1,\dots,l$),

2.2⁺ upper semi-stability, uniform with respect to t_0: 2.1⁺ with $\gamma(\beta)$ independent of t_0,

while 2.3⁺, 2.4⁺, 2.5⁺ are introduced similarly (in the case of 2.5⁺ it is assumed that $a_0=+\infty$, $A=I_0\times\mathscr{R}^k$).

Theorem 2.1. Let the Lyapunov vector function exist for eq. (4) with respect to ρ_M, $\rho^0_{M_0}$ in T. Then

1° & (e, d, f) & 3° & 2.1⁺ ⇒ 2.1 (ρ_M, $\rho^0_{M_0}$ stability in T),
1° & (c, d, f) & 4° & 2.2⁺ ⇒ 2.2,
1° & (c, d, f) & 3° & 2.3⁺ ⇒ 2.3,
1° & (c, d, f) & 4° & 2.4⁺ ⇒ 2.4,
2° & 3° & 2.5⁺ ⇒ 2.5.

Here, 1°& (c, d, f), and also 1°& (c, e, f), can be replaced by condition 2°. In the case of 2.3 the following holds for the domain of attraction $\Pi(t_0)$:

$$\{x_0\in H_0:\overline{\lim_{t\to\tau-}}\,\bar{y}^s(t,t_0,y_0)\leqslant 0\ \ (s=1,\dots,l)\ \text{ for }\ y_0=v(t_0,x_0)\}\subseteq\Pi(t_0).$$

Under particular assumptions, corollaries may be obtained, closely similar to the familiar theorems of V. I. Zubov [8], A. A. Movchan [10], N. P. Bhatia and G. P. Szego [17, 18] ($k = 1$, $f^1\leq 0$), K. Kordunyanu [47] ($k=1$, $f^1\in\mathscr{C}$, $M = 0$), V. Lakshmikantham [30, 54] ($E=H=\mathscr{R}^n$, $H_0=M_{(n-k)}$, $M(t)=M(0)$, X and f continuous) and others [45, 49, 62, 64, 69].

If the function v is known, f makes a natural appearance as a result of majorizing the right upper derivative $\bar{D}_+v(t, x(t))$ of the vector Lyapunov function along the solutions.

When constructing the vector Lyapunov function, the class of functions f may be specified by starting with the simplest, e.g., with linear or quasi-linear functions such that the question of the semi-stability of the interval of the $y = 0$ axis, $t \in T$ of the comparison system is decided immediately. Consider a theorem of this kind (cf. [49, 69]) regarding exponential ρ_M, $\rho^0_{M_0}$ stability in I_0: $\exists\lambda>0,\ \delta>0,\ \pi>0\Rightarrow\forall t_0\in T_0\ \forall x(t_0)\in M_0(t_0)^\pi(\forall x \in\mathscr{X}(t_0, x(t_0)),\ t\in I_{t_0})$.

$$d_t(x(t),M(t))\leqslant\delta\cdot d^0_{t_0}(x(t_0),M_0(t_0))e^{-\lambda(t-t_0)}.$$

Theorem 2.2. A sufficient condition for exponential $\rho_M, \rho^0_{M_0}$ stability is that there exist a Lyapunov vector function v for Eq. (4) with respect to $\rho_M, \rho^0_{M_0}$ in I_0, having the properties:

5°. $\beta\cdot[d_t(x, M(t))]^\xi \leqslant \bar{v}(t, x)\ (\forall\{t, x\} \in I_0\times E : d_t(x, M(t)) < \alpha)$.

6°. $$\|v(t_0, x_0)\| \leqslant \gamma[d^0_{t_0}(x_0, M_0(t_0))]^\xi (\forall\{t_0, x_0\} \in T_0\times H_0 : d^0_{t_0}(x_0, M_0(t_0)) < \alpha_0),\ (c, e, f) \vee (\alpha_0 = \alpha = +\infty)\ (\beta, \gamma, \xi > 0;\ \alpha, \alpha_0 \in (0 + \infty]).$$

IV. $f(t, y) = Py + \tilde{f}(t, y)$ for $\{t, y\} \in A$, where P is a constant $(k\times k)$-matrix, $p_{s\sigma} \geqslant 0\ (\forall s \neq \sigma)$,

$$(-1)^s \begin{vmatrix} p_{11} \dots p_{1s} \\ \dots \\ p_{s1} \dots p_{ss} \end{vmatrix} > 0 \quad (s = 1, \dots, k), \quad \frac{\|\tilde{f}(t, y)\|}{\|y\|} \underset{t\in I_0}{\rightrightarrows} 0 \text{ for } \|y\| \to 0. \tag{6}$$

If we have $\eta = \alpha = \alpha_0 = +\infty$, $\tilde{f}(t, y) \equiv 0$ here, we obtain exponential $\rho_M, \rho^0_{M_0}$-stability in the large ($\pi = +\infty$).

The methods for constructing the vector Lyapunov function for a composite system depend on allowing for its structural features and the possibility of decomposition. Consider an extension to systems with distributed parameters of the method for constructing the vector Lyapunov function, proposed by F. N. Bailey [66] for finite-dimensional systems and $M(t) = 0$.

Let

$$M(t) = M_1(t)\times\dots\times M_m(t) \qquad \text{for } t \in I_0,\ M_\mu(t) \subset E_\mu;$$

$$M_0(t_0) = M_{01}(t_0)\times\dots\times M_{0m}(t_0) \qquad \text{for } t_0 \in I_0,\ M_{0\mu}(t_0) \subseteq M_\mu(t_0);$$

$$d^0_{t_0}(x_0, M_0(t_0)) = \sqrt{\sum_{\mu=1}^m d^{0\mu}_{t_0}(x_{0\mu}, M_{0\mu}(t_0))^2};$$

$$d_t(x, M(t)) = \sqrt{\sum_{\mu=1}^m d^\mu_t(x_\mu, M_\mu(t))^2};$$

let interactions of the subsystems be absent when $x \in M(t)$ and let

$$X_\mu(t, x_1, \dots, x_\mu, \dots, x_m) = X_\mu{}^0(t, x_\mu) \text{ for } x_\mu \in H_\mu,\ x_\nu \in M_\nu(t) \cap H_\nu\ (\nu = \mu).$$

Consider the differential equations of the isolated subsystems

$$\frac{dx_\mu}{dt} = X_\mu(t, x_\mu), \qquad x_\mu \in E_\mu \tag{7}$$

with $X_\mu{}^0 : I_0\times H_\mu \to E_\mu$ and assume that functionals $v^\mu : I_0 \times E_\mu \to \mathcal{R}^1$, have been constructed for them, satisfying a one-sided Lipschitz condition with respect to $\{t, x_\mu\}$ in any bounded closed domain $\Psi_\mu \subset I_0\times E_\mu$ and estimates, typical of quadratic forms, but with respect to M_μ

$$\beta_\mu[d_t^\mu(x_\mu, M_\mu(t))]^2 \leqslant v^\mu(t, x_\mu)\,\forall\{t, x_\mu\} \in I_0\times E_\mu,$$

$$|v^\mu(t_0, x_0)| \leqslant \gamma_\mu[d^{0\mu}_{t_0}(x_{0\mu}, M_{0\mu}(t_0))]^2 \forall\{t_0, x_{0\mu}\} \in T_0 \times H_{0\mu},$$

$$v'^\mu_{(7)}(t, x_\mu) \leqslant -\zeta_\mu v^\mu(t, x_\mu)\,\forall t \in I_0, \quad x_\mu \in H_\mu(\beta_\mu, \gamma_\mu, \zeta_\mu > 0).$$

By majorizing, while recalling the interaction of the subsystems in (3) (assuming $l = k = m$)

$$v'^\mu_{(3)}(t, x) \leqslant \sum_{\nu=1}^m p_{\mu\nu}v^\nu(t, x_\nu) + \tilde{f}^\mu(t, v^1(t, x_1), \dots, v^m(t, x_m))$$

the problem reduces to the conditions IV. If the functionals v^μ are differentiable and $\tilde{f} = 0$, sufficient conditions for system (3) to be exponentially $\rho_M, \rho^0_{M_0}$ stable in the large are (6), $p_{\mu\nu} \geq 0\ (\nu \neq \mu)$ and

$$\frac{\partial v^{\mu}(t, x_{\mu})}{\partial x}(X_{\mu}^{0}(t, x_{\mu}) - X_{\mu}(t, x)) + \sum_{\nu=1}^{m} (p_{\mu\nu} + \delta_{\mu\nu}\zeta_{\mu}) v^{\nu}(t, x_{\nu}) \geqslant 0$$

($\delta_{\mu\nu}$ is Kronecker delta, $\mu = 1, \ldots, m$).

When constructing the functionals v^{μ} for the isolated subsystems (7), the familiar methods for constructing Lyapunov functionals may be used, specially in the case $M(t) \equiv 0$ [4-14, 25, 48, 72]. The existence of the function v^1, satisfying estimates typical of quadratic forms, was proved by N. N. Krasovskii [6] for the first (finite-dimensional) subsystem under the condition that its zero solution be exponentially stable, while $X_1^0(t, x_1)$ is continuous and has a continuous and bounded derivative $\partial X_1^0/\partial x_1$. For the second subsystem (integro-differential equations), reference may be made to the results of E. A. Barbashin [5], R. Gabasov [72], T. K. Sirazetdinov [11] and others. With regard to systems of evolutionary partial differential equations, e.g., of the type of the m-th subsystem, T. K. Sirazetdinov's method [11] etc. may be used for constructing the Lyapunov functionals in the form of "integral quadratic forms."

While it was noted in [67] that F. N. Bailey's method of obtaining sufficient conditions for stability has the merit of simplicity, which becomes increasingly more valuable with the number of the subsystem, it was also pointed out that these conditions are so to speak "over-sufficient," though this drawback is inherent to some degree in most methods for investigating the stability of nonlinear systems.

A great deal depends on how successfully the system decomposition is performed. Obviously, good stability conditions will be obtained if, in particular, the subsystems are weakly connected. Weak connections of this kind appear, for instance, in the equations of the spatial movements of certain rocket vehicles. Their stability can be successfully investigated by means of the vector Lyapunov function, formed from functionals constructed for the pitching, rolling and yawing channels.

The principle of least interaction, discovered by I. M. Gel'fand and M. L. Tsetlin [73], must be mentioned: according to this, composite controlled systems are typified by a structure which allows individual relatively autonomous subsystems to be isolated, thereby offering the advantage of minimizing the interactions between these subsystems.

Sometimes the optimal control of each subsystem (channel) of the weakly connected system is synthesized under the assumption that it is isolated. When constructing the vector Lyapunov function for the over-all composite (multichannel) system it is then natural to utilize the Lyapunov-Bellman functionals, obtained by solving "analytic construction of controlled" problems [11, 74] for the individual subsystems (channels) on the basis of some local optimality criterion.

The matrix P in the conditions for exponential stability in the large can be optimized on the basis of a min $\lambda_{max}(P)$ criterion.

A further advance may be made in constructing the vector Lyapunov function if some finite iterative process for improving it is employed, based on a hierarchy of subsystems and taking as the first approximation the function obtained by the generalized F. N. Bailey method. Turning to papers on the optimization of large systems, in which the decomposition principle is employed, we find not only a multilevel organization of the control, but also a hierarchical structure of the actual computational algorithm [75, 76].

In this context, after constructing the functionals $v_1^1(t, x_1), \ldots, v_1^m(t, x_m)$, corresponding to the isolated subsystems, it becomes necessary to estimate the strength of the connections between them and the influence of these individual connections on the comparison system. After picking out the most influential connections and allowing for their structural features, subsystems of the second level may be formed, in which two or more of the most strongly connected subsystems of the first level are combined. For these, the functionals $v_2^1(t, x_1'), \ldots, v_2^k(t, x_{k'})$ are constructed and corrected, then the vector Lyapunov function $v_2 = \{v_2^1, \ldots, v_2^k\}$ $(k < m)$ is considered (each phase coordinate appears in just one of the functionals v_2^s, $\{x_1, \ldots, x_m\} = \{x_1', \ldots, x_k'\} = x \in E$. Certain subsystems may be characterized by several functionals $v_2^s(t, x_{s'})$. If necessary, subsystems of the 3rd, 4th, etc. levels may be formed (the number of levels is finite).

Quite complicated technical complexes, including elements with distributed parameters of different types, may be analyzed by this method. Another possible approach to construction of the vector Lyapunov function involves the introduction of models describing the operation of the subsystems [69].

3. Asymptotic Behavior

The vector Lyapunov function may be used for investigating many other properties of processes with distributed parameters [62, 64, 69].

3.1. $\rho_M, \rho^0_{M_0}$-invariance (right $\rho_M, \rho^0_{M_0}$-invariance) in T: $\forall t_0 \in T_0\ (\forall x(t_0) \in H_0 : d^0_{t_0}(x(t_0), M_0(t_0)) = 0)\ (\forall x \in \mathscr{X}(t_0, x(t_0)),\ t \in T_{t_0})\ d_t(x(t), M(t)) = 0$.

Theorem 3.1. For $\rho_M, \rho^0_{M_0}$-invariance (right $\rho_M, \rho^0_{M_0}$-invariance) in T, it is sufficient that a vector Lyapunov function v exist for eq. (4), satisfying the following condition.

7°. $(\forall \{t_0, x_0\} \in T_0 \times H_0 : d^0_t(x_0, M_0(t_0)) = 0)\ v(t_0, x_0) \leqslant 0,\ \bar{v}(t, x) > 0 (\forall \{t, x\} \in T \times E : d_t(x, M(t)) > 0)$, while the corresponding comparison system (5) has an upper solution $\bar{y}(t, t_0, /0)$ (with $\bar{y}(t_0, t_0, 0) = 0 \forall t_0 \in T_0$), in which $\bar{y}^s(t, t_0, 0) = 0$ for $t \in T_{t_0}$ $(s = 1, \ldots, l)$.

3.2. $\rho_M, \rho^0_{M_0}$-boundedness (of the solutions of eq. (4)) in $T : \forall t_0 \in T_0 \exists \pi(t_0) \in (0, +\infty] \ni \forall x(t_0) \in M_0(t_0)^{\pi(t_0)} \exists \beta(t_0, x(t_0)) > 0 \ni (\forall x \in \mathscr{X}(t_0, x(t_0)),\ t \in T_{t_0})\ d_t(x(t), M(t)) < \beta(t_0, x(t_0))$.

3.3. Equi- $\rho_M, \rho^0_{M_0}$-boundedness in T: $\forall t_0 \in T_0 \exists \pi(t_0) \in (0, +\infty] \ni \forall \alpha \in (0, \pi(t_0)) \exists \beta(\alpha, t_0) > 0 \ni \forall x(t_0) \in M_0(t_0)^{\alpha} \forall x \in \mathscr{X}(t_0, x(t_0))\ (\forall t \in T_{t_0}) d_t(x(t), M(t)) < \beta(\alpha, t_0)$.

3.4. Uniform $\rho_M, \rho^0_{M_0}$-boundedness in T: 3.3 with π and $\beta(\alpha)$ independent of $t_0 \in T_0$.

3.5. $\rho_M, \rho^0_{M_0}$-boundedness in the limit in T: $\exists \beta > 0 \ni \forall t_0 \in T_0 \exists \pi(t_0) \in (0, +\infty] \ni \forall x(t_0) \in M_0(t_0))^{\pi(t_0)} (\forall x \in \mathscr{X}(t_0, x(t_0)))\ \overline{\lim_{t \to \tau-}}\, d_t(x(t), M(t)) < \beta$.

$\rho_M, \rho^0_{M_0}$-boundedness (or equal-boundedness, uniform boundedness, boundedness in the limit) in T in the large: 3.2 (or 3.3, 3.4, 3.5) with $\pi = +\infty \forall t_0 \in T_0$.

It will be assumed henceforth that $A = T \times \mathscr{R}^k$.

Theorem 3.2. Let there exist a vector Lyapunov function v for eq. (4) with respect to $\rho_M, \rho^0_{M_0}$ in T and some $\pi(t_0) \in (0, +\infty]$.

If, in the upper solutions $\bar{y}(t)$ of the comparison system (5) with $\bar{y}(t_0) \in V_{0\pi} = \{v(t_0, x_0) \in \mathscr{R}^k : \{t_0, x_0\} \in T_0 \times H_0,\quad d^q_{t_0}(x_0, M_0(t_0)) < \pi\}$ the components $\bar{y}^1(t), \ldots, \bar{y}^l(t)$ are upper-bounded in T_{t_0} (3.2+)

8°. $\bar{v}(t_i, x_i) \to +\infty$ for any sequence $\{t_i, x_i\} \in T \times E : t_i \to \tau-,\ d_{t_i}(x_i, M(t_i)) \to +\infty$, then the solutions of eq. (4) are $\rho_M, \rho^0_{M_0}$-bounded in T.

If $\bar{y}^1(t), \ldots, \bar{y}^l(t)$ are equi-upper-bounded in T with $\bar{y}(t_0) \in V_{0\pi} : \forall t_0 \in T_0,\ \delta \in (0, \pi) \exists \gamma(\delta, t_0) > 0 \ni (\forall t \in T_{t_0},\ y_0 \in V_{0\delta})$

$$\bar{y}^s(t, t_0, y_0) < \gamma(\delta, t_0) \quad (s = 1, \ldots, l) \tag{3.3+}$$

and

9°. $\lim_{d_t(x, M(t)) \to \infty} v(t, x) = +\infty$ uniformly with respect to $t \in T_{t_0} (\forall t_0 \in T_0)$, implies the solutions x(t) of eq. (4) are equi- $\rho_M, \rho^0_{M_0}$-bounded in T.

If the $\bar{y}^1(t), \ldots, \bar{y}^l(t)$ are uniformly upper-bounded in T with $\bar{y}(t_0) \in V_{0\pi}$: 3.3+ with π, $\gamma(\delta)$ independent of t_0 (3.4+), and the limit in 9° is uniform with respect to $t_0 \in T_0$ (9°(u)), then the solutions x(t) are uniformly $\rho_M, \rho^0_{M_0}$-bounded in T.

If $\bar{y}^1(t), \ldots, \bar{y}^l(t)$ are upper-bounded in the limit in T with $\bar{y}(t_0) \in V_{0\pi}$

$$\exists \gamma > 0 \ni (\forall t_0 \in T_0,\ y_0 \in V_{0\pi})$$
$$\overline{\lim_{t \to \tau-}}\, \bar{y}^s(t, t_0, y_0) < \gamma \quad (s = 1, \ldots, l) \tag{3.5+}$$

and we have

10°. $\lim\limits_{t\to\tau-,\ d_t(x,\ \mu(t))\to+\infty} \bar{v}(t, x) = +\infty$ (uniformly with respect to x), implies the solutions of Eq. (4) are ρ_M, $\rho^0_{M_0}$-bounded in the limit in T.

When 3.2^+ (3.3^+, 3.4^+, 3.5^+) is satisfied with $\pi(t_0) = +\infty$, the corresponding properties of ρ_M, $\rho^0_{M_0}$-boundedness in the large in T are obtained in a natural way. Similar assertions hold for equi-ρ_M, $\rho^0_{M_0}$-boundedness in the limit in T under condition 10°; while under condition 9°(u) also, they hold for uniform ρ_M, $\rho^0_{M_0}$-boundedness in the limit in T and for ρ_M, $\rho^0_{M_0}$-dissipativeness in T (i.e., when property 3.4 and the property of uniform ρ_M, $\rho^0_{M_0}$-boundedness in the limit in T hold simultaneously). Sufficient conditions can also be obtained for ρ_M, $\rho^0_{M_0}$-continuity of the solutions x(t) in T with respect to the initial data $x(t_0) \in H_0$ (among which are included conditions 2° and 3°).

Under special assumptions, we arrive at theorems obtained earlier by T. Yoshizawa [29], V. Lakshmikantham [30], and others [62, 64].

Take the case when the functionals ρ_M, $\rho^0_{M_0}$ represent the distance from a current point x to the coordinate origin 0, i.e., the norm in E, or the distance between points of different solutions. We shall start by examining non-local continuability of the solutions (rightwards) in the interval T, under the assumption that the solutions exist and can be continued up to the boundary of any bounded closed domain $\bar{K} \subset T \times E$, containing the initial point $\{t_0, x_0\} \in T_0 \times H_0$ [64].

Theorem 3.3. Any solution x(t) of eq. (4) with $\{t_0, x(t_0)\} \in T_0 \times H_0$ can be continued in T_{t_0} if there exists a vector Lyapunov function v for (4) (with respect to the norm) in T, with the following properties

11°. $(\forall \tau^* \in \bar{T}) \lim \bar{v}(t, x) = +\infty$ as $t \to \tau^* - \|x\| \to \infty$, and the upper solutions $\bar{y}(t, t^*, y^*)$ of the comparison system (5), passing through any point $\{t^*, y^*\}$ $ct^* \in T$, $y^* = v(t^*, x^*)$, $\|x^*\| > 0$, are defined in T.

With k = 1 this represents an extension of theorems due to M. A. Krasnosel'skii and S. G. Krein [34] $(X \in \mathscr{C}_E(T \times E),\ f^1 = k(t) g(y^1))$ and other authors [37, 64].

Now introduce Banach space B as the direct product E × E with natural linearization and the definition

$$\|\{x', x''\}\|_B = \|x'\| + \|x''\|$$

for the norm of an element $\{x', x''\} \in B(x' \in E, x'' \in E)$. A system of two identical equations of the type (4) may be considered in it as a single differential equation (*) with operator $\{X(t, x'), X(t, x'')\}: I_0 \times H \times H \to B$ and solutions $\{x'(t), x''(t)\}$, composed of the solutions x'(t), x"(t) of (4), $\{x'(t_0), x''(t_0)\} = \{x_0', x_0''\} \in H_0 \times H_0$. A metric is introduced similarly into the space of initial data $H_0 \times H_0$. As the set M(t) we take the plane $\{x' = x''\} \subset B$ (the "diagonal"), while $M_0(t_0) = \{\{x_0', x_0''\} \in H_0 \times H_0 : x_0' = x_0''\}$. Then, given any t, the distance between two solutions x'(t) and x"(t) of (4) is equal to the distance in B of the corresponding solution $\{x'(t), x''(t)\}$ of the new system from this plane $d^B(\{x'(t), x''(t)\}, M(t)) = \|x'(t) - x''(t)\|$ $(t \in T)$,

$$d^{0B}(\{x_0', x_0''\}, M_0(t_0)) = d^0(x_0'\ x_0'').$$

The vector Lyapunov function considered for eq. (*) is the function v(t, x', x") with respect to these distances in T.

Boundedness of the solutions of eq. (*) with respect to the plane M in the large (equi- or uniform boundedness or boundedness in the limit) corresponds to boundedness of the distance between any two solutions of (4) or in T. Yoshizawa's terminology, D-boundedness of the system (4) (equi- or uniform boundedness or boundedness in the limit). Stability of the plane M (uniform, asymptotic) for (*) implies stability of all the solutions x(t) with $x(t_0) \in H_0$ of eq. (4) and corresponds to equi-stability of system (4) (uniform, asymptotic) in T. Yoshizawa's sense. Using the vector Lyapunov function v(t, x', x") with respect to these distances for eq. (1), Theorems 2.1 and 3.2 yield conditions of these types for stability and boundedness of the system (4). In the case when M is asymptotically stable in the large in T, we obtain extremal stability of the system (4) in T:

$$(\forall t_0 \in T_0,\ x'(t_0),\ x''(t_0) \in H_0)\ \ \|x'(t) - x''(t)\| \to 0 \ \text{for}\ \ t \to \tau -.$$

The phenomenon of convergence in system (4) in T is defined as the existence of a solution, bounded in T_{t_0}, and asymptotically stable in the large in T. To discover this solution, we only need to construct the vector Lyapunov function v(t, x', x") with respect to the distances to the plane M for eq. (*), such that it satisfies all the conditions of Theorem 2.1 and ensures the asymptotic stability of M in the large in T, and also the vector Lyapunov function $v_1(t, x)$ for eq. (4) with respect to $\|\cdot\|$ in T, satisfying conditions similar to those of Theorem 3.2, ensuring the existence of a bounded solution [64, 69].

Right invariance of the plane $\{x' = x''\}$ corresponds to the solutions of system (3) being (right) unique in T_{t_0} for all $\{t_0, x_0\} \in T_0 \times H_0$. A sufficient condition for this is that there exist a vector Lyapunov function $v(t, x', x'')$ for eq. (*) with respect to the distances of this plane in T, satisfying the conditions of Theorem 3.1. The property whereby the solutions of (*) are ρ_M, $\overset{0}{\rho}_{M_0}$-continuous with respect to the initial data $\{x'(t_0), x''(t_0)\}$ in T, transforms into a right continuous dependence in T of the solutions $x'(t)$ of eq. (4) on the initial data; consequently, this continuity can be ensured by constructing a vector Lyapunov function $v(t, x', x'')$, satisfying conditions 2° and 3°.

4. Influence of Perturbations

The influence of perturbations is taken into account similarly. We consider the distributed equation

$$\frac{dz}{dt} = X(t, z) + R(t, z), \quad z \in E \tag{8}$$

with perturbations $R(t, z)$ from the same class of operators as X. As in [64], we introduce definitions of ρ_M, $\overset{0}{\rho}_{M_0}$-stability or ρ_M, $\overset{0}{\rho}_{M_0}$-boundedness in T when perturbations which are small at each instant are continuously operative:

$$\forall t_0 \in T_0 \exists \pi(t_0) \in (0, +\infty] \gg \forall z(t_0) \in M_0(t_0)^{\pi(t_0)}$$
$$\exists \varepsilon(t_0, \ z(t_0)) > 0, \ \ \delta(t_0, z(t_0)) > 0 \gg (\forall t \in T_{t_0}) d_t(z(t), \ \ M(t)) < \varepsilon(t_0, \ \ z(t_0))$$

for every solution $z(t)$ of any equation (8), for which

$$(\forall t \in T_{t_0}, \ z \in H : d_t(z, \ M(t)) \leqslant \varepsilon(t_0, \ z(t_0))) \ \|R(t, \ z)\| \leqslant \delta \tag{9}$$

(and similarly in the case of perturbations which are small on the average or integrally small in $T = I_0$).

We can extend to this case the relevant theorems of [64] on stability and boundedness of sets in the case of continuously operative perturbations.

Theorem 4.1. Let a vector Lyapunov function v exist for eq. (4) with respect to ρ_M, $\overset{0}{\rho}_{M_0}$ in T, satisfying a Lipschitz condition with respect to x in any domain $\{\{t, x\} \in T \times E : d_t(x, M(t)) \le \alpha\}$ and (f). If, in the upper solutions $\bar{y}(t)$ with $\bar{y}(t_0) \in V_{0\pi}$ of the comparison system (5), $\bar{y}^1(t), \ldots, \bar{y}^l(t)$ are upper bounded under perturbations which operate continuously and are small at each instant, and 10° holds, then the solutions of (4) will be ρ_M, $\overset{0}{\rho}_{M_0}$-bounded in T under perturbations which operate continuously and are small at each instant. (Here, 10° and (f) can be replaced by condition 9° (u), provided that the condition $d_t(z, M(t)) \le \varepsilon$ is discarded from (9)).

The same theorem also holds for ρ_M, $\overset{0}{\rho}_{M_0}$-continuity in T of the solutions (of (4)) with respect to the initial data and the right side of the equation. It extends the result of [64], referring to the case $M = \{x' = x''\}$, $v(t, x', x'')$. The vector Lyapunov function $v(t, x', x'')$ was used in [64] to prove theorems on the convergence of approximations and the continuous dependence of the solutions of (4) on a parameter; these theorems can be extended in a similar way.

The results obtained may thus be summarized as the following comparison principle.

If there exists for the system with distributed parameters (3) a vector Lyapunov function with respect to ρ_M, $\overset{0}{\rho}_{M_0}$ in T, satisfying the definitions of conditions 1°-11° (these definitions are in general proper to each case), then the various properties of processes in the system, such as ρ_M, $\overset{0}{\rho}_{M_0}$-uniqueness and nonlocal continuability of the solutions, ρ_M, $\overset{0}{\rho}_{M_0}$-continuity of the solutions as functions of the initial data, the right side or a parameter, ρ_M, $\overset{0}{\rho}_{M_0}$-stability and boundedness in the detailed classification, dissipativeness, and convergence, follow from the corresponding properties of the solutions of the k-dimensional comparison system.

5. Some Problems

1. Before discussing unsolved problems in connection with the use of the vector Lyapunov function for qualitative study of composite systems with distributed parameters, it must be emphasized that methods for constructing such functions still need to be developed, both for concrete cases and for general classes of system.

For properties other than stability, this topic is wide open (methods have not even been developed for constructing Lyapunov functionals for subsystems of types 2, ..., m). In the methods discussed above, for exponential stability, further progress may be gained by utilizing the results of M. G. Krein and Yu. L. Daletskii [77] regarding exponential splitting, when the Banach space has a direct decomposition into a finite number of subspaces, with disjoint intervals for the general exponents of the solutions which start in these subspaces.

2. An interesting problem concerns the simultaneous proof of different dynamic properties of a system with distributed parameters (while also obtaining quantitative estimates for the domains of attraction etc.) on the basis of the topological picture of the disposition of the integral curves of the comparison system in space $\mathscr{R}^k$, at any rate for k = 2, 3.

3. From the abstract point of view, vector Lyapunov functions define a mapping of the space of states of the initial system into the space of states of the comparison system, under which the dynamic properties of the system are in general preserved in just the backward direction: the properties of the comparison system induce the corresponding properties of the initial system. Another feature is bound up with utilizing partial ordering of the phase space of the comparison system* by means of some cone (in our treatment, the cone of nonnegative elements.)

An investigation of stability-preserving mappings under very general assumptions was commenced by W. Hahn in a short note [20] (see also [24, 25]). In a discussion of one of the present author's papers, A. D. Myshkis suggested that it would be worth developing a theory of such "one-sided" mappings of a topological vector space into a manifold with a cone, by utilizing modern topology and theory of dynamic systems on manifolds. This would enable many results to be generalized and stated in an invariant form.

The present situation in topological dynamics is said to be similar to that in functional analysis in the thirties: the potential of the subject was obvious, though little had been actually accomplished. It is to be hoped that the same is true in regard to applying the method of Lyapunov functions for qualitative analysis of composite systems.

LITERATURE CITED

1. A. M. Lyapunov, Collected Works [in Russian], Vols. I-IV, Izd. AN SSSR (1954, 1956, 1959); Vol. V, Nauka (1965).
2. A. Poincaré, On Curves Defined by Differential Equations [Russian translation], Gostekhizdat (1947).
3. N. G. Chetaev, Stability of Motion. Works on Analytic Mechanics [in Russian], Izd. AN SSSR (1962).
4. K. P. Persidskii, "On the stability of the solutions (of a denumerable system) of differential equations," Izv. AN Kazakh. SSR, Matematika i Mekhanika, 2, No. 56 (1948); 4, No. 97 (1950); 5, No. 62 (1951); 9 (13) (1960) (1961).
5. E. A. Barbashin, "The method of sections in theory of dynamic systems," Matem. Sb., 29 (71), No. 2 (1951); "On conditions for preserving stability properties of the solutions of integro-differential equations," Izv. Vuzov, Matematika, No. 1 (1957); Introduction to Theory of Stability of Motion [in Russian], Nauka (1967); Lyapunov Functions [in Russian], Nauka (1970).
6. N. N. Krasovskii, "Converses of theorems in Lyapunov's second method and aspects of stability of motion with respect to the first approximation," Prikl. Matem. i Mekh., 20, No. 2 (1956), 20, No. 4 (1956); Some Problems in Theory of Stability of Motion [in Russian], Fizmatgiz (1959).
7. V. V. Rumyantsev, "Stability of rotation of a rigid body with a liquid-filled cavity," Prikl. Matem. i Mekh., 21, No. 6 (1957); 23, No. 6 (1959); 24, No. 1 (1960); 24, No. 4 (1960); 26, No. 6 (1962); 28, No. 6 (1964); 33, No. 6 (1969); "Stability of motion of a rigid body with liquid-filled cavities," Proceedings of the First and Second All-Union Meetings on Theoretical and Applied Mechanics [in Russian], Izd. AN SSSR (1962); Nauka (1965).
8. V. I. Zubov, Lyapunov's Methods and Their Applications [in Russian], Izd. LGU (1957).
9. D. M. Volkov, "An analog of Lyapunov's second method in boundary value problems for nonlinear hyperbolic equations," Dokl. AN BSSR, 2, No. 8 (1958); Uch. Zap. LGU, No. 271 (1958).
10. A. A. Movchan, "On Lyapunov's direct method in problems of stability of elastic systems," Prikl. Matem. i Mekh., 23, No. 3 (1959); "Stability of processes with respect to two metrics," Prikl. Matem. i Mekh., 24, No. 6 (1960); 29, No. 4 (1965); "On stability of processes of deformation of continuous bodies," Archiv. Mech. Stosowanej, 15, No. 5 (1963); Inzh. Sb., 29, No. 1 (1960); Inzh. Zhurnal, 5, No. 4 (1965).

*As distinct from introducing partial ordering directly into the phase space E of the initial system (4), as is usually done.

11. T. K. Sirazetdinov, "On stability of processes with distributed parameters," Tr. Kazansk. Aviats. In-ta, Matem. i Mekh., No. 83 (1964); No. 109 (1969); Prikl. Matem., 31, No. 1 (1967); "On stability of stochastic processes with distributed parameters," Prik . Matem. i Mekh., 28, No. 6 (1964); Tr. Kazansk. Aviats. In-ta, Matem. i Mekh., No. 97 (1968); Stability of Systems with Distributed Parameters [in Russian], Izd. Kazansk. In-ta, Kazan' (1971).
12. P. K. C. Wang, "Stability analysis of simplified flexible vehicle via Lyapunov's direct method," AIAA Journ., 3, No. 9 (1965); Raketn. Tekhn. i Kosmonavtika, No. 9 (1965); "Stability analysis of elastic and aeroelastic systems via Lyapunov's direct method," Journ. Franklin Inst., 281, No. 1 (1966); "On the feedback control of distributed parameter systems," Internat. Journ. Control, 3, No. 3 (1966); IEEE Trans. Automat. Control, Vol. AC-11, No. 1 (1966); Journ. Appl. Mech. Trans. ASME, Ser. E., 33, No. 1 (1966).
13. Extension of Lyapunov's second method to partial differential equations, Bibliographical Index 1957-67, Sib. NII Energetiki, Izd. IGiG SO AN SSSR (1969).
14. M. G. Krein, Lectures on Theory of Stability of Solutions of Differential Equations in Banach Space (1947-48) [in Russian], Izd. In-ta Matem. AN USSR, Kiev (1964); Usp. Matem. Nauk, 33 (25) (1948).
15. J. L. Massera, "Contributions to stability theory," Annals of Math., 64, No. 1 (1956).
16. J. Auslender and P. Seibert, "Prolongations and generalized Lyapunov functions," Proc. Internat. Sympos. Nonlin. Diff. Equations and Nonlinear Mechanics, Academic Press (1963); Ann. Inst. Fourier, 14, No. 2 (1964).
17. N. P. Bhatia, Stability and Lyapunov functions in dynamical systems," Contrib. Diff. Equations, 3 (1964); Math. Syst. Theory, 1, No. 2 (1967).
18. N. P. Bhatia and G. P. Szego, Dynamical Systems: Stability Theory and Applications, Springer, Berlin (1967).
19. N. P. Bhatia and O. Hajck, Local Semi-Dynamical Systems, Springer, Berlin (1969).
20. W. Hahn, Theory of Stability of Motion, Springer, Berlin (1967); Journ. Reine und Angew. Math., 228 (1967).
21. E. Roksin, Asymptotic Foundations of Theory of Control Systems [in Russian], Proceedings of Second IFAK Congress; Theory of Continuous Automatic Systems [in Russian], Nauka (1965).
22. A. A. Kayande and V. Lakshmikantham, "General dynamical systems and conditional stability," Proc. Cambridge Philos. Soc., 63, No. 1 (1967).
23. K.S. Sibirskii, Introduction to Topological Dynamics [in Russian], Red.-Izd. Otdel AN Mold. SSR (1970).
24. D. Bushaw, "A stability criterion for general systems," Math. System Theory, 1, No. 1 (1967).
25. I. E. Gilbert and R. J. Knops, "Stability of general systems," Arch. Rat. Mech. Anal., 25, No. 4 (1967).
26. M. S. Izman, "Application of Lyapunov's second method to investigation of stability and asymptotic stability of sets in dispersive dynamic systems," Differ. Uravn., 5, No. 7 (1969); 7, No. 4 (1971); "On stable and attracting sets in dynamic systems," in; Mathematical Studies [in Russian], Red.-Izd. Otd. AN Mold, SSR, 3, No. 3 (1968); 3, No. 4 (1968).
27. J. Nagy, "Lyapunov's direct method in abstract local semi-flows," Comment. Math. Univ. Carolinae, 8, No. 2 (1967); "Lyapunov's direct method in abstract control processes," Časopis pro Pětováni Mat., 93, No. 3 (1968).
28. J. K. Hale, "Dynamical systems and stability," Journ. Mat. Anal. and Appl., 26, No. 1 (1969); Proceedings of International Symposium on Nonlinear Oscillations, Vol. II, Izd. AN USSR (1963).
29. T. Yoshizawa, "Stability theory by Lyapunov's second method," Math. Soc. Japan (1966).
30. V. Lakshmikantham, "Differential equations in Banach spaces and the extension of Lyapunov's methods," Proc. Cambridge Phil. Soc., 59, No. 2 (1963); 60, No. 4 (1964); J. Math. Anal. and Appl., 8 (1964); 9 (1964).
31. E. Tumajek, "Omezenost abstraktnich procesu," Caspis pro Pestovani Mat., 95, No. 2 (1970).
32. V. M. Alekseev, "A Theorem on an integral inequality and some applications," Matem. Sb., 68 (110), No. 2 (1965).
33. A. A. Kayande and D. B. Muley, "Lyapunov functions and a control problem," Proc. Cambridge Philos. Soc., 63, No. 2 (1967).
34. M. A. Krasnosel'skii and S. G. Krein, "On theory of ordinary differential equations in Banach spaces," Proc. of seminar on functional analysis, Izd. Voronezhsk. Un-ta, No. 2 (1956); Proceedings of Third All-Union Mathematical Meeting, Vol. 3, Izd. AN SSSR (1958).
35. A. I. Perov, "On integral inequalities," Proc. of seminar on functional analysis, Izd. Voronezhsk. Un-ta, No, 5 (1957); "On uniqueness theorems for ordinary differential equations," Dokl. AN SSSR, 120, No. 4 (1958).
36. M. A. Krasnosel'skii and Ya. D. Mamedov, "A note on the application of differential and integral inequalities in problems on correct posing of the Cauchy problem for ordinary differential equations in Banach spaces," Nauchn. Dokl. Vyssh. Shkoly, Fiz.-Matem. N., No. 2 (1959).
37. A. V. Kibenko, "On theory of ordinary differential equations in Banach space," Dokl. AN SSSR, 135, No. 5 (1961).

38. A. V. Kibenko, M. A. Krasnosel'skii, and Ya. D. Mamedov, "One-sided estimates in the existence conditions for solutions of differential equations in functional spaces," Uch. Zap. Azerb. Gos. Un-ta, Fiz.-Mat. i Khim. N., No. 3 (1961).
39. H. A. Antosiewicz, "An inequality for approximate solutions of ordinary differential equations," Math. Zeitschr., 78 (1962); Proceedings of International Symposium on Nonlinear Oscillations, Vol. 1, Izd. AN SSSR (1963); "Recent contributions to Lyapunov's second method," Colloq. Internat. Centre Nat. Res. Scient., 148 (1965).
40. Ya. D. Mamedov, "One-sided estimates in conditions for investigating the solutions of differential equations with a bounded operator and delayed argument in Banach space," in: Aspects of Computational Mathematics [in Russian], Izd. AN Az. SSR (1967); Proc. of Seminar on Theory of Differential Eqs. with a Deviating Argument, Un-t Dryzhby Narodov im. P. Lumumba, 5 (1967); 6 (1968).
41. S. G. Krein, Linear Differential Equations in Banach Space [in Russian], Nauka (1967).
42. S. G. Krein and V. B. Osipov, "Lyapunov functions and Cauchy problems for some systems of partial differential equations," Differents. Uravnen., 6, No. 11 (1970).
43. S. A. Chaplygin, Selected Works on Mechanics and Mathematics [in Russian], GITTL (1954).
44. Z. B. Tsalyuk, "A note on application of solvability conditions for Chaplygin's problem to aspects of qualitative theory of equations," Dokl. AN SSSR, 134, No. 1 (1960).
45. V. Lakshmikantham and S. Leela, Differential and Integral Inequalities. Theory and Applications, Vol. 1, Ordinary Differential Equations, Vol. 2, Functional, Partial, Abstract and Complex Differential Equations, Acad. Press, New York (1969).
46. M. A. Krasnosel'skii, Operators of Displacement over the Trajectories of Differential Equations [in Russian], Nauka (1966).
47. K. Kordunyanu, "Application of differential inequalities to theory of stability," Analese Stiintifice ale Univ., "Al. I. Cuza," Iasi. Ser. Noua, Sec. 1, Math. Fiz., Chimie, 6, No. 1 (1960); 7, No. 2 (1961); Boll. Acad. Polon. Sci., 10 (1962); Rev. Roumaine Math. Pures et Appl., 7, No. 3 (1964); Compt. Rend. Acad. Sci., 256, No. 1 (1963); Abhandl. Dtsch. Acad. Wiss., Berlin, Kl. Math. Phis und Techn., No. 1 (1965).
48. H. Leipholz, Stabilitätstheorie. Eine Einfuhrung in die Theorie der Stabilität dynamischer Systeme und fester Körper, Taubner, Stuttgart (1968).
49. V. M. Matrosov, "On theory of stability of motion," Prikl. Matem. i Mekh., 26, No. 6 (1962); J. Appl. Math. and Mech., 26 (1963); Tr. Kazansk. Aviats. In-ta, Matem. i Mekh., 80 (1963); Proceedings of Inter-Collegiate Conference on Applications of Theory of Stability of Motion and Analytic Mechanics [in Russian], Kazan', Izd. KAI (1964); "Development of the method of Lyapunov functions in stability theory," Proc. of Second All-Union Meeting on Theoretical and Applied Mechanics [in Russian], 1, Nauka (1965).
50. R. Bellman, "Vector Lyapunov functions," Journ. Soc. Industr. and Appl. Math., Ser. A, Control, 1, No. 1 (1962).
51. T. Wazewski, "Systemes des equations et des inequalities differentielles ordinaires aux deuxiemes membres montones et leurs applications," Ann. Soc. Polon. Mat., 23 (1950).
52. W. Walter, Differential- und Integralungleichungen und ihre Anwendung bei Abschätzungs- und Eindeutigkeits-Problemen, Springer, Berlin (1964).
53. J. Szarski, Differential Inequalities, Warsaw, PWN (1965).
54. V. Lakshmikantham, "Vector functions and conditional stability," J. Math. Anal. and Appl., 10, No. 2 (1965).
55. S. K. Persidskii, "On some theorems of Lyapunov's second method," Differents. Uravn., 5, No. 4 (1969); Izv. AN KazSSR, Ser. Fiz.-Mat., No. 1 (1967).
56. A. A. Kayande and V. Lakshmikantham, "Conditionally invariant sets and vector Lyapunov functions, Journ. Math. Anal. and Appl., 19, No. 3 (1967).
57. G. R. Verma and V. Lakshmikantham, Several Lyapunov functions and mixed stability. Bull. Math. Soc. Sci. Math. RS Romania, 11, No. 2 (1967), (1968).
58. V. Lakshmikantham, S. Leela, and T. Sastri, Converse theorems for conditional stability of motion, J. Math. Anal. and Appl., 19, No. 3 (1967).
59. S. K. Persidskii and V. F. Kondratenko, "On stability of the solutions of linear systems of differential equations," Izv. AN KazSSR, Ser. Fiz.-Mat., No. 5 (1968).
60. G. A. Egorov, "On investigation of the stability of integral manifolds," Tr. Kazansk. Aviats. In-ta, Matem. i Mekh., No. 107 (1968).
61. V. D. Irtegov, "On the stability of the solutions of difference equations," Tr. Kazansk. Aviats. In-ta, Matem. i Mekh., No. 125 (1970).
62. V. Lakshmikantham and S. Leela, "Parabolic differential equations and conditional stability," Rend. Circolo Matem. Palermo, Ser. II, 15, No. 2 (1966); Conditional invariant sets and functional differential equations," Rend. Circolo Matem. Palermo, Ser. II, 15, No. 3 (1966).

63. P. Hartman, "The existence and stability of stationary points," Duke Math. J., 33, No. 2 (1966).
64. V. M. Matrosov, "The principle of comparison with a Lyapunov vector function," Differents. Uravn., 4, No. 8 (1968); 4, No. 10 (1968); 5, No. 7 (1969); 5, No. 12 (1969).
65. G. Kron, Diakoptics, Macdonald, London (1963).
66. F. N. Bailey, "The application of Lyapunov's second method to interconnected systems," Journ. Soc. Industr. and Appl. Math., Control, Ser. A, 3, No. 3 (1965) (1966); Proc. Nat. Electron. Conference, Chicago, 21 (1965).
67. A. A. Piontkovskii and L. D. Rutkovskaya, "Study of some problems in stability theory by the method of the vector Lyapunov function," Avtomat. i Telemekhan., No. 10 (1967).
68. A. I. Kukhtenko, "On theory of composite systems with hierarchical control structure," in: Composite Control Systems [in Russian], Izd. Naukova Dumka, No. 1 (1966); No. 4 (1968); Proc. of Seminar on Composite Control Systems, Izd. In-ta Kibern. AN USSR, No. 1 (1968).
69. V. M. Matrosov, "Vector Lyapunov functions in the analysis of nonlinear interconnected systems," Symposia Matematica, Vol. VI, Nation. Inst. Alta Matem. Ital., Acad. Pres (1971).
70. V. M. Matrosov, "On differential equations and inequalities with discontinuous right sides," Differents. Uravn., 3, No. 3 (1967); 3, No. 5 (1967).
71. N. V. Azbelev and Z. B. Tsalyuk, "On integral and differential inequalities," Proc. of Fourth All-Union Mathematical Meeting, Vol. II, Nauka (1964).
72. R. Gabasov, "On the stability of the solutions of differential-operator equations under continuously operative perturbations," Izv. Vuzov, Matematika, No. 5(30) (1962).
73. I. M. Gel'fand and M. L. Tsetlin, On Mathematical Modeling of the Mechanisms of the Central Nervous System. Models of the Structural-Functional Organization of Some Biological Systems [in Russian], Nauka (1966).
74. A. M. Letov, "Analytic construction of control devices," Avtomat. i Telemekhan., Nos. 4, 5, 6 (1960); No. 4 (1961); No. 11 (1962); Dynamics of Flight and Control [in Russian], Nauka (1969).
75. J. I. Sanders, IEE Trans. Appl. and Ind., 83, No. 75 (1964).
76. D. D. Pirson, "Systems of multi-level control. Theory of self-adaptive control systems," Proceedings of the Second International Symposium IFAK on Self-Adaptive Systems, Nauka (1969).
77. Yu. L. Daletskii and M. G. Krein, Stability of the Solutions of Differential Equations in Banach Space [in Russian], Nauka (1970).

On the Stabilization of Decentralized Control Systems

SHIH-HO WANG AND EDWARD J. DAVISON

Abstract—This paper considers the problem of stabilizing a linear time-invariant multivariable system by using several local feedback control laws. Each local feedback control law depends only on partial system outputs. A necessary and sufficient condition for the existence of local control laws with dynamic compensation to stabilize a given system is derived. This condition is stated in terms of a new notion, called "fixed modes," which is a natural generalization of the well-known concept of uncontrollable modes and unobservable modes that occur in centralized control system problems. A procedure that constructs a set of stabilizing feedback control laws is given.

I. Introduction and Problem Statement

There is an increasing interest in the study of decentralized control systems [1]–[4]. When control theory is applied to solve problems of large systems, e.g., electric power systems, socioeconomic systems, etc, an important feature called decentralization often arises. Such systems have several local control stations; at each station, the controller observes only local system outputs and controls only local inputs. All the controllers are involved, however, in controlling the same large system. It is obviously important to know under what conditions there exists a set of appropriate local feedback control laws that will stabilize the complete system. Several different versions of this problem have been formulated and examined in [1]. In the present paper, this problem is treated intensively. Section II contains the main result (Theorem 1) of this paper—a necessary and sufficient condition for the existence of local output feedback control laws with dynamic compensation to stabilize a given system. This condition is stated in terms of a new notion, called "fixed modes" of a decentralized control system. In Section III, some preliminary results on the stabilization of a usual centralized control system via output feedback with dynamic compensation are given. It is also shown that the "fixed modes" introduced in Section II is a natural generalization of the well-known concept of uncontrollable modes and unobservable modes that occur in the usual centralized control system problems. Section IV contains the proof of Theorem 1.

The formal problem statement is as follows

Consider a linear time-invariant multivariable system with ν local control stations described by

$$\dot{x}(t) = Ax(t) + \sum_{i=1}^{\nu} B_i u_i(t) \tag{1a}$$

$$y_i(t) = C_i x(t), \qquad (i = 1,\cdots,\nu), \tag{1b}$$

where $x(t) \in \mathcal{R}^n$ is the state, $u_i(t) \in \mathcal{R}^{m_i}$ and $y_i(t) \in \mathcal{R}^{p_i}$ are the input and output, respectively, of the ith local control station $(i = 1,\cdots,\nu)$. The matrices A, B_i, and C_i, $(i = 1,\cdots,\nu)$, are real, constant, and of appropriate size. The problem is to find ν local output feedback control laws with dynamic compensation in order to stabilize the whole system. The set of local feedback laws are assumed to be generated by the following feedback controllers:

$$\dot{z}_i(t) = S_i z_i(t) + R_i y_i(t) \tag{2a}$$

$$u_i(t) = Q_i z_i(t) + K_i y_i(t) + v_i(t), \qquad i = 1,\cdots,\nu, \tag{2b}$$

where $z_i(t) \in \mathcal{R}^{\eta_i}$ is the state of the ith feedback controller, $v_i(t) \in \mathcal{R}^{m_i}$ is the ith local external input, and S_i, R_i, Q_i, and K_i are real constant matrices of appropriate size. Equation (2) can be written more compactly as follows:

$$\dot{z}(t) = Sz(t) + Ry(t) \tag{3a}$$

$$u(t) = Qz(t) + Ky(t) + v(t), \tag{3b}$$

where

$$\begin{aligned} S &\triangleq \text{block diag } [S_1,\cdots,S_\nu] \\ R &\triangleq \text{block diag } [R_1,\cdots,R_\nu] \\ Q &\triangleq \text{block diag } [Q_1,\cdots,Q_\nu] \\ K &\triangleq \text{block diag } [K_1,\cdots,K_\nu] \\ z^T(t) &\triangleq [z_1^T(t) \vdots \cdots \vdots z_\nu^T(t)] \\ y^T(t) &\triangleq [y_1^T(t) \vdots \cdots \vdots y_\nu^T(t)] \\ u^T(t) &\triangleq [u_1^T(t) \vdots \cdots \vdots u_\nu^T(t)] \end{aligned}$$

and

$$v^T(t) \triangleq [v_1^T(t) \vdots \cdots \vdots v_\nu^T(t)]. \tag{4}$$

When the feedback control law generated by (3) is applied to system (1), the closed-loop system is described by

$$\begin{bmatrix} \dot{x}(t) \\ \dot{z}(t) \end{bmatrix} = \begin{bmatrix} A + BKC & BQ \\ RC & S \end{bmatrix} \begin{bmatrix} x(t) \\ z(t) \end{bmatrix} + \begin{bmatrix} B \\ 0 \end{bmatrix} v(t), \tag{5}$$

where

$$B = [\underbrace{B_1}_{m_1} \vdots \cdots \vdots \underbrace{B_\nu}_{m_\nu}] \text{ and } C = \begin{bmatrix} C_1 \\ \cdot \\ \cdot \\ \cdot \\ C_\nu \end{bmatrix} \begin{matrix} \}p_1 \\ \\ \\ \\ \}p_\nu \end{matrix}. \tag{6}$$

Manuscript received December 4, 1972. Paper recommended by I. B. Rhodes, Chairman of the IEEE S-CS Technical Committee. This work was supported by the National Research Council of Canada under Grant A 4396.

S. H. Wang was with the Department of Electrical Engineering, University of Toronto, Toronto, Ont., Canada. He is now with the National Aeronautics and Space Administration, Langley Research Center, Hampton, Va. 23365.

E. J. Davison is with the Department of Electrical Engineering, University of Toronto, Toronto, Ont., Canada.

Reprinted from *IEEE Trans. Automat. Contr.*, vol. AC-18, pp. 473–478, Oct. 1973.

The set of local feedback control laws of (3) is to be chosen so that the overall system (5) is asymptotically stable, (i.e., all poles of system (5) are in the open left-half complex plane $\mathcal{C}^-$). This then is the problem of stabilizing a decentralized control system via local output feedback with dynamic compensation.

II. Main Result

The following definitions are required to state the main theorem of this paper.

Definition 1

Consider the triple $(C,A,B) \in \mathcal{R}^{p\times n} \times \mathcal{R}^{n\times n} \times \mathcal{R}^{n\times m}$ and the two sets of integers $m_1,\cdots,m_\nu$ and $p_1,\cdots,p_\nu$, with $m = \sum_{i=1}^{\nu} m_i$ and $p = \sum_{i=1}^{\nu} p_i$, which specifies system (1). Let K be the set of block diagonal matrices as follows:

$$\mathrm{K} = \{K | K = \text{block diag } [K_1,\cdots,K_\nu], \quad K_i \in \mathcal{R}^{m_i\times p_i}, \quad i = 1,\cdots,\nu\}. \tag{7}$$

Then the greatest common divisor of the set of characteristic polynomials of $(A + BKC)$, for all $K \in \mathrm{K}$, is called the *fixed polynomial* of the triple (C,A,B) with respect to K, denoted by $\psi(\lambda; C,A,B,\mathrm{K})$, i.e.,

$$\psi(\lambda; C,A,B,\mathrm{K}) = \gcd_{K\in \mathrm{K}}\{\det(\lambda I - A - BKC)\}.$$

Comment: The map $\psi(\lambda; C,A,B,\cdot):\mathrm{K} \mapsto \gcd_{K\in\mathrm{K}}\{\det(\lambda I - A - BKC)\}$ is well defined for any subset $\mathrm{K} \subset \mathcal{R}^{m\times p}$ and not just for the set K of block diagonal matrices in (7).

Definition 2

For a given triple $(C,A,B) \in \mathcal{R}^{p\times n} \times \mathcal{R}^{n\times n} \times \mathcal{R}^{n\times m}$ and the given set K of block diagonal matrices in (7), the set of *fixed modes* of (C,A,B) with respect to K is defined as follows:

$$\Lambda(C,A,B,\mathrm{K}) = \bigcap_{K\in\mathrm{K}} \sigma[A + BKC],$$

where $\sigma[A + BKC]$ denotes the set of eigenvalues of $(A + BKC)$. Equivalently, $\Lambda(C,A,B,\mathrm{K})$ can be defined as follows:

$$\Lambda(C,A,B,\mathrm{K}) = \{\lambda | \lambda \in \mathcal{C} \text{ and } \psi(\lambda; C,A,B,\mathrm{K}) = 0\}.$$

Remark 1: The set of fixed modes of a given triple (C,A,B) with respect to K can be computed as follows. Let $K \triangleq$ block diag $[K_1,\cdots,K_\nu] \in \mathrm{K}$, and let $k(i,r,s)$ denote the (r,s)th element of K_i, $(i = 1,\cdots,\nu)$, $(r = 1,\cdots,m_i)$, $(s = 1,\cdots,p_i)$. Notice that K contains the null matrix, hence $\Lambda(C,A,B,\mathrm{K}) \subseteq \sigma[A] = \{\lambda_1,\cdots,\lambda_n\}$. For each $\lambda_i \in \sigma[A]$, $\det(\lambda_i I - A - BKC)$ is a polynomial in $k(i,r,s)$, $(i = 1,\cdots,\nu)$, $(r = 1,\cdots,m_i)$, $(s = 1,\cdots,p_i)$. If $\det(\lambda_i I - A - BKC)$ is identically zero, then clearly $\lambda_i \in \Lambda(C,A,B,\mathrm{K})$. On the other hand, if $\det(\lambda_i I - A - BKC)$ is not identically zero, then there exists $\hat{K} \in \mathrm{K}$, such that $\det(\lambda_i I - A - B\hat{K}C) \neq 0$, hence $\lambda_i \notin \Lambda(C,A,B,\mathrm{K})$. This procedure determines the set of fixed modes $\Lambda(C,A,B,\mathrm{K})$. The computation of the fixed polynomial $\psi(\lambda; C,A,B,\mathrm{K})$ can be performed in a similar fashion.

The following theorem is the main result of this paper.

Theorem 1

Consider the given system (C,A,B) of (1), with B and C defined in (6). Let K be the set of block diagonal matrices defined in (7). Then a necessary and sufficient condition for the existence of a set of local feedback control laws (3) such that the closed-loop system (5) is asymptotically stable is that

$$\Lambda(C,A,B,\mathrm{K}) \subset \mathcal{C}^-,$$

where $\mathcal{C}^-$ denotes the open left-half complex plane.

III. Preliminary Developments

In this section, some preliminary results on the stabilization of usual centralized linear control systems with dynamic compensation are given. These results are required in the proof of Theorem 1.

Lemma 1 (Kalman Canonical Structure Theorem [5])

Consider a linear time-invariant multivariable system specified by

$$\dot{x}(t) = Ax(t) + Bu(t) \tag{8a}$$

$$y(t) = Cx(t), \tag{8b}$$

where $x(t) \in \mathcal{R}^n$ is the state, $u(t) \in \mathcal{R}^m$ is the input, $y(t) \in \mathcal{R}^p$ is the output, and A,B,C are real constant matrices of appropriate size. Then there exists an $n \times n$ real nonsingular matrix T such that

$$\tilde{A} \triangleq TAT^{-1} = \begin{bmatrix} \tilde{A}_{11} & 0 & \tilde{A}_{13} & 0 \\ \tilde{A}_{21} & \tilde{A}_{22} & \tilde{A}_{23} & \tilde{A}_{24} \\ 0 & 0 & \tilde{A}_{33} & 0 \\ 0 & 0 & \tilde{A}_{43} & \tilde{A}_{44} \end{bmatrix} \qquad \tilde{B} \triangleq TB = \begin{bmatrix} \tilde{B}_1 \\ \tilde{B}_2 \\ 0 \\ 0 \end{bmatrix} \tag{9}$$

$$\tilde{C} \triangleq CT^{-1} = [\tilde{C}_1 \; 0 \; \tilde{C}_3 \; 0],$$

where

$$\left\{\begin{bmatrix} \tilde{A}_{11} & 0 \\ \tilde{A}_{21} & \tilde{A}_{22} \end{bmatrix}, \begin{bmatrix} \tilde{B}_1 \\ \tilde{B}_2 \end{bmatrix}\right\}$$

is a completely controllable pair and

$$\left\{[\tilde{C}_1 \; \tilde{C}_3], \begin{bmatrix} \tilde{A}_{11} & \tilde{A}_{13} \\ 0 & \tilde{A}_{33} \end{bmatrix}\right\}$$

is a completely observable pair. The set of eigenvalues of $\tilde{A}_{11}$ is said to be the *controllable and observable modes* of (C,A,B). Similarly, the union of the eigenvalues of $\tilde{A}_{22}$, $\tilde{A}_{33}$, and $\tilde{A}_{44}$ (i.e., $\sigma[\tilde{A}_{22}] \cup \sigma[\tilde{A}_{33}] \cup \sigma[\tilde{A}_{44}]$) is said to be the set of *uncontrollable modes and unobservable modes* of (C,A,B). Furthermore, $\tilde{C}_1(\lambda I - \tilde{A}_{11})^{-1}\tilde{B}_1 = C(\lambda I - A)^{-1}B$.

For a proof and some related discussions of Lemma 1, see [5], [6]. The following lemma relates the uncontrollable modes and unobservable modes of a triple (C,A,B) to its fixed modes as defined in Section II.

Lemma 2

Consider a given triple $(C,A,B) \in \mathcal{R}^{p\times n} \times \mathcal{R}^{n\times n} \times \mathcal{R}^{n\times m}$ and its Kalman canonical structure form $(\tilde{C},\tilde{A},\tilde{B})$ in (9), then

$$\psi(\lambda;\, C,A,B,\mathcal{R}^{m\times p}) = \prod_{i=2}^{4} \det(\lambda I - \tilde{A}_{ii}). \tag{10}$$

The above lemma can be easily proved by using [7, theorem 2] (see also [8, theorem 2]); the details are omitted.

Remark 2: Lemma 2 clearly implies that the notion of fixed modes, i.e., $\Lambda(C,A,B,\mathcal{R}^{m\times p})$, in the usual centralized control case reduces to the well-known concept of the uncontrollable modes and unobservable modes of a triple (C,A,B).

The following result is a simple consequence of a result due to Brasch and Pearson (see [9, theorem 2]); its proof is omitted here.

Lemma 3

Consider a given triple $(C,A,B) \in \mathcal{R}^{p\times n} \times \mathcal{R}^{n\times n} \times \mathcal{R}^{n\times m}$, which is not necessarily completely controllable or completely observable. Assume that the triple is transformed into its Kalman canonical structure form $(\tilde{C},\tilde{A},\tilde{B})$ in (9), and suppose the submatrix $\tilde{A}_{11}$(in $\tilde{A}$) is of size $n_1 \times n_1$, $(0 \leq n_1 \leq n)$. Then there exists an integer $\tilde{n}$, $(0 \leq \tilde{n} \leq n_1)$, such that for any polynomial $\phi(\lambda) = \lambda^{\tilde{n}+n_1} + a_{\tilde{n}+n_1-1}\lambda^{\tilde{n}+n_1-1} + \cdots + a_0$ of degree $\tilde{n} + n_1$ with real coefficients, there exists a real constant matrix K^e of size $(m + \tilde{n}) \times (p + \tilde{n})$, such that

$$\det(\lambda I - A^e - B^e K^e C^e) = \phi(\lambda)\cdot \prod_{i=2}^{4} \det(\lambda I - \tilde{A}_{ii}), \tag{11}$$

where A^e, B^e, and C^e are shown by the following:

$$A^e = \underbrace{\left[\begin{array}{c|c} A & 0 \\ \hline 0 & 0 \end{array}\right]}_{n\ \tilde{n}} \begin{array}{l}\}n \\ \}\tilde{n}\end{array} \qquad B^e = \underbrace{\left[\begin{array}{c|c} B & 0 \\ \hline 0 & I \end{array}\right]}_{m\ \tilde{n}} \begin{array}{l}\}n \\ \}\tilde{n}\end{array} \qquad C^e = \underbrace{\left[\begin{array}{c|c} C & 0 \\ \hline 0 & I \end{array}\right]}_{n\ \tilde{n}} \begin{array}{l}\}p \\ \}\tilde{n}\end{array}. \tag{12}$$

Remark 3: The augmented triple (C^e,A^e,B^e) of (12) represents the given system (C,A,B) plus $\tilde{n}$ integrators, see [9] and [10]. Lemma 3 states that with a suitable number of integrators, the controllable and observable modes of (C,A,B) can be arbitrarily assigned via output feedback. It can also be shown that the remaining uncontrollable modes and unobservable modes of (C,A,B) are invariant under output feedback even with dynamic compensation.

IV. Proof of the Main Result

In this section, the main result of the paper (Theorem 1) will be proved. For its proof, it is more convenient to restate the stabilization problem of decentralized control systems described in Section I in the following equivalent form.

Consider the given system $(C,A,B) \in \mathcal{R}^{p\times n} \times \mathcal{R}^{n\times n} \times \mathcal{R}^{n\times m}$ of (1), where B,C are shown in (6). The problem of stabilizing the system (1) via local output feedback with dynamic compensation is to find a set of integers $\eta_1 \geq 0,\cdots,\eta_\nu \geq 0$ and a real constant matrix K^e, of size $(m + \eta) \times (p + \eta)$ where $\eta \triangleq \sum_{i=1}^{\nu} \eta_i$, with the special structure given in (14), such that all zeros of the polynomial

$$\det(\lambda I - A^e - B^e K^e C^e)$$

are in the open left-half complex plane, where

$$A^e = \underbrace{\left[\begin{array}{c|c} A & 0 \\ \hline 0 & 0 \end{array}\right]}_{n\ \eta} \begin{array}{l}\}n \\ \}\eta\end{array} \qquad B^e = \underbrace{\left[\begin{array}{c|c} B & 0 \\ \hline 0 & I \end{array}\right]}_{m\ \eta} \begin{array}{l}\}n \\ \}\eta\end{array} \qquad C^e = \underbrace{\left[\begin{array}{c|c} C & 0 \\ \hline 0 & I \end{array}\right]}_{n\ \eta} \begin{array}{l}\}p \\ \}\eta\end{array},$$

$$\eta = \textstyle\sum_{i=1}^{\nu} \eta_i, \tag{13}$$

and

$$K^e = \left[\begin{array}{cccc|cccc} K_1 & & & 0 & Q_1 & & & 0 \\ & K_2 & & & & Q_2 & & \\ & & \ddots & & & & \ddots & \\ 0 & & & K_\nu & 0 & & & Q_\nu \\ \hline R_1 & & & 0 & S_1 & & & 0 \\ & R_2 & & & & S_2 & & \\ & & \ddots & & & & \ddots & \\ 0 & & & R_\nu & 0 & & & S_\nu \end{array}\right] \begin{array}{l} \}m_1 \\ \}m_2 \\ \vdots \\ \}m_\nu \\ \}\eta_1 \\ \}\eta_2 \\ \vdots \\ \}\eta_\nu \end{array} \tag{14}$$

$$\quad p_1 \quad p_2 \ \cdots\ p_\nu \qquad \eta_1\ \eta_2\ \cdots\ \eta_\nu$$

where K_i, Q_i, R_i, and S_i, $(i = 1,\cdots,\nu)$, are any real constant matrices of appropriate size.

It can be easily verified that the above problem is solvable iff the stabilization problem of Section I is solvable.

In order to prove the necessity part of Theorem 1, it is enough to prove the following proposition.

Proposition 1

Consider a given triple $(C,A,B) \in \mathcal{R}^{p\times n} \times \mathcal{R}^{n\times n} \times \mathcal{R}^{n\times m}$, where B and C are partitioned in (6), and the set K of block diagonal matrices defined in (7). Then for any given set of integers $\eta_1 \geq 0,\cdots,\eta_\nu \geq 0$,

$$\psi(\lambda;\, C,A,B,\mathrm{K}) = \psi(\lambda;\, C^e,A^e,B^e,\mathrm{K}^e)$$

where C^e, A^e, and B^e are shown in (13), and K^e is the set of $(m + \eta) \times (p + \eta)$ real constant matrices shown in (14), with $\eta = \sum_{i=1}^{\nu} \eta_i$.

Proof: First, consider the special case where $\eta_1 = 1$ and $\eta_2 = \cdots = \eta_\nu = 0$. Then the general case can be easily proved by inductive arguments.

In the case $\eta_1 = 1, \eta_2 = \cdots = \eta_\nu = 0$, K^e has the following form:

$$K^e = \left[\begin{array}{cc:c} \multicolumn{2}{c:}{K} & q_1 \\ & & 0 \\ \hdashline r_1 & 0 & s_1 \end{array}\right] \begin{array}{l} \}m_1 \\ \}m_2 + \cdots + m_\nu, \\ \}1 \end{array}$$
$$\qquad\; p_1 \;\; p_2 + \cdots + p_\nu \;\; 1$$

with $K \in \mathrm{K}$.

From the definition of ψ,

$$\begin{aligned} &\psi(\lambda;\, C^e,A^e,B^e,\mathrm{K}^e) \\ &= \gcd_{K^e \in \mathrm{K}^e} \{\det(\lambda I - A^e - B^e K^e C^e)\} \\ &= \gcd_{K \in \mathrm{K}} \left\{ \begin{array}{l} \gcd \\ q_1 \in \mathcal{R}^{m_1} \\ r_1{}' \in \mathcal{R}^{p_1} \\ s_1 \in \mathcal{R} \end{array} \left(\det \left[\begin{array}{c:c} \lambda I - A - BKC & -B_1 q_1 \\ \hdashline -r_1 C_1 & \lambda - s_1 \end{array} \right] \right) \right\} \\ &= \gcd_{K \in \mathrm{K}} \left\{ \begin{array}{l} \gcd[(\lambda - s_1) - r_1 C_1 (\lambda I - A - BKC)^{-1} B_1 q_1] \\ \qquad \cdot \det(\lambda I - A - BKC) \\ q_1 \in \mathcal{R}^{m_1},\, r_1{}' \in \mathcal{R}^{p_1},\, s_1 \in \mathcal{R} \end{array} \right\}. \end{aligned} \tag{15}$$

In the last step, the following formula is used. Let F be a square matrix partitioned into four blocks:

$$F = \left[\begin{array}{c:c} A & B \\ \hdashline C & D \end{array}\right] \text{ and } \det A \neq 0,$$

then $\det F = \det(D - CA^{-1}B) \det A$, see [11].

For each fixed $K \in \mathrm{K}$, consider the triple $(C_1, A + BKC, B_1)$. From Lemma 1, there exists an $n \times n$ real nonsingular matrix T that transforms $(C_1, A + BKC, B_1)$ into its Kalman canonical structure form as follows:

$$T(A + BKC)T^{-1} = \left[\begin{array}{c:c:c:c} \tilde{A}_{11} & 0 & \tilde{A}_{13} & 0 \\ \hdashline \tilde{A}_{21} & \tilde{A}_{22} & \tilde{A}_{23} & \tilde{A}_{24} \\ \hdashline 0 & 0 & \tilde{A}_{33} & 0 \\ \hdashline 0 & 0 & \tilde{A}_{43} & \tilde{A}_{44} \end{array}\right], \qquad TB_1 = \left[\begin{array}{c} \tilde{B}_1 \\ \hdashline \tilde{B}_2 \\ \hdashline 0 \\ \hdashline 0 \end{array}\right]$$

$$C_1 T^{-1} = [\tilde{C}_1 \,\vdots\, 0 \,\vdots\, \tilde{C}_3 \,\vdots\, 0].$$

Also from Lemma 1, $C_1(sI - A - BKC)^{-1}B_1 = \tilde{C}_1(sI - \tilde{A}_{11})^{-1}\tilde{B}_1$, and $\det(\lambda I - A - BKC) = \prod_{i=1}^{4} (\lambda I - \tilde{A}_{ii})$. Then the following identity is easily derived.

$$\begin{aligned} &\{(\lambda - s_1) - r_1 C_1 (\lambda I - A - BKC)^{-1} B_1 q_1\} \\ &\quad \cdot \det(\lambda I - A - BKC) = \{(\lambda - s_1) \det(\lambda I - \tilde{A}_{11}) \\ &\quad - r_1 \tilde{C}_1 [\operatorname{adj}(\lambda I - \tilde{A}_{11})] \tilde{B}_1 q_1\} \times \prod_{i=2}^{4} (\lambda I - \tilde{A}_{ii}). \end{aligned} \tag{16}$$

On the other hand, for any $K \in \mathrm{K}$,

$$\begin{aligned} \psi(\lambda;\, C,A,B,\mathrm{K}) &= \psi(\lambda;\, C, A + BKC, B, \mathrm{K}) \\ &\triangleq \gcd_{\hat{K} \in \mathrm{K}} \{\det[\lambda I - (A + BKC) - B\hat{K}C]\} \\ &= \gcd_{\hat{K}_i \in \mathcal{R}^{m_i \times p_i}} \left\{ \det\left[\lambda I - (A + BKC) - \sum_{i=1}^{\nu} B_i \hat{K}_i C_i \right] \right\}, \end{aligned}$$

which divides

$$\begin{aligned} &\gcd_{\substack{\hat{K}_1 \in \mathcal{R}^{m_1 \times p_1} \\ \hat{K}_i = 0, \quad (i = 2, \cdots, \nu).}} \left\{ \det\left[[\lambda I - (A + BKC) - \sum_{i=1}^{\nu} B_i \hat{K}_i C_i \right] \right\} \\ &= \gcd_{\hat{K}_1 \in \mathcal{R}^{m_1 \times p_1}} \{\det[\lambda I - (A + BKC) - B_1 \hat{K}_1 C_1]\} \\ &\triangleq \psi(\lambda;\, C_1, A + BKC, B_1, \mathcal{R}^{m_1 \times p_1}) \\ &= \prod_{i=2}^{4} \det(\lambda I - \tilde{A}_{ii}). \end{aligned}$$

The last step follows from Lemma 2. Hence for any $K \in \mathrm{K}$, $\psi(\lambda;\, C,A,B,\mathrm{K})$ divides $\prod_{i=2}^{4} \det(\lambda I - \tilde{A}_{ii})$. From (15) and (16), for any $K \in \mathrm{K}$, $\prod_{i=2}^{4} \det(\lambda I - \tilde{A}_{ii})$ divides

$$\left\{ \begin{array}{l} \gcd\limits_{q_1 \in \mathcal{R}^{m_1},\; r_1{}' \in \mathcal{R}^{p_1},\; s_1 \in \mathcal{R}} [(\lambda - s_1) - r_1 C_1 (\lambda I - A \\ \qquad - BKC)^{-1} B_1 q_1] \times \det(\lambda I - A - BKC) \end{array} \right\}.$$

Then from (15), it is clear that $\psi(\lambda;\, C,A,B,\mathrm{K})$ divides $\psi(\lambda;\, C^e,A^e,B^e,\mathrm{K}^e)$.

It remains to show that $\psi(\lambda;\, C,A,B,\mathrm{K})$ is a multiple of $\psi(\lambda;\, C^e,A^e,B^e,\mathrm{K}^e)$. Consider the following set of matrices

$$\tilde{\mathrm{K}}^e \triangleq \left\{ \left[\begin{array}{c:c} K & 0 \\ \hdashline 0 & s \end{array}\right] \middle| \quad K \in \mathrm{K}, \quad s \in \mathcal{R} \right\}.$$

It is clear that $\psi(\lambda;\, C^e,A^e,B^e,\tilde{\mathrm{K}}^e)$ is a multiple of $\psi(\lambda;\, C^e,A^e,B^e,\mathrm{K}^e)$, since $\tilde{\mathrm{K}}^e$ is a subset of K^e. Moreover,

$$\begin{aligned} &\psi(\lambda;\, C^e,A^e,B^e,\tilde{\mathrm{K}}^e) \\ &= \gcd_{\substack{K \in \mathrm{K} \\ s \in \mathcal{R}}} \{\det(\lambda I - A - BKC) \times (\lambda - s)\} \\ &= \psi(\lambda;\, C,A,B,\mathrm{K}). \end{aligned}$$

Hence $\psi(\lambda;\, C,A,B,\mathrm{K})$ is a multiple of $\psi(\lambda;\, C^e,A^e,B^e,\mathrm{K}^e)$. So far it has been proved that

$$\psi(\lambda;\, C,A,B,\mathrm{K}) = \psi(\lambda;\, C^e,A^e,B^e,\mathrm{K}^e)$$

for the special case $\eta_1 = 1$ and $\eta_2 = \cdots = \eta_\nu = 0$. The general case is proved by the repeated application of the above result. Q.E.D.

Proof of the Necessity Part of Theorem 1

Suppose that the system (1) can be stabilized by using local output feedback with dynamic compensation (as according to the new problem formulation given at the beginning of this section), then there exists a matrix K^e such that the polynomial $\det(\lambda I - A^e - B^e K^e C^e)$ has only stable zeros, where A^e, B^e, C^e, and K^e are shown in (13) and (14). By definition, $\psi(\lambda;\, C^e,A^e,B^e,\mathrm{K}^e)$ is a divisor of $\det(\lambda I - A^e - B^e K^e C^e)$. From Proposition 1, $\psi(\lambda;\, C,A,B,\mathrm{K}) = \psi(\lambda;\, C^e,A^e,B^e,\mathrm{K}^e)$. Hence $\psi(\lambda;\, C,A,B,\mathrm{K})$ has only stable zeros, i.e., $\Lambda(C,A,B,\mathrm{K}) \subset \mathcal{C}^-$. This completes the proof of the necessity part of Theorem 1.

In order to prove the sufficiency part of Theorem 1, the following lemmas are required.

Lemma 4

Consider a given triple $(C,A,B) \in \mathbb{R}^{p\times n} \times \mathbb{R}^{n\times n} \times \mathbb{R}^{n\times m}$ and the set K of block diagonal matrices in (7). For each $K \in$ K, the characteristic polynomial of $(A + BKC)$ can be factored as

$$\det(\lambda I - A - BKC) = \psi(\lambda; C,A,B,\mathrm{K})\phi(\lambda; K), \quad (17)$$

where $\psi(\lambda; C,A,B,\mathrm{K})$ is the fixed polynomial of (C,A,B) with respect to K, and $\psi(\lambda; C,A,B,\mathrm{K})$ is independent of K. Write $\phi(\lambda; 0)$ as follows:

$$\phi(\lambda; 0) = (\lambda - \lambda_1)^{\mu_1}(\lambda - \lambda_2)^{\mu_2} \cdots (\lambda - \lambda_k)^{\mu_k}, \quad (18)$$

where λ_j, $(j = 1,\cdots,k)$, are distinct zeros of $\phi(\lambda; 0)$ with multiplicity μ_j. Among the set of zeros λ_j, $(j = 1,\cdots,k)$, assume that $\mathrm{Re}\lambda_j \geq 0$, for all $j \in \{\alpha,\beta,\cdots,\delta\}$, and the remaining zeros have negative real parts. Under these conditions the following hold.

Part I: There exists $\epsilon > 0$, such that $B(\lambda_j; \epsilon) \triangleq \{\lambda | \lambda \in \mathbb{C}, |\lambda - \lambda_j| < \epsilon\} \subset \mathbb{C}^-$, for all $j \in \{1,\cdots,k\} \setminus \{\alpha,\beta,\cdots,\delta\}$.

Part II: For all $\epsilon > 0$, there exists $\gamma > 0$ such that for all $K \in$ K with $\|K\| < \gamma$, there are exactly μ_j zeros of $\phi(\lambda; K)$ in $B(\lambda_j; \epsilon)$, for all $j \in \{1,\cdots,k\}$.

Part III: For all $\gamma > 0$, there exists $K \in$ K with $\|K\| < \gamma$, such that $\phi(\lambda_j,K) \neq 0$, for all $j \in \{1,\cdots,k\}$.

Proofs:

Part I: This part simply follows from the fact that the open left-half complex plane $\mathbb{C}^-$ is an open set.

Part II: Write $\phi(\lambda; K)$ in the following form:

$$\phi(\lambda; K) = \lambda^{\mu} + a_{\mu-1}(K)\lambda^{\mu-1} + \cdots + a_0(K), \quad (19)$$

where $a_\tau(K)$, $(\tau = 0,\cdots,\mu - 1)$, are polynomials in $k(i,r,s)$, and $k(i,r,s)$ are elements of K (see Remark 1), and $\mu = \sum_{j=1}^{k} \mu_j$. From Rouché's theorem (see [12, p. 218]), if f and g are two analytic functions defined on the complex plane $\mathbb{C}$, and

$$|f(\lambda) - g(\lambda)| < |f(\lambda)|, \qquad \text{if } |\lambda - \lambda^*| = \epsilon',$$

then f and g have the same number of zeros in $B(\lambda^*; \epsilon')$ (if they are counted according to their multiplicities). Since $\phi(\lambda; 0)$ is a polynomial in λ and has only a finite number of zeros, then for any $\epsilon > 0$ and any complex number λ^*, there exist ϵ', $(0 < \epsilon' \leq \epsilon)$, and $\eta > 0$, such that $|\phi(\lambda; 0)| \geq \eta$ if $|\lambda - \lambda^*| = \epsilon'$. From (19),

$$|\phi(\lambda; 0) - \phi(\lambda; K)| \leq \sum_{\tau=0}^{\mu-1} |a_\tau(0) - a_\tau(K)| \cdot |\lambda|^\tau \leq \sum_{\tau=0}^{\mu-1} |a_\tau(0) - a_\tau(K)| \cdot (|\lambda^*| + \epsilon')^\tau. \quad (20)$$

Since $a_\tau(K)$, $(\tau = 0,\cdots,\mu - 1)$, are continuous functions in $k(i,r,s)$ (the elements of K), there exists $\gamma > 0$, such that if

$$\|K\| \triangleq \max_i \left(\max_r \sum_{s=1}^{p_i} |k(i,r,s)| \right) < \gamma,$$

then

$$|a_\tau(0) - a_\tau(K)| < \frac{\eta}{\mu \times (|\lambda^*| + \epsilon')^\tau}. \quad (21)$$

From (20), (21), if $\|K\| < \gamma$, then

$$|\phi(\lambda; 0) - \phi(\lambda; K)| < \eta \leq |\phi(\lambda; 0)|,$$

if $|\lambda - \lambda^*| = \epsilon'$. Hence from Rouché's theorem, $\phi(\lambda; 0)$ and $\phi(\lambda; K)$ have the same number of zeros in $B(\lambda^*; \epsilon') \subset B(\lambda^*; \epsilon)$. When this argument is applied to $\lambda^* = \lambda_j$, $(j = 1,\cdots,k)$, Part II of the lemma is proved.

Part III: In order to prove Part III, first define the following set:

$$\mathcal{L}_j \triangleq \{K | K \in \mathrm{K} \text{ and } \phi(\lambda_j; K) = 0\}, \qquad j = 1,\cdots,k. \quad (22)$$

From (17), (18), and the definition of $\psi(\lambda; C,A,B,\mathrm{K})$,

$$\gcd_{K\in\mathrm{K}}\{\phi(\lambda; K)\} = 1.$$

Hence for any $\lambda_j \in \mathbb{C}$, there exists $\hat{K} \in$ K, such that

$$\phi(\lambda_j; \hat{K}) = \lambda_j^{\mu} + a_{\mu-1}(\hat{K})\lambda_j^{\mu-1} + \cdots + a_0(\hat{K}) \neq 0. \quad (23)$$

From (18) and (23), $\phi(\lambda_j; K)$ is a nonconstant polynomial in $k(i,r,s)$ (i.e., the elements of K), so the set $\mathcal{L}_j$ of (22) is a hypersurface in the parameter space of K with dimension $\sum_{i=1}^{\nu} m_i \times p_i$ (see [8] and [13]). In any nonempty open set of the parameter space of K, there exists $K \notin \bigcup_j \mathcal{L}_j$. This completes the proof of Part III of this lemma.

Q.E.D.

Proof of the Sufficiency Part of Theorem 1

Suppose $\Lambda(C,A,B,\mathrm{K}) \subset \mathbb{C}^-$, then $\psi(\lambda; C,A,B,\mathrm{K})$ has only stable zeros. Write

$$\det(\lambda I - A - BKC) = \psi(\lambda; C,A,B,\mathrm{K})\phi(\lambda; K), \qquad K \in \mathrm{K} \quad (17')$$

and

$$\phi(\lambda; 0) = (\lambda - \lambda_1)^{\mu_1}(\lambda - \lambda_2)^{\mu_2} \cdots (\lambda - \lambda_k)^{\mu_k}. \quad (18')$$

The set of unstable zeros of $\phi(\lambda; 0)$ is $\{\lambda_\alpha,\lambda_\beta,\cdots,\lambda_\delta\}$. From Lemma 4, there exist $\gamma > 0$ and $\hat{K} \triangleq$ block diag $[\hat{K}_1,\cdots,\hat{K}_\nu] \in$ K with $\|\hat{K}\| < \gamma$ such that the following statements hold.

Statement I: $\phi(\lambda_j; \hat{K}) \neq 0$, for all $j \in \{1,\cdots,k\}$.

Statement II: For all $K \in$ K with $\|K\| < \gamma$, the number of unstable zeros of $\phi(\lambda; K)$ is no greater than the number of unstable zeros of $\phi(\lambda; 0)$. This implies that the number of unstable eigenvalues of $(A + BKC)$ is no greater than the number of unstable eigenvalues of A (if they are counted according to their multiplicities).

From the above matrix $\hat{K} \triangleq$ block diag$[\hat{K}_1,\cdots,\hat{K}_\nu]$, define the following set of matrices:

$$\hat{K}(i) \triangleq \text{block diag}[0,\cdots,0,\hat{K}_i,\cdots,\hat{K}_\nu] \in \mathrm{K}, \qquad i = 1,\cdots,\nu$$

and

$$\hat{K}(\nu + 1) = 0 \in \mathbb{R}^{m\times p}.$$

Let $\pi \in \{1,\cdots,\nu\}$ be an integer with the following property:

$$\phi(\lambda_j,\hat{K}(\pi)) \neq 0, \qquad \text{for all } j \in \{\alpha,\beta,\cdots,\delta\}$$

and

$$\phi(\lambda_j,\hat{K}(\pi + 1)) = 0, \qquad \text{for some } j \in \{\alpha,\beta,\cdots,\delta\}. \quad (24)$$

The existence of π follows from the fact that $\phi(\lambda_j;\hat{K}(1)) = \phi(\lambda_j;\hat{K}) \neq 0$ and $\phi(\lambda_j;\hat{K}(\nu+1)) = \phi(\lambda_j;0) = 0$ for all $j \in \{\alpha,\beta,\cdots,\delta\}$.

Let $\hat{A}$ denote $A + B\hat{K}(\pi+1)C$. Then (24) is equivalent to saying that $\hat{A}$ has an unstable eigenvalue λ_j, and that there exists $\hat{K}_\pi \in \mathbb{R}^{m_\pi \times p_\pi}$ such that $\lambda_j \notin \sigma[\hat{A} + B_\pi \hat{K}_\pi C_\pi]$. Hence from Lemma 2, λ_j is a controllable and observable mode of the triple $(C_\pi,\hat{A},B_\pi)$. Then from Lemma 3 (see also Remark 3), the controllable and observable mode λ_j can be stabilized via output feedback with dynamic compensation (at the πth control station). Hence the number of unstable eigenvalues of $\hat{A}$ can be reduced at least by one via the above local feedback.

Also notice that the number of unstable eigenvalues of $\hat{A} \triangleq A + B\hat{K}(\pi+1)C$ is no greater than the number of unstable eigenvalues of A. This follows from the fact that $\|\hat{K}(\pi+1)\| \leq \|\hat{K}\| < \gamma$ [see Statement II following (18′)]. There is only a finite number of unstable eigenvalues $\{\lambda_\alpha,\lambda_\beta,\cdots,\lambda_\delta\}$ of the given matrix A. By repeated application of the above procedures, the given system (C,A,B) can therefore be stabilized by using local feedback with dynamic compensation. Q.E.D.

By using the same arguments as in the proof of Theorem 1, one can easily obtain the following more general result.

Corollary 1

Under the same assumptions as in Theorem 1, the necessary and sufficient condition for the existence of a set of local feedback control laws in (3) such that all poles of the closed-loop system of (5) are in S is that

$$\Lambda(C,A,B,\mathrm{K}) \subset S,$$

where S is any nonempty symmetric open subset of the complex plane $\mathbb{C}$ (i.e., if $\lambda \in S$ then its complex conjugate $\lambda^* \in S$).

V. Conclusions

The problem of stabilizing a decentralized linear time-invariant multivariable system via local output feedback with dynamic compensation is investigated in this paper. A necessary and sufficient condition for the existence of a solution is given in Theorem 1. This condition is stated in terms of a new concept called "fixed modes." In general, the construction of the local dynamic controllers given in the proof of Theorem 1 requires more integrators than necessary. The problem of constructing a set of minimal-order local dynamic controllers to stabilize a given system is still open. Some related topics can be found in [14].

It should be remarked that other classes of feedback control configurations, which would correspond to a broader interpretation of the term decentralized control, may be imposed in the problem formulation; for example, if we assume that some local control stations cannot observe their outputs in the usual decentralized sense, then this can be trivially considered in the problem formulation of Theorem 1, by letting K be the class of matrices with some zero blocks in the diagonal. Another example is the case where the given system consists of an ordered set of subsystems in which all output information preceding any given subsystem is available to that subsystem. This constraint can be dealt with by letting K in Theorem 1 be the class of matrices in block triangular form. In any of these cases, Theorem 1 is still valid.

Acknowledgment

The authors are grateful to Professor W. M. Wonham and Dr. J. Pauksens for their many stimulating discussions.

References

[1] M. Aoki, "On feedback stabilizability of decentralized dynamic systems," *Automatica*, vol. 8, pp. 163–173, Mar. 1972.
[2] Y. C. Ho and K. C. Chu, "Team decision theory and information structures in optimal control problems Parts I," *IEEE Trans. Automat. Contr.*, vol. AC-17, pp. 15–21, Feb. 1972.
[3] R. Lau, R. M. Persiano, and P. Varaiya, "Decentralized information and control: A network flow example," *IEEE Trans. Automat. Contr.*, vol. AC-17, pp. 466–473, Aug. 1972.
[4] P. Varaiya, "Trends in the theory of decision-making in large systems," *Ann. Economic and Social Measurement*, vol. 1/4, pp. 493–500, 1972.
[5] R. E. Kalman, "Canonical structure of linear dynamical systems," *Proc. Nat. Acad. Sci., U.S.A.*, vol. 48, pp. 596–600, 1962.
[6] C. A. Desoer, *Notes for A Second Course on Linear Systems.* New York: Van Nostrand Reinhold, 1970, pp. 189–192.
[7] S. H. Wang and E. J. Davison, "A note on the simple modes of multivariable systems," *IEEE Trans. Automat. Contr.* (Tech. Notes and Corresp.), vol. AC-17, p. 548, Aug. 1972.
[8] E. J. Davison and S. H. Wang, "Properties of linear time-invariant multivariable systems subject to arbitrary output and state feedback," *IEEE Trans. Automat. Contr.*, vol. AC-18, pp. 24–32, Feb. 1973.
[9] F. M. Brasch, Jr., and J. B. Pearson, "Pole placement using dynamic compensators," *IEEE Trans. Automat. Contr.*, vol. AC-15, pp. 34–43, Feb. 1970.
[10] A. S. Morse and W. M. Wonham, "Decoupling and pole assignment by dynamic compensation," *SIAM J. Contr.*, vol. 8, pp. 317–337, Aug. 1970.
[11] F. R. Gantmacher, *The Theory of Matrices*, vol. 1. New York: Chelsea, 1960, pp. 45–46.
[12] W. Rudin, *Real and Complex Analysis.* New York: McGraw-Hill, 1966.
[13] J. Fogarty, *Invariant Theory.* New York: Benjamin, 1969.
[14] S. H. Wang and E. J. Davison, "A minimization algorithm for the design of linear multivariable systems," *IEEE Trans. Automat. Contr.*, vol. AC-18, pp. 220–225, June 1973.

New Results on the Controllability and Observability of General Composite Systems

E. J. DAVISON AND S. H. WANG

Abstract—**Using Rosenbrock's frequency domain approach [1], necessary and sufficient conditions are obtained for a general series-parallel or feedback composite system to be controllable or/and observable. The conditions are given in terms of the state equations of the composite system and are very simple to compute. Using these results, it is shown that a general series-parallel feedback cascade system is controllable or/and observable for "almost all" interconnection matrices, provided some mild conditions on the subsystems are imposed.**

I. Introduction

There has been much interest recently in determining conditions which guarantee that a composite system is either controllable or observable in terms of the parameters of the subsystems. See, for example, [1, p. 125]–[7]. The present development obtains necessary and sufficient conditions for a general series-parallel or feedback composite system to be either controllable or observable in terms of the state equations of the composite system. The conditions obtained are very simple to compute.

II. Main Results

Series-Parallel Composite Systems

Given the general series-parallel composite system

$$\begin{bmatrix} \dot{x}_1 \\ \dot{x}_2 \end{bmatrix} = \begin{bmatrix} A_1 & E \\ 0 & A_2 \end{bmatrix} \begin{bmatrix} x_1 \\ x_2 \end{bmatrix} + \begin{bmatrix} B_1 \\ B_2 \end{bmatrix} u$$

$$y = (C_1 \quad C_2) \begin{bmatrix} x_1 \\ x_2 \end{bmatrix} + Du \tag{1}$$

with input u and output y, where $x_1 \in R^{n_1}$, $x_2 \in R^{n_2}$, $u \in R^m$, $y \in R^r$, let the eigenvalues of A_1 be given by λ_i^1, $i = 1,2,\cdots,n_1$ and the eigenvalues of A_2 be given by λ_i^2, $i = 1,2,\cdots,n_2$. The following result is obtained.

Theorem 1: The series-parallel composite system (1) is controllable if and only if the following two conditions both hold:

a) (A_2, B_2) is controllable (2)

b) $$\text{rank}\begin{bmatrix} A_1 - \lambda_i^1 I & E & B_1 \\ 0 & A_2 - \lambda_i^1 I & B_2 \end{bmatrix} = n_1 + n_2, \qquad i = 1,2,\cdots,n_1 \tag{3}$$

and the system (1) is observable if and only if the following two conditions both hold:

c) (C_1, A_1) is observable (4)

d) $$\text{rank}\begin{bmatrix} A_1 - \lambda_i^2 I & E \\ 0 & A_2 - \lambda_i^2 I \\ C_1 & C_2 \end{bmatrix} = n_1 + n_2, \qquad i = 1,2,\cdots,n_2. \tag{5}$$

Proof of Theorem 1: Consider the proof of the controllability results initially. It is desired to determine when the pair

$$\left\{ \begin{bmatrix} A_1 & E \\ 0 & A_2 \end{bmatrix}, \begin{bmatrix} B_1 \\ B_2 \end{bmatrix} \right\}$$

is controllable. From Rosenbrock [1, p. 72] the pair is controllable if and only if the polynomial matrices

$$\left(sI - \begin{bmatrix} A_1 & E \\ 0 & A_2 \end{bmatrix} \right), \begin{bmatrix} B_1 \\ B_2 \end{bmatrix}$$

are relatively left prime, or alternately, the pair is controllable if and only if

$$\text{rank}\left\{ \left(sI - \begin{bmatrix} A_1 & E \\ 0 & A_2 \end{bmatrix} \right), \begin{bmatrix} B_1 \\ B_2 \end{bmatrix} \right\} = n_1 + n_2, \qquad \forall s \in C \tag{6}$$

which becomes

$$\text{rank}\left\{ \begin{bmatrix} A_2 - sI & E & B_1 \\ 0 & A_2 - sI & B_2 \end{bmatrix} \right\} = n_1 + n_2, \qquad \forall s \in C. \tag{7}$$

If it is assumed now that (A_2, B_2) is controllable, this implies that rank $(A_2 - sI, B_2) = n_2$, $\forall s \in C$, which means that condition (7) simplifies to condition (3), thereby establishing the proof of the sufficiency of the result. Since it is obvious that (A_2, B_2) being controllable is also a necessary condition that must hold, then the proof of the necessity of the results immediately follows.

The proof of the observability result follows immediately by duality and so is omitted. Q.E.D.

Feedback Composite Systems

Given the feedback composite system

$$\begin{bmatrix} \dot{x}_1 \\ \dot{x}_2 \end{bmatrix} = \begin{bmatrix} A_1 & -B_1 C_2 \\ B_2 C_1 & A_2 - B_2 D_1 C_2 \end{bmatrix} \begin{bmatrix} x_1 \\ x_2 \end{bmatrix} + \begin{bmatrix} B_1 \\ B_2 D_1 \end{bmatrix} u$$

$$y = (C_1, -D_1 C_2) \begin{bmatrix} x_1 \\ x_2 \end{bmatrix} + D_1 u \tag{8}$$

with input u and output y, where $x_1 \in R^{n_1}$, $x_2 \in R^{n_2}$, $u \in R^m$, $y \in R^r$, let the eigenvalues of A_2 be given by λ_i^2, $i = 1,2,\cdots,n_2$. The preceding composite system arises from the following feedback system:

$$\dot{x}_1 = A_1 x_1 + B_1 e \quad \dot{x}_2 = A_2 x_2 + B_2 y \quad e = u - z$$

$$y = C_1 x_1 + D_1 e \quad z = C_2 x_2. \tag{9}$$

The following result is obtained.

Theorem 2: The feedback composite system (8) is controllable if and only if the following two conditions both hold:

a) (A_1, B_1) is controllable (10)

Manuscript received March 1, 1974. Paper recommended by A. S. Morse, Chairman of the IEEE S-CS Linear Systems Committee. This work was supported by the National Research Council of Canada under Grant A4396.

E. J. Davison is with the Department of Electrical Engineering, University of Toronto, Toronto, Ont., Canada.

S. H. Wang is with the College of Engineering and Applied Science, University of Colorado, Colorado Springs, Colo.

Reprinted from *IEEE Trans. Automat. Contr.*, vol. AC-20, pp. 123–128, Feb. 1975.

$$\text{b)}\quad \operatorname{rank}\begin{bmatrix} A_1-\lambda_i^2 I & 0 & B_1 \\ B_2C_1 & A_2-\lambda_i^2 I & B_2D_1 \end{bmatrix}=n_1+n_2,\quad i=1,2,\cdots,n_2 \tag{11}$$

and the system (8) is observable if and only if the following two conditions both hold:

$$\text{c)}\quad (C_1,A_1) \text{ is observable} \tag{12}$$

$$\text{d)}\quad \operatorname{rank}\begin{bmatrix} A_1-\lambda_i^2 I & -B_1C_2 \\ 0 & A_2-\lambda_i^2 I \\ C_1 & -D_1C_2 \end{bmatrix}=n_1+n_2,\qquad i=1,2,\cdots,n_2. \tag{13}$$

Proof of Theorem 2: Consider the controllability result initially. The composite system (8) is controllable if and only if

$$\left\{\begin{bmatrix} A_1 & -B_1C_2 \\ B_2C_1 & A_2-B_2D_1C_2 \end{bmatrix}, \begin{bmatrix} B_1 \\ B_2D_1 \end{bmatrix}\right\}$$

is controllable, or alternately if and only if

$$\left\{\begin{bmatrix} A_1 & 0 \\ B_2C_1 & A_2 \end{bmatrix}, \begin{bmatrix} B_1 \\ B_2D_1 \end{bmatrix}\right\}$$

is controllable, or if and only if

$$\left\{\begin{bmatrix} A_2 & B_2C_1 \\ 0 & A_1 \end{bmatrix}, \begin{bmatrix} B_2D_1 \\ B_1 \end{bmatrix}\right\}$$

is controllable. Theorem 1 can now be applied to establish the necessary and sufficient conditions for the controllability of this later pair which results in conditions a) and b) of Theorem 2. The proof of the observability result follows by similar arguments and so is omitted. Q.E.D.

Remarks:

1) The composite systems (1), (8) considered include as special cases all other previous results treated before in [1]–[7].

2) It follows by the development of the proof of Theorem 1 that condition a) of Theorem 1 can be replaced by the equivalent condition

$$\text{a}')\quad \operatorname{rank}(A_2-\lambda_i^2 I,\ B_2)=n_2,\qquad i=1,2,\cdots n_2 \tag{14}$$

and that condition c) of Theorem 1 can be replaced by the equivalent condition

$$\text{c}')\quad \operatorname{rank}\begin{bmatrix} A_1-\lambda_i^1 I \\ C_1 \end{bmatrix}=n_1,\qquad i=1,2,\cdots,n_1 \tag{15}$$

and similar substitutions can be made for Theorem 2.

3) It should be noted that the conditions obtained in Theorems 1 and 2 and (14), (15) are remarkably easy to check, it just being a matter of determining eigenvalues and finding the rank of a small dimension matrix. In this regard, the new conditions obtained should be much more advantageous to use computationally than the standard conditions obtained by checking the rank of the controllability matrix of the augmented system. This is because the new conditions do not require the calculation of A^jB, $j=0,1,2,\cdots$, type terms, which, in the case of the standard controllability matrix test, are known to produce numerical difficulties in high order systems, due to the finite word length of a digital computer.

4) Many special results of interest can be obtained from Theorems 1 and 2; for example, the following problem arises in disturbance observation [8].

Definition [9]: Let the Jordan canonical form of the $n\times n$ matrix A be given by $A=\Gamma\Lambda\Gamma^{-1}$, where $\Lambda=\operatorname{diag}({}_1\Lambda,{}_2\Lambda,\cdots,{}_q\Lambda)$ where ${}_i\Lambda$ consists of all the Jordan blocks associated with the eigenvalue λ_i, $i=1,2,\cdots,q$, where λ_i are all distinct. Let these eigenvalues λ_i, $i=1,2,\cdots,q$, be called the *modal eigenvalues* of A, or, alternately, the *modes* of the system. The system $\dot{x}=Ax$, $y=Cx$ is said to be *observable with respect to the modes* $\lambda_r,\lambda_s,\cdots$ if there exists an initial state $x(0)$ such that $y(t)=Ce^{At}x(0)$ only contains terms containing $\exp(\lambda_r t)p_r(t)$, $\exp(\lambda_s t)p_s(t),\cdots$, where $p_r(t)$, $p_s(t),\cdots$ are nonzero polynomial vectors of t.

Theorem 3 (Disturbance Observation): Consider the system (1) and assume that A_2 is completely unstable, i.e., $\operatorname{Re}\lambda_i^2>0$, $i=1,2,\cdots,n_2$. Then the unstable modes of A_1 and the unstable modes of A_2 are observable from y if and only if (C_1,A_1) is detectable and condition d) of Theorem 1 holds.

Proof: Let a coordinate transformation be applied to (C_1, A_1, E) so that

$$(C_1,A_1,E)\sim\left\{(\bar{C}_1,0),\ \begin{pmatrix} \bar{A}_1^1 & 0 \\ \bar{A}_1^2 & \bar{A}_1^3 \end{pmatrix},\ \begin{pmatrix} E_1 \\ E_2 \end{pmatrix}\right\}$$

where $(\bar{C}_1,\bar{A}_1^1)$ is observable and $\bar{A}_1^1\in R^{\bar{n}_1^1\times\bar{n}_1^1}$; then the unstable modes of A_1 and the unstable modes of A_2 are observable from y if and only if

$$\left\{(\bar{C}_1,C_2),\ \begin{pmatrix} \bar{A}_1^1 & E_1 \\ 0 & A_2 \end{pmatrix}\right\}$$

is observable and $\bar{A}_1^3$ is stable, or alternately if and only if

1) (C_1,A_1) is detectable

$$2)\quad \operatorname{rank}\begin{bmatrix} \bar{A}_1^1-\lambda_i^2 I & E_1 \\ 0 & A_2-\lambda_i^2 I \\ \bar{C}_1 & C_2 \end{bmatrix}=\bar{n}_1^1+n_2,\qquad i=1,2,\cdots n_2 \tag{16}$$

or alternately if and only if

1) (C_1,A_1) is detectable

$$2)\quad \operatorname{rank}\begin{bmatrix} \bar{A}_1^1-\lambda_i^2 I & 0 & E_1 \\ \bar{A}_1^2 & \bar{A}_1^3-\lambda_i^2 I & E_2 \\ 0 & 0 & A_2-\lambda_i^2 I \\ \bar{C}_1 & 0 & C_2 \end{bmatrix}=n_1+n_2,\quad i=1,2,\cdots,n_2 \tag{17}$$

since $\operatorname{rank}(\bar{A}_1^3-\lambda_i^2 I)=n_1-\bar{n}_1^1$, $i=1,2,\cdots n_2$. This later condition is condition d) of Theorem 1. Q.E.D.

The following theorems show that a general series-parallel feedback cascade system is controllable or/and observable for "amost all" [9] interconnection matrices, provided some mild conditions on the subsystems are imposed.

Definition: If $f=f(x_1,\cdots x_n)$ is a polynomial in the n variables $x_1,\cdots,x_n$ with coefficients in R, then the point $\hat{x}=(\hat{x}_1,\cdots,\hat{x}_n)$ in R^n is said to be a zero of f if $f(\hat{x}_1,\cdots,\hat{x}_n)=0$. The set of zeros of f is called the locus of f, and is denoted by $V(f)$. A proper subset V of R^n is called a *hypersurface* in R^n if it is the locus of a nonconstant polynomial.

Theorem 4: Assume that

$$\left\{(C_1,C_2),\ \begin{bmatrix} A_1 & 0 \\ 0 & A_2 \end{bmatrix},\ \begin{bmatrix} B_1 \\ B_2 \end{bmatrix}\right\}$$

is controllable and observable; then the following result is obtained.

a) The class of all interconnection matrices E which makes (1) not both controllable and observable is either an empty set or a hypersurface in the parameter space of E.

b) The class of all interconnection matrices (E_1, E_2) which makes

$$\left\{(C_1, C_2), \begin{bmatrix} A_1 & E_1 \\ E_2 & A_2 \end{bmatrix}, \begin{bmatrix} B_1 \\ B_2 \end{bmatrix}\right\}$$

not both controllable and observable is either an empty set or a hypersurface in the parameter space of (E_1, E_2).

Proof: Consider the proof of a). Let

$$\mathcal{E} = \left\{ E \mid E \in R^{n_1 \times n_2} \text{ and } \left\{(C_1, C_2), \begin{bmatrix} A_1 & E \\ 0 & A_2 \end{bmatrix}, \begin{bmatrix} B_1 \\ B_2 \end{bmatrix}\right\} \text{ is not both controllable and observable} \right\} \tag{18}$$

and let the controllability matrix of the pair

$$\left\{\begin{bmatrix} A_1 & E \\ 0 & A_2 \end{bmatrix}, \begin{bmatrix} B_1 \\ B_2 \end{bmatrix}\right\}$$

be denoted by $R(E)$. Also, let $g(E)$ denote the polynomial which is equal to the sum of squares of all minors of order $(n_1 + n_2)$ in $R(E)$. Then, since the system is assumed to be controllable when $E = 0$, it follows that $g(0) \neq 0$, and from this, it follows that either $\mathcal{E}$ is an empty set or that $\mathcal{E}$ is a hypersurface in the parameter space of E, thereby establishing the result. Q.E.D.

Remark 5: The requirement that

$$\left\{(C_1, C_2), \begin{bmatrix} A_1 & 0 \\ 0 & A_2 \end{bmatrix}, \begin{bmatrix} B_1 \\ B_2 \end{bmatrix}\right\}$$

be controllable and observable is a fairly strong condition; a weaker condition is assumed in the following result.

Theorem 5: Assume that (A_2, B_2) is controllable, that (C_1, A_1) is observable, and that A_1, A_2 each have distinct eigenvalues; then the following result is obtained.

a) The class of all interconnnection matrices E which makes

$$\left\{(C_1, 0), \begin{bmatrix} A_1 & E \\ 0 & A_2 \end{bmatrix}, \begin{bmatrix} 0 \\ B_2 \end{bmatrix}\right\}$$

not both controllable and observable is either an empty set or a hypersurface in the parameter space of E.

b) The class of all matrices (E, C_2, B_1) which makes

$$\left\{(C_1, C_2), \begin{bmatrix} A_1 & E \\ 0 & A_2 \end{bmatrix}, \begin{bmatrix} B_1 \\ B_2 \end{bmatrix}\right\}$$

not both controllable and observable is either an empty set or a hypersurface in the parameter space of (E, C_2, B_1).

c) The class of all interconnection matrices (E_1, E_2) which makes

$$\left\{(C_1, 0), \begin{bmatrix} A_1 & E_1 \\ E_2 & A_2 \end{bmatrix}, \begin{bmatrix} 0 \\ B_2 \end{bmatrix}\right\}$$

not both controllable and observable is either an empty set or a hypersurface in the parameter space of (E_1, E_2).

d) The class of all matrices (E_1, E_2, B_1, C_2) which makes

$$\left\{(C_1, C_2), \begin{bmatrix} A_1 & E_1 \\ E_2 & A_2 \end{bmatrix}, \begin{bmatrix} B_1 \\ B_2 \end{bmatrix}\right\}$$

not both controllable and observable is either an empty set or a hypersurface in the parameter space of (E_1, E_2, B_1, C_2).

Proof: Let the Jordan canonical form of A_1 and A_2 be given respectively, by $A_1 = \Gamma_1 \lambda_1 \Gamma_1^{-1}$, $A_2 = \Gamma_2 \lambda_2 \Gamma_2^{-1}$ where $\lambda_1 = \text{diag}(\lambda_1^1, \lambda_2^1, \cdots, \lambda_{n_1}^1)$ and $\lambda_2 = \text{diag}(\lambda_1^2, \lambda_2^2, \cdots, \lambda_{n_2}^2)$. Let $B_2^* = \Gamma_2^{-1} B_2$, $C_1^* = C_1 \Gamma_1$. It can now be observed that since (A_2, B_2) is controllable and since A_2 has distinct eigenvalues, then rank $(\lambda_2 - \lambda I, B_2^*) = n_2$, $\forall \lambda \in C$, and similarly since (C_1, A_1) is observable and since A_1 has distinct eigenvalues, then

$$\text{rank}\begin{bmatrix} \lambda_1 - \lambda I \\ C_1^* \end{bmatrix} = n_1, \qquad \forall \lambda \in C.$$

Now let $E = \Gamma_1 E^* \Gamma_2^{-1}$ where E^* is an $n_1 \times n_2$ matrix given by

$$E^* = \theta \gamma' \tag{19}$$

where $\theta = (\theta_1, \theta_2, \cdots, \theta_{n_1})'$ and $\gamma' = (\gamma_1, \gamma_2, \cdots, \gamma_{n_2})$ and where θ_j, γ_j are real nonzero scalars chosen (nonuniquely) so that

$$\text{rank}\begin{bmatrix} \gamma' & 0 \\ \lambda_2 - \lambda_i^1 I & B_2^* \end{bmatrix} = n_2 + 1, \qquad i = 1, 2, \cdots, n_1 \tag{20}$$

$$\text{rank}\begin{bmatrix} \lambda_1 - \lambda_i^2 I & \theta \\ C_1^* & 0 \end{bmatrix} = n_1 + 1, \qquad i = 1, 2, \cdots, n_2. \tag{21}$$

Note that since $\text{rank}(\lambda_2 - \lambda_i^1 I, B_2^*) = n_2$, $i = 1, 2, \cdots n_1$ and

$$\text{rank}\begin{bmatrix} \lambda_1 - \lambda_i^2 I \\ C_1^* \end{bmatrix} = n_1, \qquad i = 1, 2, \cdots, n_2$$

this is always possible to do. (In fact, (20) and (21) will be satisfied for almost any $\{\gamma_j\}$, $\gamma_j \in R^1$, $\gamma_j \neq 0$, $j = 1, 2, \cdots, n_2$ and for almost any $\{\theta_j\}$, $\theta_j \in R^1$, $\theta_j \neq 0$, $j = 1, 2, \cdots, n_1$.)

Now

$$\left\{(C_1, 0), \begin{bmatrix} A_1 & E \\ 0 & A_2 \end{bmatrix}, \begin{bmatrix} 0 \\ B_2 \end{bmatrix}\right\}$$

is controllable and observable if and only if

$$\left\{(C_1^*, 0), \begin{bmatrix} \lambda_1 & E^* \\ 0 & \lambda_2 \end{bmatrix}, \begin{bmatrix} 0 \\ B_2^* \end{bmatrix}\right\}$$

is controllable and observable, and the later triple is controllable and observable, from Theorem 1, if and only if the following two conditions both hold:

$$\text{rank}\begin{bmatrix} \lambda_1 - \lambda_i^1 I & E^* & 0 \\ 0 & \lambda_2 - \lambda_i^1 I & B_2^* \end{bmatrix} = n_1 + n_2, \qquad i = 1, 2, \cdots, n_1 \tag{22}$$

$$\text{rank}\begin{bmatrix} \lambda_1 - \lambda_i^2 I & E^* \\ 0 & \lambda_2 - \lambda_i^2 I \\ C_1^* & 0 \end{bmatrix} = n_1 + n_2, \qquad i = 1, 2, \cdots, n_2. \tag{23}$$

It can now be observed that the preceding two conditions will always be satisfied; in particular,

$$\operatorname{rank}\begin{bmatrix} \lambda_1-\lambda_i^1 I & E^* & 0 \\ 0 & \lambda_2-\lambda_i^1 I & B_2^* \end{bmatrix} = n_1 - 1 + \operatorname{rank}\begin{bmatrix} \theta_i \gamma' & 0 \\ \lambda_2-\lambda_i^1 I & B_2^* \end{bmatrix}$$

$$= n_1 - 1 + \operatorname{rank}\begin{bmatrix} \gamma' & 0 \\ \lambda_2-\lambda_i^1 I & B_2^* \end{bmatrix}$$

$$= n_1 + n_2, \qquad i = 1,2,\cdots,n_1 \tag{24}$$

$$\operatorname{rank}\begin{bmatrix} \lambda_1-\lambda_i^2 I & E^* \\ 0 & \lambda_2-\lambda_i^2 I \\ C_1^* & 0 \end{bmatrix} = n_2 - 1 + \operatorname{rank}\begin{bmatrix} \lambda_1-\lambda_i^2 I & \gamma_i \theta \\ C_1^* & 0 \end{bmatrix}$$

$$= n_2 - 1 + \operatorname{rank}\begin{bmatrix} \lambda_1-\lambda_i^2 I & \theta \\ C_1^* & 0 \end{bmatrix}$$

$$= n_1 + n_2, \qquad i = 1,2,\cdots,n_2. \tag{25}$$

It has now been shown that there exists an E matrix, namely, $E = \Gamma_1 E^* \Gamma_2^{-1}$, which makes

$$\left\{ (C_1, 0), \begin{bmatrix} A_1 & E \\ 0 & A_2 \end{bmatrix}, \begin{bmatrix} 0 & B_2 \end{bmatrix} \right\}$$

controllable and observable. The same type of argument as used in the proof of Theorem 4 can now be immediately applied to prove parts a)–d) of Theorem 5. Since there are no new features in this part of the argument, it will be omitted. Q.E.D.

Remarks:

6) The statements of Theorems 4 and 5 are true if the assumption on observability (controllability) is removed and if the corresponding observability (controllability) statements in the theorems are deleted.

7) Theorem 5 is most interesting; it implies, for example, that if there exist two subsystems with distinct eigenvalues, then if one subsystem is controllable, almost all interconnection matrices between the two subsystems will result in the composite system being controllable.

8) The following example,

$$A_1 = \begin{pmatrix} 0 & 0 \\ 0 & 0 \end{pmatrix}, \quad A_2 = (1), \quad B_2 = (1), \quad C_1 = \begin{pmatrix} 1 & 0 \\ 0 & 1 \end{pmatrix}$$

shows that Theorem 5 is not necessarily true unless A_1 and A_2 each have distinct eigenvalues.

The following remark is made for completeness.

9) If the added assumption is made in Theorem 5 that the eigenvalues of A_1 and A_2 are disjoint from each other, then the class of all matrices (B_1, C_2) which makes

$$\left\{ (C_1, C_2), \begin{pmatrix} A_1 & 0 \\ 0 & A_2 \end{pmatrix}, \begin{pmatrix} B_1 \\ B_2 \end{pmatrix} \right\}$$

not both controllable and observable is either an empty set or a hypersurface in the parameter space of (B_1, C_2). The proof of this observation is trivial and is omitted. It is to be noted that this result is not necessarily true if the distinct and disjoint assumption is relaxed.

III. Extension of Previous Results to Multicomposite Systems

Given the general series-parallel multicomposite system

$$\dot{x} = \begin{bmatrix} A_1 & E_{11} & E_{12} & E_{13} & \cdots & E_{1,\gamma-1} \\ 0 & A_2 & E_{22} & E_{23} & \cdots & E_{2,\gamma-1} \\ 0 & 0 & A_3 & E_{33} & \cdots & E_{3,\gamma-1} \\ 0 & 0 & 0 & A_4 & \cdots & E_{4,\gamma-1} \\ \vdots & \vdots & \vdots & \vdots & & \vdots \\ 0 & 0 & 0 & 0 & \cdots & E_{\gamma-1,\gamma-1} \\ 0 & 0 & 0 & 0 & \cdots & A_\gamma \end{bmatrix} x + \begin{bmatrix} B_1 \\ B_2 \\ B_3 \\ B_4 \\ \vdots \\ B_{\gamma-1} \\ B_\gamma \end{bmatrix} u$$

$$y = (C_1 \;\; C_2 \;\; C_3 \;\; C_4 \;\; \cdots \;\; C_\gamma) x + Du \tag{26}$$

with input u and output y, where $x_i \in R^{n_i}$, $i = 1,2,\cdots,\gamma$, $u \in R^m$, $y \in R^r$. Let the eigenvalues of A_i, $i = 1,2,\cdots,\gamma$, be given by λ_j^i, $j = 1,2,\cdots,n_i$; $i = 1,2,\cdots,\gamma$.

The following result is immediately obtained by applying Theorem 1 to (26) in a sequential manner.

Theorem 6: The series-parallel multicomposite system (26) is controllable if and only if the following conditions all hold.

C1) $\operatorname{rank}(A_\gamma - \lambda_j^\gamma I, B_\gamma) = n_\gamma, \quad j = 1,2,\cdots,n_\gamma$

C2)
$$\operatorname{rank}\begin{bmatrix} A_{\gamma-1} - \lambda_j^{\gamma-1} I & E_{\gamma-1,\gamma-1} & B_{\gamma-1} \\ 0 & A_\gamma - \lambda_j^{\gamma-1} I & B_\gamma \end{bmatrix} = n_{\gamma-1} + n_\gamma, \quad j = 1,2,\cdots,n_{\gamma-1}$$

C3)
$$\operatorname{rank}\begin{bmatrix} A_{\gamma-2} - \lambda_j^{\gamma-2} I & E_{\gamma-2,\gamma-2} & E_{\gamma-2,\gamma-1} & B_{\gamma-2} \\ 0 & A_{\gamma-1} - \lambda_j^{\gamma-2} I & E_{\gamma-1,\gamma-1} & B_{\gamma-1} \\ 0 & 0 & A_\gamma - \lambda_j^{\gamma-2} I & B_\gamma \end{bmatrix} = n_{\gamma-2} + n_{\gamma-1} + n_\gamma, \quad j = 1,2,\cdots,n_{\gamma-2} \tag{27}$$

$\vdots$

Cγ)
$$\operatorname{rank}\begin{bmatrix} A_1 - \lambda_j^1 I & E_{11} & E_{12} & E_{13} & \cdots & E_{1,\gamma-1} & B_1 \\ 0 & A_2 - \lambda_j^1 I & E_{22} & E_{23} & \cdots & E_{2,\gamma-1} & B_2 \\ 0 & 0 & A_3 - \lambda_j^1 I & E_{33} & \cdots & E_{3,\gamma-1} & B_3 \\ 0 & 0 & 0 & A_4 - \lambda_j^1 I & \cdots & E_{4,\gamma-1} & B_4 \\ \vdots & \vdots & \vdots & \vdots & & \vdots & \vdots \\ 0 & 0 & 0 & 0 & & E_{\gamma-1,\gamma-1} & B_{\gamma-1} \\ 0 & 0 & 0 & 0 & & A_\gamma - \lambda_j^1 I & B_\gamma \end{bmatrix}$$

$$= \sum_{i=1}^{\gamma} n_i, \qquad j = 1,2,\cdots,n_1$$

and the system (26) is observable if and only if the following conditions all hold:

$$\text{C1)}' \quad \text{rank}\begin{bmatrix} A_1-\lambda_j^1 I \\ C_1 \end{bmatrix} = n_1, \qquad j=1,2,\cdots,n_1$$

$$\text{C2)}' \quad \text{rank}\begin{bmatrix} A_1-\lambda_j^2 I & E_{11} \\ 0 & A_2-\lambda_j^2 I \\ C_1 & C_2 \end{bmatrix} = n_1+n_2, \qquad j=1,2,\cdots,n_2$$

$$\text{C3)}' \quad \text{rank}\begin{bmatrix} A_1-\lambda_j^3 I & E_{11} & E_{12} \\ 0 & A_2-\lambda_j^3 I & E_{22} \\ 0 & 0 & A_3-\lambda_j^3 I \\ C_1 & C_2 & C_3 \end{bmatrix} = n_1+n_2+n_3, \qquad j=1,2,\cdots,n_3 \tag{28}$$

$$\vdots$$

$$\text{C}\gamma)' \quad \text{rank}\begin{bmatrix} A_1-\lambda_j^\gamma I & E_{11} & E_{12} & E_{13} & \cdots & E_{1,\gamma-1} \\ 0 & A_2-\lambda_j^\gamma I & E_{22} & E_{23} & \cdots & E_{2,\lambda-1} \\ 0 & 0 & A_3-\lambda_j^\gamma I & E_{33} & \cdots & E_{3,\gamma-1} \\ 0 & 0 & 0 & A_4-\lambda_j^\gamma I & \cdots & E_{4,\gamma-1} \\ \vdots & \vdots & \vdots & \vdots & & \vdots \\ 0 & 0 & 0 & 0 & \cdots & E_{\gamma-1,\gamma-1} \\ 0 & 0 & 0 & 0 & \cdots & A_\gamma-\lambda_j^\gamma I \\ C_1 & C_2 & C_3 & C_4 & \cdots & C_\gamma \end{bmatrix} = \sum_{i=1}^{\gamma} n_i, \qquad j=1,2,\cdots,n_\gamma.$$

Remarks:

10) Similar results can likewise be obtained for the corresponding feedback multicomposite system (as was done in Theorem 2).

11) It is clear that the results of Theorems 4 and 5 can trivially be extended to deal with the above multicomposite system.

IV. Conclusions

Necessary and sufficient conditions have been obtained for a general series-parallel or feedback composite system to be controllable or/and observable. The conditions obtained are expressed in terms of the parameters of the state equations of the subsystems and in terms of the interconnecting parameters of the composite system and are extremely simple to compute. Using these results it is shown that a composite system is controllable or/and observable for "almost all" interconnection parameters, provided some mild conditions on the system are imposed. An application of the results obtained is made to the problem of observing an unstable disturbance in a system.

References

[1] H. H. Rosenbrock, *State-space and Multivariable Theory*. Camden, N. J.: Nelson, 1970.

[2] E. G. Gilbert, "Controllability and observability in multivariable control systems," *SIAM J. Contr.*, vol. 1, pp. 128–151, 1963.

[3] C. T. Chen and C. A. Desoer, "Controllability and observability of composite systems," *IEEE Trans. Automat. Contr.*, vol. AC-12, pp. 402–409, Aug. 1967.

[4] M. V. Bhandarkar and M. M. Fahmy, "Controllability of tandem connected systems," *IEEE Trans. Automat. Contr.*, vol. AC-17, pp. 150–151, 1972.

[5] S. H. Wang and E. J. Davison, "On the controllability and observability of composite systems," *IEEE Trans. Automat. Contr.*, vol. AC-18, pp. 74–75, 1973.

[6] W. A. Wolovich and H. L. Hwang, "Composite system controllability and observability," *Automatica*, vol. 10, no. 2, pp. 209–212, 1974.

[7] M. L. J. Hautus, "Input regularity of cascaded systems", *IEEE Trans. Automat. Contr.*, to be published.

[8] E. J. Davison, "The feedforward, feedback, and robust control of a general servomechanism problem—parts I, II," Dep. Elec. Eng., Univ. Toronto, Toronto, Canada, Control System Reps. 7305, 7306, Apr. 1973; presented at the 11th Allerton Conf. Signals and Systems, Oct. 1973.

[9] E. J. Davison and S. H. Wang, "Properties of linear time-invariant multivariable systems subject to arbitrary output and state feedback," *IEEE Trans. Automat. Contr.*, vol. AC-18, pp. 24–32, Feb. 1973.

Stability of Interconnected Dynamical Systems Described on Banach Spaces

ROBERT D. RASMUSSEN AND ANTHONY N. MICHEL, MEMBER, IEEE

Abstract—**New stability results for a large class of interconnected dynamical systems (also called composite systems or large-scale systems) described on Banach spaces are established. In the present approach, the objective is always the same: to analyze large-scale systems in terms of their lower order and simpler subsystems and in terms of their interconnecting structure. The present results provide a systematic procedure of analyzing *hybrid dynamical systems* (i.e., systems that are described by a mixture of different types of equations). To demonstrate the method of analysis advanced, two specific examples are considered.**

I. Introduction

THE continuing use of Lyapunov's direct method as an important tool in the qualitative analysis of dynamical systems has been augmented in recent years by the extension of this method to systems defined on abstract spaces (see, e.g., [1], [2]). However, as is well known, one of the major difficulties involved in the use of the direct method is in the construction of appropriate Lyapunov functionals. As the size and complexity of systems increases, this problem acquires formidable dimensions, and therefore a method is called for which makes the analysis tractable.

Many high-dimensional systems with complex structure, which are of practical interest, may be viewed as consisting of several lower order (and simpler) subsystems which when interconnected in an appropriate fashion yield the original *large-scale systems*. This viewpoint motivates also the terms *composite system* and *interconnected system*. Frequently, it is possible to analyze such systems in terms of their lower order subsystems and in terms of their interconnecting structure. In this way, difficulties that arise when the Lyapunov method is applied to complex high-dimensional systems may often be circumvented. This method of analysis has been considered by several workers. Specifically, the Lyapunov stability of composite systems has been considered, e.g., in [3]–[13]; the input-output stability of interconnected systems (which in the case of L_2- and l_2-stability is often equivalent to Lyapunov stability) is considered, e.g., in [14], [15]; and estimates for trajectory behavior and trajectory bounds (using the Lyapunov-type approach) are established, e.g., in [12], [16].

Manuscript received September 23, 1975; revised February 18, 1976 and April 8, 1976. Paper recommended by G. L. Blankenship, Chairman of the IEEE S-CS Stability, Nonlinear, and Distributed Systems Committee. This work was supported by the National Science Foundation under Grant Eng 75-14093 and by the Engineering Research Institute, Iowa State University.

R. D. Rasmussen is with the Jet Propulsion Laboratory, Pasadena, CA.

A. N. Michel is with the Department of Electrical Engineering and Engineering Research Institute, Iowa State University, Ames, IA 50011.

In the present paper new stability results (all in the sense of Lyapunov) for large-scale dynamical systems described on Banach spaces are established. The present results are in the spirit of those cited in [3]–[16]. That is, the analysis is accomplished in terms of the lower order subsystems and in terms of their interconnecting structure.

The present results include most of those cited in [3]–[7], [9]–[13] (which deal with dynamical systems defined on finite-dimensional spaces) and differ significantly from those presented in [8]. The results presented herein are general enough to allow stability analysis of dynamical systems described by ordinary differential equations, differential-difference equations, functional differential equations, integro-differential equations, certain classes of partial differential equations, and the like. In addition, the present results are flexible enough to allow stability analysis of *hybrid dynamical systems* (i.e., dynamical systems described by a mixture of equations) in a systematic fashion.

To demonstrate the usefulness of the results presented, two specific examples are considered. The first example, representing a *hybrid system*, consists of an interconnection of two subsystems, one of these being described by a partial differential equation, while the other subsystem is represented by an ordinary differential equation. In the second example, the *point kinetics model of a coupled core nuclear reactor*, described by a set of functional differential equations, is analyzed.

II. Notation

Let $R=(-\infty,\infty)$ and $R^+=[0,\infty)$. A function $\psi: R^+\to R^+$ is said to belong to *class K* (i.e., $\psi\in K$) if it is continuous, strictly increasing, and if $\psi(0)=0$. If $\psi\in K$ and if $\lim_{r\to\infty}\psi(r)=\infty$, then ψ is said to belong to *class KR*. The inverse of $\psi\in KR$ is denoted by ψ^{-1}. Two functions $\psi_1,\psi_2\in K$ are said to be of the *same order of magnitude* if there exist positive constants k and l such that $k\psi_1(r)\leqslant\psi_2(r)\leqslant l\psi_1(r)$ for all $r\in R^+$. (For a discussion of the properties of functions of class K and KR, refer, e.g., to [1].)

The notation $0(r)$ denotes terms of second or higher order, so that $\lim_{r\to 0^+}0(r)/r=0$.

Let $C=((c_{ij}))$, $i=1,\cdots,m$, $j=1,\cdots,n$ denote an $m\times n$ matrix and let $x=(x_i)$, $i=1,\cdots,l$ denote an $l\times 1$ column

Reprinted from *IEEE Trans. Automat. Contr.*, vol. AC-21, pp. 464–471, Aug. 1976.

vector. The transpose of C and x is denoted by C' and x', respectively. For a square matrix D, $\lambda(D)$ denotes an eigenvalue of D. If all eigenvalues of D are real, then $\lambda_{\max}(D)$ and $\lambda_{\min}(D)$ denote the largest and smallest eigenvalues of D, respectively.

Banach spaces are denoted by X and Z (with appropriate subscripts, if necessary) and norms are denoted by $\|\cdot\|$ (with appropriate subscripts, if necessary). In particular, $X=L_p[a,b]$ is the set of all Lebesgue measurable functions $f:[a,b]\to R$ with $\|f\|_p=[\int_a^b|f(t)|^p\,dt]^{1/p}<\infty$, $1\leqslant p<\infty$.

An n-dimensional Euclidean space is denoted by R^n with the norm denoted by $|\cdot|$. If $x\in R^n$, then $x=(x_i)>0$ indicates that $x_i>0$, $i=1,\cdots,n$.

III. Systems Described on Banach Spaces

Systems are considered which may be described by differential equations of the form

$$\dot{x}_t=Ax_t, \qquad x_0=a \tag{1}$$

where $t\in R^+$ and A is an operator (possibly nonlinear) with domain $D(A)$ dense in a Banach space X. A function $x_t, t\in R^+$, is said to be a *solution* of (1) if $x_t\in D(A)$ and if x_t possesses a derivative which satisfies (1) for all $t\in R^+$. Henceforth, system (1) is assumed to be *well posed* in the sense that 1) it possesses a unique solution for each $a\in D(A)$, and 2) solutions depend continuously on $x_0=a$ for all $t\in R^+$. It is also assumed that $A0=0$, so that (1) admits the trivial solution $x_t\equiv 0$.

Stability results for system (1) involve the existence of Lyapunov functionals and their time derivatives along solutions of (1).

Definition 1: A function $V:X\to R$ is said to be an *admissible Lyapunov functional* if 1) $V(0)=0$, 2) V is continuous, 3) the closure of the set

$$Q_m=\{a\in X: V(a)<m\},$$

denoted by $\overline{Q}_m$, is bounded for all $m>0$, and 4) there exists a function $\nabla V: X\times X\to R$ such that for all $a,x\in X$,

1) $V(a+x)-V(a)\leqslant \nabla V(a,x)+0(\|x\|)$
2) $\nabla V(a,x)$ is linear in x, and
3) $\nabla V(a,x)$ is continuous in x, uniformly with respect to $a\in Q_m$, for each $m>0$. ■

Under the assumptions of Definition 1, one has for all $a\in\overline{Q}_m\cap D(A)$ and any $m>0$, along solutions of (1),

$$\dot{V}(a)=\overline{\lim_{t\to 0^+}}\, t^{-1}[V(x_t)-V(a)]\leqslant \nabla V(a,Aa).$$

Differential equations defined on Banach spaces are treated in [2], and their stability properties (in the sense of Lyapunov) are investigated in [1] and [2].

IV. Interconnected Dynamical Systems

In the present section large-scale dynamical systems described on Banach spaces are considered which may be viewed as an interconnection of lower order subsystems. When disconnected from the system structure, these subsystems are called *isolated subsystems*. First the isolated subsystems $\mathcal{S}_i$ are introduced and then the interconnected system $\mathcal{S}$ is described. Finally, several results for asymptotic stability and exponential stability of interconnected systems are established. The proofs of these results are given in Section VI.

Consider systems of the form

$$\mathcal{S}_i: \dot{z}_t^i=F_i z_t^i,\ z_0^i=a^i, \qquad i=1,\cdots,l \tag{2}$$

where $z_t^i\in Z_i$ for all $t\in R^+$ and Z_i is a Banach space with norm $\|\cdot\|_i$. Henceforth it is assumed that each isolated subsystem $\mathcal{S}_i$ possesses all properties imposed on system (1).

In order to characterize the stability properties of isolated subsystems $\mathcal{S}_i$, the following conventions are employed.

Definition 2: Isolated subsystem $\mathcal{S}_i$ is said to possess *Property A* if there exist an admissible Lyapunov functional V_i, three functions $\psi_{i1},\psi_{i2}\in KR, \psi_{i3}\in K$, and a real constant σ_i such that for some $m_i>0$, 1) $\psi_{i1}(\|a^i\|_i)\leqslant V_i(a^i)\leqslant\psi_{i2}(\|a^i\|_i)$, $a^i\in\overline{Q}_{m_i}$ and 2) $\nabla V_i(a^i,F_ia^i)\leqslant\sigma_i\psi_{i3}(\|a^i\|_i)$, $a^i\in\overline{Q}_{m_i}\cap D(F_i)$. ■

Definition 3: If in Definition 2 $\psi_{i1},\psi_{i2},\psi_{i3}\in KR$ and if $\psi_{i1},\psi_{i2},\psi_{i3}$ are of the same order of magnitude, then isolated subsystem $\mathcal{S}_i$ is said to possess *Property B*. ■

If in Definitions 2 and 3, $\sigma_i<0$, then the trivial solution of isolated subsystem $\mathcal{S}_i$ is uniformly asymptotically stable and exponentially stable, respectively. If in Definitions 2 and 3 $\overline{Q}_{m_i}\cap D(F_i)=Z_i$, then the trivial solution of $\mathcal{S}_i$ is uniformly asymptotically stable in the large and exponentially stable in the large, respectively. (For proofs of these statements, refer to [1] and [2].)

The parameter σ_i in these definitions is a measure of the degree of stability of isolated subsystem $\mathcal{S}_i$ and will be used in studying the qualitative effects of subsystem $\mathcal{S}_i$ on the behavior of the entire interconnected system, which is characterized next.

Let $X=Z_1\times Z_2\times\cdots\times Z_l=\times_{i=1}^l Z_i$. If $x\in X$, then $x=(z^i)$, $i=1,\cdots,l$. Let

$$\|x\|=\max_i\|z^i\|_i. \tag{3}$$

Since each Z_i is a Banach space, it follows that X is also a Banach space with respect to the norm (3).

Interconnected dynamical systems are considered which are of the form

$$\mathcal{S}: \dot{z}_t^i=F_iz_t^i+G_ix_t, \qquad i=1,\cdots,l \tag{4}$$

with $x_0=a=(a^i)$, $i=1,\cdots,l$. The operator G_i has domain $D(G_i)\subset X$ and maps $D(G_i)$ into Z_i.

Letting $Ax=(F_iz^i+G_ix)$, $i=1,\cdots,l$ (4) can equivalently be expressed as

$$\mathcal{S}: \dot{x}_t=Ax_t,\ x_0=a, \tag{5}$$

which is clearly of the same form as (1).

Henceforth, it is assumed that (4), or equivalently (5), possesses all the properties imposed on (1). Thus, system (4) is well posed.

System (5) is called a *large-scale system*, or a *composite system*, or an *interconnected system* with *decomposition* (4). It may be viewed as an interconnection of l isolated subsystems $\mathcal{S}_i$ characterized by (2). The operators G_i, $i=1,\cdots,l$ in (4) comprise the *interconnecting structure* of system $\mathcal{S}$.

The proofs of the following results are given in Section VI.

Theorem 1: Assume that for interconnected system $\mathcal{S}$ the following conditions hold:

1) Each isolated subsystem $\mathcal{S}_i$ possesses Property A.

2) Given the functionals V_i and the functions ψ_{i3}, $i=1,\cdots,l$ of hypothesis 1) there exist real constants b_{ij}, $i,j=1,\cdots,l$, such that

$$\nabla V_i(a^i, G_i a) \leqslant \left[\psi_{i3}(\|a^i\|_i)\right]^{1/2} \sum_{j=1}^{l} b_{ij}\left[\psi_{j3}(\|a^j\|_j)\right]^{1/2} \quad (6)$$

for all $a \in \times_{i=1}^{l} \bar{Q}_{m_i} \cap D(A)$.

3) There exists a vector $\alpha=(\alpha_i)\in R^l$, $\alpha_i>0$, $i=1,\cdots,l$, such that the *test matrix* $S=((s_{ij}))$, $i,j=1,\cdots,l$, defined by

$$s_{ij}=\begin{cases}\alpha_i(\sigma_i+b_{ii}), & i=j\\ \frac{1}{2}(\alpha_i b_{ij}+\alpha_j b_{ji}), & i\neq j\end{cases}$$

is negative definite.

Then the trivial solution of $\mathcal{S}$ is *asymptotically stable.* ■

Theorem 2: Assume that for interconnected system $\mathcal{S}$ the following conditions hold:

1) Each isolated subsystem $\mathcal{S}_i$ possesses Property B.

2) All functions ψ_{ij}, $i=1,\cdots,l$, $j=1,2,3$ in hypothesis 1) are of the same order of magnitude.

3) Hypotheses 2) and 3) of Theorem 1 hold.

Then the trivial solution of $\mathcal{S}$ is *exponentially stable.* ■

Remark 1: It is emphasized that in Theorems 1 and 2 the stability analysis is accomplished in terms of the lower order (and hopefully simpler) subsystems and in terms of the system interconnecting structure. ■

Remark 2: Concerning the test matrix S of Theorems 1 and 2, some observations are in order.

1) The negative definiteness condition on S requires that

$$\sigma_i < -b_{ii}, \qquad i=1,\cdots,l.$$

Thus, each subsystem must either possess a certain degree of stability, or the interconnecting structure must provide stabilizing effect (i.e., stabilizing feedback), associated with each unstable subsystem.

2) The nature of the bound on G_i is such as to restrict the strength of interconnection between subsystems. Thus, the present results take the form of weak coupling conditions.

3) Bounds of form (6) are not particularly hard to satisfy. (See, e.g., the examples in Section V.)

4) Stability criteria for which the conditions in item 1) are not satisfied have not appeared in the literature, using the present approach of analysis.

5) The observations in item 1) suggest a systematic stabilization procedure of unstable interconnected dynamical systems, using *local feedback* for each unstable $\mathcal{S}_i$.

6) It is emphasized that there are no restrictions in sign of the off-diagonal terms in the S matrices of Theorems 1 and 2. This is in contrast with related results (e.g., [3]–[9]) where it is *always* required that the off-diagonal terms of S be of the same sign. In this *special* case, the observations in the next remark are applicable. ■

Remark 3: If in Theorems 1 and 2, $b_{ij}\geqslant 0$, $i,j=1,\cdots,l$, $i\neq j$, as is the case in many applications, then the theory of M-matrices (see, e.g., [17], [18]) can be used to advantage.

1) An $(n\times n)$ matrix $A=((a_{ij}))$, $i,j=1,\cdots,n$, is said to be an M-matrix if $a_{ij}\leqslant 0$ for all $i\neq j$, and if one of the following equivalent statements is satisfied:

a) All successive principal minors of A are positive.

b) There is a vector $x>0$ such that $Ax>0$.

c) There is a vector $y>0$ such that $A'y>0$.

d) A is nonsingular and all elements of A^{-1} are nonnegative.

e) The real parts of the eigenvalues of A are all positive.

2) The applicability of M-matrices to Theorems 1 and 2 lies in the following result, which is a consequence of the preceding statements (see, e.g., [7]): let $\Sigma=((\delta_{ij}\sigma_i))$ and $B=((b_{ij}))$, $i,j=1,\cdots,l$ where δ_{ij} denotes the Kronecker delta. If $D=-(\Sigma+B)$ is an M-matrix, then there exists a diagonal matrix $F=((\delta_{ij}\alpha_i))$, $\alpha_i>0$, $i,j=1,\cdots,l$ such that

$$S=\tfrac{1}{2}\left[F(\Sigma+B)+(\Sigma+B)'F\right] \quad (7)$$

is negative definite. ■

Identifying (7) with matrix S of Theorems 1 and 2 yields the following result.

Corollary 1: Assume that conditions 1) and 2) of Theorem 1 hold. Then the trivial solution of system $\mathcal{S}$ is *asymptotically stable* if $b_{ij}\geqslant 0$ for all $i\neq j$, and if the successive principal minors of the test matrix $D=((d_{ij}))$ are positive, where

$$d_{ij}=\begin{cases}-(\sigma_i+b_{ii}), & i=j\\ -b_{ij}, & i\neq j.\end{cases}$$ ■

Remark 4: Results such as Corollary 1 have been reported for finite-dimensional systems described by ordinary differential equations (see, e.g., [3]–[9]). Note that whereas it is always possible to deduce such results (as Corollary 1) from Theorems 1 and 2, the converse is in general not true [see Remark 2-6)]. Thus, even when specialized to finite-dimensional systems, Theorems 1 and 2 appear to be more general than corresponding results involving M-matrices. ■

Theorem 1 and item b) of Remark 3-1) yield the following result.

Corollary 2: Assume that conditions 1) and 2) of Theorem 1 hold. Then the trivial solution of system $\mathcal{S}$ is *asymptotically stable* if $b_{ij} \geqslant 0$ for all $i \neq j$, and if there exist positive constants λ_i, $i=1,\cdots,l$, such that

$$(\sigma_i + b_{ii}) + \sum_{j=1}^{l} \frac{\lambda_j}{\lambda_i} b_{ij} < 0, \qquad i=1,\cdots,l. \qquad \blacksquare$$

Remark 5: It can be shown (see, e.g., [7]) that if a matrix A is an M-matrix, then $A-\mu I$ is an M-matrix if and only if $\mu < \min \operatorname{Re}(\lambda(A))$, where I denotes the identity matrix. ■

Corollaries 1 and 2 and Remark 5 yield the following.

Corollary 3: Let D denote the matrix defined in Corollary 1 and assume that system $\mathcal{S}$ has been shown to be asymptotically stable either by Corollary 1 or Corollary 2. Then any modification of isolated subsystems $\mathcal{S}_i$ (or their local feedback) which increases each $(\sigma_i + b_{ii})$, $i=1,\cdots,l$ by an amount less than $\mu = \min_k \operatorname{Re}[\lambda_k(D)]$, $k=1,\cdots,l$, will leave system $\mathcal{S}$ asymptotically stable. ■

Corollary 3 provides a bound on the degree of stability of the subsystems (together with their local feedback) which is a possible criterion for the degree of stability of interconnected system $\mathcal{S}$.

Corollaries 1 and 2 yield sufficient conditions for exponential stability if the comparison functions $\psi_{ij} \in KR$, $i=1,\cdots,l$, $j=1,2,3$ are of the same order of magnitude.

Remark 6: If in Theorems 1 and 2 $\overline{Q}_{m_i} \cap D(F_i) = Z_i$, $i=1,\cdots,l$ and $\times_{i=1}^{l} \overline{Q}_{m_i} \cap D(A) = X$, then these results yield conditions for uniform asymptotic stability in the large and exponential stability in the large, respectively. Similar statements are true for Corollaries 1 and 2. ■

V. Applications

Presently two specific examples are considered.

Example 1: Consider the *hybrid system* described by the equations

$$\left.\begin{aligned} \dot{z}_t^1 &= A z_t^1 + b \int_0^L f(y) z_t^2(y)\,dy \\ \dot{z}_t^2 &= \alpha \frac{\partial^2}{\partial y^2} z_t^2(y) - \beta z_t^2(y) + g(y) c' z_t^1 \\ \frac{\partial}{\partial y}\left[z_t^2(y)\right]^2\Big|_0^L &\triangleq \frac{\partial}{\partial y}\left[z_t^2(y)\right]^2\Big|_L - \frac{\partial}{\partial y}\left[z_t^2(y)\right]^2\Big|_0 < 0 \end{aligned}\right\} \tag{8}$$

where $z_t^1 \in R^n$, A is a stable $n \times n$ matrix, $b, c \in R^n$ are constant n-vectors, $z_t^2 \in L_2[0,L]$, and α, β, and L are positive constants. Also, f and g are real functions such that $\|f\|_2 < \infty$ and $\|g\|_2 < \infty$, where

$$\|f\|_2 = \left[\int_0^L f(y)^2\,dy\right]^{1/2}, \quad \|g\|_2 = \left[\int_0^L g(y)^2\,dy\right]^{1/2}.$$

System (8) may be viewed as an interconnection of isolated subsystems

$$\left.\begin{aligned} \mathcal{S}_1 &: \dot{z}_t^1 = A z_t^1, & z_0^1 &= a_1 \\ \mathcal{S}_2 &: \dot{z}_t^2 = \alpha \frac{\partial^2}{\partial y^2} z_t^2 - \beta z_t^2, & z_0^2 &= a_2 \end{aligned}\right\}. \tag{9}$$

Since A is stable there exists a positive definite symmetric matrix P such that $A'P + PA \triangleq -Q$ is negative definite. Choose $V_1(a_1) = a_1' P a_1$ and $V_2(a_2) = \frac{1}{2}\|a_2\|_2^2$. Using the notation of Definition 3, one obtains

$$\psi_{11}(|a_1|) \triangleq \lambda_{\min}(P)|a_1|^2 \leqslant V_1(a_1) \leqslant \lambda_{\max}(P)|a_1|^2 \triangleq \psi_{12}(|a_1|)$$

$$\nabla V_1(a_1, F_1 a_1) = a_1'(A'P + PA)a_1 = -a_1' Q a_1 \leqslant -\lambda_{\min}(Q)|a_1|^2 = \sigma_1 \psi_{13}(|a_1|)$$

$$\psi_{21}(\|a_2\|_2) \triangleq \tfrac{1}{2}\|a_2\|_2^2 \leqslant V_2(a_2) \leqslant \tfrac{1}{2}\|a_2\|_2^2 \triangleq \psi_{22}(\|a_2\|_2), \quad \text{and}$$

$$\begin{aligned} \nabla V_2(a_2, F_2 a_2) &= \int_0^L a_2(y)\left[\alpha \frac{\partial^2}{\partial y^2} a_2(y) - \beta a_2(y)\right] dy \\ &= \int_0^L \left[-\alpha\left(\frac{\partial}{\partial y} a_2(y)\right)^2 - \beta a_2(y)^2\right] dy + \frac{\alpha}{2}\frac{\partial}{\partial y} a_2(y)^2\Big|_0^L \\ &\leqslant -\beta\|a_2\|_2^2 \triangleq \sigma_2 \psi_{23}(\|a_2\|_2). \end{aligned}$$

Clearly, isolated subsystems $\mathcal{S}_1$ and $\mathcal{S}_2$ possess Property B.

Also, using the notation of Theorems 1 and 2 one has

$$\begin{aligned} \nabla V_1(a_1, G_1 a) &= 2a_1' P b \int_0^L f(y) a_2(y)\,dy \\ &\leqslant 2\lambda_{\max}(P)|b|\,\|f\|_2 |a_1|\,\|a_2\|_2 \\ &\triangleq [\psi_{13}(|a_1|)]^{1/2} \sum_{j=1}^{2} b_{1j}[\psi_{j3}(\|a_j\|_j)]^{1/2} \end{aligned}$$

and

$$\begin{aligned} \nabla V_2(a_2, G_2 a) &= \int_0^L a_2(y) g(y) c' a_1\,dy \leqslant |c|\,\|g\|_2 \|a_2\|_2 |a_1| \\ &\triangleq [\psi_{23}(\|a_2\|_2)]^{1/2} \sum_{j=1}^{2} b_{2j}[\psi_{j3}(\|a_j\|_j)]^{1/2}. \end{aligned}$$

Since $\psi_{13}(r) = \psi_{23}(r) = r^2$, it follows that $b_{11} = b_{22} = 0$, $b_{12} = 2\lambda_{\max}(P)|b|\,\|f\|_2$ and $b_{21} = |c|\,\|g\|_2$.

Choosing $\alpha_1 = 2/b_{12}$ and $\alpha_2 = 2/b_{21}$, matrix S of Theorem 2 assumes the form

$$S = \begin{bmatrix} \dfrac{-\lambda_{\min}(Q)}{\lambda_{\max}(P)|b|\,\|f\|_2} & 2 \\ 2 & \dfrac{-2\beta}{|c|\,\|g\|_2} \end{bmatrix}.$$

Matrix S is negative definite if and only if

$$\beta\left[\frac{\lambda_{\min}(Q)}{\lambda_{\max}(P)}\right] > 2|b|\,|c|\,\|f\|_2 \|g\|_2. \tag{10}$$

Thus, if (10) is satisfied, the trivial solution of system (8) is exponentially stable in the large. ■

Example 2: Consider the *point kinetics model* of a *coupled core nuclear reactor* with l cores (see [19], [20]) described by the set of equations

$$\left.\begin{aligned} \Lambda_i\dot{p}_i(t)&=[\rho_i(t)-\epsilon_i-\beta_i]p_i(t)+\rho_i(t)+\sum_{k=1}^{6}\beta_{ki}c_{ki}(t)\\ &\quad+\sum_{j=1}^{l}\epsilon_{ji}\frac{P_{j0}}{P_{i0}}\int_{-\infty}^{t}h_{ji}(t-s)p_j(s)\,ds\\ \dot{c}_{ki}(t)&=\lambda_{ki}[p_i(t)-c_{ki}(t)],\qquad i=1,\cdots,l,\ k=1,\cdots,6 \end{aligned}\right\} \tag{11}$$

where $p_i:R\to R$ and $c_{ki}:R\to R$ represent power in the ith core and concentration of the kth precursor in the ith core, respectively. The constants $\Lambda_i,\epsilon_i,\beta_i,\beta_{ki},\epsilon_{ji},P_{i0},\lambda_{ki}$ are all positive, where $\beta_i=\sum_{k=1}^{6}\beta_{ki}$ and $h_{ij}:R\to R$ determine the coupling via neutron migration from the jth core to the ith core. The reactivity $\rho_i(t)$ of the ith core is expressed by the relation

$$\rho_i(t)=\int_{-\infty}^{t}w_i(t-s)p_i(s)\,ds,\qquad i=1,\cdots,l \tag{12}$$

where $w_i:R\to R$. Making the physically realistic assumption

$$\lim_{t\to-\infty}c_{ki}(t)e^{\lambda_{ki}t}=0,\qquad k=1,\cdots,6,\ i=1,\cdots,l \tag{13}$$

one obtains

$$c_{ki}(t)=\int_{-\infty}^{t}\lambda_{ki}e^{-\lambda_{ki}(t-s)}p_i(s)\,ds,\qquad k=1,\cdots,6,\ i=1,\cdots,l. \tag{14}$$

To put (11) into the appropriate form, let the state be $z(\tau)$, where

$$z_t^i(\tau)=p_i(t+\tau),\qquad i=1,\cdots,l,\ \tau<0 \tag{15}$$

denotes the entire past history of p_i at time t, and eliminate $\rho_i(t)$ and $c_{ki}(t)$ from (11) using (12) and (14). Let

$$f_i(t)\triangleq\Lambda_i^{-1}\left[w_i(t)+\sum_{k=1}^{6}\beta_{ki}\lambda_{ki}e^{-\lambda_{ki}t}+\epsilon_{jj}h_{jj}(t)\right], \tag{16}$$

$$K_i\triangleq\Lambda_i^{-1}(\epsilon_j+\beta_j) \tag{17}$$

$$n_i(t)\triangleq\Lambda_i^{-1}w_i(t),\qquad i=1,\cdots,l \tag{18}$$

and

$$g_{ij}(t)\triangleq\frac{\epsilon_{ji}P_{j0}}{\Lambda_iP_{i0}}h_{ji}(t),\qquad i\neq j,\ i,j=1,\cdots,l. \tag{19}$$

Making the change of variable $s=t+\tau$ in the integrals and noting that $\dot{p}_i(t)=\dot{z}_t^i(0)$, one obtains $\dot{x}_t=Ax_t$, where $x_t=(z_t^i)$, $i=1,\cdots,l$ and

$$\dot{z}_t^i(\tau)=A_ix_t(\tau)=\left\{\begin{array}{ll} -K_iz_t^i(0)+\int_{-\infty}^{0}f_i(-u)z_t^i(u)\,du & \\ +z_t^i(0)\int_{-\infty}^{0}n_i(-u)z_t^i(u)\,du+\sum_{\substack{j=1\\ i\neq j}}^{l}\int_{-\infty}^{0}g_{ij}(-u)z_t^i(u)\,du, & \tau=0\\ \left.\dfrac{\partial z_t^i(u)}{\partial u}\right|_{u=\tau}, & \tau<0 \end{array}\right\} \tag{20}$$

(Note that the running time variable is still t and not τ, and $\dot{z}_t^i(\tau)$ signifies $\partial z_t^i(\tau)/\partial t$.) The initial condition is therefore $z_0^i(\tau)=a^i(\tau)=p_i(\tau)$, $\tau\leqslant0$, $i=1,\cdots,l$ the past history of $p_i(t)$ at $t=0$.

For some $L_i>0$, define

$$m_i(u)=\left\{\begin{array}{ll} L_i^{-1}e^{L_iu}, & u<0,\ i=1,\cdots,l\\ L_i^{-1}+1, & u=0,\ i=1,\cdots,l \end{array}\right\} \tag{21}$$

and let Z_i be the function space on $(-\infty,0]$, with norm defined, using the Lebesgue–Stieltjes integral, as

$$\|z^i\|_i=\left[\int_{-\infty}^{0}z^i(u)^2\,dm_i(u)\right]^{1/2}=\left[z^i(0)^2+\int_{-\infty}^{0}z^i(u)^2e^{L_iu}\,du\right]^{1/2}, \tag{22}$$

$i=1,\cdots,l$. Then Z_i is an L_2-space with the above norm and $z_t^i\in Z_i$.

System (20) may be viewed as an interconnection of l subsystems $\mathcal{S}_i$ described by

$$\dot{z}_t(\tau)=F_iz_t^i(\tau)=\left\{\begin{array}{ll} -K_iz_t^i(0)+\int_{-\infty}^{0}f_i(-u)z_t^i(u)\,du+z_t^i(0)\int_{-\infty}^{0}n_i(-u)z_t^i(u)\,du, & \tau=0\\ \left.\dfrac{\partial z_t^i(u)}{\partial u}\right|_{u=\tau}, & \tau<0 \end{array}\right\}. \tag{23}$$

For $\mathcal{S}_i$ choose

$$V_i(a^i)=a^i(0)^2+K_i\int_{-\infty}^{0}a^i(u)^2e^{L_iu}\,du,\qquad i=1,\cdots,l. \tag{24}$$

Then

$$\min(1,K_i)\|a^i\|_i^2 \leqslant V_i(a^i) \leqslant \max(1,K_i)\|a^i\|_i^2 \quad (25)$$

and

$$\nabla V_i(a^i,F_ia^i)=2a^i(0)F_ia^i(0) + 2K_i\int_{-\infty}^{0} a^i(u)F_ia^i(u)e^{L_iu}\,du. \quad (26)$$

For initial conditions satisfying

$$\lim_{t\to-\infty} a^i(\tau)^2e^{L_i\tau}=0, \qquad i=1,\cdots,l, \quad (27)$$

integration by parts yields

$$\int_{-\infty}^{0} a^i(u)F_i(u)e^{L_iu}\,du=\int_{-\infty}^{0} a^i(u)\frac{\partial a^i(u)}{\partial u}e^{L_iu}\,du$$
$$=\tfrac{1}{2}a^i(0)^2-\tfrac{1}{2}L_i\int_{-\infty}^{0} a^i(u)e^{L_iu}\,du, \qquad i=1,\cdots,l. \quad (28)$$

Let

$$b_i \triangleq \left[\int_{-\infty}^{0} a^i(u)^2e^{L_iu}\,du\right]^{1/2}, \qquad i=1,\cdots,l\ ; \quad (29)$$

choose L_i in such a fashion that the positive constants

$$c_{ii} \triangleq \left[\int_{0}^{\infty} f_i(u)^2e^{L_iu}\,du\right]^{1/2} \quad (30)$$

and

$$d_i \triangleq \left[\int_{0}^{\infty} n_i(u)^2e^{L_iu}\,du\right]^{1/2}, \quad (31)$$

$i=1,\cdots,l$ are finite. Applying (28) and the Cauchy–Schwartz inequality to (26) one obtains

$$\nabla V_i(a^i,F_ia^i) \leqslant -K_ia^i(0)^2+2c_{ii}a_i(0)b_i - K_iL_ib_i^2+2d_ia^i(0)^2b_i, \quad (32)$$

$i=1,\cdots,l$, a polynomial in $a^i(0)$ and b_i. The first three terms are a quadratic form which is negative definite if

$$K_i\sqrt{L_i} > c_{ii}, \qquad i=1,\cdots,l. \quad (33)$$

The fourth term in (32) is of third order and thus, if (33) is satisfied, then $\nabla V_i(a^i,F_ia^i)$ is negative definite in a neighborhood of the origin. That is, for any σ_i such that

$$0>\sigma_i \geqslant -\tfrac{1}{2}K_i(L_i+1)+\left[\tfrac{1}{4}K_i(L_i-1)^2+c_{ii}^2\right]^{1/2} \quad (34)$$

where the right-hand side of (34) is the maximum eigenvalue of the quadratic terms, one has

$$\nabla V_i(a^i,F_ia^i) \leqslant \sigma_i\left(a^i(0)^2+b_i^2\right)=\sigma_i\|a^i\|_i^2, \qquad i=1,\cdots,l \quad (35)$$

for $\|a^i\|_i$ sufficiently small. It follows from (25) and (35) that subsystems $\mathcal{S}_i$ possess Property B.

For the interconnections, let

$$c_{ij}=\left[\int_{0}^{\infty} g_{ij}(u)^2e^{L_ju}\,du\right]^{1/2} \geqslant 0, \qquad i\neq j, \quad (36)$$

$i,j=1,\cdots,l$. Then

$$\nabla V_i(a^i,G_ia)=2a^i(0)\sum_{\substack{j=1\\ i\neq j}}^{l}\int_{-\infty}^{0} g_{ij}(-u)a^j(u)\,du$$
$$\leqslant a^i(0)\sum_{\substack{j=1\\ i\neq j}}^{l} 2c_{ij}b_j \leqslant \|a^i\|_i\sum_{\substack{j=1\\ i\neq j}}^{l} 2c_{ij}\|a^j\|_j. \quad (37)$$

Since $\psi_{i3}(r)=r^2$ it follows that

$$b_{ii}=0, \quad b_{ij}=2c_{ij}>0, \qquad i\neq j,\ i,j=1,\cdots,l. \quad (38)$$

Corollaries 1–3 may now be applied with test matrix $D=((d_{ij}))$ characterized by (34) and (38), i.e.,

$$d_{ij}=\left\{\begin{array}{ll} -\sigma_i, & i=j \\ -2c_{ij}, & i\neq j \end{array}\right\}. \quad (39)$$

For example, if $l=2$, then system (11) is exponentially stable if

$$\sigma_1\sigma_2>4c_{12}c_{21}. \quad (40)$$

Thus, if in a two-core coupled nuclear reactor, each core is exponentially stable and if the coupling between cores via neutron migration is sufficiently weak, then the reactor is exponentially stable. ∎

VI. Proofs of Theorems 1 and 2

The proofs of Theorems 1 and 2 are combined.

To prove *Theorem 1*, note that by hypothesis 1) there exists for each isolated subsystem $\mathcal{S}_i$ an admissible Lyapunov functional $V_i(a^i)$ on the closed and bounded set $\overline{Q}_{m_i}$. Choose the vector $\alpha>0$ as in hypothesis 3) and let

$$V(a)=\sum_{i=1}^{l}\alpha_iV_i(a^i) \quad (41)$$

where $a\in \times_{i=1}^{l}\overline{Q}_{m_i}$. Let $m=\min_i(\alpha_im_i)$ and $Q_m=\{a\in X: V(a)<m\}$. Then, if $a\in Q_m$,

$$V_i(a^i) \leqslant V(a)/\alpha_i < \frac{m}{\alpha_i} \leqslant m_i,$$

which implies that $a^i\in Q_{m_i}$, $i=1,\cdots,l$. Thus,

$$\overline{Q}_m \subset \mathop{\times}_{i=1}^{l} \overline{Q}_{m_i}, \quad m=\min_i(\alpha_im_i) \quad (42)$$

where $\times_{i=1}^{l}\overline{Q}_{m_i}$ is closed and bounded in X, making $\overline{Q}_m$ closed and bounded also. Furthermore, it is easily shown that $V(a)$ is continuous. Moreover, applying Definition 1 to (41), we have

$$\nabla V(a,x)=\sum_{i=1}^{l}\alpha_i\nabla V_i(a^i,z^i) \quad (43)$$

which is linear in x, continuous in x, uniformly with

respect to $a \in Q_m$. Therefore $V(a)$ is an admissible Lyapunov functional (in the sense of Definition 1) for interconnected system (4).

Since each isolated subsystem $\mathbb{S}_i$ possesses Property A, one has for $a \in \overline{Q}_m$,

$$\psi_1(\|a\|) \leq \sum_{i=1}^{l} \alpha_i \psi_{i1}(\|a^i\|_i) \leq V(a)$$

$$\leq \sum_{i=1}^{l} \alpha_i \psi_{i2}(\|a^i\|_i) \leq \psi_2(\|a\|) \quad (44)$$

where ψ_1 and ψ_2 can always be chosen such that $\psi_1, \psi_2 \in KR$.

Next, $\nabla V(a, Aa)$ for (4) is evaluated. By linearity and hypotheses 1) and 2) one has

$$\nabla V(a,Aa) = \sum_{i=1}^{l} \alpha \nabla V_i(a^i, F_i a^i + G_i a)$$

$$= \sum_{i=1}^{l} \alpha_i \left[\nabla V_i(a^i, F_i a^i) + \nabla V_i(a^i, G_i a) \right]$$

$$\leq \sum_{i=1}^{l} \alpha_i \left\{ \sigma_i \psi_{i3}(\|a^i\|_i) + \left[\psi_{i3}(\|a^i\|_i)\right]^{1/2} \cdot \sum_{j=1}^{l} b_{ij} \left[\psi_{j3}(\|a^j\|_j)\right]^{1/2} \right\}$$

$$= \sum_{i=1}^{l} \alpha_i(\sigma_i + b_{ii}) \left\{ \left[\psi_{i3}(\|a^i\|_i)\right]^{1/2} \right\}^2 + \sum_{\substack{i,j=1 \\ i \neq j}}^{l} \tfrac{1}{2}(\alpha_i b_{ij} + \alpha_j b_{ji}) \left[\psi_{i3}(\|a^i\|_i)\right]^{1/2} \left[\psi_{j3}(\|a^j\|_j)\right]^{1/2}$$

$$= u' S u$$

where $u' = ([\psi_{13}(\|a^1\|_1)]^{1/2}, \cdots, [\psi_{l3}(\|a^l\|_l)]^{1/2})$ and S is the matrix of hypothesis 3). Since S is a negative definite symmetric matrix, the largest eigenvalue of S, $\lambda_{\max}(S)$, is a negative real number. Therefore,

$$\nabla V(a,Aa) \leq u'Su \leq \lambda_{\max}(S)|u|^2 = \lambda_{\max}(S) \sum_{i=1}^{l} \psi_{i3}(\|a^i\|_i)$$

$$\leq -\psi_3(\|a\|), \quad (45)$$

where ψ_3 can always be chosen such that $\psi_3 \in K$.

The conclusion for Theorem 1 now follows from (44) and (45). ∎

To prove *Theorem 2*, it remains to be shown that hypothesis 2) of Theorem 2 implies that ψ_1, ψ_2, ψ_3 in the preceding proof of Theorem 1 can always be chosen to be of the same order of magnitude (see [1]).

By hypothesis 2), all functions $\psi_{ij} \in KR$, $i = 1, \cdots, l$, $j = 1,2,3$ are of the same order of magnitude. Therefore there is a function $\psi \in KR$, e.g., $\psi = \psi_{11}$ and positive constants d_{ij}, $i = 1, \cdots, l$, $j = 1,2,3$ such that

$$\psi_{i1}(r) \geq d_{i1}\psi(r), \; \psi_{i2}(r) \leq d_{i2}\psi(r), \; \psi_{i3}(r) \geq d_{i3}\psi(r)$$

for all $r \geq 0$, $i = 1, \cdots, l$. Now choose the functions ψ_1, ψ_2, and ψ_3 as

$$\psi_1(r) = \min_i (\alpha_i d_{i1})\psi(r), \quad \psi_2(r) = \left(\sum_{i=1}^{l} \alpha_i d_{i2} \right) \psi(r),$$

$$\psi_3(r) = -\lambda_{\max}(S) \min_i (d_{i3}) \psi(r),$$

each of class KR. Then one obtains

$$\sum_{i=1}^{l} \alpha_i \psi_{i1}(\|a^i\|_i) \geq \sum_{i=1}^{l} \alpha_i d_{i1} \psi(\|a^i\|_i)$$

$$\geq \min_i (\alpha_i d_{i1}) \sum_{i=1}^{l} \psi(\|a^i\|_i)$$

$$\geq \min_i (\alpha_i d_{i1}) \max_i \psi(\|a^i\|_i)$$

$$= \min_i (\alpha_i d_{i1}) \psi(\max \|a^i\|_i) = \psi_1(\|a\|). \quad (46)$$

Also,

$$\sum_{i=1}^{l} \alpha_i \psi_{i2}(\|a^i\|_i) \leq \sum_{i=1}^{l} \alpha_i d_{i2} \psi(\|a^i\|_i)$$

$$\leq \left(\sum_{i=1}^{l} \alpha_i d_{i2} \right) \max_i \psi(\|a^i\|_i)$$

$$= \left(\sum_{i=1}^{l} \alpha_i d_{i2} \right) \psi\left(\max_i \|a^i\|_i \right) = \psi_2(\|a\|). \quad (47)$$

Finally, since $\lambda_{\max}(S) < 0$, one has

$$\lambda_{\max}(S) \sum_{i=1}^{l} \psi_{i3}(\|a^i\|_i) \leq \lambda_{\max}(S) \sum_{i=1}^{l} d_{i3} \psi(\|a^i\|_i)$$

$$\leq \lambda_{\max}(S) \min_i (d_{i3}) \sum_{i=1}^{l} \psi(\|a^i\|_i)$$

$$\leq \lambda_{\max}(S) \min_i (d_{i3}) \max_i \psi(\|a^i\|_i)$$

$$= \lambda_{\max}(S) \min_i (d_{i3}) \psi\left(\max_i \|a^i\|_i \right) = -\psi_3(\|a\|). \quad (48)$$

Since the functions ψ_1, ψ_2, and ψ_3 are each multiples of ψ, they are of the same order of magnitude. They also satisfy (44) and (45). Therefore, the conclusion of Theorem 2 follows. ∎

VII. Concluding Remarks

At this point the following comments are in order.

1) Sufficient conditions for asymptotic stability and exponential stability of a large class of interconnected dynamical systems were established (Theorems 1 and 2).

2) Properties of M-matrices were employed to obtain several corollaries to the results cited in item 1 (Corollaries 1, 2, and 3).

3) The present results, which constitute a generalization to previous ones, dealing with systems described on finite-dimensional spaces, are general enough to allow stability analysis of systems described by ordinary differential equations, certain classes of partial differential

equations, integro-differential equations, differential difference equations, functional differential equations, stochastic differential equations, and other types of evolutionary systems. In addition, *hybrid systems*, described by a mixture of different types of equations, may be analyzed in a systematic fashion by the present method (see Example 1).

4) Although the present results deal with strong dynamical systems, the present approach may be applied to other types (e.g., weak dynamical systems, extended dynamical systems, and limit dynamical systems) with similar results.

5) It is emphasized that the present results differ significantly from those obtained in [8]. Specifically, in [8] vector Lyapunov functionals are employed, a comparison principle is utilized, and a test matrix (similar to matrix S of Theorems 1 and 2) is obtained which is *always* an M-matrix. In the present approach, scalar Lyapunov functions are used (consisting of a weighted sum of Lyapunov functions for the isolated subsystems), explicit usage of a comparison principle is avoided, and the test matrix S need not be an M-matrix. Also, in the present approach, the interconnecting structure of composite systems is exploited to a greater degree than in [8]. The present set of results and those of [8] are *not* equivalent, nor does one imply the other.

6) The purpose of Example 2 is to present a noncontrived example. No claims of a significant contribution in the area of nuclear reactors are intended.

Acknowledgment

The authors are indebted to Prof. R. K. Miller of the Department of Mathematics, Iowa State University, for numerous constructive suggestions and criticism. They would also like to thank Prof. P. J. Moylan of the University of Newcastle, Australia, for comments which improved the original manuscript.

References

[1] W. Hahn, *Stability of Motion*. New York: Springer, 1967.
[2] G. E. Ladas and V. Lakshminkantham, *Differential Equations in Abstract Spaces*. New York: Academic, 1972.
[3] F. N. Bailey, "The application of Lyapunov's second method to interconnected systems," *J. SIAM Contr.*, ser. A, vol. 3, pp. 443–462, 1966.
[4] A. A. Piontkovskii and L. D. Rutkovskaya, "Investigation of stability theory problems by the vector Lyapunov function method," *Automat. Remote Contr.*, vol. 10, pp. 1422–1429, 1967.
[5] W. E. Thompson, "Exponential stability of interconnected systems," *IEEE Trans. Automat. Contr.*, vol. AC-15, pp. 504–506, Aug. 1970.
[6] A. N. Michel and D. W. Porter, "Stability analysis of composite systems," *IEEE Trans. Automat. Contr.*, vol. AC-17, pp. 222–226, Apr. 1972.
[7] M. Araki and B. Kondo, "Stability and transient behavior of composite nonlinear systems," *IEEE Trans. Automat. Contr.*, vol. AC-17, pp. 537–541, Aug. 1972.
[8] V. M. Matrosov, "The method of vector Lyapunov functions in analysis of composite systems with distributed parameters," *Automat. Remote Contr.*, vol. 34, pp. 5–22, 1973.
[9] L. T. Grujić and D. D. Šiljak, "Asymptotic stability and instability of large-scale systems," *IEEE Trans. Automat. Contr.*, vol. AC-18, pp. 636–645, Dec. 1973.
[10] A. N. Michel, "Stability analysis of interconnected systems," *SIAM J. Contr.*, vol. 12, pp. 554–579, Aug. 1974.
[11] ——, "Stability analysis of stochastic composite systems," *IEEE Trans. Automat. Contr.*, pp. 246–250, Apr. 1975.
[12] ——, "Stability analysis and trajectory behavior of composite systems," *IEEE Trans. Circuits Syst.*, vol. CAS-22, pp. 305–312, Apr. 1975.
[13] A. N. Michel and R. D. Rasmussen, "Stochastic stability analysis of interconnected systems," in *Advances in Control and Dynamical Systems: Theory and Applications*, vol. 13, C. T. Leondes, Ed. New York: Academic, 1976.
[14] D. W. Porter and A. N. Michel, "Input-output stability of time-varying nonlinear multiloop feedback systems," *IEEE Trans. Automat. Contr.*, vol. AC-19, pp. 422–427, Aug. 1974.
[15] P. A. Cook, "On the stability of interconnected systems," *Int. J. Contr.*, vol. 20, pp. 407–415, 1974.
[16] A. N. Michel, "Quantitative analysis of simple and interconnected systems: Stability, boundedness and trajectory behavior," *IEEE Trans. Circuit Theory*, vol. CT-17, pp. 292–301, Aug. 1970.
[17] A. Ostrowski, "Determinanten mit überwiegender Hauptdiagonale und die absolute Konvergenz von linearen Iterationprozessen," *Comentarii Math. Helvetici*, vol. 30, pp. 175–210, 1956.
[18] M. Fiedler and V. Ptak, "On matrices with non-positive off-diagonal elements and positive principal minors," *Czech. Math. J.*, vol. 12, pp. 382–400, 1962.
[19] Z. Akcasu, G. S. Lillouche, and L. M. Shotkin, *Mathematical Methods in Nuclear Reactor Dynamics*. NY: Academic, 1971.
[20] H. Plaza and W. H. Kohler, "Coupled reactor kinetics equations," *Nucl. Sci. Eng.*, vol. 22, 1966.

Decentralized Control of Linear Multivariable Systems*

J. P. CORFMAT† and A. S. MORSE‡

Concepts from graph theory and from the geometric theory of linear systems are used to derive explicit conditions for determining when the closed-loop spectrum of a multi-channel linear system can be freely assigned or stabilized with decentralized control.

Summary—This paper studies the effect of decentralized feedback on the closed-loop properties of jointly controllable, jointly observable k-channel linear systems. Channel interactions within such systems are described by means of suitably defined directed graphs. The concept of a complete system is introduced. Complete systems prove to be precisely those systems which can be made both controllable and observable through a single channel by applying nondynamic decentralized feedback to all channels. Explicit conditions are derived for determining when the closed-loop spectrum of a k-channel linear system can be freely assigned or stabilized with decentralized control.

INTRODUCTION

It is sometimes desirable or required to construct a system's feedback control or decision strategy based on a constrained measurement or information pattern. If the information pattern consists of disjoint subsets of 'local' measurement variables, one subset for each local input variable, and if the information fed back to each local input depends only on the input's corresponding subset of measurement variables, then the information pattern is decentralized and the overall feedback strategy is called a decentralized control.

Decentralization arises in many contexts. Examples can be found of team decision and adjustment processes [2], [3] where the objective is for an organization of n members to optimize the expected value of a cost or payoff functional while each member is allowed to observe only part of a changing or random environment. Much work dealing with decentralization in economic systems centers around linear programming and the decomposition principle of [4]. There is a growing literature concerned with applications of decentralized control to frequency and voltage regulation problems associated with power sharing in multi-area electrical power pools [5, 6]. A conventional multiloop industrial controller not involving 'cross-couplings' can be viewed as a form of decentralized control, although the term decentralization is rarely used in this case.

Recent research concerned with decentralization has led to a number of interesting and useful results. Gersho and Karafin [7] and later Sandberg [8] proposed a form of decentralized control to synchronize data transmissions in a network of remotely located transmitting and receiving stations. Decentralization has been successfully applied to a single-commodity network flow problem by Lau, Persiano and Varaiya [9]. Aoki and Li [10] have studied the extent to which the state of a linear system can be recovered from information available to local control agents. Chu [11] has considered a decentralized version of the linear regulator problem for a system consisting of a string of identical subsystems.

McFadden [12] and later the present authors [13] studied the problem of stabilizing a specially structured linear system using a simple decentralized decision strategy appropriate to economic systems; similar problems have recently been discussed in [14]. Another study of the stabilization question for more general linear systems was undertaken by Aoki [15] who developed a number of interesting but partial results. A solution to the stabilization problem for a large class of linear systems was recently obtained by Wang and Davison [16].

In the present paper we study the effects of

*Received 7 August 1975; revised 25 February 1976. The original version of this paper, based on the principal results of [1], was presented at the *6th IFAC Congress* which was held in Boston/Cambridge, Mass. during August 1975. The published Proceedings of this IFAC Meeting may be ordered from: Instrument Society of America, 400 Stanwix Street, Pittsburgh PA 15222. This paper was recommended for publication in revised form by associate editor E. J. Davison.

†Department of Engineering and Applied Science Yale University, New Haven, CT 06520, U.S.A.

‡The work of the authors was supported by the U.S. Air Force Office of Scientific Research (Grant No. 72–2211) and by the National Research Council of Canada, (Grant No. 28140).

Reprinted with permission from *Automatica*, vol. 12, pp. 479–495, Sept. 1976.

decentralized feedback on the closed-loop properties of k-channel, jointly controllable, jointly observable, linear systems (§1). Channel interactions within such systems are usefully characterized in qualitative terms by means of suitably defined directed graphs (§2). The concept of a complete system is introduced in §3. Complete systems prove to be precisely those systems which can be made both controllable and observable through a single channel by application of nondynamic decentralized feedback to all channels (§3).

It is well-known[17] that the closed-loop spectrum of a linear system can be freely adjusted or assigned with centralized control provided the system is both controllable and observable. However with the added constraint of decentralization, controllability and observability are no longer sufficient as pointed out in[15]. In the sequel (§4) it is shown that if the graph of a k-channel system Σ is strongly connected, then the closed-loop spectrum of Σ can be freely assigned just in case Σ is complete. If the graph of Σ is not strongly connected, then spectrum assignment is possible if and only if (1) the dimension of Σ equals the sum of the dimensions of a set of strongly connected subsystems of Σ and (2) each subsystem is complete (§5).

If for a particular system, free spectrum assignment is not possible, then the spectrum of the closed-loop system which results from application of any specific linear decentralized control, contains a uniquely determined subset which cannot be shifted no matter what control is applied. By characterizing this subset directly in terms of the system's parameters (§4, 5), a complete solution is obtained to the problem of stabilizing a k-channel linear system with decentralized control. These results significantly extend the arthors' earlier work[18], and encompas the contributions of[16].

Notation

In the sequel all vector spaces, maps and linear transformations are defined over the reals; $\mathcal{R}^n$ is the real n-space; $\mathrm{d}(\mathcal{X})$ is the dimension of a vector space $\mathcal{X}$. Both linear maps and matrices are denoted by capital Roman letters $A, B, \ldots$; prime (′) means transpose.

An m-input, p-output, n-dimensional linear system $\dot{x} = Ax + Bu$, $y = Cx$, is often represented by the triple (C, A, B). We write $\mathcal{B}$ for image B, $\langle A|\mathcal{B}\rangle \equiv \mathcal{B} + A\mathcal{B} + \cdots + A^{n-1}\mathcal{B}$ for the controllable space of (A, B) and $[C|A] \equiv \bigcap_{i=1}^{n}$ kernel CA^{i-1} for the unobservable space of (C, A). The uncontrollable polynomial (u.p.) of (A, B) is the characteristic polynomial of A mod $\langle A|\mathcal{B}\rangle$.

If $\mathcal{R}$ is A-invariant, $A|\mathcal{R}$ is the restriction of A to $\mathcal{R}$. If $\mathcal{S} \supset \mathcal{R}$ and $\mathcal{S}$ is A-invariant, $A\|(\mathcal{S}/\mathcal{R})$ denotes the map induced by A in $\mathcal{S}/\mathcal{R}$.

The root set of a real polynomial $\alpha(\lambda)$ is the symmetric set of complex roots of $\alpha(\lambda)$ repeated according to multiplicity. The spectrum of (C, A, B) (i.e. the spectrum of A) is the root set of the characteristic polynomial (c.p.) of A. A set of complex numbers is stable, if all elements in the set are points in the open left half complex plane; $\alpha(\lambda)$ is a stable polynomial if its root set is stable.

The (unique) monic least common multiple of two polynomials $\alpha(\lambda)$ and $\beta(\lambda)$ is written as $\alpha(\lambda) \vee \beta(\lambda)$; the monic greatest common divisor of $\alpha(\lambda)$ and $\beta(\lambda)$ is denoted by $\alpha(\lambda) \wedge \beta(\lambda)$. The complement of a subset $\mathbf{s}$ in a set $\mathbf{k}$ is written as $\mathbf{k} - \mathbf{s}$.

The complement of a proper variety in a vector space is called a robust set. Properties of robust sets are discussed in the appendix.

1. FORMULATION

The model to be considered is a k-channel linear system $\Sigma \equiv \{C_i, A, B_i; \mathbf{k}\}$ described by the equations

$$\left.\begin{aligned} \dot{x}(t) &= Ax(t) + \sum_{i=1}^{k} B_i u_i(t) \\ y_i(t) &= C_i x(t), \qquad i \in \mathbf{k} \equiv \{1, 2, \ldots, k\} \end{aligned}\right\} \quad (1)$$

where $u_i(t)$ and $y_i(t)$ are vectors of control input and measured output variables associated with channel i; the time functions $u_i(\cdot)$, $x(\cdot)$ and $y_i(\cdot)$ take values in $\mathcal{U}_i(=\mathcal{R}^{m_i})$, $\mathcal{X}(=\mathcal{R}^n)$ and $\mathcal{Y}_i(=\mathcal{R}^{p_i})$ respectively, and $A, B_1, \ldots, B_k, C_1, \ldots, C_k$ are constant matrices or linear maps. We deal primarily with systems for which $k \geq 2$, for if $k = 1$ decentralization is not an issue.

To rule out trivialities, only those systems for which

$$\left.\begin{aligned} C_i &\neq 0 \\ B_i &\neq 0 \end{aligned}\right\}, \qquad i \in \mathbf{k} \quad (2)$$

are considered. We usually assume that Σ is a jointly controllable, jointly observable system; i.e. $\langle A|\sum_{i=1}^{k} \mathcal{B}_i\rangle = \mathcal{X}$ and $\bigcap_{i=1}^{k} [C_i|A] = 0$. Note however, that this assumption does not necessarily mean that Σ is controllable or observable through a single channel; e.g. joint controllability of Σ does not imply controllability of, for example, (A, B_1).

Roughly speaking, a feedback control for Σ is decentralized if each local control u_i

depends only on local measurement y_i for its implementation. A local controller for the ith channel of Σ is a linear dynamical system Σ_i forced by y_i and possibly a command input v_i; the output of Σ_i then serves as the control input u_i to the channel i. Thus Σ_i admits a model of the form

$$\left.\begin{aligned}\dot{z}_i(t) &= H_i z_i(t) + L_i y_i(t) + R_i v_i(t)\\ u_i(t) &= M_i z_i(t) + F_i y_i(t) + G_i v_i(t)\end{aligned}\right\} \qquad (3)$$

where $z_i(\cdot)$ takes values in $\mathscr{X}_i(=\mathscr{R}^{n_i})$. Note that the nondynamic local feedback law $u_i(t) = F_i y_i(t) + v_i(t)$ is a specialization of (3) obtained by setting $n_i = 0$ and G_i equal to the $p_i x p_i$ identity.

If the local controllers (3) are applied to (1), there results a closed-loop system (1), (3) of dimension $n + \sum_{i=1}^{k} n_i$ with spectrum (i.e., eigenvalue set) depending on the properties of the local controllers Σ_i. We say that the closed-loop spectrum of Σ is *freely assignable* with decentralized control if for some fixed set of positive integers $\{q_1, q_2, \ldots, q_r\}$ and for each symmetric set of complex numbers Λ, compatible with $\{q_1, q_2, \ldots, q_r\}$,† there exist local controllers Σ_i which, when applied to Σ, result in a closed-loop system (1), (3) with spectrum Λ. The main objective of this paper is to develop explicit conditions in terms of the matrices which characterize Σ, for determining when the closed-loop spectrum of Σ is freely assignable with decentralized control. This is the *spectrum assignment problem* for decentralized feedback systems.

One possible approach to the spectrum assignment problem is to first apply to Σ, local nondynamical controls of the form

$$u_i(t) = F_i y_i(t) + v_i(t), \qquad i \in \mathbf{k}, \qquad (4)$$

the objective being to select the F_i so that the resulting closed-loop system

$$\left.\begin{aligned}\dot{x}(t) &= \left(A + \sum_{j=1}^{k} B_j F_j C_j\right)x(t) + \sum_{i=1}^{k} B_i v_i(t)\\ y_i(t) &= C_i x(t), \qquad i \in \mathbf{k}.\end{aligned}\right\} \qquad (5)$$

is both controllable and observable through a single channel, say j. For if this can be accomplished, then standard techniques[17] can be used to construct a second local control system

$$\dot{z}_j(t) = H_j z_j(t) + L_j y_j(t)$$
$$v_j(t) = M_j z_j(t) + N_j y_j(t)$$

which, when applied to channel j of (5), results in a closed-loop system with prescribed spectrum. In the sequel we develop conditions in terms of the parameter matrices of Σ which tell when this program can be carried out. To simplify notation we henceforth write F for the matrix diag. $[F_1, F_2, \ldots, F_k]$, $\mathscr{F}$ for the vector space of all such matrices, A_F for $A + \sum_{i=1}^{k} B_i F_i C_i$, and $\Sigma_F \equiv \{C_i, A_F, B_i; \mathbf{k}\}$ for the k-channel system represented by (5).

Closely related to the spectrum assignment problem is the problem of determining when Σ can be stabilized with decentralized control. The stabilization problem has been studied[16] where it is shown that Σ can be stabilized with decentralized control just in case the polynomial

$$\alpha_\Sigma(\lambda) \equiv \bigwedge_{F \in \mathscr{F}} \det[\lambda I - A_F], \qquad (6)$$

called the *fixed polynomial* of Σ, has all its roots in the open left-half complex plane. The necessity of this condition is a direct consequence of the following proposition which can be deduced directly from the results of[16].

Proposition 1. *Let* $\Lambda(\Sigma_i, \mathbf{k})$ *denote the spectrum of the closed-loop system described by* (1) *and* (3); *and write* Λ_Σ *for the root set of* $\alpha_\Sigma(\lambda)$. *Then*

$$\Lambda_\Sigma \subset \Lambda(\Sigma_i, \mathbf{k})$$

for all linear controls Σ_i, i $\in \mathbf{k}$.

In the sequel, Proposition 1 will be used to arrive at an alternative solution to the stabilization problem with solvability conditions expressed directly in terms of the problem data. Our approach will, in addition, enable us to illustrate when the stabilization problem is well-posed, and when it is not.

2. CONTROLLABLE, SINGLE-CHANNEL SUBSYSTEMS

The principal problem to be considered here and in §3 is as follows: For given Σ and fixed $j \in \mathbf{k}$, find conditions for the existence of an $F \in \mathscr{F}$ such that (C_j, A_F, B_j) is a controllable, observable system. If an F with the required property exists for some j, then it turns out that such an F exists for each and every $j \in \mathbf{k}$; and when this is so we say that Σ can be made controllable and observable through a single channel with nondynamic decentralized control.

We begin by studying the simpler problem of determining when an $F \in \mathscr{F}$ can be found so

†A set of complex numbers Λ is compatible with $\{q_1, q_2, \ldots, q_r\}$ if Λ can be expressed as a disjoint union of r symmetric subsets, the number of elements in the ith subset equalling q_i.

that (A_F, B_j) is a controllable pair. This problem has in effect been solved in[19] for the case $k = 2$. Central to the problem's solution is the concept of a remnant polynomial. The *remnant polynomial* of a (single-channel) linear system (C, A, B) is defined as

$$\rho(C, A, B) = \begin{Bmatrix} 1 & \text{if } r \geq t \\ \prod_{i=r+1}^{t} \tau_i(\lambda) & \text{if } r < t \end{Bmatrix}, \qquad (7)$$

where r is the rational function rank of the transfer matrix $C(\lambda I - A)^{-1}B$ and $\tau_1(\lambda)$, $\tau_2(\lambda), \ldots, \tau_t(\lambda)$ are the transmission polynomials of (C, A, B)[20], ordered so that $\tau_i(\lambda)$ divides $\tau_{i-1}(\lambda)$, $i \in \{2, 3, \ldots, t\}$; (C, A, B) is called a *complete* (single-channel) system if its transfer matrix is nonzero and if $\rho(C, A, B) = 1$.

The main result of[19] is as follows.

Proposition 2. *Let* (C, A, B) *and* $D: \mathcal{V} \to \mathcal{X}$ *be fixed with* $\langle A|\mathcal{B} + \mathcal{D}\rangle = \mathcal{X}$ *and* $C(\lambda I - A)^{-1}B \neq 0$. *For all* $K: \mathcal{Y} \to \mathcal{V}$, $\rho(C, A, B)$ *divides the uncontrollable polynomial of* $(A + DKC, B)$, *the word 'divides' being replaceable by 'equals' just in case* K *is selected so that* $d(\langle A + DKC|\mathcal{B}\rangle)$ *is as large as possible.*

By making use of the following lemma, it is possible to restate Proposition 2 in a more convenient way.

Lemma 1.† *For the two-channel system* $\Sigma \equiv \{C_i, A, B_i; 2\}$ *and any decentralized feedback matrix* $F = \text{diag}.\ [F_1, F_2]$, $\langle A_F|\mathcal{B}_1\rangle = \langle A + B_2F_2C_2|\mathcal{B}_1\rangle$ *and* u.p. $(A_F, B_1) =$ u.p. $(A + B_2F_2C_2, B_1)$.

Lemma 1 and Proposition 2 clearly imply that

Proposition 2a. *If* $\Sigma \equiv \{C_i, A, B_i; 2\}$ *is jointly controllable and if* $C_2(\lambda I - A)^{-1}B_1 \neq 0$, *then for all* $F \in \mathcal{F}$, $\rho(C_2, A, B_1)$ *divides the uncontrollable polynomial of* (A_F, B_1), *the work divides being replaceable by equals just in case* $F \in \mathcal{F}$ *is selected so that* $d(\langle A_F|\mathcal{B}_1\rangle)$ *is as large as possible.*

From Proposition 2a it follows that if Σ is jointly controllable and if $C_2(\lambda I - A)^{-1}B \neq 0$, then F can be selected so that $\langle A_F|\mathcal{B}_1\rangle = \mathcal{X}$ if and only if (C_2, A, B_1) is complete.

In order to extend these results to k-channel systems, it is necessary to take into account which transfer matrices within Σ are zero and which ones are not. This is a topological matter and is therefore best treated in graph-theoretic terms.

We define the *graph* of Σ, written G_Σ, to be a pair $(\mathbf{k}, \Gamma)$ consisting of k nodes labeled 1, 2, . . , k, with node i representing the ith channel of Σ, and a function Γ mapping $\mathbf{k}$ into its power set according to the rule $\Gamma(j) = \{i\colon\ H_{ij} \neq 0\}$; $H_{ij} \equiv C_i(\lambda I - A)^{-1}B_j$ is the transfer matrix from channel input u_j of Σ to channel output y_i. The graph may be represented by a diagram consisting of a set of k nodes, labeled $1, 2, \ldots, k$, with a directed arc drawn from node j to node i just in case u_j influences y_i in the sense that $H_{ij} \neq 0$. The graph of Σ thus represents the open-loop pattern of signal flow between the channels of Σ.

It is perhaps natural to expect that if (A_F, B_j) is controllable for some F, then each node of G_Σ must be reachable from node j along a directed path† in G_Σ. This is, in fact, the case.

Proposition 3. *If for some* $j \in \mathbf{k}$ *and* $F \in \mathcal{F}$, (A_F, B_j) *is controllable, then for each* $i \in \mathbf{k}$, *with* $i \neq j$, *there exists a path in* G_Σ *from node* j *to node* i.

While this property of G_Σ is necessary for (A_F, B_j) to be controllable for some F, it is not sufficient for as we have already noted, when $k = 2$ completeness of a certain subsystem of Σ is also required. To describe analogous subsystems for k-channel systems, we introduce the following notation: If $\mathbf{s}$ is a nonempty subset of $\mathbf{k}$, with elements $i_1, i_2, \ldots, i_p$ ordered so that $i_1 < i_2 < \ldots < i_p$, then we define $B_\mathbf{s}$ and $C_\mathbf{s}$ so that

$$C_\mathbf{s} = \begin{bmatrix} C_{i_1} \\ C_{i_2} \\ \vdots \\ C_{i_q} \end{bmatrix} \qquad B_\mathbf{s} = [B_{i_1}, B_{i_2}, \ldots, B_{i_q}]$$

Thus $(C_{\mathbf{s}_1}, A, B_{\mathbf{s}_2})$ is a subsystem of Σ modeling the relationship between channel inputs u_i, $i \in \mathbf{s}_2$ and channel outputs y_j, $j \in \mathbf{s}_1$. We call $(C_{\mathbf{s}_1}, A, B_{\mathbf{s}_2})$ a *complementary subsystem* of Σ if $\mathbf{s}_2$ is a proper subset of $\mathbf{k}$ and if $\mathbf{s}_1 = \mathbf{k} - \mathbf{s}_2$; a complementary subsystem $(C_{\mathbf{k}-\mathbf{s}}, A, B_\mathbf{s})$ is said to *contain* input (resp. output) channel j if j is an element of $\mathbf{s}$ (resp. $\mathbf{k} - \mathbf{s}$).

The following lemma provides a useful relationship between path properties of G_Σ and transfer matrices of complementary subsystems of Σ. Below, $\mathbf{k}(j)$ denotes the class of all proper subsets of $\mathbf{k}$ which contain j, and $\bar{\mathbf{k}}$ is the class of all nonempty subsets of $\mathbf{k}$.

†The simple proof of this lemma is omitted.

†A *path* from node i_1 to node i_p of G_Σ can be interpreted as a sequence of nodes $i_1, \ldots, i_p$ satisfying $i_q \in \Gamma(i_{q-1})$, $q \in \{2, 3, \ldots, p\}$.

Lemma 2. *Let* i *and* j *be distinct integers in* **k**. *There exists a path in* G_Σ *from node* j *to node* i *if and only if*

$$C_{\mathbf{k}-\mathbf{s}}(\lambda I - A)^{-1}B_{\mathbf{s}} \neq 0, \qquad \mathbf{s} \in (\bar{\mathbf{k}} - \mathbf{k}(i)) \cap \mathbf{k}(j) \tag{8}$$

Proof. The existence of a path in G_Σ from node j to node i is equivalent to the requirement that for each proper subset $\mathbf{s} \subset \mathbf{k}$ satisfying $j \in \mathbf{s}$ and $i \in \mathbf{k} - \mathbf{s}$, there exist integers $i_1 \in \mathbf{s}$ and $i_2 \in \mathbf{k} - \mathbf{s}$ such that $i_2 \in \Gamma(i_1)$. But the class of proper subsets $\mathbf{s}$ satisfying $j \in \mathbf{s}$ and $i \in \mathbf{k} - \mathbf{s}$ coincides with $(\bar{\mathbf{k}} - \mathbf{k}(i)) \cap \mathbf{k}(j)$ and the condition $i_2 \in \Gamma(i_1)$ is equivalent to $C_{i_2}(\lambda I - A)^{-1}B_{i_1} \neq 0$. Furthermore, $C_{\mathbf{k}-\mathbf{s}}(\lambda I - A)^{-1}B_{\mathbf{s}} \neq 0$ if and only if $C_{i_2}(\lambda I - A)^{-1}B_{i_1} \neq 0$ for some $i_1 \in \mathbf{s}$ and $i_2 \in \mathbf{k} - \mathbf{s}$. Thus there exists a path in G_Σ from node j to node i just in case (8) is true. □

From the easily verified identity

$$\mathbf{k}(j) = \bigcup_{\substack{i=1 \\ i \neq j}}^{k} ((\bar{\mathbf{k}} - \mathbf{k}(i)) \cap \mathbf{k}(j))$$

and Lemma 2 follows that for each $i \in \mathbf{k}$, with $i \neq j$, there exists a path in G_Σ from node j to node i if and only if

$$C_{\mathbf{k}-\mathbf{s}}(\lambda I - A)^{-1}B_{\mathbf{s}} \neq 0, \qquad \mathbf{s} \in \mathbf{k}(j)$$

Since $\{(C_{\mathbf{k}-\mathbf{s}}, A, B_{\mathbf{s}}) : \mathbf{s} \in \mathbf{k}(j)\}$ is the set of all complementary subsystems of Σ containing input channel j, we can state

Lemma 3. *The transfer matrices of all complementary subsystems of* Σ, *which contain input channel* j, *are nonzero if and only if for each* $i \in \mathbf{k}$, *with* $i \neq j$, *there exists a path in the graph of* Σ *from node* j *to node* i.

Proof of Proposition 3. Let (A_F, B_j) be controllable. Fix $\mathbf{s} \in \mathbf{k}(j)$; then $\langle A + \sum_{i \in \mathbf{k}-\mathbf{s}} B_iF_iC_i | \mathcal{B}_{\mathbf{s}} \rangle \supset \langle A + \sum_{i \in \mathbf{k}} B_iF_iC_i | \mathcal{B}_j \rangle$. Since $(A + \sum_{i \in \mathbf{k}} B_iF_iC_i,\ B_j)$ is controllable, it follows that $\langle A + \sum_{i \in \mathbf{k}-\mathbf{s}} B_iF_iC_i | \mathcal{B}_{\mathbf{s}} \rangle = \mathcal{X}$.

If $C_{\mathbf{k}-\mathbf{s}}(\lambda I - A)^{-1}B_{\mathbf{s}} = 0$, then $C_{\mathbf{k}-\mathbf{s}}A^iB_{\mathbf{s}} = 0$, $i \geq 0$; thus $\langle A | \mathcal{B}_{\mathbf{s}} \rangle \subset \text{kernel } C_{\mathbf{k}-\mathbf{s}}$ and $\langle A | \mathcal{B}_{\mathbf{s}} \rangle = \langle A + \sum_{i \in \mathbf{k}-\mathbf{s}} B_iF_iC_i | \mathcal{B}_{\mathbf{s}} \rangle$. Therefore $\mathcal{X} \subset \text{kernel } C_{\mathbf{k}-\mathbf{s}}$ or equivalently, $C_{\mathbf{k}-\mathbf{s}} = 0$. This clearly contradicts (2); therefore $C_{\mathbf{k}-\mathbf{s}}(\lambda I - A)^{-1}B_{\mathbf{s}} \neq 0$. Since $\mathbf{s} \in \mathbf{k}(j)$ is arbitrary, it must be true that the transfer matrices of all complementary subsystems of Σ containing input channel j, are nonzero. It follows from Lemma 3 that Proposition 3 is true. □

As noted earlier, Proposition 3 provides only necessary conditions for (A_F, B_j) to be controllable for some F. A complete solution to the problem is provided by the following theorem.

Theorem 1. *Let* $\Sigma \equiv \{C_i, A, B_i; \mathbf{k}\}$ *and* $j \in \mathbf{k}$ *be fixed with* Σ *jointly controllable. There exists an* $F \in \mathcal{F}$ *such that*

$$\langle A_F | \mathcal{B}_j \rangle = \mathcal{X}, \tag{9}$$

if and only if each complementary subsystem of Σ *which contains input channel* j *is complete.* For fixed $j \in \mathbf{k}$, let $\mathcal{F}_j$ denote the set of all $F \in \mathcal{F}$ for which $d(\langle A_F | \mathcal{B}_j \rangle)$ is as large as possible. As defined, $\mathcal{F}_j$ is clearly robust {cf. Appendix} in $\mathcal{F}$. This implies that if $F \in \mathcal{F}$ is selected at random, then F is almost certain to be in $\mathcal{F}_j$. Therefore, if (9) holds for some F, then it holds for 'almost every' $F \in \mathcal{F}$.

The following theorem further characterizes the pair (A_F, B_j) for the case when the conditions of Theorem 1 fail to hold.

Theorem 2. *Let* Σ *and* j *satisfy the hypotheses of Theorem* 1, *and write* $\alpha_F(\lambda)$ *for the uncontrollable polynomial of* (A_F, B_j).
i. *For all* $F \in \mathcal{F}$,

$$\bigvee_{\mathbf{s} \in \mathbf{k}(j)} \rho(C_{\mathbf{k}-\mathbf{s}}, A, B_{\mathbf{s}}) \text{ divides } \alpha_F(\lambda) \tag{10}$$

ii. *If, in addition, for each* $i \in \mathbf{k}$ *with* $i \neq j$ *there is a path in* G_Σ *from node* j *to node* i, *then* $\alpha_F(\lambda)$ *is independent of* $F \in \mathcal{F}_j$ *and*

$$\alpha_F(\lambda) \text{ divides } \prod_{\mathbf{s} \in \mathbf{k}(j)} \rho(C_{\mathbf{k}-\mathbf{s}}, A, B_{\mathbf{s}}) \tag{11}$$

Theorem 2 implies that the uncontrollable spectrum of (A_F, B_j), namely the root set of $\alpha_F(\lambda)$, contains the root set of $\bigvee_{\mathbf{s} \in \mathbf{k}(j)} \rho(C_{\mathbf{k}-\mathbf{s}}, A, B_{\mathbf{s}})$ as a fixed subset for all $F \in \mathcal{F}$. The theorem also implies that if node j of G_Σ is connected by a directed path to each other node of G_Σ, then the uncontrollable spectrum of (A_F, B_j) is contained in another fixed subset, namely the root set of $\prod_{\mathbf{s} \in \mathbf{k}(j)} \rho(C_{\mathbf{k}-\mathbf{s}}, A, B_{\mathbf{s}})$, whenever F is selected so that $d(\langle A_F | \mathcal{B}_j \rangle)$ is as large as possible. From the obvious relationship between the two fixed subsets bounding the root set of $\alpha_F(\lambda)$, it follows that if node j of G_Σ is connected to each other node of G_Σ, then F can be selected so that (A_F, B_j) is stabilizable just in case each polynomial $\rho(C_{\mathbf{k}-\mathbf{s}}, A, B_{\mathbf{s}})$, $\mathbf{s} \in \mathbf{k}(j)$, is stable.

In the sequel, use will be made of the following remarks, which pertain to any linear system (C, A, B).

Remark 1. If D is the identity on $\mathscr{X}$, then $\langle A|\mathscr{B}+\mathscr{D}\rangle=\mathscr{X}$. From this and Proposition 2 it follows that if $C(\lambda I-A)^{-1}B\neq 0$, then u.p. $(A+KC, B)=\rho(C, A, B)$ whenever K is selected so that $d(\langle A+KC|\mathscr{B}\rangle)$ is as large as possible.

Remark 2. It is easy to verify that the rational function rank of the transfer matrix $C(\lambda I-A-KC)^{-1}B$ is independent of K. Since the transmission polynomials of $(C, A+KC, B)$ are known to be independent of K {cf. [20], Corollary 4.1}, it follows from the definition of $\rho(\cdot)$, that $\rho(C, A+KC, B)$ is independent of K as well.

The proofs of Theorems 1 and 2 are based on the following lemmas.

Lemma 4. *Let* (C, A, B) *and* $D:\mathscr{V}\to\mathscr{X}$ *be fixed with* $C(\lambda I-A)^{-1}B\neq 0$. *Then* u.p. (A + DKC, B) *divides* (ρ(C, A, B)) (u.p. (A, [B, D])) *whenever* K *is selected so that* $d(\langle A+DKC|\mathscr{B}\rangle)$ *is as large as possible.*

Proof. Let $(\bar{C}, \bar{A}, \bar{B})$ denote the system obtained by restricting the maps of (C, A, B) to $\bar{\mathscr{X}}\equiv\langle A|\mathscr{B}+\mathscr{D}\rangle$; and write $\bar{D}$: $\mathscr{V}\to\bar{\mathscr{X}}$ for the restricted map $v\mapsto Dv$. Then for any K, u.p. $(A+DKC, B)=$ (u.p. $(\bar{A}+\bar{D}K\bar{C}, \bar{B})$) (u.p. $(A+DKC, [B, D])$); but since u.p. $(A+DKC, [B, D])$ is independent of K,

$$\text{u.p.}(A+DKC, B)=(\text{u.p.}(\bar{A}+\bar{D}K\bar{C}, \bar{B}))\times(\text{u.p.}(A, [B, D])) \quad (12)$$

Now fix K_0 so that $d(\langle A+DK_0C|\mathscr{B}\rangle)$ is as large as possible; since the relation $\langle A+DKC|\mathscr{B}\rangle=\langle\bar{A}+\bar{D}K\bar{C}|\bar{\mathscr{B}}\rangle$ holds for all K, it follows that $K\equiv K_0$ maximizes $d(\langle\bar{A}+\bar{D}K\bar{C}|\bar{\mathscr{B}}\rangle s$ as well. But $\langle\bar{A}|\bar{\beta}+\bar{\mathscr{D}}\rangle=\bar{\mathscr{X}}$ and the hypothesis $C(\lambda I-A)^{-1}B\neq 0$ implies $\bar{C}(\lambda\bar{I}-\bar{A})^{-1}\bar{B}\neq 0$. It therefore follows from Proposition 2 applied to the data $(\bar{C}, \bar{A}, \bar{B})$ and $\bar{D}$, that u.p. $(\bar{A}+\bar{D}K_0\bar{C}, \bar{B})=\rho(\bar{C}, \bar{A}, \bar{B})$. This and (12) evaluated at $K=K_0$ the yield u.p.$(A+DK_0C, B)=(\rho(\bar{C}, \bar{A}, \bar{B}))$(u.p. $(A, [B, D])$. To complete the proof it is therefore enough to show that $\rho(\bar{C}, \bar{A}, \bar{B})$ divides $\rho(C, A, B)$.

Let $\mathscr{S}$ and $\mathscr{R}$ denote the largest (A, B)-invariant and controllability subspaces in kernel C [22]. Since $\mathscr{R}\subset\langle A|\mathscr{B}+\mathscr{D}\rangle=\bar{\mathscr{X}}$ and kernel $\bar{C}=\bar{\mathscr{X}}\cap$ (kernel C), $\mathscr{R}$ is also the largest $(\bar{A}, \bar{B})$-invariant subspace in kernel $\bar{C}$. Furthermore, $\mathscr{B}\subset\bar{\mathscr{X}}$ and $A\bar{\mathscr{X}}\subset\bar{\mathscr{X}}$; thus $\mathscr{S}\cap\bar{\mathscr{X}}$ is (A, B)-invariant {cf. [22], Lemma 7.1}. Clearly $\mathscr{S}\cap\bar{\mathscr{X}}$ is the largest (A, B)-invariant subspace in kernel $\bar{C}$. Therefore if $M:\mathscr{U}\to\mathscr{X}$ is selected so that both $\mathscr{S}$ and $\mathscr{S}\cap\bar{\mathscr{X}}$ are $(A+BM)$-invariant, then

$$\text{transmission pol.}(C, A, B)=\text{invariant factors }(A+BM)\|(\mathscr{S}/\mathscr{R})$$
$$\text{transmission pol.}(\bar{C}, \bar{A}, \bar{B})=\text{invariant factors }(A+BM)\|((\mathscr{S}\cap\bar{\mathscr{X}})/\mathscr{R})$$

Since $(\mathscr{S}\cap\bar{\mathscr{X}})/\mathscr{R}\subset\mathscr{S}/\mathscr{R}$, it follows {cf. [19], Lemma 1} that if $\{\tau_1, \tau_2, \ldots, \tau_t\}\equiv$ trans. pol. (C, A, B) and $\{\bar{\tau}_1, \bar{\tau}_2, \ldots, \bar{\tau}_{\bar{t}}\}\equiv$ trans. pol. $(\bar{C}, \bar{A}, \bar{B})$, then $\bar{t}\leq t$ and $\bar{\tau}_i$ divides τ_i, $i\in\{1, 2, \ldots, \bar{t}\}$. From this, the definition of a remnant polynomial and the fact that the rank of $C(\lambda I-A)^{-1}B$ equals the rank of $\bar{C}(\lambda\bar{I}-\bar{A})^{-1}\bar{B}$, it follows that $\rho(\bar{C}, \bar{A}, \bar{B})$ divides $\rho(C, A, B)$. □

Lemma 5. *Let* $\Sigma=\{C_i, A, B_i; 2\}$ *be a fixed two-channel system.*

i. *There exists a map* $F\in\mathscr{F}$ *such that*

$$C_1(\lambda I-A_F)^{-1}\neq 0 \quad (13)$$

if and only if

$$C_1(\lambda I-A)^{-1}[B_1, B_2]\neq 0 \quad (14a)$$

and

$$\begin{bmatrix}C_1\\C_2\end{bmatrix}(\lambda I-A)^{-1}B_1\neq 0 \quad (14b)$$

ii. *If* (14) *holds, then there exists a robust set* $\mathscr{F}^*\in\mathscr{F}$ *with the property that if* $F\in\mathscr{F}^*$, *then* (13) *is satisfied and*

$$\rho(C_1, A_F, B_1) \textit{ divides } \rho\left(\begin{bmatrix}C_1\\C_2\end{bmatrix}, A, B_1\right)\rho(C_1, A, [B_1, B_2]). \quad (15)$$

Proof. (i) If (14a) fails to hold, then $C_1A^i[B_1, B_2]=0$ $(i\geq 0)$ or equivalently, $\langle A|\mathscr{B}_1+\mathscr{B}_2\rangle\subset$ kernel C_1. Since $\langle A_F|\mathscr{B}_1\rangle\subset\langle A|\mathscr{B}_1+\mathscr{B}_2\rangle$, it follows that $\langle A_F|\mathscr{B}_1\rangle\subset$ kernel C_1; hence $C_1A_F^iB_1=0$ $(i\geq 0,\ F\in\mathscr{F})$ which is contrary to (13). Therefore (14a) and (by duality) (14b) are necessary.

For sufficiency, suppose (14) holds. If $C_1(\lambda I-A)^{-1}B_1\neq 0$, then (13) is satisfied with $F=0$. If $C_1(\lambda I-A)^{-1}B_1=0$, then (14) implies that $C_1(\lambda I-A)^{-1}B_2\neq 0$ and that $C_2(\lambda I-A)^{-1}B_1\neq 0$. Hence in this case there must be least positive integers q_2 and q_1 such that $C_1A^{q_2-1}B_2\neq 0$ and that $C_2A^{q_1-1}B_1\neq 0$. But by assumption $C_1A^iB_1=0$ $(i\geq 0)$; it follows that $A_F^{i-1}B_1=A^{i-1}B_1$ $(i\in\{1, 2, \ldots, q_1\})$ and that $C_1A_F^{i-1}=C_1A^{i-1}$ $(i\in\{1, 2, \ldots, q_2\})$ for any matrix $F=$ diag. $[F_1,$

$F_2]$ in $\mathscr{F}$. Thus

$$C_1 A_F^{(q_1+q_2-1)} B_1 = C_1 A_F^{q_2} A^{q_1-1} B_1 = C_1 A^{q_2-1}(A + B_1 F_1 C_1 + B_2 F_2 C_2) A^{q_1-1} B_1 = C_1 A^{q_2-1} B_2 F_2 C_2 A^{q_1-1} B_1$$

Since $C_1 A^{q_2-1} B_2 \neq 0$ and $C_2 A^{q_1-1} B_1 \neq 0$, F (i.e. F_2) can now be chosen so that $C_1 A_F^{(q_1+q_2-1)} B_1 \neq 0$; and (13) clearly holds with this choice of F.

(ii) Assume that (14) is true and let $\hat{\mathscr{F}}$ denote the set of $F \in \mathscr{F}$ for which (13) is satisfied; *i* above implies that $\hat{\mathscr{F}}$ is nonempty. Since (13) fails to hold only a proper variety in $\mathscr{F}$, $\hat{\mathscr{F}}$ is robust in $\mathscr{F}$.

Let $\hat{\mathscr{K}}$ denote the set of maps K in the linear space $\mathscr{K}$ of maps $\mathscr{Y}_1 \to \mathscr{X}$ for which $d(\langle A + KC_1 | \mathscr{B}_1 + \mathscr{B}_2\rangle)$ is maximal; $\hat{\mathscr{K}}$ is clearly robust in $\mathscr{K}$. Therefore the product set $\hat{\mathscr{F}} \times \hat{\mathscr{K}}$ is robust in $\mathscr{F} \oplus \mathscr{K}$. Write $\mathscr{M}$ for the set of pairs (F, K) in $\mathscr{F} \oplus \mathscr{K}$ for which $d(\langle A_F + KC_1 | \mathscr{B}_1\rangle)$ is maximal. Then $\mathscr{M}$ is robust in $\mathscr{F} \oplus \mathscr{K}$ as is the intersection $\mathscr{M} \cap (\hat{\mathscr{F}} \times \hat{\mathscr{K}})$.

Let $(F_0, K_0) \in \mathscr{M} \cap (\hat{\mathscr{F}} \times \hat{\mathscr{K}})$ be fixed. Define $\mathscr{F}^*$ to be the set of $F \in \hat{\mathscr{F}}$ such that $d(\langle A_F + K_0 C_1 \| \mathscr{B}_1\rangle)$ is maximal; then $\mathscr{F}^*$ is robust in $\mathscr{F}$. It will now be shown that (13) and (15) hold for any $F \in \mathscr{F}^*$.

Fix $F^* \in \mathscr{F}^*$; then F^* maximizes $d(\langle A_F + K_0 C_1 | \mathscr{B}_1\rangle)$ with respect to $F \geqslant \hat{\mathscr{F}}$ as does F_0 (since $(F_0, K_0) \in \mathscr{M}$ and $F_0 \in \hat{\mathscr{F}}$). But $d(\langle A_F + K_0 C_1 | \mathscr{B}_1\rangle)$ can have only one maximal value with respect to $F \in \hat{\mathscr{F}}$ so $d(\langle A_{F^*} + K_0 C_1 | \mathscr{B}_1\rangle) = d(\langle A_{F_0} + K_0 C_1 | \mathscr{B}_1\rangle)$. Since $(F_0, K_0) \in \mathscr{M}$, it must be true that $(F^*, K_0) \in \mathscr{M}$; in other words

$$d(\langle A_{F^*} + K_0 C_1 | \mathscr{B}_1\rangle) = \max_{F, K} d(\langle A_F + KC_1 | \mathscr{B}_1\rangle) \tag{16}$$

Equation (16) implies that K_0 maximizes $d(\langle A_{F^*} + KC_1 | \mathscr{B}_1\rangle)$ with respect to $K \in \mathscr{K}$. In addition, $C_1(\lambda I - A_{F^*})^{-1} B_1 \neq 0$ since $F^* \in \hat{\mathscr{F}}$. Therefore, by Remark 1,

$$\text{u.p.}\,(A_{F^*} + K_0 C_1, B_1) = \rho(C_1, A_{F^*}, B_1) \tag{17}$$

Write $F = \text{diag.}\,[F_1, F_2]$. Then (16) implies that (F_1^*, F_2^*, K_0) maximizes $d(\langle A + B_1 F_1 C_1 + B_2 F_2 C_2 + KC_1 | \mathscr{B}_1\rangle)$ with respect to (F_1, F_2, K). Since $\langle A + B_1 F_1 C_1 + B_2 F_2 C_2 + KC_1 | \mathscr{B}_1\rangle$ is independent of F_1 {cf. Lemma 1}, it must be that (F_2^*, K_0) maximizes $d(\langle A + B_2 F_2 C_2 + KC_1 | \mathscr{B}_1\rangle)$ with respect to (F_2, K). Furthermore since the substitution $K = B_2 \bar{F} + \bar{K}$ is reversible, $d(\langle A + B_2 F_2 C_2 + B_2 \bar{F} C_1 + \bar{K} C_1 | \mathscr{B}_1\rangle)$ attains it maximum with respect to $(\bar{F}, F_2, \bar{K})$ at $(0, F_2^*, K_0)$. It follows that

$$d\left(\left\langle A + B_2[0, F_2^*]\begin{bmatrix} C_1 \\ C_2 \end{bmatrix} + K_0 C_1 \middle| \mathscr{B}_1\right\rangle\right) = \max_{[\bar{F}, F_2]} d\left(\left\langle A + B_2[\bar{F}, F_2]\begin{bmatrix} C_1 \\ C_2 \end{bmatrix} + K_0 C_1 \middle| \mathscr{B}_1\right\rangle\right) \tag{18}$$

Remark 2 and (14b) imply that $\begin{bmatrix} C_1 \\ C_2 \end{bmatrix} (\lambda I - A - K_0 C_1)^{-1} B_1 \neq 0$. From this, (18) and Lemma 4 applied to the system $\left(\begin{bmatrix} C_1 \\ C_2 \end{bmatrix}, A + K_0 C_1, B_1\right)$ with $D \equiv B_2$, it follows that u.p. $\left(A + B_2[0, F_2^*]\begin{bmatrix} C_1 \\ C_2 \end{bmatrix} + K_0 C_1, B_1\right)$ divides $\left(\rho\left(\begin{bmatrix} C_1 \\ C_2 \end{bmatrix}, A + K_0 C_1, B_1\right)\right)$ (u.p. $(A + K_0 C_1, [B_1, B_2])$); but by Lemma 1, u.p. $\left(A + B_2[0, F_2^*]\begin{bmatrix} C_1 \\ C_2 \end{bmatrix} + K_0 C_1, B_1\right) =$ u.p. $(A_{F^*} + K_0 C_1, B_1)$ and by Remark 2, $\rho\left(\begin{bmatrix} C_1 \\ C_2 \end{bmatrix}, A + K_0 C_1, B_1\right) = \rho\left(\begin{bmatrix} C_1 \\ C_2 \end{bmatrix}, A, B_1\right)$. Thus u.p. $(A_{f^*} + K_0 C_1, B_1)$ divides $\left(\rho\left(\begin{bmatrix} C_1 \\ C_2 \end{bmatrix}, A, B_1\right)\right)$ (u.p. $(A + K_0 C_1, [B_1, B_2])$). This and (17) show that

$$\rho(C_1, A_{F^*}, B_1) \text{ divides } \left(\rho\left(\begin{bmatrix} C_1 \\ C_2 \end{bmatrix}, A, B_1\right)\right) \times (\text{u.p.}(A + K_0 C_1, [B_1, B_2]) \tag{19}$$

Since K_0 maximizes $d(\langle A + KC_1 | \mathscr{B}_1 + \mathscr{B}_2\rangle)$ with respect to $K \in \mathscr{K}$ (i.e. $K_0 \in \hat{\mathscr{K}}$), it follows from (14a) and Remark 1 that u.p. $(A + K_0 C_1, [B_1, B_2]) = \rho(C_1, A, [B_1, B_2])$. Substitution into (19) yields (15) which is the desired result. □

Lemma 6. *For given* $\Sigma \equiv \{C_i, A, B_i; k\}$, *suppose that* u.p. (A_F, B_1) *divides a fixed polynomial* $\delta(\lambda)$ *for all* F *in some robust subset of* $\mathscr{F}$. *Then there exists a monic polynomial* $\mu(\lambda)$ *which divides both* $\delta(\lambda)$ *and* u.p. (A_F, B_1) *for all* $F \in \mathscr{F}$, *and which equals* u.p. (A_F, B_1) *whenever* F *is selected so that* $d(\langle A_F | \mathscr{B}_1\rangle)$ *is as large as possible.*

Proof. Let $\mathscr{F}^*$ denote the set of $F \in \mathscr{F}$ for which $d(\langle A_F | \mathscr{B}_1\rangle)$ is as large as possible. Since $d(\langle A_F | \mathscr{B}_1\rangle)$ is independent of $F \in \mathscr{F}^*$, the integer $n^* \equiv d(\mathscr{X} / \langle A_F | \mathscr{B}_1\rangle)$ is uniquely determined by any $F \in \mathscr{F}^*$; clearly

$$\text{degree}\,(\text{u.p.}\,(A_F, B_1)) = n^*, \qquad F \in \mathscr{F}^* \tag{20}$$

Let $\mathscr{K}$ denote the vector space of maps $\mathscr{X} \to \mathscr{U}_1$, and write $\alpha_{F,K}(\lambda) \equiv$ c.p. $(A_F + B_1K)|\langle A_F + B_1K|\mathscr{B}_1\rangle$. Since u.p. $(A_F + B_1K, B_1)$ is clearly independent of K, we can write c.p. $(A_F + B_1K) = \alpha_{F,K}(\lambda)$(u.p. (A_F, B_1)). But c.p. $(A_F + B_1K)$ can also be written in factored form as $\mu(\lambda)\beta_{F,K}(\lambda)$ where $\mu(\lambda) \equiv \bigwedge_{F,K}$ c.p. (A_F+B_1K); thus

$$\alpha_{F,K}(\lambda)(\text{u.p.}\,(A_F, B_1)) = \mu(\lambda)\beta_{F,K}(\lambda),$$
$$(F, K) \in \mathscr{F}\oplus\mathscr{K} \qquad (21)$$

Since $\alpha_{F,K}(\lambda) =$ c.p. $(A_F + B_1K)$ restricted to the controllable space of (A_F, B_1), it follows that for each fixed $F \in \mathscr{F}$, a $K \in \mathscr{K}$ can be found for which $\alpha_{F,K}(\lambda)$ and $\mu(\lambda)$ are coprime. This and (21) clearly imply that

$$\mu(\lambda) \text{ divides u.p. } (A_F, B_1), \qquad F \in \mathscr{F} \qquad (22)$$

In addition, since by hypothesis u.p. (A_F, B_1) divides $\delta(\lambda)$ for F in some robust set $\underline{\mathscr{R}} \subset \mathscr{F}$, $\mu(\lambda)$ must divide $\delta(\lambda)$.

By Lemma A.1, the set $\underline{\mathscr{S}}$ of pairs $(F, K) \in \mathscr{F}\oplus\mathscr{K}$ such that $\beta_{F,K}(\lambda)$ and $\delta(\lambda)$ are coprime, is robust in $\mathscr{F}\oplus\mathscr{K}$. Since $\underline{\mathscr{F}}^*$ is robust in $\mathscr{F}$, so is $\underline{\mathscr{R}} \cap \underline{\mathscr{F}}^*$. Therefore $\underline{\mathscr{S}} \cap ((\underline{\mathscr{R}} \cap \underline{\mathscr{F}}^*) \times \mathscr{K})$ is robust in $\mathscr{F}\oplus\mathscr{K}$. This means that there exists a pair (F_0, K_0) such that $F_0 \in \underline{\mathscr{R}} \cap \underline{\mathscr{F}}^*$ and $\beta_{F_0,K_0}(\lambda)$ and $\delta(\lambda)$ are coprime; but u.p. (A_F, B_1) divides $\delta(\lambda)$ for $F \in \underline{\mathscr{R}}$, so $\beta_{F_0,K_0}(\lambda)$ and u.p. (A_{F_0}, B_1) are coprime. It follows from (21) that u.p. (A_{F_0}, B_1) divides $\mu(\lambda)$; hence by (22), u.p. $(A_{F_0}, B_1) = \mu(\lambda)$. But $F_0 \in \underline{\mathscr{F}}^*$ and from (20), degree (u.p. (A_F, B_1)) is independent of $F \in \underline{\mathscr{F}}^*$. Therefore degree (u.p. (A_F, B_1)) $=$ degree $\mu(\lambda)$, $F \in \underline{\mathscr{F}}^*$. But this and (22) can be true only if u.p. $(A_F, B_1) = \mu(\lambda)$ for $F \in \underline{\mathscr{F}}^*$. □.

Lemma 7. *Let* $\Sigma \equiv \{C_i, A, B_i; \mathbf{k}\}$ *be fixed and suppose that for each* $i \in \mathbf{k}$, *with* $i \neq 1$, *there exists a path in* G_Σ *from node* 1 *to node* i. *Then*

$$\text{u.p.}\,(A_F, B_1) \textit{ divides } (\text{u.p.}\,(A, b_k))\left(\prod_{s \in \mathbf{k}(1)} \rho(C_{\mathbf{k}-s}, A, B_s)\right)$$

whenever $F \in \mathscr{F}$ *is selected so that* $d(\langle A_F|\mathscr{B}_1\rangle)$ *is as large as possible.*

Proof. According to Lemma 3, the hypothesis of the present lemma is equivalent to

$$C_{\mathbf{k}-s}(\lambda I - A)^{-1}B_s \neq 0, \qquad s \in \mathbf{k}(1) \qquad (23)$$

Suppose $k = 2$, in which case (23) becomes $C_2(\lambda I - A)^{-1}B_1 \neq 0$. If F^* (i.e. diag. $[F_1^*, F_2^*]$) maximizes $d(\langle A_F|\mathscr{B}_1\rangle)$, then by Lemma 1, F_2^* maximizes $d(\langle A + B_2F_2C_2|\mathscr{B}_1\rangle)$ and u.p. $(A_{F^*}, B_1) =$ u.p. $(A + B_2F_2^*C_2, B_1)$. From this and Lemma 4 applied to the system (C_2, A, B_1) with $D = B_2$, it follows that u.p. (A_{F^*}, B_1) divides (u.p. (A, B_2))$(\rho(C_2, A, B_1))$. Hence the lemma is true for $k = 2$.

Let $p \geq 2$ be fixed and set $q = p + 1$. Suppose that the lemma is true for $k = p$. Assume that the lemma's hypothesis holds for $k = q$; i.e. by Lemma 3,

$$C_{\mathbf{q}-s}(\lambda I - A)^{-1}B_s \neq 0 \qquad s \in \mathbf{q}(1) \qquad (24)$$

Write $\mathscr{F}^p$ (resp. $\mathscr{F}_q$) for the linear space of all matrices $F^p \equiv$ diag. $[F_1, F_2, \ldots, F_p]$ (resp., F_q).

Fix $s^* \in \mathbf{p}(1)$; then $s^* \in \mathbf{q}(1)$. Since

$$C_{\mathbf{q}-s^*} = \begin{bmatrix} C_{\mathbf{p}-s^*} \\ C_q \end{bmatrix}, \qquad (25)$$

it follows from (24) that

$$\begin{bmatrix} C_{\mathbf{p}-s^*} \\ C_q \end{bmatrix}(\lambda I - A)^{-1}B_{s^*} \neq 0 \qquad (26)$$

Since $s^* \cup \{q\} \in \mathbf{q}(1)$, (24) holds for $s = s^* \cup \{q\}$. Using the fact that $\mathbf{q} - (s^* \cup \{q\}) = \mathbf{p} - s^*$, we can therefore write

$$C_{\mathbf{p}-s^*}(\lambda I - A)^{-1}[B_{s^*}, B_q] \neq 0 \qquad (27)$$

From (26), (27) and Lemma 5, it follows that there exists a robust set $\underline{\mathscr{F}}(s^*) \subset \mathscr{F}_q$ such that with $F_q \in \underline{\mathscr{F}}(s^*)$,

$$C_{\mathbf{p}-s^*}(\lambda I - A - B_qF_qC_q)^{-1}B_{s^*} \neq 0 \qquad (28)$$

and

$$\rho(C_{\mathbf{p}-s^*}, A + B_qF_qC_q, B_{s^*}) \text{ divides}$$
$$\rho\left(\begin{bmatrix} C_{\mathbf{p}-s^*} \\ C_q \end{bmatrix}, A, B_{s^*}\right)\rho(C_{\mathbf{p}-s^*}, A, [B_{s^*}, B_q]) \qquad (29)$$

Since s^* is arbitrary, the preceding statement is valid for all $s^* \in \mathbf{p}(1)$. Thus (28) and (29) hold for all $s^* \in \mathbf{p}(1)$ and all F_q in the robust set $\bigcap_{s^* \in \mathbf{p}(1)} \underline{\mathscr{F}}(s^*)$.

Let $\underline{\mathscr{F}}_q^*$ denote the robust set obtained by intersecting $\bigcap_{s^* \in \mathbf{p}(1)} \underline{\mathscr{F}}(s^*)$ with the set of all $F_q \in \mathscr{F}_q$ which maximize $d(\langle A + B_qF_qC_q|\mathscr{B}_p\rangle)$. Then for arbitrary $F_q^* \in \underline{\mathscr{F}}_q^*$,

$$d(\langle A + B_qF_q^*C_q|\mathscr{B}_p\rangle) = \max_{F_q} d(\langle A + B_qF_qC_q|\mathscr{B}_p\rangle) \qquad (30)$$

$$C_{\mathbf{p}-s}(\lambda I - A - B_qF_q^*C_q)^{-1}B_s \neq 0, \quad s \in p(1) \qquad (31)$$

$$\rho(C_{\mathbf{p}-\mathbf{s}}, A + B_qF^*_qC_q, B_\mathbf{s}) \text{ divides}$$

$$\rho(C_{\mathbf{q}-\mathbf{s}}, A, B_\mathbf{s})\rho(C_{\mathbf{q}-\mathbf{s}\cup\{q\}}, A, B_{\mathbf{s}\cup\{q\}}). \qquad \mathbf{s} \in \mathbf{p}(1) \tag{32}$$

In writing the right side of (32), we have made use of (29), (25) and the fact that $\mathbf{q} - \mathbf{s} \cup \{q\} = \mathbf{p} - \mathbf{s}$ for $\mathbf{s} \in \mathbf{p}(1)$.

Write $\mathcal{F}^q$ for the linear space of all matrices $F^q \equiv \text{diag.}\ [F_1, F_2, \ldots, F_q]$ and let $\mathcal{F}^q_*$ denote the set of all $F^q \in \mathcal{F}^q$ for which $d(\langle A_{F^q}|\mathcal{B}_1\rangle)$ is maximal. Then $\mathcal{F}^q_*$ is robust in $\mathcal{F}^q$ as is the set $\mathcal{R} \equiv \mathcal{F}^q_* \cap (\mathcal{F}^p \times \mathcal{F}^*_q)$. Let $F^q_* \equiv (F^p_*,\ F^*_q) \in \mathcal{R}$ be fixed. Then $(F^p_*,\ F^*_q) \in \mathcal{F}^q_*$ so F^p_* maximizes $d(\langle A_{F^p} + B_qF^*_qC_q|\mathcal{B}_1\rangle)$ with respect to F^p. In addition, since $F^*_q \in \mathcal{F}^*_q$, (31) holds. It follows from Lemma 3 that $\Sigma^* \equiv \{C_i, A + B_qF^*_qC_q, B_i\,; \mathbf{p}\}$ satisfies the hypothesis of the present lemma at $k = p$. Since by assumption the lemma is valid at $k = p$, it must be that u.p. $(A_{F^p_*} + B_qF^*_qC_q, B_1)$ divides (u.p. $(A + B_qF^*_qC_q,\ B_\mathbf{p})$) $\prod_{\mathbf{s}\in\mathbf{p}(1)} \rho(C_{\mathbf{p}-\mathbf{s}}, A + B_qF^*_qC_q, B_\mathbf{s})$. From this, (32) and the fact that $A_{F^q_*} = A_{F^p_*} + B_qF^*_qC_q$, it follows that

$$\text{u.p.}\ (A_{F^q_*}, B_1) \text{ divides (u.p. } (A + B_qF^*_qC_q, B_\mathbf{p})) \prod_{\mathbf{s}\in\mathbf{m}} \rho(C_{\mathbf{q}-\mathbf{s}}, A, B_\mathbf{s}) \tag{33}$$

where $\mathbf{m} = \{\mathbf{s} \cup \{q\}\colon \mathbf{s} \in \mathbf{p}(1)\} \cup \mathbf{p}(1)$.

From (30) and (24) evaluated at $\mathbf{s} = \{q\}$, it follows that Lemma 4 can be applied to the system $(C_q, A, B_\mathbf{p})$ with $D = B_q$ to yield

$$\text{u.p.}\,(A + B_qF^*_qC_q, B_\mathbf{p}) \text{ divides}$$

$$(\text{u.p.}\,(A, [B_\mathbf{p}, B_q]))\rho(C_q, A, B_\mathbf{p}) \tag{34}$$

Observe that $[B_\mathbf{p},\ B_q] = B_\mathbf{q}$, $C_q = C_{\mathbf{q}-\mathbf{p}}$ and $\mathbf{m} \cup \{\mathbf{p}\} = \mathbf{q}(1)$; these relations, together with (33) and (34) imply that

$$\text{u.p.}\ (A_{F^q_*}, B_1) \text{ divides u.p.}\,(A, B_\mathbf{q}) \prod_{\mathbf{s}\in\mathbf{q}(1)} \rho(c_{\mathbf{q}-\mathbf{s}}, A, B_\mathbf{s}) \tag{35}$$

Since $F^q_* \in \mathcal{R}$ is arbitrary, (35) holds for all $F^q_* \in \mathcal{R}$. Hence by Lemma 6, u.p. (A_{F^q}, B_1) is fixed and independent of $F^q \in \mathcal{F}^q_*$. Since $\mathcal{F}^q_* \cap \mathcal{R}$ cannot be empty, (35) holds for all $F^q_* \in \mathcal{F}^q_*$. Thus the lemma is true at $k = q(= p + 1)$. By induction, the proof is complete. □

Proof of Theorem 2

i. Fix $\mathbf{s} \in \mathbf{k}(j)$ and $F = \text{diag.}\ [F_1, F_2, \ldots, F_k] \in \mathcal{F}$. Since $\langle A_F|\mathcal{B}_1\rangle \subset \langle A_F|\mathcal{B}_\mathbf{s}\rangle$, it must be that u.p. $(A_F,\ B_\mathbf{s})$ divides $\alpha_F(\lambda)$. For suitably defined block diagonal matrices K_1 and K_2, it is possible to write $A_F = A + B_\mathbf{s}K_1C_\mathbf{s} + K_{\mathbf{k}-\mathbf{s}}K_2C_{\mathbf{k}-\mathbf{s}}$; thus by Lemma 1, u.p. $(A_F, B_\mathbf{s}) =$ u.p. $(A + B_{\mathbf{k}-\mathbf{s}}K_2C_{\mathbf{k}-\mathbf{s}}, B_\mathbf{s})$. It follows that

$$\text{u.p.}\,(A + B_{\mathbf{k}-\mathbf{s}}K_2C_{\mathbf{k}-\mathbf{s}}, B_\mathbf{s}) \text{ divides } \alpha_F(\lambda) \tag{36}$$

We wish to show that

$$\rho(C_{\mathbf{k}-\mathbf{s}}, A, B_\mathbf{s}) \text{ divides } \alpha_F(\lambda) \tag{37}$$

For the case when $C_{\mathbf{k}-\mathbf{s}}(\lambda I - A)^{-1}B_\mathbf{s} \neq 0$, Proposition 2 applied to the system $(C_{\mathbf{k}-\mathbf{s}},\ A,\ B_\mathbf{s})$ with $D = B_{\mathbf{k}-\mathbf{s}}$, asserts that $\rho(C_{\mathbf{k}-\mathbf{s}},\ A,\ B_\mathbf{s})$ divides u.p. $(A + B_{\mathbf{k}-\mathbf{s}}K_2C_{\mathbf{k}-\mathbf{s}},\ B_\mathbf{s})$; this and (36) therefore imply (37).

Now suppose that $C_{\mathbf{k}-\mathbf{s}}(\lambda I - A)^{-1}B_\mathbf{s} = 0$, or equivalently that $\langle A|\mathcal{B}_\mathbf{s}\rangle \subset [C_{\mathbf{k}-\mathbf{s}}|A]$. Set $\mathcal{R} = \langle A|\mathcal{B}_\mathbf{s}\rangle$ and $\mathcal{S} = [C_{\mathbf{k}-\mathbf{s}}|A]$. Since both $\mathcal{R}$ and $\mathcal{S}$ are subspaces of kernel $C_{\mathbf{k}-\mathbf{s}}$, it must be that $\mathcal{R} = \langle A + B_{\mathbf{k}-\mathbf{s}}K_2C_{\mathbf{k}-\mathbf{s}}|\mathcal{B}_\mathbf{s}\rangle$, that $(A + B_{\mathbf{k}-\mathbf{s}}K_2C_{\mathbf{k}-\mathbf{s}})\mathcal{S} \subset \mathcal{S}$ and thus that $(A + B_{\mathbf{k}-\mathbf{s}}K_2C_{\mathbf{k}-\mathbf{s}})\|(\mathcal{S}/\mathcal{R}) = A\|(\mathcal{S}/\mathcal{R})$. But $\mathcal{S}/\mathcal{R} \subset \mathcal{X}/\mathcal{R}$ implying that c.p. $(A + B_{\mathbf{k}-\mathbf{s}}K_2C_{\mathbf{k}-\mathbf{s}})\|(\mathcal{S}/\mathcal{R})$ divides c.p. $(A + B_{\mathbf{k}-\mathbf{s}}K_2C_{\mathbf{k}-\mathbf{s}})\|(\mathcal{X}/\mathcal{R})$; therefore c.p. $A\|(\mathcal{S}/\mathcal{R})$ must divide c.p. $(A + B_{\mathbf{k}-\mathbf{s}}K_2C_{\mathbf{k}-\mathbf{s}})\|(\mathcal{X}/\mathcal{R})$ which in turn equals u.p. $(A + B_{\mathbf{k}-\mathbf{s}}K_2C_{\mathbf{k}-\mathbf{s}}, B_\mathbf{s})$. From this and (36) it follows that

$$\text{c.p.}\ A\|(\mathcal{S}/\mathcal{R}) \text{ divides } \alpha_F(\lambda) \tag{38}$$

Since $\mathcal{B}_\mathbf{s}$ is a subspace of both $\mathcal{R}$ and $\mathcal{S}$, $\mathcal{R}$ and $\mathcal{S}$ must be the largest $(A, B_\mathbf{s})$-controllability and $(A, B_\mathbf{s})$-invariant subspaces respectively in kernel $C_{\mathbf{k}-\mathbf{s}}$. But $\mathcal{R}$ and $\mathcal{S}$ are both A-invariant and rank $(C_{\mathbf{k}-\mathbf{s}}(\lambda I - A)^{-1}B_\mathbf{s}) = 0$. It follows from the definition of a remnant polynomial that $\rho(C_{\mathbf{k}-\mathbf{s}},\ A,\ B_\mathbf{s}) =$ c.p. $A\|(\mathcal{S}/\mathcal{R})$. This and (38) imply that (37) holds. Thus (37) is true in general.

Since $\mathbf{s} \in \mathbf{k}(j)$ is arbitrary, (37) must be valid for all $\mathbf{s} \in \mathbf{k}(j)$. It follows that assertion i of the theorem is true.

ii. Without loss of generality, assume $j = 1$. By hypothesis Σ is jointly controllable, so u.p. $(A, B_\mathbf{k}) = 1$. From this and Lemma 7, it follows that assertion ii of the theorem is true. □

Proof of Theorem 1. The requirement that each complementary subsystem of Σ containing input channel j be complete, is equivalent to the conditions

$$C_{\mathbf{k}-\mathbf{s}}(\lambda - A)^{-1}B_\mathbf{s} \neq 0, \qquad \mathbf{s} \in \mathbf{k}(j) \tag{39a}$$

$$\rho(C_{\mathbf{k}-\mathbf{s}}, A, B_\mathbf{s}) = 1, \qquad \mathbf{s} \in \mathbf{k}(j) \tag{39b}$$

Suppose that for fixed $F \in \mathcal{F}$, (A_F, B_j) is

controllable. Then (39a) follows from Proposition 3 and Lemma 3, while (39b) is a consequence of i/Theorem 2 and the fact that u.p. $(A_F, B_j) = 1$.

Assume (39) is true. Then Lemma 3 and (39a) insure that the hypothesis of ii/Theorem 2 is satisfied; (39b) and (11) thus imply for suitable $F \in \mathscr{F}$, that u.p. (A_F, B_j) must divide 1 or equivalently that (A_F, B_j) is controllable. □

3. CONTROLLABLE AND OBSERVABLE SINGLE-CHANNEL SUBSYSTEMS

By making use of Theorem 1 and duality it is possible to derive conditions for the existence of an $F \in \mathscr{F}$ which makes the jth channel of Σ_F both controllable and observable. First note from Theorem 1 (dualized), that if Σ is a jointly observable system, then there exists an $F \in \mathscr{F}$ such that $[C_j|A_F] = 0$ just in case each complementary subsystem of $\Sigma' \equiv \{B'_i, A', C'_i; \mathbf{k}\}$ containing input channel j, is complete; since completeness is invariant under transposition {cf. [19]}, this condition can be replaced with the equivalent requirement that each complementary subsystem of Σ containing output channel j, be complete. Thus if (C_j, A_F, B_j) is both controllable and observable for some F, then it is necessary that all complementary subsystems of Σ containing either input or output channel j, be complete. We claim that these conditions are sufficient as well—assuming, of course, that Σ is both jointly controllable and jointly observable. If the conditions hold, then the sets of F and $\bar{F}$ for which (A_F, B_j) is controllable and $(C_j, A_{\bar{F}})$ is observable are each robust in $\mathscr{F}$, as is their intersection $\mathscr{F}^*$; this being so, the set of $F \in \mathscr{F}$ for which (C_j, A_f, B_j) is both controllable and observable, namely $\mathscr{F}^*$, is nonempty.

The preceding argument shows that there exists an $F \in \mathscr{F}$ such that (C_j, A_F, B_j) is controllable and observable if and only if each complementary subystem of Σ containing either input channel j or output channel j, is complete. But from the definition of a complementary subsystem it directly follows that the union of all complementary subsystems containing either input or output channel j equals the set of all complementary subsystems of Σ. Hence if we now extend the previous definition of completeness to k-channel systems by calling such systems *complete* whenever all of their complementary subsystems are complete, then we can consisely summarize what has been accomplished so far.

Theorem 3. *Let* $\Sigma \equiv \{C_i, A, B_i; \mathbf{k}\}$ *be a jointly controllable jointly observable system and let* $j \in \mathbf{k}$ *be fixed. There exists an* $F \in \mathscr{F}$ *such that*

$$\langle A_F|\mathscr{B}_j\rangle = \mathscr{X}$$

and

$$[C_j|A_F] = 0$$

if and only if Σ *is complete.*

Since completeness is a property not depending on j, it must be true that if Σ is complete, then for any $j \in \mathbf{k}$, F_j can be selected so that (C_j, A_{F_j}, B_j) is controllable and observable. Furthermore, since for each j the set of all such F_j is robust in $\mathscr{F}$, the set of all F for which (C_j, A_F, B_j) is controllable and observable for all $j \in \mathbf{k}$ must also be robust in $\mathscr{F}$ and therefore nonempty. In other words, if Σ_F can be made controllable and observable through one channel, then it can be made controllable and observable through every channel. More precisely,

Corollary 1. *If* Σ *is a complete, jointly controllable, jointly observable system, then there exists an* $F \in \mathscr{F}$ *such that for all* $j \in \mathbf{k}$, (C_j, A_F, B_j) *is controllable and observable.*

In order for Σ to be complete it is necessary that the transfer matrices of each of its complementary subsystems be nonzero. But from Lemma 3, it is obvious that

Lemma 8. *The transfer matrices of all complementary subsystems of* Σ *are nonzero if and only if the graph of* Σ *is strongly connected.*†

Therefore Σ_F can be made controllable and observable through a single channel, only if the graph of Σ is strongly connected; in this case we call Σ a *strongly connected* system.

Remark. It is *not* true that the graph of Σ_F is independent of F. However, if the graph is strongly connected for some F, then it remains strongly connected for all F. It is also true that completeness of Σ_F is independent of F.

Suppose we write $\mathscr{S}$ for the linear space of all n-dimensional, single-channel systems $(C^{(p\times n)}, A^{(n\times n)}, B^{(n\times m)})$. It is shown in [19] that $(C, A, B) \in \mathscr{S}$ is complete just in case

$$C(\lambda I - A)^{-1}B \neq 0$$

and the rank of the matrix

$$M(\lambda) = \begin{bmatrix} \lambda I - A & B \\ C & 0 \end{bmatrix}$$

†A directed graph is strongly connected if for each distinct pair of nodes i and j, the graph contains a path from i to j.

is no less than n for all real and complex values of λ. Since rank $(\lambda I - A) = n$ for all λ not in the spectrum of A, it is easy to see that the rank condition on $M(\lambda)$ is equivalent to the requirement that no prime factor of c.p. A be a factor of every nth order minor of $M(\lambda)$, the minors of $M(\lambda)$ being viewed as polynomials in λ. Since the coefficients of each of these polynomials depend algebraically on the parameters of (C, A, B), it must be that the resultant of each prime factor of c.p. A together with each nth order minor of $M(\lambda)$, is a polynomial in the parameters of (C, A, B). From this it directly follows that the class of systems in $\mathcal{S}$ which are not complete, is a variety in $\mathcal{S}$. Since $\mathcal{S}$ always contains at least one complete system (e.g. select $(C, A, B) \in \mathcal{S}$ with (A, B) controllable and $C(\lambda I - A)^{-1}B \neq 0$), the variety is proper. The complement $\mathcal{S}^*$ of this variety, which is the set of complete systems in $\mathcal{S}$, is therefore robust. This implies that $\mathcal{S}^*$ is open and dense in $\mathcal{S}$ (with respect to the standard Euclidean metric), and thus that completeness is a generic property of single-channel systems in $\mathcal{S}$.

Suppose we now write $\mathscr{C}$ for the linear vector space of all k-channel systems $\{C_i^{(p_i \times n)}, A^{(n \times n)}, B_i^{(n \times m_i)}; \mathbf{k}\}$. Using essentially the same arguments as above, it is quite straightforward to show that if $\mathscr{C}^*$ is the set of jointly controllable, jointly observable, complete systems in $\mathscr{C}$, then $\mathscr{C}^*$ is robust in $\mathscr{C}$. Thus one can also state that joint-controllability, joint-observability and completeness are generic properties of systems in $\mathscr{C}$.

From a practical point-of-view, such an assertion can be very misleading in that it is based on hypotheses which fail to take into account possible dependencies between model parameters. For example, if there really is no internal physical connection between the ith channel input and the jth channel output of the process modelled by Σ, then the transfer matrix between channel input i and channel output j of Σ must be exactly, not approximately, zero, and the class of admissible model parameter perturbations should preserve this property, if conclusions based on perturbation analysis are to make sense.

With this in mind, suppose we call two systems in $\mathscr{C}$, *topologically equivalent* if both systems have the same graph. For the topogical equivalence class E_Σ of $\Sigma \in \mathscr{C}$ to contain at least one complete system, it is necessary (cf. Lemma 8) that Σ be strongly connected. Assume, therefore, that Σ is strongly connected; and write E_Σ^* for the set of all jointly controllable, jointly observable, complete systems in E_Σ^*. Since E_Σ^* coincides with the intersection of $\mathscr{C}^*$ and E_Σ, and since $\mathscr{C}^*$ is open in E, with respect to the standard Euclidean metric, it follows that E_Σ^* is open in E_Σ, with respect to the restriction of this metric to E_Σ. Thus if Σ can be made single channel controllable and observable with decentralized feedback then so can any system $\bar{\Sigma} \in E_\Sigma$, provided $\bar{\Sigma}$ is sufficiently close to Σ. In other words, the problem of finding an $F \in \mathscr{F}$ for which Σ_F is single channel controllable and observable, is well-posed at any point in E_Σ^*.

We would like to be able to assert that 'almost every system' in E_Σ is in E_Σ^*, i.e. that E_Σ^* is dense in E_Σ, but to prove this seems to be a difficult matter. Note that the usual approach, i.e. show that E_Σ^* is the complement of a proper variety in E_Σ, will not work here since in general E_Σ is a proper variety in $\mathscr{C}$ rather than a linear subspace. Thus the problem of deciding whether or not joint controllability, joint observability and completeness are generic properties of systems in E_Σ, remains unsolved.

Remark. The topological constraints imposed on the graph of Σ can lead to some unusual consequences. For example, if $\Sigma \equiv \{C_i, A, B_i; \mathbf{3}\}$ is a two-dimensional, three-channel, jointly controllable, jointly observable, strongly connected system possessing a transfer matrix with zero diagonal blocks (i.e. the transfer matrix from channel input i to channel output i is zero, $i \in \mathbf{3}$), then the two eigenvalues of A must be equal.

4. STRONGLY CONNECTED SYSTEMS

By making use of the preceding results it is now possible to obtain complete and constructive solutions to the spectrum assignment and stabilization problems for strongly connected systems. We begin by reviewing some basic properties of linear systems.

It is well known that the closed-loop spectrum of a linear system (C, A, B), resulting from the application of (centralized) linear control, consists of two subsets: one is controller dependent and freely assignable with dynamic controllers of sufficiently high order[17]; the other, called the *unassignable spectrum* of (C, A, B), is controller independent and is therefore a fixed subset of the closed-loop system's spectrum, no matter what the controller design. It is also known that if $\alpha(\lambda)$ is the characteristic polynomial of A mod $\langle A|\mathscr{B}\rangle$ and if $\beta(\lambda)$ is the characteristic polynomial of A restricted to $\langle A|\mathscr{B}\rangle \cap [C|A]$, then the root set of $\alpha(\lambda)\beta(\lambda)$ coincides with the unassignable spectrum of (C, A, B).† It is natural to

†Note that the roots of $\alpha(\lambda)$ and $\beta(\lambda)$ are the 'uncontrollable' and 'controllable but unobservable' eigenvalues of (C, A, B), respectively.

call $\alpha(\lambda)\beta(\lambda)$ the *unassignable polynomial* of (C, A, B).

For fixed $\Sigma \equiv \{C_i, A, B_i; \mathbf{k}\}$, it is possible to characterize the unassignable polynomial of (C_j, A_F, B_j) as a function of $j \in \mathbf{k}$ and $F \in \mathcal{F}$. For this, let $\mathbf{k}^*$ denote the set of all proper subsets of $\mathbf{k}$, and write $\mathcal{F}_j$ for the set of $F \in \mathcal{F}$ for which the rank of the nth order Hankel matrix

$$H_F(j) \equiv \begin{bmatrix} C_jB_j & C_jA_FB_j & \dots & C_j(A_F)^{n-1}B_j \\ \vdots & & & \\ C_j(A_F)^{n-1}B_j & \dots\dots & & C_j(A_F)^{(2n-1)}B_j \end{bmatrix}$$

is as large as possible. Clearly each $\mathcal{F}_j$, $j \in \mathbf{k}$, is robust in $\mathcal{F}$ as is the intersection $\mathcal{F}^* \equiv \bigcap_{j\in\mathbf{k}} \mathcal{F}_j$.

Theorem 4. *Let* $\Sigma \equiv \{C_i, A, B_i; \mathbf{k}\}$ *be a jointly controllable, jointly observable, strongly connected system. Let* $\mu_F{}^j(\lambda)$ *denote the unassignable polynomial of* (C_j, A_F, B_j) *and write* $\alpha_\Sigma(\lambda)$ *for the fixed polynomial of* Σ *as defined by* (6). *Then*

i. $\bigvee_{s\in\mathbf{k}^*} \rho(C_{\mathbf{k}-s}, A, B_s)$ *divides* $\mu_F{}^j(\lambda)$ $\quad j \in \mathbf{k}, \quad F \in \mathcal{F}$

ii. $\mu_F{}^j(\lambda)$ *divides* $\prod_{s\in\mathbf{k}^*} \rho(C_{\mathbf{k}-s}, A, B_s)$ $\quad j \in \mathbf{k}, \quad F \in \mathcal{F}^*$

iii. $\mu_F{}^j(\lambda) = \alpha_\Sigma(\lambda)$ $\quad j \in \mathbf{k}, \quad F \in \mathcal{F}^*$

Note that condition iii asserts that $\mu_F{}^j(\lambda)$ is independent of $F \in \mathcal{F}^*$ *and* $j \in \mathbf{k}$. Thus for fixed $F \in \mathcal{F}^*$, the set of possible closed-loop spectra which can be achieved by applying dynamic feedback to a single channel of Σ_F, is the same for all channels of Σ_F.

We remark that condition iii is true *only* under the hypotheses of the theorem. In particular, if Σ is not strongly connected, then iii need not necessarily hold.

Recall that Proposition 1 in effect states that $\alpha_\Sigma(\lambda)$ must divide the characteristic polynomial of the closed-loop system which results from application of any linear decentralized control to Σ. It follows that for stabilization, $\alpha_\Sigma(\lambda)$ must be a stable polynomial, while for spectrum assignment, $\alpha_\Sigma(\lambda)$ must equal one. In view of iii/Theorem 4, these conditions are also sufficient for stabilization and spectrum assignment respectively. But conditions i and ii of Theorem 4 imply that $\alpha_\Sigma(\lambda)$ is stable {resp., equals one} just in case for each $s \in \mathbf{k}^*$, $\rho(C_{\mathbf{k}-s}, A, B_s)$ is stable {resp. equals one}. From this, Lemma 8 and Theorem 3, we arrive at the following theorem which provides explicit conditions for the spectrum assignment and stabilization problems to be solvable for strongly connected systems.

Theorem 5. *Let* $\Sigma \equiv \{C_i, A, B_i; \mathbf{k}\}$ *be a jointly controllable, jointly observable, strongly connected system. Then with decentralized control*

i. *the spectrum of* Σ *can be freely assigned if and only if* Σ *is complete.*

ii. Σ *can be stabilized if and only if for each* $s \in \mathbf{k}^*$, $\rho(C_{\mathbf{k}-s}, A, B_s)$ *is a stable polynomial.*

Remark. In view of the discussion in §3, it is evident that the spectrum assignment problem is well-posed at any solution point within the topological equivalence class of Σ, provided Σ is strongly connected; using similar arguments it can easily be shown that this is also the case for the stabilization problem. However, for systems which are not strongly connected, quite the opposite can be true as will be illustrated by example at the end of the next section.

Example 1. The two channel system described by the matrices

$$C_1 = \begin{bmatrix} 1 & 0 & 0 & 0 \\ 0 & 1 & 0 & 0 \end{bmatrix}$$

$$C_2 = [1 \quad 0 \quad 1 \quad 0]$$

$$A = \begin{bmatrix} 0 & 1 & 1 & 0 \\ 0 & 0 & 0 & 1 \\ 0 & 0 & 1 & 0 \\ 0 & 0 & 0 & 2 \end{bmatrix} \quad B_1 = \begin{bmatrix} 1 \\ 0 \\ 0 \\ 0 \end{bmatrix} \quad B_2 = \begin{bmatrix} 0 \\ 0 \\ 1 \\ 1 \end{bmatrix}$$

is jointly controllable, jointly observable and complete. Spectrum assignment can be achieved by first applying nondynamic feedback to channel 2, to make channel 1 controllable and observable, and then applying appropriate dynamic feedback to channel 1 using standard techniques [17].

Example 2. Although the system with matrices

$$C_1 = \begin{bmatrix} 1 & 0 & 0 & 0 \\ 0 & 1 & 0 & 0 \\ 0 & 0 & 0 & 1 \end{bmatrix}$$

$$C_2 = [0 \quad 1 \quad 1 \quad 0]$$

$$A = \begin{bmatrix} 0 & 1 & 0 & 0 \\ 0 & 1 & 0 & 0 \\ 0 & 0 & 1 & 0 \\ 0 & 0 & 0 & 1 \end{bmatrix} \quad B_1 = \begin{bmatrix} 0 \\ 1 \\ 0 \\ 0 \end{bmatrix}, \quad B_2 = \begin{bmatrix} 0 & 0 \\ 0 & 0 \\ 1 & 0 \\ 0 & 1 \end{bmatrix}$$

is jointly controllable, jointly observable and strongly connected, it fails to satisfy condition ii of Theorem 5, and thus cannot be stabilized. For this example, the maximal rank of the Hankel matrix associated with either channel of Σ_F is three, and there is an unassignable eigenvalue at $\lambda = 1$ which cannot be shifted with decentralized control.

Proof of Theorem 4. Set $\bar{\mathbf{k}}(j) \equiv \{\mathbf{k}-s : s \in \mathbf{k}(j)\}$: then

$$\mathbf{k}^* = \mathbf{k}(j) \cup \bar{\mathbf{k}}(j) \qquad (40)$$

i. Let $j \in \mathbf{k}$ and $F \in \mathcal{F}$ be fixed. By i/Theorem 2, $\bigvee_{s \in \mathbf{k}(j)} \rho(C_{\mathbf{k}-s}, A, B_s)$ divides u.p. (A_F, B_j), while by the dual of i/Theorem 2, $\bigvee_{s \in \bar{\mathbf{k}}(j)} \rho(B'_s, A', C'_{\mathbf{k}-s})$ divides u.p. (A'_F, C'_j). Since u.p. (A_F, B_j) and u.p. (A'_F, C'_j) each divide $\mu^j_F(\lambda)$, and since the remnant polynomial of a system is invariant under system transposition[19], it must be that $\bigvee_{s \in (\mathbf{k}(j) \cup \bar{\mathbf{k}}(j))} \rho(C_{\mathbf{k}-s}, A, B_s)$ divides $\mu^j_F(\lambda)$. From this and (40) it follows that assertion i is true.

ii. Let $\mathcal{F}_1$ {resp. $\mathcal{F}_2$} denote the set of $F \in \mathcal{F}$ for which $d(\langle A_F|\mathcal{B}_j\rangle)$ {resp. $d(\langle A'_F|C'_j\rangle)$} is as large as possible. Then $\mathcal{F}_1$ and $\mathcal{F}_2$ are each robust in $\mathcal{F}$ as is the set $\mathcal{F}_1 \cap \mathcal{F}_2 \cap \mathcal{F}^*$. Let $F \in \mathcal{F}_1 \cap \mathcal{F}_2 \cap \mathcal{F}^*$ and $j \in \mathbf{k}$ be fixed. Since Σ is strongly connected, G_Σ must satisfy the hypothesis of ii/Theorem 2; therefore

$$\text{u.p.}\,(A_F, B_j) \quad \text{divides} \quad \prod_{s \in \mathbf{k}(j)} \rho(C_{\mathbf{k}-s}, A, B_s) \qquad (41)$$

Similarly, by the dual of ii/Theorem 2 and the fact that $F \in \mathcal{F}_2$,

$$\text{u.p.}\,(A'_F, C'_j) \quad \text{divides} \quad \prod_{s \in \bar{\mathbf{k}}(j)} \rho(C_{\mathbf{k}-s}, A, B_s) \qquad (42)$$

But u.p. (A'_F, C'_j) equals c.p. $(A_F \| [C_j|A_F])$ which in turn is divisible by c.p. $(A_F|(\langle A_F|\mathcal{B}_j\rangle \cap [C_j|A_F]))$. From this, (41), (42) and the definition of $\mu^j_F(\lambda)$, it follows that $\mu^j_F(\lambda)$ divides $\prod_{s \in (\mathbf{k}(j) \cup \bar{\mathbf{k}}(j))} \rho(C_{\mathbf{k}-s}, A, B_s)$. This and (40) imply that

$$\mu^j_F(\lambda) \quad \text{divides} \quad \prod_{s \in \mathbf{k}^*} \rho(C_{\mathbf{k}-s}, A, B_s),$$

$$F \in \mathcal{F}_1 \cap \mathcal{F}_2 \cap \mathcal{F}^* \qquad (43)$$

Since (43) holds for at least one point in $\mathcal{F}^*$, to prove assertion ii, it is enough to show that assertion iii is true.

iii. For fixed $j \in \mathbf{k}$ and $F \in \mathcal{F}$, set $\mathcal{R}_F \equiv \langle A_F|\mathcal{B}_j\rangle$ and $\mathcal{N}_F \equiv [C_j|A_F]$. Note that degree $\alpha^j_F(\lambda) = d(\mathcal{X}) - d(\mathcal{R}_F/(\mathcal{R}_F \cap \mathcal{N}_F))$; in addition, from realization theory, $d(\mathcal{R}_F/(\mathcal{R}_F \cap \mathcal{N}_F)) =$ rank $H_F(j)$. Since rank $H_F(j)$ is independent of $F \in \mathcal{F}^*$, it follows that for some integer n^*,

$$\text{degree } \mu^j_F(\lambda) = n^*, \qquad F \in \mathcal{F}^* \qquad (44)$$

Let $\mathcal{K}$ denote the space of linear maps $\mathcal{Y}_j \to \mathcal{U}_j$. Since $A_F + B_jKC_j = A + \sum_{i=1, i \neq j}^{k} B_iF_iC_i + B_j(F_j + K)C_j$, it must be that $\alpha_\Sigma(\lambda) = \bigwedge_{F,K}$ c.p. $(A_F + B_jKC_j)$ {cf. equation (6)}. Thus for any F and K, c.p. $(A_F + B_jKC_j)$ can be written in factored form as $\alpha_\Sigma(\lambda)\beta_{F,K}(\lambda)$. In addition, since the controllable space, unobservable space and unassignable polynomial of $(C_j, A_F + B_jKC_j, B_j)$ are clearly independent of K, it follows that c.p. $(A_F + B_jKC_j)$ can also be written as $\mu^j_F(\lambda)\gamma_{F,K}(\lambda)$ where $\gamma_{F,K}(\lambda) =$ c.p. $(A_F + B_jKC_j)\|(\mathcal{R}_F/(\mathcal{R}_F \cap \mathcal{N}_F))$. Therefore

$$\alpha_\Sigma(\lambda)\beta_{F,K}(\lambda) = \mu^j_F(\lambda)\gamma_{F,K}(\lambda), \quad (F, K) \in \mathcal{F} \oplus \mathcal{K} \qquad (45)$$

For arbitrary but fixed F, let $(\bar{C}, \bar{A}, \bar{B})$ be the controllable, observable system induced by (C_j, A_F, B_j) in $\mathcal{R}_F/\mathcal{R}_F \cap \mathcal{N}_F)$. Using the results of [17], it is possible to find vectors $g \in \mathcal{U}_j$ and $h \in \mathcal{Y}_j$ and a matrix $K_0 \in \mathcal{K}$ such that $(h'\bar{C}, \bar{A} + \bar{B}K_0\bar{C}, \bar{B}g)$ is controllable and observable. Hence by the root-locus technique, a scalar δ can be found for which c.p. $(\bar{A} + \bar{B}K_0\bar{C} + \bar{B}g\delta h'\bar{C})$ is coprime with $\alpha_\Sigma(\lambda)$. Setting $K \equiv K_0 + \delta gh'$ and recognizing that c.p. $(\bar{A} + \bar{B}K\bar{C}) = \gamma_{F,K}(\lambda)$, we see that $\gamma_{F,K}(\lambda)$ and $\alpha_\Sigma(\lambda)$ are coprime. From this, (45) and the fact that F was selected arbitrarily, it follows that

$$\alpha_\Sigma(\lambda) \quad \text{divides} \quad \mu^j_F(\lambda) \qquad F \in \mathcal{F} \qquad (46)$$

By Lemma A.1, the set $\mathcal{G}$ of pairs $(F, K) \in \mathcal{F} \oplus \mathcal{K}$ such that $\beta_{F,K}(\lambda)$ and $\prod_{s \in \mathbf{k}^*} \rho(C_{\mathbf{k}-s}, A, B_s)$ are coprime, is robust in $\mathcal{F} \oplus \mathcal{K}$. Since $\mathcal{F}_1 \cap \mathcal{F}_2 \cap \mathcal{F}^*$ is robust in $\mathcal{F}$, it must be that $\mathcal{G} \cap ((\mathcal{F}_1 \cap \mathcal{F}_2 \cap \mathcal{F}^*) \times \mathcal{K})$ is robust in $\mathcal{F} \oplus \mathcal{K}$. This means that there exists a pair (F_0, K_0) such that $F_0 \in \mathcal{F}_1 \cap \mathcal{F}_2 \cap \mathcal{F}^*$ and $\beta_{F_0,K_0}(\lambda)$ and $\prod_{s \in \mathbf{k}^*} \rho(C_{\mathbf{k}-s}, A, B_s)$ are coprime; thus from (43), it must be true that $\beta_{F_0,K_0}(\lambda)$ and $\mu^j_{F_0}(\lambda)$ are coprime. This and (45) imply that $\mu^j_{F_0}(\lambda)$ divides $\alpha_\Sigma(\lambda)$. Therefore from (46) $\mu^j_{F_0}(\lambda) = \alpha_\Sigma(\lambda)$. But $F_0 \in \mathcal{F}^*$ and from (44), degree $\mu^j_F(\lambda)$ is independent of $F \in \mathcal{F}^*$. Therefore degree $\mu^j_F(\lambda) =$ degree $\alpha_\Sigma(\lambda)$, $F \in \mathcal{F}^*$. But this and (46) can be true only if $\mu^j_F(\lambda) = \alpha_\Sigma(\lambda)$ for $F \in \mathcal{F}^*$; since $j \in \mathbf{k}$ is arbitrary and since $\alpha_\Sigma(\lambda)$ is not a function of j, this relationship remains true for all $j \in \mathbf{k}$. □

5. *k*-CHANNEL SYSTEMS

We now examine the spectrum assignment and stabilization problems for systems which are not necessarily strongly connected. For such systems the methods appropriate to strongly connected systems cannot be directly applied; e.g. if Σ is not strongly connected, it is impossible to find an F which will make a single channel of Σ_F controllable and observable {cf. Theorem 3}. However it still may be possible to assign or stabilize the system's spectrum with decentralized control. To determine under what conditions this can be done we use standard techniques from graph theory[21] to decompose G_Σ into a number—say p—of strongly connected components $G_i \equiv (\mathbf{q}_i, \Gamma_i)$; the subsets $\mathbf{q}_i$ form a partition of $\mathbf{k}$ (i.e. $\mathbf{k}$ can be written as the disjoint union of the $\mathbf{q}_i$); Γ_i is a function mapping $\mathbf{q}_i$ into its power set according to the rule $\Gamma_i(j) \equiv \{r: r \in (\Gamma(j)) \cap \mathbf{q}_i\}$, Γ having already been defined in §2. The *reduced graph* of Σ, written G^R, is the pair $(\mathbf{q}^R, \Gamma^R)$ where $\mathbf{q}^R \equiv \{\mathbf{q}_1, q_2, \ldots, \mathbf{q}_p\}$ and Γ^R is the mapping from $\mathbf{q}^R$ into its power set, defined in such a way that $\mathbf{q}_i \in \Gamma^R(\mathbf{q}_j)$ if and only if there exist nodes $i_0 \in \mathbf{q}_i$ and $j_0 \in \mathbf{q}_j$ such that $i_0 \in \Gamma(j_0)$. We say that a component G_i is *isolated* if $\Gamma^R(\mathbf{q}_i)$ is the empty set and $\mathbf{q}_i \notin \bigcup_{\substack{j \in \mathbf{p} \\ j \neq i}} \Gamma^R(\mathbf{q}_j)$. Thus G_i is isolated if no arc enters or leaves the set $\mathbf{q}_i$. Observe that the reduced graph G^R contains no circuits (i.e. if $\mathbf{q}_i \in \Gamma^R(\mathbf{q}_j)$, then $\mathbf{q}_j \notin \Gamma^R(\mathbf{q}_i)$; thus we may assume that the set $\mathbf{q} \equiv \{\mathbf{q}_1, \mathbf{q}_2, \ldots, \mathbf{q}_p\}$ is ordered in such a way that

$$\Gamma^R(\mathbf{q}_i) \subset \{q_1, q_2, \ldots, q_{i-1}\}, \qquad i = 2, 3, \ldots, p \tag{47}$$

Associated with Σ, consider the $k \times k$ matrix T_Σ defined by

$$t_{ij} = \begin{cases} 0 \text{ if } i \notin \Gamma(j) \text{ \{i.e. if } C_i(\lambda I - A)^{-1}B_j = 0\} \\ 1 \text{ otherwise} \end{cases}$$

We call T_Σ the *topology* matrix of Σ; clearly T_Σ is the transpose of the incidence matrix[21] of G_Σ, and it completely characterizes G_Σ.

Let k_i, $i \in \mathbf{p} \equiv \{1, 2, \ldots, p\}$, be the number of elements in $\mathbf{q}_i$. Observe that is always possible to relabel the channels of Σ in such a way that the set $\mathbf{q}_1$ contains the k_1 first integers, $\mathbf{q}_2$ the next k_2, etc.; to this relabelling there corresponds a permutation of the rows of T_Σ, together with the *same* permutation of the columns, and when this is carried out there results a matrix T_Σ^* which, because of (47), has the block triangular form

$$T_\Sigma^* = \begin{bmatrix} T_{11} & & \ldots T_{1p} \\ 0 & T_{22} & \ldots T_{2p} \\ \vdots & & \vdots \\ 0 & & \ldots T_{pp} \end{bmatrix} \tag{48}$$

Without loss of generality, it will be assumed in the sequel that the relabelling has been done so that the topology matrix of Σ has a block triangular structure of the form (48). Correspondingly, the overall transfer matrix of Σ admits a similar block triangular structure; the ith diagonal block in the latter is the transfer matrix of the k_i-channel strongly connected system $\{C_j, A, B_j; \mathbf{q}_i\}$ described by the equations

$$\left.\begin{aligned} \dot{x}_i(t) &= Ax_i(t) + \sum_{j \in \mathbf{q}_i} B_j u_j(t) \\ y_j(t) &= C_j x_i(t), \qquad j \in \mathbf{q}_i \end{aligned}\right\} \tag{49}$$

Note that in general this system is neither jointly controllable nor jointly observable. Nevertheless realization theory insures the existence of a jointly controllable, jointly observable system $\Sigma^i \equiv \{\bar{C}_j, \bar{A}_i, \bar{B}_j; \mathbf{q}_i\}$ with the same overall transfer matrix as (49). We call Σ^i the ith *strongly connected subsystem* of Σ.

We wish to relate the closed-loop spectrum of Σ to the closed-loop spectrum of the k-channel system determined by the Σ^i, when *identical* local controllers Σ_j are applied to both systems. For this write $\Lambda(\Sigma_j, \mathbf{k})$ for the closed-loop spectrum of Σ, and with identical controls applied to the Σ^i, let $\Lambda_i(\Sigma_j, \mathbf{q}_i)$ denote the closed-loop spectrum of Σ^i.

Observe that the block-triangular structure of the transfer matrix of Σ, with the transfer matrix of Σ^i being the ith diagonal block, implies that the parameter matrices describing Σ can be written as

$$C_j = [\hat{C}_j, x, \ldots, x], \quad (j \in \mathbf{q}_1)$$
$$C_j = [0, \hat{C}_j, x, \ldots, x], \quad (j \in \mathbf{q}_2)$$
$$\vdots$$
$$C_j = [0, \ldots, 0, \hat{C}_j], \quad (j \in \mathbf{q}_p)$$

$$A = \begin{bmatrix} \hat{A}_1 & x & \cdot & \cdot & \cdot & x \\ 0 & \hat{A}_2 & x & \cdot & \cdot & x \\ \cdot & 0 & \cdot & & & \cdot \\ \cdot & \cdot & & \cdot & & \cdot \\ \cdot & \cdot & & & \cdot & x \\ 0 & \cdot & \cdot & \cdot & 0 & \hat{A}_p \end{bmatrix}$$

$$B_j = \begin{bmatrix} \hat{B}_j \\ 0 \\ \vdots \\ 0 \end{bmatrix} \quad (j \in \mathbf{q}_1),$$

$$B_j = \begin{bmatrix} x \\ \hat{B}_j \\ 0 \\ \vdots \\ 0 \end{bmatrix} \quad (j \in \mathbf{q}_2), \ldots,$$

$$B_j = \begin{bmatrix} x \\ \vdots \\ \cdot \\ x \\ \hat{B}_j \end{bmatrix} \quad (j \in \mathbf{q}_p)$$

where x-marks denote matrices of no importance to the present analysis, and $\hat{\Sigma}^i \equiv \{\hat{C}_j, \hat{A}_i, \hat{B}_j; \mathbf{q}_i\}$ is a system with the same overall transfer matrix as $\Sigma^i \equiv \{\bar{C}_j, \bar{A}_i, \bar{B}_j; \mathbf{q}_i\}$. Since Σ^i is jointly controllable and jointly observable, the parameter matrices of $\hat{\Sigma}^i$ can be written as

$$\hat{C}_j = [0, \bar{C}_j, x], (j \in \mathbf{q}_i)$$

$$\hat{A}_i = \begin{bmatrix} A_{i1} & x & x \\ 0 & \bar{A}_i & x \\ 0 & 0 & A_{i2} \end{bmatrix} \quad \begin{bmatrix} x \\ \bar{B}_j \\ 0 \end{bmatrix} = \hat{B}_j, \qquad (j \in \mathbf{q}_i)$$

Note that the set $\Lambda_0 \equiv \bigcup_{i \in \mathbf{p}} (\sigma(A_{i1}) \cup \sigma(A_{i2}))$ is the complement of the union of the open-loop spectra of the Σ^i in the spectrum of Σ.

For each $j \in \mathbf{k}$, let $\dot{z}_j = H_j z_j + L_j y_j$, $u_j = M_j z_j + F_j y_j$ denote a local control system Σ_j to be applied to the jth channel of Σ. Write $\bar{A}^*_i$ for the system matrix describing the overall closed-loop system which results when controllers $\Sigma_j (j \in \mathbf{p}_i)$ are applied to Σ^i. It is easy to see that the system matrix A^* associated with the closed-loop system resulting from application of the Σ_j to Σ, can be written as

$$A^* = \begin{bmatrix} \hat{A}^*_1 & x & \cdots & x \\ 0 & \hat{A}^*_2 & x & x \\ \vdots & 0 & \ddots & \vdots \\ \vdots & \vdots & \ddots & x \\ 0 & \cdot & \cdots & \hat{A}^*_p \end{bmatrix}$$

where

$$\hat{A}^*_i = \begin{bmatrix} A_{i1} & x & x \\ 0 & \bar{A}^*_i & x \\ 0 & 0 & A_{i2} \end{bmatrix}$$

Clearly $\sigma(A^*) = \Lambda_0 \cup_{i \in \mathbf{p}} \sigma(\bar{A}^*_i)$. We are led to the following proposition which characterizes the closed-loop spectrum of Σ in terms of the closed-loop spectra of its strongly connected subsystems.

Proposition 4. *For all linear controls* Σ_j, $j \in \mathbf{k}$,

$$\Lambda(\Sigma_j, \mathbf{k}) = \Lambda_0 \bigcup_{i \in \mathbf{p}} \Lambda_i(\Sigma_j, \mathbf{q}_i)$$

Proposition 4 clearly implies that the closed-loop spectrum of Σ can be freely assigned (resp. stabilized) just in case Λ_0 is empty (resp. stable) and spectrum assignment (resp. stabilization) is possible for each strongly connected subsystem of Σ. If these conditions are satisfied, then spectrum assignment or stabilization can be achieved by applying the methods of §4 to each one of the strongly connected subsystems of Σ. If the resulting decentralized control is then applied to the original system, the spectrum of the resulting closed-loop system will coincide with the disjoint union of Λ_0 and the assigned or stabilized spectra of each of the p strongly connected subsystems. A short calculation shows that the total dimension of the decentralized control system needed to achieve stability or to assign a prescribed spectrum, never need exceed n, the dimension of Σ.

The preceding observations can be summarized in the following way.

Theorem 6. *Let* $\Sigma \equiv \{C_i, A, B_i; \mathbf{k}\}$ *be a jointly controllable, jointly observable system. Then with decentralized control, the closed-loop spectrum of* Σ *can be*

i. *freely assigned if and only if the sum of the dimensions of* Σ*'s strongly connected subsystems is equal to the dimension of* Σ *and spectrum assignment is possible for each strongly connected subsystem.*

ii. *stabilized if and only if the complement of the disjoint union of the spectra of* Σ*'s strongly connected subsystems in the spectrum of* Σ *is stable and stabilization is possible for each strongly connected subsystem.*

Theorem 6 together with its predecessor, Theorem 5, thus provide explicit conditions for the solvability of both the spectrum assignment and stabilization problems for k-channel linear systems.

So far in studying the spectrum assignment and stabilization problems, we have implicitly assumed the system data to be exact. However, difficulties may arise if the preceding results are applied without qualification. To illustrate this, consider the two-channel system with overall transfer matrix

$$\begin{bmatrix} \dfrac{1}{s-1} & \dfrac{1}{s-1-\epsilon} \\ 0 & \dfrac{1}{s} \end{bmatrix}$$

The system is not strongly connected. With $\epsilon = 0$, stabilization can be achieved with the decentralized control $u_1 = -2y_1 + v_1$, $u_2 = -y_2 + v_2$. However, if the same control is applied with $\epsilon \neq 0$, there results the closed-loop transfer matrix

$$\begin{bmatrix} \dfrac{1}{s+1} & \dfrac{(s-1)s}{(s-1-\epsilon)(s+1)^2} \\ 0 & \dfrac{1}{s+1} \end{bmatrix}$$

and it is apparent that a right-half plane pole-zero cancellation takes place when $\epsilon = 0$. Thus although for $\epsilon = 0$, stabilization (and even spectrum assignment) is possible, for arbitrary small ϵ stabilization is impossible and the pole at $1+\epsilon$ cannot be shifted with any decentralized control.

From this we may draw the following conclusions, which apply in general to systems which are not strongly connected, but which will be discussed in terms of a simple example: Suppose that it is found that a physical process can be represented by a transfer matrix of the form

$$\begin{bmatrix} \dfrac{1}{s-a_1} & \dfrac{1}{s-a_2} \\ 0 & \dfrac{1}{s} \end{bmatrix} \tag{50}$$

with the a_i positive. Then two possibilities arise:

i. The parameter a_1 exactly equals a_2, the equality remaining valid for all physically meaningful perturbations of the process data. In this case the process is truly of dimension two and it makes sense to apply the theory developed above to stabilize the system: the closed-loop system will remain stable for small perturbations of the process data.

ii. The parameters a_1 and a_2 are equal to within computer tolerances, but are not exactly equal. It then makes little sense to represent the process by a model of dimension two and apply the theory to that model; for if a decentralized control is designed on such a basis and then applied to the process, instability will inevitably occur. This, of course, is because the process really requires a model of dimension three, and the theory should be applied to such a model to correctly predict that the pole at a_2 cannot be shifted and therefore that the process cannot be stabilized with decentralized control.

Now suppose that in place of (50), the process is initially modeled in state space form; i.e. Σ is a minimal realization of (50). Then Σ will be of dimension two if $a_1 = a_2$, and dimension three if $a_1 \neq a_2$. Suppose now that *all* parameters of Σ are perturbed subject only to the constraint that the perturbed system has the same topology (graph) as Σ. Then it is easy to check that by perturbing the parameters of Σ ever so slightly, it is possible to obtain a system for which stabilization *is* possible, *even for case* ii when $a_1 \neq a_2$. Thus because of computer inaccuracy, for example, one might well conclude that stabilization is possible when in fact it is not. Of course the problem here is that the class of admissible perturbations is too large.

CONCLUSIONS

In this paper we have developed explicit conditions for determining when the closed-loop spectrum of a linear system can be freely assigned or stabilized with decentralized control; and we have described in reasonably complete terms what sort of spectra one might hope to achieve. We have made use of some ideas from graph theory to obtain certain results, e.g. single channel controllability and observability imply strong connectedness, which no doubt have implications for more general, non-linear systems. While our synthesis technique is constructive, it may not always prove to be the best way to construct decentralized controls, particularly for systems known to possess 'weak couplings' between channels. On the other hand, both the stability and dynamic response of an entire decentralized system are determined only after *all* local controllers have been applied, at least for strongly connected systems, and so if channel couplings are 'strong', final system tuning from just one channel would appear to make sense.

Although some of the preceding solvability conditions undoubtedly hold generically, the corresponding nongeneric cases cannot be ignored. For example, if a strongly connected system is in some sense 'close' to being incomplete, then it may be better to adopt an incomplete model for purposes of design.

The subject of decentralized dynamic control is quite new and it is evident from the preceding discussion, that there is much work to be done.

REFERENCES

[1] J. P. CORFMAT: Decentralized Control of Linear Multivariable Systems. Yale University Becton Center Technical Report CT–67, October, 1974; also Yale University Doctoral Dissertation, December, 1974.

[2] T. A. MARSHAK: Computation in Organizations: The Comparison of Price Mechanisms and Other Adjustment Processes. *Decision and Organization*, 237–281. North Halland, Amsterdam (1972).

[3] J. MARSHAK and R. RADNER: *The Economic Theory of Teams*, Yale University Press, New Haven (1971).

[4] G. B. DANTZIG and P. WOLFE: Decomposition principle for linear programs. *Operations Res.* **8**(1), 1960.

[5] O. I. ELGERD: *Electric Energy Systems Theory: An Introduction.* McGraw-Hill, New York (1971).

[6] C. W. SANDERS, E. C. TACKER and T. D. LINTON: A new class of decentralized filters for interconnected systems. *IEEE Trans. Aut. Control* **AC–19**, 259–262 June (1974).

[9] A. GERSHO and B. J. KARAFIN: Mutual synchronization of geographically separated oscillators. *Bell Systems Tech. J.* 1689–1701, December (1966).

[8] I. W. SANDBERG: On conditions under which it is possible to synchronize digital transmission systems. *Bell Systems Tech. J.* **46**(6), 1999–2020 (1969).

[9] R. LAU, R. C. M. PERSIANO and P. P. VARAIYA: Decentralized information and control: a network flow example, *IEEE Trans. Aut. Control* **AC-17**, 446–473.

[10] M. AOKI and M. T. LI: Partial reconstruction of state vectors in decentralized dynamic systems. *IEEE Trans. Aut. Control.* **AC-18**(3), 289–294 (1973).

[11] K. C. CHU: Optimal decentralized regulation for a strong of coupled systems. *IEEE Trans. Aut. Control* **AC-19**(3), 243–246 (1974).

[12] D. MCFADDEN: On the controllability of decentralized macroeconomic systems: the assignment problem. *Mathematical Systems Theory and Economics* **1**, 221–240. Springer–Verlag (1967).

[13] J. P. CORFMAT and A. S. MORSE: Stabilization with decentralized feedback control. *IEEE Trans. Aut. Control* **AC-18**(6), 673–682 (1973).

[14] M. AOKI: On dynamic assignment problems and decentralized stabilization policies. UCLA, Dept. of System Science Technical Report, 1974.

[15] M. AOKI: On feedback stabilizability of decentralized dynamic systems. *Automatica* **8**(2), 163–173 (1972).

[16] S. H. WANG and E. J. DAVISON: On the Stabilization of decentralized control systems. *IEEE Trans. Aut. Control* **AC–18**(5), 473–478 (1973).

[17] F. M. BRASCH, JR. and J. B. PEARSON: Pole placement using dynamic compensators. *IEEE Trans. Aut. Control* **AC–15**(1), 34–43 (1970).

[18] J. P. CORFMAT and A. S. MORSE: Spectrum assignment with decentralized feedback control. *Proc. 1973 Princeton Conference on Information Science and Systems*, 228–231, March (1973).

[19] J. P. CORFMAT and A. S. MORSE: Control of linear systems through specified input channels. *SIAM J. Control and Optimization* **14**(1), 163–175, January (1976).

[20] A. S. MORSE: Structural invariants of linear multivariable systems. *SIAM J. Control* **11**(3), 446–465. August (1973).

[21] R. G. BUSACKER and T. L. SAATY: *Finite Graphs and Networks.* McGraw-Hill, New York (1965).

[22] W. M. WONHAM and A. S. MORSE: Decoupling and pole assignment in linear multivarible systems: A geometric approach. *SIAM J. Control* **8**(1), 1–18, February (1970).

APPENDIX

Properties of robust sets

In this appendix we discuss certain properties of robust subsets which are used throughout the paper. The assertions which follow can be easily justified with very elementary arguments. With the exception of Lemma A.1, proofs will therefore not be given.

For our purposes, a subset V of a real n-dimensional vector space $\mathcal{F}$, is a *variety* in $\mathcal{F}$, if the coordinates of each element of V, relative to a fixed basis for $\mathcal{F}$, coincide with a zero of a finite system of (real coefficient) polynomial equations in n-indeterminates, and visa versa. If at least one polynomial in the system is not the zero polynomial, then $V \neq \mathcal{F}$; in this case, V is a *proper* variety. If V is a proper variety, then its complement $\underline{\mathcal{V}}$ is open and dense in $\mathcal{F}$ with respect to the standard Euclidean metric; it is natural to call $\underline{\mathcal{V}}$ a *robust* subset of $\mathcal{F}$. Note that a robust subset is never empty.

If $\underline{\mathcal{U}}$ and $\underline{\mathcal{V}}$ are robust in $\mathcal{F}$, then so is their intersection $\underline{\mathcal{U}} \cap \underline{\mathcal{V}}$. If $\underline{\mathcal{U}}$ and $\underline{\mathcal{W}}$ are robust in vector spaces $\mathcal{F}$ and $\mathcal{G}$ respectively, then the Cartesian product $\underline{\mathcal{U}} \times \underline{\mathcal{W}}$ is robust in the 'outer' direct sum $\mathcal{F} \oplus \mathcal{G}$.

If $\alpha_F: \mathcal{F} \to$ reals is a polynomial function in the coordinates of $F \in \mathcal{F}$, relative to a fixed basis, we sometimes say that α_F is *algebrically dependent* on F. If M_F is a matrix with all entries algebrically dependent on F and if rank $M_F \geq r$ for some $F \in \mathcal{F}$, then the set of F for which rank $M_F \geq r$ is robust in $\mathcal{F}$. From this it follows that if (C_F, A_F, B_F) is a linear system with coefficients algebraically dependent on F, then the set of F for which $d(\langle A_F | \mathcal{B}_F \rangle)$ is maximal, is robust in $\mathcal{F}$. It also follows that if the transfer matrix of (C_F, A_F, B_F) is nonzero for some F, then the set of F with this property is robust in $\mathcal{F}$.

Suppose that $\alpha_F(\lambda)$ is a real monic polynomial with coefficients depending algebraically on $F \in \mathcal{F}$. It is always possible to write $\alpha_F(\lambda) = \mu(\lambda)\beta_F(\lambda)$ where $\mu(\lambda) \equiv \bigwedge_{F \in \mathcal{F}} \alpha_F(\lambda)$; $\beta_F(\lambda)$ is a polynomial, with coefficients depending algebraically on F, satisfying

$$\bigwedge_{F \in \mathcal{F}} \beta_F(\lambda) = 1 \tag{A.1}$$

Lemma A.1. Let $\gamma(\lambda)$ be any fixed polynomial. The set of $F \in \mathcal{F}$ for which $\gamma(\lambda)$ and $\beta_F(\lambda)$ are coprime is robust in $\mathcal{F}$.

Proof. Write $\gamma(\lambda) = \prod_{i=1}^{m} \gamma_i^{n_i}(\lambda)$ where the $\gamma_i(\lambda)$ are the prime factors of $\gamma(\lambda)$. For fixed i, there must exist $F \in \mathcal{F}$ for which $\beta_F(\lambda)$ and $\gamma_i(\lambda)$ are coprime; if this were false, $\gamma_i(\lambda)$ (being a prime) would have to divide $\beta_F(\lambda)$ for all $F \in \mathcal{F}$, contradicting (A.1). Therefore, the set $\mathcal{R}_i$ of F for which the resultant of $\gamma_i(\lambda)$ and $\beta_F(\lambda)$ is nonzero, is nonempty and therefore robust in $\mathcal{F}$. It follows that the set of F for which $\gamma(\lambda)$ and $\beta_F(\lambda)$ are coprime, namely $\bigcap_{i=1}^{m} \mathcal{R}_i$, is robust in $\mathcal{F}$. □

State Estimation for Discrete Systems With Switching Parameters

C.B. CHANG, Member, IEEE
M.I.T. Lincoln Laboratory
M. ATHANS, Fellow, IEEE
M.I.T. Electronic Systems Laboratory

Abstract

The problem of state estimation for discrete systems with parameters which may be switching within a finite set of values is considered. In the general case it is shown that the optimal estimator requires a bank of elemental estimators with its number growing exponentially with time. For the Markov parameter case, it is found that the optimal estimator requires only N^2 elemental estimators where N is the number of possible parameter values.

Manuscript received June 14, 1977; revised October 17, 1977

This work was supported by the U.S. Department of the Army. The work of M. Athans was also supported in part by the Air Force Office of Scientific Research under AFOSR Grant 77-3281.

The views and conclusions contained in this document are those of the contractor and should not be interpreted as necessarily representing the official policies, either expressed or implied, of the United States Government.

Authors' addresses: C.B. Chang, Lincoln Laboratory, Massachusetts Institute of Technology, Lexington, MA 02173; M. Athans, Electronic Systems Laboratory, Massachusetts Institute of Technology, Cambridge, MA 02139.

I. Introduction

In most practical applications of recursive estimation theory, there are difficulties in obtaining an exact mathematical model of the physical dynamic process. The uncertain parts of the system are sometimes represented by an unknown parameter vector. When the state estimation for this type of system has to be carried out, the variations of these parameters and their identification play a critical role.

Many approaches have been proposed in attempting to perform state estimation together with parameter identification. There is one class of adaptive estimation scheme which calls for the construction of a bank of elemental estimators with each matched to a possible parameter value [1]-[7]. The optimal state estimate is obtained via a weighted sum over the elemental estimates with the a posteriori hypothesis probabilities as weighting factors. Based upon this idea, algorithms for both adaptive estimation and control have been designed. These algorithms are optimum if 1) one of the assumed models matches the physical process and 2) the unknown parameter is a constant vector. Various arbitrary modifications have been suggested to alleviate the above restrictions.

The purpose of this paper is to present the extension of the above adaptive estimation scheme to the case when the parameter vector is assumed to vary within a finite number of values. For a general parameter process when the present parameter value can depend on the past history of the parameter values, the optimum state estimate is found to be the weighted sum of elemental estimates obtained by estimators matched to all possible parameter histories. The weighting factor is the a posteriori hypothesis probability determined by 1) residual processes of all elemental estimators and 2) the conditional probability characterizing properties of the parameter process. Since each elemental estimator has to match to a given parameter history and the number of possible histories grows with time, the optimal estimator requires exponentially growing memory. When the parameter follows a Markov process, i.e., the present parameter value depends *only* on the previous parameter value, this great computational and memory requirement is relaxed. It is shown that the required number of elemental estimator stays at N^2 for the Markov parameter case where N is the number of possible parameter values. Since the estimator for the Markov parameter allows the parameter to change with time and its computational requirement is quite modest (when compared with the exponentially growing memory requirement), it may be used as a feasible suboptimal estimator design for the general time-varying parameter estimation problem.

There are a number of potential application areas of the above proposed scheme. For example, the constant-parameter multiple-elemental- estimator algorithm was applied in the F-8C aircraft real-time control problem [6]. In that application, the aerodynamic constants were

Reprinted from *IEEE Trans. Aerosp. Electron. Syst.*, vol. AES-14, pp. 418–425, May 1978.

discretized into several sets of values. A bank of Kalman filters were constructed to match each set of parameter values. The switching parameter algorithm discussed here will allow a more rapid switching among different sets of parameters when the true aerodynamic constants have indeed changed. Another potential application of this scheme is in re-entry vehicle (RV) tracking [8]-[10]. A RV re-enters the atmosphere following a ballistic trajectory. It may perform aerodynamic maneuvers at a medium altitude region and return to ballistic at a low altitude region. A detection scheme was suggested in [8] and [9] allowing the detection of a RV maneuver. This scheme suffers from detection delay and filter transients in the switch-over region. In addition, it cannot identify the change to a trajectory from maneuver to ballistic. Another solution is to construct two filters, a ballistic filter and a maneuvering filter. If one chooses the constant parameter algorithm, then a large bias error will develop in the ballistic filter when the RV initiates a maneuver. It therefore cannot identify the maneuver quench. The proposed algorithm allowing transition probabilities to exist between ballistic and maneuvering filters thus allows the rapid identification of a maneuvering as well as a ballistic trajectory.

The optimal estimator for continuous systems with continuously time-varying parameters was considered before, e.g., see [2]-[4]. A representation theorem was given in [2] and [3] and the differential equation for the a posteriori hypothesis probability density function was given in [4]. It is, however, very difficult to derive the algorithm for the discrete case by directly discretizing the results of [2]-[4]. The derivation used in this paper is that of Bayesian approach [11]. Similar derivations were used in [5] and [7] where the adaptive control problem was discussed in [5] and a tutorial treatment of the multiple elemental estimator approach to adaptive estimation was presented in [7].

This paper is organized as follows. The state estimation problem considered in this paper is stated in Section II. The optimal state estimators for 1) a general parameter process and 2) a Markov parameter process are presented in Section III. The time evolution of the parameter process is characterized by a set of hypothesis processes which is also discussed in the Section III. The implementation of the optimal state estimator depends upon the elemental estimators. This subject is briefly discussed in Section IV. Derivations of the estimation algorithms are given in the Appendix.

II. Problem Statement

Consider the following discrete system and measurement equations:

$$\mathbf{x}(t+1) = \mathbf{f}(\mathbf{x}(t), \gamma(t), \mathbf{u}(t), \xi(t)) \tag{1}$$

$$\mathbf{z}(t+1) = \mathbf{h}(\mathbf{x}(t+1), \theta(t+1)) \tag{2}$$

where $\mathbf{x}(t)$ is the state vector, $\mathbf{u}(t)$ is the known control vector, $\mathbf{z}(t)$ is the measurement vector, $\xi(t)$ is the system noise vector, $\boldsymbol{\theta}(t)$ is the measurement noise vector, $\boldsymbol{\gamma}(t)$ is the random, unknown, and time-varying parameter vector, and t is the discrete time index.

The parameter vector is restricted to take values from a finite set. Define a finite-dimensional parameter vector set R_q containing N distinct parameter vectors, i.e.,

$$R_q = \{\gamma : \gamma = \gamma_i ; i = 1, \ldots, N\}. \tag{3}$$

The parameter may switch to any parameters in R_q at the next discrete time, i.e.,

$$\gamma = \gamma_i \text{ at } t \text{ for } \gamma_i \in R_q$$

and

$$\gamma = \gamma_j \text{ at } \tau \text{ for } \gamma_j \in R_q$$

when $t \neq \tau$, i may or may not be equal to j. Due to this property, γ is called a stochastic jump parameter. A similar parameter process was considered earlier in a control design problem [12].

Given the problem defined above, we would like to find the minimum variance estimator (true conditional mean) of the state

$$\hat{\mathbf{x}}(t/t) = E[\mathbf{x}(t)/Z(t)] \tag{4}$$

and the conditional covariance matrix

$$\Sigma(t/t) = \operatorname{cov}[\mathbf{x}(t); \mathbf{x}(t)/Z(t)] \tag{5}$$

where $Z(t)$ is the set of all past measurements and controls, i.e.,

$$Z(t) = \{\mathbf{u}(0), \mathbf{u}(1), \ldots, \mathbf{u}(t-1), \mathbf{z}(1), \ldots, \mathbf{z}(t)\}. \tag{6}$$

This is clearly a discrete nonlinear estimation problem. With the parameter vector defined above, it can be shown that the optimal estimator has a fixed structure. That is, the optimal estimate is a weighted sum of elemental estimates with each matched to a possible parameter history. Furthermore, when the system becomes linear even though the parameter process is still related to the state vector nonlinearly, the optimal estimates can be computed for a finite discrete time t.

III. Problem Solution

In this section we present solutions to the described estimation problem for 1) a general parameter process, and 2) a Markov parameter process. In order to mathematically characterize the time evolution of the parameter process, we first define a set of hypothesis processes in the following subsection.

A. Hypothesis Processes

The parameter vector may switch to any vectors in R_q

at the next discrete time, the number of parameter permutation therefore grows exponentially. A set of indices is defined to represent the time evolution of the parameter permutation.

1) I is an index set denoting

$$I = \{i : i = 1, ..., N\}. \tag{7}$$

2) I_t is a t-tuple index set, i.e.,

$$I_t = \{i : i = i_t \times i_{t-1} \times ... \times i_1 ; i_t, ..., i_1 \in I\}. \tag{8}$$

3) $I_{t+1} = i_{t+1} \times I_t$ for $i_{t+1} \in I$.

Notice that I is the index for parameter vectors in R_q and I_t is the index for the permutation of the parameter vector from the initial time to time t. In the following, two hypothesis processes representing the parameter time history are defined.

1) $H_i(t)$, $i \in I$: The hypothesis that $\gamma = \gamma_i$ at t. It is a "local" hypothesis since it only concerns the status of γ at t.

2) $\overline{H}_i(t)$, $i \in I_t$: The hypothesis that a given history of γ for up to time t is true. This history is indexed by a $i \in I_t$. It is a "global" hypothesis since it concerns the time history of γ. For all $i \in I_t$, it defines all possible γ sequences for up to time t.

Clearly, the global hypothesis describes the time evolution of the parameter process. It is specified by local hypotheses at each discrete time instance, i.e.,

$$\overline{H}_i(t) = (H_{i_t}(t), H_{i_{t-1}}(t-1), ..., H_{i_1}(1)) \tag{9}$$

where $i \in I_t$, $i_t, ..., i_1 \in I$, and i is determined by a given sequence of $i_t, ..., i_1$.

In the discussions to follow, we will frequently refer to the a posteriori hypothesis probability. It is the probability of a given hypothesis being true conditioned upon the past measurements, i.e.,

$$P_i(t) = \Pr(\text{the } i\text{th hypothesis is true } /Z(t)). \tag{10}$$

There is also a conditional probability which is specified by the nature of the parameter process. In the general case, the present parameter value may depend upon the values of its entire past history. This is mathematically characterized as

$$\Pr(H_i(t) \text{ is true } /H_{i_{t-1}}(t-1), H_{i_{t-2}}(t-2), ..., H_{i_1}(1))$$
$$\cdot \triangleq P(H_i(t)/\overline{H}_j(t-1)), \quad i \in I; j \in I_{t-1}. \tag{11}$$

If, however, the parameter follows a Markov process, its present value will depend only on its immediate past value, i.e.,

$$P(H_i(t)/\overline{H}_j(t-1)) = P(H_i(t)/H_j(t-1)) \triangleq P_{ij} = \text{the transition probability between parameters } \gamma_i(t) \text{ and } \gamma_j(t-1). \tag{12}$$

These probabilities play an important role in the optimal state estimators to be discussed in the next two subsections.

B. General Parameter Case

The algorithms for computing the state estimates, the covariance matrix, and the a posteriori hypothesis probability are stated below. The derivations are included in the Appendix.

1) For state estimate and covariance,

$$\hat{x}(t/t) = \sum_{i \in I_t} P_i(t)\, \hat{x}_i(t/t) \tag{13}$$

$$\Sigma(t/t) = \sum_{i \in I_t} P_i(t)[\Sigma_i(t/t) + (\hat{x}_i(t/t) - \hat{x}(t/t))(\hat{x}_i(t/t) - \hat{x}(t/t))^T] \tag{14}$$

where $\hat{x}_i(t/t) = E[x(t)/Z(t), \overline{H}_i(t)]$ and $\Sigma_i(t/t) = \operatorname{cov}[x(t); x(t)/Z(t), \overline{H}_i(t)]$.

2) For the a posteriori hypothesis probability,

$$P_j(t+1) = \{[p(z(t+1)/H_i(t+1), \overline{H}_k(t), Z(t))]/[p(z(t+1)/Z(t))]\} \cdot P(H_i(t+1)/\overline{H}_k(t), Z(t))\, P_k(t) \tag{15}$$

where 1) $i \in I$, $k \in I_t$, and $j = i \times k$, and 2) $P(H_i(t+1)/\overline{H}_k(t), Z(t)) = \Pr(\gamma(t+1) = \gamma_i/\overline{H}_k(t), Z(t))$. We make the following remarks:

1) The elemental estimate $\hat{x}_i(t/t)$ is obtained by constructing a minimum variance estimator based upon systems and measurement equations defined by (1) and (2) with the parameter process matched to the ith global hypothesis $\overline{H}_i(t)$.

2) The elemental estimates $\hat{x}_i(t/t)$ and covariances $\Sigma_i(t/t)$ are combined to obtain the optimal estimate $\hat{x}(t/t)$ and covariance $\Sigma(t/t)$ by using a weighted sum. The ith weighting factor $P_i(t)$ is the a posteriori hypothesis probability of the ith global hypothesis being true conditioned upon all the past measurements and controls.

3) The above equations are independent of whether the system is linear or nonlinear. When the system is nonlinear, the elemental estimates can, of course, only be obtained approximately. The problem of realization will be discussed in the next section.

4) The probability density functions of (15) are computed by using residual processes of elemental estimators. That is,

a) $p(z(t+1)/H_i(t+1), \overline{H}_k(t), Z(t))$ = the density of the residual process of the elemental estimator matching to $\overline{H}_k(t)$ and $H_i(t+1)$.

b)

$$p(\mathbf{z}(t+1)/Z(t)) = \sum_{m \in I_t} p(\mathbf{z}(t+1)/\bar{H}_m(t), Z(t))\, P_m(t)$$

$$= \sum_{m \in I_t} \Big[\sum_{n \in I} p(\mathbf{z}(t+1)/\bar{H}_m(t), H_n(t+1), Z(t)) \cdot p(H_n(t+1)/\bar{H}_m(t), Z(t)) \Big] P_m(t). \tag{16}$$

5) The conditional hypothesis probability, $P(H_i(t+1)/\bar{H}_k(t), Z(t))$, represents the property of the hypothesis process. It characterizes the evolution of γ in R_q. Notice that we do not restrict the behavior of γ variations. In the most uncertain situation such that the parameter may change to any parameters in R_q with equal probability, one obtains

$$P(H_i(t+1)/\bar{H}_k(t), Z(t)) = 1/N, \quad \text{for all } i \in I. \tag{17}$$

If the parameter is a Markov process, it becomes the transition probability

$$P(H_i(t+1)/\bar{H}_k(t), Z(t)) = P(H_i(t+1)/H_k(t)), \quad \text{for } i, k \in I. \tag{18}$$

6) Clearly, the difficulty of implementing this estimator lies in the fact that the number of elemental estimators grows exponentially with time. The time evolution of elemental estimators for $N = 2$ is shown in Fig. 1. Notice that at time t, the number of elemental estimators grows to 2^t.

This difficulty is, however, alleviated when the parameter follows a Markov process. It can be shown that for a Markov parameter process, the number of elemental estimator is limited to N^2. This case is discussed in the next section.

C. Markov Parameter Case

In presenting the algorithm for the Markov parameter case, we use the following simplified notations:

$\hat{\mathbf{x}}_i(t/t) = E[\mathbf{x}(t)/H_i(t), Z(t)]$

$\Sigma_i(t/t) = \text{cov}[\mathbf{x}(t); \mathbf{x}(t)/H_i(t), Z(t)]$

$\hat{\mathbf{x}}_{ij}(t/t-1) = E[\mathbf{x}(t)/H_i(t), H_j(t-1), Z(t-1)]$

= the predicted state

$\hat{\mathbf{x}}_{ij}(t/t) = E[\mathbf{x}(t)/H_i(t), H_j(t-1), Z(t)]$

= the updated state

$P_i(t) = \Pr(H_i(t) \text{ is true} / Z(t))$

$P_{ij} = \Pr(H_i(t) \text{ is true}/H_j(t-1))$

= the transition probability

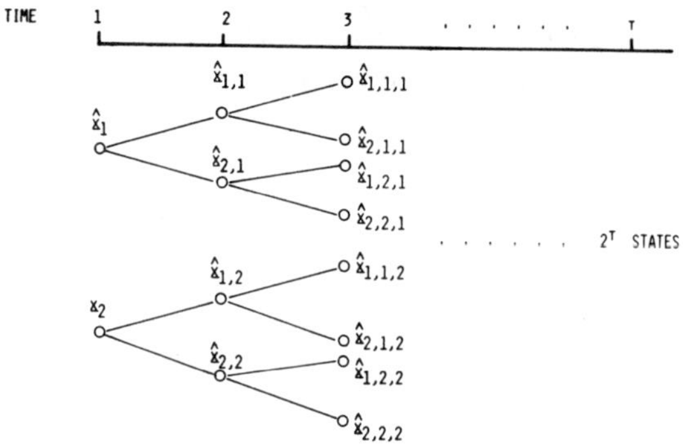

Fig. 1. Time evolution of elemental estimates for $N = 2$. (Underlined symbols correspond to boldface symbols in the text.)

$p_{ij}(t) = p(\mathbf{z}(t)/H_i(t), H_j(t-1), Z(t-1))$

= the residual density of the (i, j)th elemental estimator.

We now proceed to state the estimator. The derivation is included in the Appendix.

1) For state estimate and covariance calculations:

a) At time $t-1$, we have

$$\hat{\mathbf{x}}_i(t-1/t-1), \Sigma_i(t-1/t-1), P_i(t-1), \text{ for } i = 1, \ldots, N.$$

b) Compute

$$\hat{\mathbf{x}}_{ij}(t/t-1), \Sigma_{ij}(t/t-1), \quad \text{for } i, j = 1, \ldots, N$$

using N^2 state and covariance predictors.

c) Compute

$$\hat{\mathbf{x}}_{ij}(t/t), \Sigma_{ij}(t/t), \quad \text{for } i, j = 1, \ldots, N$$

using N^2 state and covariance updators and the new measurement $\mathbf{z}(t)$.

d) Compute $\hat{\mathbf{x}}_i(t/t)$ and $\Sigma_i(t/t)$ for $i = 1, \ldots, N$ using

$$\hat{\mathbf{x}}_i(t/t) = \sum_{j=1}^{N} \hat{\mathbf{x}}_{ij}(t/t)\, P(H_j(t-1)/H_i(t), Z(t)) \tag{19}$$

$$\Sigma_i(t/t) = \sum_{j=1}^{N} P(H_j(t-1)/H_i(t), Z(t)) \,[\Sigma_{ij}(t/t) + (\hat{\mathbf{x}}_{ij}(t/t) - \hat{\mathbf{x}}_i(t/t))(\hat{\mathbf{x}}_{ij}(t/t) - \hat{\mathbf{x}}_i(t/t))^T] \tag{20}$$

where

$$P(H_j(t-1)/H_i(t), Z(t)) = p_{ij}(t) P_{ij} P_j(t-1) \Big/ \sum_{j=1}^{N} p_{ij}(t) P_{ij} P_j(t-1). \tag{21}$$

e) The minimum variance (true conditional mean) state estimate $\hat{\mathbf{x}}(t/t)$ and covariance $\Sigma(t/t)$ are obtained by

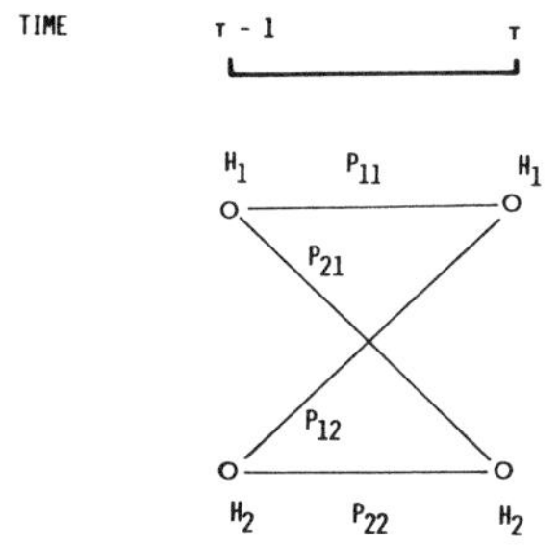

Fig. 2. Time evolution of hypothesis processes for Markov parameters with N = 2.

Fig. 3. State estimator for Markov parameters with N = 2. (Underlined symbols correspond to boldface symbols in the text.)

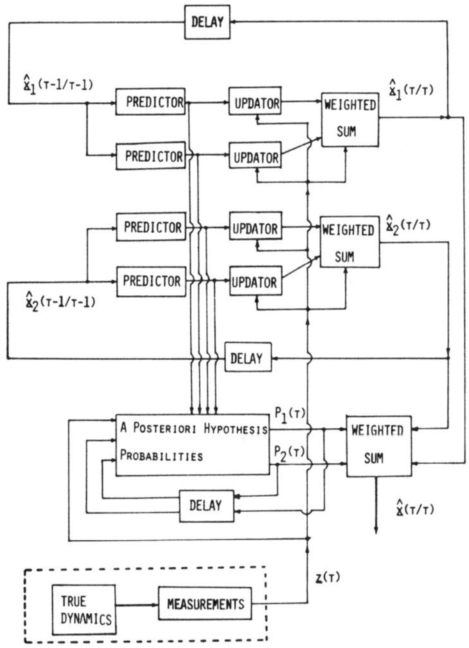

$$\hat{\mathbf{x}}(t/t) = \sum_{i=1}^{N} P_i(t)\,\hat{\mathbf{x}}_i(t/t) \tag{22}$$

$$\Sigma(t/t) = \sum_{i=1}^{N} P_i(t)[\Sigma_i(t/t) + (\hat{\mathbf{x}}_i(t/t) - \hat{\mathbf{x}}(t/t))(\hat{\mathbf{x}}_i(t/t) - \hat{\mathbf{x}}(t/t))^{\mathrm{T}}]. \tag{23}$$

The above completes one cycle of the state estimate and covariance computation.

2) For the a posteriori hypothesis probability calculations:

$$P_i(t) = \sum_{j=1}^{N} p_{ij}(t)\,P_{ij}P_j(t-1) \Big/ \sum_{i=1}^{N}\sum_{j=1}^{N} p_{ij}(t)\,P_{ij}P_j(t-1). \tag{24}$$

We make the following remarks:

1) The above algorithm requires implementing N^2 elemental estimators. If the elemental estimator is a minimum variance estimator, then the state estimate $\hat{\mathbf{x}}(t/t)$ is optimum in the minimum variance sense.

2) Equation (24) may be rewritten in a matrix form: Let $\mathbf{q}(t)$ be an $N \times 1$ column vector consisting of all a posteriori local hypothesis probabilities at time t

$$\mathbf{q}(t) = \begin{bmatrix} p(H_1(t)/Z(t)) \\ \vdots \\ p(H_N(t)/Z(t)) \end{bmatrix}$$

and let $S(t+1/t)$ be an $N \times N$ matrix with the (i, j)th element s_{ij} being equal to the product of the residual density of the elemental estimator matched to $H_i(t+1)$ and $H_j(t)$ and the probability of $H_i(t+1)$ being true conditional upon $H_j(t)$.

$$s_{ij} = [p(z(t+1)/H_i(t+1), H_j(t), Z(t))\,P(H_i(t+1)/H_j(t))].$$

Then

$$\mathbf{q}(t+1) = S(t+1/t)\mathbf{q}(t)/[p(\mathbf{z}(t+1)/Z(t))] \tag{25}$$

The time evolution of the hypothesis process for a Markov parameter process is illustrated in Fig. 2 for N = 2. A detailed flowchart illustrating the state estimation process for N = 2 is shown in Fig. 3.

A special case of the above occurs when the parameter is known to be time invariant. Under this condition, the above estimator is reduced to the estimator derived by Magill [1]. This can be obtained by substituting

$$P(H_i(t)/H_j(t-1)) = \begin{cases} 1, & \text{when } i = j \\ 0, & \text{otherwise} \end{cases} \tag{26}$$

into the estimator equations.

IV. Elemental Estimator Implementation

The significance of the estimator presented in the previous section is that the estimator structure is the same for both linear and nonlinear systems. This conclusion was also found in the continuous case [2]-[4]. The estimator realization therefore depends on 1) the realization of the elemental estimator and 2) the practicality of constructing a large number of elemental estimators. Three cases can be considered.

1) *General Case.* For the system and measurement equations considered in (1) and (2), $\hat{\mathbf{x}}_i(t/t)$ and $\Sigma_i(t/t)$ cannot be computed exactly. Suboptimal nonlinear filters such as the extended Kalman filter [13] or the second-order filter [14] must be used. These filters have found many practical applications. They can be applied to approximate the optimal estimators.

2) *Linear System Case.* When the system is linear and noise vectors are additive, the optimal estimate can be

computed exactly. In this case, the elemental estimator is constructed by using the discrete Kalman filter.

3) *Linear System and Constant Parameter Case.* This is the case considered by Magill [1]. The estimator presented in this paper can be reduced to the Magill's estimator by properly setting the conditional probabilities.

V. Discussion and Conclusions

In this paper we have presented the optimal estimator for discrete systems with finite-valued switching parameters. For a general parameter process in which the present parameter value can depend on the past history of the parameter values, the memory requirement of the estimator grows exponentially with time. When the present parameter value depends only on the previous parameter value and this dependence can be characterized by the transition probability, the required number of elemental estimator is found to be equal to N^2 where N is the number of possible parameter vectors.

We conclude by making the following comments:

1) We have derived algorithms for the general case (i.e., the parameter depends upon its entire past history) and the Markov parameter case (the parameter only depends upon its immediate past history). Using the same approach, one can derive the algorithm for parameters depending upon a finite past history. That is, the parameter process evolves as

$$P(H_i(t)/\bar{H}_j(t-1)) = P(H_i(t)/H_{j_1}(t-1), H_{j_2}(t-2), \ldots, H_{j_n}(t-n)) \tag{27}$$

where n is a constant integer.

2) Although the algorithm for the Markov process does not require a growing number of elemental estimators, the filter complexity can still be severe when N is large. The feasibility of applying this estimator is therefore constrained by the available capability of the real-time processor.

3) The asymptotic properties of this estimator are not addressed in this paper. This is a complicated problem and is still an open area for further research. An attempt was made in [15] to address the asymptotic property of the Magill's estimator. Extending the approach of that reference to the estimator discussed in this paper is nontrivial. In order to address this problem, one may have to first investigate all pertinent properties of mismatched Kalman filters [16].

4) A useful study could be made comparing this algorithm with several other adaptive estimation algorithms by using a significant physical problem (e.g., the aircraft control problem, vehicle tracking problems, etc.). The comparison should be based upon the estimation accuracy (covariance), speed of convergence (i.e., identifying the correct system), computational burdens, among others.

Appendix

Derivations

A. Derivation of (13) and (14)

This derivation involves a straightforward application of the definitions of $\hat{x}(t/t)$ and $\Sigma(t/t)$ and

$$p(x(t)/Z(t)) = \sum_{i \in I_t} P_i(t)\, p(x(t)/\bar{H}_i(t), Z(t)). \tag{A1}$$

The details are omitted here.

B. Derivation of (15)

Using the conditional probability relation yields

$$P_j(t+1) = P(H_i(t+1)/\bar{H}_k(t), Z(t+1))\, P(\bar{H}_k(t)/Z(t+1)) \tag{A2}$$

where $i \in I$ and $k \in I_t$.

Using Bayes' rule one obtains

$$P(H_i(t+1)/\bar{H}_k(t), Z(t+1)) = [p(z(t+1)/H_i(t+1), \bar{H}_k(t), Z(t))/p(z(t+1)/\bar{H}_k(t), Z(t))]\, P(H_i(t+1)/\bar{H}_k(t), Z(t)) \tag{A3}$$

and

$$P(\bar{H}_k(t)/Z(t+1)) = [p(z(t+1)/\bar{H}_k(t), Z(t))/ p(z(t+1)/Z(t))]\, P(\bar{H}_k/Z(t)). \tag{A4}$$

Substituting (A3) and (A4) into (A2) one obtains (15). This completes the derivation.

C. Derivation of (19), (20), and (21)

Using the definitions of $\hat{x}_i(t/t)$ and $\Sigma_i(t/t)$, i.e.,

$$\hat{x}_i(t/t) = \int x(t)\, p(x(t)/H_i(t), Z(t))\, dx \tag{A5}$$

$$\Sigma_i(t/t) = \int (x(t) - \hat{x}_i(t/t))\, (x(t) - \hat{x}_i(t/t))^T\, p(x(t)/H_i(t), Z(t))\, dx \tag{A6}$$

and the relation

$$p(x(t)/H_i(t), Z(t)) = \sum_{j=1}^{N} p(x(t)/H_i(t), H_j(t-1), Z(t))\, P(H_j(t-1)/H_i(t), Z(t)) \tag{A7}$$

(19) and (20) follow immediately. It remains to show (21). Using the conditional probability and Bayes' rule yield

$$P(H_j(t-1)/H_i(t), Z(t)) = P(H_j(t-1), H_i(t)/Z(t))/P_i(t) \tag{A8}$$

and

$$P(H_j(t-1), H_i(t)/Z(t)) = [p(z(t)/H_i(t), H_j(t-1), Z(t-1))/ p(z(t)/Z(t-1))]\ P(H_i(t)/H_j(t-1))\ P_j(t-1). \tag{A9}$$

Notice that $p(z(t)/Z(t-1))$ is computed by using

$$p(z(t)/Z(t-1)) = \sum_{i=1}^{N} \sum_{j=1}^{N} p(z(t)/H_i(t), H_j(t-1), Z(t-1))\, P(H_i(t)/H_j(t-1))\, P_j(t-1). \tag{A10}$$

Substituting (A9) and (A10) into (A8) and using (24), one obtains (21). This completes the derivation.

D. Derivation of (22) and (23)

This derivation involves a straightforward application of the definitions of $\hat{x}(t/t)$ and $\Sigma(t/t)$ and

$$p(x(t)/Z(t)) = \sum_{i=1}^{N} P_i(t)\, P(x(t)/H_i(t), Z(t)). \tag{A11}$$

The details are omitted here.

E. Derivations of (24)

Using the conditional probability yields

$$P(H_i(t)/Z(t)) = \sum_{j=1}^{N} P(H_i(t)/H_j(t-1), Z(t))\, P(H_j(t-1)/Z(t)). \tag{A12}$$

Using Bayes' rule one obtains

$$P(H_i(t)/H_j(t-1), Z(t)) = [p(z(t)/H_i(t), H_j(t-1), Z(t-1))/p(z(t)/H_j(t-1), Z(t-1))]\ P(H_i(t)/H_j(t-1), Z(t-1)) \tag{A13}$$

and

$$P(H_j(t-1)/Z(t)) = [p(z(t)/H_j(t-1), Z(t-1))/ p(z(t)/Z(t-1))]\ P(H_j(t-1)/Z(t-1)). \tag{A14}$$

Substituting (A13) and (A14) into (A12) and using (A10) one obtains (24). This completes the derivation.

Acknowledgment

The authors would like to thank Dr. K.P. Dunn of M.I.T. Lincoln Laboratory for many stimulating discussions during the course of this work.

References

[1] D.T. Magill, "Optimal adaptive estimation of sampled stochastic processes," *IEEE Trans. Automat. Contr.*, vol. AC-10, Oct. 1965.

[2] D.G. Lainiotis, "Optimal adaptive estimation: Structure and parameter adaptation," *IEEE Trans. Automat. Contr.*, vol. AC-16, Apr. 1971.

[3] D.G. Lainiotis, "Partitioning: A unifying framework for adaptive systems, I: Estimation," *Proc. IEEE*, vol. 64, Aug. 1976.

[4] K.P. Dunn, "Measure transformation, estimation, detection, and stochastic control," D.Sc. Thesis, Washington Univ., St. Louis, Mo., May 1974.

[5] D. Willner, "Observation and control of partially unknown systems," Ph.D. Thesis, Dep. Electrical Engineering, M.I.T., Cambridge, Mass., and M.I.T. Electronic Systems Lab., Rep. ESL-R-496, June 1973.

[6] M. Athans et al., "The stochastic control of the F-8C aircraft using the multiple model adaptive control (MMAC) method," *Proc. 1975 IEEE Conf. Decision and Control* (Houston, Tex.), pp. 217-228.

[7] M. Athans and C.B. Chang, "Adaptive estimation and parameter identification using multiple model estimation algorithms," M.I.T. Lincoln Lab., Lexington, Mass., Rep. TN-1976-28, June 23, 1976.

[8] C.B. Chang, R.H. Whiting, and M. Athans, "On the state and parameter estimation for maneuvering re-entry vehicles," *IEEE Trans. Automat. Contr.*, vol. AC-22, Feb. 1977.

[9] ———, "The application of adaptive filtering methods to maneuvering trajectory estimation," M.I.T. Lincoln Lab., Lexington, Mass., Rep. TN-1975-59, Nov. 24, 1975.

[10] M. Athans, R.H. Whiting, and M. Gruber, "A suboptimal estimation algorithm with probabilistic editing for false measurements with applications to target tracking with wake phenomena," *IEEE Trans. Automat. Contr.*, vol. AC-22, June 1977.

[11] Y.C. Ho and R.C.K. Lee, "A Bayesian approach to problems in stochastic estimation and control," *IEEE Trans. Automat. Contr.*, vol. AC-9, Oct. 1964.

[12] B.D. Pierce and D.D. Sworder, "Bayes and minimax controllers for a linear system with stochastic jump parameters," *IEEE Trans. Automat. Contr.*, vol. AC-16, Aug. 1971.

[13] A.H. Jazwinski, *Stochastic Processes and Filtering Theory.* New York: Academic Press, 1970.

[14] M. Athans, R.P. Wishner, and A. Bertolini, "Suboptimal state estimation for continuous time nonlinear systems from discrete noisy measurements," *IEEE Trans. Automat. Contr.*, vol. AC-13, Oct. 1968.

[15] Y. Baran and N.R. Sandell, "An information theoretic approach to dynamical systems modeling and identification," *IEEE Trans. Automat. Contr.*, vol. AC-23, Feb. 1978.

[16] C.B. Chang and K.P. Dunn, "Kalman filter compensation for a special class of systems," *IEEE Trans. Aerosp. Electron. Syst.*, vol. AES-13, Nov. 1977.

A Distributed System Adaptive Control Strategy

C. RICHARD JOHNSON, JR., Member, IEEE
Virginia Polytechnic Institute and State University
RAYMOND C. MONTGOMERY
NASA Langley Research Center

Abstract

One attitude control device being studied for large spacecraft consists of two counter-rotating rings, each designated as an annular momentum control device (AMCD), that are attached to a spacecraft using several magnetic bearings distributed along the circumference of the rings. For large spacecraft large rings are desirable. Unfortunately, for large rings flexibility is appreciable and it becomes necessary to account for the distributed nature of the rings in the design of the magnetic bearing controllers. Also ring behavior is unpredictably sensitive to ring termperature, spin rate, manufacturing imperfections, and other variables. For that reason a distributed adaptive microcomputer-based control system is being sought for ring stabilization and maneuvering. An original adaptive-control methodology for distributed-parameter systems is detailed and application to spinning ring, i.e., AMCD, stabilization is used as an illustration. The proposed methodology, presented as a step-by-step procedure, combines a lumped-parameter expansion description of distributed parameter systems with a fundamental simultaneous identification and control strategy. Simulations are presented providing preliminary evidence of the capabilities of the proposed procedure.

Manuscript received December 12, 1978; revised February 21, 1979.

This work was supported in part by NASA-ASEE Summer Faculty Research Fellowship and NASA Langley Research Center Grant NSG-1527.

Authors' addresses: C. R. Johnson, Jr., Department of Electrical Engineering, Virginia Polytechnic Institute and State University, Blacksburg, VA 24061; R. C. Montgomery, Flight Dynamics and Control Division, NASA Langley Research Center, Hampton, VA 23665.

I. Introduction

Distributed systems are described by partial differential equations (PDE). For mechanical systems the equations often take the form

$$\ddot{w}(s,t) + K(w) = f(s,t). \tag{1}$$

In (1) the variable w is vector valued with $\ddot{w}$ indicating the second partial derivative of $w(s,t)$ with respect to time t. The operator $K(w)$ is a partial differential operator and $f(s,t)$ is the force input. The independent variable s is a naming variable that labels an infinitesimal system particle, the motion of which from (1) is illustrated if Fig. 1. If the operator resulting from linearization of (1) is autonomous, the solutions of (1) take the form [1, 2]

$$w(s,t) = \sum_{i=1}^{\infty} W_i(t)\,\phi_i(s) \tag{2}$$

where $\phi_i(s)$ is an eigenvector of K and $W_i(t)$ is the amplitude of the ith eigenvector at time t. The "mode" shapes ϕ_i are determined so as to satisfy the homogeneous PDE (1) and the boundary conditions of the problem. A convenient representation of the solutions in (2) will involve eigenvector arguments that explicitly depend on s and t, so that the modal amplitudes W_i may be determined by substituting (2) into (1). Using the orthonormal properties of the modes this substitution yields a set of ordinary differential equations

$$M_i\ddot{W}_i(t) + K_iW_i(t) = F_i(t) \tag{3}$$

where the M_i are the effective modal masses

$$M_i = \int \phi_i^2(s)ds \tag{4}$$

K_i the linearized model operators

$$K_i = \int \phi_i(s)\,K[\phi_i(s)]\,ds \tag{5}$$

and F_i the modal forces

$$F_i(t) = \int \phi_i(s) f(s,t)\,ds \tag{6}$$

with each integral taken over all system particles.

A priori prediction of the K_i in (5) is very difficult for large scale distributed systems such as flexible spacecraft. Simulation results to be presented will indicate that errors in prediction of the K_i can produce instability in a fixed control system designed to be stable. Simulations will also indicate that the discretized spatial distribution of force actuators implementing the modal forces in (3) via (6) can require continual active stabilization. Unfortunately this re-

Reprinted from *IEEE Trans. Aerosp. Electron. Syst.*, vol. AES-15, pp. 601–612, Sept. 1979.

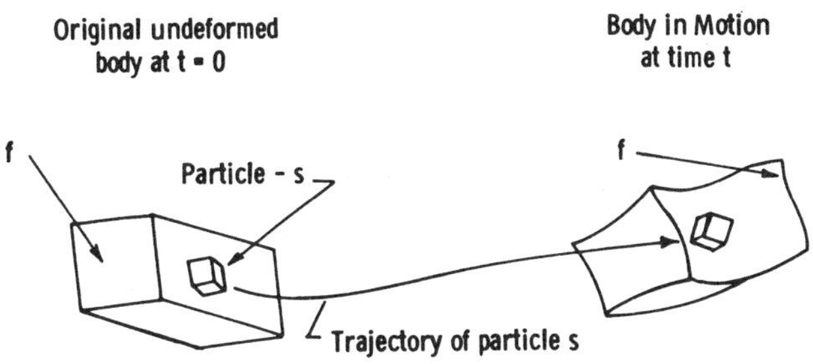

Fig. 1. Dynamics of distributed structural systems.

quirement of active stabilization may be difficult to meet with a fixed-gain controller if modeling is inadequate. Approaches to resolving this dichotomy include improving structural modeling capabilities, designing robust fixed-gain controllers tolerant of limited modeling uncertainties, and/or developing adaptive (or learning) control systems. This paper takes the last approach and details a real-time adaptive control methodology for distributed systems initially proposed in [3] and [4], which is similar to that independently developed in [5]. The adaptive control strategy in [5] is similar to that developed here in that both propose finite lumped-parameter expansion descriptions of distributed parameter systems, which are adaptively controlled by separate simultaneous identification and control of the expansion modes via a lumped-parameter plant adaptive control concept initially proposed in [6] and achieving wide acceptance as self-tuning regulation [7]. The expansions chosen and control objectives pursued are different, however, in [5] and the present work. This difference has an immediate effect on case-by-case utility. The presentation here, with deference to [5], attempts to more clearly state the generic quality of the lumped-parameter expansion/simultaneous identification and control approach to distributed parameter system control in the presence of PDE parameter uncertainty. These two works represent the birth of adaptive distributed-parameter system control.

The new adaptive-control methodology, the centerpiece of this paper, is outlined in Section II. Section III discusses the dual momentum control vector concept, which provides the test example (and original motivation) for the proposed adaptive-control strategy. Section IV follows the steps of the proposed procedure in developing the algorithms for the stabilization of a flexible, spinning ring, which represents an annular momentum control device (AMCD) suggested in [8] for attitude control of large space systems via a dual momentum vector configuration. Preliminary simulation results are presented in Section V. The conclusion briefly summarizes the questions to be addressed in further evaluation of the proposed approach to the adaptive control of distributed-parameter systems via the test problem.

II. Adaptive-Control Methodology for Distributed Systems

An adaptive-control methodology for distributed systems utilizing simultaneous identification and control techniques developed for lumped-parameter systems with a lumped-parameter expansion description of a distributed parameter system can be divided into two stages. The first segment of controller synthesis is done prior to controller activation and focuses on appropriate expansion selection and actuator/sensor distribution. The second design stage formulates the real-time sequence of signal processing and control formation tasks from sensor measurement to actuator activation.

The recommended stages of *preactivation analysis* are

1) Choose expansion descriptor $\phi_i(s)$ in (2).
2) Select finite-expansion upper limit to approximate (2).
3) Specify sensor locations and relate distributed measurments $w(s,t)$ to the modal amplitudes $W_i(t)$ in (3) via (2).
4) Determine actuator distribution and formulate effect on modal forces $F_i(t)$ in (3) via (6).
5) Establish modal control objectives.

Remarks. 1) The choice for the ϕ_i is dependent on the system PDE and the boundary conditions as well as the types of output $w(s,t)$.

2) The upper limit in an approximation of (2) represents a tradeoff between the accuracy of the approximation and the tractability of performing required real-time calculations, the complexity of which increases with the order of the approximate expansion, within the digital control system sample period.

3) The sensor spacing is highly dependent on ϕ_i and the anticipated form of w. For example, if the mode shapes represent incremental spatial frequencies then the Nyquist criterion limits mode recoverability via (2) for equally spaced measurement points. The sensor type and therefore placement is also clearly problem/cost dependent.

4) Actuator type varies between force input, boundary value alteration, and plant reconfiguration, i.e., PDE parameter variation. Boundary value control is a tool added to the arsenal versus lumped-parameter systems control due to the added complexity of the problem [9]. The distribution of actuators varies from multiple points of interaction to continuously-distributed control effort with obvious increases in analytical complexity and construction difficulty.

5) Since the adaptive-control strategy focuses on identifying the modal difference equation parameters and then solving for a modal control action, as is clear from the following real-time procedure, the more common statement of overall system performance objectives must be reduced to a modal level.

The combination of modal performances via (2) though independently satisfactory runs the risk of overall undesirable behavior, as can occur, for example, in the unstabilizability of an overall system via decentralized state feedback despite the local controllability of each subsystem [10].

The recommended steps of real-time, active, adapting, sampled data control formation are

1) Process sensor data to determine modal amplitudes $W(t)$ in (3).
2) Select the modes requiring control.
3) Improve the identification of the discretization of (3).
4) Design modal controllers on-line with current parameter estimates to meet modal performance objectives.
5) Convert desired modal controls to actual actuator commands.
6) Repeat (1)—(5) at the next sample instant.

Remarks. 1. Multiple repetition of (2) over various s will permit formation of a solvable set of linear equations in w and W_i if ϕ_i is an explicit function of neither. Such an equation set could be solved by pseudoinversion, Gaussian elimination, or other appropriate linear equations solution techniques [11]. Efficient, fast solutions may utilize the particular form of ϕ_i and the sensor distribution to advantage.

2. If more modes can be observed than controlled due to the greater ability of the sensor versus actuator distribution, then some automatic selection procedure must be formulated. The behaviors of certain structural modes, possibly exhibiting instability if sufficiently excited, are commonly recognized as being practically more critical than others. Such distinctions coupled with limited computational capability require this step, which is unfamiliar to lumped-parameter controllers.

3. A vast array of recursive parameter identification techniques exists for different equation coefficient estimation [12]. The algorithm selected must be appropriate for the circumstance, i.e., incorporating a sufficient noise model [13] and permitting real-time updating. Since each modal model is noninteracting, the separate parameter sets can be computed in parallel, an attractive attribute for a time-limited operation.

4. For simultaneous real-time plant identification and control determination the calculation of control law parameters from current plant-parameter estimates must commonly be in closed form. Such a restriction typically excludes the iterations necessary for high-order, distant-horizon quadratic cost-function minimization except as suggested in [14], where only one iteration of the Ricatti equation solution is made per sample instant, an unproven procedure. This requirement of explicit controller parameterization from plant-parameter estimates has resulted in the use of simplistic control objectives in the more widely utilized simultaneous identification and control schemes, e.g. single-stage cost-function minimization [15, 16], inverse control [17], and minimum variance regulation [18]. Again, as with the parameter identification, these calculations and the subsequent modal control formations can be computed in parallel.

5. Depending on the actuator type and distribution, this step is effectively the recombinant reverse of the initial stage of modal amplitude calculation from actual output measurement.

The designer's ultimate possibility of adapting the preactivation decisions, which determine the rigid structure within which the real-time processing is done, suggests the consideration of the formalism of layerable learning system models [19].

III. Dual Momentum Vector Control Device

One area in which large space systems will require new technology is attitude control. Large space antennas, as an example, will require establishing and maintaining a fixed attitude over long periods of time. Reaction control systems are unsuitable for this application because of the severe logistics problem associated with fuel resupply. Furthermore, many large space structures are inherently flexible and the motions set up by the on-off activity of reaction control systems results in acute structural dynamic problems. An approach which provides the necessary control forces while avoiding the above problems is to magnetically couple to the spacecraft to be controlled a pair of very large, flexible counterrotating rings. The rings nominally spin in the plane of a large area making this dual momentum vector device particularly suitable for large area structures. The resupply problem is largely alleviated since the primary requirement is for electric power which can be obtained from a solar cell array in combination with long lifetime batteries. Reaction jets required for desaturation of the rings after long periods of use would involve small quanties of fuel. Magnetic suspension provides for control smoothness not obtainable from reaction jets.

An early analysis of this dual momentum vector control concept for all axis attitude control was presented in [20]. The concept involves independently orienting the spin axis of two identical wheels. The wheels nominally have a net zero angular momentum. This is in contrast to the control moment gyro concept which uses a single wheel or a cluster of nominally orthogonal wheels to achieve attitude control. The dual momentum vector control process can be visualized in the following way. Suppose that the spin axes of the two identical wheels are aligned and that the wheels are spinning with equal but opposite spin rate as shown in Fig. 2(a). The net angular momentum of the pair is zero and, if they are constrained (by bearings and a gimbal assembly) to maintain the

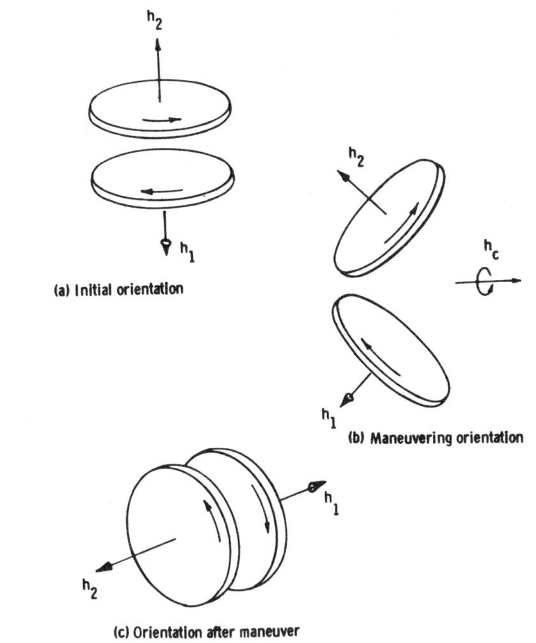

Fig. 2. Dual momentum vector control concept.

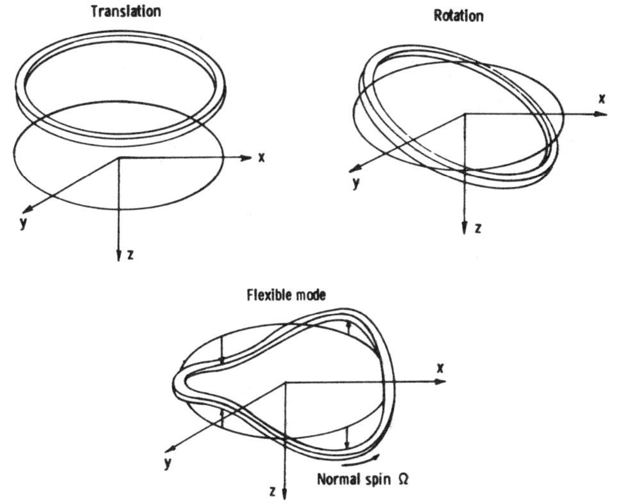

Fig. 3. Modes of motion of a spinning ring.

same net angular momentum condition, then any applied external moment will cause the pair to tumble as if the wheels had no individual momentum. If the momentum vectors are skewed by some mechanical means without applying any external moments as in Fig. 2(b), then, as conservation of momentum demands, there exists a momentum component, $\bar{h}_c = -\bar{h}_1 - \bar{h}_2$ called the control momentum. Therefore, the assembly (the wheels, suspension, and spacecraft) must rotate with an angular velocity ω_c. A large maneuver may be accomplished by skewing the momentum vector about an appropriate axis to create a control angular velocity, waiting until the assembly has rotated to the desired orientation, and unskewing the momentum vector to stop the rotation as in Fig. 2(c).

Large area space systems are expected to be relatively flexible and possess large inertias. For the application of a dual momentum vector control device to a large (e.g. 800-m diameter) antenna the wheel momentum required is huge and careful design of the wheels must be made to optimize the momentum-to-mass ratio and lower the ultimate cost of the device. Concentrating the mass of the wheels at the periphery leads to the most efficient momentum/mass ratio [8]. Also, it is desirable to maximize the speed of rotation of the wheels to maximize momentum storage. The speed is limited by allowable material stresses, but is still quite large, making the use of mechanical bearings inappropriate. Hence, the wheels should be like rings as large as the spacecraft can accommodate which are attached to the spacecraft with a suspension that is noncontacting.

One of the first noncontacting annular suspension devices was a centrifuge [21]. For space control applications, the first noncontacting annular device was the AMCD described in [8]. The device is a spinning ring that is magnetically suspended. In concept, the ring can be of arbitrary size. However, currently the technology has been studied only for rings of about 2-m in diameter. For very large rings of 800-m diameter, for example, such as are of interest for controlling large platforms, a ring behaves like a lariat and problems exist in stabilizing the ring during spin-up and during spacecraft maneuvers. Also, problems exist in the a priori modeling of the dynamics of a spinning ring. They are the result of the scale of the problem as well as the uncertain influence of spin rate, temperature, unknown material characteristics, and a host of other factors exerted on the dynamics of the ring.

IV. Adaptive Stabilization of a Spinning Ring

The adaptive-control methodology proposed in Section II is followed by the development in this section of an adaptive stabilizer for the large AMCD ring discussed in Section III.

Preactivation Analysis. 1) Because of the spatial repetition caused by ring closure, the linearized out-of-plane deformation w of a spinning, flexible ring can be decomposed as a function of s and t using an infinite sine/cosine series.

2) Assuming that the higher spatial frequency deformations will exhibit lower amplitudes due to the resilience of the ring material, this frequency series can be truncated with arbitrary accuracy at a finite number of modes simulated (NMS) so that

$$w(s,t) = W_1(t) + \sum_{j=1}^{NMS} [W_{j+1}(t) \cos(js) + W_{j+\mathrm{NMS}+1}(t) \sin(js)] \tag{7}$$

where W_j represents the modal deflection amplitudes and s the angular location in a ring particle reference frame. Note that W_1 corresponds to the translational mode in Fig. 3, W_2 and W_{NMS+2} to the rotational mode, and W_3 and W_{NMS+3} to the deformation mode.

3) Assume that the number of sensors (NS) measure the instantaneous deflection of the ring normal to a desired inertially fixed reference plane and that all sensors can be interrogated simultaneously. This could be accomplished by a single rapidly rotating scanner located at the ring center. The sensor measurements are to be taken at NS points located along the periphery of the ring at angles $s = \mathrm{LS}_i$ (location of sensor point i), $i = 1, 2, \ldots, \mathrm{NS}$, which do not change with respect to the reference plane. At time t, therefore, deflection measurements are made at particles $\{s = \mathrm{SS}_i\}$, $i = 1, \ldots, \mathrm{NS}$. The decomposition of these deflection measurements into a modal representation W_i can be made using a multiple formulation of (7)

$$\begin{bmatrix} w(\mathrm{SS}_1,t) \\ w(\mathrm{SS}_2,t) \\ \vdots \\ w(\mathrm{SS}_{\mathrm{NS}},t) \end{bmatrix} = \begin{bmatrix} 1 & \cos \mathrm{SS}_1 & \cdots & \cos(\mathrm{NMM}\cdot\mathrm{SS}_1) & \sin \mathrm{SS}_1 & \cdots & \sin(\mathrm{NMM}\cdot\mathrm{SS}_1) \\ 1 & \cos \mathrm{SS}_2 & \cdots & \cos(\mathrm{NMM}\cdot\mathrm{SS}_2) & \sin \mathrm{SS}_2 & \cdots & \sin(\mathrm{NMM}\cdot\mathrm{SS}_2) \\ \vdots & \vdots & & \vdots & \vdots & & \vdots \\ 1 & \cos \mathrm{SS}_{\mathrm{NS}} & \cdots & \cos(\mathrm{NMM}\cdot\mathrm{SS}_{\mathrm{NS}}) & \sin \mathrm{SS}_{\mathrm{NS}} & \cdots & \sin(\mathrm{NMM}\cdot\mathrm{SS}_{\mathrm{NS}}) \end{bmatrix} \cdot \begin{bmatrix} W_1(t) \\ W_2(t) \\ \vdots \\ W_{2\mathrm{NMM}+1}(t) \end{bmatrix} \quad (8)$$

where NMM (number of modes measured) is the number of modes used by the controller to represent the measurement data. Clearly, an exact representation of the NS measurements may exist only if NMS $\leqslant$ NMM. Note that no assumption has been made concerning the values of SS_i (sensor i at location s) which may differ from sample to sample. The inertia reference frame location of the measurement can be related to the spinning particle reference frame using the equation $\mathrm{SS}_i = \mathrm{LS}_i - \Omega t$, $i = 1, 2, \ldots, NS$, assuming that $\mathrm{SS}_i = \mathrm{LS}_i$ at $t = 0$. This is required since the ring particles move with angular velocity Ω with respect to the reference frame.

4) Also assume that there is a numbr of actuators (NA), each of which is located in the reference plane at angles $s = \mathrm{LA}_i$ (location of actuator i), $i = 1, 2, \ldots, \mathrm{NA}$, along the periphery of the ring and instantaneously produces a magnetic force constant over the same interval on the ring at the inertially fixed-point location of the magnet. The modal forces are functions from (6) of both the mode shapes in (7) and the actual forces f applied to the ring.

$$F_j(t) = \begin{cases} \int_0^{2\pi} f(s,t)\, ds, & j = 1 \\ \int_0^{2\pi} \cos[(j-1)s]\; f(s,t)\, ds, & j = 2, \ldots, \mathrm{NMS} + 1 \\ \int_0^{2\pi} \sin[(j-1)s]\; f(s,t)\, ds, & j = \mathrm{NMS} + 2, \ldots, 2\mathrm{NMS} + 1. \end{cases} \quad (9)$$

The point distribution of magnetic actuators selected converts, (9) to

$$F_j(t) = \begin{cases} \sum_{i=1}^{\mathrm{NA}} f_i(t), & j = 1 \\ \sum_{i=1}^{\mathrm{NA}} f_i(t) \cos[(j-1)\,\mathrm{Sf}_i(t)], & j = 2, \ldots, \mathrm{NMS} + 1 \\ \sum_{i=1}^{\mathrm{NA}} f_i(t) \sin[(j-1)\,\mathrm{Sf}_i(t)], & j = \mathrm{NMS} + 2, \ldots, \mathrm{NMS} + 1 \end{cases} \quad (10)$$

where

$$Sf_i(t) = LA_i\,(t) - \Omega t$$

and Sf_1 denotes the location s of the ith magnetic force f. Multiple concantenation of (10) once the modal forces are selected results in

$$\begin{bmatrix} F_1(t) \\ \cdot \\ \cdot \\ \cdot \\ F_{2NMS+1}(t) \end{bmatrix} = \begin{bmatrix} 1 & 1 & \cdots & 1 \\ \cos(Sf_1) & \cos(Sf_2) & \cdots & \cos(Sf_{NA}) \\ \cos(2Sf_1) & \cos(2Sf_2) & \cdots & \cos(2Sf_{NA}) \\ \cdot & & & \\ \cdot & & & \\ \cdot & & & \\ \cos(NMS\cdot Sf_1) & \cos(NMS\cdot Sf_2) & \cdots & \cos(NMS\cdot Sf_{NA}) \\ \sin(Sf_1) & \sin(Sf_2) & \cdots & \sin(Sf_{NA}) \\ \sin(2Sf_1) & \sin(2Sf_2) & \cdots & \sin(2Sf_{NA}) \\ \cdot & & & \\ \cdot & & & \\ \cdot & & & \\ \sin(NMS\cdot S_1) & \sin(NMS\cdot(Sf_2) & \cdots & \sin(NMS\cdot Sf_{NA}) \end{bmatrix} \begin{bmatrix} f_1(t) \\ \cdot \\ \cdot \\ \cdot \\ f_{NA}(t) \end{bmatrix} \tag{11}$$

5) For this ring stabilization problem an appropriate control objective is to damp the motions of the ring. This will be attempted by separate modal difference equation pole placement achieved via dynamic output feedback. The low order of the modal difference equations allows a closed-form solution for the controller parameters from the plant parameter estimates.

Real-Time Processing. 1) The NS × (2NMM + 1) matrix of (8) relating $w(s,t)$ and $W_i(t)$ without the far right column will be denoted as MADCM (modal-to-actual deflection conversion matrix). Clearly the inverse of MADCM can exist only if it is square, thereby uniquely relating w and W_j. Note that for 2NMM equally spaced measurement locations the sensed behavior of sin [(s)(NMM)] and the translation mode W_1 will be indistinguishable requiring the removal of the rightmost column and the reduction of NS to 2NMM to retain invertibility. The number of sensor measurements may be greater than twice the number of modes processed by the controller, thus requiring that the detected-to-modal amplitude conversion matrix (DMACM) be the pseudoinverse of MADCM for a least squares solution

$$\text{DMACM} = [(\text{MADCM})^T(\text{MADCM})]^{-1}(\text{MADCM})^T \tag{12}$$

which may exist only if NS ⩾ 2NMM.

2) If the number of modes controlled (NMC) is less than that measured (NMM) then some modes will have to be deleted from consideration. However, at the current state of this work NMC and NMM have been chosen equal not requiring a difficult decision.

3) For large, lightweight rings the $W_i(t)$ of (3) will effectively be undamped such that for small amplitude motion

$$(d/dt^2)\,[W_j(t)] + \omega_j^2(\Omega,t)\,W_j(t) = (1/M_j)\,F_j(t) \tag{13}$$

where ω_j represents the modal time frequency, which can be a function of both time and ring spin rate Ω. For measurements of W_j taken at uniform sample intervals of T seconds, the discretized equivalent of (13) can be written as

$$W_j(k) = A_{1j}W_j(k-1) + A_{2j}W_j(k\text{-}2) + B_{1j}F_j(k-1) + B_{2j}F_j(k-2) \tag{14}$$

where $A_{1j} = 2\cos(\omega_j T)$, $A_{2j} = -1$ and $B_{1j} = B_{2j} = (1 - \cos\omega_j T)/(M_j\omega_j^2)$. Note that by L'Hospital's rule for $\omega_j = 0$, $B_{1j} = B_{2j} = T^2/2M_j$. The effective modal masses M_j are given by

$$M_j = \begin{cases} \int_0^{2\pi} ds, & j = 1 \\ \int_0^{2\pi} \cos^2[(j-1)s]\, ds, & j = 2, \ldots, \text{NMS} + 1 \\ \int_0^{2\pi} \sin^2[(j-1)s]\, ds, & j = \text{NMS} + 2, \ldots, 2\text{NMS} + 1. \end{cases} \tag{15}$$

The assumption here requiring adaptivity is that the modal frequencies ω_j (and perhaps the actual damping coefficient assumed zero) have unknown infrequently varying values.

Identifying the difference equation descriptions of (14) via an equation error formulation yields an adaptive algorithm of the form [22]

$$\begin{bmatrix} \hat{A}_{1j}(k) \\ \hat{A}_{2j}(k) \\ \hat{B}_{1j}(k) \\ \hat{B}_{2j}(k) \end{bmatrix} = \begin{bmatrix} \hat{A}_{1j}(k-1) \\ \hat{A}_{2j}(k-1) \\ \hat{B}_{1j}(k-1) \\ \hat{B}_{2j}(k-1) \end{bmatrix} + e_j(k-1) \begin{bmatrix} \text{ASSW}_{1j} W_j(k-2) \\ \text{ASSW}_{2j} W_j(k-3) \\ \text{ASSW}_{3j} F_j(k-2) \\ \text{ASSW}_{4j} F_j(k-3) \end{bmatrix} \tag{16}$$

where e_j is the modal output prediction error defined by

$$e_j(k-1) = W_j(k-1) - [A_{1j}(k-1)\, W_j(k-2) + A_{2j}(k-1)\, W_j(k-2) + B_{1j}(k-1)\, F_j(k-2) + B_{2j}(k-1)\, F_j(k-3)]. \tag{17}$$

If each adaptive step-size weight ASWW_{ij} is positive and appropriately bounded, e.g.,

$$[\text{ASSW}_{1j} W_j^2(k-2) + \text{ASSW}_{2j} W_j^2(k-3) + \text{ASSW}_{3j} F_j^2(k-2) + ASSW_{4j} F_j^2(k-3)] < 2 \tag{18}$$

and W_j and F_j are "persistently exciting" then convergence of $[A_{1j}(k)\ A_{2j}(k)\ B_{1j}(k)\ B_{2j}(k)]$ to $[A_{1j}\ A_{2j}\ B_{1j}\ B_{2j}]$ as $k \rightarrow \infty$ can be guaranteed. This identifier can be used even if the assumption of no damping is violated since (16) assumes only the second-order form of (14). During actual operation some modes may not possess sufficient excitation to guarantee the consistency of the parameter estimates required for appropriate controller parameterization. In that event tests for convergence and acceptability of the data as in [23] may be required. Such identifiability tests could be incorporated in the mode selection procedures, if used, of step 2).

4) The modal pole placement objective corresponding to a specific fourth-order characteristic equation

$$z^4 + \text{DCEC}_{1j} z^3 + \text{DCEC}_{2j} z^2 + \text{DCEC}_{3j} z + \text{DCEC}_{4j} = 0 \tag{19}$$

(with DCEC signifying desired characteristic equation coefficient) can be achieved by utilizing the parameter estimates of (16) in a second order, dynamic-output feedback element as illustrated in Fig. 4. This controller does not explicitly require rate information, a quantity anticipated to be difficult to sense due to extremely low modal frequencies inherent in such large structures. The configuration of Fig. 4 yields

$$\begin{aligned} &z^4 + z^3[-\hat{A}_{1j}(k) - \hat{D}_{1j}(k)] \\ &+ z^2[\hat{D}_{1j}(k)\hat{A}_{1j}(k) - \hat{A}_{2j}(k) - \hat{D}_{2j}(k) - \hat{B}_{1j}(k)\hat{C}_{1j}(k)] \\ &+ z[\hat{D}_{1j}(k)\, \hat{A}_{2j}(k) + \hat{D}_{2j}(k)\hat{A}_{1j}(k) - \hat{B}_{1j}(k)\hat{C}_{2j}(k) - \hat{B}_{2j}(k)\hat{C}_{1j}(k)] \\ &+ [\hat{D}_{2j}(k)\hat{A}_{2j}(k) - \hat{B}_{2j}(k)\hat{C}_{2j}(k)] = 0 \end{aligned} \tag{20}$$

as its characteristic equation. Equating coefficients in (19) and (20) permits solution for the controller parameter estimates

$$\hat{D}_{1j}(k) = -\text{DCEC}_{1j} - \hat{A}_{1j}(k) \tag{21}$$

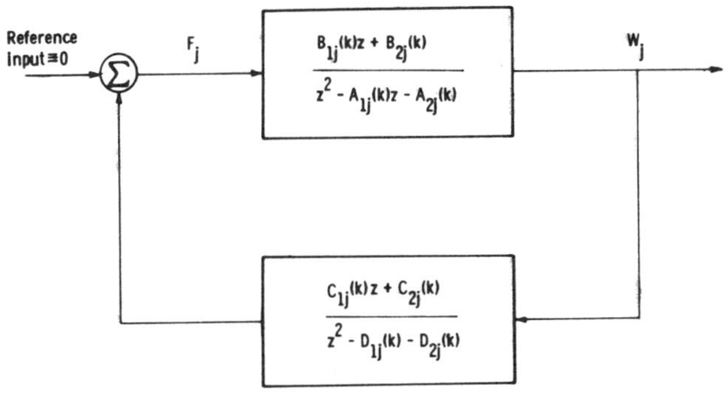

Fig. 4. Dynamic output feedback configuration.

$$\hat{C}_{1j}(k) = \{\hat{D}_{1j}(k)\hat{A}_{2j}(k) - \mathrm{DCEC}_{3j} + [\hat{D}_{1j}(k)\hat{A}_{1j}(k) - \hat{A}_{2j}(k) - \mathrm{DCEC}_{2j}] \cdot [\hat{A}_{1j}(k) - \hat{A}_{2j}(k)\hat{B}_{1j}(k)/\hat{B}_{2j}(k)] + \mathrm{DCEC}_{4j}\hat{B}_{1j}(k)/\hat{B}_{2j}(k)\} / [\hat{B}_{1j}(k)\hat{A}_{1j}(k) + \hat{B}_{2j}(k) - \hat{A}_{2j}(k)\hat{B}_{1j}(k)/\hat{B}_{2j}(k)] \tag{22}$$

$$\hat{D}_{2j}(k) = \hat{A}_{1j}(k)\hat{D}_{1j}(k) - \mathrm{DCEC}_{2j} - \hat{B}_{1j}(k)\hat{C}_{1j}(k) - \hat{A}_{2j}(k) \tag{23}$$

$$\hat{C}_{2j}(k) = [\hat{D}_{2j}(k)\hat{A}_{2j}(k) - \mathrm{DCEC}_{4j}]/\hat{B}_{2j}(k). \tag{24}$$

The modal forces provided by decentralized dynamic-output feedback as in Fig. 4 are calculable given (21) – (24) via

$$F_j(k) = \hat{C}_{1j}(k)W_j(k-1) + \hat{C}_{2j}(k)W_j(k-2) + \hat{D}_{1j}(k)F_j(k-1) + \hat{D}_{2j}(k)F_j(k-2). \tag{25}$$

5) The selection of actual forces to be applied to the ring once desired modal forces have been selected requires pseudoinversion of the (2NMS + 1) × NA actual to modal forces conversion matrix (AMFCM) from (11) resulting in a desired modal to actual forces conversion matrix (MAFCM)

$$\mathrm{MAFCM} = [(\mathrm{AMFCM})^T(\mathrm{AMFCM})]^{-1}(\mathrm{AMFCM})^T \tag{26}$$

with NMC replacing NMS in AMFCM, which can only exist if NMC is not less than half of one less than the number of actuators, i.e., NMC $\nless$ (NA-1)/2 or 2NMC + 1 $\nless$ NA. Note, however, that since NMC is typically less than NMS, forces entering "uncontrolled" modes need not be zero thereby exciting the uncontrolled modes. This undesirable effect has been designated "control spillover" [24].

Therefore, simultaneous identification and control of a flexible ring is achieved by a) using DMACM to convert an instantaneous spatial disturbance distribution to modal amplitudes, b) performing the correction of the estimates of the A_{ij}, B_{ij} coefficients of (13) via (16) to overcome uncertainty in ω_j, c) solving for the controller parameters in (14), via (21) – (24), d) inserting them in (25) to select the desired modal forces, and e) utilizing MAFCM to determine the actual forces to be applied. Note that the current modal control effort $F_j(k)$ is calculable entirely from past measurements of modal amplitudes and forces. This allows control calculation during the sample interval, thereby requiring no degrading delay between deflection measurement and control application, which could be appreciable due to the necessary pseudoinversions. Furthermore, except for the matrix operations in a) and e) which can be quickened by exploiting their special structure, as noted in the third and fourth remarks following the real-time sequencing recommended in Section II, steps b)-d) can be computed in parallel making the objective of a multiple microcomputer-based implementation possible.

V. Simulation Studies

A digital computer program has been written [3] to simulate out-of-plane motions of the ring and the adaptive controller described in Section IV. Fig. 5 is a block diagram indicating the simulation flow and capability. To observe the behavior of the ring alone the uncompensated step response and uncompensated step torque response of an initially unperturbed spinning, flexible ring with

$$\begin{aligned}
\mathrm{NMS} &= \mathrm{NMM} = \mathrm{NMC} = 4\\
\omega_j^2 &= 0, 1, 4, 5, 10, 1, 4, 5, 10\\
M_j &= 1, 1, 1, 1, 1, 1, 1, 1, 1 \quad \text{(in violation of (15))}\\
\Omega &= 60\ \mathrm{deg/s}\\
\mathrm{NS} &= 8\\
\mathrm{LS}_i &= 0, 45, 90, 135, 180, 225, 315\ \mathrm{deg}\\
\mathrm{NA} &= 9\\
\mathrm{LA}_i &= 0, 20, 60, 90, 135, 180, 225, 270, 315\ \mathrm{deg}\\
T &= 0.1\ \mathrm{s}
\end{aligned} \tag{27}$$

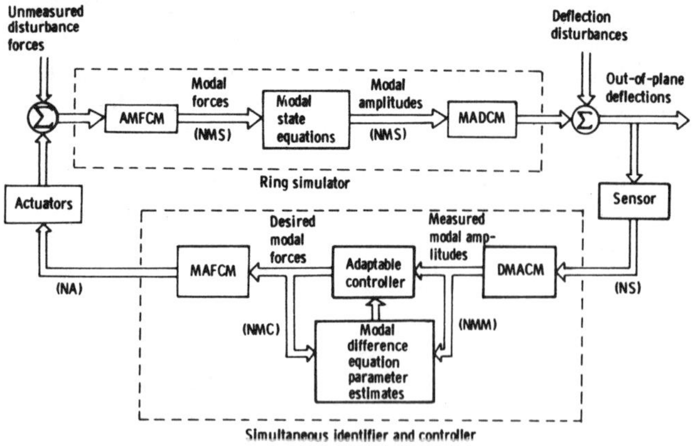

Fig. 5. Ring control simulator.

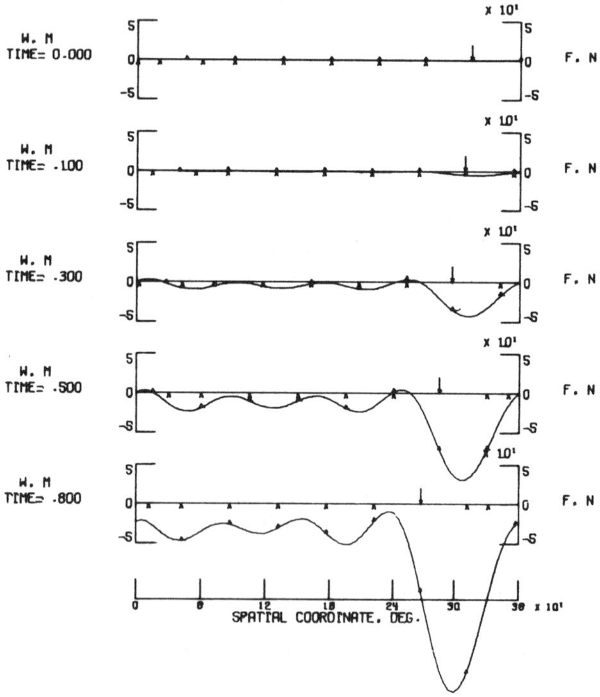

Fig. 6. Uncompensated step response.

are illustrated in Figs. 6 and 7, respectively. The numerical data in (27) do not correspond to any particular physical ring but were selected as representative merely to illustrate the simulation capability and the subsequent behavior of the distributed adaptive control laws. Note that the response of the ring to a single step input includes translation while the response to a balanced torque does not, as expected. The triangles on the displacement curves mark measurement points. The arrowheads along the spatial coordinate axis show actuator locations with the length of the arrowshaft proportional to the applied force according to the right-hand scale. The plots are based in the ring particle reference frame so the sensor and actuator locations appear to regress for a progressing ring. Note the movement of the two actuator locations 20° apart. For the large scale rings

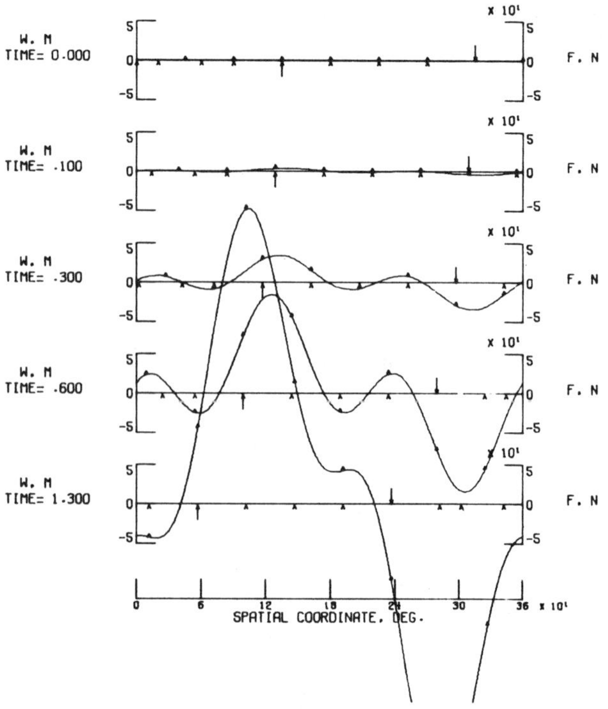

Fig. 7. Uncompensated step torque response.

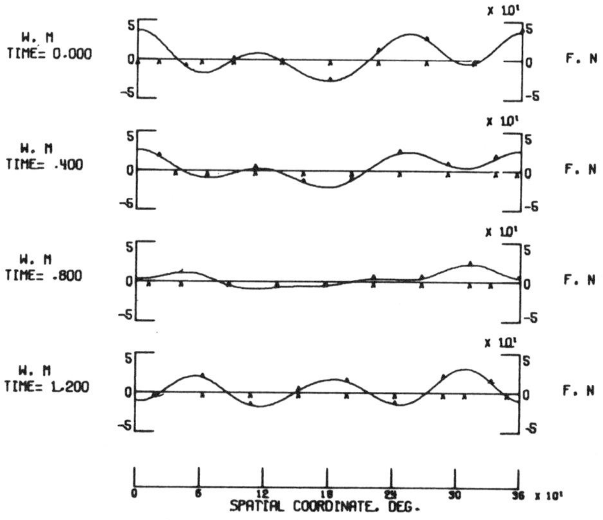

Fig. 8. Free response to an initial deformation.

currently under consideration the x axis covers over 2 km. Therefore the out-of-plane deflections, approximately 100 times smaller, are well within the region of linear behavior. More realistically, the modal frequencies would be divided by 1000 for the structures involved thereby multiplying the stated times by 1000. Note that the uncompensated response to magnetic force inputs of only 20 N produces severe excitation of the flexible modes in addition to the rigid body modes. This implies a requirement for damping the flexible modes.

The undamped behavior of the same ring is illustrated in Fig. 8 by release from the initial deformation disturbance of

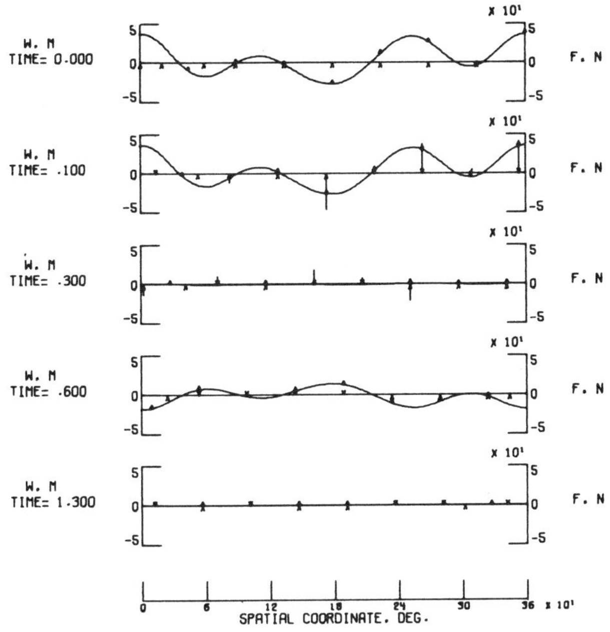

Fig. 9. Desired damping of an initial deformation.

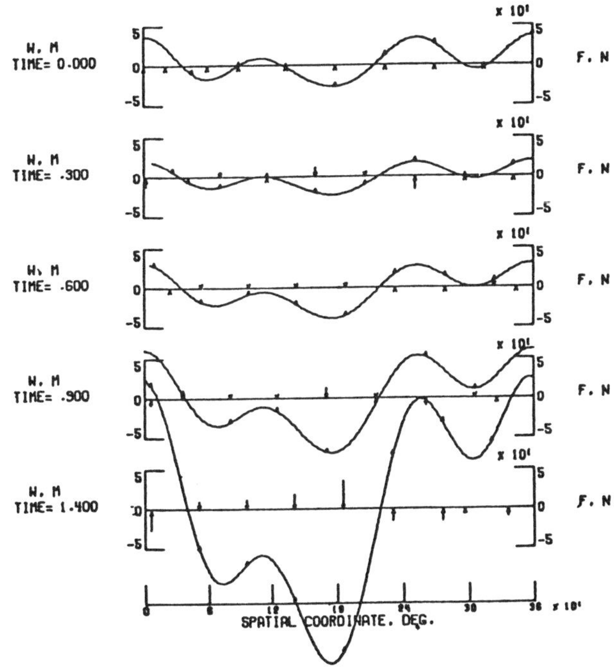

Fig. 10. Fixed incorrect parameter control response.

$$w(LS_i,0) = 3.7,\ -0.96,\ 0,\ 0.23,\ -2.7,\ 1.16,\ 2.6,\ -0.37\ \text{m}. \tag{28}$$

The predictable oscillatory behavior occurs as the ring appears nearly flat at 0.8 s but near maximum deflection only 0.4 s later. The damping chosen for each mode implicit in

$$DCEC_{1j} = -1.7,\ DCEC_{2j} = 1.06,\quad DCEC_{3j} = -0.272,\ DCEC_{4j} = 0.024 \tag{29}$$

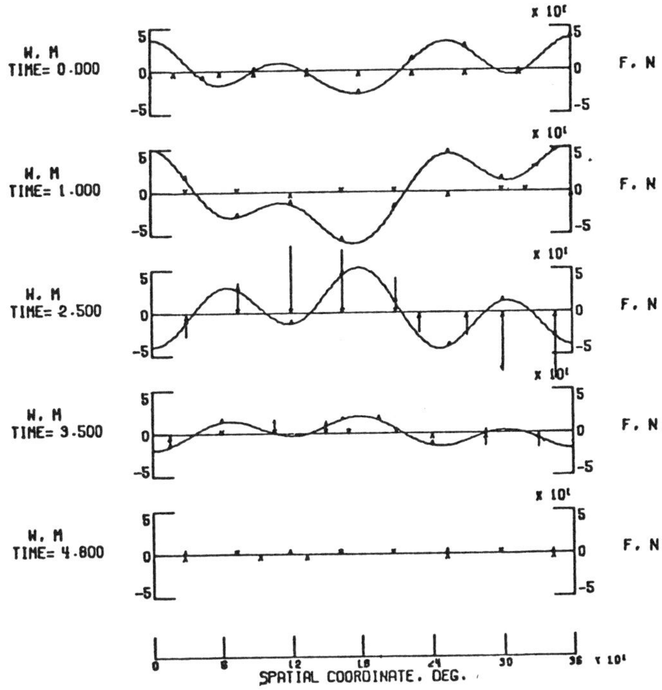

Fig. 11. Adaptive control response despite initial control law instability.

removed the disturbance within 1.3 s as shown in Fig. 9. The required modal forces were calculated using (25) with the difference equation parameter estimates equaling their actual values. Note that they remain within the ± 50 N range. For computational simplicity the applied forces from solution of (11) were assumed spatially fixed with respect to the ring during the sample interval, i.e., ring movement was inappropriately modeled to occur instantaneously at the sample instants. Correction of this procedure can be achieved by appropriate actuator windowing, therefore not affecting the generality of these simulation results.

Assuming that the modal difference equation parameter estimates are less than 10 percent in error on A_{1j}, exact for A_{2j} and within 20 percent for B_{1j} and B_{2j}, which corresponds to a 20 percent estimate error in M_j and 50 percent to 350 percent over-estimation of the structural frequencies, i.e.,

$$\begin{aligned}\hat{A}_{1j} &= 2,\ 1.8,\ 1.8,\ \ldots,\ 1.8\\ \hat{A}_{2j} &= -1,\ -1,\ \ldots,\ -1\\ \hat{B}_{1j} &= B_{2j} = 0.004,\ 0.004,\ \ldots,\ 0.004\end{aligned} \tag{30}$$

results in an unstable controller unless adapted. Fixed use of these erroneous plant parameter estimates leads, as shown in Fig. 10, to over 20 m (and growing) maximum disturbance within the desired damping interval. Alternatively, adaptive identification with $ASSW_{ij} = 1/[W_j^2(k-2) + W_j^2(k-3) + F_j^2(k-2) + F_j^2(k-3)]$ in satisfaction of (18) and controller recalculation via (25) bounds the deformation within the ±8 m maximum deflection range within the same interval in Fig. 11. Maximum forces are applied by the adaptive controller at 2.5 s and the deformation has disappeared within 5 s. Though convergence to proper time-invariant parameter values from rough initial values does not represent the practical circumstances of unpredictable time variations in the modal plant parameters, the success in such a case of the adaptive controller versus the fixed controller clearly illustrates the circumstantial superiority of the adaptive scheme.

Note that in these simulations since NMS = NMC = NMM both AMFCM and MADCM are square, making (12) and (26) exact inverses and avoiding control and observation spillover. If NMS > NMC and NMS > NMM, i.e., AMFCM and MADCM are noninvertible, then the desired modal forces will not match the applied modal forces and the measured modal amplitudes will not match the actual values. Therefore the parameter estimator in the identifier/controller will be using invalid information causing prediction error even if given correct parameter estimates. This problem briefly examined in [25] can be tackled in the case of control spillover by filtering the actuator commands through a (2NMC + 1) × NA AMFCM before providing the parameter estimator with the desired modal forces. However, inaccurate modal amplitude measurements cannot be corrected by the same strategy due to the receipt of the signals by the identifier/controller rather than their provision. This illustrates that observation spillover is a more severe problem than control spillover for the adaptive component of the control methodology as it also is for the nonadaptive modal control stategy [26].

VI. Conclusion

An original adaptive-control methodology applicable to the control of distributed systems has been presented. It has been applied to the stabilization of the flexible modes of a spinning ring. Simulation studies presented confirm the need for active stabilization of rings if they are to be used to conduct spacecraft maneuvers. A fixed-point controller design that gives desirable damping if the ring dynamics are as prespecified was shown to possibly be unstable when the assumed parameter values are in error. The adaptive controller, however, did produce stable results by quickly identifying the plant parameters and simultaneously adjusting the feedback controller gains. Therefore, the distributed system adaptive-control methodology presented can be declared feasible for control of a flexible, spinning ring. Future efforts improving these preliminary results will be directed at evaluation of controller objective specifications and requirements, sensor noise effects, criteria for selection of the modes to be controlled, and robustness despite control and observation spillover. Furthermore, the direct approach to adaptive control [26] of lumped-parameter plants will be utilized in replacement of the simultaneous identification and control strategy imbedded in the espoused procedure thereby bypassing identifiability constraints cited in Section IV. A long-range goal is to conduct studies evaluating the proposed adaptive-control philosophy for distributed systems with more than one spatial coordinate.

References

[1] A. Bradshaw and B. Porter, "Modal control of a class of distributed-parameter systems: multi-eigenvalue assignment," *Int. J. Contr.*, vol. 16, pp. 277-285, Aug. 1972.

[2] L. Meirovitch, H.F. VanLandingham, and H. Oz, "Control of spinning flexible spacecraft by modal synthesis," *Acta Astronautica*, vol. 4, pp. 985-1010, 1977.

[3] C.R. Johnson, Jr., "An approach to adaptive simultaneous identification and control of a large dual momentum ring control device," NASA-ASEE summer faculty fellowship with FDCD of NASA, Langley Research Center, Project Rep., Aug. 1978.

[4] R.C. Montgomery and C.R. Johnson, Jr., "The dual momentum ring attitude control device for large space systems: An example of distributed system adaptive control," *Proc. 12th Asilomar Conf. Circuits, Systems, and Computers*, Pacific Grove, CA, Nov. 1978.

[5] M.H. Hamza and M.A. Sheirah, "A self-tuning regulator for distributed parameter systems," *Automatica*, vol. 14, pp. 453-463, Sept. 1978.

[6] R.E. Kalman, "Design of a self-optimizing control system," *Trans. ASME*, vol. 80, pp. 468-478, Feb. 1958.

[7] K.J. Astrom, U. Borisson, L. Ljung, and B. Wittenmark, "Theory and applications of self-tuning regulators," *Automatica*, vol. 13, pp. 457-476, Sept. 1977.

[8] W.W. Anderson and N.J. Groom, "The annular momentum control device (AMCD) and potential applications," NASA Tech. Note D-7866, Mar. 1975.

[9] P.C. Parks, "On how to shake a piece of string to a standstill," in *Recent Mathematical Developments in Control*, D.J. Bell, Ed. London: Academic Press, 1973.

[10] S.-H. Wang, "An example in decentralized control systems," *IEEE Trans. Automat. Contr.*, vol. AC-23, p. 938, Oct. 1978.

[11] B. Noble, *Applied Linear Algebra*. Englewood Cliffs, NJ: Prentice-Hall, 1969.

[12] P. Eykhoff, *System Identification: Parameter and State Estimation*. London: Wiley-Interscience, 1974.

[13] L. Ljung, "Convergence analysis of parametric identification methods," *IEEE Trans. Automat. Contr.*, vol. AC-23, pp. 770-783, Oct. 1978.

[14] W.R.E. Wouters, personal communication of adaptive breeder reactor control research conducted at Koninklijke/Shell-Laboratorium, Amsterdam, Oct. 1978.

[15] G.N. Saridis and R.N. Lobbia, "Parameter identification and control of linear discrete-time systems," *IEEE Trans. Automat. Contr.*, vol. AC-17, pp. 52-60, Feb. 1972.

[16] D.W. Clarke and P.J. Gawthorp, "Self-tuning controller," *Proc. IEE*, vol. 122, pp. 929-934, Sept. 1975.

[17] J.M. Martin-Sanchez, "A new solution to adaptive control," *Proc. IEEE*, vol. 64, pp. 1209-1218, Aug. 1976.

[18] K.J. Astrom and B. Wittenmark, "On self-tuning regulators," *Automatica*, vol. 9, pp. 185-199, Mar. 1973.

[19] B.G. Buchanan, T.M. Mitchell, R.G. Smith, and C.R. Johnson, Jr., "Models of learning systems," in *The Encyclopedia of Computer Science and Technology*, J. Belzer,

A.G. Holzman, and A. Kent, Eds. New York: Marcel Dekker, vol. 11, 1978, pp. 24-51.

[20] H.B. Kennedy, "A gyro momentum exchange device for space vehicle attitude," *AIAA J.*, vol. 1, pp. 1110-1118, May 1963.

[21] J.W. Beams, "Early history of the gas centrifuge work in the U.S.A.," Department of Physics and School of Engineering and Applied Science, University of Virginia, Charlottesville, VA, May 1975.

[22] J.M. Mendel, *Discrete Techniques of Parameter Estimation: The Equation Error Formulation.* New York: Marcel Dekker, 1973, ch. 4.

[23] H.J. Dunn and R.C. Montgomery, "A moving window parameter adaptive control system for the F8-DFBW Aircraft," *IEEE Trans. Automat. Contr.*, vol. AC-22, pp. 788-795, Oct. 1977.

[24] M.J. Balas, "Active control of flexible systems," *Proc. AIAA Symp. Dyn. and Control of Large Flexible Spacecraft,* Blacksburg, VA, June 1977, pp. 217-236.

[25] A.L. Hamm and C.R. Johnson, Jr., "Indirect adaptive stabilization of a large, flexible, spinning spacering: Simulation studies," *Proc. 1979 IEEE Southeastcon,* Roanoke, VA.

[26] M.J. Balas, "Feedback control of flexible systems," *IEEE Trans. Automat. Contr.*, vol. AC-23, pp. 673-679, Aug. 1978.

[27] C.R. Johnson, Jr., and E. Tse, "Adaptive implementation of one-step-ahead optimal control via input matching," *IEEE Trans. Automat. Contr.*, vol. AC-23, pp. 865-872, Oct. 1978.

Part III

Networks
(Complexity, Partitioning, Layout)

An Efficient Heuristic Procedure for Partitioning Graphs

By B. W. KERNIGHAN and S. LIN

(Manuscript received September 30, 1969)

We consider the problem of partitioning the nodes of a graph with costs on its edges into subsets of given sizes *so as to minimize the sum of the costs on all edges cut. This problem arises in several physical situations—for example, in assigning the components of electronic circuits to circuit boards to minimize the number of connections between boards.*

This paper presents a heuristic method for partitioning arbitrary graphs which is both effective in finding optimal partitions, and fast enough to be practical in solving large problems.

I. INTRODUCTION

1.1 *Definition of the Problem*

This paper deals with the following combinatorial problem: given a graph G with costs on its edges, partition the nodes of G into subsets no larger than a *given maximum size,* so as to minimize the total cost of the edges cut.

One important practical example of this problem is placing the components of an electronic circuit onto printed circuit cards or substrates, so as to minimize the number of connections between cards. The components are the nodes of the graph, and the circuit connections are the edges. There is some maximum number of components which may be placed on any card. Since connections between cards have high cost compared to connections within a board, the object is to minimize the number of interconnections between cards.

This partitioning problem also arises naturally in an attempt to improve the paging properties of programs for use in computers with paged memory organization. A program (at least statically) can be thought of as a set of connected entities. The entities might be subroutines, or procedure blocks, or single instruction and data items, depending on viewpoint and the level of detail required. The connections between the entities might represent possible flow or transfer of control, or references from one entity to another. The problem is to assign the objects to "pages" of a given size so as to minimize the number of references between objects which lie on different pages.

To pose the partitioning problem mathematically, we shall need the following definitions. Let G be a graph of n nodes, of sizes (weights) $w_i > 0$, $i = 1, \cdots, n$. Let p be a positive number, such that $0 < w_i \leq p$ for all i. Let $C = (c_{ij})$, $i, j = 1, \cdots, n$ be a weighted connectivity matrix describing the edges of G.

Let k be a positive integer. A *k-way partition* of G is a set of nonempty, pairwise disjoint subsets of G, $v_1, \cdots, v_k$ such that $\cup_{i=1}^{k} v_i = G$. A partition is *admissible* if

$$|v_i| \leq p \quad \text{for all} \quad i,$$

where the symbol $|x|$ stands for the *size* of a set x, and equals the sum of the sizes of all the elements of x. The *cost* of a partition is the summation of c_{ij} over all i and j such that i and j are in different subsets. The cost is thus the sum of all external costs in the partition.

The partitioning problem we consider here is to find a minimal-cost admissible partition of G.

There are three other problems which are equivalent to this one. First, minimizing external cost is equivalent to maximizing internal cost because the total cost of all edges is constant. Further, by changing the signs of all c_{ij}'s, we can maximize external cost, or minimize internal cost.

1.2 *Exact Solutions*

A strictly exhaustive procedure for finding the minimal cost partition is often out of the question. To see this suppose that G has n nodes of size 1 to be partitioned into k subsets of size p, where $kp = n$. Then there are $\binom{n}{p}$ ways of choosing the first subset, $\binom{n-p}{p}$ ways for the second, and so on. Since the ordering of the subsets is immaterial, the number of cases is

$$\frac{1}{k!}\binom{n}{p}\binom{n-p}{p}\cdots\binom{2p}{p}\binom{p}{p}.$$

For most values of n, k, and p, this expression yields a very large number; for example, for $n = 40$ and $p = 10$ ($k = 4$), it is greater than 10^{20}.

Formally the problem could also be solved as an integer linear programming problem, with a large number of constraint equations necessary to express the uniformity of the partition.

Because it seems likely that any direct approach to finding an optimal solution will require an inordinate amount of computation, we turn to an examination of heuristics. Heuristic methods can produce good solutions (possibly even an optimal solution) quickly. Often in practical applications, several good solutions are of more value than one optimal one.

The first and foremost consideration in developing heuristics for combinatorial problems of this type is finding a procedure that is powerful and yet sufficiently fast to be practical. A process whose running time grows exponentially or factorially with the number of vertices of the graph is not likely to be practical. In most cases, a growth rate of more than the square of the number of vertices is still not too practical. (If the running time of a procedure grows as $f(n)$, where n is the number of vertices involved, we shall refer to it as an $f(n)$-procedure.)

1.3 *False Starts*

To point out a few pitfalls, we mention some unsuccessful attempts at heuristic solutions to the partitioning problem.

1.3.1 *Random Solutions*

One tactic is simply to generate random solutions, keeping the best seen to date, and terminating after some predetermined time or value is reached. This is quite fast, although actually an n^2-procedure. Unfortunately, this approach is unsatisfactory for problems of even moderate size, since there are generally few optimal or near-optimal solutions, which thus appear randomly with very low probabilities. Experience with 2-way partitions for a class of 0–1 matrices of size 32×32, for example, has indicated that there are typically 3 to 5 optimal partitions, out of a total of $\frac{1}{2}\binom{32}{16}$ partitions, giving a probability of success on any trial of less than 10^{-7}.

1.3.2 *Max Flow-Min Cut*

Another partitioning method is the Ford and Fulkerson max flow-min cut algorithm[1]. The graph is treated as a network in which edge costs correspond to maximum flow capacities between pairs of nodes. A cut is a separation of the nodes into two disjoint subsets. The max flow-min cut theorem states that the maximal flow values between any pair of nodes is equal to the minimal cut capacity of all cuts which separate the two nodes. In our terminology, a cut is a 2-way partition, and the cut capacity is the cost of the partition. The Ford and Fulkerson algorithm finds a cut with maximal flow, which is thus a minimal cost cut; this represents a minimum cost partition of the graph into two subsets of unspecified sizes.

There are several difficulties involved in using the Ford and Fulkerson algorithm for our partitioning problem. The most severe of these is the fact that the algorithm has no provision for constraining the sizes of the resultant subsets, and there seems to be no obvious way to extend it to include this. Thus if flow methods are used to perform a split, then further processing is necessary to make the resulting subsets the correct size. If the subsets are greatly different in size, then use of this algorithm will have produced essentially no benefit. Hence in spite of its theoretical elegance, the Ford and Fulkerson algorithm is not suitable for this application. (Note however, that since it does find the minimal cost unconstrained 2-way partition, the value it produces is a lower bound for solutions produced by any method.)

Reprinted with permission from *Bell Syst. Tech. J.,* vol. 49, pp. 291–307, Feb. 1970.

1.3.3 *Clustering*

A class of much more intuitive methods is based on identifying "natural clusters" in the given cost matrix—that is, groups of nodes which are strongly connected in some sense. For example, one can use very simple heuristics for building up clusters, based on collecting together elements corresponding to large values in the cost matrix. But again these methods do not in general include much provision for satisfying constraints on the sizes of the subsets, nor do they provide for systematic assignment of "stragglers" (nodes which do not obviously belong to any particular subset).

1.3.4 *λ-Opting*

Lin, working on the Traveling Salesman Problem, [See Ref. 2] categorized a set of methods of improving given solutions by rearranging single links, double links, triplets, and in general, λ links. He referred to a change involving the movement of λ links as a λ-change. If a configuration of the system is reached in which no λ-change can be made which results in a decrease in cost, the configuration is said to be "λ-opt."

For the partitioning problem, an analogous operation is the interchange of groups of λ points between a pair of sets. Thus a 1-change is the exchange of a single point in one set with a single point in another set. A configuration is then said to be "1-opt" if there exists no interchange of two points which decreases the cost of the partition. Experiments to evaluate 1-opting for 2-way partitions of 0-1 matrices (32 × 32) within which about one-half of the elements were nonzero, show that apparently optimal values can be achieved in about 10 percent of the trials; values within 1 or 2 of the optimal can be achieved in about 75 percent of cases.

It appears fruitless to extend λ beyond 1 (1-opting is already an n^2-procedure), or to extend 1-opting experiments to partitions into more than two subsets, since more powerful methods have been developed. These methods are the topic of the next sections.

II. TWO-WAY UNIFORM PARTITIONS

2.1 *Introduction*

The simplest partitioning problem which still contains all the significant features of larger problems is that of finding a minimal-cost partition of a given graph of $2n$ vertices (of equal size) into two subsets of n vertices each. The solution of the 2-way partitioning problem is the subject of this section. The solution provides the basis for solving more general partitioning problems. In Section 2.6, we discuss 2-way partitions into sets of unequal size.

Let S be a set of $2n$ points, with an associated cost matrix $C = (c_{ij})$, $i, j = 1, \cdots, 2n$. We assume without loss of generality that C is a symmetric matrix, and that $c_{ii} = 0$ for all i. There is no assumption about nonnegativity of the c_{ij}'s. We wish to partition S into two sets A and B, each with n points, such that the "external cost" $T = \sum_{A \times B} c_{ab}$ is minimized.

In essence, the method is this: starting with any arbitrary partition A, B of S, try to decrease the initial external cost T by a series of interchanges of subsets of A and B; the subsets are chosen by an algorithm to be described. When no further improvement is possible, the resulting partition A', B' is locally minimum with respect to the algorithm. We shall indicate that the resulting partition has a fairly high probability of being a globally minimum partition.

This process can then be repeated with the generation of another arbitrary starting partition A, B, and so on, to obtain as many locally minimum partitions as we desire.

Given S and (c_{ij}), suppose A^*, B^* is a minimum cost 2-way partition. Let A, B be any arbitrary 2-way partition. Then clearly there are subsets $X \subset A$, $Y \subset B$ with $|X| = |Y| \leq n/2$ such that interchanging X and Y produces A^* and B^* as shown below.

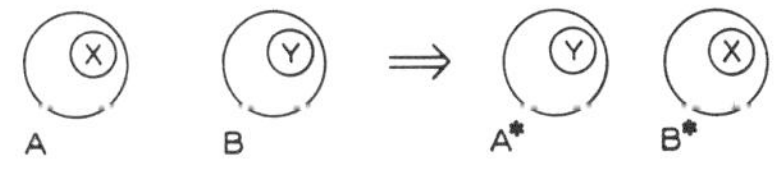

$A^* = A - X + Y$
$B^* = B - Y + X$

The problem is to identify X and Y from A and B, without considering all possible choices. The process we describe finds X and Y approximately, by sequentially identifying their elements.

Let us define for each $a \in A$, an *external cost* E_a by

$$E_a = \sum_{y \in B} c_{ay}$$

and an *internal cost* I_a by

$$I_a = \sum_{x \in A} c_{ax}.$$

Similarly, define E_b, I_b for each $b \in B$. Let $D_z = E_z - I_z$ for all $z \in S$; D_z is the difference between external and internal costs.

Lemma 1: Consider any $a \in A$, $b \in B$. If a and b are interchanged, the gain (that is, the reduction in cost) is precisely $D_a + D_b - 2c_{ab}$.

Proof: Let z be the total cost due to all connections between A and B that do not involve a or b. Then

$$T = z + E_a + E_b - c_{ab}.$$

Exchange a and b; let T' be the new cost. We obtain

$$T' = z + I_a + I_b + c_{ab}$$

and so

$$\text{gain} = \text{old cost} - \text{new cost} = T - T' = D_a + D_b - 2c_{ab}.$$

2.2 *Phase 1 Optimization Algorithm*

In this subsection we present the algorithm for 2-way partitioning.

First, compute the D values for all elements of S. Second, choose $a_i \in A$, $b_j \in B$ such that

$$g_1 = D_{a_i} + D_{b_j} - 2c_{a_i b_j}$$

is maximum; a_i and b_j correspond to the largest possible gain from a single interchange. (We will return shortly to a discussion of how to select a_i and b_j quickly.) Set a_i and b_j aside temporarily, and call them a_1' and b_1', respectively.

Third, recalculate the D values for the elements of $A - \{a_i\}$ and for $B - \{b_j\}$, by

$$D_x' = D_x + 2c_{xa_i} - 2c_{xb_j} \qquad x \in A - \{a_i\},$$
$$D_y' = D_y + 2c_{yb_j} - 2c_{ya_i} \qquad y \in B - \{b_j\}.$$

The correctness of these expressions is easily verified: the edge (x, a_i) is counted as internal in D_x, and it is to be external in D_x', so c_{xa_i} must be added twice to make this correct. Similarly, c_{xb_j} must be subtracted twice to convert (x, b_j) from external to internal.

Now repeat the second step, choosing a pair a_2', b_2' from $A - \{a_1'\}$ and $B - \{b_1'\}$ such that $g_2 = D_{a_2'} + D_{b_2'} - 2c_{a_2'b_2'}$ is maximum (a_1' and b_1' are *not* considered in this choice). Thus g_2 is the additional gain when the points a_2' and b_2' are exchanged as well as a_1' and b_1'; this additional gain is maximum, given the previous choices. Set a_2' and b_2' aside also.

Continue until all nodes have been exhausted, identifying $(a_3', b_3'), \cdots, (a_n', b_n')$, and the corresponding maximum gains $g_3, \cdots, g_n$. As each (a', b') pair is identified, it is removed from contention for further choices so the size of the sets being considered decreases by 1 each time an (a', b') is selected.

If $X = a_1', a_2', \cdots, a_k'$, $Y = b_1', b_2', \cdots, b_k'$, then the decrease in cost when the sets X and Y are interchanged is precisely $g_1 + g_2 + \cdots + g_k$. Of course $\sum_1^n g_i = 0$. Note that some of the g_i's are negative, unless all are zero.

Choose k to maximize the partial sum $\sum_{i=1}^k g_i = G$. Now if $G > 0$, a reduction in cost of value G can be made by interchanging X and Y. After this is done, the resulting partition is treated as the initial partition, and the procedure is repeated from the first step.

If $G = 0$, we have arrived at a locally optimum partition, which we shall call a *phase 1 optimal partition*. We now have the choice of repeating with another starting partition, or of trying to improve the phase 1 optimal partition. We shall discuss the latter option shortly. Figure 1 is a flowchart for the phase 1 optimization procedure.

2.3 *Effectiveness of the Procedure*

One general approach to solving problems such as this one is to find the *best* exchange involving say λ pairs of points, for some λ specified in advance[2]. The difficulty encountered is that use of a small value of λ is not sufficient to identify good exchanges, but the computational effort required grows rapidly as λ increases.

The procedure we have described *sequentially* finds an approximation to the best exchange of λ pairs. λ is not specified in advance, but rather is chosen to make the improvement as large as possible. This technique sacrifices a certain amount of power for a considerable gain in speed.

Since we construct a sequence of gains g_i, $i = 1, \cdots, n$, and find the

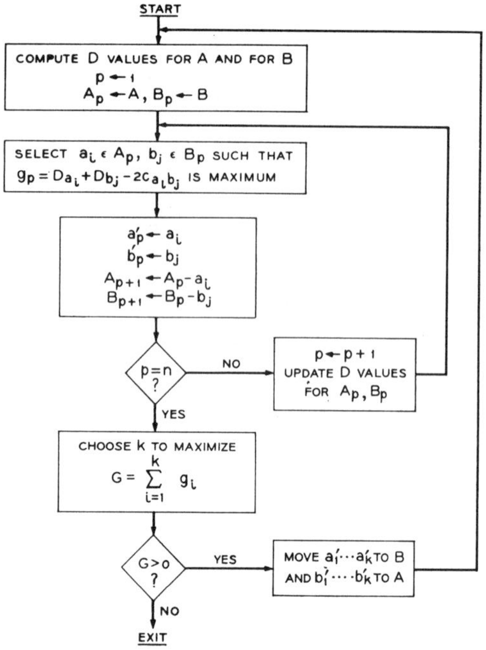

Fig. 1 — Flowchart of phase 1 optimization procedure.

maximum partial sum, *the process does not terminate immediately when some g_i is negative*. This means that the process can sequentially identify sets for which the exchange of only a few elements would actually increase the cost, while the interchange of the entire sets produces a net gain.

Numerous experiments have been performed to evaluate the procedure on different types of cost matrices. The matrices used have included *(i)* 0-1 matrices, with density of nonzero elements ranging from 5 percent to 50 percent, *(ii)* integer matrices with elements uniformly distributed on $[0, k]$, $k = 2, \cdots, 10$, *(iii)* matrices with clusters of known sizes and binding strength. Results on all of these matrices have been similar, so we shall only summarize them here. A more extended discussion may be found in Ref. 3.

A useful measure of the power of a heuristic procedure is the probability that it finds an optimal solution in a single trial. Suppose that p is the probability that a phase 1 optimal solution found using a random starting partition is globally optimal. We have examined the behavior of this probability as the size of the matrices involved is varied. Experiments show p is around 0.5 for matrices of size 30×30, 0.2 to 0.3 for 60×60, and 0.05 to 0.1 for 120×120. The functional behavior of p is approximately $p(n) = 2^{-n/30}$.

These values are derived primarily from 0-1 matrices having about 50 percent 1's (randomly placed). Experiments on matrices with lower densities of 1's yield larger variances, but substantially identical mean values for p.

2.4 *Running Time of the Procedure*

Let us define a *pass* to be the operations involved in making one cycle of identification of $(a'_1, b'_1), \cdots, (a'_n, b'_n)$, and selection of sets X and Y to be exchanged. The total time for a pass can be estimated this way. First, the computation of the D values initially is an n^2-procedure, since for each element of S, all the other elements of S must be considered. The time required for updating the D values is proportional to the number of values to be updated, so the total updating time in one pass grows as

$$(n - 1) + (n - 2) + \cdots + 2 + 1$$

which is proportional to n^2.

The dominant time factor is the selection of the next pair a'_i, b'_i to be exchanged. The method we have used to perform this searching is to sort the D values so that

$$D_{a_1} \geqq D_{a_2} \geqq \cdots \geqq D_{a_n}$$

and

$$D_{b_1} \geqq D_{b_2} \geqq \cdots \geqq D_{b_n} .$$

When sorting is used, only a few likely contenders for a maximum gain need be considered. This is because when scanning down the set of D_a's and D_b's, if a pair D_{a_i}, D_{b_j} is found whose sum does not exceed the maximum improvement seen so far in this pass, then there cannot be another pair a_k, b_l with $k \geqq i$, $l \geqq j$, with a greater gain, (assuming $c_{ij} \geqq 0$) and so the scanning can be terminated. Thus the next pair for interchange is found rapidly. Sorting is an $n \log n$ operation, so in this method, the total time required to sort D values in a pass will be approximately

$$n \log n + (n - 1) \log (n - 1) + \cdots + 2 \log 2$$

which grows as $n^2 \log n$.

To reduce the time for selection of an (a, b) pair, it is possible to use techniques which are faster than sorting, but which do not necessarily always give the maximum gain at each stage. For example, one method is to scan for the largest D_a and the largest D_b, and use the corresponding a and b as the next interchange. This method is essentially linear-time and would probably be implemented as part of the recomputation of the D values. It is best suited for sparse matrices, where the probability that $c_{ab} > 0$ is small. A slight extension, involving negligible extra cost, is to save the largest two or three D_a's and D_b's, so that if the largest pair does not give the maximum gain (because c_{ab} is too large), then another can be tried. Experience indicates that three values are sufficient in virtually all cases, even for matrices with a relatively high percentage of nonzero entries. Use of this method reduces running time by about 30 percent in the present implementation, with very small degradation of power.

The number of passes required before a phase 1 optimal partition is achieved is small. On all matrix sizes tested at the time of writing (up to 360 points), it has been almost always from 2 to 4 passes. On the basis of this experimental evidence, the number of passes is not strongly dependent on the value of n.

From the foregoing observations, it is possible to estimate the total running time of the procedure. If we use a method which sorts the D values at each stage (time proportional to $n^2 \log n$), then the running time should grow as $n^2 \log n$. If a fast-scan method is used, and the number of passes is constant, the running time should have an n^2 growth rate; this is a lower bound.

For comparison, examination of all pairs of sets X and Y, and evaluation of the costs would require time proportional to

$$n^2 \sum_{k=1}^{n/2} \binom{n}{k}^2 \sim \frac{n^2}{2} \sum_{k=0}^{n} \binom{n}{k}^2$$
$$= \frac{n^2}{2} \binom{2n}{n}$$
$$\sim \frac{n^2}{2} 4^n \left(\frac{1}{\pi n}\right)^{\frac{1}{2}}$$

for large n. This function grows as $n^{3/2} 4^n$.

Running times have been plotted in Fig. 2. The observed times have an apparent growth rate of about $n^{2.4}$, which is reasonably close to n^2. Although on the logarithmic plot this curve is close to linear over the range $n = 20$ to $n = 130$, it may actually be $n^2 \log n$; insufficient data is available to check this. All times are based on an implementation in FORTRAN G on an IBM System 360 Model 65.

2.5 *Improving the Phase 1 Optimal Partition*

In this section, we discuss a method which might be used to improve the partition produced by the phase 1 procedure, which may not be globally optimum. The method suggested in this section is based heavily on experimental evidence, although there are quite plausible reasons for performing the particular set of operations. The basic idea is to perturb the locally optimal solution in what we hope is an enlightened manner, so that an iteration of the process on the perturbed solution will yield a further reduction in the total cost. If this tactic fails, nothing has been lost except some computation time, since the best solution seen so far is always saved.

Computer results for problems with up to 64 points suggest that whenever a phase 1 optimal solution is not globally optimal, $|X| = |Y| \approx n/2$. Roughly, this implies that if $|X|$ and $|Y|$ had been small compared to $n/2$, they would have been found by the process; it is only larger sets which are not identified all the time.

A successful heuristic to find the correct X and Y in this case is to find a phase 1 optimal partition for each of the sets A and B, say $A \to [A_1, A_2]$ and $B \to [B_1, B_2]$. (That is, find near-optimal partitions of A and of B separately.) Recombine the 4 sets into 2, say $A_0 = A_1 \cup B_1$ and $B_0 = A_2 \cup B_2$, and continue with phase 1 optimization. If our expectation is correct, the new X and Y will be small, and thus readily identified by the phase 1 process.

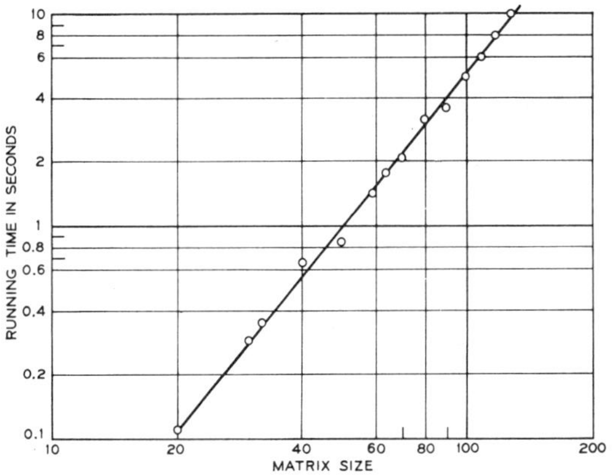

Fig. 2 — Running time.

When A is split into A_1, A_2 and B into B_1, B_2 there are two ways in which the smaller sets can be recombined. A series of tests was made on matrices of moderate size (up to 64×64), in which both possible recombinations were done, generating three phase 1 optimal values for each starting partition. For matrices of size 32×32, the apparent optimal value was observed at least once in each triple of values, for a large number of cases. With matrices of size 64×64, there were occasional failures.

It might be noted that the extra time involved for the recombination approach is three times that required to do a completely new partition from a random start, assuming an n^2-procedure.

It is possible to estimate whether a particular improvement tactic is profitable or not in the following way. Suppose that some method increases the probability of finding an optimal partition from p to p', while it increases the running time from t to t'. Then in a fixed amount of time, it is possible to do k trials of the basic procedure, and kt/t' trials of the improved method. The corresponding probabilities of achieving an optimal solution are $1 - (1 - p)^k$ and $1 - (1 - p')^{kt/t'}$ respectively. The improved method is then desirable if the second expression is greater than the first; by simple manipulation, this condition becomes

$$1 - p' < (1 - p)^{t'/t}.$$

On the basis of the numerical values in this section, it may be useful to try the recombination method.

2.6 *Partitioning into Unequal-Sized Sets*

It is simple to modify the procedure to partition a set S with n elements into two sets of specified sizes n_1 and $n_2 (n_1 + n_2 = n)$. Assume $n_1 < n_2$. Then restrict the maximum number of pairs that can be exchanged in one pass of the procedure to n_1. All other operations are performed on all elements of each set. (The starting partition is into two sets, of n_1 and n_2 elements respectively.)

Suppose we wish to partition S into two sets, such that there are at least n_1 elements and at most n_2 elements in each subset; $n_1 + n_2 = n$, but they are not specified further.

The procedure is easily modified to handle this sort of constraint by the addition of "dummy" elements. These are elements which have no connections whatsoever; that is, they have zero entries in the cost matrix wherever they appear. Add $2n_2 - n$ dummies so S has $2n_2$ elements, and perform the procedure on it. The resulting partition will assign the dummy elements to the two subsets so as to minimize the external cost; at this point the dummies are discarded, leaving a partition into two subsets that satisfy the size constraints given.

2.7 *Elements of Unequal Sizes*

We have made the assumption so far that the elements (vertices) of the graph are all of the same size. This requirement may be relaxed to a large extent by converting any node of size $k > 1$ to a cluster of k nodes of size 1, bound together by edges of appropriately high cost. The size of the problem will obviously increase proportionally to the value of k, so it may be necessary to sacrifice some accuracy to keep the number of generated nodes within reasonable bounds.

III. MULTIPLE-WAY PARTITIONS

3.1 *Reduction to 2-Way Partitioning Problem*

So far, the discussion has been concerned exclusively with the basic problem of performing a 2-way partition on a set of $2n$ objects. In this section we extend the technique to perform k-way partitions on a set of kn objects, using the 2-way procedure as a tool.

The essential idea is to start with some partition into k sets of size n and by repeated application of the 2-way partitioning procedure to pairs of subsets, make the partition as close as possible to being pairwise optimal. (Section 3.2 treats the question of what starting sets to use.) Of course pairwise optimality is only a necessary condition for global optimality. There may be situations where some complex interchange of three or more items from three or more subsets is required to reduce a pairwise optimal solution to globally optimum; at the moment, no reasonable method for identifying such sets is known.

There are $\binom{k}{2}$ pairs of subsets to consider, so the time for one pass through all pairs is (assuming an n^2-procedure) $\binom{k}{2}n^2 \approx (kn)^2/2 =$ (number of points)$^2/2$. In general, more passes than this will actually be required, since when two sets are made optimal, this may change their optimality with respect to other sets.

Experience indicates that the number of passes is small and the process converges quickly. For example, our algorithm selects (i, j) as the next pair of sets to be optimized, where either i or j has been changed since the last time the pair (i, j) was selected. Using this selection process, the average number of passes through each pair of sets is a slowly growing function of both k and n. For matrices of size 100 or less and $k < 6$, the number of passes has been less than 5. [The average number of passes is computed as the average number of pairs considered to reach pairwise optimality, normalized by $\binom{k}{2}$.]

In any particular trial, there is a correlation between the number of pairs selected and the quality of the final partition. To get a better solution requires more work.

Convergence is rapid: two passes account for more than 95 percent of the improvement in most cases; the remaining passes contribute only small further reductions. Let $p(n, k)$ be the proportion of minimum cost solutions found for a particular n and k. For k fixed and small compared to n, the functional behavior of $p(n, k)$ is similar to the case $k = 2$, but the actual values are lower. Roughly, we observe $p(n, k + 1) \approx \frac{1}{2}p(n, k)$ for k in the range 2–4, and n up to 100, with considerable variation depending on the matrix being tested. For instance, for matrices of size about 40, $p(40, 2) \approx 0.4$, $p(42, 3) \approx 0.2$, and $p(40, 4) \approx 0.1$.

Another interesting question is measurement of how close to optimum the partitions found are. The solutions obtained by pairwise optimization have values concentrated in a narrow range. In almost all cases, the largest value found by the procedure is within 4–5 percent of the smallest. As another measure, if c is the mean cost of random partitions and b is the cost of the best partition observed, then virtually all partitions found have values v such that

$$v - b \leqq 0.1(c - b).$$

For instance, one test case was a series of 4-way partitions of a 0–1 matrix of size 80. This matrix had 1278 nonzero entries (a density of 0.2), corresponding to 639 edges in the graph. The mean value of randomly chosen partitions was 480.6. Twenty-four partitions of this matrix were found using the method described above. The lowest value encountered was 352 (1 time), the highest 365 (1 time); the mean value was 359.5, the median 360.

3.2 *Starting Partition*

In this subsection we discuss various methods of generating good starting partitions, based on modifications of the basic procedure.

The primary reason for choosing good starting partitions is that this particular form of preprocessing reduces the amount of work required to make the system pairwise optimal. It may also make the probability of an optimal solution higher, although this tendency is very difficult to evaluate.

Several methods for finding good multi-way starting partitions which are based on repeated application of the procedure itself have been investigated. The essential idea is to generate a k-way starting partition by first forming an r-way partition, then an s-way partition on each of the resulting subsets, and so on, up to t-way. (Here $k = rs \cdots t$.) The partitions found this way will in general be better than those which are completely arbitrary. A pairwise optimization stage is applied to the final set of subsets.

For example, if k is a power of 2, then perform a 2-way split, then a

2-way split on each of these subsets, and so on until the desired size of subsets is found.

This general approach is prone to the following difficulty: the first split divides the original set into r subsets by trying to make the internal connections in each subset as large as possible. Obviously this may conflict directly with the next stage, which is to try to divide each subset further. Carried to several levels, it can lead to a relatively poor overall solution. In experiments with 4-way partitions of matrices of sizes up to 64×64, this method yields optimal solutions approximately as often as does starting with a 4-way partition in the first place. In addition, this method will be effective if the matrix happens to have natural clusters of approximately the correct size (that is, equal to the final subset size).

A second method which can be used is to partition the set of kn elements into a set of n and a set of $(k - 1)n$, using the slightly modified version of the basic procedure discussed in the first part of Section 2.6. The set of n elements is set aside, and the next n elements from the remaining $(k - 1)n$ are identified. This continues until k subsets have been formed; again the pairwise optimization technique is used to improve on this partition.

This method can make an error in the identification of the first set which will bias the choice of the second, and so on; the effect is most severe for the case where k is large, so each set is small.

The method of breaking off subsets sequentially has another potential flaw: regardless of the starting configuration, it will identify approximately the same set each time it is used on a particular problem, and hence little is gained by using it twice on one cost matrix. However variations in the order of performing pairwise optimizations can still produce different final partitions in general.

Limited computational experience with sequential break-off followed by pairwise optimization suggests that it yields solutions which are on the average at least as good as (and sometimes slightly better than) those provided by pairwise optimization applied to an arbitrary k-way starting partition. Pairwise optimization yields the optimum with a higher probability, however, because it is less susceptible to error caused by a bad choice made early. For instance, in tests on the 80 point matrix mentioned previously, sequential break off yielded 4-way solutions with a mean value of 358.6, but the lowest value found was 355. (The highest was 363.) These may be compared to 359.5, 352 and 365 for the standard partitioning method.

Running time for the sequential break-off method is lower than for straight pairwise optimization.

Insufficient data is available for a direct comparison between sequential break off and the method of repeated subdivision.

In all cases, the original process, be it a completely random generation of some initial configuration, or the production of a good starting partition, is followed by a pairwise optimizing phase. It is unlikely that using better starting partitions will lead to worse results than random starts, on the average. Whether the possible improvement in results and running times will justify the extra computational effort required to generate the starting partition depends on the characteristics of the particular class of matrices being studied.

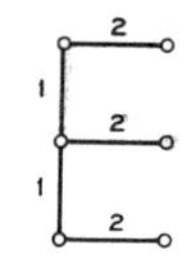

Fig. 3 — Cost reduction by expansion.

Some limited experiments were performed to compare the present procedure with a multi-dimensional scaling technique[4], on a Boolean matrix of 316 points, with about 1400 nonzero entries. The results indicated that the procedure identifies clusters well, even when no attempt is made to provide a good starting partition.

3.3 *Expansion Factor*

The introduction of dummy elements was mentioned in Section 2.6 as a method of handling partitioning into subsets of unequal sizes. This can be viewed equally well as a means of introducing "slack" into a solution, in an attempt to get a lower overall cost by allowing "expansion." That is, so far we have treated the problem of finding a partition with a constraint on the sizes of the subsets, *and* on the number of subsets, since given kn points, we have tried to find the best partitions into exactly k subsets of n points each. Suppose we now relax this second constraint by permitting the addition of dummy elements to increase the size of the problem, and attempt to find the best solution involving *any number* (greater than or equal to k) of subsets, with *at most* n points in each. This solution with k or greater subsets will in general have a lower cost than the constrained solution.

Figure 3 shows an example in which introducing slack permits a lower overall cost. Assume n is 3 and all nodes are size 1. The vertical edges have cost 1 and the horizontal ones cost 2. Any partition into 2 equal subsets has a cost of at least 3, but there is an obvious partition into 3 subsets with cost 2. Any nontrivial partition into 4 or more subsets has a cost greater than 2, so 3 subsets represents the optimal expansion. It is possible to find the minimal cost solution and the corresponding optimal amount of expansion as follows. Suppose the problem has kn points to be partitioned into k sets of n points each. Starting with no slack (kn points), the optimal assignment is found. Then n dummies, enough to create one extra subset, are added, making a $(k + 1)n$ problem, and so on. Eventually, one subset is produced which consists entirely of dummies. When this occurs, we take the partition with this set of dummies removed as our optimum solution.

REFERENCES

1. Ford, L. R., and Fulkerson, D. R., *Flows in Networks*, Princeton, New Jersey: Princeton University Press, 1962, p. 11.
2. Lin, S., "Computer Solutions of the Traveling Salesman Problem," B.S.T.J., *44*, No. 10 (December 1965), pp. 2245–2269.
3. Kernighan, B. W., "Some Graph Partitioning Problems Related to Program Segmentation," Ph.D. Thesis, Princeton University, January 1969, pp. 74–126.
4. Kruskal, J. B., Multi-Dimensional Scaling by Optimizing Goodness of Fit to a Non-Metric Hypothesis," *Psychometrika, 29,* No. 1 (March 1964), pp. 1–27, and No. 2 (June 1964), pp. 115–129.

DEPTH-FIRST SEARCH AND LINEAR GRAPH ALGORITHMS*

ROBERT TARJAN†

Abstract. The value of depth-first search or "backtracking" as a technique for solving problems is illustrated by two examples. An improved version of an algorithm for finding the strongly connected components of a directed graph and an algorithm for finding the biconnected components of an undirect graph are presented. The space and time requirements of both algorithms are bounded by $k_1V + k_2E + k_3$ for some constants k_1, k_2, and k_3, where V is the number of vertices and E is the number of edges of the graph being examined.

Key words. Algorithm, backtracking, biconnectivity, connectivity, depth-first, graph, search, spanning tree, strong-connectivity.

1. Introduction. Consider a graph G, consisting of a set of vertices $\mathscr{V}$ and a set of edges $\mathscr{E}$. The graph may either be directed (the edges are ordered pairs (v, w) of vertices; v is the *tail* and w is the *head* of the edge) or undirected (the edges are unordered pairs of vertices, also represented as (v, w)). Graphs form a suitable abstraction for problems in many areas; chemistry, electrical engineering, and sociology, for example. Thus it is important to have the most economical algorithms for answering graph-theoretical questions.

In studying graph algorithms we cannot avoid at least a few definitions. These definitions are more-or-less standard in the literature. (See Harary [3], for instance.) If $G = (\mathscr{V}, \mathscr{E})$ is a graph, a *path* $p: v \overset{*}{\Rightarrow} w$ in G is a sequence of vertices and edges leading from v to w. A path is *simple* if all its vertices are distinct. A path $p: v \overset{*}{\Rightarrow} v$ is called a *closed path*. A closed path $p: v \overset{*}{\Rightarrow} v$ is a *cycle* if all its edges are distinct and the only vertex to occur twice in p is v, which occurs exactly twice. Two cycles which are cyclic permutations of each other are considered to be the same cycle. The *undirected version* of a directed graph is the graph formed by converting each edge of the directed graph into an undirected edge and removing duplicate edges. An undirected graph is *connected* if there is a path between every pair of vertices.

A (directed rooted) *tree* T is a directed graph whose undirected version is connected, having one vertex which is the head of no edges (called the *root*), and such that all vertices except the root are the head of exactly one edge. The relation "(v, w) is an edge of T" is denoted by $v \rightarrow w$. The relation "There is a path from v to w in T" is denoted by $v \overset{*}{\rightarrow} w$. If $v \rightarrow w$, v is the *father* of w and w is a *son* of v. If $v \overset{*}{\rightarrow} w$, v is an *ancestor* of w and w is a *descendant* of v. Every vertex is an ancestor and a descendant of itself. If v is a vertex in a tree T, T_v is the subtree of T having as vertices all the descendants of v in T. If G is a directed graph, a tree T is a *spanning tree* of G if T is a subgraph of G and T contains all the vertices of G.

If R and S are binary relations, R^* is the transitive closure of R, R^{-1} is the inverse of R, and

$$RS = \{(u, w) \mid \exists v((u, v) \in R \ \& \ (v, w) \in S)\}.$$

If $f, f_1, \cdots, f_n$ are functions of $x_1, \cdots, x_n$, we say f is $O(f_1, \cdots, f_n)$ if

$$|f(x_1, \cdots, x_n)| \leq k_0 + k_1|f_1(x_1, \cdots, x_n)| + \cdots + k_n|f_n(x_1, \cdots, x_n)|$$

for all x_i and some constants $k_0, \cdots, k_n$. We shall assume a random-access computer model.

2. Depth-first search. Backtracking, or depth-first search, is a technique which has been widely used for finding solutions to problems in combinatorial theory and artificial intelligence [2], [11] but whose properties have not been widely analyzed. Suppose G is a graph which we wish to explore. Initially all the vertices of G are unexplored. We start from some vertex of G and choose an edge to follow. Traversing the edge leads to a new vertex. We continue in this way; at each step we select an unexplored edge leading from a vertex already reached and we traverse this edge. The edge leads to some vertex, either new or already reached. Whenever we run out of edges leading from old vertices, we choose some unreached vertex, if any exists, and begin a new exploration from this point. Eventually we will traverse all the edges of G, each exactly once. Such a process is called a *search* of G.

There are many ways of searching a graph, depending upon the way in which edges to search are selected. Consider the following choice rule: when selecting an edge to traverse, always choose an edge emanating from the vertex most recently reached which still has unexplored edges. A search which uses this rule is called a *depth-first search.* The set of old vertices with possibly unexplored edges may be stored on a stack. Thus a depth-first search is very easy to program either iteratively or recursively, provided we have a suitable computer representation of a graph.

DEFINITION 1. Let $G = (\mathscr{V}, \mathscr{E})$ be a graph. For each vertex $v \in \mathscr{V}$ we may construct a list containing all vertices w such that $(v, w) \in \mathscr{E}$. Such a list is called an *adjacency list* for vertex v. A set of such lists, one for each vertex in G, is called an *adjacency structure* for G.

If the graph G is undirected, each edge (v, w) is represented twice in an adjacency structure; once for v and once for w. If G is directed, each edge (v, w) is represented once: vertex w appears in the adjacency list of vertex v. A single graph may have many adjacency structures; in fact, each ordering of the edges around the vertices of G gives a unique adjacency structure, and each adjacency structure corresponds to a unique ordering of the edges at each vertex. Using an adjacency structure for a graph, we can perform depth-first searches in a very efficient manner, as we shall see.

Suppose G is a connected undirected graph. A search of G imposes a direction on each edge of G given by the direction in which the edge is traversed when the search is performed. Thus G is converted into a directed graph G'. The set of edges which lead to a new vertex when traversed during the search defines a spanning tree of G'. In general, the arcs of G' which are not part of the spanning tree interconnect the paths in the tree. However, if the search is depth-first, each edge (v, w) not in the spanning tree connects vertex v with one of its ancestors w.

DEFINITION 2. Let P be a directed graph, consisting of two disjoint sets of edges, denoted by $v \rightarrow w$ and $v \dashrightarrow w$ respectively. Suppose P satisfies the following properties:

(i) The subgraph T containing the edges $v \rightarrow w$ is a spanning tree of P.

(ii) We have $\dashrightarrow \subseteq (\overset{*}{\rightarrow})^{-1}$, where "$\rightarrow$" and "$\dashrightarrow$" denote the relations defined by the corresponding set of edges. That is, each edge which is not in the spanning tree T of P connects a vertex with one of its ancestors in T.

Then P is called a *palm tree*. The edges $v \dashrightarrow w$ are called the *fronds* of P.

THEOREM 1. *Let P be the directed graph generated by a depth-first search of a connected graph G. Then P is a palm tree. Conversely, let P be any palm tree. Then P is generated by some depth-first search of the undirected version of P.*

Proof. Consider the program listed below, which carries out a depth-first search of a connected graph, starting at vertex s, using an adjacency structure of the graph to be searched. The program numbers the vertices of the graph in the order they are reached during the search and constructs the directed graph (P) generated by the search.

```
BEGIN
    INTEGER i;
    PROCEDURE DFS(v, u); COMMENT vertex u is the father of
                    vertex v in the spanning tree being constructed;
        BEGIN
            NUMBER (v) := i := i + 1;
            FOR w in the adjacency list of v DO
                BEGIN
                    IF w is not yet numbered THEN
                        BEGIN
                            construct arc v → w in P;
                            DFS(w, v);
                        END
                    ELSE IF NUMBER (w) < NUMBER (v) and w ¬= u
                        THEN construct arc v -→ w in p;
                END;
        END;
    i := 0;
    DFS(s, 0);
END;
```

Figure 1 gives an example of the application of DFS to a graph. Suppose $P = (\mathscr{V}, \mathscr{E})$ is the directed graph generated by a depth-first search of some connected graph G, and assume that the search begins at vertex s. Examine the procedure DFS. The algorithm clearly terminates because each vertex can only be

* Received by the editors August 30, 1971, and in revised form March 9, 1972.

† Department of Computer Science, Cornell University, Ithaca, New York 14850. This research was supported by the Hertz Foundation and the National Science Foundation under Grant GJ-992.

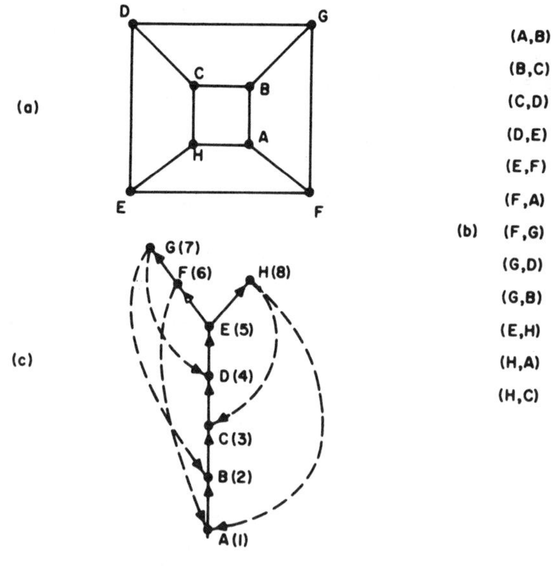

FIG. 1. *Application of depth-first search to a graph*

(a) *The graph*
(b) *Order of edge exploration*
(c) *Generated palm tree, with numbers of vertices in ( )*

numbered once. Furthermore, each edge in the graph is examined exactly twice. Therefore the time required by the search is linear in V and E.

For any vertices v and w, let $d(v, w)$ be the length of the shortest path between v and w in G. Since G is connected, all distances are finite. Suppose that some vertex remains unnumbered by the search. Let v be an unnumbered vertex such that $d(s, v)$ is minimal. Then there is a vertex w such that w is adjacent to v and $d(s, w) < d(s, v)$. Thus w is numbered. But v will also be numbered, since it is adjacent to w. This means that all vertices are numbered during the search.

The vertex s is the head of no edge $w \rightarrow s$. Each other vertex v is the head of exactly one edge $w \rightarrow v$. The subgraph T of P defined by the edges $v \rightarrow w$ is obviously connected. Thus T is a spanning tree of P.

Each arc of the original graph is directed in at least one direction; if (v, w) does not become an arc of the spanning tree T, either $v \dashrightarrow w$ or $w \dashrightarrow v$ must be constructed since both v and w are numbered whenever edge (v, w) is inspected and either NUMBER (v) < NUMBER (w) or NUMBER (v) > NUMBER (w).

The arcs $v \rightarrow w$ run from smaller numbered points to larger numbered points. The arcs $v \dashrightarrow w$ run from larger numbered points to smaller numbered points. If arc $v \dashrightarrow w$ is constructed, arc $w \rightarrow v$ is not constructed later because v is numbered. If arc $w \rightarrow v$ is constructed, arc $v \dashrightarrow w$ is not later constructed, because of the test "$w \neg = u$" in procedure DFS. Thus each edge in the original graph is directed in one and only one direction.

Consider an arc $v \dashrightarrow w$. We have NUMBER (w) < NUMBER (v). Thus w is numbered before v. Since $v \dashrightarrow w$ is constructed and not $w \rightarrow v$, v must be numbered before edge (w, v) is inspected. Thus v must be numbered during execution of DFS $(w, -)$. But all vertices numbered during execution of DFS $(w, -)$ are descendants of w. This means that $w \overset{*}{\rightarrow} v$, and G is a palm tree.

Let us prove the converse part of the theorem. Suppose that P is a palm tree, with spanning tree T and undirected version G. Construct an adjacency structure of G in which all the edges of T appear before the other edges of G in the adjacency lists. Starting with the root of T, perform a depth-first search using this adjacency structure. The search will traverse the edges of T preferentially and will generate the palm tree P; it is easy to see that each edge is directed correctly. This completes the proof of the theorem.

We may state Theorem 1 nonconstructively as the following corollary.

COROLLARY 2. *Let G be any undirected graph. Then G can be converted into a palm tree by directing its edges in a suitable manner.*

3. Biconnectivity. The value of depth-first search follows from the simple structure of a palm tree. Let us consider a problem in which this structure is useful.

DEFINITION 3. Let $G = (\mathscr{V}, \mathscr{E})$ be an undirected graph. Suppose that for each triple of distinct vertices v, w, a in $\mathscr{V}$, there is a path $p: v \overset{*}{\Rightarrow} w$ such that a is not on the path p. Then G is *biconnected.* (Other equivalent definitions of biconnectedness exist.) If, on the other hand, there is a triple of distinct vertices v, w, a in $\mathscr{V}$ such that a is on any path $p: v \overset{*}{\Rightarrow} w$, and there exists at least one such path, then a is called an *articulation point* of G.

LEMMA 3. *Let $G = (\mathscr{V}, \mathscr{E})$ be an undirected graph. We may define an equivalence relation on the set of edges as follows: two edges are equivalent if and only if they belong to a common cycle. Let the distinct equivalence classes under this relation be $\mathscr{E}_i$, $1 \leqq i \leqq n$, and let $G_i = (\mathscr{V}_i, \mathscr{E}_i)$, where $\mathscr{V}_i$ is the set of vertices incident to the edges of $\mathscr{E}_i$; $\mathscr{V}_i = \{v | \exists w((v, w) \in \mathscr{E}_i)\}$. Then:*

(i) *G_i is biconnected, for each $1 \leqq i \leqq n$.*
(ii) *No G_i is a proper subgraph of a biconnected subgraph of G.*
(iii) *Each articulation point of G occurs more than once among the $\mathscr{V}_i$, $1 \leqq i \leqq n$. Each nonarticulation point of G occurs exactly once among the $\mathscr{V}_i$, $1 \leqq i \leqq n$.*
(iv) *The set $\mathscr{V}_i \cap \mathscr{V}_j$ contains at most one point, for any $1 \leqq i, j \leqq n$. Such a point of intersection is an articulation point of the graph.*

The subgraphs G_i of G are called the *biconnected components of G*.

Suppose we wish to determine the biconnected components of an undirected graph G. Common algorithms for this purpose, for instance, Shirey's [14], test each vertex in turn to discover if it is an articulation point. Such algorithms require time proportional to $V \cdot E$, where V is the number of vertices and E is the number of edges of the graph. A more efficient algorithm uses depth-first search. Let P be a palm tree generated by a depth-first search. Suppose the vertices of P are numbered in the order they are reached during the search (as is done by the procedure DFS above). We shall refer to vertices by their numbers. If u is an ancestor of v in the spanning tree T of P, then $u < v$. For any vertex v in P, let LOWPT (v) be the smallest vertex in the set $\{v\} \cup \{w | v \overset{*}{\rightarrow} \dashrightarrow w\}$. That is, LOWPT (v) is the smallest vertex reachable from v by traversing zero or more tree arcs followed by at most one frond. The following results form the basis of an efficient algorithm for finding biconnected components. This algorithm was discovered by Hopcroft and Tarjan [4]. Paton [12] describes a similar algorithm.

LEMMA 4. *Let G be an undirected graph and let P be a palm tree formed by directing the edges of G. Let T be the spanning tree of P. Suppose $p: v \overset{*}{\Rightarrow} w$ is any path in G. Then p contains a point which is an ancestor of both v and w in T.*

Proof. Let T_u with root u be the smallest subtree of T containing all vertices on the path p. If $u = v$ or $u = w$, the lemma is immediate. Otherwise, let T_{u_1} and T_{u_2}

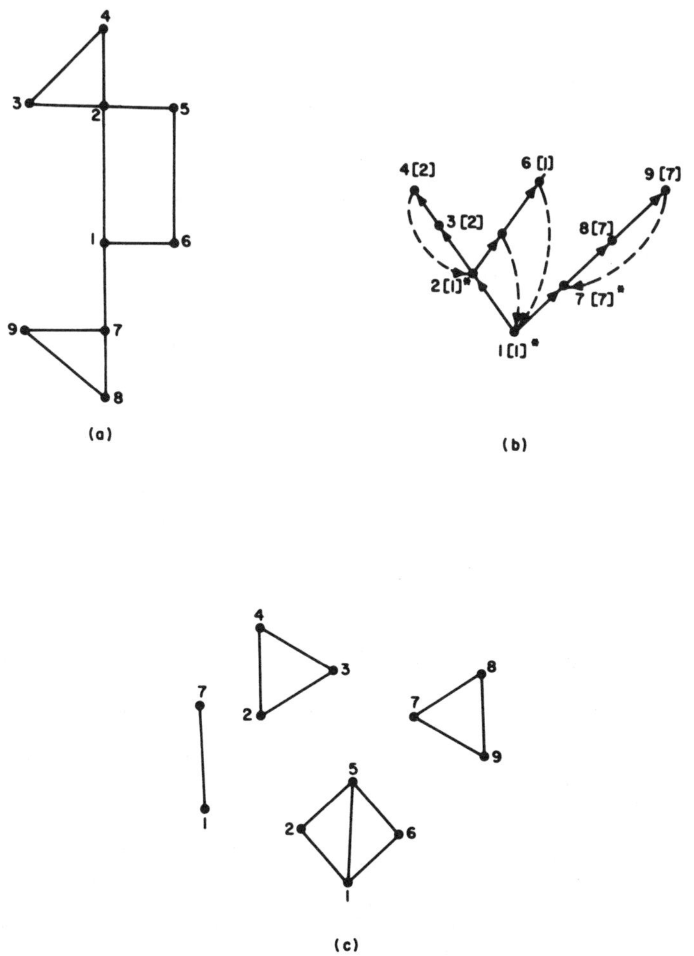

FIG. 2. *A graph and its biconnected components*

(a) *Graph*
(b) *A palm tree with LOWPT values in [ ], articulation points marked with **
(c) *Biconnected components*

be two distinct subtrees containing points on p such that $u \to u_1$ and $u \to u_2$. If only one such subtree exists, then u is on p since T_u is minimal. If two such subtrees exist, path p can only get from T_{u_1} to T_{u_2} by passing through vertex u, since no point in one of these trees is an ancestor of any point in the other, while both $\to$ and $-\!\to$ connect only ancestors in a palm tree. Since u is an ancestor of both v and w, the lemma holds.

LEMMA 5. *Let G be a connected undirected graph. Let P be a palm tree formed by directing the edges of G, and let T be the spanning tree of P. Suppose a, v, w are distinct vertices of G such that $(a, v) \in T$, and suppose w is not a descendant of v in T. (That is, $\neg(v \overset{*}{\to} w)$ in T.) If LOWPT $(v) \geq a$, then a is an articulation point of P and removal of a disconnects v and w. Conversely, if a is an articulation point of G, then there exist vertices v and w which satisfy the properties above.*

Proof. If $a \to v$ and LOWPT $(v) \geq a$, then any path from v not passing through a remains in the subtree T_v, and this subtree does not contain the point w. This gives the first part of the lemma.

To prove the converse, let a be an articulation point of G. If a is the root of P, then at least two tree arcs must emanate from a. Let v be the head of one such arc and let w be the head of another such arc. Then $a \to v$, LOWPT $(v) \geq a$, and w is not a descendant of v. If a is not the root of P, consider the connected components formed by deleting a from G. One component must consist only of descendants of a. Such a component can contain only one son of v by Lemma 4. Let v be such a son of a. Let w be any proper ancestor of a. Then $a \to v$, LOWPT $(v) \geq a$, and w is not a descendant of v. Thus the converse part of the lemma is true.

COROLLARY 6. *Let G be a connected undirected graph, and let P be a palm tree formed by directing the edges of G. Suppose that P has a spanning tree T. If C is a biconnected component of G, then the vertices of C define a subtree of T, and the root of this subtree is either an articulation point of G or is the root of T.*

Figure 2 shows a graph, its LOWPT values, articulation points, and biconnected components. The LOWPT values of all the vertices of a palm tree P may be calculated during a single depth-first search, since

$$\text{LOWPT}(v) = \min(\{\text{NUMBER}(v)\} \cup \{\text{LOWPT}(w) | v \to w\} \cup \{\text{NUMBER}(w) | v -\!\to w\}).$$

On the basis of such a calculation, the articulation points and the biconnected components may be determined, all during one search. The biconnectivity algorithm is presented below. The program will compute the biconnected components of a graph G, starting from vertex s.

```
BEGIN
   INTEGER i;
   procedure BICONNECT (v, u);
      BEGIN
         NUMBER (v) := i := i + 1;
         LOWPT (v) = NUMBER (v);
         FOR w in the adjacency list of v DO
            BEGIN
               IF w is not yet numbered THEN
                  BEGIN
                     add (v, w) to stack of edges;
                     BICONNECT (w, v)
                     LOWPT (v) := min (LOWPT (v), LOWPT (w));
                     IF LOWPT (w) ≥ NUMBER (v) THEN
                        BEGIN
                           start new biconnected component;
                           WHILE top edge e = (u1, u2) on edge
                                 stack has NUMBER (u1)
                                 ≥ NUMBER (w) DO
                              delete (u1, u2) from edge stack and
                                 add it to current component;
                           delete (v, w) from edge stack and add
                              it to current component;
                        END;
                  END
               ELSE IF (NUMBER (w) < NUMBER (v)) and
                  (w ¬= u) THEN
                  BEGIN
                     add (v, w) to edge stack;
                     LOWPT(v) := min(LOWPT(v), NUMBER(w));
                  END;
            END;
      END;
   i := 0;
   empty the edge stack;
   FOR w a vertex DO IF w is not yet numbered THEN BICONNECT (w, 0);
END;
```

The edges of G are placed on a stack as they are traversed; when an articulation point is found the corresponding edges are all on top of the stack. (If $(v, w) \in T$ and LOWPT $(w) \geq$ LOWPT (v), then the corresponding biconnected component contains the edges in $\{(u_1, u_2) | w \overset{*}{\to} u_1\} \cup \{(v, w)\}$ which are on the edge stack.) A single search on each connected component of a graph G will give us all the connected components of G.

THEOREM 7. *The biconnectivity algorithm requires O(V, E) space and time when applied to a graph with V vertices and E edges.*

Proof. The algorithm clearly requires space bounded by $k_1V + k_2E + k_3$, for some constants k_1, k_2, and k_3. The algorithm is an elaboration of the depth-first search procedure DFS. During the search, LOWPT values are calculated and each edge is placed on the edge stack once and removed from the edge stack once. The amount of extra time required by these operations is proportional to E. Thus BICONNECT has a time bound linear in V and E.

THEOREM 8. *The biconnectivity algorithm correctly gives the biconnected components of any undirected graph G.*

Proof. The actual depth-first search undertaken by the algorithm depends on the adjacency structure chosen to represent G; we shall prove that the algorithm is correct for all adjacency structures. Notice first that the biconnectivity algorithm analyzes each connected component of G separately to find its biconnected components, applying one depth-first search to each connected component. Thus we need only prove that the biconnectivity algorithm works correctly on connected graphs G.

The correctness proof is by induction on the number of edges in G. Suppose G is connected and contains no edges. G either is empty or consists of a single point. The algorithm will terminate after examining G and listing no components. Thus the algorithm operates correctly in this case. Now suppose that the algorithm works correctly on all connected graphs with $E - 1$ or fewer edges. Consider applying the algorithm to a connected graph G with E edges.

Each edge placed on the stack of edges is eventually removed and added to a component since everything on the edge stack is removed whenever the search returns to the root of the palm tree of G. Consider the situation when the first component G' is formed. Suppose that this component does not include all the edges of G. Then the vertex v currently being examined is an articulation point of the graph and separates the edges in the component from the other edges in the graph by Lemma 5.

Consider only the set of edges in the component. If BICONNECT $(v, 0)$ is executed, using the graph G' as data, the steps taken by the algorithm are the same as those taken during the analysis of the edges of G' when the data consists of the entire graph G. Since G' contains fewer edges than G, the algorithm operates correctly on G' and G' must be biconnected. If we delete the edges of G' from G, we get another subgraph G'' with fewer edges than G since G' is not empty. The algorithm operates correctly on G'' by the induction assumption. The behavior of the algorithm on G is simply a composite of its behavior on G' and on G''; thus the algorithm must operate correctly on G.

Now suppose that only one component is found. We want to show that in this case G is biconnected. Suppose that G is not biconnected. Then G has an articulation point a. By Lemma 5, LOWPT $(v) \geq a$ for some son v of a. But the articulation point test in the program will succeed when the edge (a, v) is examined, and more than one biconnected component will be generated. This contradiction shows that G is biconnected, and the algorithm works correctly in this case.

By induction, the biconnectivity algorithm gives the correct components when applied to any connected graph, and hence when applied to any graph.

4. Strong connectivity. The biconnectivity algorithm shows how useful depth-first search can be when applied to undirected graphs. However, when a directed graph is searched in a depth-first manner, a simple palm tree structure does not result, because the direction of search on each edge is fixed. The more complicated structure which results in this case is still simple enough to prove useful in at least one application.

DEFINITION 4. Let G be a directed graph. Suppose that for each pair of vertices v, w in G, there exist paths $p_1 : v \overset{*}{\Rightarrow} w$ and $p_2 : w \overset{*}{\Rightarrow} v$. Then G is said to be *strongly connected.*

LEMMA 9. *Let $G = (\mathscr{V}, \mathscr{E})$ be a directed graph. We may define an equivalence relation on the set of vertices as follows: two vertices v and w are equivalent if there is a closed path $p : v \overset{*}{\Rightarrow} v$ which contains w. Let the distinct equivalence classes under this relation be $\mathscr{V}_i$, $1 \leq i \leq n$. Let $G_i = (\mathscr{V}_i, \mathscr{E}_i)$, where $\mathscr{E}_i = \{(v, w) \in \mathscr{E} | v, w \in \mathscr{V}_i\}$. Then:*

(i) *Each G_i is strongly connected.*

(ii) *No G_i is a proper subgraph of a strongly connected subgraph of G.*

The subgraphs G_i are called the *strongly connected components* of G.

Suppose we wish to determine the strongly connected components of a directed graph. This problem is related to the problem of determining the ergodic subchains and transient states of a Markov chain. Fox and Landy [1] give an algorithm for solving the latter problem; Purdom [13] and Munro [10] present virtually identical methods for solving the former problem. These algorithms use depth-first search. Purdom claims a time bound of kV^2 for his algorithm; Munro claims $k \max(E, V \log V)$, where the graph has V vertices and E edges. Their algorithm attempts to construct a cycle by starting from a point and beginning a depth-first search. When a cycle is found, the vertices on the cycle are marked as being in the same strongly connected component and the process is repeated. The algorithm has the disadvantage that two small strongly connected components may be collapsed into a bigger one; the resultant extra work in relabeling may contribute V^2 steps using a simple approach, or $V \log V$ steps if a more sophisticated approach is used (see Munro [10]). In fact, the time bound may be reduced further if an efficient list merging algorithm [9] is used. However, a more careful study of what a depth-first search does to a directed graph reveals that an $O(V, E)$ algorithm which requires *no* merging of components may be devised.

Consider what happens when a depth-first search is performed on a directed graph G. The set of edges which lead to a new vertex when traversed during the search form a tree. The other edges fall into three classes. Some are edges running from ancestors to descendants in the tree. These edges may be ignored, because they do not affect the strongly connected components of G. Some edges run from descendants to ancestors in the tree; these we may call *fronds* as above. Other edges run from one subtree to another in the tree. These, we call *cross-links*. It is easy to verify that if the vertices of the tree are numbered in the order they are reached during the search, a cross-link (v, w) always has NUMBER $(v) >$ NUMBER (w). We shall denote tree edges by $v \to w$, and fronds and cross-links by $v \dashrightarrow w$.

Suppose G is a directed graph, to which a depth-first search algorithm is applied repeatedly until all the edges are explored. The process will create a set of trees which contain all the vertices of G, called the *spanning forest* F of G, and sets of fronds and cross-links. (Other edges are thrown away.) A directed graph consisting of a spanning forest and sets of fronds and cross-links is called a *jungle*. Suppose the vertices are numbered in the order they are reached during the search and that we refer to vertices by their number. Then we have the following results.

LEMMA 10. *Let v and w be vertices in G which lie in the same strongly connected component. Let F be a spanning forest of G generated by repeated depth-first search. Then v and w have a common ancestor in F. Further, if u is the highest numbered common ancestor of v and w, then u lies in the same strongly connected component as v and w.*

Proof. Without loss of generality we may assume $v \leqq w$. Let p be a path from v to w in G. Let T_u with root u be the smallest subtree of a tree in F containing all the vertices in p. There must be such a tree, since p can pass from one tree in F to another tree with smaller numbered vertices but p can never lead to a tree with larger numbered vertices. If p were contained in two or more trees of F, it could not end at w, since $v \leqq w$.

Thus T_u exists, and v and w have a common ancestor in F. In fact, p must pass through vertex u, by a proof similar to the proof of Lemma 4, and u, v, w must all be in the same strongly connected component. This gives the lemma.

COROLLARY 11. *Let C be a strongly connected component in G. Then the vertices of C define a subtree of a tree in F, the spanning forest of G. The root of this subtree is called the* root *of the strongly connected component C.*

The problem of finding the strongly connected components of a graph G thus reduces to the problem of finding the roots of the strongly connected components, just as the problem of finding the biconnected components of an undirected graph reduces to the problem of finding the articulation points of the graph. We can construct a simple test to determine if a vertex is the root of a strongly connected component. Let

$$\text{LOWLINK}(v) = \min(\{v\} \cup \{w \mid v \overset{*}{\to}\dashrightarrow w \;\&\; \exists u (u \overset{*}{\to} v \;\&\; u \overset{*}{\to} w \;\&\; u \text{ and } w \text{ are in the same strongly connected component of } G)\}).$$

That is, LOWLINK (v) is the smallest vertex which is in the same component as v and is reachable by traversing zero or more tree arcs followed by at most one frond or cross-link.

LEMMA 12. *Let G be a directed graph with* LOWLINK *defined as above relative to some spanning forest F of G generated by depth-first search. Then v is the root of some strongly connected component of G if and only if* LOWLINK $(v) = v$.

Proof. Obviously, if v is the root of a strongly connected component C of G, then LOWLINK $(v) = v$, since if LOWLINK$(v) < v$, some proper ancestor of v would be in C and v could not be the root of C.

Consider the converse. Suppose u is the root of a strongly connected component C of G, and v is a vertex in C different from u. There must be a path $p: v \overset{*}{\Rightarrow} u$. Consider the first edge on this path which leads to a vertex w not in the subtree T_v. This edge is either a vine or a cross-link, and we must have LOWLINK $(v) \leqq w < v$, since the highest numbered common ancestor of v and w is in C.

Figure 3 shows a directed graph, its LOWLINK values, and its strongly connected components. LOWLINK may be calculated using depth-first search. An algorithm for computing the strongly connected components of a directed graph in $O(V, E)$ time may be based on such a calculation. An implementation of such an algorithm is presented below. The points which have been reached during the search but which have not yet been placed in a component are stored on a stack. This stack is analogous to the stack of edges used by the biconnectivity algorithm.

```
BEGIN
    INTEGER i;
    PROCEDURE STRONGCONNECT (v);
        BEGIN
            LOWLINK (v) := NUMBER (v) := i := i + 1;
            put v on stack of points;
            FOR w in the adjacency list of v DO
                BEGIN
                    IF w is not yet numbered THEN
                        BEGIN comment (v, w) is a tree arc;
                            STRONGCONNECT (w);
                            LOWLINK (v) := min (LOWLINK (v),
                                            LOWLINK (w));
                        END
                    ELSE IF NUMBER (w) < NUMBER (v) DO
                        BEGIN comment (v, w) is a frond or cross-link;
                            if w is on stack of points THEN
                                LOWLINK (v) := min (LOWLINK (v),
                                               NUMBER (w));
                        END;
                END;
            If (LOWLINK (v) = NUMBER (v)) THEN
                BEGIN comment v is the root of a component;
                    start new strongly connected component;
                    WHILE w on top of point stack satisfies
                            NUMBER (w) ≧ NUMBER (v) DO
                        delete w from point stack and put w in
                            current component;
                END;
        END;
    i := 0;
    empty stack of points;
    FOR w a vertex IF w is not yet numbered THEN STRONGCONNECT (w);
END;
```

THEOREM 13. *The algorithm for finding strongly connected components requires $O(V, E)$ space and time.*

Proof. The algorithm clearly requires space bounded by $k_1 V + k_2 E + k_3$, for some constants k_1, k_2, and k_3. The algorithm is an elaboration of the depth-first search procedure DFS, modified to apply to directed graphs. During the search, LOWLINK values are calculated, each point is placed on the stack of

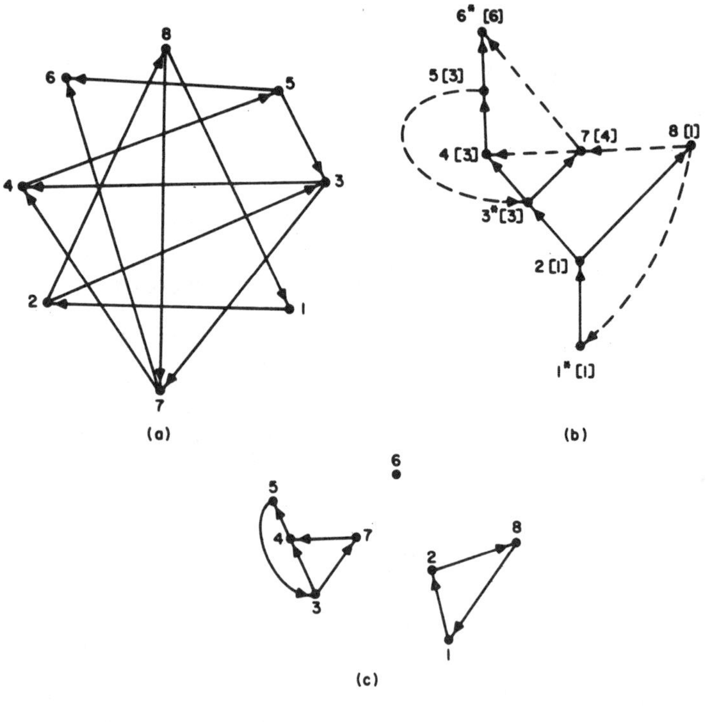

FIG. 3. *A graph and its strongly connected components*
(a) *Graph*
(b) *Jungle generated by depth-first search with* LOWLINK *values in* [ ], *roots of components marked with* *
(c) *Strongly connected components*

points once, and each point is removed from the stack of points once. Testing to see if a vertex is on the point stack can be done in a fixed time if a Boolean array is kept which answers this question for each vertex. The amount of extra time required by these operations is linear in V and E. Thus STRONGCONNECT has a time bound linear in V and E.

THEOREM 14. *The algorithm for finding strongly connected components works correctly on any directed graph G.*

Proof. We prove by induction that the calculation of LOWLINK (v) is correct. Suppose as the induction hypothesis that for all vertices v such that v is a proper descendant of vertex k or $v < k$, LOWLINK (v) is computed correctly. This means that the test to determine if v is the root of a component is performed correctly for all such vertices v. The reader may verify that this somewhat strange induction hypothesis corresponds to considering vertices in the order they are examined for the *last* time during the depth-first search process.

Consider vertex $v = k$. Let $v \overset{*}{\rightarrow} w_1$ and let $w_1 \dashrightarrow w_2$ be a vine or cross-link such that $w_2 < v$. If vertices v and w_2 have no common ancestor, then before vertex v is reached during the search, vertex w_2 must have been removed from the stack of points and placed in a component. (The smallest numbered ancestor of vertex w_2 must be a component root.) Thus edge $w_1 \dashrightarrow w_2$ does not enter into the calculation of LOWLINK (v).

Otherwise, let u be the highest common ancestor of v and w_2. Vertex v is also the highest common ancestor of w_1 and w_2. If u is not in the same strongly connected component as w_2, then there must be a strongly connected component root on the tree path $u \overset{*}{\rightarrow} w_2$. Since $w_2 < v$, this root was discovered and w_2 was removed from the stack of points and placed in a component before the edge $w_1 \dashrightarrow w_2$ is traversed during the search. Thus $w_1 \dashrightarrow w_2$ will not enter into the calculation of LOWLINK (v). (This can only happen if $w_1 \dashrightarrow w_2$ is a cross-link.) On the other hand, if u is in the same strongly connected component as w_2, there is no component root $r \neg = u$ on the branch $u \overset{*}{\rightarrow} w_2$, and $v \dashrightarrow w_2$ will be used to calculate LOWLINK (w_2), and also LOWLINK (v), as desired. Thus LOWLINK (v) is calculated correctly, and by induction LOWLINK is calculated correctly for all vertices.

Since the algorithm correctly calculates LOWLINK, it correctly identifies the roots of the strongly connected components. If such a root u is found, the corresponding component contains all the descendants of u which are on the stack of points when u is discovered. These vertices are all on top of the stack of points, and are all put into a component by STRONGCONNECT. Thus STRONGCONNECT works correctly.

5. Further applications. We have seen how the depth-first search method may be used in the construction of very efficient graph algorithms. The two algorithms presented here are in fact optimal to within a constant factor, since every edge and vertex of a graph must be examined to determine a solution to one of the problems. (Given a suitable theoretical framework, this statement may be proved rigorously.) The similarity between biconnectivity and strong connectivity revealed by the depth-first search approach is striking. The possible uses of depth-first search are very general, and are certainly not limited to the examples presented. Hopcroft and Tarjan have constructed an algorithm for finding triconnected components in $O(V, E)$ time by extending the biconnectivity algorithm [8]. An algorithm for testing the planarity of a graph in $O(V)$ time [15] is also based on depth-first search. Combining the connectivity algorithms, the planarity algorithm, and an algorithm for testing isomorphism of triconnected planar graphs [7], we may construct an algorithm to test isomorphism of arbitrary planar graphs in $O(V \log V)$ time [8]. Depth-first search is a powerful technique with many applications.

REFERENCES

[1] B. L. FOX AND D. M. LANDY, *An algorithm for identifying the ergodic subchains and transient states of a stochastic matrix*, Comm. ACM, 11 (1968), pp. 619–621.

[2] S. W. GOLOMB AND L. D. BAUMERT, *Backtrack programming*, J. ACM, 12 (1965), pp. 516–524.

[3] F. HARARY, *Graph Theory*, Addison-Wesley, Reading, Mass., 1969.

[4] J. HOPCROFT AND R. TARJAN, *Efficient algorithms for graph manipulation*, Tech. Rep. 207, Computer Science Department, Stanford University, Stanford, Calif., 1971.

[5] ———, *A V^2 algorithm for determining isomorphism of planar graphs*, Information Processing Letters, 1 (1971), pp. 32–34.

[6] ———, *Planarity testing in V log V steps: Extended abstract*, Tech. Rep. 201, Computer Science Department, Stanford University, Stanford, Calif., 1971.

[7] J. HOPCROFT, *An N log N algorithm for isomorphism of planar triply connected graphs*, Tech. Rep. 192, Computer Science Department, Stanford University, Stanford, Calif., 1971.

[8] J. HOPCROFT AND R. TARJAN, *Isomorphism of planar graphs*, IBM Symposium on Complexity of Computer Computations, Yorktown Heights, N.Y., March, 1972, to be published by Plenum Press.

[9] J. HOPCROFT AND J. ULLMAN, *A linear list-merging algorithm*, Tech. Rep. 71-111, Cornell University Computer Science Department, Ithaca, N.Y., 1971.

[10] I. MUNRO, *Efficient determination of the strongly connected components and transitive closure of a directed graph*, Department of Computer Science, University of Toronto, 1971.

[11] N. J. NILSON, *Problem Solving Methods in Artificial Intelligence*, McGraw-Hill, New York, 1971.

[12] K. PATON, *An algorithm for the blocks and cutnades of a graph*, Comm. ACM, 14 (1971), pp. 468–475.

[13] P. W. PURDOM, *A transitive closure algorithm*, Tech. Rep. 33, Computer Sciences Department, University of Wisconsin, Madison, 1968.

[14] R. W. SHIREY, *Implementation and analysis of efficient graph planarity testing algorithms*, Doctoral thesis, Computer Sciences Department, University of Wisconsin, Madison, 1969.

[15] R. TARJAN, *An efficient planarity algorithm*, Tech. Rep. 244, Computer Science Department, Stanford University, Stanford, Calif., 1971.

Topological Optimization of Computer Networks

HOWARD FRANK AND WUSHOW CHOU

Invited Paper

***Abstract*—Modeling, analysis, and design problems and methodologies for centralized and distributed computer-communication networks are discussed. The basic problem is to specify the location and capacity of each communication link within the network. The design objective is to provide a low-cost network which satisfies constraints on response time, throughput, reliability, and other parameters. Fundamental network models for queuing and reliability analysis are described, as are basic properties of various network structures. A number of approaches to the topological design problem are presented; areas where further research is needed are indicated; possible approaches to several unsolved problems are suggested.**

I. Introduction

IN THIS PAPER we cover both centralized and distributed computer networks. For illustration, we give specific examples of two highly successful networks—the NASDAQ Over-the-Counter Quotation System (see Fig. 1) and the ARPA Computer Network (see Fig. 2).

"NASDAQ" stands for the National Association of Securities Dealers Automated Quotation System. It is a system of computers and communication and terminal devices, designed primarily for the collection and distribution of real-time quotations for the over-the-counter securities market. The NASDAQ System is composed of low-, medium-, and high-speed communication lines (from 1600–50 000 bits/s) tied to the NASDAQ Central Processing Center in Trumbull, Conn., through control units and data concentrators. Ten concentrators are utilized in the NASDAQ System–two each in Chicago, Ill., San Francisco, Calif., and Atlanta, Ga., and four in New York, N. Y. For reliability, there are two lines from the central processor to each concentrator facility. The regional circuits are dedicated leased-line multipoint data circuits. Design constraints for the system include small response times for several types of messages, extremely high reliability and availability specifications, low error rates on circuits, and the ability of the network to handle a large number of calls without loss of data.

The ARPA Network [19], [44]–[48] provides store-and-forward communications between a set of computers distributed across the United States. The message handling at each center in the network is performed by a special-purpose interface message processor (IMP). The centers are connected through the IMP's by full duplex high-speed communication lines. Messages are broken up into sets of packets each with appropriate header information. Each packet independently makes its way through the network to its destination. A goal of the system is to achieve an average delay of less than 0.2 s for single packets. The network must accommodate variations in traffic flow without degradation in performance. For reliability, the network is designed so that at least two IMP's and/or lines must fail before it can become disconnected. Control of the system is distributed throughout the network with each IMP providing routing and flow control on the basis of its current knowledge of the network status.

Fig. 1. The NASD automated quotation system.

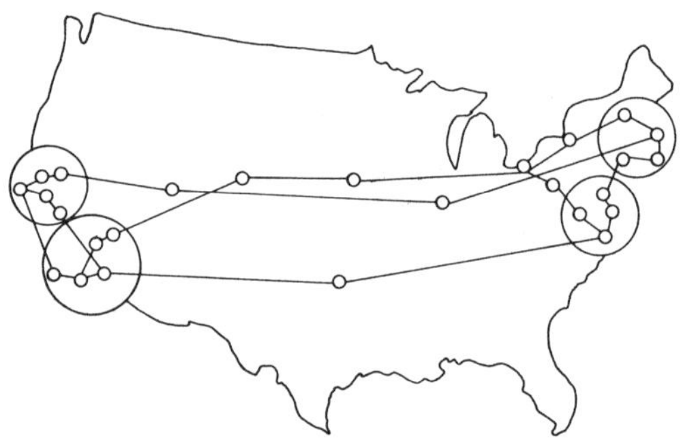

Fig. 2. The ARPA computer network (March 1972).

Although the NASDAQ and ARPA networks differ significantly in their philosophy of design and their operation, the development of networks of both types involves three similar principal activities: 1) the design of the IMP's, concentrators, and multiplexers (hereafter called *nodes*) to act as store-and-forward switches; 2) the topological modeling and design to specify the capacity and location of each communication circuit within the network; 3) the design of protocols to meet user requirements. We will concentrate on item 2), the topological modeling and design, although this area clearly interacts with the other problems.

Many formidable theoretical problems are encountered during the topological optimization of any network. These problems include:

Network Choice: In general, there are $[N(N-1)/2)!]/\{[(N(N-1)/2-M)!]M!\}$ ways of arranging M links among N nodes. Considering all possible designs by computer is out

Manuscript received July 19, 1972. This work was supported by the Advanced Research Projects Agency of the Department of Defense under Contract DAHC15-70-C-0120.

The authors are with the Network Analysis Corporation, Glen Cove, N. Y. 11542.

Reprinted from *Proc. IEEE*, vol. 60, pp. 1385–1397, Nov. 1972.

of the question, and no known computationally feasible method exists for finding an optimal computer-communication layout for a system with a sufficiently large number of nodes.

Discrete Elements: Components usually are available in discrete sizes. Thus line speeds can be at 2000, 2400, 3600, · · · , 9600, · · · , 50 000, 240 000 bits/s, etc. This means an integer optimization problem must be solved. Except in special cases for centralized tree design, no theoretical methods now exist for problems of practical size.

Nonlinearities: Component cost structures, time-delay functions, and reliability functions are all nonlinear. Typical cost functions are neither "concave" nor "convex" and no analytical methods are available to obtain optimal solutions for networks containing such elements.

However, in spite of these difficulties, enormous progress has been made in the understanding of computer-communication networks and in the development of effective analysis and design procedures. In this paper, we will describe a number of approaches to these problems, indicate areas where additional research is needed, and suggest possible approaches to several outstanding problems.

II. Performance Criteria and System Models and Computations

To specify the topology of a computer-communication network, the following questions must be answered.

Design Constraints

What average, maximum, or distribution of time delays will be tolerated? (The ARPA Network requires average response times of less than 0.2 s, while NASDAQ requires that over 50 percent of all messages be answered by the system in less than 5 s and 90 percent in less than 7 s.)

What average and peak throughputs are required? (ARPA requirements are generally on the order of 10 kbits/s/node on the average and peak rates of 85 kbits/s/node pair in an unloaded network. NASDAQ specifies busy-hour traffic in excess of 100 000 messages (each about 900 bits long) per busy hour.)

How should reliability be defined and what degree of reliability is necessary? (Minimum ARPA Network reliability is specified by a two-connectivity constraint, while NASDAQ System requirements are specified for each component within the system.)

What types of computing and terminal equipment are available for the system? (Within the ARPA Network, there are a variety of computers and terminals while the centralized NASDAQ design contains only one main computer center and common concentrators, control devices, and I/O terminals.)

What types of communication modes, such as leased, point-to-point lines, direct-dial, multipoint polled, etc., are feasible? What are the characteristics, such as transmission rates, average downtimes, and error rates, of feasible communication lines? (NASDAQ utilizes both point-to-point and multipoint lines with transmission rates of from 1600–50 000 bits/s. The ARPA Network uses primarily 50 000-bit/s lines, although other speeds are assigned in a few selected places. In both networks, line speeds were chosen to yield required response times at minimum cost.)

Object of Network Optimization

The goal of the network optimization is to provide a "robust" network design which meets all design constraints and has the lowest possible cost. The network design must provide solutions to the following problems.

1) What are the best message formats and transmission schemes?

2) What kind of concentrator, multiplexer, and front-end configurations are needed, and where should they be located?

3) What combination of leased lines, dial-up, point-to-point, multipoint, etc., is to be used, and what are the best line speeds and locations?

4) What is the sensitivity of the design's performance to variations in traffic loads, equipment availabilities, and other parameter changes?

The two most important design constraints for both the NASDAQ System and the ARPA Network are the time-delay–throughput contraints and the network reliability constraints. These can be expected to be major considerations in the design of most significant systems. The time-delay and reliability analyses of networks are difficult problems for which only recently have adequate approaches been developed. We, therefore, summarize some of the pertinent work on these subjects.

Time Delay Models and Computations [31]–[33]

Queuing models for computer networks evolved from similar models for message-switched communication systems which are generalized from the well-established theory of single-server queues. The approach to be discussed is oriented towards the analysis of store-and-forward systems to point-to-point communication lines. Analytic approaches are also possible for multipoint lines.

Queues on a single communication link are characterized by the distribution of times between consecutive arrivals of messages and the distribution of transmission times. The *average delay time T* is most often taken as the measure of performance. The major analytical difficulty of *network* queuing problems is that flows throughout the network are correlated to circumvent this problem; the *length* of a given message can be considered as an independent random variable as it passes from node to node. Although this independence assumption is not physically realistic, it results in a mathematically tractable model which does not seem to affect the accuracy of the predicted time delays [31].

With the independence assumption, the network queuing analysis can be decomposed into solvable single-server problems on a link-by-link basis. Assume Poisson arrival processes for all messages with an average of $\gamma_{j,k}$ messages per second and an exponential distribution of message lengths with an average of $1/\mu$ bits per message. Then, if γ is the sum of the quantities $\gamma_{j,k}$, C_i is the capacity (bits/second) of the ith link, λ_i is its average message traffic (messages/second), and B is the number of links in the network,

$$T = \sum_{i=1}^{B} \frac{\lambda_i}{\gamma} \frac{1}{\mu C_i - \lambda_i} \cdot$$

Moreover, the effect of propagation time and overhead traffic in the actual network can be included to obtain a more detailed equation for average delay [33]. Thus

$$T = K + \sum_{i=1}^{B} \frac{\lambda_i}{\gamma}\left[\frac{1}{\mu' C_i} + \frac{\lambda_i/\mu C_i}{\mu C_i - \lambda_i} + P_i + K\right]$$

where $1/\mu'$ represents the average number of bits in an information message, and $1/\mu$ represents the average length of all messages (including acknowledgments, headers, requests for next messages, parity checks, etc.). Finally, K is the nodal processing time, assumed constant (for the ARPA Network about 0.35 ms), and P_i is the propagation time on the ith channel (about 20 ms for a 3000-mi link).

The main difficulty in applying the above queuing approach is the calculation of the link flows $\lambda_1, \cdots, \lambda_B$. For a tree network, this is no problem since there is one unique path between any pair of points. However, for a distributed design, there may be many alternatives, and hence the queuing analysis can only be applied in conjunction with a routing strategy. We will discuss this problem later in the paper.

Reliability Models and Computations

A simple and natural characterization of network reliability is the ability of the network to sustain communication between all operable pairs of nodes. For example, for design purposes, a network may be required to have at least two independent paths between all pairs of nodes. This insures that at least two nodes and/or links must fail before any pair of operable nodes cannot communicate. This reliability criterion, while conceptually appealing, is independent of the properties of the nodes and links and does not take into account the "degree" of disruption that may occur. Hence it does not reflect the actual availability of resources in the network.

A more refined measure (among others) is the average fraction of node pairs that cannot communicate because of node and link failures. To calculate this measure, knowledge of the element failure rates must be available or estimated. If desired, the availability of specific nodes and the existence of specified communicating node pairs can also be considered using this formulation.

The most detailed level of analysis of reliability incorporates element failures, flow requirements, routing strategies, acceptable delays, and other pertinent network characteristics. At this level of analysis, an appropriate criterion is the expected throughputs of the intact portions of the overall network. Only "brute-force" procedures are presently available at this level of analysis.

A variety of approaches to the analysis problem are possible [20], [52], [53], [57]. The authors believe that purely analytical approaches are computationally intractable for large networks, and, therefore, the best presently available methods employ a combination of analysis and simulation to obtain estimates of parameters such as the probability of the network not being connected or the average fraction of noncommunicating node pairs. Computations for tree networks can be *analytically* performed [30]. Hence we outline several approaches which apply to distributed networks.

Consider the basic analysis problem given below. Suppose a network has N nodes and B links, and each element has probability p of failing. Estimate $h(p)$, the probability of the network being disconnected by the failing components. Alternatively, estimate $n(p)$, the expected number of node pairs not able to communicate, for P values of p. We discuss three approaches to this problem [52]–[54].

The first approach, which we shall call the *naive approach,* is first to generate a random number for each element. If the number is less than p, the corresponding element is considered failed and removed from the net; otherwise it is left in. Then the resulting network is checked for connectivity or the number of node pairs disconnected depending on whether $h(p)$ or $n(p)$ is desired. This computation is repeated enough times to obtain a sufficiently accurate estimate according to the usual standards of Monte Carlo simulation. This entire procedure must then be repeated for each of the P values of p of interest.

The second approach, which we will call the *functional-simulation approach,* is based on a technique used for simulation approaches to percolation problems. Here, as in the naive method, a random number is generated for each element. Now, however, we place these numbers and their corresponding nodes and links in decreasing order. Let r_i be the random number generated for the ith element, and suppose r_{i_1}, $r_{i_2}, \cdots, r_{i_{N+B}}$ are the random numbers in decreasing order. Then, in the naive method, if $r_{i_k} \geq p > r_{i_{k+1}}$, we would analyze for connectivity the subnetwork consisting of the elements $i_1, i_2, \cdots, i_k$. We can evaluate one point in the sample for each value of p for one set of random numbers using the following procedure: for $1 \geq p > r_{i_1}$ the network has no links or nodes; for $r_{i_1} \geq p > r_{i_2}$ the network consists of the element i_1 only; for $r_{i_2} \geq p > r_{i_3}$ the network consists of elements i_1 and i_2, etc. What makes this technique especially effective is that there exist connectivity algorithms which can be used to "update" the analysis of a given network when an additional element is added. Thus for example, if the network with all the components present is analyzed for connectivity, introducing the components in the order $i_1, i_2, \cdots, i_{N+B}$ using one of these techniques, we achieve the analysis for the subnetworks consisting of $i_1, i_2, \cdots, i_k$ for $k=1, \cdots, B$. In particular, we can determine $h(p)$ or $n(p)$ for all P values of p at once using a connectivity algorithm *once* for each sample point rather than P times, as would be required in the naive method.

By a theorem due to Kruskal, each connectivity calculation is equivalent to determining a maximum total length spanning forest where the length of the ith link is r_i. This relation is discussed in detail in a paper by Kershenbaum and Van Slyke [29].

The final technique, which we call the *Moore–Shannon approach,* is based on the use of the following equation (with $q=1-p$):

$$h(p) = \sum_{k=0}^{N+B} C(k) p^{N+B-k} q^k$$

where $C(k)$ is the number of disconnected subnetworks with exactly k elements. This relation is similar to one due to Moore and Shannon [39] for analyzing the reliability of switching networks. A similar relation can be used for $n(p)$:

$$n(p) = \sum_{k=0}^{N+B} D(k) p^{N+B-k} q_k$$

where $D(k)$ is the average number of node pairs not communicating over all subnetworks with exactly k elements. In this approach, each $C(k)$ or $D(k)$ is estimated by simulation rather than attempting to estimate directly the sums $h(p)$ or $n(p)$. It is shown in [53] that this is especially effective for $h(p)$ because all but $B-N+2$ of the $C(k)$ are known *a priori.* There are $\binom{B}{k}$ subnetworks with exactly k elements. If this number is reasonably small, $C(k)$ and $D(k)$ can be calculated by enumeration, and if the number is too large, they can be estimated by simulation. In either case it is useful to use a connectivity algorithm by which, after the connectivity of one subnetwork is determined, subnetworks similar to the first can be easily analyzed using the results of the first analysis. Such an algorithm is described in [54].

The Moore–Shannon approach can be made especially effective for reliability analysis by using stratified random sampling. The strata are defined by the number of links failing *and* the number of nodes failing. Proportional random sampling is then used to divide the effort devoted in the simulation according to the probability of each stratum.

Basic Design Techniques for Centralized Networks

Basic Properties and Heuristic Structures

A good "heuristic structure" may be thought of as any topology which a designer thinks is both "low" cost and feasible. Such a structure may be the result of a "back-of-the-envelope" design or the output of a sophisticated computer-design program. When a designer follows the conventional but naive approach of placing a link wherever there is enough traffic to warrant it, he is creating heuristic structure. Usually, a good designer incorporates whatever theoretical results are available from simpler network models. For example, simple structures for the design of a centralized network are often based on a shortest tree (often called a minimal spanning tree). Such a tree has the smallest total number of miles of communication lines of all possible trees. In general, the use of a tree in a centralized design may not provide the minimum cost solution. However, it can be shown that if all link costs are *convex* functions of capacity (this implies continuous functions), then some tree solution is optimal [58]. On the other hand, the use of a shortest tree connecting all communicating points even under the simplest and most ideal conditions may still not be optimal. This is because one can add a set of S additional points, called Steiner points, to an N-node system and then find the shortest tree connecting the $N+S$ points [25]. If the new points are chosen appropriately, the resulting tree is called a *Steiner minimal tree* and the total length of all lines in the Steiner tree is less than the total length of all lines in the ordinary shortest tree without Steiner points. A simple example of a shortest tree and a Steiner tree is shown in Fig. 3. Physically, a Steiner point corresponds to the location of a multiplexer or concentrator, and although some procedures are available for finding good locations, this problem is presently tractable only by using heuristic procedures. An effective heuristic approach for the problem is described in [6].

As already stated, the most economical topology, even for a centralized network, may not be a tree. In practice, there is a finite set of choices for line capacities, and thus line costs are not convex functions of capacity. In this case, a nontree topology will often be more economical than any tree topology. Fig. 4 illustrates this point with a simple example in which we must connect two points (1 and 2) to a central node 3 with either 19.2- or 50-kbit/s lines. For the flow requirements given, the most economical configuration connecting only the three points shown is not a tree. (This assumes that the facilities at nodes 1 and 2 are capable of relaying the flow.) The nontree configuration is also more reliable since, if any line should fail, the network can still transmit some traffic from both nodes 1 and 2 to the central site.

An interesting and important property for centralized nontree structures is that, given any optimal topology, there exists

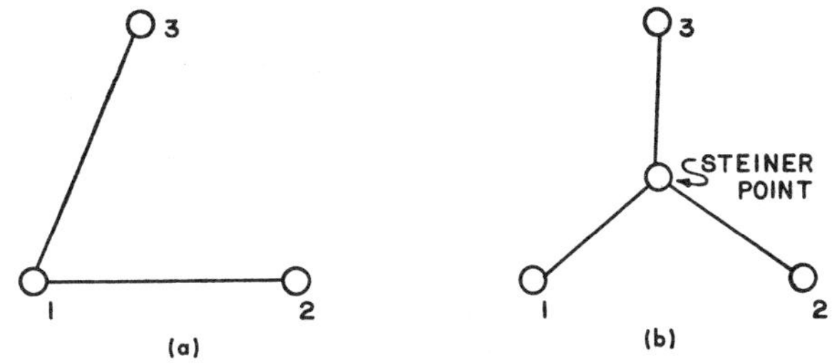

Fig. 3. (a) Shortest tree (minimal spanning tree). (b) Steiner minimal tree.

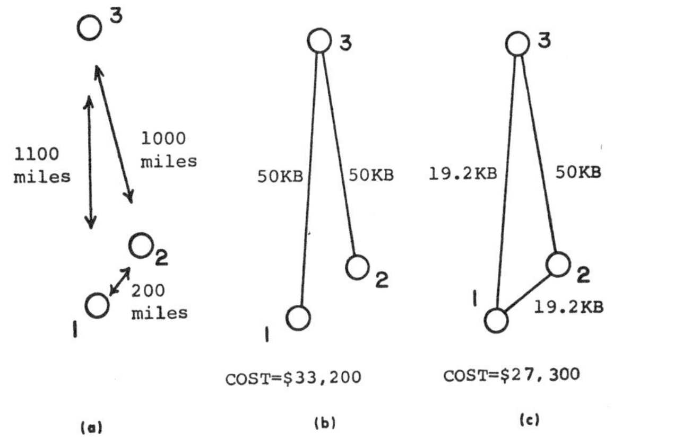

FLOW MATRIX (KBPS)

$$\begin{bmatrix} 0 & 0 & 25 \\ 0 & 0 & 40 \\ 25 & 40 & 0 \end{bmatrix}$$

MONTHLY LINE COSTS ($)

Capacity	Data Set	Per Mile
19.2	850.	7.50
50.0	850.	15.00

Fig. 4. Example of utility of nontree structure for centralized network.

a tree within the network such that all *unsaturated* links *are in* the tree, and, furthermore, all flows to the central node in each link are in the same direction (and vice versa). In a sense this means that the topology departs from a tree structure only when completely utilized links can be added. A complete discussion of this property is given in [49]. Several interesting design procedures could be derived from this work, but these have not yet been investigated.

A consideration in the design of a nontree structure such as the one shown in Fig. 4(c) is the use of split routing. The network shown satisfies flow requirements by sending up to 10 Kbits/s along the path from node 1 to node 2 to node 3 and the remainder of the 25-kbit/s requirement directly from node 1 to node 3 through the direct 19.2-kbit/s link. In a general network, with flow requirements between many pairs of nodes (rather than to and from only the central node), a distributed design may be best, but split routing may not be any more economical than sending all flow between any pair of nodes over a single path. If the cost of installing channel capacity is convex, this can be shown to be the case [23]. (A special case of this is when the cost of installing C units of channel capacity is linear.) Thus if $f(C)$ is the cost of installing C units of channel capacity, split routing gains nothing if

$$f(C+2) - 2f(C+1) + f(C) \leq 0.$$

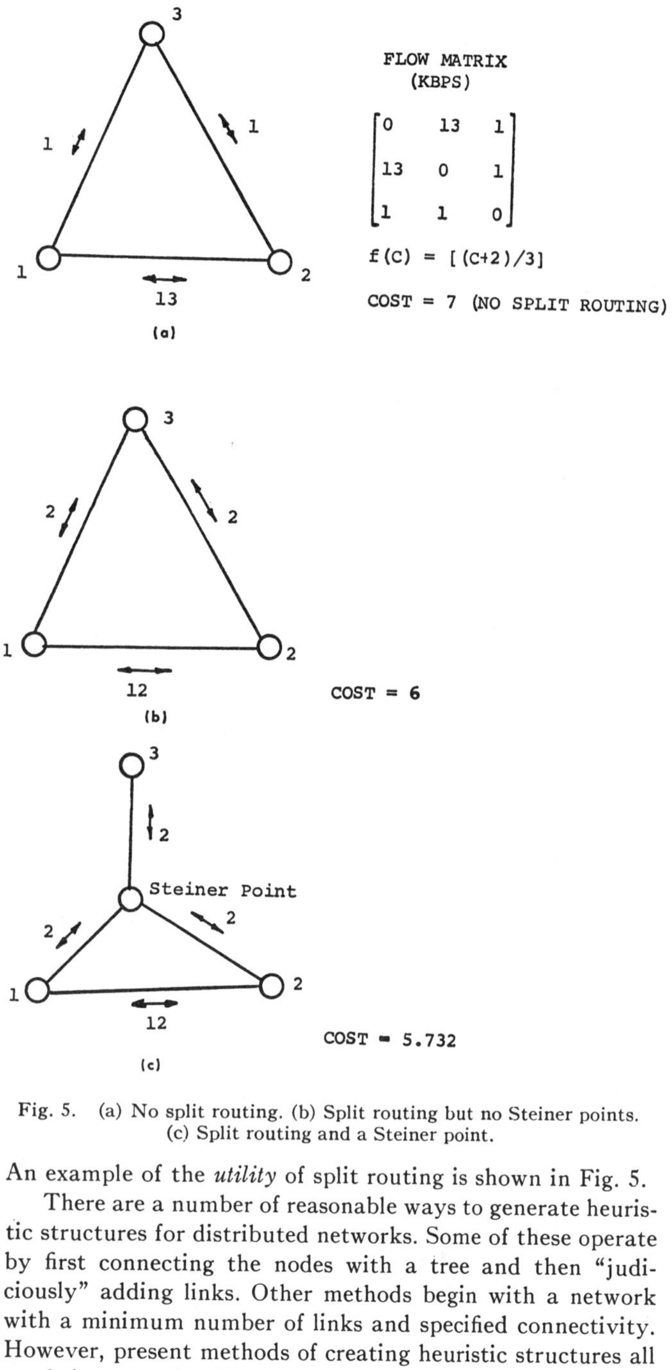

Fig. 5. (a) No split routing. (b) Split routing but no Steiner points. (c) Split routing and a Steiner point.

An example of the *utility* of split routing is shown in Fig. 5.

There are a number of reasonable ways to generate heuristic structures for distributed networks. Some of these operate by first connecting the nodes with a tree and then "judiciously" adding links. Other methods begin with a network with a minimum number of links and specified connectivity. However, present methods of creating heuristic structures all result in networks that can be considerably improved by the use of iterative optimization techniques, and hence we will postpone our discussion of distributed network design until the next section.

Design via an Interactive Terminal [3], [4], [7], [12], [42]

One of the most routine ways of enhancing human design capabilities has been the development of numerous "design" programs which enable a design to interact with a time-sharing system via a Teletype or graphics terminal. Although the computer *per se* does only a minimal amount of design, the human effort is reduced by the designer rapidly being able to test various alternatives.

A typical interactive design program accepts as input parameters such as the node locations and names, the traffic requirements and message-length statistics, and, optionally, line locations or capacities. If line locations are not input, the computer might generate a shortest tree. In application-oriented packages, the program will allow the designer to include concentrators and multiplexers in his design.

The designer would then use his terminal to modify the initial configuration. After each modification the computer would recalculate the cost for the current design. With a graphics terminal and light pen, the typical program gives the designer the following capabilities.

1) A link can be added by indicating with the light pen the two nodes to be joined.

2) A link's parameters can be changed by using the light pen to indicate a parameter set to be changed and using the keyboard to type new parameter values.

3) A link can be deleted by using the light pen.

4) The position of a node can be changed with the light pen, first indicating a new position, and then the node to be moved.

Most interactive design programs are proprietary, and, therefore, details are not available. However, nearly all presently available interactive programs known to us perform virtually no significant analysis except for cost calculation during the *design* process. A few programs allow for a network simulation either before or after a design step and some calculate simple graph theoretic properties such as the radius and average path length.

There is a useful place for interactive program packages of the type described. These programs can eliminate many of the difficult jobs of finding tariffs and costing systems and finding "reasonable" networks. However, the current interactive approach is highly inferior to iterative schemes using remote batch techniques. Some simple iterative methods are described next.

Iterative Techniques [8], [9], [16], [17], [24], [50], [56]

Iterative techniques for computer-network design involve the heuristic application of a family of optimization procedures called "node" or "link" exchanges. Exchange methods are search procedures which optimize network structure by changing small sections of a large network at a time. The methods begin with a network satisfying all constraints either selected by a human or generated by a different computer-design program. Exchange methods are then applied to the starting network until a new network which both satisfies all constraints and has lower cost is found. The process is then repeated until cost can no longer be lowered. The network resulting is then said to be "locally optimal." The process may then be repeated using either a new starting network, a different class of exchange methods, or the same methods in a different order. To illustrate the types of exchange methods that can be employed, we discuss the design of centralized networks using several different methods. We defer our discussion of distributed nets until we examine several key problems in the design of such networks. In addition to topological specification, link capacities must be selected for the network design. If time delay is the dominant constraint in capacity selection, a method for optimally choosing from an available set of options exists [16]. This procedure can be used in conjunction with any of the following methods.

Multipoint Path Creation: One of the earliest exchange methods, developed by Esau and Williams [9], forms the basis of the IBM proprietary program CNDP. This method, as initially developed, applies to a specialized class of centralized networks which has links with only one line-transmission

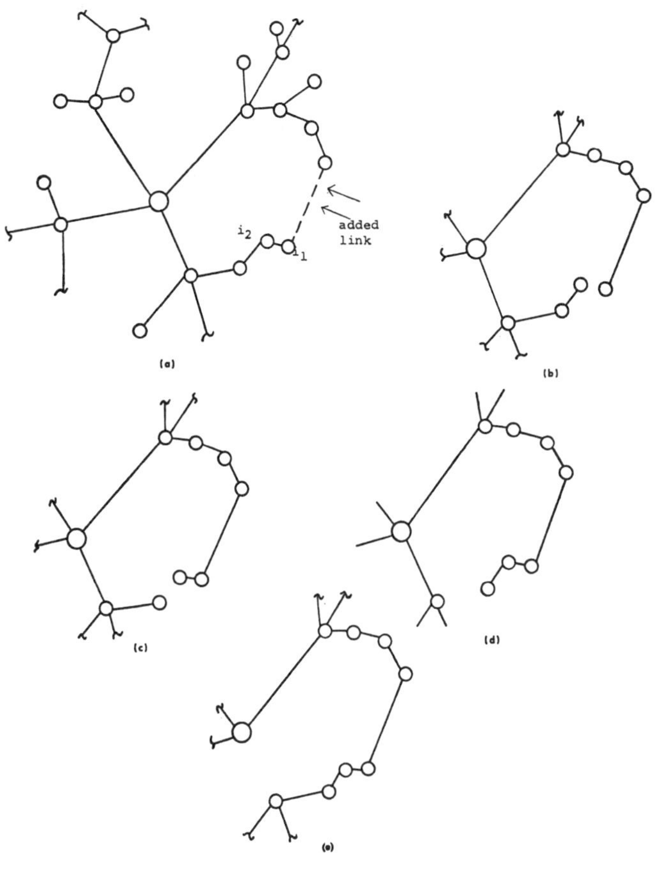

Fig. 6. Illustration of a simple branch exchange procedure.

rate. Initially, a star network of links from every terminal to the central node is chosen. At successive iterations, one of these links is removed and the terminal thus disconnected is placed on an established path to reduce costs by creating a multipoint circuit. To determine allowable link exchanges, the amount of traffic to be accommodated is checked first. The tradeoff value for performing each feasible exchange is calculated, and the one with the greatest gain is chosen.

A Simple Path Exchange [50]: The procedure begins with any tree, and thus each node has a unique path to the central node. Improvements are attempted by exchanging nodes on different paths. The algorithm first uniquely orders the paths as $\pi_1, \pi_2, \cdots$, and then begins by considering the farthest node, say n_1, from the central site on the first path π_1. The closest node to n_1 on a different path, say π_j, is added to the first path if the total cost is reduced and the exchange is feasible. If infeasible, then the first combination of nodes on π_1 closest to the central node whose total number of data terminals is equal to the number of data terminals of the added node is removed from π_1 and added to the path π_j. The next farthest node from the center on the first path is then considered in the same manner. When all nodes on the first path have been examined, that path is excluded from consideration. The second path π_2 is then treated, and so on.

Circuit Deletions [16]: An effective local optimization procedure operates by starting with any tree and then applying a special kind of elementary tree transformation. It can be shown that any tree can be obtained from any other tree by a sequence of such transformations. The transformation used is as follows: For a given tree, choose a node i and find the node i_1 closest to i but not already connected to i. Add link (i_1, i) to the tree and identify the circuit formed. Suppose that this circuit consists of links (i_1, i), (i, i_2), (i_2, i_3), $\cdots$, (i_j, i_1). New trees are formed by deleting in turn links (i, i_2), (i_2, i_3), $\cdots$, and finally (i_j, i_1) as illustrated in Fig. 6. Each time a link is deleted, the optimal-link-capacities algorithm is applied to determine network cost. Each node i is scanned in turn; link additions from i to its d nearest neighbors are considered. Whenever L lower cost trees have been generated, the node scan is begun again. (For most problems, $d=3$ and $L=1$ yield excellent results.) This procedure has proven to be extremely powerful in finding low-cost trees. Whenever the problem has been small enough to exhaustively find optimal trees, the method has converged to the optimum solution. In addition, tests in which link cost is set equal to link length have shown that the procedure can readily converge to the shortest tree solution.

The Capacity-Optimization Problem

One of the most difficult of all network design problems is the selection of link capacities from a *finite* set of options. Although there are many heuristic approaches to this problem, rigorous results are scarce. For most existing work, the objective is either to minimize cost subject to a fixed average time-delay constraint or minimize average time delay (or some function of the average delay) subject to fixed cost constraint.

If the object is to minimize cost subject to a specified *maximum* allowable average time delay, an *optimum* procedure exists for centralized tree networks [16]. If the object is to minimize average time delay at a fixed cost (or the sum of the kth power of average delay on each link), analytic results are available if link capacities are continuous variables [33], [38]. For distributed networks the problem is far more complex than for tree networks since the major variable affecting a link capacity choice is the link flow. A tree has only one path between any pair of points, so that link flow is known *a priori*. For a distributed design, link capacities and link flows interact since a good routing algorithm will tend to send flow around saturated portions of the network. Thus link capacity selection and the network routing should be simultaneously considered. Because of this difficulty, there are no *practical optimal* results for distributed networks, although many seemingly "good" heuristics are available. Thus to illustrate the capacity selection problem, we examine centralized nets where we can compare heuristics to known optimum solutions.

The tree in Fig. 7 has had its capacities optimally selected from a finite set of options with realistic data set and line costs. The system was designed to produce a *maximum* one-way average delay of 0.3 s from any node to the central node (node 1). The overall monthly network line cost is 21 775 dollars. The capacities resulting from the best continuous linear approximation are indicated in the same figure by encircled numbers. These correspond to the capacities chosen to minimize average delay for an expenditure of 21 775 dollars. For both sets of capacities, the average delay is 0.19 s.

The limitations of the continuous selection method based on *average delay* can be seen by examining the decisions made for several links by each algorithm. Thus for the link (7, 15) the optimal discrete choice was 50 kbits/s, while the continuous choice was 37.8 kbits/s. (This would become 50 kbits/s if capacity were rounded up to the nearest available capacity.) Both methods recognized that this link capacity could be inexpensively made very large to make its delay contribution negligible. However, for another similar situation, link (2, 16), the optimal discrete capacity was chosen to be 4.8 kbits/s,

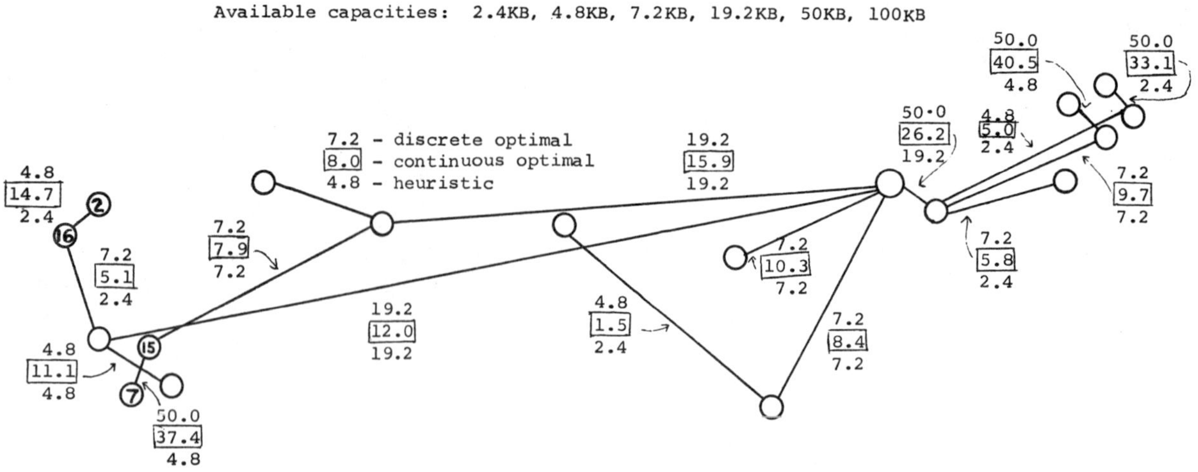

Fig. 7. Discrete optimal: Maximum average delay = 0.3 s. Discrete and continuous optimal: Average delay = 0.9 s. Cost = 21 775 dollars. Heuristic: Maximum average delay = 1.4 s. Average delay = 0.5 s. Cost = 17 692 dollars.

while the continuous capacity selected was 14.7 kbits/s. (Rounded up, this becomes 19.2 kbits/s.) The optimum discrete decision recognized that this link delay contribution was not critical, even though its capacity could be inexpensively increased. The continuous choice was not able to recognize this fact. Finally, to contrast the results of optimization to the results of the application of simple heuristic, we assign discrete capacities according to the following rule: Set the capacity of a link equal to the smallest capacity that can accommodate the flow in the link. With this decision rule, the network line cost is reduced to 17 692 dollars per month. (This selection criterion produces a *stable* system whose set of capacities has the smallest possible cost.) However, the maximum average delay from any node to the central node is now 1.4 s, and the overall average delay is 0.5 s.

IV. Distributed Network Design Problems

Imbedded in the centralized network problem are a number of simplifications to the general network design problem. In a centralized network, after one specifies line speeds, transmission schemes, and traffic statistics, time-delay analyses can be performed. However, in a distributed design, these data are not sufficient since we must also specify how flow is to be routed through the net. Using current heuristic exchange techniques for minimum-cost design, the major analysis problems one encounters in feasibility analysis are routing (and delay) and reliability analysis.

The Routing Problem [1], [13], [18], [21], [22], [27], [58]

A good routing procedure must 1) make full use of available line capacities; 2) be computationally efficient for repeated use during an iterative design process; and 3) exhibit the same characteristics as the one implemented in the operating network.

If the objective is to minimize average delay, the routing problem is a nonlinear multicommodity flow problem [15]. This problem [13], [22], [58] can be formulated as a separable convex programming problem, and either minimum delay or maximum ideal throughput can be achieved if the routing procedure follows the solution of the programming problem. However, for networks with more than a few nodes, this approach is prohibitively expensive for repeated applications of the routing algorithm during the design stage.

A proven heuristic procedure is to route flow over the least utilized paths with the minimum number of intermediate nodes. This approach generally gives a result within 5–20 percent of the optimum. In addition to being fast (over three orders of magnitude faster than the programming approach) specialized procedures can be derived with this heuristics that facilitate local optimization of the network structure. However, the basic method does not take full advantage of split routing and significant variations in link capacities, and leaves room for considerable improvements especially in the case when the network contains a wide distribution of different line capacities (an apparently desirable characteristic of very large networks).

A more efficient routing procedure, called *cut saturation*, is a generalization of the above basic method which allows split routing and the utilization of paths with more than the minimum number of nodes [18]. As much flow as possible between all node pairs is first simultaneously routed over the paths with minimum numbers of nodes. (These paths are "shortest paths" using a simple unit metric for each line.) When no shortest path with excess capacity is available, the "saturated" lines are deleted from the network representation and flows are then routed over the shortest paths of the remaining subnetwork. The process is continued until the network representation has had enough lines removed to become disconnected. By this technique, a message is sent down a path with fewest intermediate nodes and excess capacity or, when that path is filled, the one with next fewest intermediate nodes and excess capacity, etc. The computational complexity of this procedure is of the same order of magnitude as the basic heuristic approach. Yet its results are extremely close to the optimal solution during heavy traffic. Furthermore, the routing strategy is very similar to physically implementable and desirable routing schemes for distributed networks such as the ARPA Network, and, both experimentally and by simulation, leads to realizable network performance.

As an example, the 23-node network shown in Fig. 8 has five saturated links. Five iterations were required to obtain these flows before a cut was saturated and the network was disconnected. The performance of the network after each iteration is illustrated by a graph of throughput versus delay in Fig. 9. Graphs illustrating computation time per iteration as a function of network size are shown in Fig. 10. The rela-

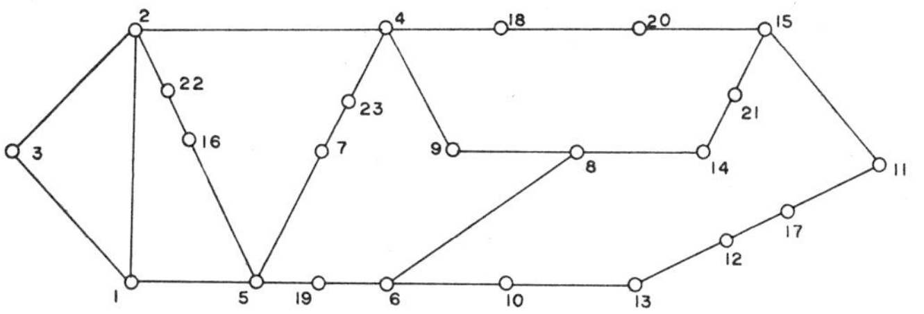

Fig. 8. Network use for example (all links have 50 000-bit/s capacity).

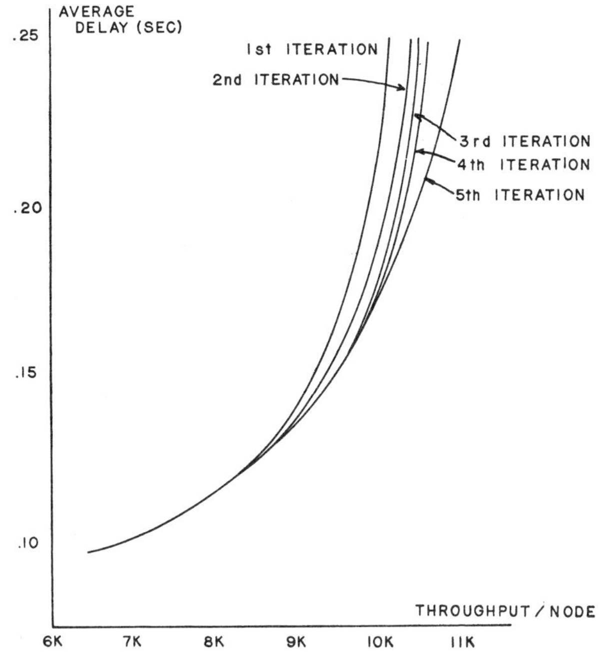

Fig. 9. Throughput versus delay using cut saturation (routing for network shown in Fig. 8).

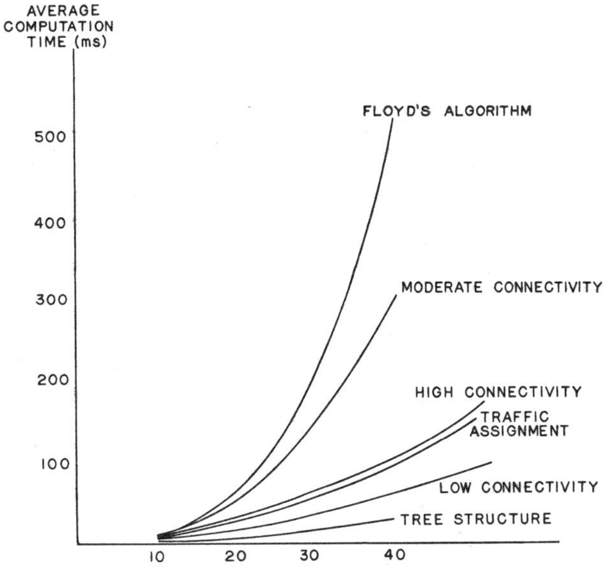

Fig. 10. Computation time versus network size and connectivity for new shortest-route algorithm and for Floyd's algorithm.

tionships shown are for specially devised algorithms described in detail in [41].

The basic limitation, and a fundamental open problem of present routing procedures, is that the performance of a particular routing strategy is critically dependent on the capacities assigned to the links. In many procedures either these capacities are assigned before the routing is performed, or a routing is first performed and capacities then assigned. The difficulty with such approaches is that *a priori* the routing-capacity assignment combination is biased, and often any chances of affecting economies are reduced. Thus a major outstanding problem is that no general procedures are yet known for assigning capacities from a wide range of options while simultaneously optimizing the routing used in the design procedure.

The Reliability Problem

An essential characteristic of any good network design is that it should not suffer significant degradation in performance if some elements fail. With the state of current technology, both nodes and communication links have nontrivial downtimes. Therefore, the network design must provide for these failures by having sufficient alternate paths to satisfy flow requirements and time-delay constraints. For example, the NASDAQ Network is made reliable by installing parallel lines and extra concentrators, while the ARPA design utilizes distributed control and at least two independent paths between each pair of nodes.

The network reliability problem has two aspects—analysis and design. Reliability analysis for centralized networks is routine. For distributed networks, it has reached the stage where large networks can be handled. Preliminary studies indicate that for large networks, reliability may be the dominant design constraint. For example, the graphs in Fig. 11 correspond to reliability analyses of 2-connected low-cost networks of 20, 40, 60, 80, 100, and 200 nodes. These networks meet all delay and throughput constraints. However, as the number of nodes increases, the probability that all nodes can simultaneously communicate continuously decreases, while the average fraction of communicating node pairs decreases for nets with more than 80 nodes.

The relationship between distributed network connectivity and availability criteria such as "average fraction of communicating node pairs" is not yet completely understood, although the problem is being studied intensively. The design problem is to identify link locations so that given reliability constraints as well as throughput and delay are satisfied. If one were to conclusively relate network connectivity to the availability constraints, then a family of proven link exchange procedures could be applied to the problem. Thus a new problem with the following formulation could be solved.

Assume we are given N geographically located nodes and a cost matrix $C=[c_{i,j}]$, where $c_{i,j}$ is the cost of constructing link $[i, j]$. We would like to construct a minimum-cost network such that there are at least $r_{i,j}$ node disjoint paths between nodes n_i and n_j, where $R=[r_{i,j}]$ is a specified redundancy matrix.

This design problem can be formulated as an integer-programming problem, but the number of constraints is astronomical even for relatively small networks. A tractable approach is to search for low-cost rather than minimum-cost networks, using link-exchange methods like the ones previously described. Initially, we generate a network satisfying all connectivity constraints using either deterministic procedures such as the ones described in [20, section IV.A.3] or randomized procedures such as those described in [51].

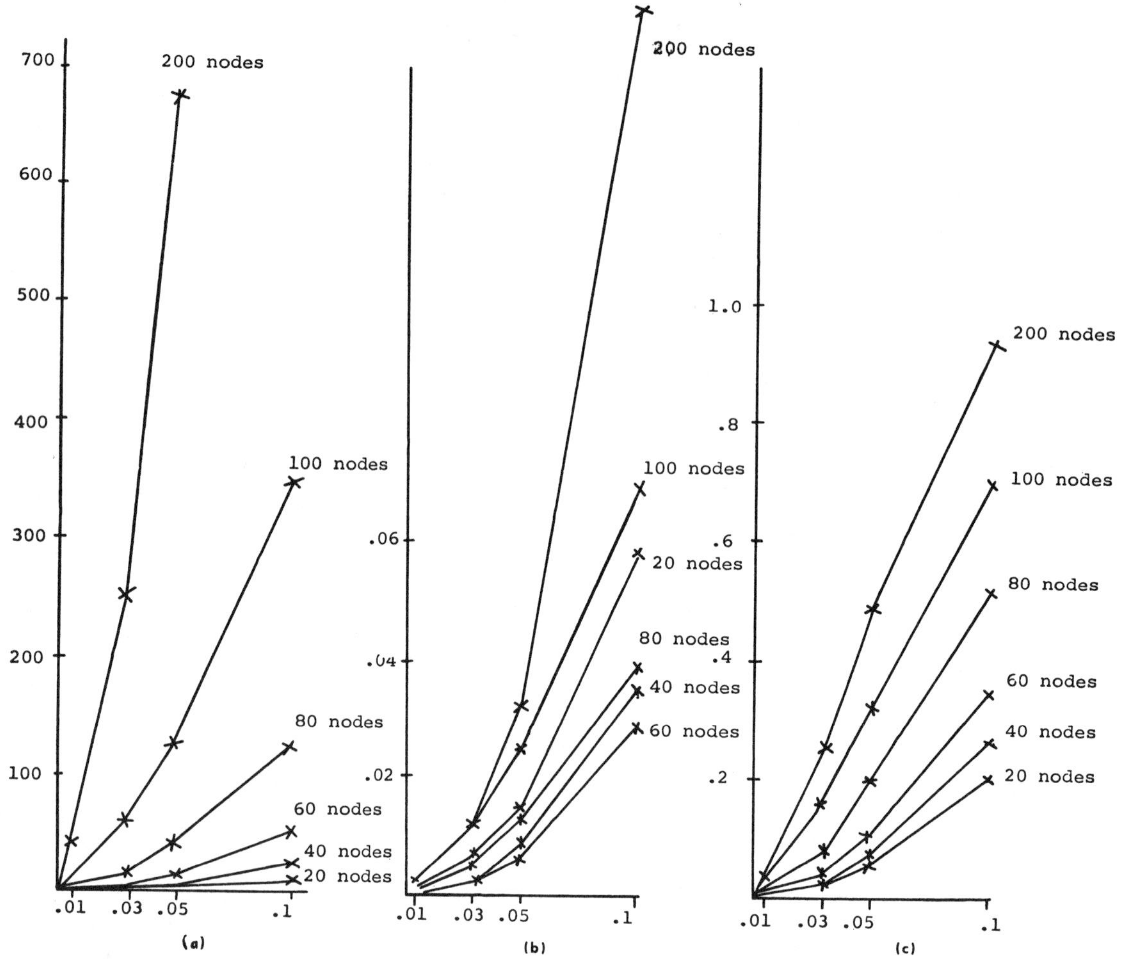

Fig. 11. Network reliability as a function of number of nodes. (a) Average number of node pairs not communicating versus probability of link failing. (b) Average fraction of node pairs not communicating versus probability of link failing. (c) Probability of net being disconnected versus probability of link failing.

The optimizing routine searches for networks generated by link exchanges of two links which have lower cost and are also feasible. Feasibility analysis now requires general connectivity analysis as well as the already required throughput and time-delay analyses. A powerful local transformation for this problem is a link exchange such that if the net has two links $[i, k]$ and $[j, h]$ such that $[i, h]$ and $[j, k]$ are not in the net, a new network can be generated by removing $[i, k]$ and $[j, h]$ and adding $[i, h]$ and $[j, k]$. If

$$c_{i,h} + c_{j,k} < c_{i,k} + c_{j,h}$$

the new network has lower cost than the original one and it must be analyzed for feasibility. Testing connectivity feasibility can be economically done because of the following fact: If a link exchange on a feasible network which results in a network G^* destroys feasibility by reducing the redundancy between nodes n_a and n_b below the requirement $r_{a,b}$, and if the link exchange resulted in the removal of links $[i, k]$ and $[j, h]$, then either $w_{i,k}(G^*) < r_{a,b}$ or $w_{j,h}(G^*) < r_{a,b}$. This reduces the number of computations required, and, in addition, highly efficient methods for performing this computation are available [20]. For example, if one wishes to check a network for k-connectivity, it is only necessary to make connectivity computations at k nodes in a sequence of progressively simpler network structures [34]. For example, to test a network for 4-connectivity, we would only have to perform the following operations:

Choose any node and verify that there are at least four node-disjoint paths from this node to every other node. Remove the node and all the branches connected to it. From the resulting network choose another node and verify that there are at least three node-disjoint paths connecting this node to every other node. Remove this node and its branches, choose another, and then verify that there are at least two node-disjoint paths from it to every other node. Choose a fourth node. Verify that there is at least one path from it to every other node. These are the only calculations that are required. If the number of disjoint paths between each pair of nodes had been only three, then at the last stage of analysis it would have been impossible to verify that the required number of paths existed. If the new network has lower cost and is feasible, the link exchange is adopted and the search for additional improvements continues. When no additional favorable exchanges can be found, the process terminates. After a locally optimum network is obtained, a new starting network is generated (if one can be found), and the optimization procedure is repeated.

Iterative Techniques for Distributed Network Design [13], [17], [24]

The basic methods for design of a distributed network follow the same patterns as the iterative centralized net-design techniques. After specifying a feasible "starting" network, the computer iteratively modifies the network in simple steps

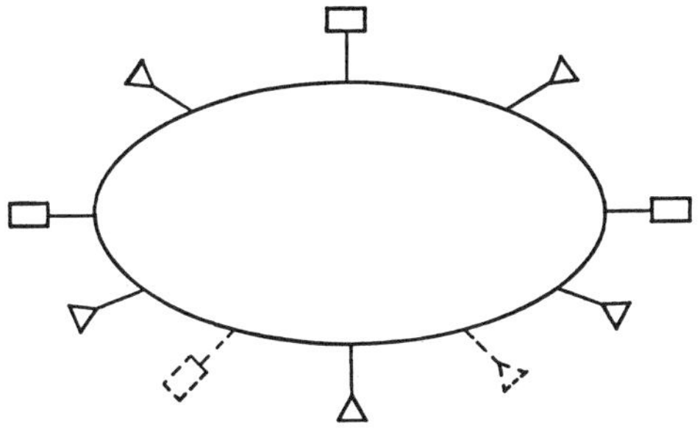

Fig. 12. A looped communication system.

until a lower cost network satisfying all constraints is found. This network becomes a new starting network and the process is continued.

A successful set of modifications used for the ARPA Network design has been a class of link exchanges involving the removal or addition of one or two links at a time. Links may be added or deleted anywhere in the networks. However, regardless of the number of links simultaneously involved in a local modification, all transformations can be formulated in terms of the removal or addition of a *single* link at a time; that is, if a modification adds two links and removes three, an equivalent set of modifications adds two links one at a time and then deletes three links, also, one at a time. Thus any set of link modifications can be considered to be composed of these two *elementary* modifications. The importance of considering such elementary transformations stems from the increased efficiency that can be obtained during each step of the optimization process. Whenever links are added or removed, traffic must be rerouted in order to reevaluate network throughput and delay. During the optimization process, it is usually adequate to route all flow over paths containing the minimum number of nodes, leaving further increases in throughput to later design stages. In this case, after a link addition or deletion, it is not necessary to reroute all traffic completely to compute the new network throughput. Note that in this case, the capacity of a link is either preassigned or set equal to the smallest capacity that will accommodate the link flow required to meet the network throughput specifications.

A simplification of the general design problem occurs in the case of *looped* networks [10], [11], [43], [55], [59]. Such a system usually consists of a set of buffered terminals and one or more computers linked by means of a single circuit, as shown in Fig. 12. The topological design problem for such a system consists in finding a minimum cost routing for the communication circuit. If the number of terminals and computers on the circuit is fixed, then design may be approached in two independent steps: 1) selection of the transmission rate of the circuit based on queuing considerations, and 2) optimization of the layout of the circuit. In this case, the layout problem is equivalent to the "traveling salesman problem" [35], [36], and can be effectively solved by using link exchanges. On the other hand, multiple loop systems in which the locations of the devices on each loop are to be selected represent "clustering" problems similar to those discussed later in this paper. Complete solutions for these problems are not yet available.

An example of distributed network optimization is shown in Fig. 13(a) and (b). The starting network is the one given in

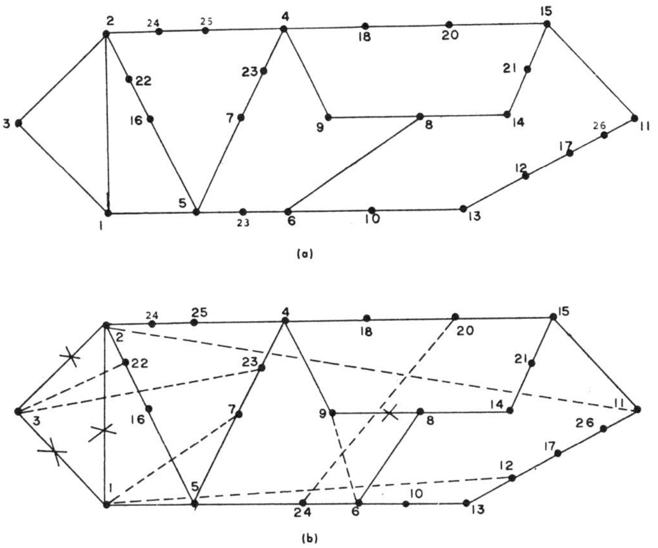

Fig. 13. Example of optimization procedure for a 26-node network. (a) Initial network. (b) Dashed lines indicate added links; "x-ed" lines indicate deleted links.

TABLE I

26-Node Optimization

Throughput (kbits/s/node) (Minimum Node Routing)	Yearly Line Cost (thousands of dollars)	Line Cost per Node	Line Cost per Megabit (cents)	Network Description	
				Added	Removed
8.6	883	34.0	12.48	network shown in Fig. 13	
8.6	885	32.9	12.20		(1,2)
10.5	991	38.0	11.60	(1,12)	
13.0	1129	43.5	10.07	(2,11)	
14.0	1144	44.0	10.00	(3,23)	
16.0	1173	45.1	9.00	(1,7) (3,22) (9,6) (24,20)	(1,3) (3,2) (9,8)

Fig. 13(a). Link additions and deletions are indicated in Fig. 13(b), and the associated throughputs and cost per megabit are listed in Table I. Note that, as described before, minimum node routing is used instead of the more optimal routing discussed in the last section. If this routing strategy had been used, for example, for the network in Fig. 13(a), a throughput of 9.9 kbits/s/node (i.e., about 15 percent higher) would result.

As the size of the network increases, the greater is the cost to perform optimization by link exchanges. Decomposition of the network design into regions becomes necessary, and additional heuristics are needed to determine effective decompositions. It presently appears that these methods can be used to design relatively efficient networks with general structures containing a few hundred nodes, while substantially different procedures will be necessary for networks of greater size.

An alternative iterative approach which only uses link exchanges in a limited sense can be developed from some basic work developed for the optimization of communication networks for "Telpak" and private line tariffs. This approach is described below.

Telpak Network Design [24]

A Telpak is an AT&T tariff that allows a user to purchase either 60 or 240 service-equivalent lines at substantially reduced rates. The user must pay for all 60 or 240 lines even if he requires fewer lines to meet his needs. The Telpak design problem is then to find the most economical arrangement of Telpaks and private lines to meet a set of flow requirements. The following method was derived for optimization of Telpaks in large (3000–4000 points) networks. However, the principles are directly applicable to computer networks. The first step to achieve an economical overall design is to generate as a starting network a simple structure connecting all nodes. A subset of key points spread over the entire country is then chosen, and a series of operations involving a relatively small number of nodes at a time are then performed to improve routing for requirements including a number of key points. After this has been accomplished, sections of the structure are sequentially treated in the same way.

The starting network for each section is a tree (derived from but not identical to a shortest tree). This establishes a unique route between each node pair. Links are then added to provide alternate paths, and all requirements are then routed along the path of shortest length. Additional links, called *longlines*, are then added. For these links, flows are only allowed to be a multiple of the capacity of the highest rate line available (for Telpak, a multiple of 240 voice-equivalent circuits). The longlines are thus all fully utilized links. (For a data link, we might define "full utilization" to be, say, 95 percent of line capacity.) After adding longlines, there are many alternate paths between each pair of points, although most requirements are routed along the shortest ones. For each requirement between a pair of nodes, an attempt is made to find less expensive although possibly longer paths by 1) reducing the flow along a path by the node pair requirement, 2) calculating the additional cost for this number of lines for each link in the network, and 3) using these incremental costs, finding the least cost path and then routing the requirement along it. Finally, a triangle optimization examines every pair of links adjacent at a common node. For example, for flow on links (i, j) and (j, k), it may be less expensive to route some flow directly from i to k. A simple calculation can determine the most economical routing among these points and the local optimization can then be performed.

The principal limitations of the aforementioned approach directly stem from the computation time required to optimize a large network. During the procedure, all flow paths are always known, and throughput and time delay can be routinely calculated. For a network with several hundred points, the method can be refined to produce better results, while for a network with several thousand points, careful attention must be given to the programming aspects of the basic method to keep computation times low enough for the procedure to be practical.

Large Network Design Problems

The outstanding design problems for large distributed networks is to specify their routing and topological structures. This specification must make full use of a wide variety of circuit options. Preliminary studies indicate that, initially, the most fruitful approaches will be based on the partitioning of the network into regions or, equivalently, constructing a large network by connecting a number of regional networks. For optimization of these networks, link-exchange methods are unfeasible unless such decompositions are used, since exchange methods are search procedures, and for network designs incorporating realistic constraints on routing, throughput, delay, cost, reliability, and *physical implementation*, the complexity of the computation is on the order of N^3 to N^6, where N is the number of nodes. For networks with more than a few hundred nodes, each with its own processing capability, present link-exchange procedures fail without network decomposition because of the large amounts of computer time needed to perform the optimization. In addition to reducing the computational complexity of the topological design problem, nodes may be clustered into regions for numerous reasons, such as: a) to partition status information for use in routing, flow control, and other decision processes within the operating network [27], [28]; b) to determine regions of low-, medium-, and high-speed lines in hierarchal structures; and c) to find concentrator–multiplexer locations. For example, the NASDAQ Network connects over 2000 points and is itself a decomposed system since the concentrators create a natural partition of the nodes. However, in this case, since most nodes are unintelligent *terminals* rather than computers with local control capabilities, design problems are not grave.

In distributed networks constructed by connecting together a set of regional networks, to send a message, a node might specify both the destination region and the destination node in that region. No detailed implementation of a large network has yet been specified, but early studies of their properties indicate that factors such as cost, throughput, delay, and reliability are similar to those of the present ARPA Network, if the ARPA technology is used [14].

One of the first questions that arises in the decomposition problem is how large to make the regions and how many regions to create. (In a multilooped network, each loop would define a region.) Once the number of regions has been determined, nodes must be assigned to each and central nodes selected. This problem is extremely difficult, and little is presently known. In order to create effective partitions, we must solve a variety of "clustering" problems. The literature on clustering and partitioning is large but fragmented and spread over many domains including information retrieval, taxonomy, and networks. A potentially valuable research area is the application of known clustering techniques to computer networks. For example, Bahl and Tang have effectively used such techniques for optimization of concentrator locations in centralized networks [2].

Before any substantial progress can be made in applying clustering techniques to large computer networks, an appropriate "distance" or "nearness" measure must be selected, probably on the basis of intuition and experiment. One can see the importance and difficulty of this step by considering an analogous problem in information retrieval. Suppose book titles are stored on a disk file for reference. If, after title A is selected by a reader, there is a high probability that title B will be requested, both titles should be stored in the same record so that overall average disk-access time is minimized. The difficult part of the problem is to determine the conditional probability that B will be referenced if A is referenced. In other words, we must assign a "distance" between A and B for the clustering algorithm. If this step is done poorly, any clustering algorithm subsequently applied will not operate well.

The same basic problem applies to using clustering techniques for the design of large computer networks. Here clustering requires the assignment of distance measures to take into account cost, capacity, traffic, delay, reliability, and routing. Almost no general theoretical results are presently known for this problem.

One approach to determining the size of regions for the decomposition problem is to attempt to minimize the computation requirements for the design procedure. Suppose we have a procedure to perform on a network for which the computing time can be characterized as a polynomial function of the number of nodes. Examples are link-exchange algorithms and many connectivity and reliability analysis procedures. Suppose further that the procedure can be carried out by performing the procedure separately on each region of the decomposed network and then performing the procedure once on the global network.

For simplicity, suppose we seek a two-level hierarchal decomposition. The nodes are divided into groups called *regions*. Nodes which are end nodes of links connecting two regions are called *central* nodes. A regional subnetwork or a *local network* is the nodes of a region together with the links which connect two nodes in the region. Let N be the total number of nodes, NR the number of regions, and NG the number of central nodes per region in the global network. We assume that the number of nodes is the same for each region as is the number of central nodes and that the computing time for a subnetwork with N nodes is proportional to N^v for some $v>1$. While these assumptions are not necessary, the conclusions based on them are indicative and the techniques used can also be applied to more general cases.

Under these assumptions the total computation time is proportional to

$$T = NR\left(\frac{N}{NR}\right)^v + (NR\cdot NG)^v$$

where the first term corresponds to the regional calculation and the last term to the global calculation. Assuming that NR and (N/NR) (the number of nodes per region) can take on noninteger values, we can determine the number of regions which corresponds to minimum computer time by setting $\partial T/\partial NR$ to zero. This implies

$$NR = \left(\frac{N}{NG}\right)^{v/(2v-1)} \left(\frac{v-1}{v}\right)^{1/(2v-1)}$$

The asymptotic result as $v \to \infty$ which is of interest is

$$\lim_{v\to\infty} \left(\frac{N}{NG}\right)^{v/(2v-1)} \left(\frac{v-1}{v}\right)^{1/(2v-1)} = \sqrt{\frac{N}{NG}}.$$

This result can also be obtained by noticing that for large v, total computer time is dominated by the subnetwork with the largest number of nodes; it is then easy to convince oneself that the minimum computation will occur when the global network has exactly as many nodes as the regions, i.e., when $NR\cdot NG = N/NR$. Solving this for NR yields $NR = \sqrt{N/NG}$.

Conclusion

The modeling, analysis, and topological design areas for computer networks have all made significant progress in recent years. General network structures with several hundred nodes can now be handled, while specialized structures (such as centralized trees) with thousands of nodes are tractable. Effective design for small networks of, say, less than about 25 nodes, is routine using the principles discussed in this paper, while design of networks with appreciably more nodes requires a high degree of both analytical and programming skill. Among the major open problems for large network design using present design philosophies are the problems of clustering and partitioning. It appears that these problems will be well in hand in the near future, but that new design philosophies will have to be developed for general network structures with significantly more than one thousand nodes.

References

[1] P. Baran, S. Boehm, and P. Smith, "On distributed communications," Series of 11 Rep. by Rand Corp., Santa Monica, Calif., 1964.

[2] L. R. Bahl and D. T. Tang, "Optimization of concentrator locations in teleprocessing networks," presented at the Symp. Computer Communications, Apr. 4–6, 1972.

[3] A. Battersby and M. L. Berners-Lee, "Communication through interactive diagrams," in *Computer Graphics in Management: Case Studies of Industrial Applications*, R. E. Green and R. D. Parslow, Eds. London: Gower, 1970.

[4] R. L. Brewster, "Designing optimum data networks," *Data Process.*, pp. 23–26, Jan. 1971.

[5] S. Carr, S. Crocker, and V. Cerf, "Host-host communication protocol in the ARPA network," in *Spring Joint Computer Conf., AFIPS Conf. Proc.* Washington, D. C.: Spartan, 1970, pp. 589–597.

[6] S. K. Chang, "The generation of a minimal tree with a Steiner topology," presented at the Symp. Computer Communications, Apr. 4–6, 1972.

[7] R. Citrenbaum, "An on-line model for computation-communication network analysis and modification," SDC, Santa Monica, Calif., Tech. Memo. 4743/003/000, May 11, 1971.

[8] D. R. Doll, "Topology and transmission rate considerations in the design of centralized computer-communication networks," *IEEE Trans. Commun. Technol.*, vol. COM-19, pp. 339–344, June 1971.

[9] L. R. Esau and K. C. Williams, "On teleprocessing system design—Part II: A method for approximating the optimal network," *IBM Syst. J.*, vol. 5, pp. 142–147, 1966.

[10] D. J. Farber and K. Larson, "The structure of a distributed communication system," presented at the Symp. Computer Communications, Apr. 4–6, 1972.

[11] W. D. Farmer and E. E. Newbell, "An experimental distributed switching system to handle bursty computer traffic," in *Proc. Ass. Comput. Mach. Symp.* (Pine Mountain, Ga.), Oct. 13–16, 1969.

[12] J. D. Foley and K. Lau, "Computer aided design of computer networks via computer graphics," presented at the Univ. of North Carolina, Chapel Hill, N. C., 1971.

[13] H. Frank and W. Chou, "Routing in computer networks," in *Networks*, vol 1, pp. 99–112, 1971.

[14] ——, "Cost and throughput in computer-communication networks," in "Infotech Report on the State of the Art of Computer Networks," 1972.

[15] H. Frank and I. T. Frisch, *Communication, Transmission and Transportation Networks.* Reading, Mass.: Addison-Wesley, 1972.

[16] H. Frank, I. T. Frisch, W. Chou, and R. Van Slyke, "Optimal design of centralized computer networks," in *Networks*, vol. 1. New York: Wiley, 1971, pp. 43–57.

[17] H. Frank, I. T. Frisch, and W. Chou, "Topological considerations in the design of the ARPA computer network," in *Spring Joint Computer Conf., AFIPS Conf. Proc.* Washington, D. C.: Spartan, 1970, pp. 581–587.

[18] W. Chou and H. Frank, "Routing strategies for computer network design," in *Proc. Brooklyn Polytechnic Inst. on Computer Networks and Teletraffic*, Apr. 1972.

[19] H. Frank, R. Kahn, and L. Kleinrock, "Computer communication network design—Experience with theory and practice," in *Spring Joint Computer Conf., AFIPS Conf. Proc.* Washington, D. C.: Spartan, 1972.

[20] H. Frank and I. T. Frisch, "Analysis and design of survivable networks," *IEEE Trans. Commun. Technol.*, vol. COM-18, pp. 501–519, Oct. 1970.

[21] L. Fratta, M. Gerla, and L. Kleinrock, "The flow deviation method: An approach to store-and-forward network design," to be published.

[22] G. Fultz and L. Kleinrock, "Adaptive routing techniques for store-and-forward computer-communication networks," in *Proc. Int. Conf. Communications*, pp. 39-1–39-8, 1971.

[23] E. N. Gilbert, "Minimum cost communication networks," *Bell Syst. Tech. J.*, pp. 2209–2227, 1967.

[24] M. Goldstein, "Design of long distance telecommunications networks-The Telepak problem," Office of Emergency Preparedness, Rep. R4, Jan. 1971.

[25] S. L. Hakimi, "Steiner's problem in graphs and its implications," in *Networks*, vol. 1, pp. 113–134, 1971.

[26] R. E. Kahn, "Terminal access to the ARPA computer network," in *Proc. 3rd Courant Inst. Symp.* (Nov. 1970). Englewood Cliffs, N. J.: Prentice–Hall, 1970.

[27] R. E. Kahn and W. R. Crowther, "Flow control in a resource sharing computer network," in *Proc. 2nd ACM/IEEE Symp. Problems in the Optimization of Data Communication Systems* (Palo Alto, Calif.), Oct. 1971.

[28] ——, "A study of the ARPA network design and performance," Rep. 2161, Aug. 1971.

[29] A. Kershenbaum and R. Van Slyke, "Minimum spanning trees in sparse graphs," in *Proc. ACM Special Interest Group in Mathematical Programming*, Aug. 1972.

[30] ——, "Recursive analysis of network reliability," to be published.

[31] L. Kleinrock, *Communication Nets: Stochastic Message Flow and Delay*. New York: McGraw–Hill, 1964.

[32] ——, "Models for computer networks," in *Proc. Int. Conf. Communications*, pp. 21-9 21-16, 1969.

[33] ——, "Analysis and simulation methods in computer network design," in *Spring Joint Computer Conf., AFIPS Conf. Proc.* Washington, D. C.: Spartan, 1970, pp. 569–579.

[34] D. Kleitman, "Methods of investigating connectivity of large graphs," *IEEE Trans. Circuit Theory* (Corresp.), vol. CT-16, pp. 232–233, May 1969.

[35] P. Krolak, W. Felts, and G. Marble, "A man-machine approach toward solving the traveling salesman problem," *Commun. Ass. Comput. Mach.*, vol. 14, pp. 327–335, May 1971.

[36] S. Lin, "Computer solutions of the traveling salesman problem," *Bell Syst. Tech. J.*, vol. 44, pp. 2245–2269, 1965.

[37] T. Marill and L. G. Roberts, "Toward a cooperative network of time-shared computers," in *Fall Joint Computer Conf., AFIPS Conf. Proc.* Washington, D. C.: Spartan, 1966.

[38] B. Meister, H. R. Muller, and H. R. Rudin, "New optimization criteria for message-switching networks," *IEEE Trans. Commun. Technol.*, vol. COM-19, pp. 256–260, June 1971.

[39] E. F. Moore and C. E. Shannon, "Reliable circuits using less reliable relays," *J. Franklin Inst.*, vol. 262, pp. 191–208, 1956.

[40] NAC 3rd Semiannual Tech. Rep. Project "Analysis and optimization of store-and-forward computer networks," Def. Doc. Cen., Alexandria, Va., June 1971.

[41] NAC 4th Semiannual Tech. Rep. Project "Analysis and optimization of store-and-forward computer networks," Def. Doc. Cen., Alexandria, Va., Dec. 1971.

[42] G. S. Pan, "Communications information system," *Telecommunications*, pp. 19–23, June 1970.

[43] J. Pierce, "A network for block switching of data," *Bell Syst. Tech. J.*, to be published.

[44] L. G. Roberts, "Multiple computer networks and inter-computer communications," presented at the Ass. Comput. Mach. Symp. Operating Systems, Gatlinburg, Tenn., 1967.

[45] ——, "A forward look," *Signal*, vol. 25, pp. 77–81, Aug. 1971.

[46] ——, "Resource sharing networks," presented at the IEEE Int. Conv., New York, N. Y., Mar. 1969.

[47] L. G. Roberts and B. Wessler, "The ARPA computer network development to achieve resource sharing," in *Spring Joint Computer Conf., AFIPS Conf. Proc.*, 1970, pp. 543–549.

[48] ——, "The ARPA computer network," in *Computer Communication Networks*, F. Kuo and N. Abramson, Eds. New York: Prentice–Hall, 1973, to be published.

[49] B. Rothfarb and M. Goldstein, "The one terminal Telpak problem," *Oper. Res.*, vol. 19, pp. 156–169, Jan.–Feb. 1971.

[50] R. L. Sharma and M. T. El Bardai, "Suboptimal communications network synthesis," in *Proc. Int. Conf. Communications*, vol. 7, pp. 19-11–19-16, 1970.

[51] K. Steiglitz, P. Weiner, and D. J. Kleitman, "The design of minimum cost survivable networks," *IEEE Trans. Circuit Theory*, vol. CT-16, pp. 455–460, Nov. 1969.

[52] R. Van Slyke and H. Frank, "Reliability of computer-communication networks," in *Proc. 5th Conf. Applications of Simulation* (New York, N. Y.), Dec. 1971.

[53] ——, "Network reliability analysis—I," *Networks*, vol. 1, no. 3, 1972.

[54] ——, "Network reliability analysis—II," to be published in *Networks*, 1972.

[55] L. P. West, "Loop-transmission control structures," *IEEE Trans. Commun.*, vol. COM-20, pp. 531–539, June 1972.

[56] V. K. M. Whitney, "A study of optimal file assignment and communication network configuration in remote-access computer message processing and communication systems," *Dep. Elec. Eng.*, Univ. of Michigan, Ann Arbor, Rep. 02641-2-T, SEL 48, Sept. 1970.

[57] R. S. Wilkov, "Analysis and design of reliable computer networks," *IEEE Trans. Commun.*, vol. COM-20, pp. 660–678, June 1972.

[58] B. Yaged, "Minimum cost routing for static network problems," *Networks*, vol. 1, pp. 139–172, 1971.

[59] M. L. T. Yuen *et al.*, "Traffic flow in a distributed loop switching system," presented at the Symp. Computer Communications, Apr. 4–6, 1972.

The Complexity of the Network Design Problem

D. S. Johnson
Bell Laboratories
Murray Hill, New Jersey

J. K. Lenstra
Mathematisch Centrum
Amsterdam, The Netherlands

A. H. G. Rinnooy Kan
Erasmus University
Rotterdam, The Netherlands

ABSTRACT

In the network design problem we are given a weighted undirected graph. We wish to find a subgraph which connects all the original vertices and minimizes the sum of the shortest path weights between all vertex pairs, subject to a budget constraint on the sum of its edge weights. In this note we establish NP-completeness for the network design problem, even for the simple case where all edge weights are equal and the budget restricts the choice to spanning trees. This result justifies the development of enumerative optimization methods and of approximation algorithms, such as those described in a recent paper by R. Dionne and M. Florian.

INTRODUCTION

In the network design problem we are given a weighted undirected graph. We wish to find a subgraph which connects all the original vertices and minimizes the sum of the shortest path weights between all vertex pairs, subject to a budget constraint on the sum of its edge weights. In this note we establish NP-completeness [7,8] for the network design problem. Briefly, this result implies that a polynomial-bounded method for its solution could be used to construct similar algorithms for a large number of combinatorial problems which are notorious for their computational intractability, such as the travelling salesman problem and the multicommodity network flow problem. Since none of these problems are known to be solvable in polynomial time, NP-completeness of the network design problem justifies the development of enumerative optimization methods and of approximation algorithms, such as those described by R. Dionne and M. Florian [2].

For our purposes, we formulate the problem in the following way.

NETWORK DESIGN PROBLEM (NDP): Given an undirected graph $G = (V,E)$, a weight function $L: E \to \mathbb{N}$, a budget B and a criterion threshold C $(B,C \in \mathbb{N})$, does there exist a subgraph $G' = (V,E')$ of G with weight $\sum_{\{i,j\}\in E'} L(\{i,j\}) \leq B$ and criterion value $F(G') \leq C$, where $F(G')$ denotes the sum of the weights of the shortest paths in G' between all vertex pairs?

By way of introduction to a quite involved NP-completeness proof for a simplified version of NDP, we shall first present a simple proof establishing NP-completeness for the general NDP.

Theorem 1: NDP is NP-complete.

Proof: Consider the following problem.

KNAPSACK: Given positive integers $t,a_1,\ldots,a_t,b$, does there exist a subset $S \subset T = \{1,\ldots,t\}$ such that $\sum_{i\in S} a_i = b$?

We will show that KNAPSACK is reducible to NDP, *i.e.*, that for any instance of KNAPSACK an instance of NDP can be constructed in polynomial-bounded time such that solving the instance of NDP solves the instance of KNAPSACK as well. The theorem then follows from the NP-completeness of KNAPSACK [7] and the fact that NDP belongs to *NP*, since any feasible subgraph can be recognized as such in polynomial time.

Given any instance of KNAPSACK, we write $A = \sum_{i\in T} a_i$ and define an instance of NDP as follows:

$$V = \{0\} \cup \{i,i': i \in T\},$$
$$E = \{\{0,i\},\{0,i'\},\{i,i'\}: i \in T\},$$
$$L(\{0,i\}) = L(\{0,i'\}) = L(\{i,i'\}) = a_i \quad (i \in T),$$
$$B = 2A+b,$$
$$C = 4tA-b.$$

Figure 1 illustrates this reduction. We claim that KNAPSACK has a solution if and only if $G = (V,E)$ contains a subgraph with weight at most B and criterion value at most C.

It is easily seen that any feasible NDP solution can be assumed to contain a star graph $G^* = (V,\{\{0,i\},\{0,i'\}: i \in T\})$; G^* has weight $2A = B-b$ and criterion value $4tA = C+b$. Adding an edge $\{i,i'\}$ to G^* increases the weight by a_i and decreases the criterion value by a_i, since $\{i,i'\}$ will appear only in the shortest path between i and i'. The equivalence now follows in a straightforward way. □

However, since KNAPSACK can be solved in $O(tb)$ time [1], Theorem 1 does not exclude the existence of a similar *pseudo-polynomial* algorithm [4] for NDP; the above construction crucially depends on allowing arbitrary positive integers as edge weights and budget. As a stronger result, we shall now prove that NDP is NP-complete even in the simple case where all edge weights are equal and the budget restricts the choice to spanning trees.

SIMPLE NETWORK DESIGN PROBLEM (SNDP): NDP with $L(\{i,j\}) = 1$ for all $\{i,j\} \in E$ and $B = |V|-1$.

Theorem 2: SNDP is NP-complete.

Proof: As a starting point we take the following NP-complete problem [3,6].

KNAPSACK: $t = 4$, $a_1 = 2$, $a_2 = 3$, $a_3 = 5$, $a_4 = 6$, $b = 7$.

NDP: G & L:

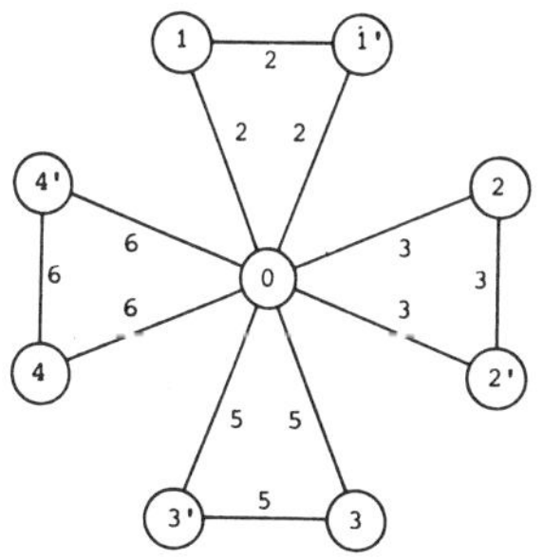

B = 39, C = 249.

Fig. 1 Equivalent instances of KNAPSACK and NDP.

Reprinted with permission from *Networks*, vol. 8, pp. 279–285, Winter 1978.

EXACT 3-COVER: $t = 2$, $s = 4$,

$S = \{\{\tau_1,\tau_2,\tau_3\},\{\tau_2,\tau_3,\tau_5\},\{\tau_2,\tau_4,\tau_5\},\{\tau_4,\tau_5,\tau_6\}\}$.

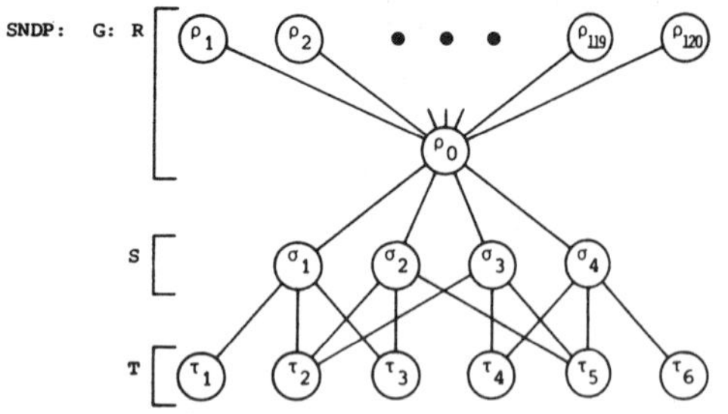

$C = 17656$.

Fig. 2 Equivalent instances of EXACT 3-COVER and SNDP.

EXACT 3-COVER: Given a family $S = \{\sigma_1,\dots,\sigma_s\}$ of 3-element subsets of a set $T = \{\tau_1,\dots,\tau_{3t}\}$, does there exist a subfamily $S' \subset S$ of pairwise disjoint sets such that $\bigcup_{\sigma\in S'} \sigma = T$?

We will show that EXACT 3-COVER is reducible to SNDP.

Given any instance of EXACT 3-COVER, we define an instance of SNDP as follows:

$$V = R \cup S \cup T,$$
$$R = \{\rho_0,\rho_1,\dots,\rho_r\},$$
$$r = C_{SS} + C_{ST} + C_{TT},$$
$$E = \{\{\rho_i,\rho_0\}: i=1,\dots,r\} \cup \{\{\rho_0,\sigma\}: \sigma \in S\} \cup \{\{\sigma,\tau\}: \tau \in \sigma \in S\},$$
$$C = C_{RR} + C_{RS} + C_{RT} + C_{SS} + C_{ST} + C_{TT},$$

where $C_{RR} = r^2$, $C_{RS} = 2rs+s$, $C_{RT} = 9rt+6t$, $C_{SS} = s^2-s$, $C_{ST} = 9st-6t$, $C_{TT} = 18t^2-12t$.

Figure 2 illustrates this reduction. We will prove that EXACT 3-COVER has a solution if and only if $G = (V,E)$ contains a spanning tree with criterion value at most C. We assume that G is connected, *i.e.*, $\bigcup_{\sigma\in S} \sigma = T$.

Let $G' = (V,E')$ be some spanning tree of G and let $F_{PQ}(G')$ denote the sum of the weights of all shortest paths in G' between vertex sets P and Q $(P,Q \subset V)$. We clearly have $\{\rho_i,\rho_0\} \in E'$ for all $i = 1,\dots,r$. If $\{\rho_0,\sigma\} \notin E'$ for some $\sigma \in S$, then

$$\begin{aligned} F(G') &> F_{RR}(G') + F_{RS}(G') + F_{RT}(G') \\ &\geq C_{RR} + C_{RS} + 2(r+1) + C_{RT} \\ &> C; \end{aligned}$$

therefore, we may assume that $\{\rho_0,\sigma\} \in E'$ for all $\sigma \in S$. It follows that in G' each vertex in T is adjacent to exactly one vertex in S. Straightforward calculations show that we now have

$$F_{PQ}(G') = C_{PQ} \quad \text{for } P = R,S \text{ and } Q = R,S,T.$$

Denoting the number of vertices in S being adjacent in G' to exactly h vertices in T by s_h $(h = 0,1,2,3)$, we have

$$\begin{aligned} F_{TT}(G') &= 4(3t(3t-1)/2) \\ &\quad - 2\left|\{\{\tau,\tau'\}: \tau \neq \tau',\ \{\{\sigma,\tau\},\{\sigma,\tau'\}\} \subset E' \text{ for some } \sigma \in S\}\right| \\ &= (18t^2-6t) - (2s_2+6s_3) \\ &= C_{TT} + 6(t-s_3) - 2s_2. \end{aligned}$$

It is easily seen that $F_{TT}(G') = C_{TT}$ if and only if $s_3 = t$, $s_2 = s_1 = 0$, $s_0 = s-t$. The first condition is now equivalent to $F(G') \leq C$, the second one to the existence of an EXACT 3-COVER solution. This completes the proof. □

Various related types of network design problems have been discussed in the literature; an excellent survey has been given by R. T. Wong [10]. For instance, the problems dealt with by A. J. Scott [9] are generalizations of NDP and hence NP-complete.

Another variation has been introduced by T. C. Hu [5]. Given a complete graph with a distance and a requirement for each vertex pair, we wish to find a spanning tree which minimizes the total cost of communication, where the cost of communication for a pair of vertices is the distance of the path between them multiplied by their requirement. The case where all distances are equal can be solved by polynomial time [5]; for the case where all requirements are equal, NP-completeness follows easily as a corollary to Theorem 2.

Finally, we mention some results with respect to the complexity of a network design problem due to F. Maffioli. Given a weighted graph with a specific vertex ρ and an integer k, we wish to find a spanning tree of minimum total weight subject to the constraint that each subtree incident with ρ contains at most k other vertices. The case $k = 2$ can be formulated as a matching problem; the case $k = 3$ can be proved NP-complete by a reduction of 3-DIMENSIONAL MATCHING, even if all edge weights are equal.

REFERENCES

1. Bellman, R. E. and S. E. Dreyfus, *Applied Dynamic Programming*, Princeton University Press, Princeton, New Jersey, 1962.

2. Dionne, R. and M. Florian, "Exact and Approximate Algorithms for Optimal Network Design," to appear in *Networks*, 9, 1, 1979.

3. Garey, M. R., R. L. Graham and D. S. Johnson, "Some NP-Complete Geometric Problems," *Proc. 8th Annual ACM Symp. Theorey Comput.*, 1976, pp. 10-22.

4. Garey, M. R. and D. S. Johnson, "Strong NP-Completeness Results: Motivation, Examples and Implications," *J. Assoc. Comput. Mach.*, 25, 1978, pp. 499-508.

5. Hu, T. C., "Optimum Communication Spanning Trees," *SIAM J. Comput.*, 3, 1974, pp. 188-195.

6. Hyafil, L. and R. L. Rivest, "Constructing Optimal Binary Decision Trees is NP-Complete," *Information Processing Letters*, 5, 1976, pp. 15-17.

7. Karp, R. M., "Reducibility Among Combinatorial Problems," in *Complexity of Computer Computations*, R. E. Miller and J. W. Thatcher, eds., Plenum Press, New York, 1972, pp. 85-103.

8. Karp, R. M., "On the Computational Complexity of Combinatorial Problems," *Networks*, 5, 1975, pp. 45-68.

9. Scott, A. J., "The Optimal Network Problem: Some Computational Procedures," *Trans. Res.*, 3, 1969, pp. 201-210.

10. Wong, R. T., "A Survey of Network Design Problems," Working Paper OR 053-76, Operations Research Center, Massachusetts Institute of Technology, Cambridge, Massachusetts, 1976.

An Automatic Layout System for Masterslice LSI: MARC

KAZUHIRO UEDA, YOSHI SUGIYAMA, AND KOU WADA

Abstract—An LSI layout software system MARC is described, which places functional blocks on a chip and wires between the blocks for a masterslice LSI. Up to a 1200-cell masterslice, which can contain 300 functional blocks, can be handled by the system. The placement procedure is based on the attractive and repulsive force method, and the routing procedure is implemented using the two-phase assignment algorithm. The efficient combination of placement and routing procedures brings about a high-quality result. Manual modification can be performed easily by the stand-alone graphic system interfaced with the main computer. It is shown that satisfactory run results have been obtained in the application to practical data.

I. Introduction

As THE SCALE and density of the large-scale integrated (LSI) circuit increases, it is estimated that more manpower and turnaround time is needed for the design and manufacturing of an LSI. Consequently, computer aids have been keenly required for each phase of LSI design.

The LSI layout design phase is said to occupy 30–50 percent of the whole LSI design cycle—from circuit specification to characteristics evaluation [1]. So, the effectiveness of design automation in the LSI layout design phase is expected to be the largest one.

A layout design software system MARC has been developed which places functional blocks on a chip and wires between these blocks for a masterslice type LSI. The system aims at a high-quality placement and wiring result by using effective algorithms in both placement and wiring procedures and combining them. Up to a 1200-cell masterslice, which is able to contain 300 functional blocks (equivalent to about 1200 logic gates) can be handled by the MARC system. Furthermore, two kinds of cells and sections can be defined on the masterslice. As a result, it is expected that the system will be applicable to the LSI in the near future as well as at the present time.

It is essentially impossible to guarantee the 100 percent wireability in the masterslice LSI routing problem in which the interconnection area is fixed and limited. From this viewpoint, it is assumed that manual modification is inevitable to complete the remaining interconnections which are left unsolved in the automated wiring process. In the MARC system, manual modification can be performed in a comparatively easy manner through the stand-alone graphic system interfaced with the main computer. Since the manually modified data may have errors, a wiring pattern check routine is provided to examine the consistency between the input logical interconnection data and the resultant physical pattern data.

Manuscript received December 20, 1977; revised March 13, 1978.

The authors are with the Bipolar IC Section, Musashino Electrical Communication Laboratory, Nippon Telegraph and Telephone Public Corporation, Tokyo, Japan.

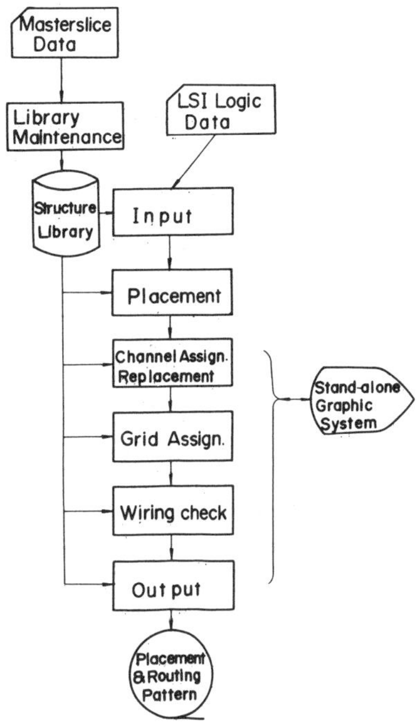

Fig. 1. Flowchart of MARC system. Manual modification is performed through the stand-along graphic system.

In this paper, the system configuration, the layout objectives, and the placement and routing algorithms are described. The run results are also discussed.

II. System Description

The MARC system software configuration is shown in Fig. 1. The system's flows are outlined as follows. The masterslice structure data library, which is provided by the library maintenance routine, are referenced by each of the following routines. The input routine converts the LSI logic circuit data from input connection format to output connection format.

Functional blocks and external terminals are allocated in the placement routine. Following the placement procedure, routing is processed in two phases: channel assignment and grid assignment. The channel assignment routine assigns the routes to channels which consist of several interconnection grids. If there are no overflows in any channel elements, the procedure goes to the following grid assignment step, otherwise, block replacement improvement is tried.

Reprinted from *IEEE J. Solid-State Circuits,* vol. SC-13, pp. 716–721, Oct. 1978.

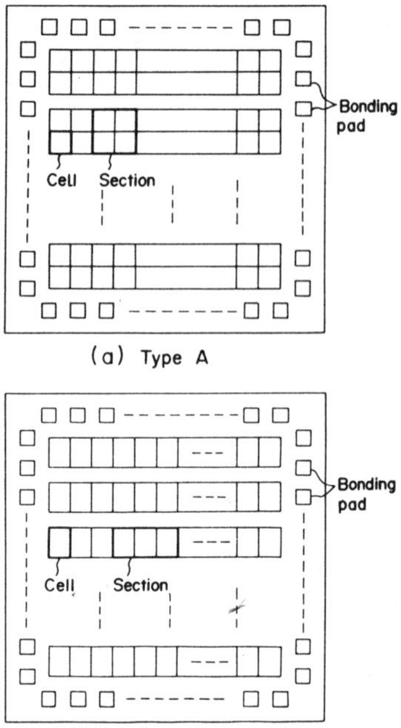

Fig. 2. Two types of masterslice structure. (a) 2 × 2–cell section. (b) 1 × n–cell section.

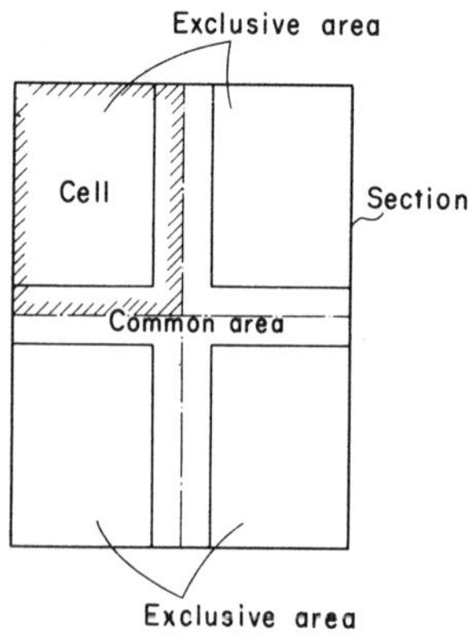

Fig. 3. Section structure containing exclusive and common area. Elements in the common area can be used with elements in exclusive area.

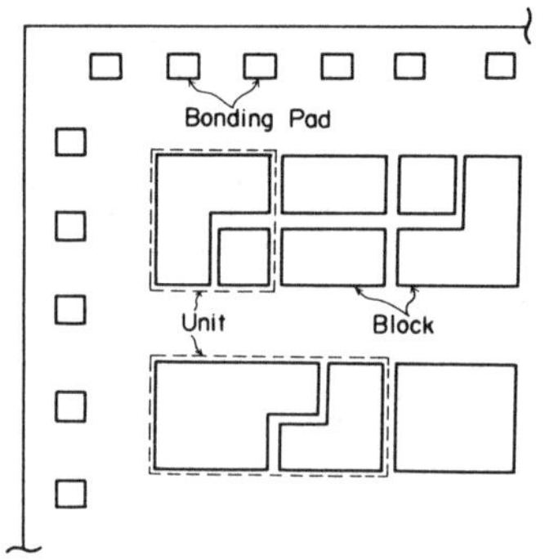

Fig. 4. Construction of "placement unit" by combining blocks. The size of unit is four or eight cells.

In the grid assignment routine, the final interconnection routes are located on the grid. If unsolved nets remain after the grid assignment, manual modification is performed through the stand-alone graphic system interfaced with the main computer, which accommodates the MARC system programs.

III. Masterslice Specification

The cells are placed in an array on a masterslice chip, as shown in Fig. 2. Either Fig. 2(a) or (b) section type is allowed. Two kinds of cells–internal cell and output cell–can be used on the same masterslice. The section may contain the common area, and, in that case, the section consists of exclusive and common areas as shown in Fig. 3. When the number of elements in the exclusive area is insufficient to construct a functional block, several elements in the common area are used in combination. The external terminals (or the bonding pads) are placed around the peripheral of a chip. The order of signal, power supply and ground pins can be arbitrary.

The program limits are 1200 cells, 300 functional blocks, and a grid 500 × 500.

IV. Placement Scheme

A. Placement Objectives

In the masterslice LSI placement problem, each functional block usually has a different number of cells depending on the function and a nonrectangular shape. To simplify the problem, several functional blocks are put together into a "unit" at the beginning of the placement phase. Thereafter, the unit, not the functional block, is handled as a "placement unit." An example of unit placement is shown in Fig. 4. Up to eight-cell functional blocks can be placed automatically in the combinational placement manner mentioned above. However, blocks having more than eight cells should be placed manually. Part or all of the external terminals can be assigned automatically.

B. Placement Algorithm

The placement process of the MARC system is divided into three phases: 1) initial placement, 2) replacement based on minimizing total-wire length, and 3) replacement based on averaging interconnection density.

The first initial placement is implemented using the algorithm called the AR method [2]. The algorithm classifies every component into two groups of "friendly" and "unfriendly" components based on connectivity: a component located near its "friendly" components and far from its "unfriendly" components results in a minimum or near minimum total-wire length placement.

In the graph of Fig. 5(a), each node represents a placement unit and an edge represents an interconnection between the units. In this example, the initial node location is given at random. First, the distance between nodes is calculated from connectivity to get the "friendly" and "unfriendly" relationship between every node.

Next, the attractive and repulsive forces based on these relationships are obtained for every node and an equilibrium node position is obtained in the procedure called the AR process.

Fig. 5(b) is a resultant node pattern after several cycles of the AR process. Then, the node pattern is transformed into the array like Fig. 5(c). This transformation process is per-

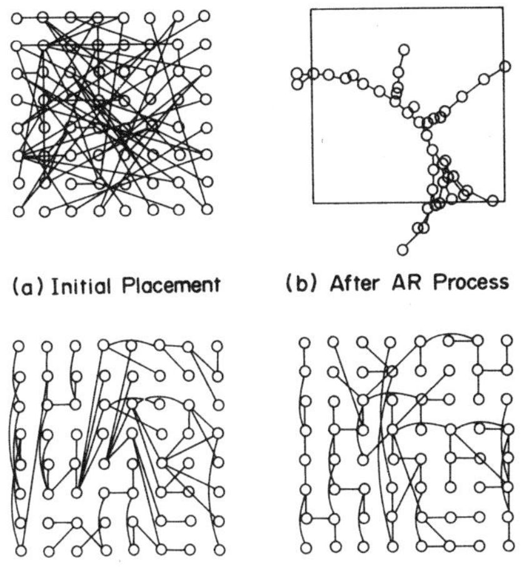

Fig. 5. Placement process by Attractive and Repulsive (AR) force method and replacement process. Criterion: Minimize total-wire length.

formed by sorting X and Y coordinates of every node. The XY sorting transformation is a simple and fast method, but it causes partial "distortions" as can been seen in Fig. 5(c). To eliminate these distortions and improve the placement, replacement of units is performed, and the final placement like Fig. 5(d) is obtained.

Following the interunit replacement, intraunit replacement is performed in which a different combination of blocks is tried in each unit so as to reduce the total-wire length. The third replacement procedure, which takes place in the channel assignment phase, is discussed later.

V. Routing Scheme

A. Routing Problems

The routing algorithms, which have been proposed for printed-circuit boards, are also applicable to the routing problem of a masterslice LSI. The maze and line-search algorithms [3], [4] are widely used, however, they have drawbacks, such as the strong dependency of wireability on the routing order and the enormous computing time in the application to the large-scale problems. One reason for the deficiencies is the fact that interconnection paths are searched sequentially on the grid level, a stroke at a time, in these routing algorithms.

The two-phase routing method, in which a routing problem is solved using two separate steps, has been proposed and implemented in a couple of systems [5], [6]. In the first phase of this algorithm, interconnection paths are assigned to the channel which consists of several wiring tracks (or grids). In the second phase, the interconnection paths are assigned to specific tracks (or grids). By this hierarchical partitioning of the routing problem, enhancement of routing efficiency can be expected.

In the MARC system, like many others, interconnection paths between blocks are automatically wired, while intraconnection paths within blocks are manually accomplished.

B. Channel Assignment

The wiring area on a masterslice consists of a grid on which interconnection paths are allowed to run. The area is divided into several horizontal and vertical channels, and the structure of the channels is modeled into a network that is used in the channel assignment procedure. Fig. 6 shows the channel partitioning of a masterslice and its network model. Since the channel may include power supply, ground, reference paths, and intraconnection paths within the functional blocks which act as obstacles to other interconnections, each channel usually has a different "capacity." The channel capacity is represented as a function proportional to the number of available tracks included in the channel. Every node in the network has the capacity and also has the value called "flow," which is incremented every time a new path is assigned to the node. Channel assignment is performed by trying to eliminate overflows in any node.

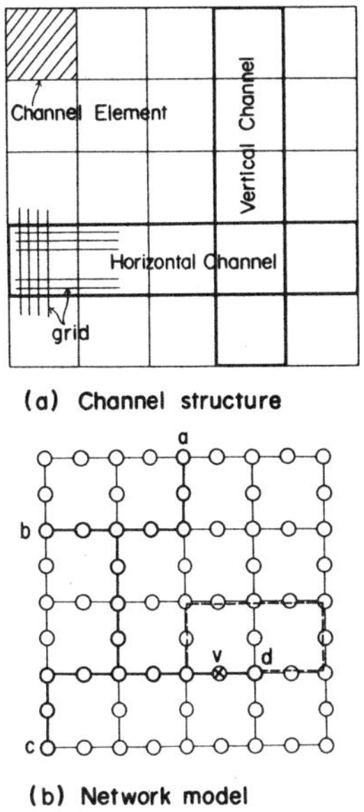

Fig. 6. Channel structure and its network model. (a) Masterslice partitioned into horizontal and vertical channels. (b) Network model used in channel assignment phase. Node v is "overflowed." Dotted-line path may be assigned to eliminate the overflow.

The channel assignment procedure is divided into two phases: initial channel assignment and reassignment phases. In the initial assignment phase, paths are assigned to the network to form quasi-minimum Steiner trees. The initial assignment is performed as follows:

1) Select terminals which bound the net (upper, right, left, bottom) and designate them $T_{i=0}$.

2) Make a minimum Steiner tree in a sense of Manhattan distance connecting each terminal in T_i (see Fig. 7).

3) Let $T \leftarrow T - T_i$ and, if T is empty, go to 5), otherwise, let $i \leftarrow i + 1$ and go to 4).

4) Select t: one of the terminals from T and select p: the nearest-to-t point on the path already obtained. Let $T \leftarrow \{t, p\}$ and go to 2).

5) Convert the quasi-minimum Steiner tree onto the network.

An attempt is made to eliminate the overflows in the nodes, which have been ignored in the initial assignment stage, in the reassignment phase. In this phase, several nets are taken from the network and assigned again to the network in order to avoid local congestions of interconnection paths.

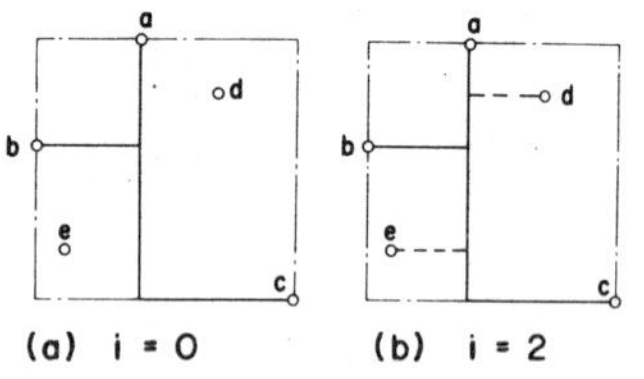

Fig. 7. Process to generate quasi-minimum Steiner tree in the initial channel assignment phase. (a) First, make a tree connecting terminals a, b, c. (b) Then, add paths from d and e.

The reassignment procedure is applied to the nets: a) passing through the overflow nodes, or b) having a comparatively long interconnection path. The quasi-minimum Steiner tree, in a sense of cost, is constructed on the network by using a labeling algorithm.

C. Replacement in Channel Assignment Phase

If overflows remain for any nodes, in spite of the improvement efforts by the reassignment procedure, it is considered that the efforts within the routing problem are insufficient and that the placement problem should be repeated at this phase. From this point of view, the replacement process, which is linked with the channel assignment, attempts to eliminate overflows in the nodes.

The replacement in this phase has the effect of averaging the wire density on the whole area of the masterslice. The replacement procedure follows:

1) Locate the blocks producing the overflows $\{a_1, a_2, \cdots, a_n\}$ and put them in descending order according to the values of their costs.

2) Exchange block a_i with one of the blocks which is surrounding a_i in a rectangular shape.

3) Reassign the paths associated with these blocks.

4) If the number of overflows is reduced, take that exchange, otherwise continue the search.

5) Continue 2)–4) until every a_i is tried.

D. Grid Assignment

Following the channel assignment and the replacement, the grid assignment is performed to fix the interconnection paths on the grid of the masterslice. The line-search algorithm, which is extensively used in PCB routing problems, is also used to assign the grid locations in the MARC system.

Since the standard line-search method is used in the MARC system, only the distinctive features are described here. Since one of the factors used in determining routing order is channel congestion, an effective ordering is expected, compared with that of the conventional programs. Secondly, the restricted search boundaries, which are determined by the channel assignment results, are employed in the first phase of the line-search process. This restriction shortens the run time for searching. The searching boundary is expanded if the interconnection paths cannot be found within the boundary and is executed in the following modes:

Mode 1: Apply the line-search method within the boundary of channels on which the net has been assigned.

Mode 2: Apply the line-search method within the rectangular area, including the unsolved terminals.

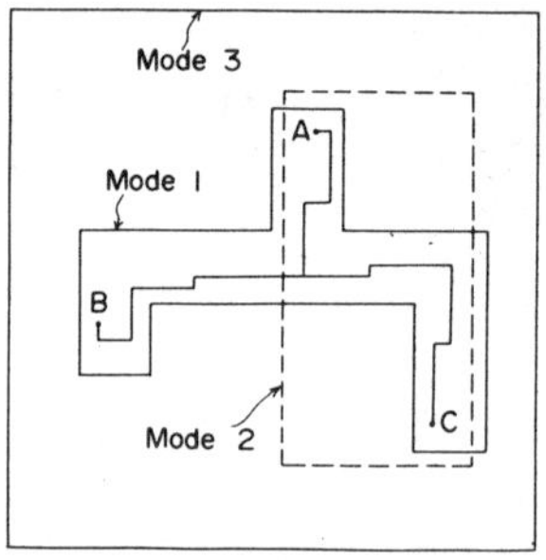

Fig. 8. Line-search boundaries enlarged in three steps: Modes 1, 2, 3 in grid assignment phase. (The mode 2 boundary is in the specific case when the path between A and C cannot be found in Mode 1.)

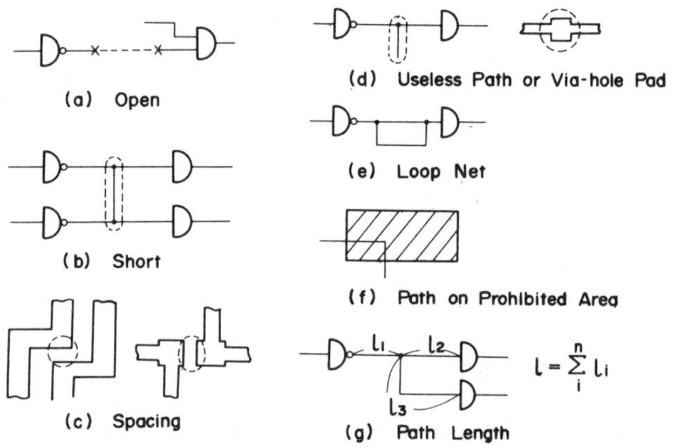

Fig. 9. Check items in wiring pattern check routine.

Mode 3: Apply the line-search method without boundary, i.e., on the whole area of the masterslice.

The boundaries of each mode are shown in Fig. 8.

VI. Wiring Pattern Check

The interconnection check routine is provided for the purpose of eliminating the data errors generated in manual modification. The check concentrates on logical short or open errors, rather than the physical. The interconnection check routine is divided into the following four phases:

1) Construct the segment and net name table, which includes all segments and net names in order, respectively, and construct a file which contains information on the interconnection prohibited area.

2) Collect all starting and ending point coordinates. Sort them in order, separately, and link the segments to each other. In this process, loop nets are detected. Insufficient spacing paths and through-hole (or via-hole) pads are also checked.

3) By investigating the coincidence between the linked segments and the net name table, short, open, useless paths, and through-holes are detected. The overlapping of each interconnection segment over the prohibited area is checked and the total path length of each net is calculated to determine whether the path length of each net exceeds the given path-length limit.

4) The number of errors and the computing time are printed. Check items are summed in Fig. 9.

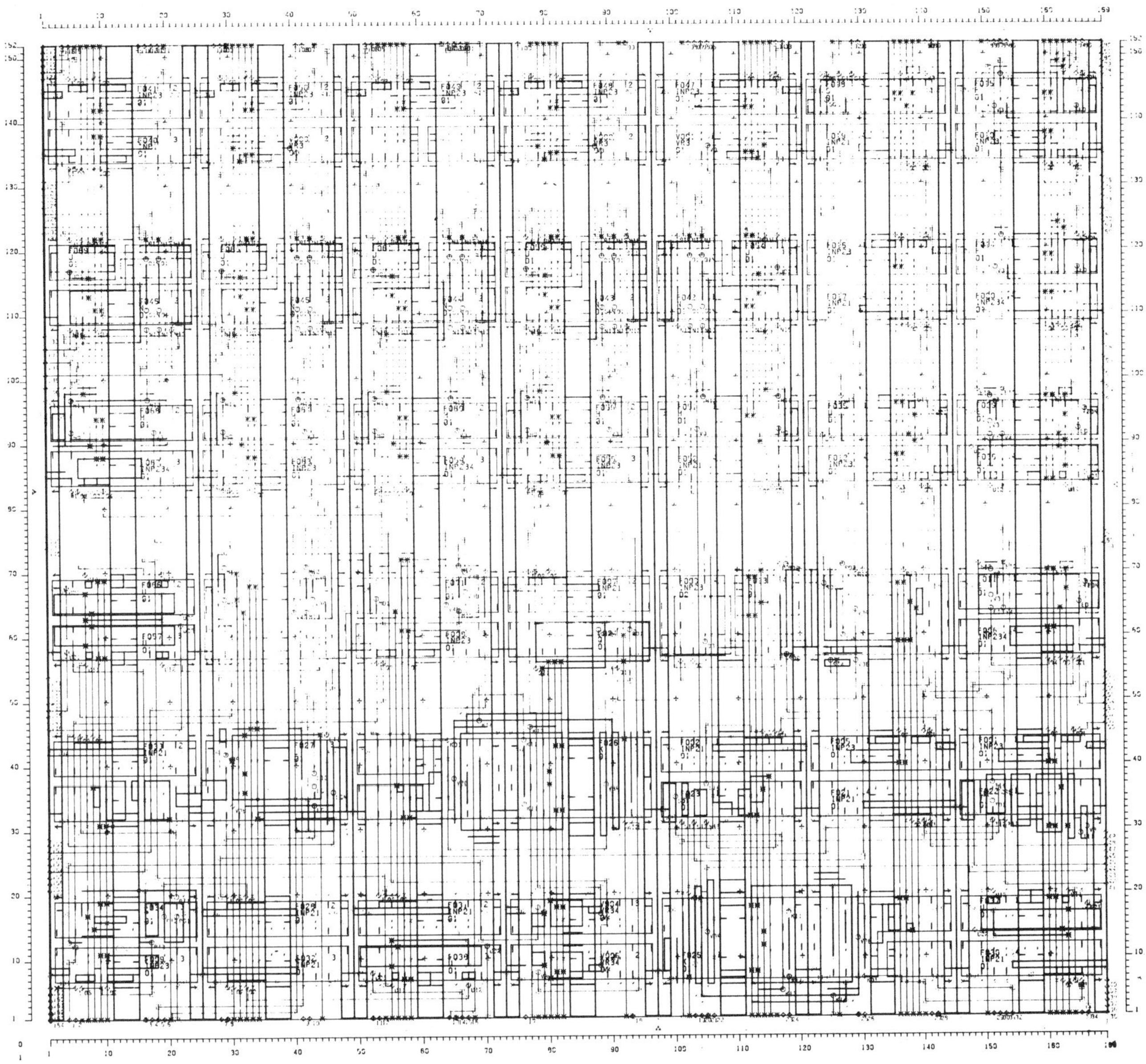

Fig. 10. Plotted pattern output for the data in Tables I and II.

TABLE I
LSI MASTERSLICE AND LOGIC DATA

Masterslice Data		LSI logic Data	
Type of Section	A	No. of Nets	200
No. of Sections	42	No. of Net Segments	358
No. of Cells	168	No. of Blocks	71
No. of Terminals	65	No. of Block Types	21

TABLE II
RUN RESULTS FOR THE DATA IN TABLE I

Process	Results			
Placement	Total-wire-length		10% less than manual	
	CPU time	AR process	19 min	Total 67 min
		Replacement	48 min	
Routing	Wireability		96 percent	
	CPU time	Channel Assign.	5 min	Total 23 min
		Grid Assign.	18 min	

VII. EXPERIMENTAL RESULTS

To evaluate the effectiveness of the system, practical data for masterslice and LSI logic circuits have been applied to the system. Typical data and run results are shown in Tables I and II. Fig. 10 shows an example of a plotted pattern. The placement result, which is about 10 percent better than that of manual performance based on total-wire-length, has been obtained, however, a comparatively large amount of computing time was required—about one hour for the 165-cell circuit data. Three quarters of the computing time for placement process is spent in the interunit and intraunit replacement phase.

A 96 percent wireability for the same circuit data has been attained. The unsolved 15 net segments were manually completed in a little more than one hour of work. Furthermore, to estimate the effectiveness of channel assignment, the following two cases were used as experiments. One is the case where the line-search method is applied within the boundary of channels in the first grid assignment phase. The other is the case where the line-search method is applied without the boundary of channels, just like with the conventional algorithms. The wireability of the former case, as expected, is higher than that of the latter.

From these results, the placement and routing algorithms used in the system have been confirmed to be quite effective. The MARC system is considered to have sufficiently good performance for application to practical data for masterslice LSI's. The programs are written in Fortran and consist of 91.6K statements, and run within 512 kbyte core memory on a DIPS-1.

VIII. Conclusion

An automatic layout system for masterslice LSI's is described. The placement based on attractive and repulsive force algorithm, and routing, based on two-phase assignment algorithm, have been confirmed to be efficient in the application to masterslice LSI layout problem. The turnaround time for LSI layout design has been reduced to about one tenth of that for manual design by utilizing the MARC system. Further improvement of the system is expected by 1) using the results of the AR method more efficiently in the unit replacement and reducing the number of exchanges and 2) optimizing the parameters in the channel and grid assignment procedures.

Acknowledgment

The authors would like to express their appreciation to H. Toyoda, M. Watanabe, H. Mukai, and K. Kataoka (all at Musashino ECL NTT) for their continuous encouragement and valuable suggestions. They also wish to thank K. Kani, H. Kawanishi, H. Yoshizawa (all at NEC Ltd.), H. Kano, Y. Ikemoto, T. Sugiyama (all at Hitachi Ltd.), Y. Ozawa, K. Ohtsu (both at Oki Electric Co.), and S. Hazama, S. Aihara (both at Fujitsu Ltd.), for their cooperation in the design and programming of the system.

References

[1] C. F. O'Donnel, "Engineering for systems using large scale integration," in *Conf. Rec., 1968 Fall Joint Comput. Conf.* Washington DC: Thompson, 1968, pp. 867–876.

[2] K. Ueda, "Placement algorithm for logic modules," *Electron. Lett.*, vol. 10, no. 10, pp. 206–207, 1974.

[3] C. Y. Lee, "An algorithm for path connections and its applications," *IRE Trans. Electron. Comput.*, vol EC-10, pp. 346–365, Mar. 1961.

[4] K. Mikami *et al.*, "A computer program for optimal routing of printed circuit connectors," in *IFIP Proc.*, H47, 1968, pp. 1475–1478.

[5] Y. Sugiyama *et al.*, "Routing program for multi-chip LSI's," in *1975 Proc. USA-Japan DA Symp.*, 1975, pp. 87–94.

[6] A. Hashimoto *et al.*, "Wire routing optimizing channel assignment within large apertures," in *Proc. 7th DA Workshop*, p. 155, 1970.

[7] Y. Ikemoto *et al.*, "Correction and wiring check system for masterslice LSI," in *Proc. 13th DA Conf.*, 1976, pp. 336–343.

[8] K. H. Khokhani *et al.*, "The chip layout problem: A placement procedure for LSI," in *Proc. 14th DA Conf.*, 1976, pp. 291–297.

[9] K. A. Chen *et al.*, "The chip layout problem: An automatic wiring procedure," in *Proc. 14th DA Conf.*, 1976, pp. 298–302.

Complexity of Network Reliability Computations

Michael O. Ball
College of Business and Management, University of Maryland, College Park, Maryland 20742

This paper considers the difficulty of computing several measures of network reliability on directed and undirected networks. Results concerning the NP-difficulty of several network reliability analysis problems are unified and in several cases generalized to wider classes of measures. Reductions are also given that relate network reliability problems on directed and undirected networks and problems with and without node failures.

1. INTRODUCTION

A large amount of work (e.g., see refs. [1-3]) has been devoted to developing algorithms to compute measures of reliability for stochastic networks. These algorithms vary widely both in terms of the measures they compute and the form of their output. This paper defines three classes of reliability analysis problems and provides several network reliability measures. We then prove in a general setting, results concerning the complexity of reliability analysis problems and apply these results to several network reliability analysis problems. This development both unifies and generalizes recent research [4-8] on the complexity of network reliability problems.

The paper is organized as follows. Section 2 places network reliability analysis problems in the more general setting of stochastic binary systems. We then define three class of reliability analysis problems: point estimate, rational, and functional. The specific network reliability measures considered are then defined. Section 3 shows that several of the problems defined in Sec. 2 are NP-hard. Section 4 gives a reduction for reliability analysis problems from undirected to directed networks. Section 5 gives reductions that eliminate the need for certain algorithms to consider node failures. Section 6 summarizes the paper.

2. PROBLEM FORMULATIONS

Standard network terminology, will be used throughout the paper (see Harary [9], for example). A stochastic network model is considered in which each node and arc may take on either of two states: operative or failed. The state of a node or arc is a random event that is independent of the state of any other node or arc. The network reliability analysis problem is, given failure probabilities for the nodes and arcs of a network, to compute a measure of network reliability. This section defines several measures of network reliability for both directed and undirected networks. Before giving these measures, a general framework for network reliability analysis problems is introduced and three classes of problems are defined.

A *stochastic binary system* (SBS) generalizes the model just introduced. It represents a system that fails randomly as a function of the random failure of its components. It is assumed that each component may take on either of two states: operative or failed and that the states of any two components are independent. An SBS may be represented by the pair (T, Γ), where T denotes the nonempty set of components and $\Gamma = \{F \subseteq T$: if all $i \in F$ fail and all $k \in T - F$ operate, then the system operates$\}$. Let $\Pr[\gamma, \{p_i\}]$ denote the probability that the system (T, Γ) operates where p_i is the failure probability of component i for $i \in T$. If all components have the same failure probability p, then p is substituted for $\{p_i\}$. If there is no ambiguity concerning the failure probabilities, then $\{p_i\}$ may be left out of the notation.

An SBS is *coherent* [14] if (1) $\phi \in \Gamma$, (2) $T \notin \Gamma$, and (3) $F \in \Gamma$, $F' \subseteq F$ imply $F' \in \Gamma$, i.e., a coherent SBS is one in which the failure of any component can only have an adverse affect on the operability of the system. For any stochastic coherent binary system (SCBS) define a *pathset* as a minimal subset of components whose operation implies system operation and a *cutset* as a minimal subset of components whose failure implies system failure. Either the set of pathsets or the set of cutsets determines Γ and thus the system.

Three classes of the reliability analysis problems are given below. Each problem class is represented by an input and an output. The input contains a representation of the system (T, Γ) and possible specifications of the component failure probabilities and of the accuracy of the answer desired. The representation of (T, Γ) could take on many forms, e.g., one of the standard representations of a network. The output is the probability that the system operates in a form compatible with the input.

(1) Point Estimate Reliability Analysis Problem

Inputs: (T, Γ), ϵ, β and $\{p_i\}_{i \in T}$, where $0 < \epsilon < 1$, $1 < \beta$, and $0 \leqslant p_i \leqslant 1$ for all $i \in T$.
Output: r, where $\Pr[\gamma, \{p_i\}]\ \epsilon < r < \Pr[\gamma, \{p_i\}]\ \beta$.

(2) Rational Reliability Analysis Problem

Inputs: (T, Γ) and $\{(d_i, e_i)\}_{i \in T}$, where e_i is a positive integer, d_i is a nonnegative integer, and $d_i \leqslant e_i$ for all $i \in T$.
Outputs: d and e, where e is a positive integer d is a non-negative integer, $d \leqslant e$, and

$$\Pr[\gamma, \{d_i/e_i\}] = d/e.$$

(3) Functional Reliability Analysis Problem

Input: (T, Γ).
Outputs: g_j for $j = 0, 1, \ldots, |T|$, where g_j for $j = 0, 1, \ldots, |T|$ are non-negative integers and

$$\Pr[\gamma, p] = \sum_{j=0}^{|T|} g_j (1-p)^j p^{|T|-j}. \tag{1}$$

The form of the input is important. It is reasonable to assume the length of the input is bounded by a polynomial in $|T|$, thus excluding, in most cases, a complete listing of Γ.

Although, most reliability analysis algorithms can compute measures in any of the three forms, the computer implementation for each differs significantly. For the point estimate problem, probabilities are represented as real numbers within a certain accuracy. For the rational problem, probabilities are represented as a pair of integers. For the functional problem, probabilities are represented as polynomials. Each case requires different techniques for handling arithmetical operations.

For applications where component failure probabilities are estimated statistically, the point estimate problem may be the most natural, since the probabilities would be specified within a certain percent accuracy and the answer would be desired in a similar form. The rational problem has the advantage that answers can be computed exactly and numerical round-off problems can be avoided. The functional problem is less general than the other two, since it requires the assumption that all components have equal failure probabilities; however, it provides a powerful sensitivity analysis capability in that once the function is obtained, the reliability can be easily computed for all values of p.

This paper considers specific network reliability analysis problems. However, some results are stated for general SBSs or SCBSs, so that the breadth of their applicability can be seen. The network reliability measures considered in the remaining three

Reprinted with permission from *Networks*, vol. 10, pp. 153–166, 1980.

sections are listed below.

For any undirected stochastic network with node set N and for any $i, j \in N, i \neq j$, let $i \sim j$ denote the random event, there exists an operative path between i and j. For any directed stochastic network with node set N and for any $i, j \in N, i \neq j$, let $i \to j$ denote the random event, there exists an operative directed path from i to j.

The following measures are defined for undirected networks:

$\Pr[h(s,t)] = \Pr[s \sim t]$, for a specified pair of nodes $\{s,t\}$ with $s, t \in N$ and $s \neq t$,
$\Pr[c_1] = \Pr[i \sim j$ for all pairs of nodes $\{i,j\}$ with $i, j \in N$ and $i \neq j]$,
$\Pr[c_2] = \Pr[i \sim j$ for all pairs of nodes $\{i,j\}$ with $i, j \in N, i \neq j$, and i, j operative$]$.

The following measures are defined for directed networks:

$\Pr[h'(s,t)] = \Pr[s \to t]$ for a specified pair of nodes $\{s,t\}$ with $s, t \in N$ and $s \neq t$,
$\Pr[c_1'(s)] = \Pr[s \to i$ for all nodes $i \in N$ with $i \neq s]$, for a specified node $s \in N$,
$\Pr[c_2'(s)] = \Pr[s \to i$ for all nodes $i \in N$, with $i \neq s$ and i operative$]$, for a specified node $s \in N$,
$\Pr[c_3'] = \Pr[i \to j$ for pairs of nodes $\{i,j\}$ with $i, j \in N$ and $i \neq j]$,
$\Pr[c_4'] = \Pr[i \to j$ for all pairs of nodes $\{i,j\}$ with $i, j \in N, i \neq j$ and i, j operative$]$.

Note that if nodes are perfectly reliable (have zero failure probability), then

$\Pr[c_1] = \Pr[c_2]$
$\quad = \Pr[$operative arcs contain a spanning tree$]$,
$\Pr[c_1'(s)] = \Pr[c_2'(s)]$
$\quad = \Pr[$operative arcs contain a spanning arborescence rooted at $s]$,
$\Pr[c_3'] = \Pr[c_4']$
$\quad = \Pr[$operative arcs contain a strongly connected spanning subnetwork$]$.

Certain measures involving expected values, e.g., Ex[number of node pairs between which operative paths exist], have also been found to be useful. These are not listed here, since in most cases they can be obtained from the measures listed above. Many other measures may be defined for stochastic networks. Those listed above arise most frequently in practice [10].

3. NP-HARD RELIABILITY ANALYSIS PROBLEMS

This section compares the complexity of reliability analysis problems with the complexity of well known combinatorial and optimization problems. In particular, it is shown that several reliability analysis problems are NP-hard. That is, the existence of a polynomially bounded algorithm to solve them implies the existence of a polynomially bounded algorithm for many classically hard combinatorial problems such as integer programming, the traveling salesman problem, and the satisfiability problem (see [11] and [12] for more details).

Theorem 1 gives the main results of this section. Its proof uses the following two lemmas. The first is given in Valiant [8], and the second generalizes a result due to Rosenthal [6].

Lemma 1. For any SBS, (T, Γ), if the functional reliability analysis problem is NP-hard, then the rational reliability analysis problem is NP-hard.

Proof: The lemma will be proved by showing that the functional reliability analysis problem can be solved by solving the rational reliability analysis problem $|T| + 1$ times.

If the rational reliability analysis problem is solved $|T| + 1$ times, where on the ith iteration all components have the common rational failure probability $p_i (p_h \neq p_k$ for $h \neq k)$, then the ith value of the solution will be $\Pr[\gamma, p_i]$. Thus,

$$\sum_{j=0}^{|T|} g_j (1 - p_0)^j p_0^{|T|-j} = \Pr[\gamma, p_0],$$

$$\sum_{j=0}^{|T|} g_j (1 - p_1)^j p_1^{|T|-j} = \Pr[\gamma, p_1],$$

$$\vdots$$

$$\sum_{j=0}^{|T|} g_j (1 - p_{|T|})^j p_{|T|}^{|T|-j} = \Pr[\gamma, p_{|T|}].$$

This is a system of $|T| + 1$ equations in $|T| + 1$ unknowns, $\{g_j\}_{j=0}^{|T|}$. The corresponding matrix is nonsingular, and an exact matrix inversion algorithm can be executed in polynomial time to solve the system of equations.

Lemma 2. For any SCBS if the problem of finding the cardinality of a minimum cardinality pathset or cutset is NP-hard then the point estimate, rational, and functional reliability analysis problems are NP-hard.

Proof: The proof of Lemma 2 draws on the following elementary properties of the polynomial (1). (For further details, see [4].)

$$g_j = |\{S \subseteq T : T - S \in \Gamma \text{ and } |S| = j\}|, \tag{2}$$

$$0 \leq g_j \leq \binom{|T|}{j}, \tag{3}$$

$$\text{if } g_j > 0, \text{ then } g_k > 0 \text{ for all } k \geq j, \tag{4}$$

if i' is the cardinality of a minimum cardinality pathset and i^* is the cardinality of a minimum cardinality cutset, then

$$i' = \min\{j : g_j > 0\}, \tag{5}$$

$$i^* = \min\left\{j : g_{|T|-j} < \binom{|T|}{j}\right\}. \tag{6}$$

As above, let i' and i^* denote the cardinality of a minimum cardinality pathset and cutset, respectively. Properties (5) and (6) show that i', and i^* can be easily determined given g_j for $j = 0, 1, \ldots, |T|$. Thus a solution to the functional reliability analysis problem gives a solution to both the minimum cardinality pathset and cutset problems. Now by Lemma 1, this also holds for the rational reliability analysis problem.

The result for the point estimate reliability analysis problem is proved by showing that for any inputs (T, Γ), ϵ, and β if p is set equal to $1 - \epsilon/(\beta 2^{|T|})$ for all $i \in T$, then the output r satisfies $i' \leq k$ if and only if $r > \epsilon (1 - p)^k$. The proof of this result is given below. The proof of an analogous result for i^* follows in a similar manner.

Properties (4) and (5) imply $g_k \geq 1$ if and only if $i' \leq k$. Thus the result will follow by showing $g_k \geq 1$ if and only if $r > \epsilon (1 - p)^k$.

$(\Longrightarrow) g_k \geq 1$ implies for $i > k$, $g_i \geq \binom{|T|-k}{i-k}$ by property (2), and since, by the definition of SCBS, when elements are added to a set of size k that makes the system operate, the system must continue to operate. This implies

$$\Pr[\gamma, p] \geq \sum_{i=k}^{|T|} \binom{|T| - k}{i - k} (1 - p)^i p^{|T|-i}$$

$$= (1 - p)^k \left(\sum_{h=0}^{|T|-k} \binom{|T| - k}{h} (1 - p)^h p^{|T|-h} \right)$$

$$= (1 - p)^k.$$

Thus, $r > \epsilon (1 - p)^k$.

$(\Longleftarrow)$ If $g_k = 0$, then property (4) implies $g_i = 0$ for all $i \leq k$, which implies

$$\Pr[\gamma, p] = \sum_{i=k+1}^{|T|} g_i (1 - p)^i p^{|T|-i},$$

$$\leq \sum_{i=k+1}^{|T|} \binom{|T|}{i} (1 - p)^i p^{|T|-i}, \tag{7}$$

$$< \sum_{i=k+1}^{|T|} \binom{|T|}{i} (1 - p)^{k+1}. \tag{8}$$

Inequality (7) follows from property (3) and inequality (8) follows by noting that since $1 - p < p$, $(1 - p)^j p^{|T|-j}$ takes on its maximum when j is minimized. Now,

$$\sum_{i=k+1}^{|T|} \binom{|T|}{i} (i - p)^{k+1} < 2^{|T|} (1 - p)^{k+1} = (1 - p)^k \epsilon / \beta.$$

Thus, $\Pr[\gamma, p] < (1 - p)^k \epsilon/\beta$, which implies $r < (1 - p)^k \epsilon$.

It should be noted that the proof for the point estimate problem made no assumptions concerning ϵ and β and assigned all components the same failure probabilities. Thus the result holds for all possible choices of ϵ and β and even when all components have the same failure probabilities.

Theorem 1 shows that a large number of reliability analysis problems are NP-hard. The first result in this area is due to Rosenthal [6], who showed that the point estimate reliability analysis problem is NP-hard for $\Pr[c_2]$. Lemma 2 generalizes his arguments and Theorem 1, part (3), applies them to a wider class of measures.

More recently Valiant [7, 8] has shown that a wide class of counting problems are NP-hard. These results indicate that both the functional and the rational reliability analysis problem for $\Pr[h(s,t)]$ and $\Pr[h'(s,t)]$ are NP-hard [Theorem 1, parts (1) and (2)]. A significant open question remaining is to determine whether or not any of the reliability analysis problems for $\Pr[c_1]$ or $\Pr[c_1']$ are NP-hard.

Theorem 1

(1) The functional reliability analysis problem is NP-hard

(a) in networks with only node failures for $\Pr[h(s, t)]$ and $\Pr[h'(s, t)]$,

(b) in networks with only arc failures for $\Pr[h(s, t)]$, $\Pr[h'(s, t)]$, and $\Pr[c_3'] = \Pr[c_4']$.

(2) The rational reliability analysis problem is NP-hard in networks with node and arc failures for $\Pr[h(s, t)]$, $\Pr[c_2]$, $\Pr[h'(s, t)]$, $\Pr[c_2'(s)]$, $\Pr[c_3']$, and $\Pr[c_4']$

(3) The point estimate reliability analysis problem is NP-hard on networks with node and arc failures for $\Pr[c_2]$, $\Pr[c_2'(s)]$, $\Pr[c_3']$, and $\Pr[c_4']$.

Proof: (1a) Valiant [8] has shown that in either a directed or undirected network, counting the number of subsets of vertices whose removal leaves an (s, t)-path in NP-hard. If

$$\Pr[h(s, t)] = \sum_{i=0}^{n} g_i (1 - p)^i p^{n-i},$$

then from property (2) it is clear that

$$\sum_{i=0}^{n} g_i = [\text{the number of subsets of vertices whose removal leaves an } (s, t)\text{-path}].$$

A similar equation holds for $\Pr[h'(s, t)]$.

(1b) Valiant [8] has shown that on either a directed or undirected network, counting the number of subsets of arcs that contain an (s, t) path is NP-hard. The proofs for $\Pr[h(s, t)]$ and $\Pr[h'(s, t)]$ follow as in (1a).

To see that the result holds for $\Pr[c_3'] = \Pr[c_4']$, note that if a directed network contains a Hamiltonian circuit, then all minimum cardinality strongly connected subnetworks (i.e., minimum cardinality pathsets) are Hamiltonian circuits. Now since the problem of determining if a directed network contains a Hamiltonian circuit is NP-hard [12], Lemma 2 implies that computing $\Pr[c_3'] = \Pr[c_4']$ is NP-hard.

(2) Lemma 1 and (1b) imply that the rational reliability analysis problem is NP-hard for $\Pr[h(s, t)]$, $\Pr[h'(s, t)]$, $\Pr[c_3']$, and $\Pr[c_4']$. To prove the result for $\Pr[c_2]$ and $\Pr[c_2'(s)]$, consider the problem of determining the reliability of a set-covering system (Fig. 1). Inputs to the problem include a ground set G and a set of covering subsets of $\overline{G}$. Now define an SCBS where the set of randomly failing elements is $\overline{G}$ and where the system operates if the set of operating members of $\overline{G}$ covers G. The minimum cardinality pathsets in this system are minimum cardinality covers. Now, by Lemma 2, since the minimum cardinality set-covering problem in NP-hard [12], the functional and rational reliability analysis problems are NP-hard for the set-covering system. Determining the reliability of a set-covering system may be reduced to determining $\Pr[c_2]$ or $\Pr[c_2(s)]$ using the transformations shown in Figs. 2 and 3.

(3) The construction given in the proofs of (1) and (2) together with Lemma 2 imply that the point estimate reliability analysis problem is NP-hard for $\Pr[c_2]$, $\Pr[c_2'(s)]$, $\Pr[c_3']$, and $\Pr[c_4']$.

4. RELATIONSHIP BETWEEN DIRECTED AND UNDIRECTED PROBLEMS

This section shows that the problems of computing $\Pr[h(s, t)]$ and $\Pr[c_1]$ may be reduced to the problems of computing $\Pr[h'(s, t)]$ and $\Pr[c_1'(s)]$, respectively. This result for $\Pr[h(s, t)]$ is given in Fratta and Montanari [5]. Theorem 2 generalizes their result to classes of SCBSs. In this context the result can easily be applied to $\Pr[c_1]$ and to other problems as well.

The proof of Theorem 2 requires the following lemma:

Lemma 3. If K and K^0 and the set of pathsets and the set of cutsets respectively for an SCBS (T, Γ), then for any $S \subseteq T$ exactly one of the following occurs:

(1) S contains a member of K,

(2) $T - S$ contains a member of K^0.

Proof: If $T - S \in \Gamma$, then S contains a member of K; and if $T - S \notin \Gamma$, then $T - S$ contains a member of K^0.

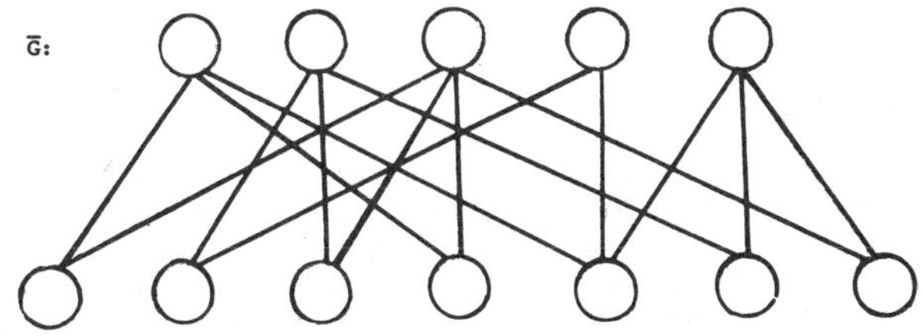

FIG. 1. Set-covering system $T = \overline{G}$, $\Gamma = \{S \subseteq T : T - S \text{ covers } G\}$.

Theorem 2. Let K_1 and K_1^0 be the set of pathsets and the set of cutsets for an SCBS (T, Γ_1). If there exist $i^0, i' \in T$ with

(1) $p_{i^0} = p_{i'}$,

(2) for all $F \in K_1$, $i^0 \in F$ implies $i' \notin F$,

(3) for all $F \in K_1^0$, $i^0 \in F$ implies $i' \notin F$,

and if K_2 and K_2^0 are the set of pathsets and the set of cutsets for SCBS $(TU\{i^*\} - \{i^0, i'\}, \Gamma_2)$ obtained by replacing each occurrence of i^0 and i' with i^* and letting $p_{i^*} = p_{i^0} = p_{i'}$, then

$$\Pr[\gamma_1] = \Pr[\gamma_2].$$

Proof: For $i \in T$, let x_i and $\bar{x}_i$ denote the events i operates and i fails, respectively. Let $\{H_k\}_{k=1}^{2^{|T|-2}}$ be all possible states of elements in $T - \{i^0, i'\}$ and let $\mathrm{OP}(H_k)$ be the set of elements H_k implies are operative and $\mathrm{FA}(H_k)$ the set of elements H_k implies are failed, i.e.,

$$H_k = \bigwedge_{i \in \mathrm{OP}(H_K)} x_i \bigwedge_{j \in \mathrm{FA}(H_k).} \bar{x}_j$$

where $\wedge$ denotes conjunction (intersection).

By the law of total probability,

$$\Pr[\gamma_1] = \sum_{k=1}^{2^{|T|-2}} \Pr[\gamma_1 | H_k] \Pr[H_k],$$

$$\Pr[\gamma_2] = \sum_{k=1}^{2^{|T|-2}} \Pr[\gamma_2 | H_k] \Pr[H_k].$$

The theorem follows by showing that for all k, $\Pr[\gamma_1 | H_k] = \Pr[\gamma_2 | H_k]$.

Case 1: $\mathrm{OP}(H_k)$ contains a member of K_1. In this case $\mathrm{OP}(H_k)$ also contains a member of K_2. Therefore,

$$\Pr[\gamma_1 | H_k] = \Pr[\gamma_2 | H_k] = 1.$$

Case 2: $\mathrm{FA}(H_k)$ contains a member of K_1^0. In this case $\mathrm{FA}(H_k)$ also contains a member of K_2^0. Therefore,

$$\Pr[\gamma_1 | H_k] = \Pr[\gamma_2 | H_k] = 0.$$

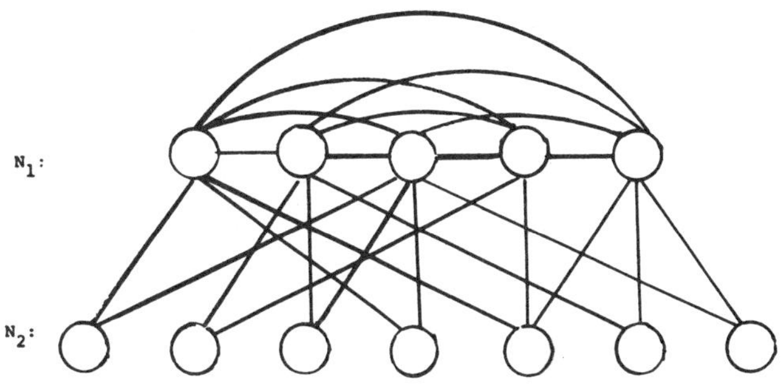

FIG. 2. Reduction for $Pr[C_2]$. All arcs are perfectly reliable; nodes in N_2 are perfectly reliable; nodes in N_1 have failure probabilities equal to the failure probabilities of the corresponding members of $\overline{G}$.

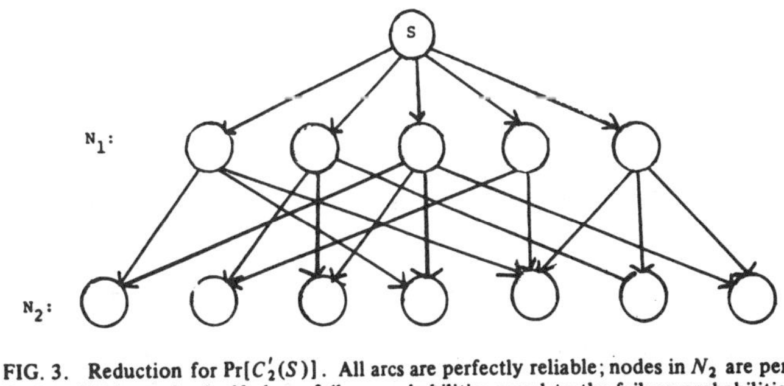

FIG. 3. Reduction for $\Pr[C_2'(S)]$. All arcs are perfectly reliable; nodes in N_2 are perfectly reliable; nodes in N_1 have failure probabilities equal to the failure probabilities of the corresponding members of $\overline{G}$.

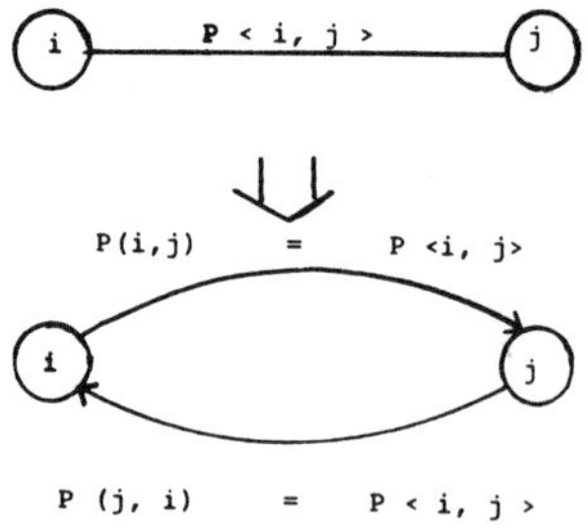

FIG. 4. Replacing an undirected arc by two directed arcs.

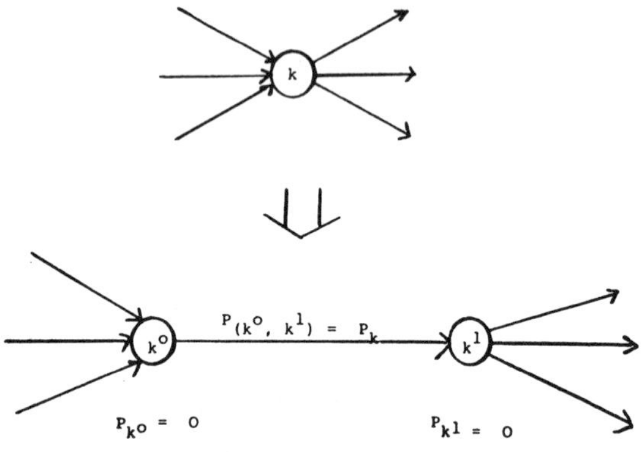

FIG. 5. Eliminating node failures.

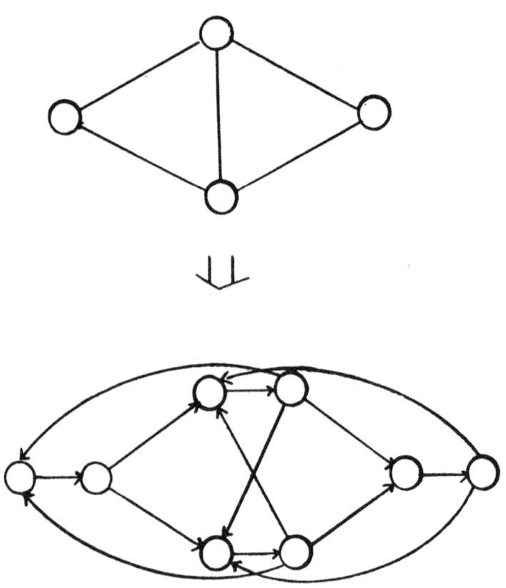

FIG. 6. Transformation from an undirected network with node failures to a directed network without node failures.

Case 3: $\mathrm{OP}(H_k)$ contains no member of K_1 and $\mathrm{FA}(H_k)$ contains no member of K_1^0. In this case, $\mathrm{OP}(H_k)$ contains no member of K_2 and $\mathrm{FA}(H_k)$ contains no member of K_2^0.

By Lemma 3, $\mathrm{OP}(H_k)\, U\{i^*\}$ contains a member of K_2 and $\mathrm{FA}(H_k)\, U\{i^*\}$ contains a member of K_2^0, so $H_k \wedge x_{i^*}$ implies γ_2 and $H_k \wedge \bar{x}_{i^*}$ implies $\bar{\gamma}_2$ (the complement of γ_2) and $P[\gamma_2 | H_k] = 1 - p_{i^*}$.

By Lemma 3, $\mathrm{OP}(H_k)\, U\{i^0, i'\}$ contains a member of K_1 and $\mathrm{FA}(H_k)\, U\{i^0, i'\}$ contains a member of K_1^0.

(I) Since i^0 and i' are in no common member of K_1, either $\mathrm{OP}(H_k)\, U\{i^0\}$ contains a member of K_1 or $\mathrm{OP}(H_k)\, U\{i'\}$ contains a member of K_1.

(II) Similarly, since i^0 and i' are in no common member of K_1^0, either $\mathrm{FA}(H_k)\, U\{i^0\}$ contains a member of K_1^0 or $\mathrm{FA}(H_k)\, U\{i'\}$ contains a member of K_1^0.

Suppose in (I), $\mathrm{OP}(H_k)\, U\{i^0\}$ contains a member of K_1. Then by Lemma 3, $T - \mathrm{OP}(H_k) - \{i^0\} = \mathrm{FA}(H_k)\, U\{i'\}$ contains no member of K_1^0, which implies by (II) that $\mathrm{FA}(H_k)\, U\{i^0\}$ contains a member of K_1^0. Thus, $H_k \wedge x_{i^0}$ implies γ_1 and $H_k \wedge \bar{x}_{i^0}$ implies $\bar{\gamma}_1$, so $\Pr[\gamma_1 | H_k] = 1 - p_{i^0}$. Similarly, if, in (I), $\mathrm{OP}(H_k)\, U\{i'\}$ contains a member of K_1, then $\Pr[\gamma_1 | H_k] = 1 - p_{i'}$. The result for case 3 now follows by the assumption that $p_{i^*} = p_{i^0} = p_{i'}$.

The impact of Theorem 2 on the measures defined in Sec. 2 is indicated by the following corollary. Figure 4 illustrates the transformation.

Corollary 1. Consider an undirected network G and a directed network G' with the properties

(1) G and G' have the same node set and node failure probabilities; and

(2) for each undirected arc $\langle i,j \rangle$ in G, there exists directed arcs (i,j) and (j,i) in G' with $p_{\langle i,j \rangle} = p_{(i,j)} = p_{(j,i)}$.

Then

(1) $\Pr[h(s,t)] = \Pr[h'(s,t)]$ for all node pairs $\{s,t\}, s \neq t$;

(2) $\Pr[c_1] = \Pr[c_1'(s)]$ for all nodes s,

where the measures on the left-hand sides are computed over G and the measures in the right-hand sides over G'.

Proof: The set of pathsets and the set of cutsets corresponding to $P[h'(s,t)]$ are directed (s,t)-paths and directed (s,t)-cuts. The set of pathsets and the set of cutsets corresponding to $\Pr[c_1'(s)]$ are spanning arborescences rooted at s and s-directed cuts. The possibility of node failures is included, so in both cases nodes are included in the appropriate structure function. It is clear that for either measure, the pair of arcs $\{(k,j), (j,k)\}$ satisfy the conditions on i^0 and i' given in Theorem 2. The undirected network may be obtained from the corresponding directed network by making a sequence of replacements of pairs of directed arcs $\{(k,j), (j,k)\}$ by undirected arcs $\langle k,j \rangle$. The theorem shows that each replacement does not alter the probability that either system operates.

Corollary 1 implies that $\Pr[h(s,t)]$ and $\Pr[c_1]$ can be transformed into corresponding problems on directed networks with twice as many arcs.

5. TRANSFORMATIONS THAT ELIMINATE NODE FAILURES

There are no known methods for replacing an undirected network with node and arc failures with an undirected network with only arc failures. However, the standard network flow technique used on directed networks also applies to reliability analysis problems (Fig. 5). This reduction applies to all directed network measures listed in Sec. 2, except $\Pr[c_2'(s)]$ and $\Pr[c_3']$. In these two cases the node states are implicit in the definition of the measures.

For $\Pr[c_1]$, $\Pr[c_1'(s)]$, and $\Pr[c_3']$ node failures can be eliminated without altering the network. If a node fails, then no path between it and any other node can exist and the events $[c_1]$, $[c_1'(s)]$, and $[c_3]$ cannot occur. Thus if $[D]$ equals $[c_2]$, $[c_1'(s)]$ or $[c_4']$, $[H]$ = [all nodes operate], and $q_i = \Pr[\text{node } i \text{ fails}]$, then

$$\Pr[D] = \Pr[D|H]\,\Pr[H] + \Pr[D|\bar{H}]\,\Pr[\bar{H}],$$

$$\Pr[D|\bar{H}] = 0, \qquad \Pr[H] = \prod_{i \in N} (1 - q_i), \text{ so that}$$

$$\Pr[D] = \Pr[D|H] \prod_{i \in N} (1 - q_i).$$

$\Pr[D|H]$ is $\Pr[D]$ computed over the network with all nodes assumed to be perfectly reliable.

The major impact of the transformations given in Secs. 4 and 5 on network reliability analysis problems may be summarized as follows. A polynomially bounded algorithm to compute $\Pr[h'(s,t)]$ ($\Pr[c_1'(s)]$) on directed networks with perfectly reliable nodes can be used to compute in polynomial time $\Pr[h'(s,t)]$ ($\Pr[c_1'(s)]$) on directed networks with node and arc failures and $\Pr[h(s,t)]$ ($\Pr[c_1]$) on undirected networks with node and arc failures. Figure 6 illustrates the transformations.

6. SUMMARY

This paper has explored the computational complexity of network reliability analysis problems. We have shown using polynomial time reductions that hard combinatorial problems can be reduced to certain reliability analysis problems and that some reliability analysis problems can be reduced to other such problems. These results can be used by algorithmic researchers in deciding how to approach certain reliability problems. In particular, the results showing that certain network reliability analysis problems are NP-hard indicate that it is unlikely that polynomially bounded algorithms exist to solve these problems. Consequently, the search for polynomially bounded algorithms should concentrate on structured network problems. For example, a polynomially bounded algorithm does exist for reliability analysis problems on tree networks [13].

I would like to thank Bruce Golden for commenting on a draft of this paper.

References

[1] M. O. Ball, "Computing network reliability," *Oper. Res.*, 27, 823–838 (1979).

[2] J. A. Buzacott, "A Recursive Algorithm for Finding the Probability that a Graph is Disconnected," Working Paper No. 76-016, Department of Industrial Engineering, University of Toronto, August, 1976.

[3] A. Rosenthal, "Computing the Reliability of Complex Networks," *SIAM J. Appl. Math.*, 32, 384–393 (1977).

[4] M. O. Ball, "Network Reliability Analysis: Algorithms and Complexity," Ph.D. thesis, Department of Operations Research, Cornell University, 1977.

[5] L. Fratta and U. Montanari, "A Recursive Method Based on Case Analysis for Computing Network Terminal Reliability," I EI-CNR Int. Rep. B-75-7, Pisa, Italy, 1976.

[6] A. Rosenthal, "A Computer Scientist Looks at Reliability Computations," in *Reliability and Fault Tree Analysis*, R. E. Barlow, J. B. Fussell and D. D. Singpurwalla, Eds., SIAM, Philadelphia, 1975, pp. 133–152.

[7] L. G. Valiant, "The Complexity of Computing the Permanent," Report CS12-14-77, University of Edinburgh, 1977.

[8] L. G. Valiant, "The Complexity of Enumeration and Reliability Problems," *SIAM J. Comput.*, 8, 410–421 (1979).

[9] F. Harary, *Graph Theory*, Addison-Wesley, Reading, MA, 1972.

[10] H. Frank, R. E. Kahn, and L. Kleinrock, "Computer Communication Network Design-Experience with Theory and Practice," *AFIPS Conference Proceedings*, **40**, 255–270 (1972).

[11] M. Garey and D. Johnson, *Computers and Intractability: A Guide to the Theory of NP-Completeness*, Freeman, San Francisco, 1979.

[12] R. M. Karp, "Reducibility Among Combinatorial Problems," in *Complexity of Computer Computations*, R. E. Miller and J. W. Thatcher, Eds., Plenum, New York, 1972.

[13] A. Kershenbaum and R. M. Van Slyke, "Recursive Analysis of Network Reliability," *Networks*, 3, 81–94 (1973).

[14] Z. W. Birnbaum, J. D. Esary, and S. C. Saunders, "Multicomponent Systems, Structures and their Reliability," *Technometrics*, 3, 55–77 (1961).

Part IV

Power Systems
(Simulation, Estimation, Planning)

Techniques for Exploiting the Sparsity of the Network Admittance Matrix

NOBUO SATO
MEMBER IEEE

W. F. TINNEY
MEMBER IEEE

Summary: This paper describes some computer programing techniques for taking advantage of the sparsity of the admittance matrix. The techniques are based on two main ideas; (1) determination of a sequence of operations which results in a near minimum of memory and computing, (2) preservation of these operations for repetition. Use of these techniques makes it possible to obtain significant reductions in memory and processing time for many network analysis programs. Claims are substantiated by actual results.

THE analysis of networks frequently involves operations with the matrix of short-circuit driving point and transfer admittances and/or its inverse, the matrix of open-circuit driving point and transfer impedances.[1] In programing digital computers to solve problems through the use of either or both of these matrices it is possible to realize very significant reductions in computing time and memory by taking advantage of properties of the admittance matrix which are usually ignored.

Emphasis on the development of efficient methods for obtaining and modifying

Paper 63-201, recommended by the AIEE Power System Engineering Committee and approved by the AIEE Technical Operations Department for presentation at the IEEE Winter General Meeting, New York, N. Y., January 27–February 1, 1963. Manuscript submitted October 15, 1962; made available for printing November 27, 1962.

NOBUO SATO and W. F. TINNEY are with the Bonneville Power Administration, Portland, Oreg.

Reprinted from *IEEE Trans. Power App. Syst.*, vol. PAS-82, pp. 944–950, Dec. 1963.

the impedance matrix has lead to its widespread use in the solution of many network problems. This emphasis has tended to obscure the possibilities inherent in the more primitive approach of solving network problems by the method of matrix reduction and back-substitution. This paper shows how the network admittance matrix can be reduced to triangular form with surprising efficiency and introduces a technique for preserving the reduction operations for repetitive application in much the same way as a matrix inverse is used. It is argued that this approach is preferable to the use of the impedance matrix in many applications. Improvements over previous methods in the order of five or ten to one are obtainable through use of the new techniques.

The main text presents the development of the techniques. Numerical examples, descriptions of practical applications, and details of programing are given in the appendixes.

Basis of Techniques

Properties of Network Matrices

For most networks the admittance matrix is relatively sparse, whereas the impedance matrix is full. Except for rather special situations which are excluded from consideration in this paper, both matrices are nonsingular and symmetric. The row and column indexes of both matrices correspond to the node numbers of the network represented. In the admittance matrix each nonzero off-diagonal element corresponds to a network branch connecting the pair of nodes indicated by the row and column indexes of the element. Since most networks exhibit irregularity in their connection arrangements, their admittance matrices exhibit a diversity in the number of nonzero elements in each row. The diversity in elements per row provides the clue to an efficient method of triangularizing the admittance matrix.

Effect of Sequence of Reduction Operations

The techniques to be described all depend in some way on the reduction or triangularization of a matrix.[2] Given a network admittance matrix, it can be shown that, generally, the number of nonzero operations and the number of additional nonzero terms introduced in the reduction to triangular form vary with the sequence in which the rows are processed. Consider the network of Fig. 1. It consists of four independent nodes numbered 1 through 4 and a reference node numbered 0. Equation 1 shows the pattern of nonzero elements of the admittance matrix Y of this network with numbering corresponding to that shown.

$$Y=\begin{Vmatrix} X & X & X & X \\ X & X & & \\ X & & X & \\ X & & & X \end{Vmatrix} \tag{1}$$

Symmetry is ignored in these displays. Equation 2 shows the pattern after processing the first row.

$$Y'=\begin{Vmatrix} 1 & X & X & X \\ & X & X & X \\ & X & X & X \\ & X & X & X \end{Vmatrix} \tag{2}$$

This step was accomplished by: (1) multiplying the first row by the reciprocal of its leading element, (2) multiplying the new first row by the negative of the leading element of the second row and adding the product to the second row, (3) repeating step 2 for the third and fourth rows. Equation 3 shows the pattern of the resultant upper triangle A after processing the second and third rows.

$$A=\begin{Vmatrix} 1 & X & X & X \\ & 1 & X & X \\ & & 1 & X \\ & & & 1 \end{Vmatrix} \tag{3}$$

Now observe the effect of processing the fourth row first or, what is equivalent, renumbering so that rows and columns 1 and 4 are interchanged. Equation 4 shows the pattern after the interchange.

$$Y=\begin{Vmatrix} X & & & X \\ & X & & X \\ & & X & X \\ X & X & X & X \end{Vmatrix} \tag{4}$$

Equation 5 shows the pattern after processing of the first (old fourth) row, and equation 6 shows the final triangle A.

$$Y'=\begin{Vmatrix} 1 & & & X \\ & X & & X \\ & & X & X \\ & X & X & X \end{Vmatrix} \tag{5}$$

$$A=\begin{Vmatrix} 1 & & & X \\ & 1 & & X \\ & & 1 & X \\ & & & 1 \end{Vmatrix} \tag{6}$$

These two examples demonstrate how the sequence of operations or arrangement of rows and columns influences the number of operations and nonzero terms. Taking full advantage of symmetry, the first example requires four divisions, six multiplications, ten multiplication-additions, and at least six words of computer memory. The second example requires four divisions, three multiplications, three multiplication-additions, and three words of computer memory. The matrices A of equations 3 and 6 are equivalent and can be used interchangeably in most applications.

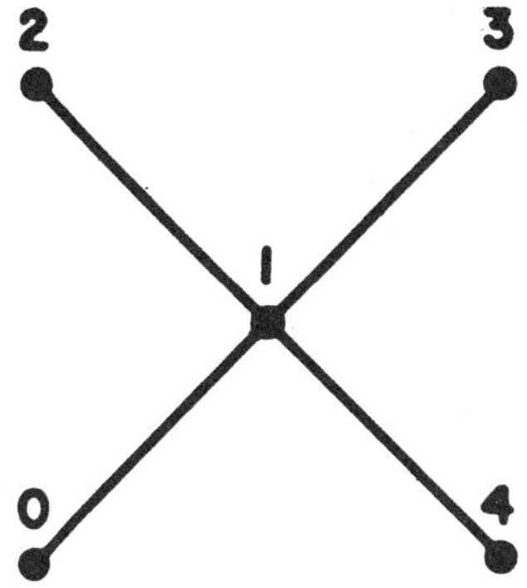

Fig. 1. Example network

The effect of varying the row and column sequence of the reduction process can be interpreted from the network viewpoint. Using the present example it is evident that a network reduction to eliminate the node at the center of the star results in an equivalent network with three independent nodes and six branches, whereas the elimination of any one of the radial nodes results in a network with three independent nodes and only three branches. Corresponding effects take place in the matrix reduction operations.

Since the sequence in which rows are processed influences the number of operations and number of nonzero terms, it is evident that for each network admittance matrix there must exist a processing sequence which requires the least operations, and a sequence, possibly but not necessarily the same, which requires the least memory. The authors have briefly investigated this problem and have tentatively concluded that any logic for determining these minima would probably be too complicated for practical application to large matrices. Nevertheless, it is possible by means of a very simple logic to establish a sequence of operations which greatly reduces the computing and memory requirements. A near optimum sequence from the standpoint of both criteria is achieved by numbering or arranging the rows so that those which initially contain the fewest terms are processed first and those with the most terms are processed last. A more detailed description of events with this condition imposed will indicate from the network viewpoint how and why the method works.

Either the sequence of operations or the arrangement of matrix rows in memory is based on the count of branches per node before the process begins. In this count, branches to the reference node are excluded because their representation in the admittance matrix does not require an off-diagonal term. First, the nodes with only one branch are assigned the lowest numbers. The effect of eliminating these nodes is to diminish the network by one branch per node eliminated. Next, the nodes with two branches are eliminated. Each node eliminated at this step deletes two existing branches and creates one new branch or, if paralleling occurs as a result of this step, modifies one old branch. At the end of these operations the reduced network will have fewer branches than before. The number of branches terminating on the remaining nodes will still agree with the original count unless paralleling occurred, in which case the count for some nodes will have diminished. An optimum logic would take these effects into account; the simple logic being used here ignores them. Next, the rows corresponding to the nodes with three branches in the original network (some of which may now have only two) are eliminated. These eliminations delete three existing branches and either create three new branches or modify and create various combinations of from one up to three branches. At the end of these eliminations the network will have no more and possibly fewer branches than before. The original branch per node count for the remaining nodes will have changed if any paralleling occurred. In an optimum logic all of these events would have to be reckoned with, but they are ignored here. From this point onward the elimination operations will necessarily introduce additional branches in the network until the process is terminated or until every node of the reduced network has a branch connecting to every other node. The process could be described analogously in terms of matrix reduction operations.

The effectiveness of this technique is a function of network configuration and size. For matrices of power system networks, improvements in the order of 4 to 1 are obtainable. The potential reduction in memory can be realized by employing a scheme of programing which stores and operates only on the nonzero elements. This is further explained in Appendix I. The computer time required to count branches and arrange data is negligible in comparison with the gains resulting therefrom.

The method of numbering given in the foregoing will be referred to as Scheme I. There is another method for sequencing reduction operations which, although considerably less efficient from the standpoint of memory and computing time, affords compensating advantages is certain applications. From the network viewpoint, node numbering is as follows: Starting with an arbitrary node as number one, the nodes one branch removed from this node are assigned the next numbers in sequence; then, the unnumbered nodes one branch removed from these are numbered next, etc. until all nodes have been numbered. The effect of this scheme on the pattern of the Y and A matrices of a 6-node network is shown in equations 7 and 8.

$$Y=\begin{array}{c|cccccc|}
 & 1 & 2 & 3 & 4 & 5 & 6 \\
1 & X & X & X & X & & \\
2 & X & X & & X & & \\
3 & X & & X & & X & \\
4 & X & X & & X & & X \\
5 & & & X & & X & X \\
6 & & & & X & X & X
\end{array} \qquad (7)$$

$$A=\begin{array}{c|cccccc|}
 & 1 & 2 & 3 & 4 & 5 & 6 \\
1 & X & X & X & X & & \\
2 & & X & X & X & & \\
3 & & & X & X & X & \\
4 & & & & X & X & X \\
5 & & & & & X & X \\
6 & & & & & & X
\end{array} \qquad (8)$$

This scheme makes it possible during the reduction process either to write out or overlay portions of the A matrix as they become unneeded in subsequent eliminations. For example, in the matrix of equations 7 and 8, it is evident that after row 4 has been processed, rows 1 and 2 are no longer needed for any subsequent operations. It will also be noted that the form of the A matrix resulting from this scheme simplifies programing. The A matrix is always full within the margin defined by the outermost elements of Y. Thus the A matrix size, pattern, and memory requirements are predictable before elimination, making it easy to establish breakpoints for writeout or overlay. Addressing is simplified because the column positions of each element of a row of A are completely specified by the position of the outermost element. This method of numbering will be referred to as Scheme II.

PRESERVING THE ELIMINATION OPERATION

In the foregoing section, techniques for determining efficient sequences of operations for the reduction of an admittance matrix have been presented. In this section, a technique for preserving these operations for repetitive application will be developed.

Frequently network problems require multisolution to equation 9:

$$I=YE \qquad (9)$$

where I is a colum vector of given node currents, Y is the network admittance matrix, and E is a column vector of node voltages to be determined. Rather than solve equation 9 repeatedly, equation 10, employing the impedance matrix Z, is usually used in order to obtain E directly.

$$E=ZI \qquad (10)$$

It will next be shown that the A matrix plus the reciprocals used in transforming Y to A contain all the information needed to transform I to E. A simple algorithm is developed to accomplish the transformation. Although the development involves some cumbersome notation, the final result is quite simple; it closely resembles the familiar process of solving simultaneous equations by reduction and back-substitution.

Equation 11 displays equation 10 in another form where the Z matrix is defined as the product of elementary matrices.[3]

$$E=(T_q \ldots T_3T_2T_1)I \qquad (11)$$

These matrices define the elementary elimination operations involved in the transformation of Y to a unit matrix U as is indicated here in the following equation:

$$U=(T_q(\ldots(T_2(T_1Y))\ldots)) \qquad (12)$$

In the transformation the elementary operations are performed in sequence as indicated by the matrix products within each pair of parentheses of equation 12. As the elementary matrices are developed successively and products are formed within each pair of parentheses, innermost pair first, the Y matrix is gradually reduced to U.

Although the product of the elementary matrices $T_1, \ldots, T_q$ of equations 11 and 12 is unique, the number and type of elementary matrices leading to this product are not. As shown in the foregoing, the number of operations and rate of accumulation of additional terms vary with the sequence in which operations are performed.

Only two types of operations are required. They are defined by the elementary matrices shown in equations 13 and 14.

$$T_{ij} = i \left\| \begin{matrix} 1 & & & & & & \\ & \cdot & & & & & \\ & & 1 & & & & \\ & & & \cdot & & & \\ & & & & c & & \\ & & & & & 1 & \\ & & & & & & \cdot \\ & & & & & & & 1 \end{matrix} \right\| \quad (i=j) \tag{13}$$

$$T_{ij} = i \left\| \begin{matrix} 1 & & & & \\ & 1 & & & \\ & c & . & 1 & \\ & & & & \cdot \\ & & & & & 1 \end{matrix} \right\| \quad (i \neq j) \tag{14}$$

Double subscript notation is now introduced to indicate the type of operation specified by an elementary matrix and the rows upon which it is to be performed. A matrix premultiplied by T_{ij} of equation 13 results in no change except in the ith row which is multiplied by the scalar c. A matrix premultiplied by T_{ij} of equation 14 results in no change except in the ith row to which is added c times the jth row. From the viewpoint of programing, the operation specified by an elementary matrix can be defined by one element and a pair of co-ordinates.

In order to illustrate the application of the ideas developed in the foregoing, the sequence of operations indicated by equations 11 and 12 is divided into two parts as shown in equation 15.

$$U=(A_{12}(\ldots(A_{n-1,n}(B_{nn}(B_{n,n-1}(\ldots \times (B_{21}(B_{11}Y))\ldots))))\ldots)) \tag{15}$$

The order of the matrix is n. The B_{ij}'s are elementary matrices defining the transformation of Y to a triangular matrix A (zeroing of columns below the diagonal) and the A_{ij}'s are elementary matrices defining the transformation of A to U (zeroing of rows to right of diagonal).

Equation 16 is equation 11 expressed in terms of the elementary matrices of equation 15.

$$E=(A_{12}(\ldots(A_{n-1,n}(B_{nn}(B_{n,n-1}(\ldots \times (B_{21}(B_{11}I))\ldots))))\ldots)) \tag{16}$$

Since I is a column matrix, each elementary matrix B_{ij} or A_{ij} defines the multiplication of one element or the multiplication of one element and the addition of this product to another element.

Equation 16 is now compared with equation 10. Assuming both Z and Y to contain all nonzero elements, the number of operations indicated by both equations is the same; for an n-th order matrix, n^2 multiplication-additions are required. However, as has already been shown, the Y matrix is sparse and by appropriate arrangement of eliminations, the triangular matrix A can also be made sparse. Although the form of equation 16 indicates n^2 elementary operations, only those operations with subscripts corresponding to the row and column positions or transpositions of the nonzero elements of A result in any change.

The essential element (nonzero, nonunity) of each matrix A_{ij} is the negative of the ijth element of A. The essential element of each matrix B_{ii} is the reciprocal of the ith pivot element in the reduction process. Each matrix B_{ij} is related to the transpose of a matrix A_{ji} as indicated in equation 17.

$$B_{ij}=B_{ii}A_{ij}B_{ii}^{-1} \quad (i \pm j) \tag{17}$$

Substitution of B_{ij} from equation 17 into equation 16 yields

$$E=(A_{12}(\ldots(A_{n-1,n}(B_{nn}(B_{n-1,n-1}, A_{n,n-1}\times B_{n-1}{}^{-1}{}_{,n-1}(\ldots(B_{11}A_{21}B_{11}^{-1}(B_{11}I))\times \ldots))))\ldots)) \tag{18}$$

Products of the form $B_{ii}^{-1}B_{ii}$ cancel giving

$$E=(A_{12}(\ldots(A_{n-1,n}(B_{nn}(B_{n-1,n-1}A_{n,n-1}\times (\ldots(A_{21}(I))\ldots))))\ldots)) \tag{19}$$

Grouping the n matrices of the form B_{ii} as indicated in equation 20 does not alter the result

$$E=(A_{12}(A_{13}(\ldots(A_{n-1,n}(B_{nn}(\ldots(B_{11}\times (A_{n,n-1}(\ldots(A_{31}(A_{21}(I)))\ldots)))\times \ldots)))\ldots))) \tag{20}$$

Equation 20 displays the operations in a form most convenient for programing. All of the required information is contained in the A matrix and the n reciprocals needed for the B_{ii} matrices. The downward operations on I are determined by following the rows of the A matrix term by term, from row 1 to row $n-1$. Then the n multiplications indicated by the B_{ii} matrices are performed. The upward operations on I (which correspond to back substitution) are determined by following the terms of the A matrix from row $n-1$ back to row 1 in reverse order. This operation is carried out in detail with a numerical example in Appendix II.

A comparison between the time and memory requirements for transforming I to E by equations 10 and 20 may now be made. For an A matrix of order n with m nonzero off-diagonal elements, the operations defined by equation 20 require n multiplications and $2m$ multiplication-additions. Memory for $m+n$ elements and an addressing scheme is required. Experience has shown that the Y matrix of a typical power system network containing 200 nodes and 400 branches and numbered according to Scheme I reduces to an A matrix of approximately 2,000 elements. Transforming a general vector I to E for such a network by the method of equation 20 would require 200 multiplications, 4,000 multiplication-additions, and memory for 2,200 elements plus an addressing scheme. Transformation by means of equation 10 would require 40,000 multiplication-additions and memory for 20,100 elements. The contrast between methods increases with the increasing of the network size.

Applications

The computing techniques which have been presented may be applied to a wide variety of network problems, only a few of which will be suggested here. Fortunately, many applications involve only small alterations in existing programs; others, however, require a completely new approach to the problems.

An obvious application is numbering by Scheme I to reduce the time and memory requirements for the single solution of a set of network equations of the form of equation 9. An example of the effectiveness of this application in connection with the power flow problem is cited in Appendix III.

Another straightforward application is the substitution of the A matrix for the impedance matrix in the many problems involving repetitive solutions of the type indicated in equation 10. The only exception to the feasibility of this substitution would be in situations where the column matrix I is not general but contains, say, only one nonzero element as in the short-circuit problem. A less obvious application which illustrates one of the possibilities for extension of the method as a substitute for the impedance matrix is discussed in Appendix IV.

Network reduction is another important application. Renumbering by Scheme I should make it possible to perform reductions on networks in the order of 1,000 nodes within a $32K$ computer memory; when the memory capacity is exceeded by employing this scheme, renumbering according to Scheme II should make it possible to accommodate network reductions of almost unlimited size.

Although in many problems the given techniques provide a superior substitute for the impedance matrix, there are problems where it is still required. In these cases, existing inversion programs[5] can be speeded up by simply applying numbering Scheme I to the order in which pivot rows are selected. Application of this scheme delays the accumulation of nonzero terms in the early stages of the inversion, thereby reducing the total number of nonzero operations. It is necessary, of course, that the program take advantage of zero terms for the technique to be effective.

Conclusions

Significant improvements in methods of solution of many network problems are obtainable through the application of the techniques presented in this paper. Heretofore, the principles on which the techniques are based have been largely ignored in methods of development and programing. The results presented suggest that further work along the same and related lines would very likely reveal more overlooked possibilities for simplifying the analysis of large networks.

Appendix I. Method of Programing Elimination Scheme

In order to obtain the reductions in memory and time made possible by a sparse matrix, it is necessary to employ a programing scheme which stores and operates only upon the nonzero elements. One such method is explained briefly in the following.

It is assumed that the Y matrix is stored compactly in some easily accessible form with a co-ordinate identification accompanying each element. The first row of Y is brought into a working area. The second and succeeding elements of this row multiplied by the reciprocal of the first (pivot) element and stored as the first entries in a table. Column positions of each element are stored in a parallel table. For some applications the reciprocal is stored in a third table. The second row of Y is then brought into the working area. If the second row contains an element in the first column, this element is immediately eliminated by appropriate combination with the modified first row. As a result of the elimination, certain elements of the second row will be modified and others added. After the elimination, each element of the modified second row except the first is multiplied by the reciprocal of the first and added to the table as before. Column positions of each element are stored in a parallel table, and, if required, the reciprocal is also stored. The third row is then brought into the working area. If it contains elements in the first or second columns, eliminations are again performed; after their completion, the modified third row is processed the same as the preceding row. If any new row does not contain elements in the columns corresponding to any of the preceding rows, it is treated the same as the first row and the process goes on to the next row. The tables produced as a result of processing the entire Y matrix in this manner form, in effect, the A matrix, stored as compactly as possible with row and column identification for each element. The reciprocals are available if needed. A program similar to this was used in the elimination method of power flow solution discussed in Appendix II.

Table I. Elements of A_{ij} Matrices

Element Values	Co-ordinates
1/2	1–4
1/3	2–3
1/3	2–4
4/5	3–4

Appendix II

The following numerical example illustrates in detail the transformation of a current vector I to a voltage vector E by the method indicated in equation 20. The Y matrix relating I and E is shown in equation 21.

$$Y=\begin{Vmatrix} 2 & 0 & 0 & -1 \\ 0 & 3 & -1 & -1 \\ 0 & -1 & 2 & -1 \\ -1 & -1 & -1 & 3 \end{Vmatrix} \qquad (21)$$

Equations 22 through 24 show successive stages of the reduction process.

$$Y'=\begin{Vmatrix} 1 & 0 & 0 & -1/2 \\ 0 & 3 & -1 & -1 \\ 0 & -1 & 2 & -1 \\ 0 & -1 & -1 & 5/2 \end{Vmatrix} \qquad (22)$$

$$Y''=\begin{Vmatrix} 1 & 0 & 0 & -1/2 \\ 0 & 1 & -1/3 & -1/3 \\ 0 & 0 & 5/3 & -4/3 \\ 0 & 0 & -4/3 & 13/6 \end{Vmatrix} \qquad (23)$$

$$Y'''=\begin{Vmatrix} 1 & 0 & 0 & -1/2 \\ & 1 & -1/3 & -1/3 \\ & & 1 & -4/5 \\ & & & 11/10 \end{Vmatrix} \qquad (24)$$

Equation 25 shows the resultant upper triangular matrix A.

$$A=\begin{Vmatrix} 1 & 0 & 0 & -1/2 \\ 0 & 1 & -1/3 & -1/3 \\ 0 & 0 & 1 & -4/5 \\ 0 & 0 & 0 & 1 \end{Vmatrix} \qquad (25)$$

According to the algorithm indicated by equation 20 the transformation of I to E is accomplished by a sequence of elementary matrix multiplications. For the given example the transformation is as indicated in equation 26.

Table II. Elements of B_{ii} Matrices

Element Values	Co-ordinates
1/2	1–1
1/3	2–2
3/5	3–3
10/11	4–4

$$E=A_{14}A_{23}A_{24}A_{34}B_{11}B_{22}B_{33}B_{44}A_{43}A_{42}A_{32}A_{41}I \qquad (26)$$

Parentheses are omitted in this equation with the understanding that operations on the column vector I are to be performed in sequence from right to left. The essential elements for the A_{ij} matrices are all obtainable from the A matrix. They are shown stored in Table I with signs reversed. The four reciprocals used to transform the diagonal elements of A to unity are needed for the B_{ii} matrices. They are shown stored in Table II.

Assume current vector I of equation 27.

$$I=\begin{Vmatrix} 1 \\ 1 \\ 1 \\ 1 \end{Vmatrix} \qquad (27)$$

The steps involved in transforming I to E are indicated in Table III. The process is guided entirely by Tables I and II. Column 0 of Table III shows I before processing. Each succeeding column shows the vector after a single step of the process. The elementary matrix affecting each step is shown at the top of the column. The affected element in each step is underlined. The vector in column 12 is E. It can be checked in equation 9 using the Y matrix of equation 21.

Column 1 of Table III is obtained from the previous (zero) column by multiplying the first element of the previous column by the first element of Table I and adding the product to the fourth element of the previous column. Note how the co-ordinates of the first element of Table I indicate the column elements to be operated upon. Column 2 is obtained from the previous (first) column by multiplying the second element of the previous column by the second element of Table II and adding the product to the third element. Columns 3 and 4 are obtained in similar manner. The process has now reached the bottom of Table I and transfers to the bottom (or top) of Table II. Column 5 is obtained from the previous column by multiplying the fourth element of the previous column by the fourth element of Table II. Note that the fourth element of E is now computed. Columns 6, 7, and 8 are obtained by similar multiplications with the third, second, and first elements of Table II. The process returns to the bottom of Table I and works in reverse order. Column 9 is obtained by multiplying the fourth element of the previous column by the fourth element of Table I and adding the product to the third element of the previous column. The process is continued in similar manner to the 12th column at which time E is completely computed.

Appendix III. Elimination Method of Power Flow Solution

When the elimination method of power flow solution was first introduced it had not been tested on large problems. During the summer of 1961, and with the assistance of one of the method's co-authors, James Van Ness, the elimination method was tested at BPA (Bonneville Power Administration). The first test results showed that the memory and time requirements increase excessively with increasing size of the problem. However, by applying the techniques described in this paper, these requirements were reduced so as to make the elimination method not only the surest but the fastest power flow solution presently known by the authors. It has now replaced the nodal iterative method for production power flow solutions at BPA

Experience has shown that the elimination method usually takes five iterations for precise solution regardless of problem size. Each iteration involves among other things the formation and solution of a set of simultaneous linear equations with real coefficients. The number of equations is twice the number of load buses plus the number of generator buses. The form of the equations corresponds to that of the network admittance matrix; therefore, the ideas expressed in this paper are applicable.

The criterion for solution in the tests required that the discrepancy between scheduled and computed quantities at all buses be less than 0.0001 per unit (0.01 mva). This requirement is approximately one hundred times more precise than that of most other programs. The effectiveness of the techniques is indicated by the results obtained with a test problem involving a network of 229 active buses (plus 73 passive buses eliminated on input), 360 branches (exclusive of 229 branches to ground). Of the 229 active buses, 194 were loads and 35 were generators. This combination resulted in a set of 423 equations. Since the coefficients of the equations change with every iteration, the method of single solution and back-substitution was used. (This is not a problem in which the elimination operations can be preserved for repetition.)

Originally the internal bus numbering corresponded to the external scheme, which, from the standpoint of minimizing operations and memory was purely random, being neither extremely favorable nor adverse. It was found with this numbering that at the end of the downward elimination process and prior to back-substitution, the triangular matrix contained 12,137 elements. Using the IBM (International Business Machines Corporation) *7090* computer, one iteration took approximately 59 seconds. (This time included the formation of coefficients, rectangular to polar to rectangular conversions, monitor outputs and other computations as well as the solution of equations.) After internal renumbering according to Scheme I, the triangular matrix contained only 4,247 elements and the time for one iteration was only 16.5 seconds. Thus, the memory requirement was reduced by almost 3 to 1 and the time by almost 4 to 1 by simply changing the order of eliminations. Actually, the time requirement for solving the equations was reduced more than this because a considerable portion of the computing included in the 16.5 seconds was unaffected by the renumbering.

The memory requirement for the triangle of a 50-equation problem was reduced from 1,264 elements before renumbering to 523 elements after renumbering. The advantage of renumbering by Scheme I is directly related to problem size but, as this example shows, it is significant even with small problems.

Table III. Transformation of I to E

0	1	2	3	4	5	6	7	8	9	10	11	12
	A_{41}	A_{32}	A_{42}	A_{43}	B_{44}	B_{33}	B_{22}	B_{11}	A_{34}	A_{24}	A_{23}	A_{14}
1	1	1	1	1	1	1	1	$\underline{1/2}$	1/2	1/2	1/2	$\underline{20/11}$
1	1	1	1	1	1	1	$\underline{1/3}$	1/3	1/3	$\underline{40/33}$	$\underline{24/11}$	24/11
1	1	$\underline{4/3}$	4/3	4/3	4/3	$\underline{4/5}$	4/5	4/5	$\underline{32/11}$	32/11	32/11	32/11
1	$\underline{3/2}$	3/2	$\underline{11/6}$	$\underline{29/10}$	$\underline{29/11}$	29/11	29/11	29/11	29/11	29/11	29/11	29/11

Appendix IV. Impedance Iterative Method of Power Flow Solution

An iterative method of power flow solution employing the impedance matrix Z has been presented,[6] but its practical value is limited because its memory and time requirements are large. Modification of the method as proposed in the following would identically duplicate the iterative results with a significant saving in memory and time.

The matrices Y, A, the n elements of the B_{ii} matrices, and the n diagonal elements of Z are required. The matrix A must be formed according to numbering Scheme II in order to duplicate the results obtained using the Z matrix alone. A node vector I is initially assumed and the downward operations indicated by the A_{ij} matrices of equation 20, with $i>j$, are performed; then, the multiplications by the B_{ii} matrices are performed. At this point the nth element of I is the nth node voltage e_n. The complex power s_n can then be calculated as

$$s_n = e_n i_n^* \tag{28}$$

where i_n^* is the conjugate of the nth element of the initial I vector prior to the downward operations.

In the impedance matrix method a correction in node current is computed which, depending on node type, makes either the complex power or the real power and voltage magnitude agree with scheduled quantities. Only the diagonal of Z is used in the correction formula. (It is assumed here that Z is Y^{-1} with branches to ground included.) The correction can also be expressed in terms of node voltage without altering the physical nature of the change. Letting node current correction be Δi_n, node voltage correction Δe_n is

$$\Delta e_n = \Delta i_n z_{nn} \tag{29}$$

where z_{nn} is the nth diagonal element of Z. The corrected node voltage e_n' is

$$e_n' = e_n + \Delta e_n \tag{30}$$

The corrected value e_n' is substituted for e_n and the upward (or back substitution) sequence of operations is initiated. The single operation indicated by the matrix $A_{n-1,n}$ yields the node voltage e_{n-1}. This voltage is corrected to e_{n-1}', substituted for e_{n-1}, and then e_{n-2} is computed, corrected, etc. When all voltages have been computed and corrected in this manner, it is necessary to obtain a new current vector I to start the second iteration. This is accomplished by substituting the vector E of corrected node voltages e_1', ..., e_n' in equation 9. The I vector obtained from this substitution is identical to the I vector which would have been obtained by using the impedance matrix method for successive current corrections from node n to node 1. The second and succeeding iterations continue in the same manner until computed and scheduled node quantities agree within set tolerances.

References

1. NETWORK ANALYSIS AND SYNTHESIS (book), **Louis Weinberg.** McGraw-Hill Book Company, Inc., New York, N. Y., 1962, chapt. 1.

2. METHODS OF APPLIED MATHEMATICS (book), **F. B. Hildebrand.** Prentice-Hall, Inc., New York, N. Y., 1952, pp. 1–4.

3. VECTOR SPACES AND MATRICES (book), **Robert M. Thrall, Leonard Tornheim.** John Wiley and Sons, Inc., New York, N. Y., 1957, pp. 82–85, 94–99.

4. ELIMINATION METHODS FOR LOAD FLOW STUDIES, **J. E. Van Ness, J. H. Griffin.** *AIEE Transactions,* pt. III (*Power Apparatus and Systems*), vol. 80, June 1961, pp. 299–304.

5. A NEW DIRECT MATRIX INVERSION METHOD, **R. B. Shipley, D. Coleman.** *Ibid.,* pt. I (*Communication and Electronics*), vol. 78, Nov. 1959, pp. 568–72.

6. POWER FLOW SOLUTION BY IMPEDANCE MATRIX ITERATIVE METHOD, **H. E. Brown, H. H. Happ, G. K. Carter, C. E. Person.** *IEEE Transactions on Power Apparatus and Systems,* vol. 82, Aug. 1963, pp. 561–64.

Discussion

A. H. El-Abiad, M. Watson, G. W. Stagg (American Electric Power Service Corporation, New York, N. Y.): In the early 1950's several papers presented the initial ideas of a number of authors in connection with the method for formulating and solving load flows on a digital computer. More recently we have witnessed the second generation of load flow papers, each publicizing improvements in the speed of solution and a reduction in storage requirements.

In 1959 Hale and Goodrich[1] introduced techniques using a hybrid matrix which reduced computation time by 80% over the previous method proposed by Ward and Hale.[2] Shortly after Van Ness and Griffin[3] proposed the elimination method, also showing 80% reduction in solution time. This was followed by a paper by Brown, Carter, Happ, and Person[4] which used the bus impedance matrix and which claimed an 80% reduction in solution time over the nodal-iterative method. The authors of the present paper, taking advantage of the sparsity of the admittance matrix claim an additional speed advantage of 75%. The question which naturally arises is how much so far have we been able to reduce computing time for the solution of load flows? Certainly the savings reported by the authors of the second generation papers could not be compounded. If they were, this would mean that a medium-sized load flow requiring 60 seconds to converge to a solution on a large-scale scientific computer would now take 0.12 second or a reduction in solution time of 99.8%.

The difficulty arises from the fact that the authors of the latter papers compare new results with their original work. This became apparent to us while doing work using the bus impedance matrix, which we presented in a conference paper, at the Fall General Meeting in Detroit in 1961. In our research we did not obtain the savings claimed by Brown, Carter, Happ, and Person.[4] We found no clear-cut speed advantage in the bus impedance method over the nodal iterative approach. We made comparisons against a different nodal iterative program than that employed by Brown, Carter, Happ, and Person[4] and used different test systems.

In an attempt to overcome some of these difficulties, the IEEE Computer Applications Subcommittee established a number of standard test systems which it made available to members working in the area of digital load flow calculations. These systems range in size from 30 to 229 buses and are available through American Electric Power Company, The Ontario Hydro-Electric System, and Bonneville Power Administration.

We believe that the value of this paper would be greatly increased if the authors could include a tabulation of computing times for their nodal iterative program, the original elimination method, and with the improved techniques described in this paper. It is hoped that they will use one or more of the standard test systems for this comparison.

REFERENCES

1. DIGITAL COMPUTATION OF POWER FLOW—SOME NEW ASPECTS, H. W. Hale, R. W. Goodrich. *AIEE Transactions*, pt. IIIA (*Power Apparatus and Systems*), vol. 78, Oct. 1959, pp. 919–24.

2. DIGITAL COMPUTER SOLUTION OF POWER FLOW PROBLEMS, J. B. Ward, H. W. Hale. *Ibid.*, vol. 75, June 1956, pp. 398–404.

3. See reference 4 of paper.

4. See reference 6 of paper.

R. B. Shipley, D. W. Coleman (Tennessee Valley Authority, Chattanooga, Tenn.): The problems of solving large systems with a limited computer memory causes us to spend a disproportionate amount of our time and initiative on computer and calculating techniques. The proposed technique appears to be extremely helpful in solving many types of network problems.

We would appreciate the authors' thoughts on a problem that we have encountered frequently: Given a network in which voltage or current, but not both, is known on each bus, the problem is to determine the remaining unknown currents and voltages. The technique presented should be helpful, but we have had only limited success in attempting to apply it.

The paper appears to be slanted toward the impedance type of load flow. Possibly the first matrix impedance type load flow was run by Tennessee Valley Authority in 1954/55. (Earlier work by Dunstan used extra mesh currents.) Convergence in a few iterations was reported in an economic dispatch paper in 1956. Other authors have presented papers confirming these results. The technique should be of great value in this type of load flow.

Our work is not primarily in the load flow area. However, we have produced a nominal 1,000 bus load flow for a *16K* computer. We accomplished this by a form of tearing. The program is slow because of input-output time of the computer. The load flow works on sections of the system individually. Then the solutions are interconnected by an iterative procedure. Accordingly, changes in one section may not affect the other sections and in these cases enormous amounts of computations are saved. This might be the key to solving extremely large load flows that cannot be accommodated in high-speed memory by any other technique.

Philosophically, do the authors expect to save machine time or merely to cause more load flow problems to be run? What per cent of presently produced digital load flow problems do they think are really essential in power system work?

Nobuo Sato, W. F. Tinney: The techniques were not intentionally slanted toward power flows, but since both discussions raise questions on this subject, we will attempt to clarify our claim about the advantages of the elimination method. Our IBM *7090* power flow program permits solution of problems using either the Gauss-Seidel or the elimination method. The Gauss-Seidel method which we employ is essentially that of Ward and Hale except for acceleration features.

We hope to be able in the near future to release a tabulation of computing times for several specific problems as suggested by El-Abiad, Watson, and Stagg. For the present we feel that a satisfactory comparison between methods can be made by discussing a typical 250-bus problem. Assuming a flat voltage start and a solution criterion that requires the largest residual to be less than 0.1 mva, the Gauss-Seidel method may require anywhere from about 50 to 200 iterations of 0.8 second each; about 100 iterations is average. The number of iterations is a widely varying function of generation, loading, and line outages as well as solution precision. There are many physically realizable conditions for which solution cannot be obtained by the Gauss-Seidel method no matter how many iterations are taken; these problems either diverge, oscillate, or converge to a certain point and go no further.

The elimination method can solve a 250-bus problem in from 3 to 5 iterations of 18 seconds each; rarely, are more than 3 iterations required for a precision of 0.1 mva. In the paper we reported that 5 iterations were required for a higher precision of 0.01 mva, but actually this precision is usually attained at the end of the fourth iteration but not measured until the end of the fifth. At the end of the fifth iteration the solution is virtually exact. If 4 iterations are assumed for the elimination method (to be on the conservative side) it is apparent that the elimination method is equal in speed to our Gauss-Seidel method when 90 iterations of the latter are required. Actually, since the average number of iterations for the elimination method for comparable precision is much nearer 3 than 4, the break-even point is closer to 70 Gauss-Seidel iterations.

Reliability, not speed, however, is the main reason for our choice of the elimination method. The elimination method can solve many problems which the Gauss-Seidel method cannot solve. It can solve problems in 3 or 4 iterations which would take several hundred iterations by Gauss-Seidel. Furthermore, by going one additional iteration beyond a relatively low precision solution such as used in the foregoing comparison, it can obtain a virtually exact solution. Formerly it was assumed that when a solution could not be obtained by the Gauss-Seidel method the system was statically unstable. We know now that in many cases this was an erroneous conclusion. We do not know when the elimination method fails to obtain a solution, as it sometimes does, if this definitely means static instability.

We have had no experience with the impedance matrix method. As originally described, the number of operations per iteration was proportional to the square of the number of buses. This would almost certainly limit its usefulness to small problems. Considerable reduction in time and storage could be obtained by the modification we suggest, but we doubt if even this would make it competitive with the elimination method. The combination of our suggested modification with the technique of network tearing as described by Shipley and Coleman might make a big difference. We think even better results would be obtained, however, by applying the tearing techniques to the elimination method.

A network with voltage or current, but not both, specified at each node can be solved as follows: Write a set of equations of the form $I=YE$ for the nodes at which current is specified and voltage is unknown. The E vector will be composed of the unknown voltages; the ith element of the I vector will be $I_i=I_i'-\Sigma Y_{ij}E_j$, where I_i' is the given ith node current and the summation contains only those $Y_{ij}E_j$ terms where j refers to a known voltage. Solve this set of equations for E by elimination and back-substitution. Then with all voltages known, the remaining unknown node currents can be calculated directly. Perhaps we have oversimplified this explanation through misunderstanding of the problem.

Direct Solutions of Sparse Network Equations by Optimally Ordered Triangular Factorization

WILLIAM F. TINNEY, SENIOR MEMBER, IEEE, AND JOHN W. WALKER, MEMBER, IEEE

***Abstract*—Matrix inversion is very inefficient for computing direct solutions of the large sparse systems of linear equations that arise in many network problems. Optimally ordered triangular factorization of sparse matrices is more efficient and offers other important computational advantages in some applications. With this method, direct solutions are computed from sparse matrix factors instead of from a full inverse matrix, thereby gaining a significant advantage in speed, computer memory requirements, and reduced round-off error. Improvements of ten to one or more in speed and problem size over present applications of the inverse can be achieved in many cases. Details of the method, numerical examples, and the results of a large problem are given.**

I. Introduction

USUALLY, the objective in the matrix analysis of networks is to obtain the inverse of the matrix of coefficients of a system of simultaneous linear network equations. However, for large sparse systems such as occur in many network problems, the use of the inverse is very inefficient. In general, the matrix of the equations formed from the given conditions of a network problem is sparse, whereas its inverse is full. By means of an appropriately ordered triangular decomposition, the inverse of a sparse matrix can be expressed as a product of sparse matrix factors, thereby gaining an advantage in computational speed, storage, and reduction of round-off error.

The method consists of two parts: 1) a scheme of recording the operations of triangular decomposition of a matrix such that repeated direct solutions based on the matrix can be obtained without repeating the triangularization, and 2) a scheme of ordering the operations that tends to conserve the sparsity of the original system. The first part of the method is not altogether new, but its computational possibilities are not widely recognized. The second part appears to be of recent origin, but it is probably only a rediscovery. Either part of the method can be applied independently, but the greatest benefit is obtained from the combined application of both parts.

The first part of the method is applicable to any matrix. Application of the second part, ordering to conserve sparsity, is limited to sparse matrices in which the pattern of nonzero elements is symmetric and for which an arbitrary

Manuscript received February 10, 1967; revised August 9, 1967. The unrevised version of this paper was published in the *Power Industry Computer Applications Conf. Rec.*, pp. 367–376, 1967. This conference was sponsored by the IEEE Power Group.

The authors are with the Bonneville Power Administration, Portland, Ore.

Reprinted from *Proc. IEEE*, vol. 55, pp. 1801–1809, Nov. 1967.

order of decomposition does not adversely affect numerical accuracy. Such matrices are usually characterized by a strong diagonal, and ordering to conserve sparsity increases the accuracy of the decomposition. A large class of network problems fulfills this condition. Generally it is not worth considering optimal ordering unless at least 80 percent of the matrix elements are zero.

At least three papers[1]–[3] have been written on various aspects of optimally ordered triangular decomposition applied to power system problems. Another paper[4] has been written on the related subject of optimally ordered Gauss-Jordan decomposition.

This paper extends previous work and presents it in a new matrix form. The method is used at Bonneville Power Administration (BPA) in many applications including power flow,[7] short circuit, transient stability, network reduction, switching transients,[8] reactive optimization,[9] tower design,[10] and others.

II. Factored Direct Solutions

This section, covering the first part of the method, shows how to derive an array of numbers from a nonsingular matrix A that can be used to obtain the effects of any or all of the following: A, A^{-1}, A^t, $(A^t)^{-1}$, and certain two-way hybrid combinations of these matrices. The superscript -1 means inverse and the superscript t means transpose. The scheme is applicable to any nonsingular matrix, real or complex, sparse or full, symmetric or nonsymmetric. Although the examples in this paper are limited to nodal equations, the method is also applicable to mesh equations. Its greatest advantage is realized in problems involving large sparse matrices.

The scheme to be described in this section is similar to those associated with the names of Gauss, Doolittle, Choleski, Banachiewicz, and others. All of these closely related schemes are computational variations of the basic process of triangularizing a matrix by equivalence transformations. They were originally developed for, and until very recently have only been described in terms of, manual computational procedures. Very little attention has been given to their special suitability for sparse matrices.

The basic scheme is first presented for the most general case, a full nonsymmetric matrix. Symmetry is then treated as a special case. Sparsity with optimal ordering, which is the primary objective of this development, is explained in Section III. Numerical examples of the basic scheme are given in Appendix I.

A. Triangular Decomposition

Triangular decomposition of a matrix by Gaussian elimination is described in many books on matrix analysis.[5],[6] Ordinarily, the decomposition is accomplished by elimination of elements below the main diagonal in successive columns. From the standpoint of computer programming for a sparse matrix, it is usually much more efficient to eliminate by successive rows. Therefore, this less familiar, but mathematically equivalent, scheme is illustrated here. The development is based on the equation

$$Ax = b \tag{1}$$

where A is a nonsingular matrix, x is a column vector of unknowns, and b is a known vector with at least one nonzero element. In the computer algorithm A is augmented by b as shown in (2) for an nth-order system.

$$\begin{bmatrix} a_{11} & a_{12} & \cdots & a_{1n} & b_1 \\ a_{21} & a_{22} & \cdots & a_{2n} & b_2 \\ \cdot & \cdot & \cdots & \cdot & \cdot \\ a_{n1} & a_{n2} & \cdots & a_{nn} & b_n \end{bmatrix} \tag{2}$$

The first step is to divide the elements of the first row by a_{11} as indicated in (3).

$$\begin{aligned} a_{1j}^{(1)} &= (1/a_{11})a_{1j} \qquad j = 2, n \\ b_1^{(1)} &= (1/a_{11})b_1. \end{aligned} \tag{3}$$

The superscripts indicate the order of the derived system.

The second step, as indicated in (4a) and (4b), is to eliminate a_{21} from the second row by linear combination with the derived first row, and then to divide the remaining derived elements of the second row by its derived diagonal element.

$$\begin{bmatrix} 1 & a_{12}^{(1)} & a_{13}^{(1)} & \cdots & a_{1n}^{(1)} & b_1^{(1)} \\ & 1 & a_{23}^{(2)} & \cdots & a_{2n}^{(2)} & b_2^{(2)} \\ \cdot & \cdot & \cdot & \cdots & \cdot & \cdot \\ a_{n1} & a_{n2} & a_{n3} & \cdots & a_{nn} & b_n \end{bmatrix} \tag{4a}$$

$$\begin{aligned} a_{2j}^{(1)} &= a_{2j} - a_{21}a_{1j}^{(1)} \qquad j = 2, n \\ b_2^{(1)} &= b_2 - a_{21}b_1^{(1)} \\ a_{2j}^{(2)} &= (1/a_{22}^{(2)})a_{2j}^{(1)} \qquad j = 3, n \\ b_2^{(2)} &= (1/a_{22}^{(1)})b_2^{(1)}. \end{aligned} \tag{4b}$$

The third step, as indicated in (5a) and (5b), is to eliminate elements to the left of the diagonal of the third row and to divide the remaining derived elements of the row by the derived diagonal element.

$$\begin{bmatrix} 1 & a_{12}^{(1)} & a_{13}^{(1)} & a_{14}^{(1)} & \cdots & a_{1n}^{(1)} & b_1^{(1)} \\ & 1 & a_{23}^{(2)} & a_{24}^{(2)} & \cdots & a_{2n}^{(2)} & b_2^{(2)} \\ & & 1 & a_{34}^{(3)} & \cdots & a_{3n}^{(3)} & b_3^{(3)} \\ \cdot & \cdot & \cdot & \cdot & \cdots & \cdot & \cdot \\ a_{n1} & a_{n2} & a_{n3} & a_{n4} & \cdots & a_{nn} & b_n \end{bmatrix} \tag{5a}$$

$$\begin{aligned} a_{3j}^{(1)} &= a_{3j} - a_{31}a_{1j}^{(1)} \qquad j = 2, n \\ b_3^{(1)} &= b_3 - a_{31}b_1^{(1)} \\ a_{3j}^{(2)} &= a_{3j}^{(1)} - a_{32}^{(1)}a_{2j}^{(2)} \qquad j = 3, n \\ b_3^{(2)} &= b_3^{(1)} - a_{32}^{(1)}b_2^{(2)} \\ a_{3j}^{(3)} &= (1/a_{33}^{(2)})a_{3j}^{(2)} \qquad j = 4, n \\ b_3^{(3)} &= (1/a_{33}^{(2)})b_3^{(2)}. \end{aligned} \tag{5b}$$

Proceeding in this manner the nth derived system is obtained.

$$\begin{bmatrix} 1 & a_{12}^{(1)} & \cdots & a_{1n}^{(1)} & b_1^{(1)} \\ & 1 & \cdots & a_{2n}^{(2)} & b_2^{(2)} \\ & & \cdots & \cdot & \cdot \\ & & \cdots & \cdot & \cdot \\ & & \cdots & \cdot & \cdot \\ & & & 1 & b_n^{(n)} \end{bmatrix} \tag{6}$$

It should be noted that at the end of the kth step, work on rows 1 to k has been completed and rows $k+1$ to n have not yet entered the process in any way.

The solution can now be obtained by back substitution.

$$\begin{aligned} x_n &= b_n^{(n)} \\ x_{n-1} &= b_{n-1}^{(n-1)} - a_{n-1,n}^{(n-1)} x_n \\ x_i &= b_i^{(i)} - \sum_{j=i+1}^{n} a_{ij}^{(i)} x_j. \end{aligned} \tag{7}$$

In programming, the x_i's replace the b_i's one by one as they are computed, starting with x_n and working back to x_1.

When A is full and n large, it can be shown that the number of multiplication-addition operations for triangular decomposition is approximately $1/3n^3$ compared with n^3 for inversion.[5]

It can easily be verified that triangularization in the same order by columns instead of rows would have produced identically the same result. Each eliminated element $a_{ij}^{(j-1)}$, $i>j$, would have been the same and the number of operations would have been the same. The back substitution also could have been accomplished by columns instead of rows in the same number of operations.

B. Recording the Operations

If the forward operations on b had been recorded so that they could be repeated, it is obvious that with this record and the upper triangle (6) for the back substitution, (1) could be solved for any vector b without repeating the triangularization. The recording of the forward operations, however, is trivial. Each forward operation is completely defined by the row and column coordinates and value of a single element $a_{ij}^{(j-1)}$, $i \geq j$, that occurs in the process. Therefore, it is unnecessary to do anything to record these elements except to leave them.

The rules for recording the forward operations of triangularization are:

1) when a term $1/a_{ii}^{(i-1)}$ is computed, store it in location ii
2) leave every derived term $a_{ij}^{(j-1)}$, $i>j$, in the lower triangle.

Since the forward as well as the back substitution operations are recorded in this scheme, it is no longer necessary to include the vector b. The final result of triangularizing A and recording the forward operations is symbolized in (8).

$$\begin{matrix} d_{11} & u_{12} & u_{13} & \cdot & \cdot & \cdot & u_{1n} \\ l_{21} & d_{22} & u_{23} & \cdot & \cdot & \cdot & u_{2n} \\ l_{31} & l_{32} & d_{33} & \cdot & \cdot & \cdot & u_{3n}. \\ \cdot & \cdot & \cdot & \cdot & \cdot & \cdot & \cdot \\ l_{n1} & l_{n2} & l_{n3} & \cdot & \cdot & \cdot & d_{nn} \end{matrix} \tag{8}$$

The elements of (8), defined in terms of the derived systems of A in (2)–(6), are:

$$\begin{aligned} d_{ii} &= 1/a_{ii}^{(i-1)} & \\ u_{ij} &= a_{ij}^{(i)} & i < j \\ l_{ij} &= a_{ij}^{(j-1)} & i > j. \end{aligned} \tag{9}$$

The matrix brackets are omitted in (8) to emphasize that the array is not strictly a matrix in the same sense as preceding examples, but only a scheme of recording. It will be referred to as the table of factors. In the literature this result is frequently shown as a factoring of the inverse matrix into the product of a lower and an upper triangular matrix, but it is more suitable for this discussion to consider it as a table of factors.

C. Computing Direct Solutions

It is convenient in symbolizing the operations for obtaining direction solutions to define some special matrices in terms of the elements of the table of factors (8). The following nonsingular matrices differ from the unit matrix only in the row or column indicated.

$$\begin{aligned} &D_i: \text{Row } i = (0, 0, \cdots 0, d_{ii}, 0, \cdots 0, 0) \\ &L_i: \text{ Col } i = (0, 0, \cdots 0, 1, -l_{i+1,i}, -l_{i+2,i}, \cdots -l_{n-1,i}, -l_{n,i})^t \\ &L_i^*: \text{Row } i = (-l_{i,1}, -l_{i,2}, \cdots -l_{i,i-1}, 1, 0, \cdots 0, 0) \\ &U_i: \text{Row } i = (0, 0, \cdots 0, 1, -u_{i,i+1}, -u_{i,i+2}, \cdots -u_{i,n-1}, -u_{i,n}) \\ &U_i^*: \text{ Col } i = (-u_{1,i}, -u_{2,i}, \cdots -u_{i-1,i}, 1, 0, \cdots 0, 0)^t. \end{aligned} \tag{10}$$

The inverses of these matrices are trivial. The inverse of the matrix D_i involves only the reciprocal of the element d_{ii}. The inverses of the matrices L_i, L_i^*, U_i, and U_i^* involve only a reversal of algebraic signs of the off-diagonal elements.

The forward and back substitution operations on the column vector b that transform it to x can be expressed as premultiplications by matrices D_i, L_i or L_i^*, and U_i or U_i^*. Thus the solution of $Ax=b$ can be expressed as indicated in (11a)–(11d).

$$U_1U_2 \cdots U_{n-2}U_{n-1}D_nL_{n-1}D_{n-1}L_{n-2} \cdots L_2D_2L_1D_1b = A^{-1}b = x \tag{11a}$$

$$U_1U_2 \cdots U_{n-2}U_{n-1}D_nL_n^*D_{n-1}L_{n-1}^* \cdots L_3^*D_2L_2^*D_1b = A^{-1}b = x \tag{11b}$$

$$U_2^*U_3^* \cdots U_{n-1}^*U_n^*D_nL_{n-1}D_{n-1}L_{n-2} \cdots L_2D_2L_1D_1b = A^{-1}b = x \tag{11c}$$

$$U_2^*U_3^* \cdots U_{n-1}^*U_n^*D_nL_n^*D_{n-1}L_{n-1}^* \cdots L_3^*D_2L_2^*D_1b = A^{-1}b = x. \tag{11d}$$

Each of these four equations describes a sequence of operations on the vector b that is equivalent to premultiplication by A^{-1}. For an nth-order system, each equation indicates n multiplications, n^2-n multiplication-additions, and n additions, excluding, of course, multiplications by unity which are only symbolic. This corresponds exactly with the n^2 multiplication-additions required for premultiplication by the inverse. Starting with D_1 and proceeding to the left, (11a) describes the forward and back substitution operations that would be performed on b if it augmented A during triangularization by columns. Equation (11b) describes the same result for triangularization by rows. Equations (11c) and (11d) describe other sequences of the same operations giving the same result. Depending on programming techniques, one of these four equivalent sequences usually will prove to be the most convenient.

Direct solutions of other systems based on A can be obtained from the table of factors by use of the following two theorems:

1) the inverse of a product of nonsingular matrix factors is the product of the inverses of the factors in reverse order
2) the transpose of a matrix product is the product of the transposes of the factors in reverse order.

Although the matrix A may have been lost in the triangularization, its effect is recoverable. Given x, the vector b can be obtained as indicated in (12a) or (12b). Two other equations analogous to (11c) and (11d) using U_i^* also could be written.

$$D_1^{-1}L_1^{-1}D_2^{-1}L_2^{-1}\cdots L_{n-1}^{-1}D_n^{-1}U_{n-1}^{-1}U_{n-2}^{-1}\cdots U_2^{-1}U_1^{-1}x = Ax = b \quad (12a)$$

$$D_1^{-1}(L_2^*)^{-1}D_2^{-1}(L_3^*)^{-1}\cdots(L_n^*)^{-1}D_n^{-1}U_{n-1}^{-1}U_{n-2}^{-1}\cdots U_2^{-1}U_1^{-1}x = Ax = b. \quad (12b)$$

Again the number of essential operations is n^2.

Direct solutions for the corresponding transpose system $A^ty=c$ can be obtained as indicated in (13) and (14).

$$D_1L_1^tD_2L_2^t\cdots L_{n-1}^tD_nU_{n-1}^tU_{n-2}^t\cdots U_2^tU_1^tc = (A^t)^{-1}c = y \quad (13)$$

$$(U_1^t)^{-1}(U_2^t)^{-1}\cdots(U_{n-2}^t)^{-1}(U_{n-1}^t)^{-1}\cdot D_n^{-1}(L_{n-1}^t)^{-1}\cdots(L_2^t)^{-1}D_2^{-1}(L_1^t)^{-1}D_1^{-1}y = A^ty = c. \quad (14)$$

Again the number of operations is n^2, and, as in the previous examples, equations could be written using L_i^* and U_i^*.

Although (11)–(14) look somewhat complicated, they represent simple operations that can be guided by the table of factors (8). Each equation indicates the following:

1) a sequence for using the elements of the array for performing operations on the independent vector
2) a rule for using the subscripts of the elements of the array to indicate the elements of the vector to be operated upon
3) an algebraic sign rule for elements l_{ij} and u_{ij}
4) a rule indicating whether to multiply or divide by elements d_{ii}.

The operations can be extended to include certain two-way hybrid solutions with the matrix partitioned at any point desired. Let the hybrid column vector g be defined as

$$g^t = (b_1, b_2, \cdots, b_k, x_{k+1}, x_{k+2}, \cdots, x_n). \quad (15)$$

If g is given, the unknown first k elements of x and the $k+1$ to nth elements of b can be obtained directly. First it is necessary to compute an intermediate vector z.

$$U_{n-1}^{-1}U_{n-2}^{-1}\cdots U_{k+2}^{-1}U_{k+1}^{-1}D_kL_k^*\cdots L_3^*D_2L_2^*D_1g = z. \quad (16)$$

Equation (16) indicates the solution of an upper triangular system to obtain the first k elements of z and the solution of an independent lower triangular system to obtain the remaining elements of z.

By using elements from z and g, the composite vector h is formed.

$$h^t = (z_1, z_2, \cdots, z_k, x_{k+1}, x_{k+2}, \cdots, x_n). \quad (17)$$

Using h, the first k unknown elements of x are obtained from (18).

$$U_1U_2\cdots U_{k-1}U_kh = x. \quad (18)$$

Equation (18) defines the back substitution of (11a) from k to 1.

The $k+1$ to nth unknown elements of b are obtained from

$$(L_{k+1}^*)^{-1}D_{k+1}^{-1}\cdots(L_{n-1}^*)^{-1}D_{n-1}^{-1}(L_n^*)^{-1}D_n^{-1}z = b'. \quad (19)$$

Equation (19) defines the back substitution of (12b) from n to $k+1$. The vector b' is composed of the first k elements of z, which were unaffected by the premultiplication on the left side of the equation, and the $k+1$ to nth elements of b. Since the first k elements of b were given, the solution is complete. Again the total number of operations is n^2.

If only the unknown elements of x are wanted, (20) should be used.

$$U_1U_2\cdots U_{k-1}U_kD_kL_k^*\cdots L_3^*D_2L_2^*D_1g = x. \quad (20)$$

This requires only kn operations. Since no elements of the table of factors below the kth row are needed in this case, it is unnecessary to compute and store them.

A hybrid solution for the equation $A^ty=c$, in which the first k elements of c and the $k+1$ to nth elements of y are given, can be developed analogous to (15)–(20). Other hybrid solutions also can be developed, but they require more than n^2 operations and will not be considered.

D. Symmetry

If A is symmetric, only the d_{ii} and u_{ij} terms of the table of factors are needed. All of the foregoing direct solutions can be obtained by noting that in the symmetric case L_i can be defined in terms of D_i and U_i:

$$L_i = D_iU_i^tD_i^{-1}. \quad (21)$$

Substituting (21) in (11a) and canceling adjacent factors of the form $D_i^{-1}D_i$ gives

$$U_1U_2\cdots U_{n-2}U_{n-1}D_nD_{n-1}U_{n-1}^t\cdots D_2U_2^tD_1U_1^tb = A^{-1}b = x. \quad (22)$$

More convenient expressions from the standpoint of operation and notation can be obtained by noting that products of the form $U_i^tD_j$, $i>j$, are commutative. Thus all of the D_i factors can be grouped into one diagonal matrix

$$D = D_1D_2\cdots D_n.$$

With this arrangement (22) becomes

$$U_1U_2\cdots U_{n-2}U_{n-1}DU_{n-1}^t\cdots U_2^tU_1^tb = A^{-1}b = x. \quad (23)$$

Similar substitutions for the symmetric case can be made in the other equations for direct solutions.

Symmetry also permits a saving of almost one half of the triangularization operations because it is unnecessary to perform any operations to the left of the diagonal in order to obtain the derived elements $a_{ij}^{(j-1)}$, $i>j$. These elements can be recovered from previously derived elements by the relationship

$$a_{ij}^{(j-1)} = (a_{jj}^{(j-1)})a_{ji}^{(j)} \quad i = 2, n; \quad j = 1, i-1. \quad \text{(24a)}$$

An alternative procedure avoids a number of multiplications equal to those indicated in (24a). In the nonsymmetric case each row was normalized by multiplying each term to the right of the diagonal by the reciprocal of the diagonal term before proceeding to the next row. If this step is deferred, leaving the terms of a row unnormalized until later, the normalizing multiplication and recovery of the symmetric term, when needed, can be combined. Thus after processing row i, each term is left in the form $a_{ij}^{(i-1)}$, $i<j$, instead of $a_{ij}^{(i)}$. When the term $a_{ji}^{(i-1)}$, $j>i$, is needed in processing row j to be used as a multiplier of row i in the same way as if row i had been normalized, it is computed from $a_{ij}^{(i-1)}$. The operations using this alternative procedure are as follows:

$$a_{jk}^{(i)} = a_{jk}^{(i-1)}[(1/a_{ii}^{(i-1)})a_{ji}^{(i-1)}]a_{ik}^{(i-1)} \quad i<j; \quad k = j, n. \quad \text{(24b)}$$

The term in brackets, which is $a_{ij}^{(i)}$ in normalized form, replaces $a_{ij}^{(i-1)}$ as soon as it has been used in (24b). The saving in operations is trivial for a full matrix, but significant for a sparse one.

If the triangularized system is left unnormalized, n additional operations are required for each direct solution. This is significant for a sparse matrix; therefore, normalization as described is recommended.

E. Modifications

If A is changed, it is necessary to modify the table of factors to reflect the change. If an element $a_{k,m}$ of A is changed, elements of the table of factors with row and column subscripts i, j are affected as follows:

1) if $k>m$, elements a_{ij} with $i=k, j\geq m$ and $i>k, j\geq k$ are affected.
2) if $k\leq m$ elements a_{ij} with $i\geq k, j=m$ and $i\geq m, j>m$ are affected.

Since most changes in a row involve a change in the diagonal term, the first case is the more important. The number of operations required to effect a change in the kth row is approximately $1/3(n-k+1)^3$. However, any number of additional changes in the lower right submatrix bounded by row k and column k can be included in the same pass without increasing the volume of computations. Changes below row k in columns 1 to $k-1$ can also be made in the same pass with only a small increase in the volume of computations. If the rows in which frequent changes will be required can be anticipated, they should be located at the bottom of the matrix.

In using the inverse, every change in A affects all n^2 elements of A^{-1}; therefore, at least n^2 operations on A^{-1} are required to account for any change in A. Whether the inverse or the table of factors is the easier to modify in a particular situation depends on the nature and number of matrix changes required and the scheme of inverse modification being used.

F. Comparative Advantages for a Full Matrix

When A is full, the advantages of the factored form of direct solution are:

1) the array of factors can be obtained in one-third the number of operations of the inverse
2) the factored form gives the effect of A and the hybrid matrix; the inverse does not.

The methods are comparable in that:

1) complete solutions normally require the same number of operations
2) the storage requirement is the same.

The advantages of the inverse are:

1) complete solutions require only kn operations when the independent vector has only k nonzero elements
2) under some circumstances the inverse can be modified to reflect changes in the original matrix more easily than the table of factors.

III. Sparsity and Optimal Ordering

When the matrix to be triangularized is sparse, the order in which rows are processed affects the number of nonzero terms in the resultant upper triangle.[11] If a programming scheme is used which processes and stores only nonzero terms, a very great savings in operations and computer memory can be achieved by keeping the table of factors as sparse as possible. The absolute optimal order of elimination would result in the least possible terms in the table of factors. An efficient algorithm for determining the absolute optimal order has not been developed, and it appears to be a practical impossibility. However, several effective schemes have been developed for determining near-optimal orders.

A. Schemes for Near-Optimal Ordering

The inspection algorithms for near-optimal ordering to be described are applicable to sparse matrices that are symmetric in pattern of nonzero off-diagonal terms; i.e., if a_{ij} is nonzero, then a_{ji} also is nonzero but not necessarily equal to a_{ij}. These are matrices that occur most frequently in network problems. From the standpoint of programming efficiency, the algorithms should be applied before, rather than during, the triangularization. It is assumed in what follows that the matrix rows are originally numbered according to some external criterion and then renumbered according to the inspection algorithm. Eliminations are then performed in ascending sequence of the renumbered system.

Following are descriptions of three schemes for renumbering in near-optimal order. They are listed in increasing order of programming complexity, execution time, and optimality.

1) Number the rows according to the number of nonzero off-diagonal terms before elimination. In this scheme the rows with only one off-diagonal term are numbered first, those with two terms second, etc., and those with the most terms, last. This scheme does not take into account

any of the subsequent effects of the elimination process. The only information needed is a list of the number of nonzero terms in each row of the original matrix.

2) Number the rows so that at each step of the process the next row to be operated upon is the one with the fewest nonzero terms. If more than one row meets this criterion, select any one. This scheme requires a simulation of the effects on the accumulation of nonzero terms of the elimination process. Input information is a list by rows of the column numbers of the nonzero off-diagonal terms.

3) Number the rows so that at each step of the process the next row to be operated upon is the one that will introduce the fewest new nonzero terms. If more than one row meets this criterion, select any one. This involves a trial simulation of every feasible alternative of the elimination process at each step. Input information is the same as for scheme 2).

The comparative advantages of these schemes are influenced by network topology and size and the number of direct solutions wanted. The only virtue of scheme 1) is its simplicity and speed. For nodal equations of power networks scheme 2) is enough better than scheme 1) to justify the additional time required for its execution. Scheme 3) does not appear to be enough better than scheme 2) to justify its use for power networks, but it is believed that it may be effective for other networks. Since the authors' experience is limited to power networks, these conclusions may not be valid for other networks. Other algorithms may need to be developed. Schemes intermediate between 1) and 2) or between 2) and 3), as well as entirely different schemes, are possible.

It can be demonstrated that there exist certain matrices for which none of the given schemes will be very effective. However, they are unlikely to occur in most network equations.

B. Other Factors Influencing Ordering

Under some conditions it may be advantageous or necessary to number certain rows last even though this adversely affects sparsity. Among these conditions are the following.

1) It is known that changes in the original matrix will occur only in a certain few rows. If these rows are numbered last, only these rows in the table of factors need to be modified to reflect the changes.

2) It is known that changes in the independent vector will occur only in a certain few elements. If rows are numbered so that no changes in the vector occur above row k, the forward operations preceding row k need be performed only once, and not repeated for each subsequent case.

3) The matrix is only slightly nonsymmetric. If the rows are numbered so that the nonsymmetric portion of the matrix is last, advantage of the symmetry can be taken down to this point.

4) The hybrid operation is to be used and it is necessary to have the appropriate rows last.

When the network on which the matrix is based is composed of subnetworks with relatively few interconnections, the matrix should be arranged as indicated schematically in (25).

$$\left[\begin{array}{c|c|c|c} A & & & X \\ \hline & B & & X \\ \hline & & C & X \\ \hline X & X & X & S \end{array}\right] \tag{25}$$

This may be interpreted to be the pattern of a nodal admittance matrix of a very large network composed of three subnetworks. The pattern of the table of factors derived from this matrix would be similar. Submatrices A, B, and C, which presumably are very sparse, represent nodes within subnetworks A, B, and C, respectively; and S represents nodes associated with their interconnections. All submatrices outside of A, B, C, and S are zero except those indicated by the X's, which contain the few off-diagonal terms interconnecting the system in S. The renumbering algorithm should be applied to each submatrix independently. Within each of the submatrices A, B, and C, the rows with nonzero terms in the columns of S should be located last. This arrangement will not ordinarily be achieved by any of the given renumbering algorithms without additional logic or external coding.

The arrangement offers the following advantages:

1) If a change is made in one of the submatrices, say B, it affects only B and S in the table of factors, but not A and C.
2) If, after obtaining a solution, a change is made in the independent vector in the rows of only one partition, say B, only the operations on the vector defined by B and S need to be repeated to obtain a new solution.
3) If computer memory is limited, the procedure can be implemented by operating on only one submatrix at a time.
4) As more subnetworks are added to a system, the relationship between system size and the table of factors tends to be almost linear.

C. Effectiveness of Optimal Ordering

The effectiveness of optimal ordering can be expressed as the ratio between the number of nonzero off-diagonal terms in the table of factors and the number of similar terms in the original matrix. The nearer this ratio is to unity, the more effective the ordering. At BPA it has been found to vary between 1.7 and 2.5 for nodal admittance matrices of power networks of up to 1000 nodes. The ratio is influenced more by network topology than by system size. Obviously, it might be different for matrices based on other kinds of networks.

When the matrix is sparse, the arithmetic operations required to compute the table of factors for an nth-order system are as follows:

$$\text{divisions} = n$$

$$\text{multiplications} = s = \sum_{i=1}^{n-1} r_i. \quad (26a)$$

The number of multiplication-additions is dependent on whether the matrix is symmetric or not. For nonsymmetric matrices with symmetric pattern of nonzero off-diagonal elements:

$$\text{multiplication-additions} = \sum_{i=1}^{n-1} r_i^2. \quad (26b)$$

For symmetric matrices:

$$\text{multiplication-additions} = \sum_{i=1}^{n-1} (r_i^2 + r_i)/2 \quad (26c)$$

where r_i is the number of nonzero terms to the right of the diagonal in row i in the table of factors and s is the total number of such terms.

Determining the operation count for changing the array of factors to reflect changes in the original matrix is quite complicated and dependent upon the actual programming scheme. In general, if the changes in A begin in row k, (26a)–(26c) may be used with summations beginning at k instead of 1 to get an approximate count.

The operations required for computing any of the possible complete direct solutions based on A are:

$$\begin{aligned} \text{multiplications or divisions} &= n \\ \text{additions} &= n \\ \text{multiplication-additions} &= 2s. \end{aligned} \quad (26d)$$

Partial solutions can be obtained in fewer operations.

D. Programming

A comprehensive presentation of the various programming techniques that have been developed for the method is beyond the scope of this paper. Some have been discussed in other papers.[1],[4],[10]

When working with a full matrix, the address in computer memory of each matrix element can be related to its row and column indices in such a way that programming is quite easy. However, in order to achieve the benefits of sparsity, the programming scheme must store and process only nonzero elements. This requires, in addition to the memory allocation for the matrix elements themselves, tables of indexing information to identify the elements and to facilitate their addressing. Programming is more difficult in this case and much of the method's potential advantage can be lost through poorly planned programming. The need for expert programming cannot be overemphasized.

With the most effective programming techniques, the operations are performed as if they are being done by visual inspection and manual computation. At the completion of the optimal ordering algorithm [scheme 2) or 3)], the exact form of the table of factors is established and this information is recorded in various tables to guide the actual elimination process. During the elimination no operation is performed that would lead to a predictable zero result and no memory allocation is made for a predictable zero element. The original matrix, the table of factors, and all indexing tables contain only nonzero elements.

Rows are transferred from the original matrix to a compact working row in which elements to the left of the diagonal are eliminated by appropriate linear combination with previously processed rows from the partially completed table of factors. When work on the row has been completed, it is added to the table of factors. The actual procedure varies depending on the nature of the application. If the matrix is nonsymmetric, the derived elements to the left of the diagonal must be stored. The arrangement of the tables depends to some extent on the subsequent needs for direct solutions.

An example of one possibility for arranging the table of factors is indicated in Table I. This represents the final result for a seventh-order symmetric matrix. The columns labeled Loc refer to relative addresses in computer memory. The ith location in the D Table contains the element d_{ii} and the location in the U Table of the first u_{ij} element of row i. The column of the U Table labeled J contains the column subscript j of u_{ij}. Thus row 2 begins in location 3 of the U Table and contains u_{23}, u_{26}, and u_{27} in locations 3, 4, and 5, respectively. Row 3 begins in location 6.

TABLE I

EXAMPLE OF STORAGE AND INDEXING SCHEME FOR TABLE OF FACTORS

D Table			*U* Table		
Loc	*D* Factors	Loc in *U* Table	Loc	*U* Factors	*J*
1	d_{11}	1	1	u_{12}	2
2	d_{22}	3	2	u_{17}	7
3	d_{33}	6	3	u_{23}	3
4	d_{44}	8	4	u_{26}	6
5	d_{55}	10	5	u_{27}	7
6	d_{66}	11	6	u_{36}	6
7	d_{77}	12	7	u_{37}	7
8	—	12	8	u_{45}	5
			9	u_{46}	6
			10	u_{56}	6
			11	u_{67}	7

E. Comparative Advantages for a Sparse Matrix

When A is sparse the advantages of the factored form in addition to those previously listed are:

1) the table of factors can be obtained in a small fraction of the time required for the inverse
2) the storage requirement is small, permitting much larger systems to be solved
3) direct solutions can be obtained much faster unless the independent vector is extremely sparse
4) round-off error is reduced
5) modifications due to changes in the matrix can be made much faster.

The only disadvantage of the method is that it requires much more sophisticated programming techniques.

The effectiveness of the method may be judged by the results of a typical large problem described in Appendix II.

IV. Conclusions

Optimally ordered triangular factorization appears to be better than any other known method for the direct solution of the sparse linear equations that arise in many network problems. The method opens the way for improvements in existing applications and offers possibilities for new applications that would not be feasible with any other approach. It eliminates the need for resorting to slowly converging iterative methods in order to solve many large network problems and the need for compromising with desired problem size in order to obtain the benefits of direct solutions. Further improvements and extensions of the method are expected.

Appendix I

Numerical Examples

The following examples based on the full nonsymmetric matrix A are given only to indicate the operations. They do not show the effects of sparsity or the methods of programming to take advantage of it.

$$A = \begin{bmatrix} 2 & 1 & 3 \\ 2 & 3 & 4 \\ 3 & 4 & 7 \end{bmatrix}. \tag{27}$$

The table of factors for A is

$$\begin{matrix} \frac{1}{2} & \frac{1}{2} & \frac{3}{2} \\ 2 & \frac{1}{2} & \frac{1}{2} \\ 3 & \frac{5}{2} & \frac{4}{5} \end{matrix}. \tag{28}$$

With b given, the equation $Ax=b$ can be solved for x using (29), which is based on (11a):

$$\overset{U_1}{\begin{bmatrix} 1 & -\frac{1}{2} & -\frac{3}{2} \\ & 1 & \\ & & 1 \end{bmatrix}} \overset{U_2}{\begin{bmatrix} 1 & & \\ & 1 & -\frac{1}{2} \\ & & 1 \end{bmatrix}} \overset{D_3}{\begin{bmatrix} 1 & & \\ & 1 & \\ & & \frac{4}{5} \end{bmatrix}} \overset{L_2}{\begin{bmatrix} 1 & & \\ & 1 & \\ & -\frac{5}{2} & 1 \end{bmatrix}}$$
$$\overset{D_2}{\begin{bmatrix} 1 & & \\ & \frac{1}{2} & \\ & & 1 \end{bmatrix}} \overset{L_1}{\begin{bmatrix} 1 & & \\ -2 & 1 & \\ -3 & & 1 \end{bmatrix}} \overset{D_1}{\begin{bmatrix} \frac{1}{2} & & \\ & 1 & \\ & & 1 \end{bmatrix}} \overset{b}{\begin{bmatrix} 6 \\ 9 \\ 14 \end{bmatrix}} = \overset{x}{\begin{bmatrix} 1 \\ 1 \\ 1 \end{bmatrix}}. \tag{29}$$

With x given, b can be obtained from (30), which is based on (12a):

$$\overset{D_1^{-1}}{\begin{bmatrix} 2 & & \\ & 1 & \\ & & 1 \end{bmatrix}} \overset{L_1^{-1}}{\begin{bmatrix} 1 & & \\ 2 & 1 & \\ 3 & & 1 \end{bmatrix}} \overset{D_2^{-1}}{\begin{bmatrix} 1 & & \\ & 2 & \\ & & 1 \end{bmatrix}} \overset{L_2^{-1}}{\begin{bmatrix} 1 & & \\ & 1 & \\ & \frac{5}{2} & 1 \end{bmatrix}}$$
$$\overset{D_3^{-1}}{\begin{bmatrix} 1 & & \\ & 1 & \\ & & \frac{5}{4} \end{bmatrix}} \overset{U_2^{-1}}{\begin{bmatrix} 1 & & \\ & 1 & \frac{1}{2} \\ & & 1 \end{bmatrix}} \overset{U_1^{-1}}{\begin{bmatrix} 1 & \frac{1}{2} & \frac{3}{2} \\ & 1 & \\ & & 1 \end{bmatrix}} \overset{x}{\begin{bmatrix} 1 \\ 1 \\ 1 \end{bmatrix}} = \overset{b}{\begin{bmatrix} 6 \\ 9 \\ 14 \end{bmatrix}}. \tag{30}$$

With c given, the equation $A^t y = c$ can be solved for y by using (31), which is based on (13):

$$\overset{D_1}{\begin{bmatrix} \frac{1}{2} & & \\ & 1 & \\ & & 1 \end{bmatrix}} \overset{L_1^t}{\begin{bmatrix} 1 & -2 & -3 \\ & 1 & \\ & & 1 \end{bmatrix}} \overset{D_2}{\begin{bmatrix} 1 & & \\ & \frac{1}{2} & \\ & & 1 \end{bmatrix}} \overset{L_2^t}{\begin{bmatrix} 1 & & \\ & 1 & -\frac{5}{2} \\ & & 1 \end{bmatrix}}$$
$$\overset{D_3}{\begin{bmatrix} 1 & & \\ & 1 & \\ & & \frac{4}{5} \end{bmatrix}} \overset{U_2^t}{\begin{bmatrix} 1 & & \\ & 1 & \\ & -\frac{1}{2} & 1 \end{bmatrix}} \overset{U_1^t}{\begin{bmatrix} 1 & & \\ -\frac{1}{2} & 1 & \\ -\frac{3}{2} & & 1 \end{bmatrix}} \overset{c}{\begin{bmatrix} 9 \\ 9 \\ 17 \end{bmatrix}} = \overset{y}{\begin{bmatrix} 2 \\ 1 \\ 1 \end{bmatrix}}. \tag{31}$$

With y given, c can be obtained from (32), which is based on (14):

$$\overset{(U_1^t)^{-1}}{\begin{bmatrix} 1 & & \\ \frac{1}{2} & 1 & \\ \frac{3}{2} & & 1 \end{bmatrix}} \overset{(U_2^t)^{-1}}{\begin{bmatrix} 1 & & \\ & 1 & \\ & \frac{1}{2} & 1 \end{bmatrix}} \overset{D_3^{-1}}{\begin{bmatrix} 1 & & \\ & 1 & \\ & & \frac{5}{4} \end{bmatrix}} \overset{(L_2^t)^{-1}}{\begin{bmatrix} 1 & & \\ & 1 & \frac{5}{2} \\ & & 1 \end{bmatrix}}$$
$$\overset{D_2^{-1}}{\begin{bmatrix} 1 & & \\ & 2 & \\ & & 1 \end{bmatrix}} \overset{(L_1^t)^{-1}}{\begin{bmatrix} 1 & 2 & 3 \\ & 1 & \\ & & 1 \end{bmatrix}} \overset{D_1^{-1}}{\begin{bmatrix} 2 & & \\ & 1 & \\ & & 1 \end{bmatrix}} \overset{y}{\begin{bmatrix} 2 \\ 1 \\ 1 \end{bmatrix}} = \overset{c}{\begin{bmatrix} 9 \\ 9 \\ 17 \end{bmatrix}}. \tag{32}$$

Given the hybrid vector g such that

$$g^t = (b_1, x_2, x_3) = (6, 1, 1) \tag{33}$$

the intermediate vector z is obtained by (34), which is based on (16):

$$\overset{U_2^{-1}}{\begin{bmatrix} 1 & & \\ & 1 & \frac{1}{2} \\ & & 1 \end{bmatrix}} \overset{D_1}{\begin{bmatrix} \frac{1}{2} & & \\ & 1 & \\ & & 1 \end{bmatrix}} \overset{g}{\begin{bmatrix} 6 \\ 1 \\ 1 \end{bmatrix}} = \overset{z}{\begin{bmatrix} 3 \\ \frac{3}{2} \\ 1 \end{bmatrix}}. \tag{34}$$

The vector h is formed from z_1, x_2, x_3 as indicated in (17) and x_1 is computed by (35), which is based on (17):

$$\overset{U_1}{\begin{bmatrix} 1 & -\frac{1}{2} & -\frac{3}{2} \\ & 1 & \\ & & 1 \end{bmatrix}} \overset{h}{\begin{bmatrix} 3 \\ 1 \\ 1 \end{bmatrix}} = \overset{x}{\begin{bmatrix} 1 \\ 1 \\ 1 \end{bmatrix}}. \tag{35}$$

The elements b_2 and b_3 are computed by (36), which is based on (18):

$$\overset{(L_2^*)^{-1}}{\begin{bmatrix} 1 & & \\ 2 & 1 & \\ & & 1 \end{bmatrix}} \overset{D_2^{-1}}{\begin{bmatrix} 1 & & \\ & 2 & \\ & & 1 \end{bmatrix}} \overset{(L_3^*)^{-1}}{\begin{bmatrix} 1 & & \\ & 1 & \\ 3 & \frac{5}{2} & 1 \end{bmatrix}}$$
$$\overset{D_3^{-1}}{\begin{bmatrix} 1 & & \\ & 1 & \\ & & \frac{5}{4} \end{bmatrix}} \overset{z}{\begin{bmatrix} 3 \\ \frac{3}{2} \\ 1 \end{bmatrix}} = \overset{b'}{\begin{matrix} 3 \\ 9 \\ 14 \end{matrix}}. \tag{36}$$

Appendix II
Example Problem

The purpose of this example problem is to demonstrate the effectiveness of the method. A program has been written using the method for solving the nodal admittance equation $YE=I$. The given data are the elements of the complex symmetric nodal admittance matrix Y and vectors of node currents I. Solutions consist of the vectors of complex node voltages E corresponding to each given vector I. The program consists of three subroutines: 1) optimal ordering, 2) formation of the table of the factors, and 3) direct solutions.

The program is dimensioned for 1000 nodes. It is written mostly in MAP assembly language for the IBM 7040, a comparatively slow computer with an eight-microsecond access time and 36-bit word size. The results of a typical power network problem using scheme 2) for optimal ordering are as follows:

number of nodes = 839
number of branches = 1122
table of factors:
d_{ii} terms = 839
u_{ij} terms = 2210.

Complex arithmetic operations for direction solutions:

additions = 839
multiplications = 839
multiplication-additions = 2 × 2210 = 4420
computing times:
optimal ordering by scheme 2 = 8.11 seconds
forming table of factors = 10.42 seconds
each direct solution = 2.80 seconds.

The total time required for optimal ordering and computing the table of factors is the time that should be compared with that of other methods for obtaining the inverse; the time required for direct solutions should be compared with that of multiplying by the inverse. The time for read-in and organization of data is not included.

The dimensions of tables for a 1000-node power network problem are as follows:

d_{ii} terms:	2 × 1000 =	2000 floating point words
u_{ij} terms:	2 × 3000 =	6000 floating point words
solution vector:	2 × 1000 =	2000 floating point words
index tables:		4000 integer words
total		14000 words.

If the matrix were real instead of complex, the time for optimal ordering would be unaffected, but the time for computing the table of factors and the direct solutions would be reduced to less than 25 percent of that for the complex matrix. The dimensions of tables with factor 2 could be reduced by one half.

If the matrix were complex and nonsymmetric, the optimal ordering and direct solution times would be unaffected, but the computation of the table of factors would take almost twice as long. An additional table of 2 × 3000 = 6000 floating point words would also be required for the l_{ij} terms.

The time of 2.8 seconds is for a complete direct solution of 839 complex node voltages. Any of the other possible complete direct solutions could be obtained in approximately the same time. Partial solutions could be obtained faster. Any row or column of the inverse, if needed, could be computed in 2.8 seconds. A complete nonsymmetric inverse could be computed in this manner in 839 × 2.8 seconds, or about 40 minutes. A symmetric inverse for which only a triangular matrix is needed would require about 70 percent as much time. Since each column of the inverse would be computed independently, the round-off error in single precision would be negligible.

The 4000 words of index tables are for unpacked integers of no more than three-decimal digits; they could be considerably compressed. The time for optimal ordering could be improved. The programs for obtaining the table of factors and direct solutions are carefully planned and probably could not be improved very much.

Acknowledgment

The authors wish to aknowledge the valuable assistance of N. R. Isseks and D. Bree, Jr., who contributed to this work by developing algorithms and programs.

References

[1] N. Sato and W. F. Tinney, "Techniques for exploiting the sparsity of the network admittance matrix," *IEEE Trans. Power Apparatus and Systems*, vol. 82, pp. 944–950, December 1963.

[2] J. Carpentier, "Ordered eliminations," *Proc. Power System Computation Conf.* (London), 1963.

[3] H. Edelmann, "Ordered triangular factorization of matrices," *Proc. Power System Computation Conf.* (Stockholm), 1966.

[4] R. Baumann, "Some new aspects of load flow calculation," *Proc. Power Industry Computer Applications Conf.* (Clearwater, Fla.), p. 91, May 1965.

[5] A. Ralston, *A First Course in Numerical Analysis*. New York: McGraw-Hill, 1965, pp. 398–415.

[6] E. Bodewig, *Matrix Calculus*. New York: Interscience, 1959, pp. 101–124.

[7] W. F. Tinney and C. E. Hart, "Power flow solution by Newton's method," presented at IEEE Winter Power Meeting, New York, N. Y., 1967.

[8] H. W. Dommel, "A method for solving transient phenomena in multi-phase systems," presented at Power Systems Computation Conf., Stockholm, 1966.

[9] J. Peschon, W. F. Tinney, D. S. Piercy, O. J. Tveit, and M. Cuenod, "Optimum control of reactive power flow," presented at IEEE Winter Power Meeting, New York, N. Y., 1967.

[10] G. H. Jensen, "Designing self-supporting transmission towers with the digital computer," presented at Power Industry Computer Applications Conf., Pittsburgh, Pa., May 15–17, 1967.

State Estimation in Power Systems
Part I: Theory and Feasibility

ROBERT E. LARSON, MEMBER, IEEE, WILLIAM F. TINNEY, SENIOR MEMBER, IEEE, AND JOHN PESCHON, SENIOR MEMBER, IEEE

***Abstract*—State estimation is a digital processing scheme which provides a real-time data base for many of the central control and dispatch functions in a power system. The estimator processes the imperfect information available and produces the best possible estimate of the true state of the system. The basic theory and computational requirements of static state estimation are presented, and their impact on the evolution of the data-acquisition, data-processing, and control subsystems are discussed. The feasibility of this technique is demonstrated on network examples.**

INTRODUCTION

THE purpose of this paper and a companion paper [1] is to present a methodology for processing measured data from electric power networks in order to obtain the best possible estimates of the system variables. The theory is discussed in this paper, and [1] summarizes extensive computational experience gained with these procedures on a large, realistic network and discusses potential applications.

The emphasis of the research was on developing efficient computational procedures capable of processing thousands of measurements per second on present-day computers. The feasibility of the method is demonstrated on relatively small network examples.

Information-processing algorithms that are mathematically optimum as well as meaningful in engineering terms were developed mostly after 1960 [2], [3]. These algorithms have since been used with great success in the defense and aerospace industry. To our knowledge, these advanced information-processing techniques have not yet been used in the power industries. Schweppe first suggested the development of power-system state estimators. He has developed static state-estimator algorithms for relatively small networks [4], [5] and outlined the computational difficulties associated with their implementation [6].

The results presented here and in [1] are based on a joint research effort by Systems Control, Inc., and the Bonneville Power Administration [7], [8]. The work was undertaken to evaluate the potential impact of state-estimation procedures on the evolution of the data-acquisition, data-processing, and control subsystems of large interconnected power systems. Interest was focused on the following specific areas:

1) new and powerful techniques for system control, monitoring, display, and alarm that might be made possible by a real-time state estimator;

2) detailed specification and design of an information system capable of supplying the required system data at the least cost in sensing, telemetry, and processing hardware.

CONCEPT OF STATE ESTIMATION IN POWER SYSTEMS

General Discussion of State Estimation

The state estimator is a data-processing scheme that computes the state of a system from the following three sources of information:

1) measurements of system variables
2) the mathematical model of the system and its instrumentation
3) prior knowledge of various system inputs and outputs, referred to as *pseudomeasurements.*

The output of the state estimator approaches the true state of the system. Discrepancies between the true and estimated states are due to the following factors:

1) noise in instruments and telemetry channels
2) incomplete instrumentation, in the sense that many variables are not measured at all
3) delayed measurements that reflect a prior system state
4) erroneous pseudomeasurements
5) inaccurate network parameters.

There are numerous approaches to the design of a state estimator. To compare the performance of various estimator designs, it is customary to define a scalar "cost" function that increases with the estimation error. An estimator that minimizes this cost function is optimal in the sense that it generates the best estimate of the state based on the measurements and pseudomeasurements received [7]. The principal papers and books on optimal estimation theory are listed by Lee [2].

The optimal estimator not only generates a state estimate but also the covariance or "error" associated with it. The covariance matrix predicts the magnitude of the estimation error and hence provides a measure of confidence for the estimated state. The predicted and the actual estimation errors should be of the same order of magnitude. Significant differences indicate that a major change, such as a fault, has occurred in the system or in the instrumentation, and an alarm signal can be given. Furthermore, the most likely of these major changes (or faults) can be determined.

Application to Power Systems

Centralized automatic control of power system dispatch requires knowledge of the steady-state performance of the system. This performance must be inferred from system measurements communicated to the control center. If enough measurements could be obtained continuously, accurately, and reliably,

Paper 69 TP 638–PWR, recommended and approved by the Power System Engineering Committee of the IEEE Power Group for presentation at the IEEE Summer Power Meeting, Dallas, Tex., June 22–27, 1969. Manuscript submitted February 20, 1969; made available for printing April 14, 1969.

R. E. Larson and J. Peschon are with Systems Control, Inc., Palo Alto, Calif. (formerly Wolf Management Services of Palo Alto, Calif.)

W. F. Tinney is with the Bonneville Power Administration, Portland, Ore.

Reprinted from *IEEE Trans. Power App. Syst.,* vol. PAS-89, pp. 345–352, Mar. 1970.

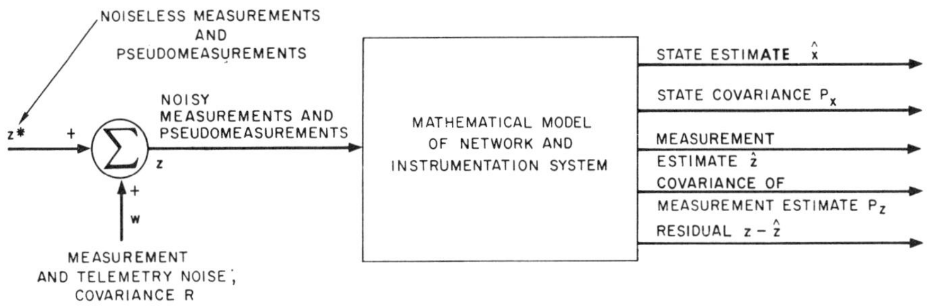

Fig. 1. Inputs and outputs of state estimator.

this would provide all the information needed for control. However, a perfect data-acquisition system is often technically and economically infeasible so that the control center must depend on measurements that are incomplete, delayed, inaccurate, and unreliable. The function of the state estimator is to process whatever information is available and to produce the best possible estimate of the true state of the system.

The operation of a power system depends upon many different control systems (e.g., turbine governors, voltage regulators, boiler controls, load-frequency control, and economic dispatching). State-estimation techniques involving the same fundamental concepts but with different equations can be applied to estimate the variables and parameters of each of these control systems. In this study, state estimation is limited to the serving of the control functions associated with the system operating under steady-state or slowly varying conditions. The corresponding mathematical model is represented by the well-known network equations; these are given in (15). The variables of concern are:

1) the injections I and K of active and reactive power at the nodes[1]
2) the voltage magnitudes and phase angles V and θ at the nodes
3) the flows T and U of active and reactive power in the branches.

The state variables are taken here to be V and θ. Once they are known, all the other electrical variables of the network can be computed directly. The measurements and pseudomeasurements could consist of any or all of the listed variables but, in practice, are predominantly of the type I, K, and T with the possibility of some additional measurements of V, U, and perhaps θ.

The State Estimator as a Data Base for Control

Fig. 1 illustrates the inputs and outputs of the state estimator, where

- z raw noisy measurements and pseudomeasurements comprising real and reactive injections I and K, real and reactive power flows T and U, voltage magnitudes and phase angles V and θ
- z^* noiseless measurements and pseudomeasurements as would be obtained from a perfect sensor
- w additive sensor and telemetry noise
- $\hat{x}$ state estimate consisting of voltage magnitude and phase angle, V and θ, at each node
- $\hat{z}$ processed measurements and pseudomeasurements; if the redundancy provided by the power-flow model is sufficient, a filtering effect occurs and the error of $\hat{z}$ is less than that of z
- $z-\hat{z}$ "residual" or difference between actual and estimated measurement; the residual is used to detect system anomalies
- R noise covariance matrix; the diagonal elements R_{ii} equal the mean noise power (or rms error) of the component w_i; the mean noise power may be defined as the time average

$$\lim_{T\to\infty} \frac{1}{T} \int_0^T w_i^2(t)dt$$

- P_x state covariance matrix; the diagonal element $(P_x)_{ii}$ predicts the mean noise power between the true state component x_i and the estimated state component $\hat{x}_i$
- P_z covariance matrix of the estimated measurement $\hat{z}$; the diagonal element $(P_z)_{ii}$ predicts the mean noise power between the noiseless measurement component z_i^* and the estimated measurement component $\hat{z}_i$.

The output of the state estimator constitutes a data base from which many of the various control, display, monitoring, alarm, logging, and interrogation functions contemplated for the advanced dispatching centers are fed. This data base is optimal in the sense that all available information is so processed as to provide the most accurate estimate of the system state and the system measurements.

Redundancy, Pseudomeasurements, and Neighboring Systems

A system of N nodes has $2N - 1$ state variables appearing in the $2N$ power flow equations,[2] namely, $N - 1$ angles θ and N voltages V. These $2N - 1$ state variables can be computed from the $2N - 1$ primary measured variables, namely, $N - 1$ active injections I and N reactive injections K.

In addition, there are numerous secondary measured variables, such as active and reactive line power flows T and U. These secondary variables can all be obtained directly from the state variables.

When only $2N - 1$ variables are measured, the state estimator reduces to a real-time power-flow solution, except for the covariance computation. When more than $2N - 1$ measurements are made, the redundancy allows filtering (or smoothing) of measurement noise, that is, reduction in estimation error and anomaly detection. These additional measurements may comprise the

[1] The term "injection" refers to the difference between local generation and local consumption.

[2] The voltage phase angle at the slack node is set to zero; it is not a state variable.

secondary variables T and U. If fewer than $2N - 1$ measurements are made, a mathematical solution remains possible but can be shown to be meaningless from an engineering point of view under most circumstances [12].

Since the number of measurements actually telemetered to the dispatching center will probably be less than $2N - 1$ in large power systems, redundancy is introduced in the form of pseudomeasurements, namely,

1) known voltage and other regulator settings
2) known generation and interchange schedules
3) short-term load forecasts broken down by nodes.

The problem of information redundancy is further complicated by the fact that few power systems are operated in isolation. They are interconnected with neighboring systems that influence them but for which measurements are frequently not available in real time. A method of taking account of the effects of neighboring systems has been devised [8]. Yet another complication can be caused by the hierarchical structuring of the information and control systems. A successful solution of this problem is possible by means of network reduction and the design of "local" estimators [8].

Theory of Least Squares State Estimation

This section summarizes least squares estimation theory as applied to a power network [7] under assumed static conditions. Modifications for updating the estimate for a continuously changing system under normal operating conditions are covered in [1].

Definition of Terms

Control and estimation theory is heavily dependent on the concept of *state* variables. For any given system the state variables are defined as a nonredundant set of variables that completely characterize the operation of the system. In power systems the state variables are usually the voltage magnitudes and phase angles at all of the nodes. The set of state variables is denoted as x, the number of state variables as n.

The quantities that can be measured are taken to be known functions of the state variables. These quantities, called the *measured* variables, can be the active and reactive power flows at the nodes, active and reactive power flows in the branches, and voltage magnitudes and phase angles at the nodes (the state variables themselves). The set of measured variables is denoted as z, the number of these variables as m.

Each measured variable is corrupted by noise. The set of measurement noises is denoted as w. The statistics of the measurement noise are assumed to be known. Specifically, each measurement noise is assumed to have zero mean. (If the noise contains a bias term, this can be accounted for [7].) Mathematically,

$$E\{w\} = 0 \tag{1}$$

where the letter E denotes expectation. The spread of the measurement noise about its mean is also assumed to be known. This can be expressed as

$$E\{ww'\} = R \tag{2}$$

where the prime denotes transposition, and R is an $m \times m$ matrix called the noise covariance matrix. If the measurements are all independent [7], then R is a diagonal matrix with diagonal terms

$$R_{ii} = \sigma_i^2 \tag{3}$$

TABLE I
LINEAR ESTIMATION EQUATIONS

Number of Measurements*	Estimator Equations
$m > n$	$x = (A'R^{-1}A)^{-1}A'R^{-1}z$ $P_x = (A'R^{-1}A)^{-1}$ $\hat{z} = A(A'R^{-1}A)^{-1}A'R^{-1}z = A\hat{x}$ $P_z = A(A'R^{-1}A)^{-1}A' = AP_xA'$
$m = n$	$\hat{x} = A^{-1}z$ $P_x = (A'R^{-1}A)^{-1}$ $\hat{z} = z$ $P_z = R$
$m < n$	$\hat{x} = A'(AA')^{-1}z$ $P_x = A'(AA')^{-1}R(AA')^{-1}A$ $\hat{z} = z$ $P_z = R$
any m	$P_x \triangleq E\{(x - \hat{x})(x - \hat{x})'\}$ $P_z \triangleq E\{(z - \hat{z})(z - \hat{z})'\}$ $R = E\{ww'\}$

* Symbols n and m denote the number of state variables and the number of measurements (including pseudomeasurements), respectively.

where σ_i is the rms value (standard deviation) of noise on the ith measurement. Numerical values for these σ_i can be calculated from knowledge of the accuracy of the instrumentation and telemetry systems. These values characterize the accuracy of the individual measurements and are fundamental in computing an optimum estimate.

The primary output of an optimum estimator is a value of the state variables called the optimum state estimate and denoted as $\hat{x}$. This estimate is optimum in the sense that it minimizes some criterion, based on knowledge of the equations relating the measured variables to the state variables, the statistics of the measurement noise, and a given set of measurements z. In least squares estimation theory, this criterion is a quadratic function of the residuals. These quantities are defined as the differences between the measurements actually made and the measurements that would have been made if the estimate were exact and there were no noise. The quadratic criterion weights the residuals according to the accuracy of the corresponding measurements. A lower penalty is assessed for a given residual in a poor measurement than is assessed in an accurate measurement.

An important by-product of the optimum estimator is the estimate covariance matrix P_x. This $n \times n$ matrix, defined as

$$E\{(x - \hat{x})(x - \hat{x})'\} = P_x \tag{4}$$

is a measure of the accuracy of the estimate. This quantity is important in a number of applications.

The optimum estimator can also be used to generate an improved estimate of the measured variables themselves. This estimate is denoted as $\hat{z}$. The covariance of this estimate, denoted as P_z, is also obtained.

Linear Estimation Theory

If the measured variables are linear functions of the state variables, then the optimum estimation equations can be solved exactly. For this case the measurement equations are written as

$$z = Ax + w \tag{5}$$

where A is the $n \times m$ measurement matrix, and the vectors z, x, and w are as previously defined. The criterion for optimum

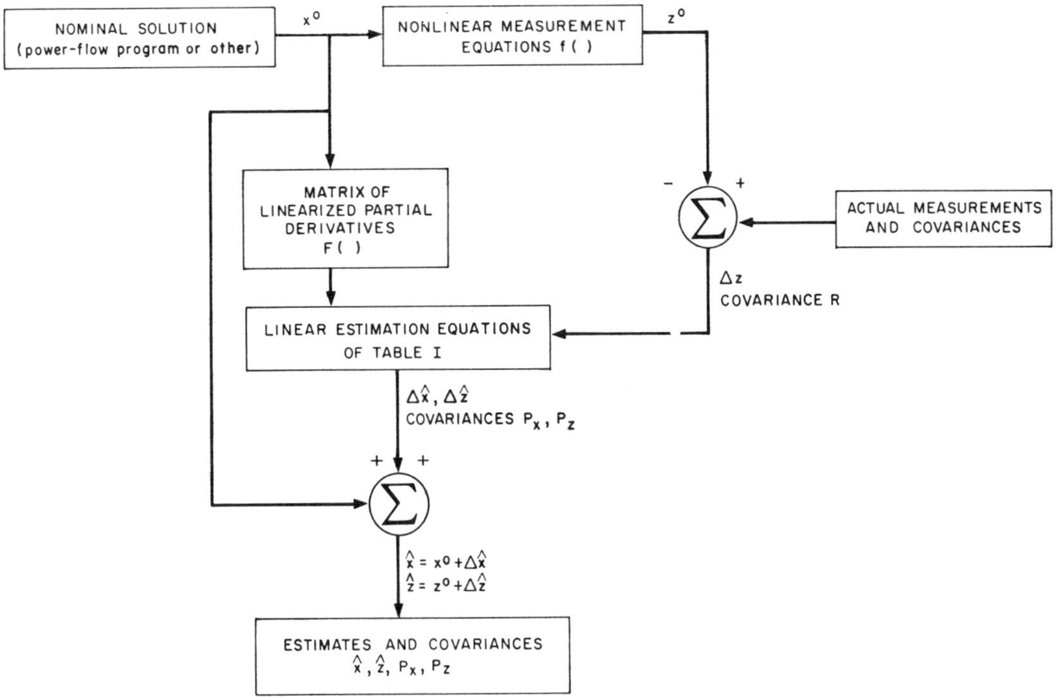

Fig. 2. Flow chart for nonlinear estimator.

estimation, denoted as J, becomes

$$J = (z - Ax)'R^{-1}(z - Ax). \tag{6}$$

The optimum estimate $\hat{x}$ is found by minimizing the scalar J. There are three separate cases to be considered, depending on the relative values of m and n. The optimum estimation equations for the three cases ($m > n$, $m = n$, $m < n$) are summarized in Table I.

Nonlinear Estimation Theory

If the measured variables are nonlinear functions of the state variables, a commonly used approach is to linearize the measurement equations about a nominal value and then to use the results of Table I. If there is an a priori estimate of the state variables that is sufficiently close to the actual values, then this approach is quite satisfactory. If the a priori estimate is not sufficiently close, then an iterative approach in which the measurement equations are relinearized about the newly obtained estimates can be applied. This approach can be justified heuristically as follows. If the nonlinear estimate equations are written

$$z = f(x) + w \tag{7}$$

where f is the nonlinear measurement function, then the least squares estimation criterion becomes

$$J = [z - f(x)]'R^{-1}[z - f(x)]. \tag{8}$$

Now, if x^0 is a nominal value of the state variables and if z^0 is defined as

$$z^0 = f(x^0) \tag{9}$$

then expansion of both $f(x)$ and z to first order yields

$$z = z^0 + \Delta z \tag{10}$$

$$f(x) = f(x^0) + F\Delta x \tag{11}$$

where F is the $n \times n$ Jacobian matrix defined by

$$F_{ij} = \frac{\partial f_i}{\partial x_j}, \; i = 1,2, \cdots, n, \quad j = 1,2, \cdots, n \tag{12}$$

and where Δz and Δx are small changes in z and x, respectively. Substitution of (9)–(11) into (8) yields

$$J = (\Delta z - F\Delta x)'R^{-1}(\Delta z - F\Delta x). \tag{13}$$

This is precisely the same quantity to be minimized as for the problem defined by

$$\Delta z = F\Delta x + w. \tag{14}$$

Thus the linear estimation equations of Table I can be applied directly to estimate the state deviations Δx based on the measurement deviations $\Delta z = z - f(x^0)$. The estimated deviations $\Delta\hat{x}$ can be added back to x^0 to obtain a new state estimate $\hat{x}$. The flow of information in this procedure is shown in Fig. 2. The accuracy of these equations has been studied extensively from both a theoretical and an empirical viewpoint [7].

For a power system, the basic ac network equations are written in the customary admittance matrix form [7], [9]

$$I_i = \sum_{j=1}^{N} V_i V_j Y_{ij} \cos(\theta_i - \theta_j - \psi_{ij})$$

$$K_i = \sum_{j=1}^{N} V_i V_j Y_{ij} \sin(\theta_i - \theta_j - \psi_{ij}), \quad i = 1,2, \cdots, N \tag{15}$$

where

- I_i real power injection at node i
- K_i reactive power injection at node i
- Y_{ij} magnitude of admittance from node i to node j
- ψ_{ij} phase angle of admittance from node i to node j.

TABLE II

MATRIX F AND VECTORS Δx AND Δz FOR MAXIMUM NUMBER OF MEASUREMENTS IN (14)

$$\Delta x = \begin{bmatrix} \theta_2 \\ \vdots \\ \theta_N \\ \hline V_1 \\ \vdots \\ V_n \end{bmatrix} \begin{matrix} \left.\vphantom{\begin{matrix}a\\b\\c\end{matrix}}\right\} N-1 \\ \left.\vphantom{\begin{matrix}a\\b\\c\end{matrix}}\right\} N \end{matrix} \qquad \Delta z = \begin{bmatrix} \Delta I_1 \\ \vdots \\ \Delta I_N \\ \hline \Delta K_1 \\ \vdots \\ \Delta K_N \\ \hline \Delta T_{ij} \\ \vdots \\ \hline U\Delta_{ij} \\ \vdots \\ \hline \Delta\theta_2 \\ \vdots \\ \Delta\theta_N \\ \hline \Delta V_1 \\ \vdots \\ \Delta V_N \end{bmatrix} \begin{matrix} \} N \\ \} N \\ \} 2M \\ \} 2M \\ \} N-1 \\ \} N \end{matrix}$$

$$F = \left[\begin{array}{c|c} a_{ij} & b_{ij} \\ i = 1,\cdots,N & i = 1,\cdots,N \\ j = 2,\cdots,N & j = 1,\cdots,N \\ \hline c_{ij} & d_{ij} \\ i = 1,\cdots,N & i = 1,\cdots,N \\ j = 2,\cdots,N & j = 1,\cdots,N \\ \hline e_{ij} & f_{ij} \\ i = 1,\cdots,2M & i = 1,\cdots,2M \\ j = 2,\cdots,N & j = 1,\cdots,N \\ \hline g_{ij} & h_{ij} \\ i = 1,\cdots,2M & i = 1,\cdots,2M \\ j = 2,\cdots,N & j = 2,\cdots,N \\ \hline \begin{matrix} 1 & & 0 \\ & \ddots & \\ 0 & & 1 \end{matrix} & 0 \\ \hline 0 & \begin{matrix} 1 & & 0 \\ & \ddots & \\ 0 & & 1 \end{matrix} \end{array}\right]$$

The corresponding real and reactive branch flow from node i to node j can be written as

$$T_{ij} = V_iV_jY_{ij}\cos(\theta_i - \theta_j - \psi_{ij}) - V_i^2Y_{ij}\cos\psi_{ij} \quad (16)$$

$$U_{ij} = V_iV_jY_{ij}\sin(\theta_i - \theta_j - \psi_{ij}) + V_i^2Y_{ij}\sin\psi_{ij}. \quad (17)$$

The measurements can include N real injections, N reactive injections, $2M$ real branch flows, $2M$ reactive branch flows, $N - 1$ voltage phase angles, and N voltage magnitudes. Because the injections and branch flows are nonlinear functions of the state variables, it is necessary to linearize the measurement equations in order to apply the estimation equations.

If all of these measurements are made, the basic F matrix and Δx and Δz vectors of (14) are as in Table II. The Δx vector contains $\Delta\theta_i$ and ΔV_i, while Δz contains ΔI_i, ΔK_i, ΔT_{ij}, ΔV_{ij}, $\Delta\theta_i$, and ΔV_i.

The linearization is performed about a nominal set of state variables θ_i^0, $i = 2, \cdots, N$ and V_i^0, $i = 1,2, \cdots, N$. The elements of the submatrices a, b, c, d, e, f, and g are the appropriate Jacobian terms of (15)–(17).

If fewer than $(4M + 4N - 1)$ measurements (the maximum) are made, then the matrix F and vector Δz have the same structure, except that the unmeasured elements of the Δz vector and the corresponding rows of A are deleted. The vector Δx is not changed.

COMPUTATIONAL CONSIDERATIONS

Straightforward Implementation of Estimation Equation

The computational requirements of directly implementing the equations in Table I are easily derived and relatively well known. However, the numerical difficulties, particularly in the case where the measurement equations must be linearized, are not generally understood or appreciated, nor are the limitations imposed by these considerations on the size of problems that can be treated. The cases $m > n$, $m = n$, and $m < n$ are treated separately.

1) *Case where $m > n$:* The basic estimation equations are

$$\hat{x} = (A'R^{-1}A)^{-1}A'R^{-1}z \quad (18)$$

$$P_x = (A'R^{-1}A)^{-1}. \quad (19)$$

These computations require basic arithmetic operations on the order of mn^2. Using one of the faster present-day computers, the computations for $n = 100$, $m > n$, would require only a few seconds; for $n = 1000$, $m > n$, the computations would require several minutes, which is too long for most practical on-line applications. The high-speed storage requirements are on the order of mn locations [7]. If $n = 100$, $m > n$, the number of locations is on the order of 10^4, which is large but manageable. However, if $n = 1000$, the number is 10^6, which is clearly infeasible for any readily available system.

Thus for present computers the limit of computational feasibility by this approach is reached for a value of n somewhere between 100 and 1000. Furthermore, unless precautions are taken, severe numerical difficulties may be encountered at values of n well below 100, and straightforward application of (18) and (19) may lead to meaningless results. For example, a problem arose in connection with a nonlinear estimation problem having 51 state variables [7]. The problem centered on the phenomenon that the matrix $(A'R^{-1}A)$ became nearly singular so that inversion was, for all practical purposes, infeasible. This difficulty occurred because some of the measurements were assumed to be orders of magnitude more accurate than others. Since most of the linearized measurement equations are proportional to differences between state variables, the situation arose where two state variables were not determined precisely, but their difference was known accurately. This problem was eliminated by placing a lower bound on the accuracy of all measurements.

A number of other difficulties also arose in this problem. However, this typical example is enough to show that a straightforward implementation of textbook equations is not sufficient for large practical systems.

2) *Case where* $m = n$: The equations for $m = n$ are

$$\hat{x} = A^{-1}z \tag{20}$$

$$P_x = (A'R^{-1}A)^{-1}. \tag{21}$$

The computational difficulties here are much the same as in the case $m > n$. Computation time is proportional to n^3, while storage requirements are on the order of n^2. Thus the limit of computational feasibility is again somewhere between $n = 100$ and $n = 1000$.

The numerical difficulties are not as severe in this case. First, since A is a square nonsingular $n \times n$ matrix, P_x can be computed as

$$P_x = A^{-1}R(A')^{-1}. \tag{22}$$

Since $(A'R^{-1}A)$ does not have to be computed or inverted, many of the problems encountered for $m > n$ are either lessened significantly or eliminated completely. Second, the computation of $\hat{x}$ requires only the inversion of A; again, this is considerably easier than the corresponding calculation for $m > n$.

3) *Case where* $m < n$: Although the optimum least-squares estimate for $m < n$ exists in principle, there are fundamental objections to its validity in a practical system. There are many estimates $\hat{x}$ for which the criterion

$$J = (z - A\hat{x})'R^{-1}(z - A\hat{x}) \tag{23}$$

is exactly zero. The optimum estimate is arbitrarily chosen to be that estimate from this set that minimizes $\hat{x}'\hat{x}$. There is no physical significance to this choice in power systems, and the expected result, that the estimate will be very poor, was confirmed in a number of experiments [7].

Improved Computational Procedures

Only the practical cases, $m > n$ and $m = n$, will be considered in this summary of some of the techniques developed [7].

1) *Use of the Power-Flow Solution for the Case* $m = n$: When $m = n$ and z consists of the n-nodal injection measurements, the problem is equivalent to the ac power-flow solution. If z is the vector of measured real and reactive injections, $\hat{x}$ is the solution vector of complex node voltages; if z is the vector of real injections only, $\hat{x}$ is the solution vector of node voltage phase angles only. In linear estimation theory the optimum estimate $\hat{x}$ is given by (20). This is in agreement with the power-flow solution where A^{-1} is the Jacobian matrix evaluated at the solution point (A is a nonsingular square matrix).

If the power flow is solved by Newton's method [9] with optimally ordered triangular factorization [10], A^{-1} is available in triangular factored form. Systems with $n = 100$ require at most a few seconds for solution, and the storage requirements are modest because the factored A^{-1} is sparse [10]. The solution of (20) for $n = 1000$ takes less than a second on a fast computer when A^{-1} is in factored form.

2) *Partitioning of* m *Measurements for* $m > n$ *into* n *Basic Measurements and* $(m - n)$ *Redundant Measurements:* For the case $m > n$ the power-flow solution cannot be used directly. However, it is possible to split the m measurements into two groups, the n basic measurements used in the power-flow solution and $(m - n)$ redundant measurements. It can be shown that the optimum estimate is the sum of the optimum estimate based on the n basic measurements and a correction term based on the $(m - n)$ redundant measurements. The optimum estimate based on the n basic measurements can be obtained using the power-flow program, exactly as before. However, in order to compute the correction term, it is necessary to process matrices of order $n \times n$ and $(m - n) \times (m - n)$. If n and $(m - n)$ are not too large, this approach is practical.

3) *Sequential Processing of the* $(m - n)$ *Redundant Measurements:* This technique is the same as in 2), except that the $(m - n)$ redundant measurements are processed one at a time. As a result, the inversion of the $(m - n) \times (m - n)$ matrix is replaced by $(m - n)$ scalar divisions. If $(m - n)$ is large, this saving can substantially increase the desirability of the approach.

4) *Sequential Processing Using Only the Diagonal Elements of the State Covariance Matrix:* The approach used in 3) still requires storage and manipulation of an $n \times n$ matrix. An approximate approach that eliminates this difficulty is the use of sequential processing, which carries only the diagonal elements of the covariance matrix in all calculations. As a result, the computation time grows as n^2 rather than n^3, and the storage grows as n rather than n^2. It is then possible to handle values of n on the order of a few thousand even with a modest computer. This approach, unlike the previous methods, does not produce an optimum estimate. However, it can be shown [7] that the difference is very slight. Furthermore, many of the numerical difficulties present when a full covariance matrix is used are eliminated, and in practice this nonoptimal estimator may actually perform better than the optimal estimator [7].

This last computational procedure forms the basis for the online implementation of the efficient state-estimator algorithm, which is discussed in [1].

Computational Results for a Static State Estimator

In order to demonstrate the performance of a power-system state estimator, a computer simulation of this process was carried out. The structure of this simulation is discussed, and numerical results are given for an 8-node network.

Structure of the Simulation

The general structure of the simulation of the static state estimator is shown in Fig. 3. The data input consists of the actual values of the system state variables, a selection of the measurements to be made, and statistics of the associated measurement noise. The actual state is specified directly, while the measurements to be made are selected from a list of allowed measurements. The standard deviation of the noise on each measurement is specified, and the measurement noise covariance matrix is constructed from this information.

The actual values of the state variables were used to generate noiseless values of the selected measurements using the measurement equations (Fig. 3). Noise was added to each measurement using a random noise generator; the noise sample for each measurement was scaled according to its standard deviation.

The optimum estimator (linear or nonlinear) operated directly on the noisy measurements z and the measurement noise covariance matrix R. The output of the estimator was the optimum state estimate $\hat{x}$, the associated estimate covariance matrix P_x, the estimate of the measured quantities $\hat{z}$, and the associated measurement covariance matrix P_z.

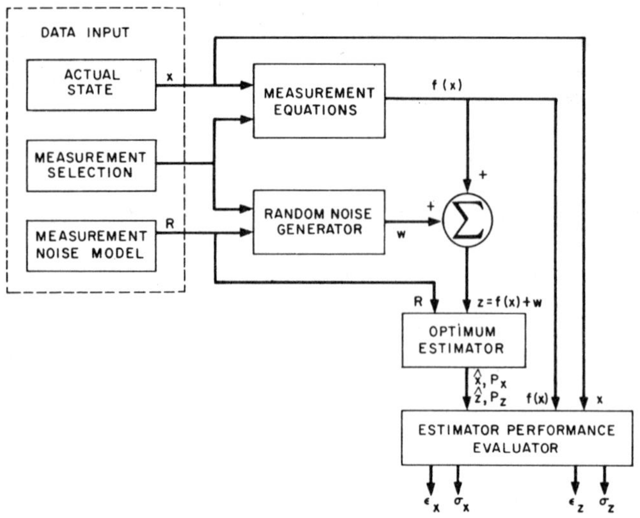

Fig. 3. Structure of the simulation for optimum estimator.

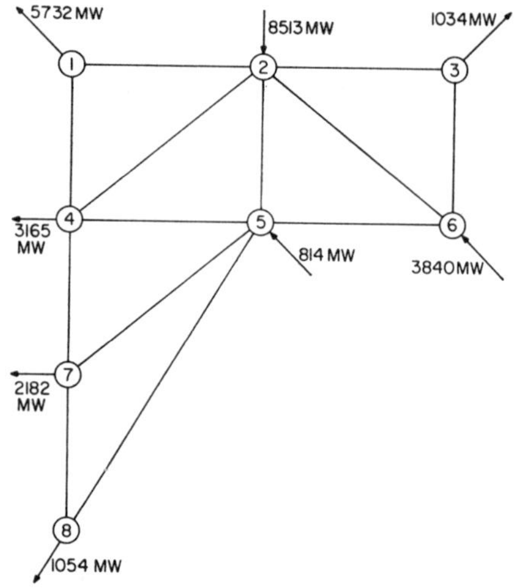

Fig. 4. 8-node example used in computational experiments.

All of these quantities were then fed into the estimator performance evaluator along with the actual values of state variables and measurements. It was then possible to evaluate estimator performance by comparing the errors between estimated and actual values of state variables and measurements; for state variable x_i this error is defined as ϵ_{xi}, where

$$\epsilon_{xi} = \hat{x}_i - x_i \tag{24}$$

while for the jth measured quantity $f_j(x)$ the error is defined as ϵ_{zj}, where

$$\epsilon_{zj} = \hat{z}_j - f_j(x). \tag{25}$$

It was also possible to compare these errors against their standard deviations; if P_{xii} denotes the ith diagonal element of P_x, then the standard deviation of the error in the estimate of x_i is given by σ_{xi}, where

$$\sigma_{xi} = \sqrt{P_{xii}} \tag{26}$$

while if P_{zjj} denotes the jth diagonal term of P_z, then the standard deviation of the error in the estimate of $f_j(x)$ is given by σ_{zj}, where

$$\sigma_{zj} = \sqrt{P_{zjj}}. \tag{27}$$

Clearly, if $|\epsilon_{xi}|$ is considerably larger than σ_{xi}, then the validity of the estimate is questionable; similarly for $|\epsilon_{zj}|$ and σ_{zj}.

Experimental Results for an 8-Node Example

The simulation was used to perform a number of computational experiments and detailed analyses [7]. An 8-node network is shown in Fig. 4 (its parameters were taken from [11]). Voltage phase angles were chosen to give the injections shown.

Case 1) State Estimation with Linear Power-Flow Model: When a linear model [11] of the network was assumed, linear estimation theory could be applied directly. The measured quantities were selected from the 8 nodal injections and the 13 branch flows. The standard deviation of each measurement was arbitrarily taken to be 2 percent of the actual value.

Different runs were made using several different sets of measurements. In all cases the actual estimation errors agreed well with the standard deviation of the estimates. Typical results are shown in Table III where the actual estimation error and the standard deviation of the estimate θ_6 are shown for several sets of measurements. These quantities are given as a percent of the actual value of θ_6 which is 30.78 degrees. The first set of measurements contains only the injections at all nodes except the slack node; there are 7 measurements and 7 state variables. The other sets of measurements contain 8, 20, and 21 measurements, respectively. This table indicates the degree to which redundant measurements improve the accuracy of the estimate.

Case 2) State Estimation with AC Power-Flow Model: The voltage-phase angles of Case 1) were used while the voltage magnitudes were selected arbitrarily. The noise model had a fixed standard deviation for each type of measurement (real injection, reactive branch flow, etc.) [7]. Because certain measurements, particularly reactive injections and branch flows, varied over a wide range, selecting the standard deviation to be a percent of the actual value led to numerical difficulties. Choosing a fixed value for each type of measurement assumes that identical meters are placed at each measuring station. It would have been just as easy to use a different standard deviation for each individual measurement.

The results in these experiments show good agreement between the estimation errors and their standard deviations. Since an ac model is used, both the voltage magnitudes and phase angles are estimated (except the phase angle at the slack node); there are a total of 15 state variables. Results for θ_6 and V_6 using 15 and 75 measurements are shown in Table IV.

Case 3) Parameter Estimation with AC Model: It is straightforward, at least in principle, to modify the estimation equations to perform parameter identification as well as state estimation [4], [7]. In one experiment the branch admittance Y_{36}, as well as the 15 state variables, was estimated. Two cases were considered. In the first case Y_{36} was at its nominal value, while in the second case Y_{36} was set to zero, corresponding to an open circuit in the branch between nodes 3 and 6. In both instances the state variables were estimated with the same accuracy as in the case where no parameter identification was attempted. Interestingly enough, the accuracy of the estimate of Y_{36} was approximately the same in both cases; these estimates are shown in Table V. Thus, although the parameter estimation accuracy is on the order of 10 percent, there is no significant loss of accuracy when the parameter varies by 100 percent. This type of performance is very desirable in fault location. The number of state variables is $n = 15$, and the number of measurements is $m = 75$. Estimation error and standard deviation are given as actual values.

TABLE III

ESTIMATION ERROR ϵ AND STANDARD DEVIATION σ FOR θ_6 IN 8-NODE EXAMPLE WITH LINEAR NETWORK MODEL USING DIFFERENT NUMBERS OF MEASUREMENTS

Number of Measurements	Estimation Error ϵ_{θ_6} (degrees)	Standard Deviation σ_{θ_6} (degrees)
7	1.844	1.037
8	0.634	0.757
20	−0.068	0.209
21	0.006	0.050

The actual value of θ_6 is 30.78 degrees. The number of state variables is $n = 7$.

TABLE IV

ESTIMATION ERROR ϵ AND STANDARD DEVIATION σ FOR θ_6 AND V_6 IN 8-NODE EXAMPLE WITH AC NETWORK MODEL USING DIFFERENT NUMBERS OF MEASUREMENTS

Number of Measurements	Estimation Error		Standard Deviation	
	ϵ_{θ_6} (degrees)	ϵ_{V_6} (kV)	σ_{θ_6} (degrees)	σ_{V_6} (kV)
15	−1.721	0.343	1.197	1.351
75	−0.083	0.252	0.292	0.312

The actual value of θ_6 is 30.78 degrees, and the actual value of V_6 is 100.8 kV. The number of state variables is $n = 15$.

TABLE V

ACTUAL AND ESTIMATED VALUES OF Y_{36} IN 8-NODE NETWORK WITH AC MODEL

Actual Y_{36}	Estimated Y_{36}	Estimation Error $\epsilon_{Y_{36}}$	Standard Deviation $\sigma_{Y_{36}}$
5419.9	4981.7	−438.2	543.6
0	−495.5	−495.5	543.6

Case 4) Outage Detection and Location with AC Model: In line outage detection the basic approach used was that of computing the difference between the estimates of the measured quantities and the measurements themselves and then comparing the differences with the standard deviation of the measurement estimate error. If the difference greatly exceeds the standard deviation, then it is likely that an outage has occurred somewhere in the network. In the numerical experiments based on the ac model of the 8-node network, various types of outages were introduced. Generally, a line outage (i.e., an open circuit) was readily diagnosed [7]. However, the detailed description of the method and the numerical results obtained are beyond the scope of this paper.

CONCLUSIONS

The least squares estimator provides the most accurate estimate of the system state since it utilizes all of the information available under the form of raw measurements, prior knowledge of certain system variables (pseudomeasurements), and the network equations. It processes this information in a manner which is statistically optimum.

Although the straightforward estimation algorithms discussed in the literature are generally not suitable for real-time power-system state estimation, several rigorous and approximate computational approaches have been developed that can serve as a point of departure for a real-time computational procedure for large networks.

The state-estimator concept is most useful when the number of measurements exceeds the number of unknown state variables (case $m > n$). The resulting redundancy of information allows filtering (reduction of estimation errors), anomaly detection, and parameter identification. Since the number of centrally available measurements is usually less than the number of unknown state variables, particularly when the state of neighboring systems is included in the estimate, it is imperative to include as many pseudomeasurements as are reasonably predictable from load forecasts, generation schedules, and regulator settings.

If the number of measurements and pseudomeasurements equals the number of unknown variables (case $m = n$), the estimator reduces to a standard load-flow solution, to which, however, a covariance computation should be added.

If the number of measurements and pseudomeasurements is less than the number of unknowns (case $m < n$), a least squares estimate can still be obtained but is of doubtful value.

The fundamental computational routine for the case $m > n$ is an efficient load-flow program which processes n of the m measurements and produces a preliminary estimate. The additional $m - n$ measurements can thereafter be processed in several ways; these additional measurements correct the preliminary estimate of the power-flow solution.

Although several state-estimation algorithms have been made to work reliably and efficiently on relatively small example problems, the difficulties of extending these algorithms to large, interconnected power systems and real-time operation should not be underrated. An efficient scheme for on-line state estimation which has a potential for real-time operation in practical power systems (with several hundred to several thousand nodes) is developed in [1].

REFERENCES

[1] R. E. Larson, W. F. Tinney, L. P. Hajdu, and D. S. Piercy, "State estimation in power systems, pt. II: implementation and applications," this issue, pp. 353–363.

[2] R. C. K. Lee, *Optimal Estimation, Identification, and Control.* Cambridge, Mass.: M.I.T. Press, 1964.

[3] R. E. Kalman and R. S. Bucy, "New results in linear filtering and prediction theory," *Trans. ASME, J. Basic Engrg.*, ser. D., vol. 83, pp. 95–108, March 1961.

[4] F. C. Schweppe and J. Wildes, "Power system static-state estimation, pt. I: exact model," *IEEE Trans. Power Apparatus and Systems*, vol. PAS-89, pp. 120–125, January 1970.

[5] F. C. Schweppe and D. Rom, "Power system static-state estimation, pt. II: approximate model," *IEEE Trans. Power Apparatus and Systems*, vol. PAS-89, pp. 125–130, January 1970.

[6] F. C. Schweppe, "Power system static-state estimation, pt. III: implementation," *IEEE Trans. Power Apparatus and Systems*, vol. PAS-89, pp. 130–135, January 1970.

[7] R. E. Larson and J. Peschon, "Feasibility study of system state estimation," Wolf Management Services, Palo Alto, Calif., Contract 14-03-79378, Bonneville Power Administration, Portland, Ore., Interim Rept., October 1968.

[8] R. E. Larson and L. P. Hajdu, "Potential applications and on-line implementation of power system state estimation," Wolf Management Services, Palo Alto, Calif., Contract 14-03-79378, Bonneville Power Administration, Portland, Ore., Final Rept., January 1969.

[9] W. F. Tinney and C. E. Hart, "Power flow solution by Newton's method," *IEEE Trans. Power Apparatus and Systems*, vol. PAS-86, pp. 1449–1460, November 1967.

[10] W. F. Tinney and J. W. Walker, "Direct solutions of sparse network equations by optimally ordered triangular factorization," *Proc. IEEE*, vol. 55, pp. 1801–1809, November 1967.

[11] J. Peschon, J. C. Kaltenbach, P. Henault, M. W. Siddiqee, and L. P. Hajdu, "Power system planning and reliability," Stanford Res. Inst., Menlo Park, Calif., Contract 14-03-72910, Project 6619, Final Rept., June 1968.

[12] R. E. Larson, "Limitations of the right pseudoinverse for state estimation," Wolf Management Services, Palo Alto, Calif., Contract 14-03-79378, Bonneville Power Administration, Portland, Ore., Project U-926, Tech. Memo 13, January 1969.

For Combined Discussion see pp. 223–226.

State Estimation in Power Systems
Part II: Implementation and Applications

ROBERT E. LARSON, MEMBER, IEEE, WILLIAM F. TINNEY, SENIOR MEMBER, IEEE, LASZLO P. HAJDU, SENIOR MEMBER, IEEE, AND DEAN S. PIERCY

***Abstract*—State estimation is a digital processing scheme which provides a real-time data base for many of the central control and dispatch functions in a power system. Its purpose is to permit improvements in system security and data accuracy and to reduce measurement and telemetry cost. The on-line implementation of an efficient state-estimator algorithm is discussed, and its feasibility is demonstrated on a 400-node network. The main motivations for and potential applications of on-line state estimation are listed, and the tradeoffs between measurement, estimation, and on-line load-flow computation are briefly discussed.**

Introduction

IN A companion paper [1] the theory of least squares state estimation for electric power systems was presented, and the feasibility of several static-estimator algorithms was established on relatively small networks.

One purpose of the present paper is to describe an improved algorithm for a practical on-line state estimator and to summarize the experimental results obtained on a 400-node network example. A second purpose is to enumerate the potential applications of state estimation in power system operation and planning.

An on-line state-estimator configuration is described, and computational results are discussed for a 400-node network example based on this on-line estimator scheme. Some potential applications for state estimation are also given, ranging from on-line monitoring for control purposes to information system planning.

On-Line Implementation of the State Estimator

A number of potential uses for state estimation were indicated in previous work [2], [3]. In most applications the state estimator(s) must be implemented on-line to process measurements as they become available. In [1], as well as in Schweppe's work [4]–[6], the state estimator was discussed primarily from the point of view of static-estimation theory in which a single batch of measurements is processed. Here, however, a state-estimator configuration suitable for on-line real-time processing of measured data is described.

Configuration of On-Line State Estimator

The basic configuration of an on-line state estimator, shown in Fig. 1, can be broken down into the five following functional groups:

Paper 69 TP 639-PWR, recommended and approved by the Power System Engineering Committee of the IEEE Power Group for presentation at the IEEE Summer Power Meeting, Dallas, Tex., June 22–27, 1969. Manuscript submitted February 20, 1969; made available for printing April 14, 1969.

R. E. Larson, L. P. Hajdu, and D. S. Piercy are with Systems Control, Inc., Palo Alto, Calif. (formerly Wolf Management Services of Palo Alto, Calif.).

W. F. Tinney is with the Bonneville Power Administration, Portland, Ore.

1) measurement and telemetry
2) input-data storage and display
3) estimate and covariance calculation
4) output-data storage and display
5) dispatching center.

Each of these groups is discussed below.

1) *Measurement and Telemetry System:* The measurement and telemetry system (block 1 of Fig. 1) collects raw measurements from various locations in the network and transmits them in digital form to the computer where the state estimator has been programmed. Any preprocessing of these measurements is included as part of this system. The set of measurements obtained is denoted as $z^{(m)}$. The measurements can be of some or all of the following types: real and reactive power injections at the nodes; real and reactive branch flows; and voltage magnitudes and phase angles at the nodes. Generally, in some parts of the network there are redundant measurements, i.e., more measurements than state variables, while in other parts of the network there are no measurements at all.

2) *Input-Data Storage and Display:* As noted previously [1] a number of quantities must be specified in order to carry out the estimate and covariance calculations. These include network configuration and parameters, measurement and telemetry system accuracy, and pseudomeasurements. This information is gathered and stored in block 2 of Fig. 1. The information is then available both for retrieval as necessary in the state-estimator calculations and for display to the dispatching center. The specification of the various types of data is discussed next.

Pseudomeasurement specification: Portions of the network may have no measurements available through the measurement and telemetry system. This can lead to a meaningless estimate [1], [2]. However, this difficulty can be resolved by defining pseudomeasurements at the nodes where actual measurements are unavailable. Pseudomeasurements are obtained from historical data, load forecasts, generation schedules, information provided by neighboring systems, operator experience, etc. They need not be very accurate, but they should be within an order of magnitude of the true value of the quantity they represent. The set of pseudomeasurements is denoted as $z^{(p)}$. These values are obtained from block 3 of Fig. 1.

Measurement accuracy specification: A critical input to the estimate and covariance calculation is the accuracy of all the measurements $z^{(m)}$ and pseudomeasurements $z^{(p)}$. This information is specified in the form of the standard deviation of each measurement and pseudomeasurement or, equivalently, the noise covariance R. These can be calculated from the characteristics of the hardware in the measurement and telemetry system and from the methods used to obtain the pseudomeasurements. These data are stored in block 4 of Fig. 1.

Network data: Network configuration and network parameters, such as branch admittance magnitudes and angles, must be known in order to carry out the estimate and covariance

Reprinted from *IEEE Trans. Power App. Syst.*, vol. PAS-89, pp. 353–363, Mar. 1970.

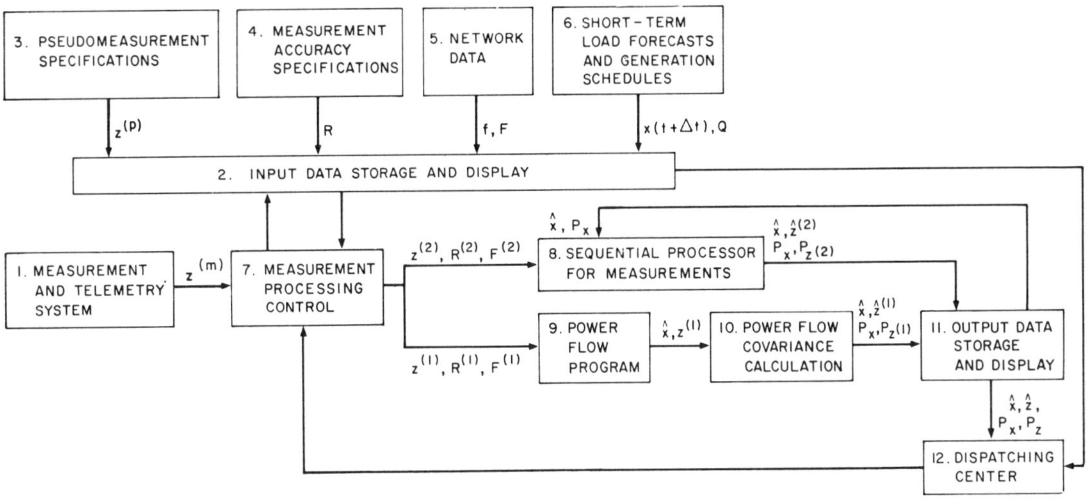

Fig. 1. On-line state-estimator block diagram.

calculations. It is also necessary to have the network equations f and the Jacobian matrix F for computing both the nonlinear and the linearized measurement equations [1], [2]. This information is provided in block 5 of Fig. 1.

Short-term load forecasts and generation schedules: In an on-line system short-term predictions of system variables play an important role [3]. Two major sources of these predictions are the short-term load forecasts and the generation schedules. This and other predictive information, such as the covariance matrix Q of the random forcing function, are available from block 6 of Fig. 1.

3) *Estimate and Covariance Calculation:* The heart of the state estimator is the estimate and covariance calculation. This function consists of the measurement processing control, which coordinates the calculations, and other blocks that carry out the specific calculations. All of these are described below.

Measurement processing control: Block 7, measurement processing control, performs the coordination and scheduling of state-estimator calculations. It controls the inflow of raw measurements and pseudomeasurements, retrieves the appropriate input data, and determines which state-estimator equations to use. It consists of both fixed logic and the capability to react to commands given at the dispatching center. This block sorts the measurements and pseudomeasurements into two sets: $z^{(1)}$ processed by the power-flow program, and $z^{(2)}$ routed to the sequential processor.

Power-flow program: One of the key elements of the static state estimator [1] is the power-flow program which computes an initial estimate x_0 based on the n measurements in $z^{(1)}$. This program is included as block 9 of the on-line state estimator. Although the power-flow program need not be used every time new measurement samples are received, it is still an important element of this state estimator, particularly in the initialization calculation.

Power-flow covariance calculation: The power-flow program provides an estimate but not its covariance. The additional calculations required to obtain the covariance P_{x_0} are performed by block 10 of Fig. 1.

Sequential processor for measurements: The most used block in the estimate and covariance calculation is the sequential processor represented by block 8 in Fig. 1. This block takes one measurement at a time from the set $z^{(2)}$ and updates the state estimate $\hat{x}$ and its covariance P_x. The updating equations are very simple, and several thousand measurements per second can be processed in this manner (using the algorithm of the next section on a CDC 6400, UNIVAC 1108, or comparable machine). Thus the sequential processor can be used to process measurements most of the time, and the power-flow program need be used only for initialization and periodic relinearization.

4) *Output-Data Storage and Display:* A prime consideration in an on-line state estimator is the continuous availability of updated estimates and corresponding covariances for both the state variables and the other measured variables. The static state estimator, which operates on a large batch of measurements, is not well-suited for this requirement, whereas the sequential processing of measurements makes this continuous availability quite straightforward to achieve. Block 11 of Fig. 1, the output-data storage and display, keeps the most recently updated values of the estimates and covariances in high-speed storage. Since a single measurement updates only a small fraction of the estimates and covariances, it is possible to read the required variables from output-data storage and display, transmit them to the sequential processor, perform the updating calculation, and then replace them with the updated values. Thus at any instant estimates and covariances of all state and measured variables are available, and all their values are completely updated, except for the small fraction affected by the single measurement currently in the sequential processor. Many of the potential applications of the state estimator require continuous display of, or access to, this updated information.

5) *Dispatching Center:* Many of the functions based on the information generated by the on-line state estimator are carried out through the dispatching center, block 12 of Fig. 1. Thus it is vital that the information in both the input-data storage and display and the output-data storage and display be available in convenient form to the dispatching center. Also, the dispatching center must have the capability to control the measurement processing. For example, the dispatcher may want to process information from a certain portion of the network in order to detect an anomaly; this capability is provided by allowing the dispatching center to send commands to the measurement processing control. In addition, the dispatching center can also modify or update some of the input data contained in blocks 3–6 of Fig. 1.

Initialization of the Estimator

In the static state estimator it is assumed that there is no information about the network state other than the measurements to be processed [1], [4]. During normal operation of the on-line estimator, however, the current estimate stored is a very accurate representation of the network state. Thus it is to be expected that initialization needs to be performed far less often in the on-line state estimator of Fig. 1 than if the static state estimator were used iteratively. Nevertheless, an initialization procedure must be available for estimator start-up and periodic realinearization. Initialization may also be necessary after a major change in the network, such as an outage, has occurred. The most promising method for initialization is the power-flow solution by Newton's method [7] using all available nodal injection measurements. At nodes from which no measurements are received, pseudomeasurements are used. This method has been used in the static state estimator with excellent results [1], [2].

Under some circumstances, it may be possible to use a priori information rather than the power-flow program to obtain the initial estimate. However, a practical on-line state estimator must have the capability of determining an initial estimate without depending on this a priori information. Furthermore, the procedure for carrying out the power-flow calculations is so fast and efficient [7] that little is to be gained computationally by processing a priori information.

The power-flow program provides the estimate but not its covariance [1]; however, the covariance matrix can be determined from the information stored in the power-flow program by a particularly efficient procedure which will be described later.

On-Line Measurement Processing

After an initial estimate and corresponding covariance matrix have been computed, the sequential processor (block 8 in Fig. 1) can be used to update these quantities on-line as the measurements are received. If the measurements were made at the same time as the measurements used in the initialization procedure, the sequential processor would perform exactly as described in [1]. If the measurements are made later, then the state estimator becomes a true dynamic estimator; three approaches for considering the dynamic effects while retaining the basic sequential processor structure have been examined elsewhere [3]. Finally, if system parameter identification is to be part of the state estimation, then special techniques are needed [1]–[4], all of which are compatible with the on-line estimator configuration described.

Computational Results for an On-Line State Estimation Algorithm Applied to a 400-Node Network

The on-line estimation procedure was tested on a 400-node example network, using a basic estimator configuration similar to that shown in Fig. 1, with blocks 6 and 12 omitted. Measurements of real powers only were considered, and the voltage phase angles were the state variables. The details of the estimation algorithm and the results are examined below.

The Power-Flow Program

The power-flow program (block 9) is used to compute the initial estimate $\hat{x}_0$ based on the n inputs $z^{(1)}$. The corresponding initial estimate covariance matrix P_{x_0} is obtained in a separate computation (block 10). The basic equations for these calculations are outlined here.

The power-flow program [7] calculates $\hat{x}_0$ as the solution to

$$z^{(1)} = f(\hat{x}_0) \tag{1}$$

where f is the nonlinear measurement function. This value of $\hat{x}_0$ is the initial state estimate [1] based on the measurements $z^{(1)}$. A by-product of this calculation is the matrix of partial derivatives

$$F_{ij}^{(1)} = \frac{\partial f_i}{\partial x_j} \tag{2}$$

evaluated at $x = \hat{x}_0$.

In this example, the measurement set $z^{(1)}$ contains the real power injections at all the nodes except the slack node, and the state variables are the voltage phase angles. Denote the injection at node i as I_i, the voltage phase angle at node j as θ_j, and the elements of the Jacobian matrix (2) as

$$H_{ij} = \frac{\partial I_i}{\partial \theta_j}. \tag{3}$$

Because of the properties of network interconnections, the matrix H is very sparse, i.e., it contains very few nonzero elements. Since the power-flow program processes and stores only the nonzero elements of H in a compact form, there is no difficulty in accommodating networks much larger than the 400-node example [7].

Once the state estimate $\hat{x}_0$ and the linearized coefficient matrix H have been obtained, the corresponding covariance matrix is calculated from

$$P_x = [H'R^{(1)^{-1}}H]^{-1} \tag{4}$$

where $R^{(1)}$ is the diagonal matrix with elements $R_{ii}^{(1)} = \sigma_i^2$, the variance of the noise in the ith measurement, and the prime denotes transposition. If only the diagonal elements of the covariance matrix are to be used [1], then only the diagonal elements of $H'R^{(1)^{-1}}H$ have to be computed.

Sequential Processing of Redundant Measurements

Once the estimate $\hat{x}_0$ and its covariance P_{x_0} have been computed, the measurements in $z^{(2)}$ are processed one at a time to obtain the final estimate $\hat{x}$ and its covariance P_x. The basic equations are

$$\hat{x}_k = \hat{x}_{k-1} + W_k[z_k^{(2)} - h_k\hat{x}_{k-1}] \tag{5}$$

$$P_k = [I - W_kh_k]P_{k-1} \tag{6}$$

$$W_k = P_{k-1}h_k'[\sigma_k^2 + h_kP_{k-1}h_k']^{-1} \tag{7}$$

where

- $\hat{x}_k$ state estimate after processing k of the measurements in $z^{(2)}$
- P_k corresponding covariance matrix
- $z_k^{(2)}$ kth measurement in $z^{(2)}$
- h_k linearized measurement equation vector for kth measurement in $z^{(2)}$
- σ_k^2 variance of noise on kth measurement in $z^{(2)}$
- W_k weighting vector for kth measurement in $z^{(2)}$.

The estimate $\hat{x}_0$ and covariance P_{x_0} are used to start the iterations. After all of the $(m - n)$ measurements in $z^{(2)}$ have been processed, the resulting estimate and covariance are $\hat{x}$ and P_x, respectively. Equations (5)–(7) are particularly easy to implement if only the diagonal elements of the covariance matrix are carried through the calculations [3].

Estimates of Measured Quantities

Estimates $\hat{z}$ of all the measured quantities in $z^{(1)}$ and $z^{(2)}$ are easily obtained from knowledge of the state estimate $\hat{x}$ after these measurements have been processed. Also, the measurement covariance matrix P_z can be obtained from the estimate covariance matrix P_x. If the composite vector z is defined to include the quantities in both $z^{(1)}$ and $z^{(2)}$, and if the measurement equation in the absence of noise is

$$z = Ax \tag{8}$$

then the relations are

$$\hat{z} = A\hat{x} \tag{9}$$

$$P_z = A(A'R^{-1}A)^{-1}A' \tag{10}$$

where R is the covariance matrix for the measurement noise on all the measurements in z. Again, the equations are very simple if only the diagonal elements of the covariance matrix are computed [3].

Data for the 400-Node Example

Network Model: The network chosen for this study is a 400-node model of the Bonneville Power Administration System and some of its connected neighboring systems. There are 610 branches connecting the 400 nodes. About half of the nodes are in the BPA (internal) system; the rest are in the neighboring (external) systems. The ac model of the network equations is used; however, only voltage phase angles are estimated, and only real power injections and branch flows are measured. Voltage magnitudes are assumed nominal. Thus the number of state variables is 399, and the maximum number of measurements is 1620.

Measurement and Pseudomeasurement Noise Variances: The noise values on the measurements and pseudomeasurements are assumed to have zero mean and to be independent. Thus the only data required are the variances of each measurement and pseudomeasurement. The estimator accepts individual variances, but for most experiments it was assumed that identical meters were used for all measurements and that all pseudomeasurements had identical accuracy. The metered quantities were injections at all nodes in the internal system and all branch flows, where the originating node was in the internal system. The meter full scale was assumed to be 1000 MW and the meter accuracy was taken to be 0.5 percent of full scale. The pseudomeasurements were assumed to be accurate to some fraction of the full scale of the meter; this fraction was varied between 10 and 40 percent.

Simulated Measurements and Pseudomeasurements: Simulated noisy measurements and pseudomeasurements were generated by adding a random number to the true value of the quantity. Each random number was scaled according to the standard deviation calculated in the previous section.

Computational Results

The estimator equations based on computing only the diagonal elements of the estimate covariance matrix were implemented in a number of computer runs. The quantities that were varied in the computations were the number of redundant measurements and σ_N, the accuracy of the pseudomeasurements expressed as a percent of full scale of the 1000-MW meter. Five cases are discussed individually.

1) σ_N = *10 Percent, All Redundant Measurements:* In the first example run the accuracy of the pseudomeasurements for the external nodes was taken to be 10 percent of full scale, or 100 MW, and all 1220 redundant measurements were made. As in all other cases, the accuracy of measurements at the internal nodes was 0.5 percent of full scale, or 5 MW.

The measurement errors at both internal and external nodes were significant fractions of the actual values of the measured quantities. Although there were injections and branch flows as large as 2150 MW, the average value of these quantities was about 100 MW. Thus the 5-MW standard deviation at internal nodes is quite substantial while the 100-MW standard deviation at external nodes in many cases is larger than the actual value of the measured quantity.

These characteristics of the measurement noise can cause difficulties with the estimator [1]. Not only is the magnitude of the noise large, but the noise on about half of the measurements is twenty times larger than the noise on the other half. Despite these difficulties, the actual errors in the state estimates were very reasonable. Naturally, the results for the internal system were far more accurate than those for the external system. Also, the results at nodes with many redundant measurements (branch flows) were more accurate than at nodes with few redundant measurements.

A quantitative measure of estimator performance is the agreement between actual estimation errors and the standard deviations of the estimates. The standard deviations are the square roots of the diagonal elements of the covariance matrices. In this regard, the performance of the estimator was extremely good. There are at least two reasons why the standard deviations might not agree with the actual estimation errors. The most important is that in the state estimator implemented here, the off-diagonal elements of the covariance matrix were not carried through the calculations. Although the validity of this approximation has been demonstrated for smaller networks [1], it is nevertheless gratifying that the results are so good in a 400-node example. A secondary reason is the large variation in the measurement accuracy at internal and external nodes; this variation has no adverse effect on the validity of the standard deviations.

Agreement between the standard deviations and estimation errors can be seen in the histogram of Fig. 2. This histogram was established by first dividing each of the 400 estimation errors by the corresponding standard deviations and taking the absolute values. These quantities were then grouped into increments of 0.08, and the number in each of the groups was computed. This number for groups 1–51 is shown by the dots in the figure. The theoretical number of quantities in each group is given by the bell-shaped curve on the figure.

2) σ_N = *10 Percent, No Redundant Measures:* The same measurement standard deviations as before were used, but none of the redundant measurements (branch flows) were made. As expected, the absence of redundant measurements detracted significantly from estimator accuracy, especially at nodes where many redundant measurements were possible. However, the estimation errors were consistent with the standard deviations, and the computational requirements were even less than in the previous case.

3) σ_N = *10 Percent, Redundant Measurements in the Internal Network:* The measurement accuracies were the same as before, but redundant measurements (branch flows) were made for originating nodes in the internal network. The estimation errors were very much the same as in Case 1 at the internal

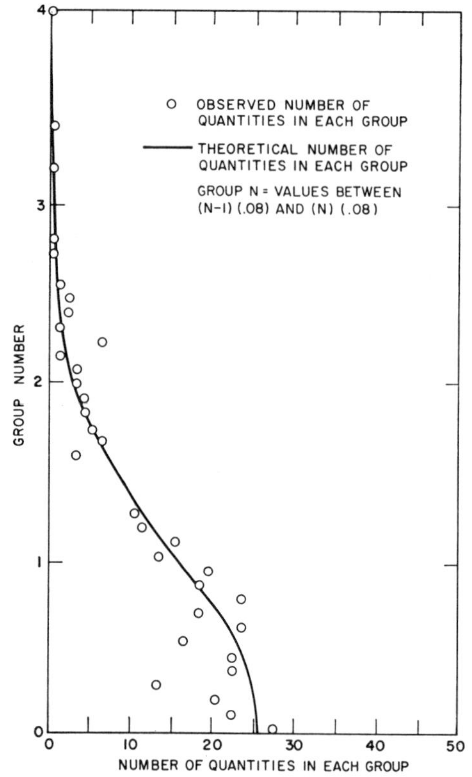

Fig. 2. Theoretical and observed normalized estimation errors for 400 states.

nodes but larger than for Case 1 at the external nodes. Again, estimation errors agreed very well with standard deviations, and computational requirements were extremely modest.

4) σ_N = *20 Percent, All Redundant Measurements:* In this case all redundant measurements were made, but the measurement standard deviations at the external nodes were increased to 20 percent of full scale on the 1000-MW meter, or 200 MW. This approximately doubled the estimation errors over those of Case 1 at the external nodes, but there was little change in the estimation errors at the internal nodes. Again, estimation errors agreed very well with their standard deviations, and computational requirements were very modest.

5) σ_N = *40 Percent, All Redundant Measurements:* In this case all redundant measurements were made, but the measurement standard deviations at the external nodes were increased to 40 percent of full scale on the 1000-MW meter, or 400 MW. This made the estimation errors approximately four times larger than for Case 1 at the external nodes. In a number of instances these errors were significantly larger than the true values of the measured quantities; nevertheless, the estimation errors and standard deviations were still in excellent agreement. As in the previous case, estimation errors at the internal nodes were nearly the same as in Case 1, and computational requirements were the same.

Computational Requirements

The computation time and storage requirements for the 400-node example were very modest. Details of the time and storage required by the power-flow program can be found elsewhere [7]; in this example, less than 5 seconds on the UNIVAC 1108 were required for the power-flow calculation, exclusive of data input and output. The experimental version of the sequential processor was capable of handling approximately 2140 measurements per second on the UNIVAC 1108, and this rate could be increased substantially by improved programming. About 100 storage locations were required for the estimator algorithm itself, 400 were required for the state estimate, and 400 for the corresponding covariances. In addition, the 1220 nonzero off-diagonal elements of the Jacobian H_{ij} that are available in the power-flow program must be retained, and as many as 3240 locations in the output-data storage and display block might be utilized for the estimates of the measured quantities and the corresponding covariances.

Potential Applications

Purposes of State Estimation

The primary reason for on-line state estimation as presently envisioned is twofold [3]:

1) to permit improvements in system security by providing a real-time data base for on-line security analysis, anomaly detection, fault diagnosis, and related display functions,

2) to facilitate the implementation of certain advanced control functions (e.g., real-time power-flow optimization) by minimizing the requirements for additional costly sensing and telemetry hardware.

A secondary purpose of on-line state estimation is to upgrade the quality of a number of control, dispatch, and record-keeping functions.

Off-line applications of the state-estimator algorithm include planning and design of the information system in order to obtain the data required for central control and dispatch in the most economical manner.

Alternative Approaches

State estimation has been considered here to mean providing estimates of the system variables (i.e., power flows and injections, voltage magnitudes and phase angles) at one central location. Three alternatives to state estimation can be envisioned that would produce a comparable data base for system control and security assessment:

1) a complete measurement and telemetry system,

2) on-line load-flow analysis,

3) some simplified, less accurate computation of the state variables (e.g., the one suggested by Siroux [8]).

It would be possible, though probably not economical, to measure all the system variables in real time; this would provide all the information needed for control. Alternatively, if enough power injection measurements were obtained in real time, a fast on-line load-flow program could provide the continuously updated set of state variables V and θ. The third approach [8] requires many more measurements than are usually available at the control center, and the results are less accurate than those obtained by the power-flow solution.

A shortcoming of the complete measurement system is the fact that if some real-time measurements are lost (or become unreliable), they must be replaced by other measurements or pseudomeasurements. This requires either operator attention (not always available) or redundancy and appropriate selection logic. Furthermore, no covariance matrices for the measurements are obtained. Hence, anomalies in the data-acquisition and transmission subsystems are not detected and identified by comparison of the covariances and residuals [1].

Simplified computations of the state variables [8] have shortcomings similar to those of a complete measurement system. Furthermore, the individual measurement and pseudomeasurement quantities are not weighted in accordance with their accuracies. Hence, a few inaccurate input quantities can degrade the accuracy of the computed state.

On-line load-flow solutions have the same difficulties as those of the complete measurement system. However, the load-flow computations could be extended to include the calculation of the state covariance matrix P_x. This approach has been discussed [1] as a special case of static state estimation where the number of measurement inputs is equal to the number of state variables ($m = n$). Each updating of the state vector would, however, require a complete power-flow computation unlike the sequential processing of measurements in the on-line state-estimator algorithm discussed here. Furthermore, on-line load-flow computation would not utilize any redundant information that might be available in the control center.

Potential Applications in System Operation

The role of the on-line state estimator within the information and control system is illustrated in Fig. 3. The estimator processes all available information about the interconnected system and provides a data base for the various control and dispatch functions. The outputs of the state estimator considered here are the estimates of the important electrical system variables. These may include the real and reactive generator outputs and loads, the real and reactive power flows in the transmission lines, the magnitudes and phase angles of node voltages, and, finally, the admittance parameters of the selected branches in the network (if parameter estimation is contemplated).

The on-line state-estimator configuration discussed in this paper is limited to driving those control and/or display functions that are associated with the power system operating under normal, slowly varying conditions (i.e., with slow changes in the state of the system, or relatively large changes on a long time scale, but not with abrupt large changes on a very short time scale).

Some of the various control and display functions that could benefit from on-line state estimation are as follows.

1) Load-frequency control [9] would benefit from receiving the optimum estimates of all power flows and injections, especially of any input quantities that are not measured in real time but are estimated and updated periodically (e.g., hourly) by the operator. State estimation appears to be the natural source for such preset input data.

2) Optimum dispatching of real and reactive power is based on the individual bus loads. The state estimator can produce the estimates of those loads whose measurements are not telemetered to the control center. Hence it is a natural adjunct of the various optimum dispatching schemes [10], [11].

3) Security monitoring, i.e., periodic checking of generation reserves, line loadings, bus voltages, transient stability properties of the network, and predicting the effects of postulated contingencies [12]–[14], could derive the necessary real-time information from the output of the state estimator.

4) Scheduled outage dispatching involves the analysis of the effects of planned outages or switching operations on system security. The estimated variables could also serve as a basis for these predictive tests.

Displays driven by the output of the on-line state estimator could provide the following information to the dispatcher:

1) estimates of line flows, generator outputs, loads, voltage levels, phase angles, etc., throughout the system, together with their respective accuracies (standard deviations),

2) anomalies in the system [1] and their geographical location to alert the dispatcher to network trouble or information-system malfunction,

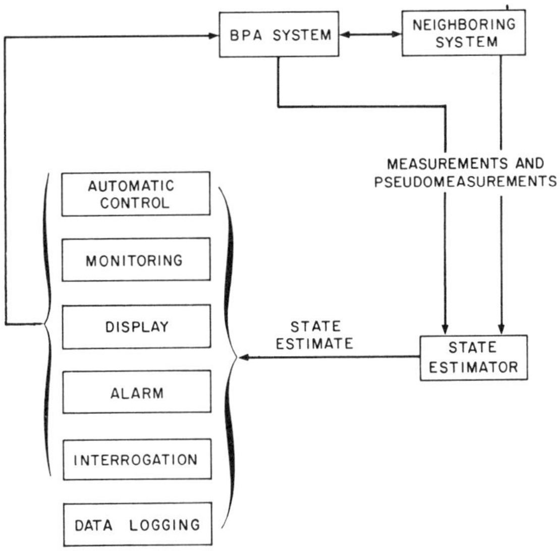

Fig. 3. Location and function of state estimator in control system.

3) unreported outages and their geographical location [1] to aid the dispatcher in diagnosing a fault,

4) marginal costs of power [10], [11] (active and reactive) at the important system nodes or interchange points,

5) indices of security [13], such as generation and transmission reserves or the transient stability margin, to alert the dispatcher whenever the system is in a vulnerable state,

6) corrective actions required to cure undesirable power-flow or voltage conditions, as well as to aid the dispatcher during system restoration procedures.

All of the information listed above is either contained in the state estimator or is easily obtainable from it. Since only a small fraction of this information needs to be displayed, the remainder could be accessible upon interrogation.

Finally, since the estimator would provide more complete and more accurate system information than is presently available, its usefulness for improved data logging is evident.

Applications in Information System Planning and Design

Information system planning and design aims to provide the information required for control, display, etc., in the most economical manner possible. Since the state estimator not only gives the best estimate $\hat{x}$ and $\hat{z}$ but also the covariances P_x and P_z for given sensor location, accuracy, and sampling rates, it can be used off-line to explore alternative configurations of the information subsystem. Some of the approaches are briefly discussed below.

1) *The Problem of Meter Placement:* In this problem it is assumed that all of the measurements made are brought to a single central processor; and it is desired to find the location, accuracy, and sampling rate of the additional sensors required to improve the accuracy of some of the state variables at the least cost in additional hardware. A solution to this problem is given elsewhere [2].

2) *Hierarchical Information Systems:* In a large power system, hierarchical structure of area and central dispatching is responsible for information collection, processing, and control. The resulting problem of information system design is more complex than that considered in the meter placement problem. The existing hierarchical structure suggests the use of area and central estimators, the feasibility of which was affirmed in the course of the study. The organization and economic tradeoffs of

information exchange between these various dispatching centers are the subjects of continuing study.

3) *Information-System Simulation:* It is believed that the analytical design relations between covariances of the state estimator that were developed for the meter placement problem [2] can be extended to this very complex information system, although the authors have not yet completed this investigation. There are excellent economic and technical reasons for relating the performance and cost of the information system to its design parameters, which can be done by means of a digital simulation. The effect of additional sensing, telemetry, and processing could be studied on such a simulator prior to any financial commitment. This simulator could subsequently be extended to analyze the performance of the various control, display, monitoring, and alarm functions contemplated for future implementation.

Use of Network Equivalents to Represent Neighboring Systems

Throughout this paper it has been assumed that pseudomeasurements are available for portions of the network where no actual measurements are made. In the case where the unobservable portion of the network corresponds to a neighboring system, an alternative procedure is available. This procedure assumes that all the branch flows between the internal system and the neighboring system are measured. It is then possible to model the influence of the neighboring system as follows.

1) Using all available measurements, obtain estimates of the system state variables for all nodes from which branches are connected to the neighboring system.

2) Using these estimates and the measured tie-line flows, determine estimates of the state variables for the boundary nodes in the neighboring system where the tie lines terminate.

3) Using a passive equivalent network of the neighboring system and the state variables of step 2), compute the equivalent injections at the boundary nodes in the neighboring system connected to the central system.

The equivalent injections of step 3) can be used to represent the influence of the neighboring system on the internal system, just as if the complete state estimate of the neighboring system were available [3]. The authors intend to discuss this technique in more detail in a future paper.

Conclusions

On-line state estimation is feasible for a practical power system with presently available analytical techniques and computer hardware.

The computational requirements of state estimation are of the same order of magnitude as those of comparable power-flow solutions.

On-line power-flow solution has been proposed to improve security and economy of operation. State estimation could replace the frequent on-line power-flow solutions in addition to its other potential applications.

On-line state estimator(s) can provide an information base for most of the conventional and advanced control and dispatch functions that are associated with operation of the power system under slowly varying conditions. However, emergency control functions that must be executed on a very short time scale can make no direct use of the output of the state-estimator algorithm described. Such functions would require a truly dynamic state estimator.

The on-line state-estimator configuration described produces the updated optimum estimates of the state and measurement variables with their corresponding covariances (or standard deviations) on a continuous basis. It can also be programmed to perform system parameter identification and short-term prediction of selected variables (e.g., loads).

Computer experiments on a 400-node example indicated that the accuracy of the estimates was always consistent with the precomputed standard deviations, even when very inaccurate pseudomeasurements were mixed with accurate real measurements. This fact has an important implication in a number of the applications contemplated, particularly in outage detection and location.

State estimation involves additional software development, which provides a trade-off possibility in terms of potential savings in telemetry hardware. Possible benefits may also be derived from the capability for anomaly detection and diagnosis, flexibility in implementing the information system, and increased accuracy.

A number of practical problems are yet to be resolved before state estimation techniques can be used in a real power system. These include the modeling of the network, the hierarchical structuring of the information system, the implementation of anomaly detection, diagnosis and prognosis functions, and methods for reading raw measurements into the computer for efficient processing.

Acknowledgment

The authors wish to thank A. R. Benson and A. W. Brooks of the Bonneville Power Administration for their valuable suggestions and stimulating discussions.

References

[1] R. E. Larson, W. F. Tinney, and J. Peschon, "State estimation in power systems, pt. I: theory and feasibility," this issue, pp. 345–352.

[2] R. E. Larson and J. Peschon, "Feasibility study of system state estimation," Wolf Management Services, Palo Alto, Calif., Contract 14-03-79378, Bonneville Power Administration, Portland, Ore., Interim Rept., October 1968.

[3] R. E. Larson and L. P. Hajdu, "Potential applications and on-line implementation of power system state estimation," Wolf Management Services, Palo Alto, Calif., Contract 14-03-79378, Bonneville Power Administration, Portland, Ore., Final Rept., January 1969.

[4] F. C. Schweppe and J. Wildes, "Power system static state estimation, pt. I: exact model," *IEEE Trans. Power Apparatus and Systems*, vol. PAS-89, pp. 120–125, January 1970.

[5] F. C. Schweppe and D. Rom, "Power system static state estimation, pt. II: approximate model," *IEEE Trans. Power Apparatus and Systems*, vol. PAS-89, pp. 125–130, January 1970.

[6] F. C. Schweppe, "Power system static state estimation, pt. III: implementation," *IEEE Trans. Power Apparatus and Systems*, vol. PAS-89, pp. 130–135, January 1970.

[7] W. F. Tinney and C. E. Hart, "Power-flow solution by Newton's method," *IEEE Trans. Power Apparatus and Systems*, vol. PAS-86, pp. 1449–1460, November 1967.

[8] J. G. Siroux and M. Adnet, "Calcul en temps réel des modules et des phrases des tensions aux noeuds du réseau de transport à partir des télémesures de puissances actives et réactives," *Rev. Gen. Elec.*, vol. 63, March/April 1967.

[9] N. Cohn, "Control of interconnected power systems," in *Handbook of Automation, Computation, and Control*, vol. 3, New York: Wiley, 1961, ch. 17.

[10] J. Peschon, D. S. Piercy, W. F. Tinney, O. J. Tviet, and M. Cuénod, "Optimum control of reactive power flow," *IEEE Trans. Power Apparatus and Systems*, vol. PAS-87, pp. 40–48, January 1968.

[11] H. W. Dommel and W. F. Tinney, "Optimal power flow solutions," *IEEE Trans. Power Apparatus and Systems*, vol. PAS-87, pp. 1866–1876, October 1968.

[12] T. E. Dy Liacco, "Control of power systems via the multi-level concept," Ph.D. dissertation, Case Western Reserve University, Cleveland, Ohio, June 1968.

[13] H. Glavitsch and R. Pichard, "Überlegungen zu einer zentralen Netzüberwachung," *Brown Boveri Mitt.*, vol. 55, July 1968.

[14] H. H. Mochon, "System security applications for the New England power exchange," presented at the 1968 Minnesota Power Systems Conf., Minneapolis, Minn.

Combined Discussion[1,2]

Gerald I. Stillman (American Electric Power Service Corporation, New York, N. Y. 10004): This paper and its companion are a pair of well-organized, well-written, and highly interesting papers on a subject that is quite new to power systems. Since this is one of the first papers to be published in this TRANSACTIONS on this subject, it might be well to differentiate, especially with the future in mind, between static state estimation, with which this paper is concerned, and dynamic state estimation. While the latter uses some of the same techniques as the former, there are important differences between the two. Perhaps the authors may wish to elaborate on these differences in the closure. Also, while state estimation is "a digital processing scheme" as applied in these two papers, it does not have to be. The important point is that state estimation is the application of statistical estimation theory, especially of weighted least-square error techniques, to a state-space formulation of power systems.

At the beginning of the paper, the authors list five factors which may cause discrepancies between true and estimated states. To these, at least two others may be added: 1) bad data points [4], i.e., gross errors in telemetered data from instrumented system locations even to the extent of receiving no data at all because of malfunction of any element of the metering channel; and 2) incorrect system model, as distinguished from "inaccurate network parameters" (this factor could include outright blunders in parameters due to incorrect input data or failure to modify the model, say, for system elements removed for maintenance). In connection with these two factors, what procedure would the authors use in determining the weighting factors to be applied to their residuals?

In the section on computational results, the authors state that "the standard deviation of each measurement was arbitrarily taken to be 2 percent of actual value." I understand the authors' desire to use a simplification of this nature in order to test the efficacy of the general method. As a practical matter, however, certain questions arise. For example, meter accuracy is generally stated in terms of full-scale reading. Consider a 500-kV circuit which has a 1000-MW full-scale meter on it to read full-load power flow. Under light-load conditions, flows as low as 100 MW might reasonably occur. Readings at such low power flows that have a standard deviation of 2 percent of actual value would require 0.2 percent accuracy at full scale, an extremely stringent requirement considering that noise from instruments, instrument transformers, analog-to-digital conversion, and telemetering channels are all included in this number. I would appreciate the authors' comments on this point.

In Tables III and IV, results are given for a number of measurements equal to the number of state variables and also to complete redundancy. If one is satisfied with the accuracy of the angle measurement of approximately 6 percent and the accuracy of the voltage measurement of approximately 1/2 percent, then the tables indicate, if I read them correctly, that a number of measurements equal to the number of state variables should be adequate. Is this a result of the assumption of 2 percent overall accuracy on all readings? If so, have the authors found a generalized approach, as distinguished from a case-by-case approach, to determine how much additional redundancy would be required if metering accuracy requirements were relaxed to 10 percent, for example, in order to obtain the same accuracy of solution?

In the example of Table V, the number of measurements was equal to complete redundancy. Would the fact that line Y_{36} was open have been nearly as obvious with a substantially smaller number of measurements?

On the matter of redundancy, particularly as applied to outage detection or to bad data points, is there any sense in talking about overall redundancy? Assuming that on a practical system complete redundancy of measurements will not even be approached for economic reasons, it would seem to be more important to guarantee a minimal amount of redundant information around each system node so as to avoid bad metering configurations, given in [4] of Part I, and the attendant numerical problems when attempting to arrive at a solution. I would appreciate the authors' comments on this point.

Once again, I should like to thank the authors for a generally lucid and thoroughly enjoyable paper.

Manuscript received July 3, 1969.

[1] R. E. Larson, W. F. Tinney, and J. Peschon, "State estimation in power systems, pt. I: theory and feasibility," this issue, pp. 345–352.

[2] R. E. Larson, W. F. Tinney, L. P. Hajdu, and D. S. Piercy, "State estimation in power systems, pt. II: implementation and applications," this issue, pp. 353–359.

F. C. Schweppe (Massachusetts Institute of Technology, Cambridge, Mass. 02139): I feel the work reported in these two papers provides a valuable forward step toward eventual implementation of a real-time monitoring and data processing system based on weighted least-squares state estimation. Most of the basic concepts and conclusions of these two papers are in close agreement with those developed during a joint research effort by the Massachusetts Institute of Technology and the American Electric Power Service Corporation (AEP) [1]–[4]. However, there are various technical areas where further discussion would be helpful.

Part I

The M.I.T.–AEP studies indicated that meter placement is a difficult problem in that the effectiveness of any given meter pattern depends heavily on the transmission network's structure and impedances. The P matrix (Σ matrix in [4]–[6] of Part I and in [1] of this discussion) or Monte Carlo trials for detection and identification can be used to evaluate the performance of any given meter placement, but our studies have not yielded any "general rules of thumb" which can be used to help minimize metering and communication costs. The problem appears to be especially acute for var and volt meters. Have the authors' developed any general meter placement rules?

The mathematical model for a state estimator depends on the measurement error covariance matrix R. For actual measurements, the values for R depend on the meter noise and bias, communication errors, system imbalances, etc. For pseudomeasurements, values for R depend on the accuracy of voltage and other regulators, load models, and forecasts, etc. Errors for certain pseudo- and actual measurements are not independent so that R theoretically contains off-diagonal terms. We feel that only an approximate R is needed and off-diagonal terms can probably be neglected. Another part of the mathematical model is the admittance matrix Y. We presently have no feel for how accurate Y has to be (or will be). What are the authors' opinions on the choice of values for R and Y?

The authors express their results in terms of the accuracy of the estimated state vector, V and θ at the buses. We have found it difficult to interpret such results and have had to look at the errors in the estimated line and bus flows to get a good physical feel. Have the authors had the same experience?

Part II

In real-time operation, measurements are repeated continually on time-varying system conditions. Nonsimultaneous meter readings are often obtained if digital telemetry is used. It is not clear exactly how the sequential processor is to operate in this situation. One approach is to go to a dynamic model and the resulting dynamic state estimator (Part II [6]). However, simpler approaches are possible. In [6] it is proposed to use a constant approximation to the P matrix (for some range of operating conditions) so that the static state estimator becomes a digital tracking loop. Do the authors have any comments on this approach?

The 400-node example uses only real power measurements. Is the resulting estimator a linear operator (as in [2] of Part II) or is it still an iterative solution of nonlinear equations? The M.I.T.–AEP studies on small networks have shown coupling between watt, var, and voltage measurements for network impedances with $X/R = 3$. Have the authors had any similar experiences with large networks and complete watt, var, and volt metering?

REFERENCES

[1] E. Hnyilicza, "A power system state estimation with applications," Power Systems Engrg. Group, Dept. of Elec. Engrg., Massachusetts Institute of Technology, Cambridge, Mass., Rept. 12, April 1969.
[4] O. J. M. Smith, A. Zarate, and G. E. Mauersberger, "State estimation of power systems," *Proc. Second Asilomar Conf. on Circuits and Systems* (October 30, 1968), pp. 260–272.
[5] O. J. M. Smith, "Power system state estimation," this issue, pp. 363–379.

Otto J. M. Smith (University of California, Berkeley, Calif. 94720): This paper utilizes the measurement noise covariance matrix R in an optimum estimation and load-flow computation. The authors' equations (6), (8), (13), and (18) show the inverse R used as a weighting matrix on the vector of measurements. If the mechanism of measurement-noise generation is time invariant, and if an infinite number of past true and corresponding noisy measurements have been analyzed to produce R, and if the minimum value of the scalar J in (6) is defined as the desired optimum, then (18) yields the optimum one-step node voltages and angles. But all of these conditional restrictions are violated in an engineering application.

Load-flow solutions are desired for different purposes. If the purpose of the load-flow computation is to reduce the errors from measured power and var node and line flows for control purposes, then one possible criterion is to minimize the squared per-unit error in every measurement. But if computer control is to be exerted on only a few of the largest machines, perhaps the desired criterion is to minimize the errors for those few machine nodes only. In other words, a use factor or engineering importance weighting should be attached to every measurement and included in R^{-1}.

Similarly, operating decisions based upon a load flow involve different costs per power and var error at each node. The correct criterion is to minimize the sum of the dollar costs of all of the resulting errors in the computed flows, starting with unknown errors in the measurements. The weighting matrix in this case is in terms of dollars per MVA.

Another function of load-flow computations is to replace missing or faulty measurements. In this case, the probability distribution of the faulty measurements is not the same as the probability distribution of the noisy measurements, and, consequently, R^{-1} is an improper weighting matrix. Furthermore, these one-step methods assume that there is no a priori information available concerning the faulty measurements, and this is not the case.

Faulty measurements can be detected by using (18) first and then recalculating the corrected measurements z' using (7). The first estimate of the measurement error vector is $(z - z')$. A plot of these errors will not necessarily have the probability distribution upon which R is based. In particular, elements in $(z - z')$ of 1 per-unit magnitude represent lost measurements, and all corresponding terms in R^{-1} should be set to zero. Also, elements in $(z - z')$ of more than three standard deviations are highly suspect, and the corresponding terms in R^{-1} should be set to the probability of the highly suspect values, which will essentially discard these measurements also.

After changing R^{-1} to discard faulty measurements and to weight the measurements in terms of importance and in terms of system costs, a second application of (18) will yield useful results in engineering terms. This method discards the highly deviant measurements which least squares improperly weighs most heavily instead of most lightly.

Newton's method and other first-order methods have difficulty in converging to the minimum of an error surface in adjustable space if the minimum is not zero, which is the usual case for real measurements. Second-order methods such as Fletcher–Powell are superior in these cases. Programs for second-order error minimization are given in [2]–[5].

REFERENCES

[2] A. R. Zarate, G. E. Mauersberger, and O. J. M. Smith, "State estimation of power systems," Electronics Research Laboratory, University of California, Berkeley, Calif., Memo. ERL-M248, June 1968.
[3] A. Bose, C. Falls, and O. J. M. Smith, "Power system post-transient steady state estimation," Electronics Research Laboratory, University of California, Berkeley, Calif., USDI Contract 14-03-77816, Rept. ERL-68-4, September 1968.

Robert E. Larson, William F. Tinney, Laszlo P. Hajdu, John Peschon, and Dean S. Piercy: We appreciate the very comprehensive discussions of our papers by Mr. Stillman, Prof. Schweppe, and Prof. Smith and would like to thank them for the many valuable comments and pertinent questions that help clarify some of the details of our work.

Mr. Stillman has raised several valid questions that pertain to the practical implementation of state estimation in electrical power systems. His remarks were directed in part toward possible future developments. In fact, since the writing of these papers we have devoted considerable effort to further developments of state-estimation techniques along the same lines. In answer to Mr. Stillman's questions, we would like to make the following statements.

Dynamic state estimation, as opposed to static state estimation discussed in the paper, is based on the principle of the "extended Kalman filter" ([2] of Part I). In this case, a dynamic model of the time behavior of the power system

$$x(k + 1) = f(x(k),u(k),w(k),k) \tag{1}$$

is considered. $x(k)$ here denotes the vector of the state variables at time k, $u(k)$ is the control vector, and $w(k)$ is a random variable vector representing the uncertainties in the model. Associated with this dynamic model is the measurement vector

$$z(k) = h(x(k)) + v(k) \tag{2}$$

where $v(k)$ is the measurement noise vector. In static state estimation, only the measurement (2) is considered, without any reference to system dynamics. The "extended Kalman filter" takes into account both the system dynamics, represented by (1), and the measurement information, given by (2). The subject of dynamic estimation has been treated in (Part II [3] and [6]) and is also a subject of a recent paper [6].

The problems arising from receiving bad data points and from "blunders" in network modeling can be readily handled by analyzing the properties of the vector $(z - \hat{z})$, which contains the differences between the true measurements and their estimates. The statistics of this vector, together with the theory of hypothesis testing, can be used to generate an algorithm for system anomaly detection and diagnosis. This has been a subject of a recent study [7], with results which we believe will have very significant applications.

In the computational example of Part I, we used standard deviations of 2 percent of the actual power flows for the sake of programming efficiency only, but this fact was not properly emphasized. However, in Part II the error was assumed to be independent of meter reading and was a fraction of the full-scale reading of every meter.

With respect to the example for parameter estimation (in the first paper), we believe that a line outage could be detected with even considerably less redundancy in measurements. How well this method works will depend not only on the amount of redundancy but also on the accuracy of the measurements, especially of those in the locality of the parameter to be estimated. At this time, no general rules are available.

We believe that questions related to tradeoff possibilities between the redundancy of the measurements, the accuracy of the estimate, and the reliability and economics of the information system represent an extremely important area of research. There is the additional problem of the sensitivity of the estimate to inaccuracies in the model, in terms of both the assumed system parameters and the entries of the noise covariance matrix R. All these questions are under investigation and a number of results have already been obtained [7]. At this stage, however, there are no definite "rules of thumb" for answering all these questions, and extensive amounts of off-line computation will have to be carried out in order to establish some fairly simple general rules.

Mr. Stillman pointed out a loosely worded statement in Part II, namely that pseudomeasurements "should be within an order of magnitude of the true value of the quantity they represent." The statement should correctly read: "on the order of the magnitude of the true value . . . ." To clarify this point, we would like to emphasize that the degree of uncertainty in pseudomeasurements is considerably higher than that for actual measurements. Uncertainties of the order of 15–30 percent are quite acceptable. We feel that good short-term load-forecasting procedures could generate much more accurate pseudomeasurements. Inaccurate pseudomeasurements will never degrade the estimate, but their usefulness in improving it may be negligible. As far as numerical difficulties are concerned, there were certain difficulties when the full covariance matrix P_x was used in the estimation procedure. Due to numerical problems this matrix tended to become singular. However, when only the diagonal elements of this matrix were used, no computational problems were observed since a diagonal matrix is always nonsingular.

There are two complementary approaches to be considered when one is faced with uncertainties in the network parameters. They are the following:

1) sensitivity analysis of the estimate of the state with respect to network parameter changes, to provide the necessary information on the possible deterioration in the performance of the estimator;

2) identification of uncertain parameters, using the results of sensitivity analysis and estimation theory, to be used on a long-range periodic basis, in order to account for the various changes in the network parameters due to external influences (such as temperature).

These two procedures are straightforward and have been a subject of recent study [7]. They are of great importance, especially when the state-estimator algorithm is implemented in an actual system. The results of off-line sensitivity analysis and parameter identification may be instrumental in the tuning of the state-estimator algorithm. (The practice of tuning of the estimator is quite standard in aerospace applications). This tuning process must also include the compensation for biasing and nonlinear effects of the meters, in addition to improving the model of the measurement-noise covariance matrix [6].

As to the question about the effect of information redundancy upon the accuracy of the estimate, we must repeat that we have no simple and general answer at this time. However, knowledge of the estimate covariance matrix is sufficient to determine the accuracy of any computed system variable since the latter is a function of the state and its variance (accuracy level) is a function of the estimate covariance matrix. This dependence will enable one to design the information system (including the extent of redundancy and meter accuracy) in such a manner that all important output variables of the state estimator (required for control or display purposes) will be estimated within a prescribed accuracy level. Our present approach is to impose these estimate accuracy requirements as a constraint on the design of the entire information system and to select from a number of admissible information system configurations (that are also practical from an engineering and reliability point of view) the most economical one [7].

Our reason for using only real-power measurements in the 400-node example was to simplify the programming effort. In order to compute the reactive power flows, the values of the line-charging capacities in the π representation of the transmission lines are needed as separate inputs for each line. For the computation of the real-power flow only, these capacitances are summed at each node and absorbed in the network admittance matrix. Thus for estimating reactive power flows, a great deal of additional information and programming is required; and this is true not only for state estimation but also for solving the load-flow program. However, we have programmed a static estimator for real and reactive powers and did not experience difficulties in obtaining the state and its covariance.

Prof. Schweppe raises three important questions with regard to Part I: 1) the problem of "meter-placement," 2) the effect of inaccuracies in the R and Y matrices, and 3) the choice of the estimated variables whose accuracy is of main concern. The first and the third questions are related to each other within the context of the design of the information system which provides measurement input data to the state estimator. We have discussed our approach to this problem in answering Mr. Stillman's questions. We would like to add here that in our opinion the problem of "meter-placement" is a subproblem of the design of the optimal (most economical) information system. The problem of optimal information system design includes many other equally important considerations, such as the cost of the communication network associated with the placed meters, the sampling rate of the information, the configuration of the computer system, the hierarchical structuring of the information system, and the problem of reliability. Our present research effort is directed along these lines. The proper choice of the R and Y matrices is also of great importance, and we have used sensitivity analysis techniques to evaluate the effects of errors in the error covariances and the network parameters upon the accuracy of the state estimate [7]. In our opinion, the off-diagonal elements of the R matrix can be neglected, and small errors in the Y matrix will not have a significant effect upon the quality of the estimate, especially if the network is large.

Prof. Schweppe raises another interesting question about the practical implementation of the sequential processing algorithm in real-time operation. The static estimator is, by assumption, time independent. In a real system, however, a set of measurements is not received simultaneously but over a finite time, and there is an additional uncertainty in the state estimate due to this effect. The static estimator does have a limited tracking capability for a slowly varying system state but needs periodic reinitialization. Since the writing of these papers, we have developed an estimator algorithm capable of tracking the state of the system without the need for initializations [6], and we believe that it provides a practical approach toward implementation in a real-time environment. The approach mentioned by Prof. Schweppe in regard to "tracking" the state using a constant P matrix seems quite reasonable. We can not offer any further comments because we feel that some numerical tests should be performed in order to verify the capability of both approaches.

It seems that many of the points raised by Prof. Smith can be clarified by thoroughly investigating the performance of the static state estimator under the following conditions: 1) the assumptions about the noise covariance matrix R are true, and 2) the assumptions about R are false.

Whenever the R matrix is known with a good degree of accuracy (through meter tests, manufacturers specifications, off-line analysis of data, etc.), the optimum estimator will minimize any of the cost functions mentioned by Prof. Smith. This statement is substantiated by a well-known result in estimation theory, noting that the variance of any linear function of the estimated state is given by an expression of the form

$$\mathrm{tr}[QP]$$

where Q is a symmetric positive semidefinite matrix, P is the covariance of the state, and $\mathrm{tr}[\cdot]$ denotes the trace operator. It is shown in [5, discussion by R. E. Larson, L. P. Hajdu, and D. G. Luenberger, pp. 376–377] that the least squares estimate $\hat{x}$ minimizes both of the following two cost functions

$$J_1 = (z - H\hat{x})^T R^{-1} (z - H\hat{x})$$

$$J_2 = E[(x - \hat{x})Q(x - \hat{x})^T] = \mathrm{tr}[QP].$$

for all matrices Q which are positive semidefinite. This means that all the various cost functions mentioned are optimized by the least-squares state estimator. For instance, if the cost criterion were to minimize the variances of some estimated injections or flows around the few largest machines, then the optimum estimator would do precisely that. Alternatively, if one attached a dollar cost of wasted energy to the error in the estimates, then the expected value of this dollar loss, being a linear function of the variances of the state variables, would also be minimized.

Now suppose that the noise covariance R is not known accurately. It can be shown [7] that the true covariance of the resulting estimate is then given by

$$P = (H^T R^{-1} H)^{-1} H^T R^{-1} R_0 R^{-1} H (H^T R^{-1} H)^{-1}$$

where R is the assumed covariance matrix and R_0 is the actual one. It can be further shown that 1) the cost $\mathrm{tr}[QP]$ is minimized when $R = R_0$ for all positive definite Q; and that 2) the first-order sensitivity matrix of the cost $\mathrm{tr}[QP]$ is given by

$$\left.\frac{\partial \mathrm{tr}[QP]}{\partial R}\right|_{R=R_0} = 0.$$

Hence small deviations in the assumed covariance R from the true covariance R_0 have only second-order effects. Consequently, if R is known within reasonable limits of its true value, then a very large class of cost functions is optimized (or very nearly so) by the optimum state estimator. This implies that the weighting matrix W should correspond as closely as possible to R^{-1} in order to optimize any one of the cost criteria suggested by Prof. Smith.

Apart from this consideration there are two major points we would like to stress.

1) Improved accuracy of the estimated state depends on the accuracy of the meters, their location in the network, and the kind of measurements they perform. The optimum state estimator will make an optimal use of the information provided by these measurements. If the accuracy of the estimate is not satisfactory, then more measurements with improved accuracy and proper location have to be used. In other words, there is a certain amount of information that one can "squeeze out" of the measurement system. The optimum estimator will "squeeze out" all that information.

2) There are certain sources of error that are far more serious than the errors resulting from an improper knowledge of the R matrix. Inaccurate network parameters will generate both bias and covariance errors. The nonlinear characteristics of meter calibration curves will also produce similar effects. A study of optimum estimation and the sensitivity of the estimate to these various effects will produce very reasonable tolerance limits on estimator performance. It will also guide the designer in identifying more accurately certain network parameters and in the design of an adequate information system.

We agree with Prof. Smith's comment that the covariance terms corresponding to lost or faulty measurements must be set to zero in the matrix R^{-1}. Faulty measurements can be readily detected by a simple algorithm [7].

Any power-flow solution method that assures fast convergence could be used to initialize the state estimator; however, initialization by power-flow solution is not likely to be required by the improved dynamic state-estimator algorithms developed for real-time implementation. An example of such an improved tracking state-estimator algorithm, which does not need initialization, is given in [6].

References

[6] A. S. Debs and R. E. Larson, "A dynamic estimator for tracking the state of a power system," presented at the IEEE Winter Power Meeting, New York, N. Y., January 25–30, 1970.

[7] L. P. Hajdu *et al.*, "New developments in system state estimation and anomaly detection," Wolf Management Services, Palo Alto, Calif., Contract 14-03-79768, Bonneville Power Administration, Portland, Ore., Final Rept., October 1969.

SPARSE SOLUTIONS USING HASH STORAGE

by Kalpana Makani

Otto J.M. Smith
Fellow IEEE

Department of Electrical Engineering and Computer Sciences
University of California, Berkeley, Calif.

Lakshmi Krishna

Abstract – An efficient computer program which permits rapid calculations for solving a large sparse system of linear equations, which arise in most power system network problems, is described here. This procedure will have an application to any system having a similar admittance matrix to the power system, such as a weather prediction or transportation system. The basic powerful tool–direct hash technique–is used to store the nonzero elements of the sparse matrix. The two main features, triangular factorization of the given sparse matrix and use of this factored form to find the direct solution to the system, are used to estimate a post-transient steady state of a power system. Updating the node admittance matrix after a change in the system structure is also performed in factored matrix form. In both factoring the Jacobian and in changing the the system structure, the introduction of additional non-zero terms in a sparse matrix is easily handled without prior knoledge of the new locations.

INTRODUCTION

In the case of a large sparse matrix where only a small fraction of the total number of elements is non-zero if the whole matrix is to be stored, then, first, the large memory space needed may not be available, and, second, even if it is available, it will be a waste of memory space if used for such a purpose. On the other hand, one can take full advantage of the sparsity of the matrix and store only nonzero elements.

For this purpose a hash technique was used, which proved to be a most powerful tool. The hash function used was a function of the matrix element position. It could be any one of several functions of the subscripts. This hash function computes a hash index from the values of the subscripts. The element value can then be accessed easily through the hash index. Though it is necessary to store the subscripts along with the element values itself, this method greatly reduced the storage requirements at the cost of some increased programming complexity and a little longer access time.

An n by n node admittance matrix can have (k + 1)n non-zero elements, where k is the average number of branch connections at each node. In general, (k + 1)n is very much less than n^2. Any method for storing these few nonzero values in a pseudorandom order can have the possibility of two different values wishing to occupy the same storage location so that an overflow vector of additional locations must be available, with a suitable pointer to direct the sequence of locations used.

Paper 71 TP 615-PWR, recommended and approved by the Power System Engineering Committee of the IEEE Power Engineering Society for presentation at the IEEE Summer Meeting and International Symposium on High Power Testing, Portland, Ore., July 18-23, 1971. Manuscript submitted September 28, 1970; made available for printing June 2, 1971.

A Scheme for Storing Nonzero Elements of the Matrix

The direct chain hash technique is used to store the nonzero elements of the matrix as follows: Given an n x n sparse matrix A, the diagonal elements are stored in a vector DR(N). The nonzero off diagonal elements in the lower and upper triangles of the matrix are stored in two value vectors JL(I) and JU(I). If the matrix is symmetric then only one value vector either JL(I) or JU(I) is needed. If the matrix is comples symmetric then again two value vectors AR(I) and AI(I) are needed for real and imaginary parts.

A common index vector is used to determine the location of a matrix element in the value vector. This index vector can be used by any number of value vectors at the same time, if the corresponding matrices have the same pattern; for example, the lower and upper triangles of a matrix having a symmetric pattern, and the real and imaginary parts of a complex matrix. This method reduces to a great extent the amount of space needed to store all of these vectors along with their separate index vectors.

The index vector called "JDEX" is divided into two parts. The lower part is used as a main index vector IA and the upper part is used as an overflow portion OA. About 25% of the total number of cells in the index vector are reserved for overflow. The contents of the first cell in JDEX is a pointer called FREE. FREE gives the displacement from JDEX of the first free cell in the overflow index OA from below.

The last 18-bit field of the second cell in index vector JDEX contains the displacement of the main index vector IA from JDEX.

Hash Code

A hash function is used to generate the index vector which is a function of the subscripts of the element of the A matrix, that is, it is a function of the row I and column J values which locate the element in the A matrix. Two possible hash functions are the following:

$$H_1(I,J) - 2 * MAX(I,J) + MOD_{BASE\ 2}(MIN(I,J)) \quad (1)$$

and

$$H_2(I,J) = 4 * MAX(I,J) + MOD_{BASE\ 4}(MIN(I,J)) \quad (2)$$

The hash function is used depending on the average number of non-zero values per row of a matrix. For example, the first hash

Reprinted from *IEEE Trans. Power App. Syst.*, vol. PAS-91, pp. 1396–1404, July/Aug. 1972.

function is used for Admittance matrix of a sparse system of linear equations which has 2 non-zero values on the average per row. The second hash function is used for Jacobian matrix which has 4 non-zero values on the average per row. This power of 2 in the hash function can be increased depending on the size of a matrix.

An example of the hash index generated is given in Table I, for the admittance matrix of a symmetric sparse system of linear equations.

Notice that the elements (7,2) and (4,7) map into the same hash-one index $H_1 = 14$. This is called a collision. All the elements having a collision are chained together. Observe that in general all the elements in a row are mapped into consecutive indices or their chains. For the hash function H_1, row elements are mapped into two consecutive indices or their chains. For the hash function H_2, all the elements in a row will be mapped into four consecutive indices or their chains. This feature is very useful in doing fast row operations, performed frequently in finding the inverse of a matrix in factored form or finding the direct solution to a sparse system of linear equations.

TABLE I

Line No.	Node	Node	Admittance Value	Hash Index $H_1(I,J)$	Hash Index $H_2(I,J)$
1	7	2	4	14	30
2	3	7	3	15	31
3	8	5	1	17	33
4	1	8	3	17	33
5	4	7	6	14	28
6	6	8	2	16	34
7	7	5	1	15	29
8	8	4	2	16	32
9	2	1	5	5	9

This hash technique described above has some advantages described below.

1. When we read the system structure from the cards and the n-node admittance matrix is assembled for all the n-nodes in the system only the non-zero values are stored using this hash technique. Regardless of the order in which we read the cards the storage order is unchanged.
2. In the process of upper triangularizing, the additional non-zero fill in is automatically handled by the program without prior knowledge of the locations of the non-zero elements of the matrix. For example, when we are eliminating the element (kl,n) in the n'th column, and let (k2,n) be a non-zero location in n'th column then there will be a new non-zero element coming into the position (k1,k2) if it were previously zero. This position is automatically handled by the hash program.
3. When the system structure changes after a transient the reduced matrix is easily updated. The changes in reduced matrix are easily computed using the information from the previous reduction. The values stored by the hash routine are easily fetched by the fetching routine.

Hash function H_1 has a total span of 13 indices for indices 5 through 17 inclusive. Hash function H_2 has a total span of 27 indices for indices 9 through 35 inclusive. H_1 has one collision each at 14, 15, 16, and 17. H_2 has only one collision at 33, due to node 8 being connected to both nodes 1 and 5, which have the same mod 4. Hash function H_2 is suitable for 4 or fewer average branch connections at each node, and was used for the power system.

The hash function value H(I,J) gives the displacement of the index of the (I,J)th element of the matrix from the base address IA. Each word in IA or OA is divided into three fields. The last 18-bit field contains the "LINK" index (displacement from FREE). The middle two 12-bit fields contain the subscript values I and J.

Initially all cells in the array JDEX are set to zero except for the first two cells. FREE points to the first cell in OA from below. To store the (I,J)th element of the matrix, H(I,J) is first calculated. Let C_o (   ) means "the contents of (   )." If C_o (JDEX+IA+H(I,J)) = 0, it means that that cell has not been used before and then I and J are packed into that word and LINK is set to zero. The value of this (I,J) th element is stored in the corresponding value vector at location VALU + H(I,J), where VALU is the name of the value vector. In case of a complex matrix both real and imaginary parts are stored with the same displacement H(I,J) for the AR and AI vectors.

If C_o(JDEX + IA + H(I,J)) ≠ 0, then i and j at that location are checked to see if I=i and J=j. If they are equal then the matrix element is added and stored in the value vector with H(I,J) as the displacement of its location from the top.

If i and j do not agree with I and J, then the overflow pointer LINK directs the program to OA, which is inspected at the location given by JDEX + LINK, where LINK is the chain displacement in the last 18 bits of JDEX + H(I,J). The process continues until correct values for I and J are found, or until LINK = 0. If LINK = 0 then LINK is replaced by FREE and FREE is decremented by 1. FREE now points to the first available cell in OA from below. I and J are then packed in OA, the overflow of the index vector, at JDEX + LINK and the value is stored in the value vector with LINK as its displacement from the top.

In order to FETCH the (I,J)th element of the matrix stored in this sparse form, the same procedure is used. If C_o(JDES + IA +H(I,J)) = 0 or LINK = 0, it indicates that the element is zero.

Inversion of a Matrix Using the Factored Form

Using the hash storage technique described earlier, the upper triangular elements of an n x n matrix V_n are stored in the array JU, the lower triangular elements of Y_n are stored in JL and the diagonal elements are stored in array JD. If the matrix Y_n is symmetric then only JL or JU needs to be stored, since the factorization process preserves symmetry. The index vector required is called JDEX. The hash function used is

$$H(I,J) = 4 * MAX(I,J) + MOD_{BASE\ 4}(MIN(I,J)) \tag{3}$$

Let i = MAX(I,J), j = MIN(I,J)

IPTR = H(I,J) + Base index of JDEX

then

JL(IPTR) = the (i,j)th element in the lower triangular matrix

JU(IPTR) = the (j,i)th element in the upper triangular matrix

The factorization scheme requires the ability to scan JL by rows and JU by columns, which is accomplished by the use of this hash function. The order in which the elements are eliminated is determined by the order in which the elements are put in the hash storage.

Since the nth row and column are the first to be eliminated, the program begins the factorization process by locating the last entry in the hash index. Suppose the storage for the nth row (column) was as follows:

In this program the matrices are either symmetric or symmetric in pattern. Hence the same index is used for JL and JU in the latter case.

The order in which the elements in the nth row will be eliminated is $(n,K_1),(n,K_2),(n,K_3),(n,K_4),(n,K_5),(n,K_6),(n,K_7),(n,K_8)$. That is all elements linked with the last entry are eliminated, then all elements linked with the next to last entry are eliminated and so on.

When element (K_1,n) in column n is eliminated, the essential element $-Y_{K1,n}/Y_{n,n}$ is stored in its place and $1/Y_{n,n}$ is stored in the corresponding diagonal location JD(n). The elements (K_2,n), (K_3,n) (K_4,n) . . . are used to modify the elements in the K_1th row, when a multiple of the nth row is added to the K_1 row to eliminate the element (K_1,n). Thus element (K_2,n) is used to modify the element (K_1,K_2). The location of the element (K_1,K_2) is found by calculating $H(K_1,K_2)$. If a new nonzero element is introduced by this process the subroutine Hash will return an index to this new nonzero element. If $K_1 > K_2$, then the element being modified is in the lower triangle, otherwise it is in the upper triangle. At the same time the element (K_1,n) in column n is eliminated, element (n,K_1) in row n can also be eliminated. This involves looking at JL(IPTR) instead of JU(IPTR); of course, this is not necessary if the matrix is symmetric. The process of elimination continues until all the elements are eliminated.

Matrix Factorization

Let Y_n be an n x n matrix

$$Y_n = \begin{bmatrix} Y_{11} & Y_{12} \cdots Y_{1n} \\ \cdots & \\ Y_{n,1} & Y_{n,2} \cdots Y_{n,n} \end{bmatrix} \tag{7}$$

The elements above the diagonal in nth column can be eliminated by premultiplying Y_n by an elemental matrix U^n, where

$$u^n = \begin{bmatrix} 1 & & & -Y_{1n}/Y_{nn} \\ & 1 & 0 & -Y_{2n}/Y_{nn} \\ & 0 & 1 & -Y_{n-1n}/Y_{n,n} \\ & & & 1 \end{bmatrix} \tag{8}$$

Similarly the elements to the left of the diagonal in the nth row can be eliminated by postmultiplying U^nY_n by an elemental row matrix L^n.

$$L^n = \begin{bmatrix} 1 & & & \\ & 1 & & \\ -\frac{Y_{n,1}}{Y_{nn}} & -\frac{Y_{n,2}}{Y_{nn}} \cdots & -\frac{Y_{n,n-1}}{Y_{nn}} & 1 \end{bmatrix} \tag{9}$$

Thus

$$U^nY_nL^n = \begin{bmatrix} Y^{(n)}_{11} & Y^{(n)}_{12} & Y^{(n)}_{1,n-1} & 0 \\ & & & \\ Y^{(n)}_{n-1,1} & Y^{(n)}_{n-1,2} & Y^{(n)}_{n-1,n-1} & \\ & 0 & & d_n \end{bmatrix} \tag{10}$$

where $d_n = Y_{n,n}$

Continuing with the (n-1)th, . . . 3rd, second column and row,

$$U^2U^3 \ldots U^n\, Y_n\, L^n\, L^{n-1} \ldots L^2 = D \tag{11}$$

where D is a diagonal matrix.

$$U^i = I + \text{col. } i: \frac{1}{Y_{ii}^{(i+1)}} \left(-Y_{1,i}^{(i+1)}, -Y_{2,i}^{(i+1)}, \ldots -Y_{i-1,i}^{(i+1)}, 0,0, \ldots 0\right)^T \tag{12}$$

$$L^i = I + \text{row } i: \frac{1}{Y_{ii}^{(i+1)}} \left(-Y_{i,1}^{(i+1)}, -Y_{i,2}^{(i+1)}, \ldots -Y_{i,i-1}^{(i+1)}, 0,0 \ldots 0\right) \tag{13}$$

$Y^{(K)}_{i,j}$ = the (i.j)th element after premultiplying by U^I and postmultiplying by L^K. (14)

Let $C = D^{-1}$

$$\text{then } C\, U^2\, U^3 \ldots U^n Y_n L^n L^{n-1} \ldots L^2 = I \tag{15}$$

$$\text{or} \quad Y_n^{-1} = L^n\, L^{n-1} \ldots L^3 L^2\, C\, U^2\, U^3 \ldots U^n \tag{16}$$

This is the expression for Y_n^{-1} in the factored form. The essential row or column of L^i, C, U^i can be stored in an n x n matrix. Thus

Y_n^{-1} = 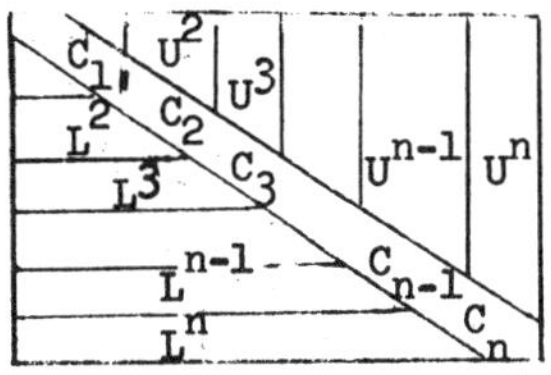

(17)

When the matrix is factored into elemental matrices, the problem of solving the matrix equation $Y_nX = b$ reduces to operation on b with the elemental matrices

$$Y_nX = b \tag{18}$$

$$X = Y_n^{-1}b \tag{19}$$

$$X = L^n\, L^{n-1} \ldots L^2\, C\, U^2\, U^3 \ldots U^n\, b \tag{20}$$

Reducing the Admittance Matrix

The admittance matrix in the power system network problem is assembled from the information about the system. The nodes in an n x n complex matrix are renumbered so that the first m nodes represent regulated voltage nodes and the last n-m nodes represent unregulated voltage nodes.

From this admittance matrix Y_n, a reduced matrix Y_m is found representing the admittance matrix for the m regulated nodes. The matrix Y_n forms the matrix equations

$$Y_n \begin{bmatrix} V_m \\ V_x \end{bmatrix} = \begin{bmatrix} I_m \\ 0 \end{bmatrix} \tag{21}$$

where V_m is m x 1, V_x is (n - m) x 1, and I_m is m x 1.

V_m stands for the voltage at the m regulated nodes.
I_m stands for the current at the m regulated nodes.

Partitioning Y_n into four parts and eliminating V_x from the system of equations,

$$\begin{bmatrix} Y_A & Y_C \\ Y_C^T & Y_D \end{bmatrix} \begin{bmatrix} V_m \\ V_x \end{bmatrix} = \begin{bmatrix} I_m \\ 0 \end{bmatrix} \tag{22}$$

$$(Y_A - Y_C Y_D^{-1} Y_C^T)(V_m) = I_m \tag{23}$$

Hence the reduced matrix for the regulated nodes is

$$Y_m = Y_A - Y_C Y_D^{-1} Y_C^T \tag{24}$$

Applying Gaussian elimination on Y_n starting from lower right-hand corner and eliminating the last n-m columns above diagonal until Y_C becomes O, the top left corner Y_A would become Y_m. Thus using an appropriate unit upper triangular matrix U,

$$\begin{bmatrix} I & U \\ 0 & V \end{bmatrix} \begin{bmatrix} Y_A & Y_C \\ Y_C^T & Y_D \end{bmatrix} = \begin{bmatrix} Y_A - Y_C Y_D^{-1} Y_C^T & 0 \\ V_C & V_D \end{bmatrix} \tag{25}$$

the required result is obtained in upper left-hand corner. Since Y_n is symmetric, Y_C^T could have been eliminated at the same time as Y_C without any extra work. Let $L = U^T$ then

$$U[Y_n]L = \begin{bmatrix} Y_m & 0 \\ 0 & D \end{bmatrix}, \quad D = \begin{bmatrix} d_{m+1} & & & \\ & \ddots & & \\ & & d_{n-1} & \\ & & & d_n \end{bmatrix} \tag{26}$$

Since Y_n is sparse, only nonzero elements in Y_n need to be operated upon. The values of U are determined by Gaussian elimination procedure.

Let

$$Y_n = \left[\begin{array}{c|c} Y_{an} & Y_{cn} \\ \hline Y_{cn}^T & Y_{nn} \end{array}\right] \begin{matrix} n-1 \\ 1 \end{matrix} \tag{27}$$

$$\qquad n-1 \quad 1$$

Premulitplying Y_n by $U_n = \begin{bmatrix} 1 & & & \bar{u}_n \\ & 1 & & \\ & & \ddots & \\ & & & 1 \end{bmatrix}$; $\bar{u}_n = -Y_{cn}/Y_{nn}$ (28)

will eliminate the nth column above the diagonal in Y_n. Similarly postmultiplying by U_n^T will eliminate the nth row.
Thus

$$U_n Y_n U_n^T = \begin{bmatrix} Y_{n-1} & 0 \\ 0 & Y_{nn} \end{bmatrix} \tag{29}$$

where $Y_{n-1} = Y_{an} - Y_{cn} Y_{cn}^T / Y_{nn}$ (30)

Continuing with the (n-1)th, (n-2)th, . . . (m+1)st columns and row we obtain

$$U_{m+1} U_{m+2} \cdots U_n Y_n U_n^T U_{n-1}^T \cdots U_{m+2}^T U_{m+1}^T = \begin{bmatrix} Y_m & 0 \\ 0 & D \end{bmatrix} \tag{31}$$

Therefore the U in equation 25 is equal to

$$U = U_{m+1} U_{m+2} \cdots U_n \tag{32}$$

The values of $U_{m+1}, U_{m+2}, \ldots U_n$ and $d_{m+1}, d_{m+2}, \ldots d_n$ should be saved after Y_n has been reduced to Y_m. They will be needed to update the matrix after a system disturbance.

Updating the Admittance Matrix after a Transient

After a system disturbance, the admittance matrix Y_n will change. Each transmission line that has been switched will contribute a change ΔY_n to the admittance matrix. Suppose the fault occurred on a line between two regulated voltage nodes a and b. Then the change in the reduced matrix is

$$\Delta Y_m = \begin{matrix} a \\ b \end{matrix} \overset{\qquad a \qquad\qquad b}{\left[\begin{array}{c|c} -Y_c - Y_{ga} & Y_c \\ \hline Y_c & -Y_c - Y_{gb} \end{array}\right]}_{m \times m} \tag{33}$$

where Y_c = admittance of line between node a and b. Y_{ga}, Y_{gb} = distributed capacitance and conductance to ground lumped at nodes a and b. If the fault occurred between nodes a and b, where one or both of the nodes are unregulated then the expression for the change is more complicated.

Using equation 24

$$Y_m + \Delta Y_m = Y_A + \Delta Y_A - (Y_C + \Delta Y_C)(Y_D + \Delta Y_D)^{-1}(Y_C + \Delta Y_C)' \tag{34}$$

$$\Delta Y_m = \Delta Y_A - (Y_C + \Delta Y_C)(Y_D + \Delta Y_D)^{-1}(Y_C + \Delta Y_C)^T + (Y_C Y_D^{-1} Y_C^T) \tag{35}$$

The following method for updating Y_m does not require a separate copy of Y_n to be stored away. This method uses information obtained from the previous reduction.

Let

$$\Delta Y_n = \overset{\quad a \qquad\qquad b}{\begin{bmatrix} \Delta Y_{aa} & \cdots & \Delta Y_{ab} \\ \Delta Y_{ba} & \cdots & \Delta Y_{bb} \end{bmatrix}} \begin{matrix} a \\ b \end{matrix} \tag{36}$$

This change has only four nonzero elements.

Let

$$C^n = \begin{bmatrix} \Delta Y_{aa} & \Delta Y_{ab} \\ \Delta Y_{ba} & \Delta Y_{bb} \end{bmatrix} \tag{37}$$

be a 2 x 2 matrix of the change. Note: $C^n = (C^n)^T$.

Let X^n be a n x 2 matrix, where the 2 columns of X^n are the a'th and b'th columns of the identity matrix.

$$X^n = \begin{bmatrix} 0 & 0 \\ 0 & 0 \\ \cdot & \cdot \\ \cdot & \cdot \\ \cdot & \cdot \\ 0 & 0 \\ 1 & 0 \\ 0 & 0 \\ 0 & 1 \\ 0 & 0 \end{bmatrix}_{n \times 2} \begin{matrix} \\ \\ \\ \\ \\ \\ \text{a'th row} \\ \\ \text{b'th row} \\ \\ \end{matrix} \tag{38}$$

Partition

$$X^n = \begin{bmatrix} X^n_t \\ \\ X^n_n \end{bmatrix} \qquad \begin{matrix} (n-1)\times 2 \\ \text{then } \Delta Y_n = X^n C^n X^{n^T} \\ \\ 1 \times 2 \end{matrix} \tag{39}$$

$$Y_n = \begin{bmatrix} Y_{an} & Y_{cn} \\ \\ Y_{cn}^T & Y_{dn} \end{bmatrix} \quad \text{and} \tag{40}$$

$$Y_n + \Delta Y_n = \begin{bmatrix} (Y_{an} + X^n_t\, C^n\, X_t^{n^T}) & (Y_{cn} + X^n_t\, C^n\, X_n^{n^T}) \\ \\ (Y_{cn} + X^n_t\, C^n\, X_n^{n^T})^T & (Y_{dn} + X^n_n\, C^n\, X_n^{n^T}) \end{bmatrix} \begin{matrix} n-1 \\ \\ 1 \end{matrix} \tag{41}$$

Restating equations 28 and 29,

$$U_n\, Y_n\, U_n^T = \begin{bmatrix} Y_{n-1} & 0 \\ \\ 0 & D \end{bmatrix} \text{ where } U_n = \begin{bmatrix} 1 & & & \bar{u}_n \\ & 1 & & \\ & & 1 \cdot & \\ & & & \cdot\ \cdot 1 \end{bmatrix} \tag{42}$$

$$\bar{u}_n = -Y_{cn}/Y_{dn} \ \ldots \tag{43}$$

$$Y_{n-1} = Y_{an} - Y_{cn} Y_{cn}^T / Y_{dn} \ \ldots \tag{44}$$

Let $d_n = Y_{dn} \cdots$ (45)

We have $U_{m+1}, U_{m+2} \ldots U_{n-1}, U_n$ and $d_{m+1}, d_{m+2} \ldots d_n$ available from previous reduction, where

$$U_{m+1}\, U_{m+2} \cdots Y_n\, U_n^T\, U_{n-1}^T \cdots U_{m+1}^T = \begin{bmatrix} Y_m & 0 \\ & d_{m+1} \\ 0 & \ddots\, d_n \end{bmatrix} \tag{46}$$

We want to calculate

$$U_{m+1}\, U_{m+2} \cdots U_n(Y_n + X^n C^n X^{n^T}) U_n^T\, U_{n-1}^T =$$

$$\begin{bmatrix} Y_m + \Delta Y_m & 0 \\ & d_{m+1} \\ 0 & \ddots\, d_n \end{bmatrix} \tag{47}$$

From 41 and 45,

$$D_n = (Y_{dn} + X^n_n\, C^n\, X_n^{n^T}) = (d_n + X^n_n\, C^n\, X_n^{n^T}) \tag{48}$$

Since everything on the right hand side is known, D_n can be calculated. From (41), (43) and (48):

$$U_n = -(Y_{cn} + X^n_n\, C^n\, X_n^{nT})/(Y_{dn} + X^n_n\, C^n\, X_n^{nT}) \tag{49}$$

$$= -(-\bar{u}_n d_n + X^n_t\, C^n\, X_n^{nT})/D_n \tag{50}$$

Again everything on the right hand side is known, and U_n can be calculated. Thus,

$$U_n(Y_n + \Delta Y_n) U_n^T = \begin{bmatrix} G_{n-1} & 0 \\ \\ 0 & D_n \end{bmatrix} \tag{51}$$

From (44)

$$G_{n-1} = Y_{an} + X^n_t\, C^n\, X^n_t - (Y_{cn} + X^n_t\, C^n\, X_n^{nT}) \cdot \\ \cdot\, (Y_{cn} + X^n_t\, C^n\, X_n^{nT})^T / (Y_{dn} + X^n_n\, C^n X_n^{nT}) \tag{52}$$

with some matrix manipulation G can be expressed as

$$Y_{n-1} + X^{n-1}\, C^{n-1}\, X^{n-1^T} \tag{53}$$

The proof of this statement is in the Appendix. Applying the same procedure to $G_{n-1} = Y_{n-1} + X^{n-1}\, C^{n-1}\, X^{n-1T}$,

U_{n-1} and D_{n-1} can be calculated. This procedure is repeated until $Y_m + X^m C^m X^{mT}$ is obtained. Hence, $\Delta Y_m = X^m C^m X^{mT}$ (54)

Let γ be a 2 x 1 vector to be used as temporary storage.

Algorithm	Comment
1. $i = n$	if $a < b < n$ then algorithm can start with $i = b$
2. $\gamma^{i^T} = X^i_i C^i$	X^i_i is 2 x 1, γ^i is 2x1
3. $D_i = d_i + X^i_i \gamma^i$	
4. $X^{i-1} = X^i_t - u_i X_i{}^i$	u_i is (i-1) x 1, X^i_i is 1 x 2
5. $U_i = u_i - X^{i-1}\, \gamma^i / D_i$	X^{i-1} is (i-1) x 2, γ^i is 2x1
6. $C^{i-1} = C^i - \gamma^i \gamma^{iT}/D_i$	
$i = i - 1$	
If $i > m$ go to 2, else go to 7.	
7. $Y_m = Y_m + X^m C^m X^{mT}$ add the change to Y_m	
STOP	

APPENDIX

The following is a detailed derivation of the matrix updating algorithm:

Given

$$Y_n = \begin{bmatrix} Y_A & Y_C \\ Y_C^T & Y_D \end{bmatrix} \begin{matrix} n-m \\ m \end{matrix}$$
$$\begin{matrix} n-m & m \end{matrix}$$

$$Y_m = Y_A - Y_C\, Y_D^{-1}\, Y_C^T$$

We want to find Y_m, the change in the admittance matrix of the regulated voltage nodes, given $Y_n = X^n\, C^n\, X^{nT}$, the change in the admittance matrix for all n nodes in the system.

1. Some relations that are useful in the derivation: Given an n x n matrix

$$Y_n = \begin{bmatrix} Y_{11} & Y_{12} & \cdots & Y_{1n} \\ Y_{21} & Y_{22} & \cdots & Y_{2n} \\ \vdots & & & \\ Y_{n1} & Y_{n2} & \cdots & Y_{nn} \end{bmatrix}$$

We can eliminate the elements above the diagonal in the n'th column by premultiplying Y by an elemental column matrix

$$U^n = \begin{bmatrix} 1 & & & -Y_{1n}/Y_{nn} \\ & 1 & 0 & -Y_{2n}/Y_{nn} \\ & & 1 & \vdots \\ & 0 & & -Y_{n-1,n}/Y_{nn} \\ & & & 1 \end{bmatrix}$$

We can similarly eliminate the elements to the left of the diagonal in the n'th row by postmultiplying $U^n Y_n$ by an elemental row matrix.

$$L^n = \begin{bmatrix} 1 & & & 0 \\ & 1 & & \\ & & \ddots & \\ \frac{-Y_{n,1}}{Y_{nn}} & \frac{-Y_{n,2}}{Y_{nn}} \cdots & \frac{-Y_{n,n-1}}{Y_{nn}} & 1 \end{bmatrix}$$

We can write the above relation in matrix notation. Since Y_n is symmetric,

$$Y_n = \begin{bmatrix} Y_{an} & Y_{cn} \\ Y_{cn}^T & Y_{dn} \end{bmatrix} \begin{matrix} n-1 \\ 1 \end{matrix} \qquad \text{A1}$$

Then $$U^n Y_n U^{n^T} = \begin{bmatrix} Y_{n-1} & 0 \\ 0 & dn \end{bmatrix} \qquad \text{A2}$$

where $$U^n = \begin{bmatrix} 1 & & & \\ & 1 \; 0 & & \bar{u}^n \\ & & \ddots & \\ 0 & & & 1 \end{bmatrix} \qquad \text{A3}$$

$$dn = Y_{dn}$$

$$\bar{u}^n = -Y_{cn} / Y_{dn} \qquad \text{A4}$$

or $$Y_{cn} = -\bar{u}^n \, dn \qquad \text{A5}$$

$$Y_{n-1} = Y_{an} - Y_{cn} Y_{cn}^T / Y_{dn} \quad \text{or}$$

$$= Y_{an} - Y_{cn} Y_{cn}^T / dn \qquad \text{A6}$$

Repeating the above operations on Y will yield d_{n-1}, $\bar{u}_{n-1}$, Y_{n-2}. Further repetitions will complete the matrix reduction process.

$$Y_n + x^n C^n x^{n^T} = \left[\begin{array}{c|c} Y_{an} + X_t^n C^n X_t^{n^T} & Y_{cn} + X_t^n C^n X_n^{n^T} \\ \hline Y_{cn} + X_n^n C^n X_t^{n^T} & Y_d + X_n^n C^n X_n^{n^T} \end{array}\right] \begin{matrix} n-1 \\ 1 \end{matrix}$$

(column sizes: $n-1$, 1) A7

$$U^n (Y_n + x^n C^n x^{n^T}) U^{n^T} = \begin{bmatrix} G_{n-1} & 0 \\ 0 & D_n \end{bmatrix} \begin{matrix} n-1 \\ 1 \end{matrix} \qquad \text{A8}$$

where $$U^n = \begin{bmatrix} 1 & & & \\ & 1 \; 0 & & \bar{u}^n \\ & & \ddots & \\ 0 & & & 1 \end{bmatrix}$$

From A4-A6 $$D_n = Y_{dn} + X_n^n C^n X_n^{n^T} \qquad \text{A9}$$

$$\bar{U}^n = -(Y_{cn} + X_t^n C^n X_n^{n^T})/(Y_{dn} + X_n^n C^n X_n^{n^T}) \qquad \text{A10}$$

$$G_{n-1} = Y_{an} + X_t^n C^n X_t^{n^T} - (Y_{cn} + X_t^n C^n X_n^{n^T}) \cdot$$

$$(Y_{cn} + X_t^n C^n X_n^{n^T})^T / V$$

where $$V = (Y_{dn} + X_n^n C^n X_n^{n^T}) \qquad \text{A11}$$

2. We note that the values of u^{m+1}, u^{m+2}, . . . u^n, d_{m+1}, d_{m+2}, . . . d_n, x^n and C^n are available. The values of Y_{an} are not available since the previous reduction process has destroyed them. From A 9 we see that D_n is easily calculated.

$$D_n = Y_{dn} + X_n^n C^n X_n^{n^T}.$$

From A4, $dn = Y_{dn}$. If we let $r^n = C^n X_n^{n^T}$ A12

This is the step 2 of the algorithm.

Then

$$D_n = dn + X_n^n r^n$$

which is step 3 in the algorithm.

From A10 $$\bar{u}^n = -(Y_{cn} + X_t^n C^n X_n^{n^T}) / (Y_{dn} + X_n^n C^n X_n^{n^T})$$

$$= \frac{Ycn}{Y_{dn}} + \frac{Ycn}{Y_{dn}} - \frac{(Y_{cn} + X_t^n C^n X_n^{n^T})}{(Y_{dn} + X_n^n C^n X_n^{n^T})}$$

$$\bar{u}^n = \frac{-Y_{cn}}{Y_{dn}} + \frac{\frac{Y_{cn}}{Y_{dn}}(Y_{dn} + X_n^n C^n X^{n^T}) - (Y_{cn} + X_t^n C^n X_n^{n^T})}{(Y_{dn} + X_n^n C^n X_n^{n^T})}$$

$$= \frac{-Y_{cn}}{Y_{dn}} + \frac{+\frac{Y_{cn}}{Y_{dn}}(X_n^n C^n X_n^{n^T}) - X_t^n C^n X_n^{n^T}}{Y_{dn} + X_n^n C^n X_n^{n^T}}$$

From A5, $\bar{u}^n = -Y_{cn}/Y_{dn}$

From A9, $D_n = Y_{dn} + X_n^n C^n X_n^{n^T}$

Factoring out $C^n X_n^{n^T}$

we get, $\bar{U}^n = \bar{u}^n - \left[u^n X_n^n + X_t^n\right] C^n X_n^{n^T} / D_n$

Set $X^{n-1} = \bar{u}^n X_n^n + X_t^n$

This is the step 4 in the algorithm.

Then $\bar{U}^n = \bar{u}^n - X^{n-1} r^n / D_n$

This is the step 5 in the algorithm.

From 11,

$$G_{n-1} = Y_{an} + X_t^n C^n X_t^{n^T} - (Y_{cn} + X_t^n C^n X_n^{n^T}) \cdot$$
$$\cdot (Y_{cn} + X_t^n C^n X_n^{n^T})^T / V$$

where $V = Y_{dn} + X_n^n C^n X_n^{n^T}$

From 10,

$$\bar{U}^n = -(Y_{cn} + X_t^n C^n X_n^{n^T}) / (Y_{dn} + X_n^n C^n X_n^{n^T})$$

Using 13, $\bar{U}^n$ may be rewritten as,

$$\bar{U}^n = \frac{-Y_{cn}}{dn} - \left[\frac{-Y_{cn}}{dn} X_n^n + X_t^n\right] C^n X_n^{n^T} / D_n$$
$$= \left(-\frac{Y_{cn}}{dn} - \frac{Y_{cn} X_n^n C^n X_n^{n^T}}{dn\, D_n} + \frac{X_t^n C^n X_n^{n^T}}{D_n}\right)$$

Substituting in A16,

$$G_{n-1} = Y_{an} + X_t^n C^n X_t^{n^T} - \left[\frac{Y_{cn}}{dn} - \frac{Y_{cn} X_n^n C^n X_n^{n^T}}{dn\, D_n} + \frac{X_t^n C^n X_n^{n^T}}{D_n}\right](Y_{cn} + X_t^n C^n X_n^{n^T})^T$$

Since $C^n = C^{n^T}$,

$$G_{n-1} = Y_{an} + X_t^n C^n X_t^{n^T} - \left[\frac{Y_{cn}}{dn} - \frac{Y_{cn} X_n^n C^n X_n^{n^T}}{dn\, D_n} + \frac{X_t^n C^n X_n^{n^T}}{D_n}\right](Y_{cn}^T + X_n^n C^n X_t^{n^T})$$

Expanding,

$$G_{n-1} = Y_{an} + X_t^n C^n X_n^{n^T} - \frac{Y_{cn} Y_{cn}^T}{dn} +$$
$$+ \frac{Y_{cn} X_n^n C^n X_n^{n^T} Y_{cn}^T}{dn\, D_n} - \frac{X_t^n C^n X_n^{n^T} Y_{cn}^T}{D_n}$$
$$- \frac{Y_{cn} X_n^n C^n X_t^{n^T}}{dn} + \frac{Y_{cn} X_n^n C^n X_n^{n^T} X_n^n C^n X_t^{n^T}}{dn\, D_n}$$

A13 $$= \frac{X_t^n C^n X_n^{n^T} X_n^n C^n X_t^{n^T}}{D_n}$$

A14 Factoring and rearranging,

$$G_{n-1} = Y_{an} - \frac{Y_{cn} Y_{cn}^T}{dn} + X_t^n \left(C^n X_t^{n^T} - \frac{C^n X_n^{n^T} Y_{cn}^T}{D_n}\right.$$

A15

$$\left. - \frac{C^n X_n^{n^T} X_n^n C^n X_t^{n^T}}{D_n}\right) - \frac{Y_{cn} X_n^n}{dn}\left[C^n X_t^{n^T} - \frac{C^n X_n^{n^T} Y_{cn}^T}{D_n}\right.$$
$$\left. - \frac{C^n X_n^{n^T} X_n^n C^n X_t^{n^T}}{D_n}\right]$$

From A6, $Y_{an} - \frac{Y_{cn} Y_{cn}^T}{dn} = Y_{n-1}$

A16

Combining terms,

$$G_{n-1} = Y_{n-1} + \left(X_t^n - \frac{Y_{cn} X_n^n}{dn}\right)\left(C^n X_t^{n^T} - \frac{C^n X_n^{n^T} Y_{cn}^T}{D_n}\right.$$
$$\left. - \frac{C^n X_n^{n^T} X_n^n C^n X_t^{n^T}}{D_n}\right)$$

factoring,

$$G_{n-1} = Y_{n-1} + \left(X_t^n - \frac{Y_{cn} X_n^n}{dn}\right)\left[\left(C^n - \frac{C^n X_n^{n^T} X_n^n C^n}{D_n}\right) X_t^{n^T}\right.$$
$$\left. - \frac{C^n X_n^{n^T} Y_{cn}^T}{D_n}\right]$$

Since G_{n-1} is symmetric, we might assume G can be written

$$G_{n-1} = Y_{n-1} + \left(X_t^n - \frac{Y_{cn} X_n^n}{dn}\right)(C^n - C^n X_n^{n^T} X_n^n C^n)$$
$$\cdot \left(X_t^{n^T} - \frac{X_n^{n^T} Y_{cn}^T}{dn}\right)$$ A17

For the assumption to be correct, we must have,

$$\left(C^n - \frac{C^n X_n^{n^T} X_n^n C^n}{D_n}\right)\left(X_t^{n^T} - \frac{X_n^{n^T} Y_{cn}^T}{dn}\right) =$$
$$\left(C^n - \frac{C^n X_n^{n^T} X_n^n C^n}{D_n}\right) X_t^{n^T} - \frac{C^n X_n^{n^T} Y_{cn}^T}{D_n}$$

$$\text{i.e. } \left(C^n - \frac{C^n X_n^{n^T} X_n^n C^n}{D_n}\right) X_t^{n^T} - \left(C^n - \frac{C^n X_n^{n^T} X_n^n C^n}{D_n}\right)$$
$$\cdot \frac{X_n^{n^T} Y_{cn}}{dn} = \left(C^n - \frac{C^n X_n^{n^T} X_n^n C^n}{D_n}\right) X_t^{n^T} - \frac{C^n X_n^{n^T} Y_{cn}^T}{D_n}$$

$$\text{i.e. } -\frac{C^n X_n^{n^T} Y_{cn}^T}{dn} + \frac{C^n X_n^{n^T} X_n^n C^n X_n^{n^T} Y_{cn}^T}{dn\, D_n} =$$
$$\frac{-C^n X_n^{n^T} Y_{cn}^T}{D_n}$$

From A9, $D_n = dn + X_n^n C^n X_n^{nT}$

Hence, $$\frac{-C^n X_n^{nT} Y_{cn}^T (dn + X_n^n C^n X_n^{nT}) + C^n X_n^{nT} X_n^n C^n X_n^{nT} Y_{cn}^T}{dn \quad D_n} = \frac{-C^n X_n^{nT} Y_{cn}^T}{D_n}$$

i.e. $$-\frac{C^n X_n^{nT} Y_{cn}^T}{D_n} - \frac{C^n X_n^{nT} Y_{cn}^T X_n^n C^n X_n^{nT}}{dn \quad D_n} + \frac{C^n X_n^{nT} X_n^n C^n X_n^{nT} Y_{cn}^T}{dn \quad D_n} = \frac{-C^n X_n^{nT} Y_{cn}^T}{D_n}$$

Since $X_n^n C^n X_n^{nT}$ is a scalar, $Y_{cn}^T (X_n^n C^n X_n^{nT}) = (X_n^n C^n X_n^{nT}) Y_{cn}^T$

then the above expression reduces to

$$\frac{-C^n X_n^{nT} Y_{cn}^T}{D_n} = \frac{-C^n X_n^{nT} Y_{cn}^T}{D_n}$$

Therefore, the assumption made in A17 is correct.

$$G_{n-1} = Y_{n-1} + \left(X_t^n - \frac{Y_{cn} X_n^n}{dn}\right)\left(C^n - \frac{C^n X_n^{nT} X_n^n C^n}{D_n}\right)\frac{\left(X_t^{nT} - X_n^{nT} - Y_{cn}^T\right)}{dn}$$

From A5, $\bar{u}^n = -Y_{cn}/dn$

From A14, $X^{n-1} = \bar{u}^n X_n^n + X_t^n$

$(X^{n-1})^T = X_n^{nT} \bar{u}^{nT} + X_t^{nT}$

Set $C^{n-1} = C^n - C^n X_n^{nT} X_n^n C^n / D_n$

Since $r^n = C^n X_n^{nT}$, $C^{n-1} = C^n - r^n r^{nT} / D_n$

This is the step 6 in the algorithm.

Therefore, $G_{n-1} = Y_{n-1} + X^{n-1} C^{n-1} (X^{n-1})^T$

Applying the same procedure to Y_{n-1}, X^{n-1}, C^{n-1} and using $\bar{u}^{n-1}$, d_{n-1} we can calculate $\bar{U}^{n-1}$ and D_{n-1}.

Further repetition of the procedure completes the reduction process.

SUMMARY

This computer program will invert a sparse matrix, and solve a system of linear equations, maintaining sparsity. It will reduce a large number of equations in many unknowns to a few equations in a few unknowns by the elimination of selected unknowns.

Applied to a power system, it generates the equivalent reduced admittance matrix for selected nodes from the admittance matrix for a larger number of nodes, and it calculates the change in the equivalent reduced admittance matrix from a change in the system configuration. Given the node voltage vector it solves for the node current vector. Given the node power vector, it solves for the node voltage vector by Newton's method. Given a transient admittance change, it solves for the new node voltage vector (new load flow) by Newton's method.

ACKNOWLEDGMENT

The authors wish to thank the Computer Center, University of California at Berkeley, for the support of this research.

BIBLIOGRAPHY

[1] Tinney, W.F. and J.W. Walker, "Direct solutions of equations by optimally ordered triangular factorizations," PICA Conference, Pittsburgh, Pa., May 15-17, 1967, pp. 367-76.

[2] Tinney, W.F. and C.E. Hart, "Power flow solutions by Newton's method," IEEE Trans. on Power Apparatus and Systems, Vol. 86, No. 11, November, 1967.

[3] Sato, N. and W.F. Tinney, "Techniques for exploiting the sparsity of the network admittance matrix." IEEE Trans. on Power Apparatus and Systems, pp. 944-950, August 2, 1963.

[4] Byrnes, F.D., "A program for the direct solution of matrix equations," Bonneville Power Administration, June, 1967.

[5] Meyer, W.S., "Notes on optimally ordered triangular factorization" March, 1968.

[6] Smith, Otto J.M., "Power system state estimation," IEEE Trans. Power Apparatus and Systems, Vol. PAS-89, No. 3, March 1970, pp. 363-379.

[7] Bose, Anjan, Charles Falls, and Otto J.M. Smith, "Power system post-transient steady state estimation," Electronics Research Laboratory, Memorandum No. ERL-M248, Contract USDI-14-03-77816, September 26, 1968.

[8] Householder, A.S., *Principles of Numerical Analysis,* McGraw-Hill Book Co., 1953.

[9] Quok, Donald and Kalpana Makani, "An application of sparse matrix techniques in power systems," M.S. Report EECS Dept., University of California, June 19, 1969.

[10] Lakshmi Krishna, "State Estimation of Power Systems applying sparse matrix techniques", M.S. Report EECS Dept., University of California, March 15, 1971.

[11] E.C. Ogbuobiri; "Dynamic Storage and Retrieval in Sparsity Programming," IEEE Trans. on Power Apparatus and Systems Vol. PAS 89, January 1970.

Discussion

N. M. Peterson and **D. W. Bree, Jr.** (Systems Control, Inc., Palo Alto, Calif.): The authors are to be commended on bringing still another powerful technique from computer science to bear upon power system analysis. There are a few points which require further comment.

Any method which exploits the sparsity of power systems should show a very great advantage when compared with methods requiring full matrix schemes. The authors indicate this fact in their introduction. However, the relevant comparison which should have been made is between the hash storage scheme and the deterministic compact storage methods currently in use in the industry as illustrated in Refs. (1)-(5) and (9). We believe that the hash method outlined in the paper will not show any advantage over these current schemes.

The hash scheme outlined also is not designed to take advantage of the specific nature of the matrix operations most frequently encountered in power system analysis. We quote from Mr. Robert Morris of Bell Telephone Labs, Ref. (12), "Observe that this (hash storage) is a useful method only when array elements are to be accessed in unpredictable order: it would not be a good method of matrix operations such as addition or multiplication were to be performed in the array" The salient characteristic of matrix operations is that an entire row or column is accessed and operated upon by another row or column. This is also characteristic of the LU factorization and direct solution. What has been effectively produced in Refs. (1)-(5) and (9) is a methodology for efficiently storing and accessing compacted rows and columns of the system matrices in the proper order to maintain sparsity throughout the calculation.

An apparent advantage for the hash code scheme of being able to directly store the input data as it is read, is offset by the fact that the system must be reordered. In order to maintain maximum system sparsity during the LU factorization, the system must be reordered, thus changing all node numbers and destroying any direct hash code correspondence.

A convincing recommendation of the hash code scheme requires a comparison of operation counts for the authors' method and the referenced direct compact storage methods. Note that all of the pointers needed for efficient row and column access are available as a byproduct of the ordering algorithm generally used to maintain sparsity during the LU factorization (1). We believe that such a comparison will show a significant advantage for the compact storage schemes currently in use, Refs. (1)-(5) and (9).

The authors' technique for updating the admittance matrix after a transient seems to assume that extensive calculations will be made using the modified matrix. We feel that in most applications, very few direct solutions will be made from the LU factorization for a particular outage. It is more likely that one may wish to examine a number of different outage conditions at a particular operating point. We, in conjunction with BPA, examined the update approach for a fast approximate solution method, and decided against it in favor of the matrix inversion lemma (also known as the matrix rank annihilation, line-at-a-time building algorithms, or compensation method).

On the average, the update of the table of factors will take about 1/2 the time required to retriangularize the complete matrix. We have found that the effect of the implicit update of the table of factor on the solution can be simulated in about 10% the time to triangularize the matrix and compute the direct solution.

We feel the value of this paper would be enhanced by references to the computer science literature the authors found helpful in selecting their hash code.

[12] R. Morris, "Scatter Storage Techniques." *Communication of the ACM,* Vol. 11, No. 1, January 1968, pp. 38-44.

Otto J. M. Smith, Kalpana Makani, and **Lakshmi Krishna:** Our hash code was invented for this application without reference to published literature.

Optimum node ordering for maintenance of sparsity should be carried out in a preliminary program using the maximum complexity system interconnection and this node order should be maintained unchanged for a large number of studies and for control computations. The node order should be changed only when additional lines are built and additional interconnections are made.

The program in this paper starts with the optimum node order. The hash code used groups all of the elements in a row together. After factorization, all of the elements of each row of the admittance matrix and the Jacobian which are processed sequentially in Newton's Method are grouped together and are read in sequence.

The special hash code which we used therefore maintains all of the computational time advantages of direct compact storage methods. Mr. Robert Morris' quote that hash storage is useful only when elements are accessed in unpredictable order is not true for our type of hash code which was designed expressly for operations by rows (or by columns).

One advantage of this storage method is that data decks can be shuffled before being read without any deleterious effects. The node numbers must be known, however. Transmission lines can be removed from a data deck without any attention to node orders.

For very long transmission lines, hyperbolic equations should be used instead of the inductance lumped into one value, and the capacitance lumped into two values. The reduced admittance matrix method permits higher accuracy for long lines. Several extra nodes are assumed at intervals along the line, and each section is modelled by the conventional lumped parameter pi Network. When the extra nodes are removed, a more accurate equivalent pi network is introduced into the admittance matrix.

Updating the matrices after a transient was included because this program was developed for on-line transient control, with machine target angles computed and transmitted to power stations during the first half-cycle of the mechanical transient. Consequently the computational time for updating the admittance matrix and the Jacobian matrix is important. We were not interested in examining a number of different outage conditions.

It would be helpful if the matrix inversion lemma is faster for updating a changed matrix, as suggested by Peterson and Bree. We have not compared the two methods to determine the minimum computational time.

ESTIMATION ALGORITHMS FOR LARGE-SCALE POWER SYSTEMS

Samir A. Arafeh *
Member IEEE

Roland Schinzinger
Member IEEE

University of California,Irvine

ABSTRACT

The operation of large-scale power systems demands that the real-time monitoring, security, and control procedures can draw on a consistent and reliable set of data representing the measurements and the network parameters. This paper considers the enhancement of this data base through advanced estimation programs for the states and parameters of the network. The role of such programs and the different forms they can take are presented in a comprehensive and unified manner. This will assist the designer in the task of selection and development of a suitable algorithm for his system from a set of possible approaches.

I INTRODUCTION

In the design of new energy control centers needed to meet the rapid expansion of electric power networks, the trend is toward more advanced engineering operation in terms of monitoring, security, and control. These requirements, so necessary to provide a reliable source of energy, have placed greater demands on the quantity and quality of the real-time data base. Limited by computational resources, the data enhancement procedures have ranged from simple validity checks on the raw measurements to complex procedures for minimizing error functions. The former approach uses the physical properties of the system and includes such techniques as equating the MVA to the square root of the real and reactive powers squared, summing the bus power injections to zero, comparing the flows at the two ends of each line, or carrying out some form of moving time-average process. The error function approach, while it does not exclude validity checks, involves more sophisticated data management and uses the mathematical model of the system cast in a statistical form for estimation of the states and, or, the parameters.[1,2,3]

Since their early development for power system applications, estimation algorithms have been the topic of a large number of papers and reports, some of which have provided assessments and comparisons of different techniques.[4,5] Nevertheless, the system's designer is still faced with a need for a unified collective source to aid him in evaluating and selecting suitable procedures. This paper attempts to fill that need. Emphasis is placed on describing the total program structure, on providing the necessary mathematical models, and on the major estimation algorithms. The observability and validity checks are outside the scope of this paper and can be found in more detail in the literature.[1,4]

The contents are organized as follows. Sections II and III present an overview of the features and fundamentals of estimation,along with the structure of an advanced estimation program. Following this, Section IV gives a comprehensive description of the mathematical model of the power network as it relates to the estimation problem. Section V emphasizes least squares methods in terms of static state estimators. Section VI discusses various tracking estimators which consider the system's time behavior. Section VII describes the various decomposition and coordination methods which allow the algorithms given earlier to be used on large-scale systems.

*) S.A.Arafeh is with Systems Control,Inc., Palo Alto, CA and a lecturer at the Univ. of California,Irvine.

Paper A 77 145-6, recommended and approved by the Power Systems Engineering Committee of the IEEE Power Engineering Society for presentation at the 1977 Winter Meeting, January 30-February 4, 1977, New York, NY. This paper was recommended for TRANSACTIONS status F 78 263-6, and presentation by title for written discussion at the 1978 Winter Meeting, January 29-February 3, 1978, New York, NY. Manuscript submitted September 16, 1976; made available for preprinting December 1, 1976. This paper has been published in 77CH1190-8-PWR IEEE 1977 Power Engineering Society Winter Meeting. Made available for publication in TRANSACTIONS December 21, 1977.

II FEATURES AND FUNDAMENTALS OF ESTIMATION

The basic function of an estimation program is to convert telemetered raw data on measurements and status into a reliable information base. This includes information on the state (typically, electrical quantities in the steady state) and network status (connectivity). When implemented, the frequency of execution of the state estimator is dictated primarily by the rate at which the user's program is to provide new information and controls. The resulting information base contains all complex bus voltages, power and current flows as well as injections, along with network status and parameter errors, if any. This information is provided with consideration to: (1) errors introduced by transducers and communication; (2) errors due to transients and telemeter failure; (3) errors in network parameter values; (4) errors due to non-simultaneity of measurements; (5) errors due to network configuration.

The fundamental assumption for successful performance of a state estimation program is that the measurement system should possess a degree of redundancy greater than unity; i.e. redundancy = (no. of measurements) / (no. of state variables) > 1.

An advanced estimation program should perform the following functions: (1) Augment the telemetered real-time data in the real-time data base to a full and reliable information base of the system. This includes the estimation of bus voltages and complex power flow of those lines and buses which are not directly telemetered, or where telemeter failure temporarily makes the data unavailable. (2) Detect, identify, and suppress bad measurements such as transients, bias, data skew errors, inaccuracy in network parameters, and unreported status changes. (3) Provide the means for a more reliable and effective closed-loop control program. Correct input data insure that the control action to be implemented is appropriate. It also prevents anomalous system response due to an inappropriate or unnecessary command. (4) Provide the means for a more reliable security analysis. An accurate and consistent set of input data results in the elimination of false alarms and insures detection of true contingencies. (5) Provide the means for more effective decision making by the dispatcher when assessing load flow solutions and recommending control actions.

Reprinted from *IEEE Trans. Power App. Syst.*, vol. PAS-98, pp. 1968–1977, Nov./Dec. 1979.

III STRUCTURE OF ESTIMATION PROGRAMS

The basic structure of an advanced state and parameter estimation program is given in Fig.1. As shown, the program starts at the INITIALIZATION module by acquiring the necessary files of raw measurements and the network model in addition to other on-line processing requirements such as detecting any change in the network status since the last processing cycle. If any change has been detected an UPDATE routine will update the NETWORK MODEL file.

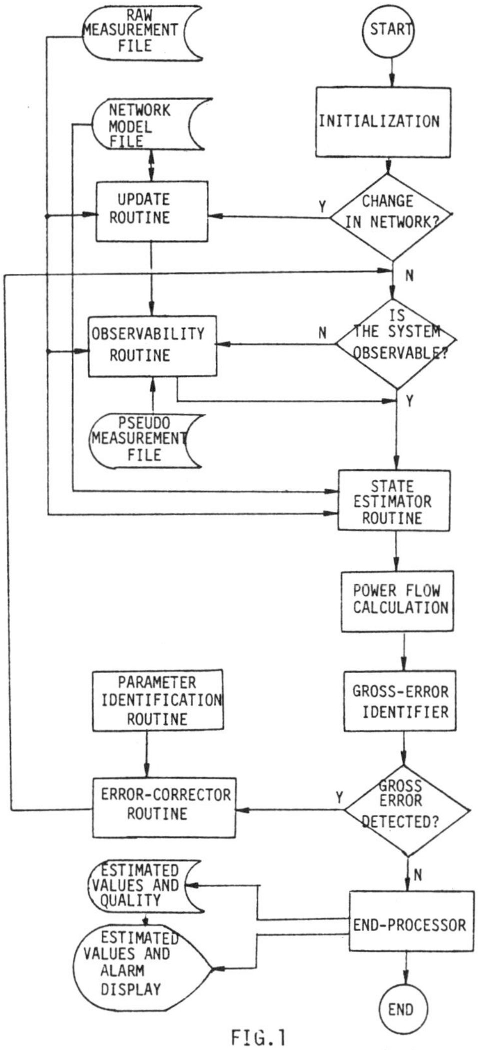

FIG.1
FUNCTIONAL FLOW DIAGRAM OF ESTIMATION PROGRAM

The INITIALIZATION module will also check for the network measurements' observability. The observability requirement is essential in order to obtain an integrated state estimation solution of the entire network. If the system is partially observable, an OBSERVABILITY routine will attempt to resolve this problem by using the RAW MEASUREMENT file or PSEUDO-MEASUREMENT file to provide an integrated state estimation solution. If the system's partial observability still persists due to lack of data and, or, there is an actual physical split in the network, the state estimator will provide independent state estimation solutions corresponding to different parts of the network.

When the network is fully observable, or partially observable, and the network model is up to date, the STATE ESTIMATOR module will make use of the raw measurements and the mathematical model in order to estimate all bus voltage magnitudes and phase angles.

Once the bus complex voltage is estimated, the POWER FLOW CALCULATION will solve for the complex power and current flow at all lines and transformers, and complex power and current injection at the load and generation buses of the entire network. This is obtained by using the familiar load flow equations; however, the solution is based on a statistically estimated input.

The GROSS-ERROR IDENTIFIER module examines the obtained results using a reasonability check and a threshold statistical test to identify any possible gross errors. If a gross error is detected as a result of bad network parameter values, transients or measurement errors, or a configuration error, an ERROR DETECTOR routine is employed. The ERROR DETECTOR routine will attempt to alleviate the error by dropping the bad measurement or by correcting the error using a PARAMETER ESTIMATION routine. The state estimation procedure is then repeated through the observability check process.

The PARAMETER ESTIMATION routine's function is to estimate both the basic model error as well as the one due to environmental effects in changing network parameter values. The method is based on a subnetwork solution to lessen the amount of compound errors of the total network and to reduce the amount of computational overhead. This is dependent on the degree of redundancy and amount of valid steady-state raw measurements which allow such a decomposition to produce valid network parameters and bias error estimaes. The procedure[6,7] is described by Fig.2. The valid, steady-state raw measurement input such as power flow, bus voltage, and line current at a specified segment of the power system is processed through the model which corresponds to a subnetwork of the system. The outputs of the real subnetwork and the model are compared according to a given statistical performance criterion. Based on a threshold test, the identifier uses the measured input signal to produce an estimate of the parameter. The sampled input values are again processed by the model and the above comparison is carried out. This procedure is continued sequentially until the discrepancy in response between the power system subnetwork and the model is minimized.

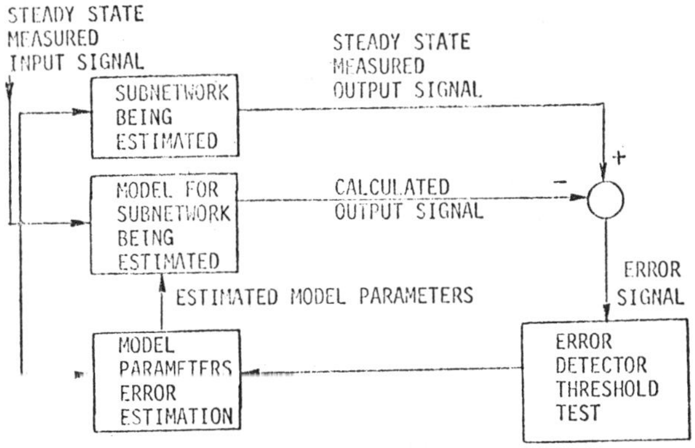

FIG.2. PARAMETER ESTIMATION ROUTINE

If no gross error is detected or gross error is detected and not corrected due to lack of data, the END PROCESSOR module will store the estimated values along with a flag indicating the quality of these values. The storage media for other power system data users is the ESTIMATED VALUES file. Also, a hardcopy

TABLE 4.1 NOTATION AND DEFINITIONS

NETWORK PARAMETERS AND VARIABLES	
$\bar{V}=e+jf=V\angle\theta$	voltage
$\bar{S}=P+jQ$	complex power
$\mathcal{Z}=R+jX=\mathcal{Z}\angle\delta$	impedance
$\mathcal{Y}=G+jB=\mathcal{Y}\angle\delta$	admittance
N_B	number of busses
SUBSCRIPTS	
r; B	bus r; set of busses B
rs; L	line between r and s; set L
n=1..N	nth component of x
m=1..M	mth component of z
i=1..	ith subsystem
SUPERSCRIPTS	
T	transpose
—	complex number
*	conjugate
i	ith iteration
^	estimate
ESTIMATION VARIABLES	
x	state variable vector
z	measurement vector
h(x)	observation fct. vector
$H=[\partial h/\partial x]$	Jacobian of h
$v=z-h(x)$	measurement error vector
b; $\mathcal{R}$	meas't noise bias & covariance
$\hat{\nu}=z-h(x)-b$	innovation vector
w	state random change vector
q; $\mathcal{Q}$	state noise bias & covariance
W	weighting matrix
J	objective function (scalar)
$\mathcal{P}$	Ricatti matrix
K	gain matrix
$\mathcal{J}$	observation matrix
t	continuous time variable
k	kth time sample
I	identity matrix
α	gradient step size

TABLE 4.2 RELATIONSHIPS BETWEEN NETWORK PARAMETERS AND VARIABLES

LINE AND BUS VOLTAGES	$\bar{V}_{rs}=\bar{V}_r-\bar{V}_s;\ \theta_{rs}=\theta_r-\theta_s$	1,2
	VECTOR TRANSFORMATION: $\bar{V}_L=T_V\bar{V}_B;\ \bar{V}_B=T_V^{-1}\bar{V}_L$	3
SERIES IMPEDANCE BETWEEN r AND s	$\mathcal{Z}_{rs}=R_{rs}+jX_{rs}=\mathcal{Z}_{rs}\angle\delta_{rs}=(G_{rs}+jB_{rs})^{-1}$	4
SHUNT ADMITTANCE OF LINE rs, ATr	THE ONE HALF OF THE SHUNT ADMITTANCE OF LINE rs ASSIGNED TO BUS r : $\mathcal{Y}r=G_r+jB_r=\mathcal{Y}_r\angle\delta_r$	5
COMPLEX POWER IN LINES AND AT BUSSES	INJECTION AT BUS r : $\bar{S}_r=\bar{S}_{r1}+\bar{S}_{r2}+..+\bar{S}_{rNB}$	6
	FLOW INTO LINE rs: $\bar{S}_{rs}=\bar{V}_r(\bar{V}_r-\bar{V}_s)^*/\mathcal{Z}_{rs}^*+V_r^2\bar{\mathcal{Y}}_r^*$	7
	VECTOR TRANSFORMATION: $\bar{S}_L=T_S\bar{S}_B;\ \bar{S}_B=T_S^{-1}\bar{S}_L$	8
LINE AND BUS CURRENTS	INJECTION AT BUS r : $\bar{I}_r=\bar{S}_r^*/\bar{V}_r^*$	9
	FLOW INTO LINE rs: $\bar{I}_{rs}=(\bar{V}_r-\bar{V}_s)/\mathcal{Z}_{rs}+\bar{V}_r\bar{\mathcal{Y}}_r$	10
COMPLEX POWER FLOW IN TERMS OF VOLTAGES	$P_{rs}=(G_{rs}+G_r)(e_r^2+f_r^2)-G_{rs}(e_re_s+f_rf_s)+B_{rs}(e_rf_s-e_sf_r)$	11
	$=V_r\{V_r\mathcal{Z}_{rs}\mathcal{Y}_r\cos\delta_r+V_r\cos\delta_{rs}-V_s\cos(\theta_{rs}+\delta_{rs})\}/\mathcal{Z}_{rs}$	12
	$Q_{rs}=-(B_{rs}+B_r)(e_r^2f_r^2)-G_{rs}(e_sf_r-e_r-e_rf_s)+B_{rs}(f_re_s+e_rf_s)$	13
	$=\{V_r\mathcal{Z}_{rs}\mathcal{Y}_r\sin\delta_r+V_r\sin\delta_{rs}-V_s\sin(\theta_{rs}+\delta_{rs})\}/\mathcal{Z}_{rs}$	14

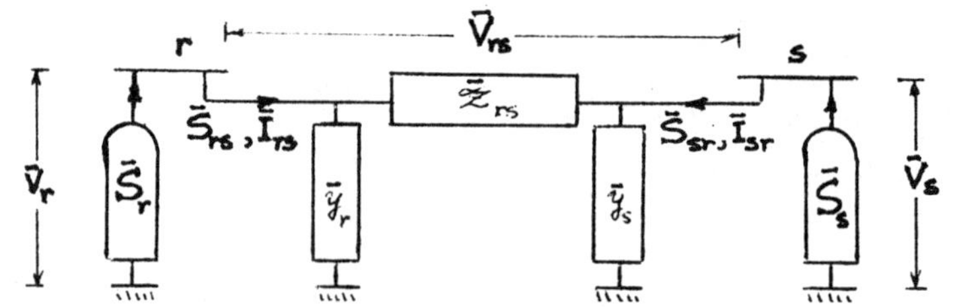

FIG.3 ELEMENTARY NETWORK MODEL

or CRT display output for values pertinent to the dispatcher can be obtained along with any alarm messages generated by the error detection and identification routine for future maintenance checking and calibration.

IV DESCRIPTION OF SYSTEM MODEL

In this section the necessary mathematical models are provided as a review and to serve in unifying notation, definitions, and relationships as used in the remaining sections. Table 4.1 lists the set of symbols and definitions associated with the network and the estimation model, the estimation model, and the observation model. These models are described next.

The Network Model

The power system network is represented by its nodes (buses) and branches (lines). The number of buses is N_B. A branch denoted by rs connects buses r and s. It is modelled as a series impedance $\bar{\mathcal{Z}}_{rs}$ representing the line and any transformers and series compensators. At any bus "r" a shunt admittance $\bar{\mathcal{Y}}_{rs}$ represents the contribution of the capacitance of line rs as seen at r. Shunt compensators are treated similarly.

The estimation problem addressed herein is concerned with the operating conditions as expressed by the complex voltages, currents, and power flows as well as the series and shunt network parameters described above. Usually the complex voltage is used as the state of the system since the total system can be solved once these values are determined. The measured quantities are the magnitudes of voltages, currents, real power, and reactive power. The relationships between these variables and parameters are given in Table 4.2 and are illustrated in Fig.3.

State and Parameter Model

For a snapshot description (i.e. at time instant t or sample k) the state and the parameters are simply described as in 4.3-1.*For a continuous or discrete time description one does not use a complete dynamic model but instead resorts to a simplified quasi-static model operating under uncertainties. This is due to the lack of prior information on the dynamic behavior of the system as a reult of random changes in load demands and operating conditions. Using the characteristics of the network under normal steady-state and slowly varying conditions, one can denote state and parameters at time $t+\Delta t$ by $x(t+\Delta t)$ in terms of x at t and the random change vector $w(t)$. The latter represents the system's random changes.[7] Thus the model can be described by $x(t+\Delta t)=x(t)+w(t)$. This can be written in discrete time form as given by 4.3-2 and in continuous time form as given by 4.3-3. The random variable w is usually represented by a Gaussian probability distribution function with mean q and covariance $\mathcal{Q}$. The corresponding network state variables and parameters are listed in the remainder of Table 4.3.

---

*) 4.3-1 stands for Table 4.3, Equation 1. - It should be noted that mathematically the state and parameters are viewed as basically the same.With reference to the network they are expressed as $x=[x_s, x_p]$. See 4.3-4.

TABLE 4.3 NETWORK AND ESTIMATION MODELS

	ESTIMATION MODEL	NETWORK MODEL (TYPICAL)
STATE & PARAMETERS	SNAPSHOT at t or k $x(t)$ or $x(k)$ (1) DISCRETE TIME $x(k+1)=x(k) + w(k)$ (2) CONTINUOUS TIME $\dot{x}(t) = w(t)$ (3)	$x \triangleq [x_s, x_p]$ (4) $x = V_L = [e_{rs}, f_{rs}]$ or $[V_{rs}, \theta_{rs}]$ (5) $x = V_B = [e_r, f_r]$ or $[V_r, \theta_r]$ (6) $x = [z_{rs}, \delta_{rs}, y_r, \delta_r, b]$ or $[B_{rs}, G_{rs}, G_r, b]$ (7)
MEASUREMENT z	$z = h(x) + v$ (8)	$z \subseteq [V_B, \theta_B, P_L, Q_L, P_B, Q_B, I_L, I_B]$ (9)

TABLE 5.1 CONDITIONS FOR AN OPTIMUM J

OBJECTIVE FUNCTION, J	$J = \sum_{m=1}^{M} \omega_m v_m^2 = v^T W v$	1
CONDITIONS FOR STATIONARY POINT ($\nabla J=0$)	$\nabla J = [\partial v/\partial x]^T W v = H^T W[z - h]$	2
FROM EQ. 2	$H^T W h = H^T W z$	3
LINEAR h, IF EXACT	$h(x) = Hx + \mathscr{b}$	4
APPROXIMATION	$h^{i+1} \cong h^i + H[x^{i+1} - x^i]$	5
	$h^{i+1} \cong Hx^{i+1} + [h^i - Hx^i]$	6
SUBSTITUTION OF h(EQ. 4 INTO 3)	$H^T W H x = H^T W[z - \mathscr{b}]$	7
SOLUTION $\hat{x}$, EXACT LINEAR CASE	$\hat{x} = [H^T W H]^{-1} H^T W[z - \mathscr{b}]$	8
LINEAR APPROXIMATION	$x^{i+1} = [H^T W H]^{-1} H^T W[z - h^i] + x^i$	9
SPECIAL CASE M = N	$x^{i+1} = H^{-1}[z - h^i] + x^i$	10
SPECIAL CASE M < N	$x^{i+1} = H^T[HH^T]^{-1} [z - h^i] + x^i$	11
HESSIAN, $\nabla^2 J$	$\nabla^2 J = [\partial^2 J/\partial x^2] = 2H^T W H$	12
USED IN EQ. 5	$x^{i+1} = [\nabla^2 J]^{-1} [\nabla J] + x^i$	13
IN TERMS OF $\Delta x = x^{i+1} - x^i$	$\Delta x = [\nabla^2 J]^{-1} [\nabla J]$	14

TABLE 5.2 AEP ALGORITHM IN OUTLINE FORM

CONCEPTUAL CRITERION,	$J = [\bar{S}_L - \hat{\bar{S}}_L]^{*T} W_L[\bar{S}_L - \hat{\bar{S}}_L]$	1
$\bar{S}_L$ AS FUNCTION OF $\bar{V}_L$: SOLVE 2.7-7 FOR $\bar{V}_{rs}=\bar{V}_r-\bar{V}_s$ IN VECTOR FORM SOLVED FOR $\bar{S}_L$	$\bar{V}_{rs} = (\bar{z}_{rs}/\bar{V}_r^*) \bar{S}_{rs}^* - \bar{z}_{rs} \bar{y}_r \bar{V}_r$ $\bar{V}_L = T_I^{-1} \bar{S}_L^* - C$ $\bar{S}_L = T_I^* \bar{V}_L^* + T_I^* C^*$	2
ERROR VECTOR, v AT ESTIMATE $\hat{x} = \hat{\bar{V}}_L$	$v = \bar{S}_L - \hat{\bar{S}}_L = \bar{S}_L - T_I^* \hat{\bar{V}}_L^* - T_I^* C^*$ or $v^* = T_I[T_I^{-1}\bar{S}_L^* - \hat{\bar{V}}_L - C] = T_I[\bar{V}_L - \hat{\bar{V}}_L]$	3 4
REFORMULATED J	$J = [\bar{V}_L - \hat{\bar{V}}_L]^T [W][\bar{V}_L - \hat{\bar{V}}_L]^*$; $W = T_I^T W T_I$	5
LINE VOLTAGES, $\bar{V}_L$, IN TERMS OF BUS VOLTAGES, $\bar{V}_B$	$\bar{V}_B$ = VECTOR OF BUS VOLTAGES, NOT INCL. $\bar{V}_b$ $\bar{V}_b$ = PRECISELY KNOWN BUS VOLTAGE $\hat{\bar{V}}_L = T_B \hat{\bar{V}}_B + T_b \bar{V}_b$	6
J IN TERMS OF $\bar{V}_L$ AND $\hat{\bar{V}}_B$ SET $\nabla J = 0$, FIND $\hat{\bar{V}}_B$	$J = [\bar{V}_L - T_B \hat{\bar{V}}_B - T_b \bar{V}_b]^T W[\bar{V}_L - T_B\hat{\bar{V}}_B - T_b\bar{V}_b]^*$ $\nabla J = -2T_B^T W[\bar{V}_L - T_B\hat{\bar{V}}_B - T_b\bar{V}_b] \equiv 0$	7 8

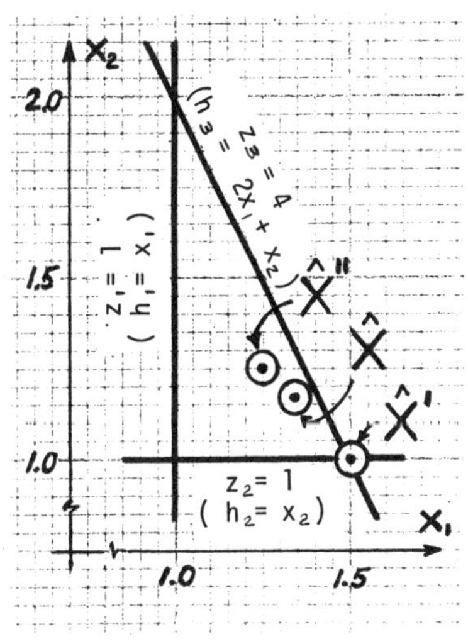

$z_1=1.00$; $h_1(x)= x_1$
$z_2=1.00$; $h_2(x)= x_2$
$z_3=4.00$; $h_3(x)=2x_1+x_2$

$v_n = z_n - h_n(x)$, $n=1,2,3$

$J = v_1^2 + v_2^2 + v_3^2$

$\hat{x} = [1.33, 1.17]$

$v = [-.33,-.17,.17]$

$J' = |v_1| + |v_2| + |v_3|$

$\hat{x}' = [1.50, 1.00]$

$v = [-.50, 0, 0]$

$J'' = \max_n \{v_n\}$

$\hat{x}'' = [1.25, 1.25]$

$v = [-.25,-.25,.25]$

FIG.5 COMPARISON OF CRITERIA J, J', AND J"

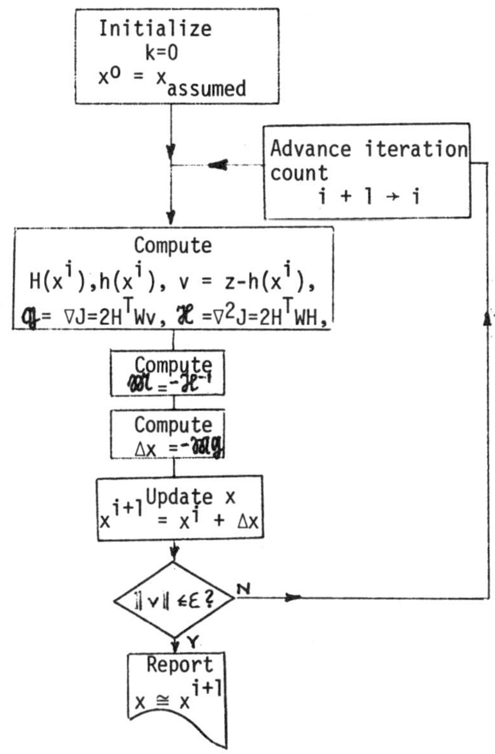

Alternative procedure:

$$\mathscr{M}^{i+1} = \mathscr{M}^i + \frac{\Delta x \Delta x^T}{\Delta x^T g} - \frac{\mathscr{M}^i \Delta g \, \Delta g^T \mathscr{M}^{iT}}{\Delta g^T \mathscr{M}^i \Delta g}$$

where

$\Delta g = g^{i+1} - g^i$

$\mathscr{M}$ = I or some typical $\mathscr{H}^{-1}$

FIG.4

FLOWCHART FOR LEAST SQUARES ALGORITHMS

Observation Model

The general observation model is described by a nonlinear function as given in 4.3-8. In this model we have: z, the collected measurement vector (see 4.3-9); $x=[x_d, x_p]$, the equivalent state vector of network state and parameters (see 4.3-1 through 4.3-7); h(.), the observation function vector which results from the relationships in Table 4.2; v, the measurement error vector, usually represented as Gaussian noise of mean (bias) b and covariance $\mathcal{R}$.

Special cases of linear observation functions also occur, but their applications are restricted by the characteristics of the measurement system. Where linear transformations or approximations are feasible they are expressed by $z' = h'(z,x) = H x$.

V WEIGHTED LEAST SQUARES METHODS (WLS)

It is appropriate to begin our discussion of estimation algorithms with least squares methods since these have been and continue to be the most widely used procedures. After briefly describing the underlying principles we will illustrate how the method can be implemented in various ways to solve the power system state estimation problem.[1 2]

Theoretical Background

Each of M measurements z_m is paired with an observation function $h_m(x)$ and is associated with an error v_m. Thus we have $z = h(x) + v$, collectively expressed in vector form. The state vector is of dimension N, and usually one works with redundant measurements such that M>N. As a criterion of the overall error one uses the sum of the measurement errors squared, with each term weighted by a factor ω_m. The objective function J appears then as given by 5.1-1. The selection of weights is discussed later. Suffice it to say for now that in matrix formulations the weights appear as the elements of W or, when statistical information is used, as the elements of $\mathcal{R}^{-1}$.

Inspection of Table 5.1 reveals how the conditions for a stationary point of J are utilized to extract the best estimate, x. When h(x) is linear we use 5.1-4 and derive a solution for x directly as given by 5.1-8.

When h(x) is nonlinear, the usual approach is to approximate it by the linear portion of a Taylor series expansion, 5.1-5, which allows one to proceed as in the linear case. This leads to an iterative process in which an update of x, say x^{i+1} for iteration i+1, is based on the preceding values of h^i and x^i. The correction is Δx as given by 5.1-14. This successive approximation scheme is carried out according to the flow chart of Fig. 5.1 until the magnitudes of the deviations v become sufficiently small.

The inversion of the Hessian, $\nabla^2 J$, requires a significant computational effort if carried out fully during every iteration. Now, this inversion needs to be carried out only once in the linear case where one cycle suffices. However, one can forego repeated evaluations of the inverse even in the nonlinear case by keeping it constant through several, or throughout all iterations. Experience with particular systems and initial values of x will dictate particular strategies. While Δx and v may be reduced less in magnitude during each of these simplified iterations, considerable time may be saved overall because each iteration requires less time. Another approach to the inversion of the Hessian is based on the successive improvement of an approximated inverse as indicated as an alternative on Fig. 4.[8] Whatever method is used to compute the inverse of the Hessian and the stepsize Δx, the deviation v is computed exactly as the difference between z and the nonlinear h(x). It should be noted that the correction Δx can be modified by inclusion of acceleration factors or other devices which take into account the trajectory followed by x during preceding iterations.

The final result, $\hat{x}$, depends on the choice of the weights used in the objective function J. The selection of identical weights for each of the M measurements, say W=I, leads to a least squares solution such as one obtains in ordinary curve fitting. If there is some knowledge of the relative significance of certain terms or of their expected instrumentation errors, then one may wish to introduce discriminatory weighting. Or one may choose a W which leads to the optimum in the statistical sense : the minimum variance estimate of the true value of x. If the v represent white Gaussian noise with zero mean, are independent, and have known spread given by the variance σ_m^2, then $\omega_m=1/\sigma_m^2$ will yield this statistically desirable answer. The σ_m^2 are usually assembled as the elements of a diagonal covariance matrix $\mathcal{R}$, which leads to $W=\mathcal{R}^{-1}$. In the absence of any detailed information about the variances it is expedient to select weights inversely proportional to the trans ducer's full scale reading.

While we restrict ourselves in this section to the weighted least squares approach we wish to point out that there exist other possible criteria. Some of these we can even derive indirectly by a judicious manipulation of weights. In each of the following cases the objective function J takes on the role of a norm of v.

Thus $\omega_m=1/|v_m|$ leads to $J' = |v_1|+..+|v_M|$, while $\omega_k=1$ for the largest of all v_m ($v_k = \max\{|v_m|\}$) and $\omega_m=0$ otherwise, would lead us to the minimax solution via $J''=\max\{v_m\}$. In the linear case the numerical answers to these objective functions are readily available through linear programming techniques. A comparison of results for a simple set of linear algebraic equations is shown in Fig.5.2. The purpose of our brief digression on objective functions is to point out that each criterion is unique.

Application to Power Systems

The application of least squares methods will be illustrated by means of some practical state estimation algorithms. Power system applications are characterized by the following: (1) $N=2N_B-1$, (2) M>N, and (3) sparse Jacob and Hessian matrices. Regarding (1) it should be noted that while system voltages constitute a convenient state vector, the direct use of power injections also has merit.[8] In either case, there are two variables associated with each of the N_B buses since we are dealing with complex state variables. But we may select an arbitrary reference for our phases (say θ=0 or f =0 at bus 1), thereby reducing our state variables by one to $N=2N_B-1$.

Note (2) refers to the fact that measurements of various types and at many locations add up to adequate redundancy. Should the case M=N be encountered, then 5.1-10 will apply whereas the underdetermined case M<N leads to the pseudoinverse incorporated in 5.1-11.

The sparsity of the Jacobian and Hessian matrices, indicated in note (3), is significant in that it allows the application of sparse matrix techniques which , along with matrix inversion by the use of the factorization method, facilitate real-time application of state estimators.[24]

The basic weighted least squares approach can be applied directly,[2] or with some approximations.[1] While the approximations do not provide the precision desired of a true estimate, they lead to fast algorithms which can be useful in other ways (fault detection, system planning). Thus the assumption of a flat voltage profile (all bus voltages equal 1 p.u., all $\theta_{rs} \cong 0$, $\sin\theta_{rs} \cong \theta_{rs}$, $\cos\theta_{rs} \cong 1$, $\delta_{rs} \cong \pi/2$, $V_r^2 \cong 2V_r - 1$) reduces power flow equations 5.2-12 and 5.2-14 to the following linear forms: $P_{rs} \cong (\theta_r - \theta_s)/z_{rs}$; $Q_{rs} \cong (V_r - V_s)/z_{rs} + y_r(2V_r - 1)$. Because these equations show P and θ to be decoupled from Q and V, it is possible to deal with two separate and reduced problems involving separate error terms and objective functions. Since $\nu_p = P - \hat{P}$ and $\nu_q = Q - \hat{Q}$ are linear functions of the state variables θ and V respectively, the objective functions are quadratic in x and the conditions of optimality are again linear. A single iteration is thus sufficient for approximately estimating the state.

As "exact" algorithms, on the other hand, we can label those procedures which call for a recalculation of the state dependent Jacobian matrix H and refactorization of H^TWH every iteration. This produces quadratic convergence, as in the Newton load flow. Coupled with the special tools of sparsity programming, these algorithms can be applied on a real-time basis to handle voltage, branch flow, and injection measurements.[24]

Accuracy and speed can be enhanced by introduction of passive nodes with identically zero injection. Such error-free "measurements" $z_f = h_f(x)$, which require no telemetry, are adjoined to the objective function by multipliers λ such that the task becomes one of iteratively solving over x and λ the following equations: $H^T(\hat{x})\, W^{-1}\, [z - h(\hat{x})] - H_f^T(\hat{x})\hat{\lambda} = 0$ and $h_f(\hat{x}) - z_f = 0$. Thus the practical redundancy of measurements is increased without additional telemetry, and the accuracy and speed of convergence are increased as well due to the inclusion of exact information.[25]

An algorithm of American Electric Power Service (AEP) uses line measurements only.[9] It is summarized in Table 5.2. The approach is conceptually based on a match between observed and calculated line power flows as opposed to bus power injections as is frequently the case in other methods. Linearization is performed by a series of transformations which permit one to rewrite the objective function in terms of complex bus voltages of which at least one must be known precisely. The weights in W, Eq. 5.2-5, are assumed to be constant.

TABLE 6.1 DISCRETE-TIME ALGORITHMS

ESTIMATED STATE, $\hat{x}$ $\hat{x}(k+1) = \hat{x}(k) + K(k+1)\hat{\nu}(k+1)$; $\hat{x}(1) = x_0$	1
(a) APPROXIMATED GAIN ALGORITHM $K(k+1) = \mathcal{P}(k+1)\, H^T(k)\mathcal{R}^{-1}$ $\mathcal{P}(k+1) = [H^T(k)\mathcal{R}^{-1}\, H(k)]^{-1}$; $\mathcal{P}(1) = \mathcal{P}_0$	2 3
(b) ESTIMATED GAIN ALGORITHM $K(k+1) = \mathcal{P}(k+1)\, H^T(k)\mathcal{R}^{-1}$ $\mathcal{P}(k+1) = [\mathcal{P}^{-1}(k) + H^T(k)\mathcal{R}^{-1}H(k)]^{-1}$; $\mathcal{P}(1) = \mathcal{P}_0$	4 5
(c) KALMAN FILTER ALGORITHM $K(k+1) = [\mathcal{P}(k)+Q(k)]H^T(k)\{\mathcal{R} + H(k)[\mathcal{P}(k)+Q(k)]H^T(k)\}^{-1}$ $\mathcal{P}(k+1) = [I - K(k+1)H(k)][\mathcal{P}(k)+Q(k)]$; $\mathcal{P}(1) = \mathcal{P}_0$	6 7
(d) SEQUENTIAL SCAN ALGORITHM $K(k+1) = \mathcal{P}(k)H_d^T\, [\mathcal{R} + H_d\mathcal{P}(k)\, H_d^T]^{-1}$ $\mathcal{P}(k+1) = [I - K(k+1)\, H_d]\mathcal{P}(k)$; $\mathcal{P}(1) = \mathcal{P}_0$	8 9
(e) ADAPTIVE ALGORITHM $\hat{x}(k+1) = \hat{x}(k) + \hat{q}(k) + K(k)\hat{\nu}(k)$ $\mathcal{P}(k+1) = [I - K(k)H(k)]\hat{Q}(k)$ $b(k+1) = [(k-1)b(k) + Z(k) - h(\hat{x},k)]/k$ $\mathcal{R}(k+1) = [(k-1)\mathcal{R}(k) + \hat{\nu}(k)\hat{\nu}^T(k) - H(k)\mathcal{P}(k)H^T(k)]/k$ $q(k+1) = [\hat{q}(k) + \hat{x}(k)]/k$ $Q(k+1) = [(k-1)\, Q(k) + K(k)\hat{\nu}(k)\hat{\nu}^T(k)K^T(k) + \mathcal{P}(k)]/k$ $K(k) \triangleq \mathcal{P}(k)H^T(k)[H(k)\mathcal{P}(k)H^T(k) + \mathcal{R}(k)]^{-1}$	10 11 12 13 14 15 16

This algorithm has been generalized by others[10] to include such additional measurements as magnitudes of currents and more bus voltages. Partition z so that $z_1 = [S_{rs}]$, $z_2 = [V_r]$ and $z_3 = [I_{rs}]$. Next, solve 2.2-7 for $V_r - V_s$ in terms of S_{rs}; write the V_r as a vector function of the V_r and θ_r; finally recast 2.2-10 as an expression for $V_r - V_s$ in terms of I_{rs}. We then have three sets of linear equations which can be further transformed and combined into a single, overdetermined set of equations $Ax = b$, with state $x = V_B$. Instead of applying the standard pseudo inverse, one can manipulate this set to advantage by premultiplying both sides by the transpose of A. In the resulting equation $A^T Ax = A^T b$ one can factor A^TA into a lower triangular matrix A_ℓ and an upper triangular matrix A_u, whence $A_\ell A_u x = A^T b$. Solving this for $\xi = A_u x = A_\ell^{-1}A^T b$ gives some computational benefits. Weighting factors can be incorporated by starting out from $WAx = Wb$.

Another approach to WLS estimation is to capitalize on the existence of load flow programs. In this method we use 2N -1 equations which balance the real and reactive power injections at the buses and are of the form 4.2-6 coupled with 4.2-12, 4.2-14.[5] The $\mathcal{M}$ measurements $z_{\mathcal{M}}$ actually undertaken will usually exceed $2N_B - 1$. One can then premultiply $z_{\mathcal{M}}$ by a transformation matrix T of dimension $(2N_B - 1) \times \mathcal{M}$ consitisting only of 0, +1, and -1 entries, such that one obtains a $(2N_B - 1)$ dimensional inferred measurement vector $z = T z_{\mathcal{M}}$. The objective function J will thus have as many terms as variables and the resulting recursion equation is the one for the special case M=N as in 5.1-10 which, in this case, takes on the form $x^{i+1} = [TH]^{-1}T[z_{\mathcal{M}} - h^i] + x^i$. The sparsity of TH is no less than that of the Jacobian H encountered in ordinary load flow work.

VI TRACKING ESTIMATOR ALGORITHMS

Tracking state estimators are used to estimate the system's steady-state time behavior based on a continuous sampling of measurements, unlike the previous WLS methods which use only one snapshot of z. Tracking estimators have been considered in lieu of an exact dynamic estimator because the complete dynamic model of the network state is not known. The tracking state estimator handles the slowly varying characteristics of the system and thus is capable of functioning properly under quasi-static operating conditions.

In the following, various tracking algorithms are described. In these algorithms the weighting matrix W is represented by the inverse of the measurement covariance matrix, R, assuming statistical data on R is available or can be estimated.

One simplified approach[11], based on assuming small observation errors so that fast algorithm convergence can be obtained, uses the basic static estimator algorithm to estimate x(k+1) for discrete time increment k+1 from z(k+1) and x(k). As described in Table 6.1-a, equations 6.1-1 through 6.1-3 follow directly from 5.1-9 by replacing the iteration index i by the time sample argument k.

Another[12] tracking algorithm accounts for the weighting

TABLE 6.2 CONTINUOUS-TIME TRACKING ALGORITHMS

(a). KALMAN FILTER ALGORITHM $\hat{x}(t) = \mathcal{P}(t)\, H^T(t)\mathcal{R}^{-1}(t)\,\hat{\nu}(t)$ $\dot{\mathcal{P}}(t) = -\mathcal{P}(t)H^T(t)\mathcal{R}^{-1}(t)H(t)\mathcal{P}(t) + Q(t)$; $\mathcal{P}(0) = \mathcal{P}_0$	1 2
(b) ADAPTIVE ALGORITHM $\dot{\hat{x}}(t) = \hat{q}(t) + K(t)\hat{\nu}(t)$; $\dot{\mathcal{P}}(t) = -K(t)H(t)\mathcal{P}(t) + \hat{Q}(t)$ $\dot{b}(t) = [Z(t) - h(\hat{x},t)]/t$; $\mathcal{R}(t) = \hat{\nu}(t)\hat{\nu}^T(t)\,/\,t$ $\dot{\hat{q}}(t) = \hat{x}(t)\,/\,t$; $\dot{Q}(t) = \mathcal{P}(t)H^T(t)\mathcal{R}^{-1}(t)\hat{\nu}(t)\hat{\nu}^T(t)/t$ $K(t) \triangleq \mathcal{P}(t)\hat{H}^T(t)\mathcal{R}^{-1}$	3;4 5;6 7;8 9

of prior estimates. It is based on the objective function $J=[x-\hat{x}]^T\mathcal{P}^{-1}[x-\hat{x}] + [z-h(x)]^T\mathcal{R}^{-1}[z-h(x)]$ where $\hat{x}$ is the prior estimate of the state and $\mathcal{P}$ is its prior covariance. The resultant sequential algorithm which minimizes this objective function is described in Table6.1-b and is seen to include a mechanism for updating at each time sample.

Tracking algorithms can also be based on the Kalman filtering approach. The continuous and discrete time algorithms which result are given in Tables 6.1-c and 6.2-a.[13]

First let us consider the state random change w and observation error v. A good and practical statistical approximation is the Gaussian distribution function. Thus, w is described to possess a mean, q, and a variance, $\mathcal{Q}$. In a similar fashion the mean and covariance of v are represented by b and $\mathcal{R}$, respectively. In comparing the Kalman filtering approach to the least squares approach it is noted that in addition to b, $\mathcal{R}$, q, and $\mathcal{Q}$ one must also provide $\mathcal{P}(0)$ or $\mathcal{P}(1)$. This has given rise to further variations among algorithms described as follows.

One approach[14] is to pre-estimate $\mathcal{P}(0)$ and to approximate $\mathcal{Q}$ by retaining the diagonal elements only, scaled by a suitable scalar factor. This factor can then be tuned by means of off-line simulated or real data.

Another, more effective approach[15] assumes $\mathcal{Q}=0$ throughout and tunes $\mathcal{P}(0)$ experimentally by off-line simulation and then processes the measurements sequentially. In this case a sequential-measurements tracking extimator is obtained; however, its response is considered slower for quasi-static changes due to the absence of $\mathcal{Q}$. In a sense the main purpose of this estimator is "to keep up with the data". Let "d" be the latest batch of data received at k+1 which, when added to the set "c" collected at time k, gives an augmented set a=c+d. Then it is possible to decompose the NxN matrix $[H^T\mathcal{R}^{-1}]_a[z_a - h_a]$ and, assuming $z_c - h_c^{\prime}$ to be negligibly small, by previous estimator action, to replace it by $H_d^T\mathcal{R}_d^{-1}[z_d - h_d]$. If we replace subscripts "c" and "a" by arguments k and k+1 respectively, then we obtain equations given in Table 6.1-d, where we should use H_d for H(k+1).

A third approach, which estimates the statistical parameters of w and v on-line from real-time data is described in Tables 6.1-e and 6.2-b. This adaptive approach is suitable only for relatively small or decomposed systems due to the algorithms' increased dimensionalities.[16]

VII DECOMPOSITION AND COORDINATION METHODS

In large scale systems, decomposition and coordination methods are essential tools[17] to ease the complexity arising from the system's high dimensionality. As a result the system's original problem can be decomposed into subproblems of lower dimensionalities, where decoupled and coordinated solutions can be obtained. In the following, several techniques are presented which can be classififed as: (1) model decomposition; (2) space decomposition; (3) time decomposition; (4) composite decomposition involving one or more of the above decompositions.

Model Decomposition (P-Q)

A state estimator algorithm based on a model decomposition method, known as the Decoupled Weighted Least Squares Estimator[18], takes advantage of a power system's weak coupling properties : While the active power is dominantly dependent on the phase angle and less on the magnitude of bus voltage, the reactive power displays the opposite relationship. Thus the P-Q decomposition illustrated in Table 7.1 can be applied. Here the elements H_{12} and H_{21} of the Jacobian in 7.1-2 may be set to zero. If in addition one assumes V=1 p.u. and θ=0, then the computational effort can be greatly reduced. Accuracy is not impaired as long as ν is calculated without resort to approximations. The algorithm is described by the flow diagram, Fig.6.

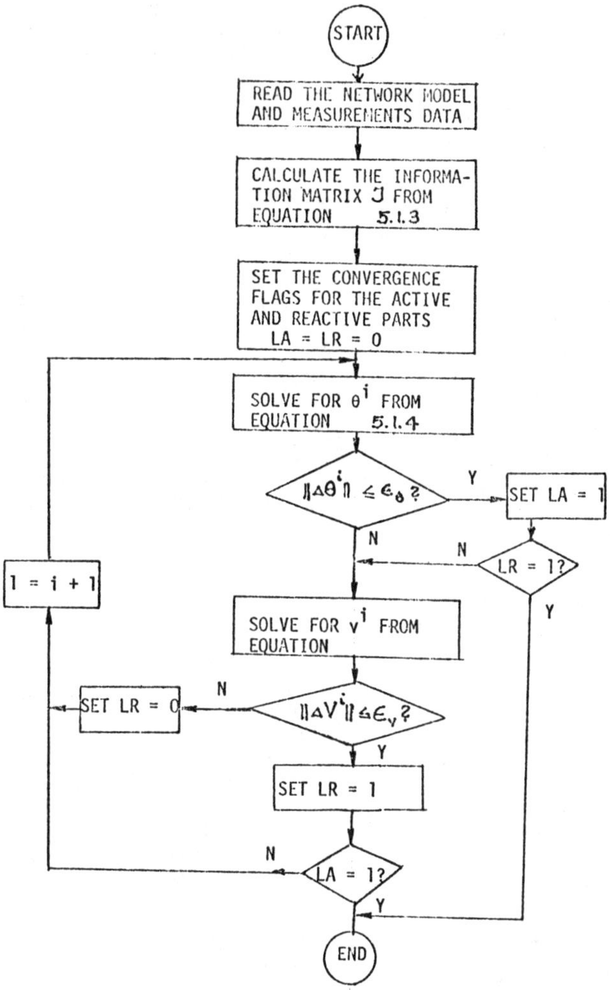

FIG.6 STATE ESTIMATOR BASED ON P-Q DECOMPOSITION

Space Decomposition

In order to describe the procedures of space decomposition in state estimation problems, it is advatageous to first consider some fundamentals of hierarchical system theory.[17, 19]

The general concept of the hierarchical system theory is to introduce a model and goal coordination variable pair denoted by π and β to the system in conjunction with an appropriate coordination procedure. The model coordination variable, π, is used to substitute for the coupling variables of the system. The original system is then decoupled to form several subsystem units with an additional set of interconnection (equality) constraints. The goal coordination variable, β, formulates these interconnection constraints into the system objective function, which results in modifying each subsystem objective to account for the

TABLE 7.1
P-Q DECOMPOSED STATIC ALGORITHM

MEASUREMENT SYSTEM PARTITIONING	$Z = \begin{bmatrix} Z_\theta \\ Z_v \end{bmatrix} \quad h(v,\theta) = \begin{bmatrix} h\,(v,\theta) \\ h_v(v,\theta) \end{bmatrix} \quad \mathcal{R} = \begin{bmatrix} \mathcal{R}_\theta & 0 \\ 0 & \mathcal{R}_v \end{bmatrix}$	(1)
JACOBIAN DEFINITION OF THE OBSERVATION FUNCTION	$\begin{bmatrix} H_{11} & H_{12} \\ H_{21} & H_{22} \end{bmatrix} = \begin{bmatrix} \frac{\partial h_P}{\partial \theta} & \frac{\partial h_P}{\partial v} \\ \frac{\partial h_Q}{\partial \theta} & \frac{\partial h_Q}{\partial v} \end{bmatrix}$	(2)
CONSTANT DECOUPLED INFORMATION MATRICES	$\mathcal{J} = \begin{bmatrix} \mathcal{A} & 0 \\ 0 & \mathcal{B} \end{bmatrix} = \begin{bmatrix} H_{11}{}^T \mathcal{R}_\theta H_{11} & 0 \\ 0 & H_{22}{}^T \mathcal{R}_v{}^{-1} H_{22} \end{bmatrix}$	(3)
DECOUPLED ESTIMATOR ALGORITHM	$[\theta^{i+1} - \theta^i] = [H_{11}{}^T(v,\theta) \;\; H_{21}{}^T(v,\theta)]\mathcal{R}^{-1} \; [Z-h(v,\theta)] \Big\vert_{\theta=\theta^i,\; v=v^i}$	(4)
	$[v^{i+1} - v^i] = [H_{12}{}^T(v,\theta) \;\; H_{22}{}^T(v,\theta)]\mathcal{R}^{-1} \; [Z-h(v,\theta)] \Big\vert_{\theta=\theta^i,\; v=v^i}$	(5)

TABLE 7.2
SPACE-DECOMPOSED STATIC ITERATIVE ALGORITHM

COORDINATOR ALGORITHM	Given x_i^{i+1}, solve for π^{i+1} and λ^{i+1} from	1
	$\pi^{i+1} = \mathcal{E}(x)^{i+1}$	
	$\beta^{i+1}\left\{ H_c^T \mathcal{R} [Z_c - h_c(\hat{x})] + \sum_{i=1}^{N} H_{c_i}^T \mathcal{R}_i^{-1} [z_i - h_i(\hat{x}_i) - h_{c_i}(\pi_i)] \right\}^{i+1}$	2
SUBSYSTEM i ALGORITHM	Given π and λ solve for x^{i+1} from	
	$\hat{x}_i^{i+1} = \hat{x}_i^i + \alpha_i^i \left\{ (H_i^T \mathcal{R}_i^{-1} H_i)^{-1} [H_i^T \mathcal{R}_i^{-1} (z_i - h_i(\hat{x}_i) + \sum_{j=1}^{N} \mathcal{E}_{ij}^T \beta_j] - \hat{x}_i \right\}^i \quad i = 1, \ldots, \mathcal{N}$	3

TABLE 7.3
SPACE-DECOMPOSED SEQUENTIAL TRACKING ALGORITHM
(DISCRETE TIME)

a. COORDINATOR ALGORITHM	Given $x(k)$, $\beta(k)$ and $\beta(k-1)$ solve for	
	$\pi(k+1)$ and $\beta(k+1)$ from	
	$\beta(k+1) = \beta(k) + \bar{\beta}(k);\; \bar{\beta} \triangleq \beta(k) - \beta(k-1)$	1
	$\pi(k+1) = \mathcal{E}\,[\hat{x}(k)]$	2
b. SUBSYSTEM i ALGORITHM	$\hat{x}_i(k+1) = \hat{x}_i(k) + K_i(k+1)\hat{\nu}_i(k+1) + \mathcal{P}_i(k)\mathcal{D}_i(k)$	3
	$K_i(k+1) = \mathcal{P}_i(k+1) H_i^T(k) \mathcal{R}_i^{-1}$	4
	$\mathcal{P}_i(k+1) = \mathcal{P}_i(k) + Q_i$	5
	$\mathcal{D}_i(k) \triangleq \frac{\partial}{\partial x_i} \sum_{j=1}^{N} \beta_j^T \mathcal{E}_{ji}\,[\hat{x}_i(k)]$	6

TABLE 7.4
SPACE-DECOMPOSED SEQUENTIAL TRACKING ALGORITHM
(CONTINUOUS TIME)

a. COORDINATOR ALGORITHM	Given $\hat{x}(t)$, $\pi(t)$ and $\beta(t)$ solve for $\pi(t+\Delta t)$ and $\beta(t+\Delta t)$ from	
	$\pi(t+\Delta t) = \mathcal{E}\,[\hat{x}(t)]$	1
	$\beta(t+\Delta t) = \beta(t) + \alpha_\beta\{\pi(t) - \mathcal{E}\,[\hat{x}(t)]\}$	2
b. SUBSYSTEM i ALGORITHM	$\dot{\hat{x}}(t) = \mathcal{P}_i \left\{ H_i^T \mathcal{R}_i^{-1} [Z_i - h_i(\hat{x}_i)] + \mathcal{D}_i(x_i) \right\}$	3
	$\dot{\mathcal{P}}_i(t) = \mathcal{P}_i \left\{ \frac{\partial}{\partial x_i} H_i^T \mathcal{R}_i^{-1} [Z_i - h_i(\hat{x}_i)] \right\} \mathcal{P}_i + \mathcal{P}_i \frac{\partial}{\partial x_i} \mathcal{D}(\hat{x}_i) \mathcal{P}_i$	4
	$\mathcal{D}_i(\hat{x}_i) \triangleq \frac{\partial}{\partial x_i} \sum_{j=1}^{\mathcal{N}} \beta_j^T \mathcal{E}_{ji}(\hat{x}_i)$	5

TABLE 7.5 SPACE-DECOMPOSED STATIC INTERLACING ALGORITHM

COORDINATOR ALGORITHM	Given Z_c^{i+1} solve for $\hat{x}_c^{i+1}$ from	
	$\hat{x}_{c_i}^{i+1} = \hat{x}_{c_i}^i + \left\{ \mathcal{P}_i K_{ji}^T [K_{ji} \mathcal{P}_i K_{ji}^T + K_{ij} \mathcal{P}_j K_{ij}^T]^{-1} + [Z_{c_j} - K_{ij} \hat{x}_{c_i}] \right\}^i$	1
	$\mathcal{P}_i^{i+1} = \left\{ I - \mathcal{P}_i K_{ji}^T [K_{ji} \mathcal{P}_i K_{ji} + K_{ij} \mathcal{P}_j K_{ij}^T]^{-1} K_{ji} \right\}^i \mathcal{P}_i^i$	2
SUBSYSTEM i ALGORITHM	Given $Z_{c_i}^{i+1} \in Z_c^{i+1}$ solve for x_i^{i+1} from	
	$\hat{x}_i^{i+1} = \hat{x}_i^i + \mathcal{P}_i H_i^T (H_i \mathcal{P}_i H_i^T + \mathcal{R}_i)^{-1} [Z_i - h(\hat{x}_i)]^i$	3
	$\mathcal{P}_i^{i+1} = [I - \mathcal{P}_i H_i^T (H_i \mathcal{P}_i H_i^T + \mathcal{R}_i)^{-1} H_i]^i \mathcal{P}_i^i$	4

TABLE 7.6 TIME-DECOMPOSED DYNAMIC ALGORITHM

$\dot{\hat{x}}_{d_i} = f(\hat{x}_{d_i}, \hat{x}_{s_i}, u_i) + \mathcal{P}_i H_i^T \mathcal{R}^{-1} [Z_{d_i} - h_{d_i}(\hat{x}_{d_i}, \hat{x}_{s_i})]$	1
$\dot{\mathcal{P}}_i = F_i \mathcal{P}_i + \mathcal{P}_i F_i^T - \mathcal{P}_i H_i^T \mathcal{R}_i^{-1} H_i \mathcal{P}_i + \mathcal{Q}_i; \quad F_i = \frac{\partial f_i}{\partial x}$	2

decomposition effect. Accordingly, if the model and goal coordination variables are known, it is possible to assign to each subsystem unit a relatively explicit subproblem which can be undertaken independently. The task of coordination is to supply and update π and β such that the overall system objectives are met. The implementation of the decoupled subsystems algorithms along with the coordination procedure are described in in the functional block diagram, Fig.7.

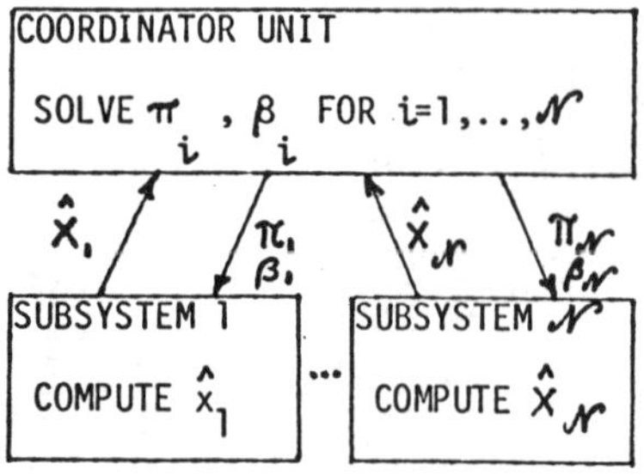

FIG. 7 HIERARCHICAL STATE ESTIMATION PROCEDURES

To formulate the space decomposed state estimator, consider that the original system, characterized by the state vector of dimension N and the measurement vector z of dimension M, is decoupled into $\mathcal{N}$ subsystems, characterized by x_i of dimension N_i and z_i of dimension M_i, such that $M_i \geq N_i$ for $i=1,\ldots,\mathcal{N}$. The coupling variables between the subsystems are defined by the model coordination vector π which results in the interconnection constraint $\pi = \mathcal{E}(x)$. By substituting the coordination vector π and the goal coordination vector β into the original estimation problem, one arrives at $\mathcal{N}$ subproblems. In this context the solution can be obtained in terms of decoupled subproblems in a two-level hierarchical structure as illustrated in Fig.7. The mathematical formulation of several algorithms are presented next.

Iterative Static Decomposition[20]

This algorithm is based on the following optimization problem:

$$\text{Minimize } J = [z_c - h_c(\pi)]^T \mathcal{R}^{-1}[z_c - h_c(\pi)] + \sum_{i=1}^{\mathcal{N}} [z_i - h_i(x_i) - h_{c_i}(\pi_i)]^T \; [z_i - h_i(x_i) - h_{c_i}(\pi_i)]$$

$$\text{Subject to } \pi_i = \mathcal{E}_i(x) = \sum_{j=1}^{N_i} \mathcal{E}_{ij}(x) \quad i = 1, \ldots, \mathcal{N}$$

where z is the measurement vector which corresponds to the coupling variables or the model coordination vector π ; h_{c_i} , h_i are the observation functions of the model coordination vector π and the ith subsystem state vector, x_i, respectively; and $\mathcal{E}$ represents the functional relationship between the coordination variables and the state variable.

By adjoining the equality constraint π to the objective function J via the goal coordination variable β and by applying nonlinear programming techniques, we arrive at the iterative algorithm described in Table 7.2.

Sequential Tracking Decomposition[21]

This algorithm is based on a maximum a posteriori criterion (MAP). The model used is generally nonlinear and cast in a continuous-time optimization problem described as follows:

$$\text{Minimize } J = \sum_{i=1}^{\mathcal{N}} J = \sum_{i=1}^{\mathcal{N}} \left\{ \frac{1}{2}\|x_i(t_o) - x_i(t_o)\|^2 + \cdot \frac{1}{2}\int_{t_o}^{t_f} \{\|z_i(t) - h_i(x_i)\|^2 + \|w_i(t)\|^2\} dt \right\}$$
$$i = 1, 2, \ldots, \mathcal{N}$$

$$\text{subject to } \dot{x}_i = w(t)$$
$$\pi_i = \mathcal{E}_i(x) = \sum_{j=1}^{\mathcal{N}} \mathcal{E}_{ij}(x_j)$$

By adjoining the interaction constraint π to the objective function J via the goal coordination vector we obtain a standard dynamic optimization problem. To get a sequential solution of this problem, the invariant embedding technique is used, which results in the sequential algorithm as described in Table 7.3 for discrete time and in Table 7.4 for continuous time.

Interlacing Coordination[22]

This algorithm uses the basic two-level structure of hierarchical solution but does not directly use the concept of model and goal coordination. The system is decomposed into subsystems such that each subsystem assumes the coupling variables x_{c_i} as a part of its internal state vector. Thus the model coordination variable is not substituted for explicitly, and the goal coordination is eliminated. Each subsystem then estimates its internal state variables based on its own measurements. To improve its performance, a second level estimator is introduced to estimate the coupling variables, based on the subsystems' estimated values. At this point, these values are viewed as coordination measurements z_c which can be expressed linearly in terms of the subsystems variables by $z_c = Hx$. The complete algorithm, which uses a linearized minimum variance solution, is described in Table 7.5.

Time Decomposition Methods

In some systems it is possible to decompose the overall problem into subproblems corresponding to elements which follow different time scales. Each subproblem can then be solved by taking advantage of the most appropriate method. Applications of this concept in power system analysis include the estimation of quasi-static and dynamic variables. A brief description of this application follows.[21, 23]

By denoting the identified slowly varying state variables by x_s and the rapidly varrying ones by x_d. The ith power plant dynamic model and observation model could be described as:

$$x_{d_i} = f(x_{d_i}, x_{s_i}, u_i) + w_i$$

$$z_{d_i} = h_{d_i}(x_{d_i}, x_{s_i}) + v_{d_i}$$

for $i=1,\ldots,\mathcal{N}$, where u_i represents the input control signal to the ith plant; $w_i(t)$ represents the plant stochatic behavior; and $\mathcal{N}$ is the total number of interconnected dynamic plants.

Typical slowly varying state variables could be identified as the voltage magnitudes and phase angles in the network. While these may undergo sudden sharp drops and rises, their values thereafter remain fairly steady. In this case x_s is also viewed as the coupling variable. Rapidly varying state variables depend on the type and particular characteristics of the plant. A typical example is a plant's frequency deviation and mechanical power which continue to vary over a period of time after perturbations. By applying time decomposition in the context of state estimation results in the two-level, tracking-dynamic estimation procedure. In this configuration, the tracking state estimator makes use of an algorithm described in

Section VI, and the dynamic state estimator makes use of an algorithm such as the one in Table 7.6 which utilizes Kalman filtering.

In the meantime, exchange of information in terms of measurements and estimates of the coupling state variables must take place to permit state estimation on the total system.

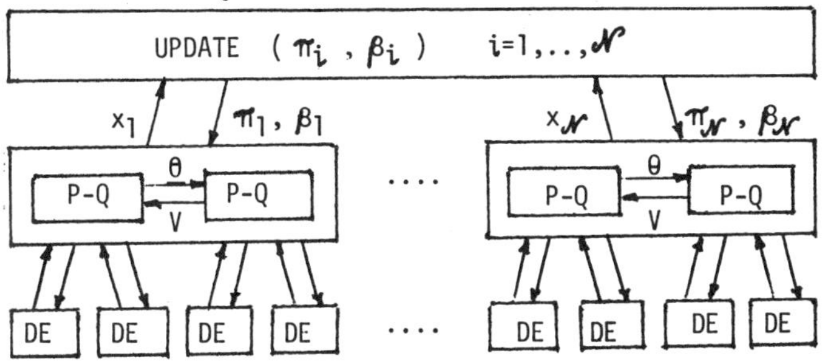

DE : Dynamic Estimator

FIG.8 COMPOSITE STATE ESTIMATOR

Composite State Estimation[19]

In many large scale systems, it is possible to apply more than one decomposition method. This could reduce complexities considerably. In terms of model, space, and time decomposition, different composite structures can be obtained by analysing the system's physical properties, its particular characteristics, and its requirements. To demonstrate these procedures in power system estimation, Fig. 8 illustrates one application. The first level represents the dynamic state estimators in each power plant. The second level makes use of the P-Q model decomposition. The third level represents the space decomposition in terms of model and goal coordination.

Requirements Placed on Algorithms

It is important to note that in order to apply decomposition methods successfully, special attention should be given to the overall system algorithm robustness, reliability, and computational efficiency. Various techniques are available in the literature; however, careful examination and evaluation of these techniques to satisfy the above conditions should be undertaken. Moreover, due to the close relationship of the decomposition methods to the physical characteristics of the problem, extended or special techniques to fit particular problems should be developed and applied.

CONCLUSIONS

The estimation procedures which are suitable for power system application can be approached in a unified way. Whether particular methods succeed or not will depend largely on a power system's peculiar characteristics. Many of the difficulties arising in the implementation of estimation algorithms are traceable to the data input rather than to the estimator. This paper stresses the differences between estimation algorithms and does not cover the data management procedures which depend so highly on the data base design of each system.

REFERENCES

1)F.C.Schweppe, J.Wildes, and D.Rom: "Power system static state estimation", partsI, II, III, IEEE Trans.Power App.Syst., v PAS-89, pp120-135, Jan 1970.

2)R.E.Larson and L.P.Hajdu: "Potential applications and on-line implementation of power system state estimation, " final report to Bonneville Power Adm., WMS Project U926, Wolf Mgt.Serv., Palo Alto, CA, 1969.

3)E.Handschin, F.C.Schweppe, J.Kohlas, and A.Fiechter: "Bad data analysis for power system state estimation, " IEEE Trans. PAS, March/April, 1975.

4)F.C.Schweppe and E.Handschin: "Static state estimation in electric power systems, " IEEE Proc., July 1974.

5)F.Arfatti, V.Castagnoli, L.Marzio, and P.Ricci: "Methods for electric power state estimation: Comparative analysis of computing simulations based upon generalized load flow, tracking, and least squares methods, "Proc.4th PSCC, Grenoble, France, 1972, paper 3.3/10.

6)S.A.Arafeh, A.P.Sage, and R.E.Kilmer: "Hierarchical decomposition of large-scale power system identification problems, "IEEE paper C74991-7, Summer Power Meeting, 1974.

7)A.S.Debs: "Estimation of steady-state power system model parameter, " IEEE Trans.PAS, Sep/Oct 1974.

8)O.J.Smith: "Power system estate estimation, " IEEE Trans. v PAS-89, pp363-379, Mar 1970.

9)J.F.Dopazo, O.A.Klitin, and L.S.VanSlyck: "State calculation of power systems from line flow measurements, Part II," IEEE Trans. v PAS-91, pp145-151, Jan/Feb 1972.

10)H.L.Fuller and T.A.Hughes: "State estimation for power systems with mixed measurements, "IEEE Summer Power Meeting, 1974, paper C74 363-8.

11)E.Handschin: "Real-time data processing using state estimation in electric power systems, " in Real-Time Control of Electric Power Systems, E.Handschin, Ed., Elsevier, Amsterdam 1972.

12)J.Bubenko and L.O.Petterson: "State estimation in electric power systems using filtering techniques, " Proc. 4th PSCC, Grenoble, France, 1972, paper 3.3/6.

13)A.P.Sage and J.L.Melsa: Estimation Theory with Application to Communication and Control, McGraw-Hill, 1971.

14)A.S.Debs and R.F.Larson: "A dynamic estimator for tracking the state of a power system, "IEEE Trans.vPAS-89, pp1670-1678, Sep/oct 1970.

15)A.S.Debs, R.F.Larson, and L.P.Hajdu: "On-line sequential state estimation for power systems, " 4th PSCC, Grenoble, France, 1972, paper 3.3/7.

16)A.P.Sage and J.L.Melsa: System Identification , Academic Press, 1971.

17)M.D.Mesarovic, D.Macko, and Y.Takahara: Theory of Hierarchical Multilevel Systems, Academic Press, 1970.

18)H.P.Horisberger, J.C.Richard, and C.Rossier: "A fast decoupled static state-estimator for electric power systems, " IEEE Summer Power Meeting, 1975, paper F75428-3.

19)S.A.Arafeh and A.P.Sage: "On the hierarchical decomposition and coordination methods in large-scale systems, "Milwaukee Symp.on Automatic Control, April 1975.

20)L.P.Hajdu et al: "New developments in system state estimation and anomaly detection, " Report prepared for Bonneville Power Adm. WMS Project U932 by Wolf Mgt.Serv., Palo Alto, CA, 1969.

21)S.A.Arafeh and A.P.Sage: "Multilevel discrete time system identification in large scale systems," Internatl. J. of Systems Science, v 5, n 8, 1974.

22)K.A.Clements, O.J.Denison, and R.J.Ringlee: "A multiarea approach to state estimation in power system networks, " IEEE Summer Power Meeting, 1972, paper C72 465-3.

23)E.Handschin and F.D.Galiana: "Hierarchical state estimation for real-time monitoring of electric power systems, " Proc.PICA, 1973.

24)F.Aschmoneit and F.Graf: "Optimaler State Estimator zur On-Line Lastflussschätzung," Z.ETG-Fachforum, Okt.1976, Munich.

25)E.Pse, R.E. Larson, and N.M. Peterson: "Algorithms for state estimation in large, interconnected power systems," Proc.1973 Internatl. Symposium on Circuit Theory.

Part V

Socioeconomic Systems (Models, Social Systems, Optimization)

Structure of the World Economy—Outline of a Simple Input-Output Formulation

WASSILY LEONTIEF

Abstract—**This invited paper presents the outline of a simple input formulation of a structure of the world economy. Such a formulation should provide a framework for assembling and organizing the mass of factual data needed to describe and understand the world economy.**

I

THE WORLD ECONOMY, like the economy of a single country, can be visualized as a system of interdependent processes. Each process, be it the manufacture of steel, the education of youth or the running of a family household, generates certain outputs and absorbs a specific combination of inputs. Direct interdependence between two processes arises whenever the output of one becomes an input of the other: coal, the output of the coal mining industry, is an input of the electric power generating sector. The chemical industry uses coal not only directly as a raw material but also indirectly in the form of electrical power. A network of such links constitutes a system of elements which depend upon each other directly, indirectly or both.

The state of a particular economic system can be conveniently described in the form of a two-way input–output table showing the flows of goods and services among its different sectors, and to and from processes or entities ("value added" and "final demand") viewed as falling outside the conventional borders of an input–output system. As the scope of the inquiry expands, new rows and columns are added to the table and some of the external inflows and outflows become internalized. Increasing the number of rows and columns that describe an economic system also permits a more detailed description of economic activities commonly described in highly aggregative terms.

Major efforts are presently underway to construct a data base for a systematic input–output study not of a single national economy but of the world economy viewed as a system composed of many interrelated parts. This global study, as described in the official document, is aimed at[1]

> "helping Member States of the United Nations make their 1975 review of world progress in accelerating development and attacking mass poverty and unemployment. First, by studying the results that prospective environmental issues and policies would probably have for world development in the absence of changes in national and international development policies, and secondly, by studying the effects of possible alternative policies to promote development while at the same time preserving and improving the environment. By thus indicating alternative future paths which the world economy might follow, the study would help the world community to make decisions regarding future development and environmental policies in as rational a manner as possible."

Preliminary plans provide for a description of the world economy in terms of 28 groups of countries, with about 45 productive sectors for each group. Environmental conditions will be described in terms of thirty principal pollutants; the use of nonagricultural natural resources in terms of some 40 different minerals and fuels.

II

The subject of this paper is the elucidation of a particular input–output view of the world economy. This formulation should provide a framework for assembling and organizing the mass of factual data needed to describe the world economy. Such a system is essential for a concrete understanding of the structure of the world economy as well as for a systematic mapping of the alternative paths along which it could move in the future.

Let us consider a world economy consisting of (1) a Developed and (2) a Less Developed region. Let us further divide the economy of each region into three productive sectors: an Extraction Industry producing raw materials; All Other Production, supplying conventional goods and services; and a Pollution Abatement Industry. In addition to these three sectors, there is also a consumption sector specified for each region. The function of the Abatement Industry is to eliminate pollutants generated by the productive sectors, consumers, and the Abatement Industry itself.

The two input–output tables displayed as Fig. 1 describe the inter-sectoral flows of goods and services within the Developed and the Less Developed economies. The flow of natural resources from the Less Developed to the Developed Countries, as well as the opposite flow of Other Goods from the Developed to the Less Developed Countries are entered in both tables: positively for the exporting region, and negatively for the importing region.

In each of the two tables the right-most entries in the first and second row represent the total domestic outputs of the Extraction Industry and of Other Production, respectively.

Each positive number along the third (Pollution) row shows the physical amount of pollutant generated by the activity named at the head of the column in which that number appears. The negative quantity shown at the intersection of the third column and the third row represents the amount of pollutant eliminated by Abatement activities.[2] Inputs such as power, chemicals, etc., purchased by the Abatement Industry from

Manuscript received July 25, 1974. Lecture delivered by W. Leontief on the occasion of his receiving the 1973 Nobel Prize for Economics, Stockholm, Sweden, December 1973.

The author is with the Department of Economics, Harvard University, Cambridge, Mass. 02138.

[1] Quoted from: *Brief Outline of the United Nations Study on the Impact of Prospective Environmental Issues and Policies on the International Development Strategy*, April 1973.

[2] All quantities are measured in billions of dollars "in current prices;" pollutants are "priced" in terms of average "per unit" abatement costs.

Reprinted from *Proc. IEEE*, vol. 63, pp. 345–351, Mar. 1975.

WORLD ECONOMY IN 1970
(Billions of 1970 dollars)

DEVELOPED COUNTRIES

	Extraction Industry	Other Production	Abatement Industry	FINAL DEMAND Domestic	FINAL DEMAND Trade	Total Output
Extraction Industry	0	76	0	2	-15	63
Other Production	21	1809	21	2414	19	4284
Pollution	5	62	-63	60	0	64
Employment	18	1372	20	287	0	
Other Value Added	21	996	22	0	0	

LESS DEVELOPED COUNTRIES

	Extraction Industry	Other Production	Abatement Industry	FINAL DEMAND Domestic	FINAL DEMAND Trade	Total Output
Extraction Industry	0	8	0	2	15	25
Other Production	7	197	0	388	-19	573
Pollution	2	8	0	11	0	21
Employment	9	149	0	99	0	
Other Value Added	8	220	0	0	0	

Fig. 1.

TECHNICAL AND CONSUMPTION COEFFICIENTS *

Developed Countries

$$A_1 = \begin{bmatrix} .0 & .0178 & .0 \\ .3372 & .4223 & .3298 \\ .0859 & .0144 & .0118 \end{bmatrix} \qquad C_1 = \begin{bmatrix} .0007 \\ .8834 \\ .0218 \end{bmatrix}$$

$$\ell_1 = \begin{bmatrix} .2867 & .3203 & .3161 \end{bmatrix} \qquad \ell_1^c = \begin{bmatrix} .1050 \end{bmatrix}$$

$$r_1 = \begin{bmatrix} .3332 & .2324 & .3482 \end{bmatrix} \qquad r_1 = \begin{bmatrix} .0 \end{bmatrix}$$

Less Developed Countries

$$A_2 = \begin{bmatrix} .0 & .0141 & .0 \\ .2934 & .3437 & .3298 \\ .0859 & .0144 & .0118 \end{bmatrix} \qquad C_2 = \begin{bmatrix} .0037 \\ .7943 \\ .0218 \end{bmatrix}$$

$$\ell_2 = \begin{bmatrix} .3729 & .2597 & .3161 \end{bmatrix} \qquad \ell_2^c = \begin{bmatrix} .2020 \end{bmatrix}$$

$$r_2 = \begin{bmatrix} .3337 & .3825 & .3541 \end{bmatrix} \qquad r_2^c = \begin{bmatrix} .0 \end{bmatrix}$$

* The coefficients in these tables do not sum to unity because the pollution generated by industry and by final demand is only partially abated in the developed countries and not abated at all in the less developed countries.

Fig. 2.

other sectors, and value added paid out by that industry are entered as positive amounts in the same third column. The difference between the total amount of pollution generated in all sectors and the amount eliminated by the Abatement Sector is represented by the *net* emission figure, the right-most entry in the third row. Finally, labor inputs used in each sector and payments made to other income-receiving agents are shown in the bottom two rows.

The numbers in these two tables are strictly speaking, fictitious. But their general order of magnitude reflects crude, preliminary estimates of intersectoral flows within and between the Developed and Less Developed Regions during the past decade.

For analytical purposes, the outputs and inputs of the Extraction Industry and Other Production, as well as the amounts of pollutants generated and abated, can be interpreted as quantities measured in the appropriate physical units (pounds, yards, kilowatts, etc.). The same is true of the services of some of the so-called primary factors: labor inputs, for example, are entered in the second to last row of each table. A similar physical measurement of the other components of value added, even if it were possible in principle, is impossible given the present state of knowledge. In pure, or should I say speculative economic theory, we can overcome this kind of difficulty by introducing some convenient albeit unrealistic assumptions. But a theoretical formulation designed to permit empirical analysis has to account for the fact that at least some components of value added cannot be interpreted as payments for measurable physical inputs, but must be treated as purely monetary magnitudes.

III

The flows described in the two input–output tables are interdependent. They have to satisfy three distinct sets of constraints. First, within each production or consumption process there exists a technological relationship between the level of output and the required quantities of various inputs. For example, if we divide each figure in the first column of the first table of Fig. 1 (the inputs of the Extraction Industry) by the total output of that sector (the last figure in the first row), we find that to produce one unit of its output this sector absorbed 0.3372 units of the output of Other Production, used 0.2867 units of Labor Services and spent 0.3332 dollars for other value added. Moreover, for each unit of useful output the Extraction Industries generated 0.0859 units of pollution. Other sets of input–output coefficients describe the technical structure of every sector of production and consumption in both groups of countries.

While statistical input–output tables continue to serve as the principal source of information on the input requirements or "cooking recipes" of various industries, increasingly we find economists using engineering data as a supplemental source. Complete structural matrices of the two groups of countries used in our example are shown in Fig. 2.

The second set of constraints that has to be satisfied by every viable system requires that the total (physical) amounts of outputs and inputs of each type of good must be in balance, i.e., total supply must equal total demand. In the case of a pollutant, *net* emission must equal the total amount generated by all sectors less the amount eliminated by the abatement process.

For example, the balance between the total output and the combined inputs of extracted raw materials can be described by the following equation:

$$\underbrace{(1-a_{11})x_1}_{\text{net output of Extraction Industry}} - \underbrace{a_{12}x_2}_{\text{amount delivered to Other Production}} - \underbrace{a_{13}x_3}_{\text{amount delivered to the Abatement Industry}} - \underbrace{c_1 y}_{\text{amount delivered to Final Users}} - \underbrace{T_1}_{\text{amount exported}} = 0. \tag{1}$$

The equation describing the balance between generation, abatement, and net emission of pollution reads as follows:

$$\underbrace{-a_{31}x_1 - a_{32}x_2}_{\text{gross amount of pollution generated by sectors 1 and 2}} + \underbrace{(1-a_{33})x_3}_{\text{amount abated by abatement activities}} - \underbrace{c_3 y}_{\text{gross amount generated by consumers and government}} + \underbrace{E}_{\text{net amount emitted into the environment}} = 0 \tag{2}$$

x_1 and x_2 represent the total outputs of the Extraction Industry and of Other Production, respectively; x_3, the level of activity of the abatement sector; y, the sum total of values added, i.e., Gross National Income. The "technical coefficient" a_{ij} represents the number of units of the product of sector i absorbed (or generated in the case of pollution) by sector j in producing one unit of its output; c_j is a "consumption coefficient" describing the number of units of the output of sector j consumed (or generated in the case of pollution) per unit of total value added, i.e., per unit of Gross National Income.

Fig. 3 displays the complete set of linear equations describing the physical balances between outputs and inputs of all sectors in both countries in terms of compact matrix notation. The last of these equations—written below in its explicit form—describes the flows of exports and imports that link the Developed and Less Developed areas into a single world economy

$$B = T_2 p_2 - T_1 p_1. \tag{3}$$

The balance of trade B, i.e., the difference between the monetary value of the two opposite trade flows, depends not only on the quantities T_1 and T_2 of traded goods but also on their prices, p_1 and p_2. The higher the price a country receives for its exports, or the lower the price it pays for imports, the better are its "terms of trade."

The last of the three sets of relationships describes the interdependence of the prices of all goods and services and the values added paid out, per unit of output, by each industry. For example, a typical equation in this set states that the price at which the Extraction Sector sells one unit of its output equals the average outlay incurred in producing it. This includes the costs (i.e., quantities X prices) of inputs purchased from other sectors, wages paid and all other value added:

$$\underbrace{p_1}_{\text{price of output}} - \underbrace{a_{11}p_1 - a_{21}p_2}_{\text{cost of material inputs}} - \underbrace{q_1 a_{31} p_3}_{\text{cost of pollution abatement}} - \underbrace{l_1 w}_{\text{cost of labor inputs}} - \underbrace{r_1}_{\text{other value added}} = 0. \tag{4}$$

The technical coefficients (a_{ij} and l_i's) appearing in this equation are the same as those appearing in the structural matrices

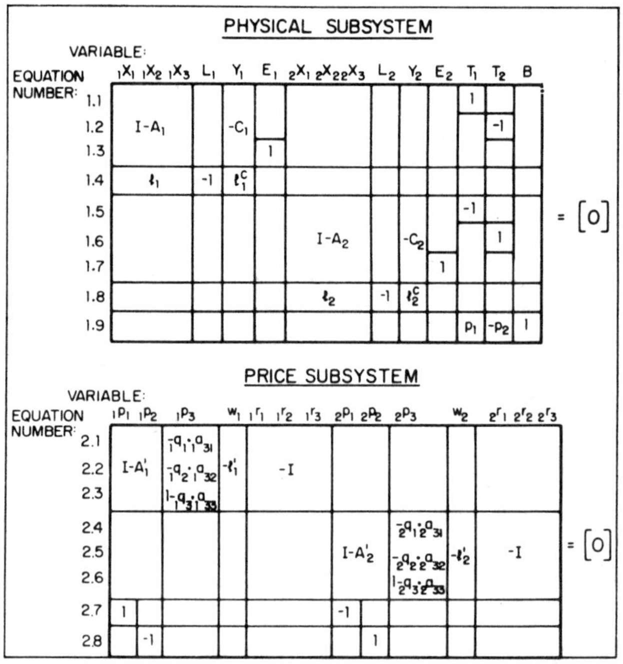

Fig. 3.

of Fig. 2. The abatement ratios q_i represent the fraction of the gross pollution emission of industry i that is eliminated (at that industry's expense)[3] by the Abatement Industry.

In this example, the system of physical balances contains 9 equations with 15 variables, while the price-values-added system has 8 equations with 14 variables. But these 14 variables are reduced to 12 and the number of equations of 6 if one assumes from the outset that the internationally traded products of the Extraction Industry and Other Production have the same price in the Developed and the Less Developed Countries. Equations

[3] This formulation is based on the assumption that the pollution generated by a particular sector is being eliminated at its own expense. In case the abatement cost is being paid out by the government out of its tax revenues, the price equations have to be modified accordingly. See W. Leontief, "Environmental repercussions and the economic structure," *Rev. Econ. Statist.*, Aug. 1970 and *The Measurement of Economic and Social Performance*, M. Moss, Ed., Income and Wealth Series, No. 38. New York: Columbia Univ. Press, 1973, pp. 565–576.

PHYSICAL SYSTEM ASSUMPTIONS							
		DEVELOPED COUNTRIES			LESS DEVELOPED COUNTRIES		
VARIABLES		Case I	Case II	Case III	Case I	Case II	Case III
Extraction output	X_1	Capacity limited to 150% of 1970 levels			Endogenous		
Other production	X_2	Endogenous			Capacity grows 6.4% per annum between 1970 and 2000		
Abatement output	X_3	Endogenous			0	Endogenous	
Employment	L	Increase proportional to population increase			Endogenous		
Final Demand	Y	Endogenous					
Net pollution emission	E	Limited to current levels assuming 1970 standards			Endogenous	Limited to twice 1970 levels	
Net trade in Extractive goods	T_1	Endogenous					
Net trade in Other goods	T_2	Endogenous					
Trade balance	B	A deficit for Less Developed Countries amounting to 1% of Developed Countries' income, reflecting capital flows and aid					
Technical Coefficients	A	Unchanged from 1970	Twice 1970 levels for Extraction Industry	Unchanged from 1970			
Labor Coefficients	ℓ	1/3 1970 levels, due to increased productivity	2/3 1970 levels for Extraction Industry	1/3 1970 levels due to increased productivity			
Consumption coefficients	C	Unchanged from 1970					
Extraction goods price	P_1	Obtained from solution of price system					
Other goods price	P_2	Obtained from solution of price system					

Fig. 4.

2.7 and 2.8 of Fig. 3 worked out explicitly read:

$$_1p_1 = {_2p_1}(\equiv p_1) \quad \text{and} \quad {_1p_2} = {_2p_2}(\equiv p_2). \tag{5}$$

The combination of both systems viewed as a whole contains 29 unknowns but only 17 equations. Thus, to arrive at a unique solution, we have to fix the values of 12 variables on the basis of some outside information, i.e., their values have to be determined exogenously.

Two types of quantitative information are required for the solution of this system. First, some data are used in the form of appropriate structural coefficients. Other kinds of factual information are introduced by assigning specific numerical values to appropriate "exogenous" variables.

In view of the uneven quality of data that will constitute the empirical basis of the present inquiry, it would be a tactical mistake to pour all the factual information we possess into the rigid mold of a single, all-embracing, inflexible explanatory scheme. The decision of which variables should be treated as dependent and which should be fixed exogenously is essentially a tactical one. The theoretical formulation is a weapon; in deciding how to use it we must take into account the nature of the particular empirical terrain.

To assess the influence of factors considered external to our theoretical description of the world economy, we earmark six physical and five value added variables as "exogenous." Figs. 4 and 5 show which variables are endogenous and assign values to all exogenous variables. These assumptions permit us to project changes in our simple world economy from a state representative of the present ("1970") to three alternative hypothetical states about thirty years hence ("2000 (I)," "2000 (II)," and "2000 (III)").

Total labor input in Developed Countries, L_1, is exogenous: under full or nearly full employment, its magnitude depends on demographic and cultural factors not accounted for within our formal theoretical system. Substantial endemic unemployment in the Less Developed Countries makes it advisable to consider the level of total employment as depending on the level of output—that is to treat L_2 as endogenous.

The output of the Extraction Industry in the Developed Countries is restricted by the limited availability of natural resources. We account for this limitation by making $_1x_1$ exogenous. In the Less Developed Countries, where natural resources are still plentiful, the output of the Extraction Industry, $_2x_1$, depends partly on a small domestic market but primarily on the import requirements of Developed Countries. Thus, $_2x_1$ can be treated as a dependent variable.

The situation is reversed in the case of Other Production. In Developed Countries the output of manufactured goods normally adjusts to the level of final demand, making $_1x_2$ a dependent variable. Yet in the Less Developed Countries the output of other production, $_1x_2$, is restricted by external factors such as weak infrastructure and limited capital. In this case rising domestic inputs usually stimulate a growing demand for imports. Hence, $_2x_2$ is treated as independent and T_1 and T_2 as dependent variables.

In the price-value-added system of equations, all money wages and other value-added payments in the Developed Countries (w, r_1, r_2, and r_3) are exogenously determined. This means

PRICE SYSTEM ASSUMPTIONS							
		DEVELOPED COUNTRIES			LESS DEVELOPED COUNTRIES		
VARIABLE		Case I	Case II	Case III	Case I	Case II	Case III
Extraction goods price	P_1	Endogenous					
Other goods price	P_2						
Abatement Price	P_3						
Wage rate	w	Kept at 1970 level (index = 1.0)					
Other value added in Extraction per unit of output	r_1	Kept at 1970 levels (index = 1.0)			Endogenous		
Other value added in Other Production	r_2						
Other value added in Abatement	r_3				Kept at 1970 level (index = 1.0)		
Technical coefficients	A	Unchanged from 1970		Twice 1970 levels for Extraction Industry	Unchanged from 1970		
Labor coefficients	ℓ	1/3 1970 levels, due to increased productivity		2/3 1970 levels for Extraction Industry	1/3 1970 levels, due to increased productivity		
Abatement coefficients	q	$q_1 = q_2 = q_3 = x_3/(x_3+E)$, that is, all Abatement coefficients of a given country are set to a value that reduces net pollution to the exogenously specified level E					

Fig. 5.

that the prices of all three products can be derived endogenously. In Less Developed Countries the situation seems to be different: since the prices of commodities produced by Extraction and Other Production are determined by the cost of their production (including the exogenous value added) in the Developed Countries, the value added that can be paid out by the two sectors producing these goods in the Less Developed Countries, ${}_2r_1$ and ${}_2r_2$, simply reflect the difference between a given price and the production costs.

Raw materials are, as a rule, relatively more abundant and more cheaply extracted in Less Developed Countries, thus the value added earned by Extraction Industries in Less Developed Countries can be expected to be relatively high. Ricardo speaks in this connection of "mining rents." On the other hand, technical input coefficients or, more properly, costs in Other Production of the Less Developed Countries can be expected to be higher than in Developed Countries. Because of this, the value added earned per unit of output in that sector tends to be relatively low.

Since a principal purpose of the aforementioned United Nations project is "a "realistic evaluation of the effects of alternative types of environmental policies on the economic prospects of Less Developed Countries," net pollution emissions E_1 and E_2 are treated as exogenously determined in two of our projections.

Assigning specific numerical magnitudes to all exogenously determined variables permits effective use of a variety of external data in arriving at a unique numerical solution of the formal input–output system. As the empirical inquiry advances, exogenous variables can be internalized through introduction of additional equations.

The most important but also the most demanding step in implementing an empirical input–output system is the determination of values of hundreds or even thousands of structural coefficients. The relevant methodologies are so varied and specialized that I abstain from discussing them in this general context.

IV

As has been explained above, three different sets of factual assumptions provided the basis for the three alternative projections of the state of one simple world economy for the year "1970" to the year "2000." Figs. 4 and 5 contain their full specification, while the results of the computations are summarized in three pairs of input–output tables presented in the Appendix.

The bar charts displayed in Figs. 6 and 7 facilitate a systematic examination of these findings. The width of each bar represents the relative size of the corresponding economic activity measured in base-year dollars. The length of each bar indicates the percentage increase or decrease in the level of each activity as the world economy passes from one state to another. Exogenous variables are identified by asterisks.

The long bars in the uppermost rows of these economic profiles indicate an upsurge in output and total consumption and a downward movement of prices: a "great leap forward" from 1970 to 2000. Case I is a projection that critically depends on two assumptions. First, the employed labor force in Developed Countries will increase with population growth. Second, labor productivity in both regions (the reciprocal of the labor coefficient) will be three times as high in 2000 as in 1970, with all other input coefficients remaining the same. Strict enforcement of standards contained in the United States Clean Air Act of 1967 (as amended in 1970) will bring about a sharp drop in unabated emissions in the Developed areas, while in Less Developed Countries the absence of any abatement activity will

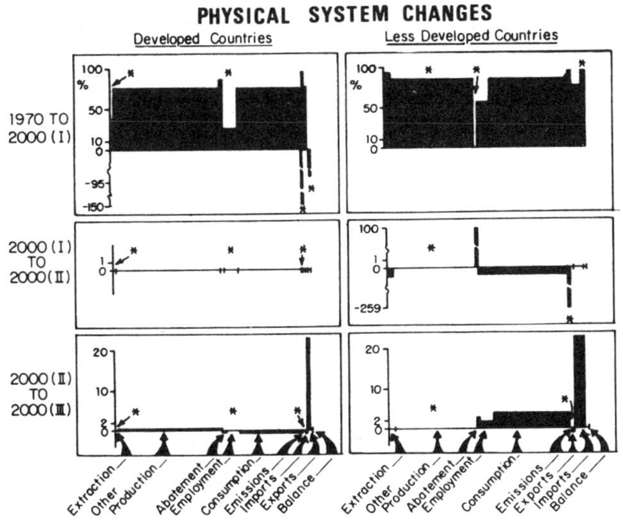

* exogenously specified variable (2000(III) values used to compute percentage changes)

Case I: Basic 2000 assumptions.
Case II: Abatement in LDC's.
Case III: Higher extraction costs in DC's.

Fig. 6.

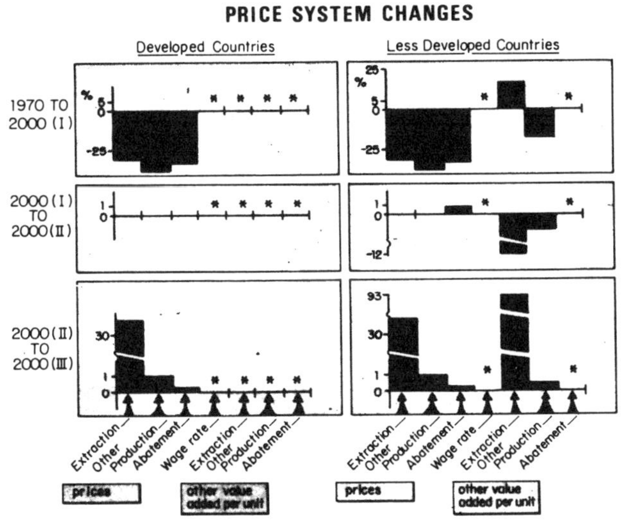

* exogenously specified variable (1970 values used to compute percentage changes)

Case I: Basic 2000 assumptions.
Case II: Abatement in LDC's.
Case III: Higher extraction costs in DC's.

Fig. 7.

force the pollution level up. International trade will expand faster than domestic economic activities. Prices (measured in wage units) will decline, while the value added in Less Developed Countries will rise in the Extraction Industry but fall in Other Production.

How would the future economic picture change if strict antipollution standards were also observed in Less Developed Countries? The answer is presented in the second row of bar graphs on Figs. 6 and 7. In the Developed Countries there will be practically no change. In Less Developed Countries the inauguration of abatement activities aimed at limiting pollution to twice its 1970 level would bring about expanded employment while requiring some sacrifices in consumption. Value added would fall sharply in the Extraction Industry and somewhat less in Other Production.

How would the situation thus attained be affected by a significant increase in the operating costs of the Extraction Industry in the Developed Countries? The bottom row of profiles in Figs. 6 and 7 shows how the conditions in both regions of the world economy would be affected if the productivity of labor in the Extraction Industry of Developed Countries rose only 1.5 rather than 3 times between 1970 and 2000 while the amounts of other Extraction inputs doubled per unit of output. The output of Other Production in the Developed Countries would register a slight increase and the level of consumption a slight decrease. Consumption in the Less Developed Countries would experience a substantial increase. The mechanism responsible for such a redistribution of income between the Developed and Less Developed Countries involves a steep increase in the price of Extraction goods compared to other prices, a corresponding rise in value added (rents yielded by the Extraction Industry of the Less Developed Countries) and, finally, a substantial increase in imports accompanied by slight reduction of exports from these countries, both reflecting a marked improvement in their "terms of trade."

I refrain from drawing any factual conclusion from the economic projections presented here. The computer received fictitious inputs and necessarily issued fictitious outputs. All theories tend to shape the facts they try to explain; any theory may thus turn into a procrustean bed. Our proposed theoretical formulation is designed to protect the investigator from this danger: it does not permit him to draw any special or general conclusions before he or someone else completes the always difficult and seldom glamorous task of ascertaining the necessary facts.

Acknowledgment

The author is indebted to P. Petri for setting up and performing all the computations, the results of which are presented here, and to D. T. Jenkins for preparing the graphs and editorial assistance.

Appendix

PROJECTED WORLD ECONOMY IN 2000 (CASE I)
(Billions of 1970 dollars)

DEVELOPED COUNTRIES

	Extraction Industry	Other Production	Abatement Industry	FINAL DEMAND Domestic	FINAL DEMAND Trade	Total Output
Extraction Industry	0	316	0	8	-226	98
Other Production	33	7502	160	9713	357	17765
Pollution	8	256	-479	240	0	25
Employment	9	1897	51	379	0	
Other Value Added	33	4129	169	0	0	

LESS DEVELOPED COUNTRIES

	Extraction Industry	Other Production	Abatement Industry	FINAL DEMAND Domestic	FINAL DEMAND Trade	Total Output
Extraction Industry	0	52	0	12	226	290
Other Production	85	1255	0	2668	-357	3650
Pollution	25	53	0	73	0	151
Employment	36	316	0	226	0	
Other Value Added	112	1143	0	0	0	

PROJECTED WORLD ECONOMY IN 2000 (CASE II)
(Billions of 1970 dollars)

DEVELOPED COUNTRIES

	Extraction Industry	Other Production	Abatement Industry	FINAL DEMAND Domestic	FINAL DEMAND Trade	Total Output
Extraction Industry	0	316	0	8	-226	98
Other Production	33	7502	160	9713	357	17765
Pollution	8	256	-479	240	0	25
Employ-ment	9	1897	51	379	0	
Other Value Added	33	4129	169	0	0	

LESS DEVELOPED COUNTRIES

	Extraction Industry	Other Production	Abatement Industry	FINAL DEMAND Domestic	FINAL DEMAND Trade	Total Output
Extraction Industry	0	52	0	12	226	290
Other Production	85	1254	36	2632	-357	3650
Pollution	25	53	-108	72	0	42
Employ-ment	36	316	12	223	0	
Other Value Added	100	1118	39	0	0	

PROJECTED WORLD ECONOMY IN 2000 (CASE III)
(Billions of 1970 dollars)

DEVELOPED COUNTRIES

	Extraction Industry	Other Production	Abatement Industry	FINAL DEMAND Domestic	FINAL DEMAND Trade	Total Output
Extraction Industry	0	315	0	8	-225	98
Other Production	66	7472	159	9678	461	17836
Pollution	8	255	-477	239	0	25
Employ-ment	19	1890	51	378	0	
Other Value Added	33	4112	168	0	0	

LESS DEVELOPED COUNTRIES

	Extraction Industry	Other Production	Abatement Industry	FINAL DEMAND Domestic	FINAL DEMAND Trade	Total Output
Extraction Industry	0	51	0	13	225	289
Other Production	85	1254	37	2735	-461	3650
Pollution	25	53	-111	75	0	42
Employ-ment	36	316	12	232	0	
Other Value Added	189	1125	40	0	0	

Social Decision Analysis

RONALD A. HOWARD, SENIOR MEMBER, IEEE

Abstract—How can a diverse free society find decision mechanisms that are logical, efficient, and timely? This is a problem that has challenged man at least from the time of Plato's *Republic*. We see today factions of society arguing over alternatives rather than over values or probabilities. Adversary proceedings encourage people to advocate extremes rather than a careful balance of several considerations.

Decision analysis, a logical procedure for balancing the many uncertain, complex, and dynamic factors that characterize a decision, offers promise of a new and valuable procedure for social decisions. The decision analyst creates an extrapersonal explicit model of the decision under consideration. Information on possible alternatives, uncertainties, relationships, or preferences can come from different groups and still be represented within the same decision model, with the implications for the decision apparent to all. One can imagine a society where decision making has become decentralized, where distinct bodies are responsible for creating social alternatives, assessing the probabilities of various outcomes for each alternative, and setting the preferences of society. Once the alternatives, information, and preferences were established, society would make the decision using only the principles of logic.

Applications to automotive pollution, hurricane seeding, and nuclear safety demonstrate the approach.

I. Introduction

IT MUST come as a surprise to most people that there is a logical way for an individual to make a decision. The essence of uncertainty seems to imply that any attempt to quantify the decision-making process must be doomed. Yet the developments of the last 10 or 20 years have shown that it is not only possible to define a good decision, but also that individual and corporate decision makers can implement a logical decision-making process in the important decisions they must face.

The question we propose to examine here is whether this process can be successfully adapted to the social decisions that we seem to encounter in increasing numbers as our society becomes more interdependent. We shall first describe the logical decision-making process we call decision analysis and show how it has been used to aid the individual decision-maker. Then we shall consider the important challenges to this procedure posed by social decisions. We shall propose a procedure for such decisions, and illustrate its use in selected social problems.

II. Decision Analysis

Good decision making is just common sense. The difficulty was that we did not know what common sense meant until we parsed the decision-making process. More complete descriptions are given elsewhere [1]–[3]; here we shall only summarize. A good decision is the logical consequence of what you can do, what you know, and what you want. "What you can do" are the alternatives available to you. Finding alternatives is the most creative part of the decision process. "What you know" means the knowledge of relationship and of magnitude, the information you bring to the decision-

Manuscript received September 19, 1974. This work was supported by the National Science Foundation under Grant GK-36491.

The author is with the Department of Engineering-Economic Systems, Stanford University, Stanford, Calif. 94305.

Reprinted from *Proc. IEEE*, vol. 63, pp. 359–371, Mar. 1975.

making process. Finally, "what you want" refers to the preference you have for the various consequences of the decision, consequences that may be distributed over time or uncertain.

The concepts we have just stated are accepted by all who work on the problem of making better decisions; however, interpretations of the concepts have varied. One group of interpreters we might call the decision theorists. The decision theorists proceed by capturing the personalistic view of the decision maker. In principle, this means having the decision maker assign a joint probability distribution on all the consequences of each alternative, and then specify a multiattribute utility function on those consequences. The best alternative is then the one with the highest expected utility. While this procedure seems simple, it rapidly exceeds the assessment capability of decision makers on any but the simplest problems.

Another group of interpreters whom we shall call the decision analysts also believe in the importance of incorporating judgment and preference. However, they spend more of their effort in constructing an extrapersonal explicit model of the decision under consideration. This model captures the important relationships of the decision problem and is open to the inspection of all who might contribute to the decision. The extrapersonalization of the decision model allows information to be collected from many disparate points of the organization: lawyers have a place to assign legal probabilities; metallurgists a place to assign distributions on material strength. While the decision maker is ultimately responsible for preferences, he can focus his attention only on the parts of the preference structure that cannot readily be delegated to others.

The creation of extrapersonal models by decision analysts was a natural consequence of applying decision analysis to the major decision problems of corporations and government agencies. Explicit models were a practical tool developed because of their usefulness in analysis and implementation of the decision. However, now that they exist, we begin to see how an entirely new structure for social decision could be built upon them. The explicit decision model provides the crucial step in dividing responsibility for the creative, informational, and preferential aspects of a decision among as many individuals or groups as required.

III. Social Decision

Now we turn to the problem of social decision and consider how decision analysis might be helpful. Most social decisions are presently made as the result of a legislative process. The legislators, primarily trained as lawyers, pursue their work through hearings intended to clarify the nature of the decision. However, these hearings often become adversary processes since the proponents of various alternatives are motivated mainly by a desire to see their favorite alternative selected rather than by a desire to illuminate the issue. Possibilities become confused with likely consequences. Bad outcomes are extensively discussed by opponents with the purpose of making their probability appear higher. From the SST to nuclear power plant siting, we have all seen and been affected by the results.

This system requires legislators to create alternatives, examine their implications, and value the outcomes for society. We can now ask whether we wish all these functions to reside in one group and, further, whether this group is properly trained to perform them. What might have been feasible in a simpler environment may no longer be adequate. The same forces that required corporations to seek more effective decision procedures are acting on society as a whole.

IV. Social Decision Analysis

Let us now assume the existence of a large supply of decision analysts capable of structuring social decision problems to the point where an explicit decision model was available. It would then be possible to constitute various bodies charged with the responsibility of providing the requisite inputs to this model.

One body of people would be responsible for the creation of new alternatives for solving society's problems. Anyone could suggest alternatives to this body for consideration. The results of its deliberations would be a set of certified alternatives for dealing with any societal problem.

A second body would be responsible for assessing the probabilities of various outcomes for each alternative. This would require capturing structural knowledge available on the alternatives and assessing probabilities on uncertainties. The work of this body would be heavily aided by decision analysts and subject-matter experts. The result would be a certified probability distribution on the outcomes of any alternative selected.

A third body would be concerned with characterizing the preferences of society. This body would be heavily influenced by the desires of all citizens. Voting procedures could be developed for determining citizen preference. However, the values of society should change slowly: major value changes would require the same care currently devoted to a constitutional amendment.

The values that would be set would be, for example, the value of a life, of health, of recreation, of cultural attainment. Outcomes extending over many years would require setting a social time preference, which would be based strongly on intergenerational considerations. While there are good arguments that society as a whole should be risk indifferent, other risk attitudes could be specified. The net product of this third body would then be a certified set of preferences to be used in evaluating outcomes.

Once the certified alternatives, information, and preferences were established, society would make the decision using only the principles of logic. If the resulting decision appeared incorrect, it could only be changed by a change in the certified inputs.

It would be possible to have different certifying bodies for different areas of social decision making. However, the fundamental nature of the deliberation of the preference body make it likely that only one such body should be constituted.

We shall now consider examples of situations where several interrelated factors, uncertainties, possibly high risks, and social concern can combine to produce a social problem worthy of the type of approach we have been discussing. We shall present these analyses in very brief summary form. Readers who desire a detailed examination of information sources and modeling choices should consult the references.

V. Automotive Emissions—The Problem of Externalities

Automotive emissions provide an excellent example of what economists call externalities.

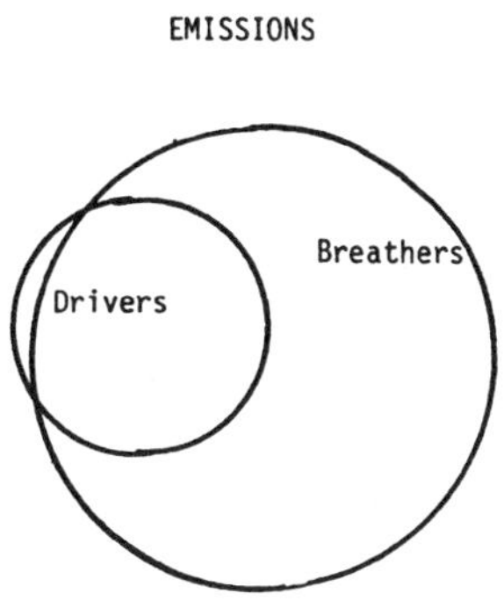

Fig. 1. The single basin model.

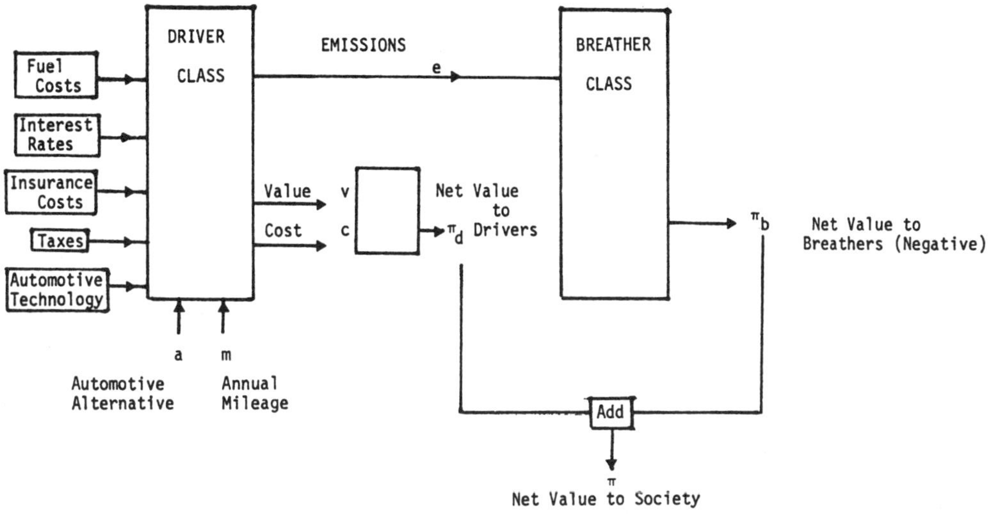

Fig. 2. The driver-breather model.

A. Externalities

An externality arises when the actions of one individual have effects on other individuals for which the other individuals are not economically charged or compensated. Externalities can be favorable, as when a world-famous violinist practices in his backyard on a Sunday afternoon and his playing is enjoyed by his neighbors, or detrimental if the same violinist practices at three o'clock in the morning.

B. Drivers and Breathers

Drivers of automotive vehicles create an externality when the emissions from their vehicles are noxious either directly or indirectly to the other individuals in the society. We shall restrict our attention to a single topographical basin where virtually all the automotive emissions are caused by people who live in the same region. Now we consider two classes, the "drivers" who create the emissions and the "breathers" who are affected by them. The term "breathers" may emphasize the human physiological effects of the emissions, but we also include individuals who are additionally concerned by aesthetic and more indirect effects such as the influence of emissions on plants; thus breathers are virtually all the human inhabitants of the region.

Fig. 1 shows the essence of the single-basin model. Drivers create emissions affecting breathers. Note that most of the drivers are breathers themselves; we are characterizing roles rather than individuals. However, typically there are many nondriving breathers, including children and the very elderly. There are also drivers who are not, or only partially, breathers, for example, those who live outside the region but who drive through it.

C. The Driver-Breather Model

Our first task is to formulate a model of externality that will allow us to see the effects of policy changes. Such a model for the case of emissions appears in Fig. 2. We imagine each member of the driver class to be faced by an economic environment described by fuel costs, insurance costs, interest costs, taxes, and automotive technology. In the face of this environment, the driver must select an automotive alternative a and an annual mileage m. We shall consider these decisions to be made on an annual basis.

Automotive Alternatives: The automotive alternative is a specification of which automotive system the driver will use for the coming year. It includes the complete spectrum of automotive conveyances each driver might buy or rent, physical modifications to those conveyances, and maintenance policies for the conveyances. For example, one alternative might be to buy an old car and do as little maintenance as possible. Another might be to buy a new no-lead car using a catalytic converter.

Cost: When the automotive alternative is selected and the annual mileage m is specified, the annual cost of operation c can be computed. Capital costs of purchase are annualized by simple or sophisticated methods.

Emissions: Associated with the (m, a) choice will be an emissions level e. While there is controversy over the consequences of various types of emissions, we shall simplify that

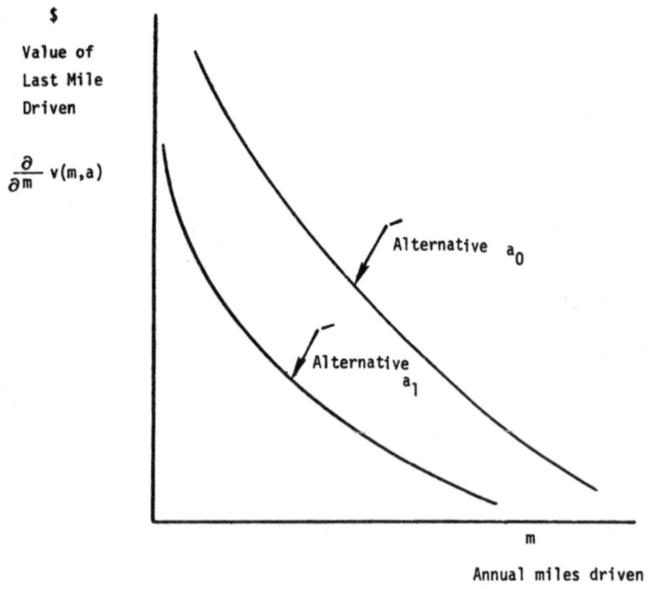

Fig. 3. The driver's marginal value of additional annual miles driven.

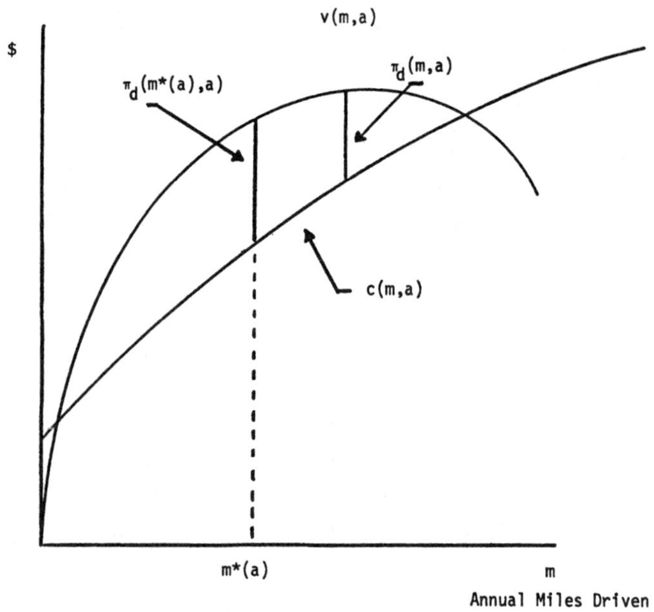

Fig. 4. Value and cost of mileage for a given alternative a.

question here by measuring the total weight of all pollutants characterized as harmful. More detailed treatment is, of course, possible.

The Driver's Decision: How does a driver decide which (m, a) choice to make? We assume that he selects the choice that maximizes the difference between his value for the choice $v(m,a)$ and the cost of the choice $c(m,a)$. This difference we call the net value to the driver $\pi_d(m,a)$.

Marginal Value of Annual Mileage: To determine $v(m,a)$ we begin with the driver's marginal value of additional annual mileage driven shown in Fig. 3. In theory, we construct this value for each automotive alternative by asking the driver how much he would pay to be able to drive 1 mi during the year, then another mile, and so forth. Presumably, the first miles are very valuable indeed since they would be used for emergency trips. When a large number of miles was already available, the values of succeeding miles should become very small. Thus we expect the marginal value of additional miles to decrease continually.

The marginal value curve depends on the automotive alternative primarily through what we could summarize as "performance." For example, the marginal value of high annual mileage in a standard car could be quite high because the driver contemplates using it for vacations. A small electric car could have a low marginal value under these conditions if it were not suitable for vacation travel. Similarly, a car with very poor acceleration might make almost all driving miles less valuable and consequently lower the curve.

Of course, it is practically almost impossible to develop the information of Fig. 3 for every driver and every alternative. However, once the concept is clear, only a few inputs may be necessary to determine the level of detail necessary for the social decision to be made.

The Net Value Calculation: Now we are in a position to determine the driver's decision. First we integrate (or sum) the marginal value curves of Fig. 3 with respect to m to obtain the driver's total value of driving m annual miles using alternative a, $v(m,a)$. This value is shown for a given alternative a in Fig. 4. The total value of driving m miles always increases with m, but with a steadily decreasing slope.

Also shown in Fig. 4 is the cost of driving m annual miles with alternative a, $c(m,a)$. For most alternatives, $c(m,a)$ will have a nonzero intercept $c(0,a)$ caused by the fixed costs of ownership such as capital costs, property and license taxes, and insurance. The annual cost $c(m,a)$ should be almost linear in m, but the figure allows for a lower marginal cost at high mileages.

The difference between the value and cost curves is the net value to the driver $\pi_d(m,a)$. For each a he should select the mileage $m^*(a)$ that maximizes this value,

$$m^*(a) = \max_m{}^{-1} \pi_d(m,a).$$

Once he knows the mileage he would drive given each automotive alternative, he would select the alternative a^* with the highest net value at its optimum mileage,

$$a^* = \max_a{}^{-1} \pi_d(m^*(a),a).$$

Some of the automotive alternatives may have negative net values, others positive. However, our range of alternatives can be quite broad from "call a taxi" to "rent a sports car." We would expect almost every member of the driver class to find an alternative of positive net value. The net value to each driver is then

$$\pi_d^* = \pi_d(m^*(a^*), a^*).$$

When this net value is summed over all drivers, we obtain the total net value of the driver class π_d.

The Breather's Consequences: The drivers as a result of their decisions have collectively produced emissions for the breathers. Since presumably no breather likes the emissions, we would expect that we would have to pay each breather a sum of money to achieve a situation where he was as happy to have the emissions and the money as he would be to have completely clean air. If we determine this sum for each level of emissions and sum the results over all breathers, we might obtain the net value to breathers curve of Fig. 5. The net value $\pi_b(e)$ is, of course, negative. We have chosen to make the curve almost linear, but with a slightly increasing slope to

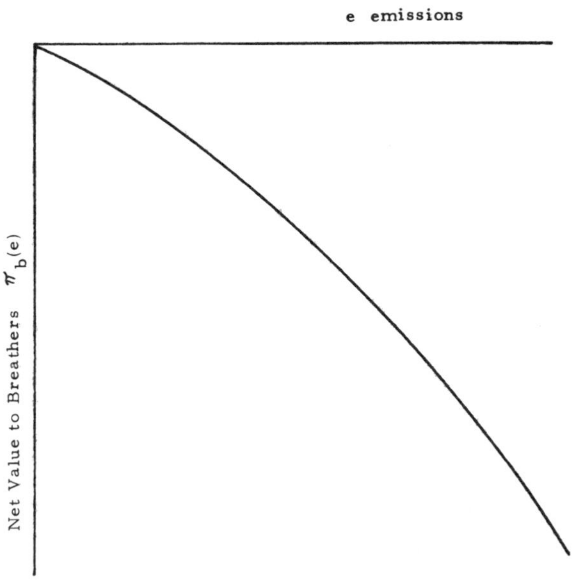

Fig. 5. Net value to breathers.

indicate that high levels of emissions may be considered to be even more undesirable than the linear model would specify.

Net Social Value: We can now evaluate any social alternative A_i for the treatment of the emission externality problem using the model of Fig. 1. We determine how drivers would make their decisions under this alternative and find the associated net drivers' value $\pi_d(A_i)$ by summing the optimum net values of each driver. Of course, it is possible that certain alternatives would cause nondrivers to become drivers or vice versa. If this effect is important, it can be treated by considering the driver class to be the union of driver classes for each alternative.

Then we would find the emissions $e(A_i)$ generated by the driver class under this alternative. The net value to breathers of this level of emissions is then $\pi_b(e(A_i)) = \pi_b(A_i)$, as computed from the net value to breathers curve. Finally the net value to society of alternative A_i, $\pi(A_i)$, is given by

$$\pi(A_i) = \pi_d(A_i) + \pi_b(A_i).$$

The alternative with the highest net value to society would be the best alternative for implementation, if alternative A_i did not in itself have associated administrative costs.

Administrative Costs: Most social alternatives A_i will, in fact, impose additional costs to the society as a result of their implementation. If these costs are $k(A_i)$, then the net social value will be

$$\pi(A_i) = \pi_d(A_i) + \pi_b(A_i) - k(A_i).$$

Change in Net Social Value: We shall now find the best social alternative by determining which alternative has the largest net social value including implementation costs. However, it serves as well to designate some particular alternative as a base alternative and then to determine how the net social value will change if any other alternative is used. Thus if we use A_0 to designate the base alternative, conveniently a continuation of present policy, then $\Delta\pi(A_i)$, the increase in social value caused by switching to social alternative A_i, is given by the following

$$\Delta\pi(A_i) = \pi(A_i) - \pi(A_0).$$

The best social alternative is then the one with the largest $\Delta\pi(A_i)$. Finally, using a parallel notation, we can write $\Delta\pi(A_i)$ in terms of the components of social value as

$$\Delta\pi(A_i) = \Delta\pi_d(A_i) + \Delta\pi_b(A_i) - \Delta k(A_i).$$

The advantage of this formulation is that we need deal only with the changes in driver net value, breather net value, and implementation cost when considering a new alternative rather than measuring the magnitude of any of these quantities. This approach simplifies considerably the input questions we discussed previously.

D. Social Alternatives

Now let us consider various social alternatives that could be employed in treating the emission externality.

Creating an Emission Market: Perhaps the simplest conceptual approach to the emission externality is to create a real emission market. Consider an agency that would charge drivers for the emissions they create and then use the money to compensate the breathers. If there were some way to discriminate among drivers so that those to whom the rights of emission were most valuable paid the highest tax per unit or if there were a way to discriminate among breathers so that those most affected by emissions received the highest compensation, then it could be possible for the revenue received by the agency in the form of taxes to be different from the compensation costs it incurs. It is likely that the agency could run at a profit even after administration costs were paid.

A Practical Pollution Market: However, for simplicity and practicality, we shall assume that the market functions by charging drivers r for each unit of emission and then compensating the breathers equally according to the total amount of emissions produced. For example, if each breather would pay $\$b$ for clean air, if there were n_b breathers, and if the total emissions were e kilograms, then the charge per kilogram (r) for emissions would be $r = bn_b/e$.

The Proper Signal: The importance of this approach is that it would reflect to the creators of the externality the costs that they are imposing on those they affect. Instituting such a charge would not drive emissions to zero, but would assure that each unit of emissions created by drivers was worth at least as much to them as the cost they were imposing on the breathers. Thus a proper "signal" is created to illustrate to those who emit just what is in the social interest. Note that this effect is preserved even if the compensation is not actually paid directly to the breathers but simply to the general funds of the community. Of course, if the emission level falls, the amount b the breathers are willing to pay for the clean air will probably fall even more so that we would expect the emission charge to fall as the air contains less emissions.

System Costs: We have not yet faced up to the question of how to implement even this simplified market. What would be required is a device to measure emissions directly, for example, a device that could be attached to the tailpipe to experience chemical changes as emissions are created. Perhaps the device could change color when a certain quantity of emissions had passed, thus requiring the owner to buy a new one at a service station. Perhaps he would have to exchange it periodically for a new one and send the old one to a center to determine the emissions he had created and consequently the tax he owed. The point is that such a method would make it the owner's responsibility to determine what kind of car he drove, how he drove it, and how he maintained it. "The more you emit, the more you pay."

TABLE I
DRIVER CHARACTERISTICS

MILEAGE CLASS (THOUSAND MILES PER YEAR)	%	VARIABLE COST PER MILE (DOLLARS)
> 30	.03	.040
20 - 29	.06	.048
15 - 20	.12	.056
10 - 15	.25	.064
5 - 10	.34	.072
< 5	.20	.080

TABLE II
ANTIPOLLUTION DEVICE CHARACTERISTICS

Abbreviation	Descriptions	Exhaust Emission Rate (g/mi) HC	NO_x	Sum	Initial Cost (dollars)	Variable Cost (dollars/1000 mi)
None	Uncontrolled[2]	11	4	15	0	0.00
CAP 68	1968 clear air package[1] Changes in the settings of carburetor and distribution parameters and some engine design modifications	4.4	6	10.4	21	0.00
LLO 71	1971 low lead, low octane engine[1] Engine modified to run well on low lead fuel	3.5	4.5	8	35	1.40
TUNE	Minor engine tune-up[1] Adjustment of idle speed and mixture	4	6.4	10.4	0	3.00
KIT	Used car kit[1] A General Motors package of parts and instructions for older cars with ignition and carburetion changes	5.1	2.8	7.9	35	0.00
CAT	Catalytic converter[1] A double catalyst bed with both oxidizing and reducing agents (low-lead gas required)	0.9	2	2.9	135	4.40
EXR	Exhaust gas recirculation[1] Some exhaust routed back to the carburetor with engine modified to run on unleaded fuel	3.4	2.1	5.5	60	1.40
VSAD	Vacuum spark advance disconnect[2] Restrictions placed on the spark advance in some driving modes	5.5	1.6	7.1	35	3.00
FLAME	Flame-afterburner[2] A system designed to complete the burning of exhaust gasses by means of a sustained flame in the exhaust system	0.3	4	4.3	200	3.00

[1] D. N. Dewees, "Automobile air pollution: An economic analysis," Ph.D. dissertation, Dep. Economics, Harvard University, Cambridge, Mass., Sept. 1971.

[2] P. B. Downing and L. Stoddard, "Benefit/cost analysis of air pollution control devices for used cars," Project Clear Air, Univ. California, Los Angeles, Sept. 1970.

Such a system could be quite expensive. The capital and service charges for the device would have to be paid. An enforcement mechanism would be necessary to assure that emissions were being propertly recorded; however, this enforcement mechanism might be self-supporting. Finally, the effectiveness of the system would be diminished if it is subverted by drivers; technology and enforcement will affect rates of subversion. Thus, when we consider the emission market solution, we must include in the net social value what could possibly be very high system costs of administration, enforcement, and subversion. It is possible that these system costs could make the emission market have a negative net social value compared with the present social policy.

Other Social Alternatives: Other social alternatives are easy to evaluate within the driver–breather model. For example, instituting a gas tax on gasoline sold, or having the tax depend on the class of car driven and on its time of last emission checkup. Other alternatives could be a mileage tax for each type of car collected according to odometer readings reported regularly by drivers. All of these are surrogates for the direct measurement of emissions. They lack the very attractive signaling effects of the emission tax, but their system costs could be sufficiently low to make them a better social policy.

The Present Alternative: Finally, we come to the present social alternative: requiring vehicles to be factory-modified with emission-control equipment. We observe first that this alternative has extremely remote signaling effects. Given that you have purchased one of these cars, there is little incentive to reduce emissions by curtailing driving or installing additional equipment. Nor is there any evidence that the present alternative is best when the effects on drivers, breathers, and

TABLE III
DRIVERS' DECISIONS AS A FUNCTION OF TAX RATE

TAX RATE	DEVICE SELECTED BY DRIVER CLASS > 30K	20-29K	15-20K	10-15K	5-10K	< 5K	
0.0000	PAYTAX	PAYTAX	PAYTAX	PAYTAX	PAYTAX	PAYTAX	
0.1000	CAP 68	CAP 68	CAP 68	CAP 68	CAP 68	PAYTAX	
0.2000	CAP 68	CAP 68	CAP 68	CAP 68	CAP 68	PAYTAX	
0.3000	KIT	KIT	KIT	CAP 68	CAP 68	PAYTAX	
0.4000	KIT	KIT	KIT	KIT	KIT	CAP 68	BASE
0.5000	KIT	KIT	KIT	KIT	KIT	CAP 68	CASE
0.6000	KIT	KIT	KIT	KIT	KIT	CAP 68	
0.7000	KIT	KIT	KIT	KIT	KIT	KIT	
0.8000	KIT	KIT	KIT	KIT	KIT	KIT	
0.9000	KIT	KIT	KIT	KIT	KIT	KIT	
1.0000	KIT	KIT	KIT	KIT	KIT	KIT	
1.1000	KIT	KIT	KIT	KIT	KIT	KIT	
1.2000	KIT	KIT	KIT	KIT	KIT	KIT	
1.3000	FLAME	KIT	KIT	KIT	KIT	KIT	
1.4000	FLAME	FLAME	FLAME	KIT	KIT	KIT	
1.5000	FLAME	FLAME	FLAME	FLAME	KIT	KIT	
1.6000	FLAME	FLAME	FLAME	FLAME	KIT	KIT	
1.7000	FLAME	FLAME	FLAME	FLAME	CAT	KIT	
1.8000	FLAME	CAT	CAT	CAT	CAT	KIT	
1.9000	CAT	CAT	CAT	CAT	CAT	KIT	
2.0000	CAT	CAT	CAT	CAT	CAT	KIT	
2.1000	CAT	CAT	CAT	CAT	CAT	KIT	
2.2000	CAT	CAT	CAT	CAT	CAT	KIT	
2.3000	CAT	CAT	CAT	CAT	CAT	KIT	
2.4000	CAT	CAT	CAT	CAT	CAT	KIT	
2.5000	CAT	CAT	CAT	CAT	CAT	KIT	
2.6000	CAT	CAT	CAT	CAT	CAT	KIT	
2.7000	CAT	CAT	CAT	CAT	CAT	KIT	
2.8000	CAT	CAT	CAT	CAT	CAT	KIT	
2.9000	CAT	CAT	CAT	CAT	CAT	CAT	
3.0000	CAT	CAT	CAT	CAT	CAT	CAT	
$/KG							

TABLE IV
BASE CASE RESULTS

TAX RATE R: $ 0.50/kg

DRIVER CLASS (Thousands of Miles Driven)	DEVICE	ANNUALIZED FIXED COST	VARIABLE COST ($/yr)	TAX ($/yr)	CHANGE IN ANNUAL MILES DRIVEN
> 30	KIT	9	29	150	-2095
20-29	KIT	7	17	94	-1091
15-20	KIT	5	12	67	- 655
10-15	KIT	4	9	48	- 409
5-10	KIT	4	5	29	- 218
< 5	CAP 68	4	0	13	- 74

ALL DRIVERS:

Average mileage	9962
Average emissions	8.53 g/Mi
Average tax/year	$43

system costs are considered. In fact, the wisdom of the present alternative can be evaluated only within the type of structure we have developed.

E. Numerical Results

Some preliminary insights into wise social choices for emission control can be gained from a numerical model developed by Judd to investigate the usefulness of the evaluation structure we have presented [4]. The model is applied to Los Angeles County in 1965, a reference year before extensive automotive modification and for which data are readily available. The 2.5 million households are assumed to be willing to pay $100 each to eliminate the 500 million kg of emissions generated annually by 3.3 million autos driving 10 000 mi/year and averaging 15 g of emissions per mile. From these data we find r = $0.50/kg is the nominal charge that should be placed on emissions.

Drivers were divided into six classes according to the number of miles they would normally drive by the partition of 5, 10, 15, 20, and 30 thousand miles per year. Each driver faced economics according to his class. Table I shows the class, the percentage of drivers in each class, and the variable cost per mile. Table II shows the various modifications to a standard car considered to be available to each driver. Table III displays how the drivers would respond in terms of these modifications if they faced various levels of emission tax. If the tax is small, up to and including $0.30/kg, low mileage drivers simply pay the tax, while high mileage drivers install CAP 68. As the tax increases beyond $0.30/kg, installation of KIT begins to drive out CAP 68 for all but the lowest mileage class. The nominal charge of $0.50 finds KIT used by all drivers except those driving less than 5000 mi/year; they use CAP 68. Increasing the charge above $0.60 drives everyone to KIT. When the charge reaches $1.30, FLAME becomes increasingly desirable for high mileage drivers. The charge has to reach 1.70 before CAT becomes desirable and it does not drive out KIT entirely until the charge reaches $2.90.

Table IV shows the detailed results of the nominal case where the emission charge is $0.50. We observe here how the drivers' choices have affected their costs, taxes, and mileage driven. We also observe that the effect of the $0.50 charge is to cut automotive emissions approximately in half.

TABLE V
SOCIAL VALUE INFERRED FROM REQUIRED DEVICES

DEVICE	SOCIAL VALUE RANGE FOR WHICH DEVICE IS OPTIMAL $/KG	IMPLIED ANNUAL VALUE OF ELIMINATING AIR POLLUTION (PER HOUSEHOLD) $/YEAR
None	0 - 0.05	0 - 10
CAP 68	.05 - 0.30	10 - 60
KIT	.30 - 1.35	60 - 270
FLAME	1.35 - 1.75	270 - 350
CAT	> 1.75	> 350
OTHERS	NEVER OPTIMAL	

Table V shows how the emission charge would have to be changed to make the various devices optimal, and then illustrates what this would mean in terms of the value to each household of eliminating automotive air pollution. The catalytic converter CAT is being installed on most 1975 cars in this region. Each household in Los Angeles County would have to be willing to pay at least $350/year, in 1965 dollars, if this is to be a wise choice for society. Since this figure represents about 5 percent of 1965 household annual income there is reason to reexamine the advisability of requiring the catalytic converter.

F. Concluding Observations

We should not take the results of the numerical model too seriously for it is only a pilot analysis of the emission situation in one region. Yet it could be expanded to comprise as many additional aspects of the problem as we wish to include, from nonautomotive pollution to the use of public transit. Our goal should be a balanced analysis comprehensive enough to investigate most concerns or suggestions regarding emissions.

It should provide a framework large enough to allow varying views to be assessed against a common background, so that information and reason will be more valuable tools of advocacy than rhetoric and histrionics.

Within the model several specific issues can be examined, such as uncertainty in the effectiveness or cost of any alternative, the choice of criteria for evaluating social desirability, or the value of emission measurement devices. The construction of a realistic model with this capability would require expertise in law, engineering, economics, psychology, and environmental medicine, to name but a few. The cost might be several hundred thousand dollars. But even this cost is a fraction of 1 percent of the total annual cost of automotive emissions estimated for Los Angeles County alone. The question is not when we can afford such models but rather whether we can afford to be without them.

VI. Hurricane Seeding—Uncertainty and Responsibility

As a further example of social decision analysis we can cite a recent study on hurricane seeding [5]. Seeding hurricanes with silver iodide crystals in an attempt to mitigate their destructive effects was a promising idea experimentally tested during the 1960's. The first experiments on Hurricanes Esther (1961) and Beulah (1963) were encouraging, but the experiment on Hurricane Debbie (1969) was the most impressive experimental result. Massive seedings on August 18 and 20, 1969, were followed by reductions of 31 and 15 percent in peak wind speed.

Hurricanes are costly to the nation. Annual property damage averages over $400 million dollars; individual Hurricanes Betsy (1965) and Camille (1969) each cost about 1.5 billion dollars of property damage. Seeding is cheap: about $0.25 million dollars. If the results of the Debbie experiment are applicable to even a few hurricanes, the savings could be very significant.

A. The Decision

The U.S. Hurricane Modification Program at the time of the study was exclusively scientific in its mission. Any seeding of hurricanes that could threaten coastal areas was prohibited. The decision that had to be faced was whether this prohibition should be removed. If it were, then study could begin on whether any specific hurricane threatening the U.S. should be seeded.

B. The Effect of Seeding

The characteristic of hurricanes of greatest economic consequence is their maximum sustained surface wind speed, since this is the characteristic that is primarily responsible for property damage. Happily, this is just the characteristic that seeding promises to reduce. However, even if seeding were certain to reduce wind speed, it still would not be clear that it should be performed. For by the very act of seeding, the Government stamps its name on the subsequent damage caused by the hurricane—"acts of God" become "acts of Uncle Sam." Even though scientists may state that the damage would probably have been worse without the seeding, the government would be in a difficult legal position.

The position is aggravated because the natural behavior of hurricanes is notoriously unpredictable. They can spontaneously intensify or diminish in their destructive power. If a hurricane were seeded when its peak winds were 100 miles per hour and it crossed the coast with peak winds of 125 miles per hour there will be few who pause to think that the winds might have been 150 miles per hour at land encounter if the seeding had not been performed. The public may well note only that a government-seeded hurricane intensified.

C. Characterizing Information on Seeding Effects

Three hypotheses were used to describe possible beliefs about the effect of seeding:

1) H_1, the "*beneficial*" hypothesis: the average effect of seeding is to reduce the maximum sustained wind speeds;
2) H_2, the "*null*" hypothesis: seeding has no effect on hurricanes;
3) H_3, the "*detrimental*" hypothesis: the average effect of seeding is to increase the maximum sustained wind speed.

After discussions with the scientists concerned, hypothesis H_1 was taken to mean a reduction of 15 percent in wind speed over a 12-h period; hypothesis H_3, a 10-percent increase. The actual wind speed after the 12-h period as a percentage of the initial wind speed is assumed to have a normal distribution in all cases. If seeding has no effect, natural variability is described by a standard deviation of 15.6 percent. If seeding has an effect, the standard deviation of its average effectiveness (7 percent) and the standard deviation of its effectiveness in any particular case (7 percent) combine to produce a 10-percent standard deviation for the effect of seeding. When this is superimposed on the natural variability, we obtain a standard deviation of 18.6 percent for the percentage change in wind speed of a seeded hurricane. Thus the 12-h percentage wind change distributions are

$$P(w|H_1) = f_n(85 \text{ percent}, 18.6 \text{ percent})$$
$$P(w|H_2) = f_n(100 \text{ percent}, 15.6 \text{ percent})$$
$$P(w|H_3) = f_n(110 \text{ percent}, 18.6 \text{ percent}).$$

It was now necessary to assign probabilities to the hypotheses based on meteorological judgment. The Debbie experiment produced what were considered to be two independent observations of 31- and 15-percent reductions in wind speed. In other words, on one occasion, the wind was 69 percent of its original speed; in the other, 85 percent. The relative likelihood of these observations under each hypothesis can be computed by multiplying together the ordinates of the corresponding normal density function at the values 69 percent and 85 percent. If we let D be the Debbie observations, the result is:

$$P(D|H_1) \sim 3.21$$
$$P(D|H_2) \sim 0.61$$
$$P(D|H_3) \sim 0.173.$$

The pre-Debbie hypothesis probabilities $P(H_1)$ must be related to the post-Debbie probabilities $P(H_i D)$ by Bayes' equation,

$$P(H_i|D) = \frac{P(D|H_i)P(H_i)}{\sum_{i=1}^{3} P(D|H_i)P(H_i)}.$$

Of course, since no one recalled the probabilities before the experiment, they had to be reconstructed. Meteorologists believed that before Debbie, H_1 was more likely than H_3 if seeding had any effect on hurricanes. They also believed that after Debbie they saw H_1 and H_2 as equally likely. Prior and posterior probabilities that meet these conditions and also satisfy Bayes' equation are

$$P(H_1) = 0.15 \qquad P(H_2) = 0.75 \qquad P(H_3) = 0.10$$

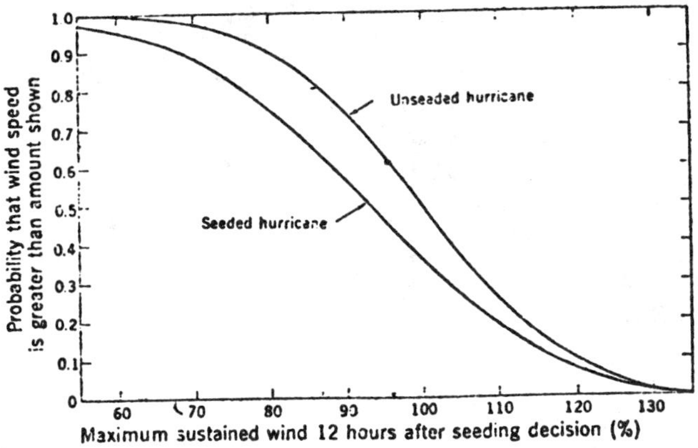

Fig. 6. Probability distribution on 12-h wind changes for the seeded and unseeded hurricane.

and

$$P(H_1 | D) = 0.49 \qquad P(H_2 | D) = 0.49 \qquad P(H_3 | D) = 0.02.$$

Note that before Debbie these assignments revealed three to one odds that seeding would have no effect, and that there was one chance in ten that it would have a detrimental effect. The Debbie experiment reduced the probability of a detrimental effect to only one in fifty, and made a beneficial and no effect equally likely. Sensitivity analysis showed there was little incentive in refining the state of knowledge beyond this specification.

D. The Wind Speed Probability Distribution

Having the posterior probabilities on the hypotheses and the percentage change in wind speed distributions given the hypotheses, we can multiply and sum to obtain the probability distribution on percentage change in wind if a hurricane is seeded. This distribution is plotted in complementary cumulative form in Fig. 6. It is also shown for a nonseeded hurricane; this is simply the complementary cumulative of $P_n(w | H_2)$. Note that for any wind speed you choose, the probability that a hurricane will exceed that wind speed is always greater for an unseeded than for a seeded hurricane. We say that the seeding alternative stochastically dominates the alternative of leaving the hurricane alone. As long as higher winds are worse, the logical man would prefer seeding to not seeding.

We must note, however, that seeded hurricanes can still intensify. For example, the probability that a seeded hurricane will intensify by 10 percent or more is 0.18, almost one chance in five. We choose seeding because an unseeded hurricane has an even higher probability of such intensification: 0.26.

E. Property Damage

To make the case complete we must examine property damage to find the economic consequences of wind speed reduction. Fig. 7 shows 21 instances of residential property damage caused by hurricanes with the associated wind speed. A least-squares fit produced an exponent of 4.363 in relating damage to wind speed. This implies that a 15-percent reduction in wind speed will cause a 51-percent reduction in damage. If seeding is beneficial, it can have a major economic impact.

Using the wind-damage function of Fig. 7, we can transform the distribution on wind speed of Fig. 6 into the distribution on property damage shown in Fig. 8. Once again we see that the seeding alternative stochastically dominates the alternative of not seeding. The probability of property damage exceeding

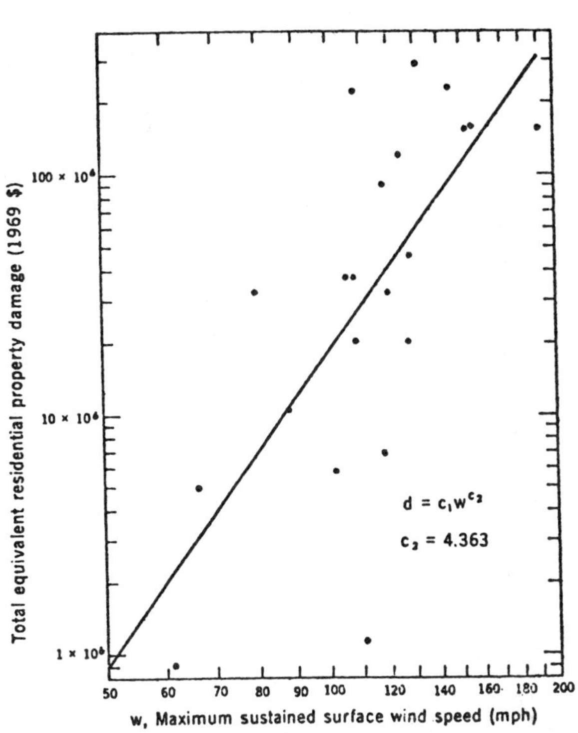

Fig. 7. Property damage plotted against maximum sustained wind speed.

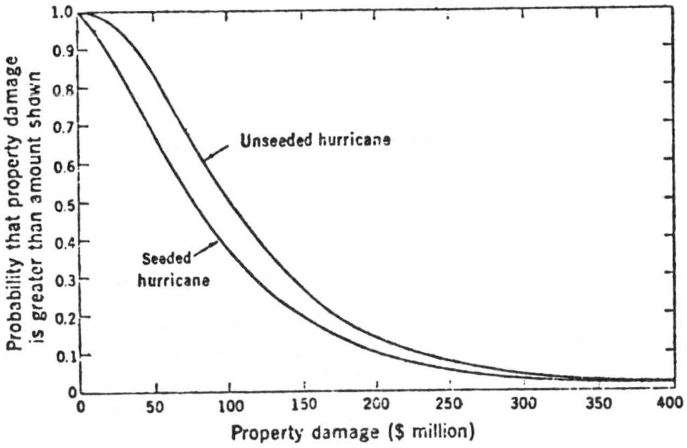

Fig. 8. Probability distributions on property damage for the seeded and unseeded hurricane.

any given level is always higher for the unseeded hurricane. If our objective is to minimize property damage, then seeding is the better alternative.

F. The Seeding Decision

A detailed description of the seeding decision for a nominal hurricane appears in Fig. 9. Here the two alternatives of seeding and not seeding lead to five-branch chance nodes with probabilities on change in maximum sustained wind obtained by discretizing the distributions of Fig. 6. The property damage associated with each change in maximum wind speed is shown at the tips of the tree. The expected loss for the seeding alternative is $94.33 million including the $0.25 million cost of seeding. The expected loss of not seeding is $116 million, or $21.67 million higher. Not only is seeding the better alternative, it is in an absolute sense significantly better.

But perhaps this result is very sensitive to the assumptions made. It does not appear to be. The probability of H_3 must

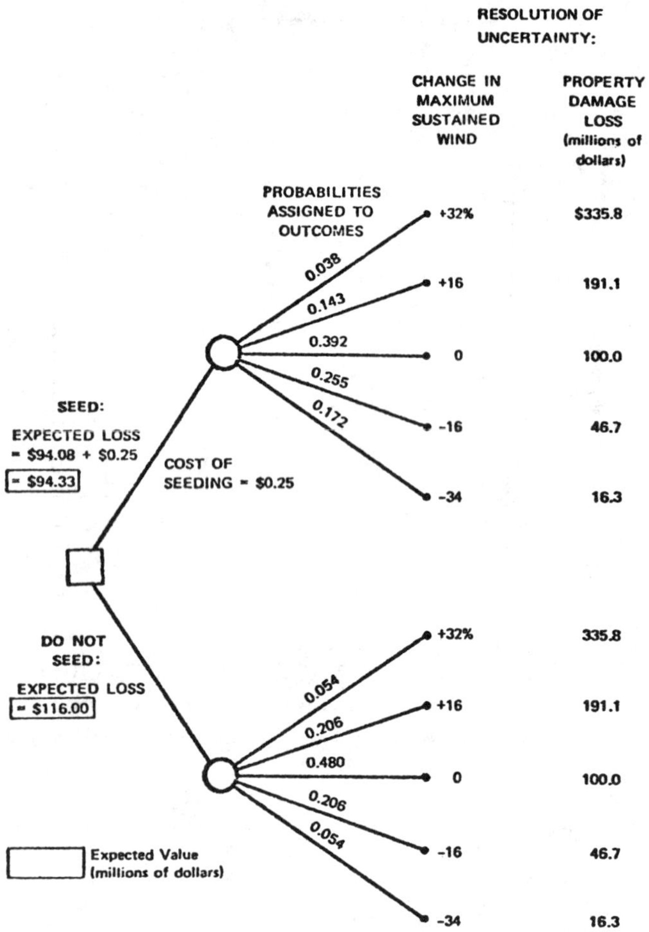

Fig. 9. The seeding decision for the nominal hurricane.

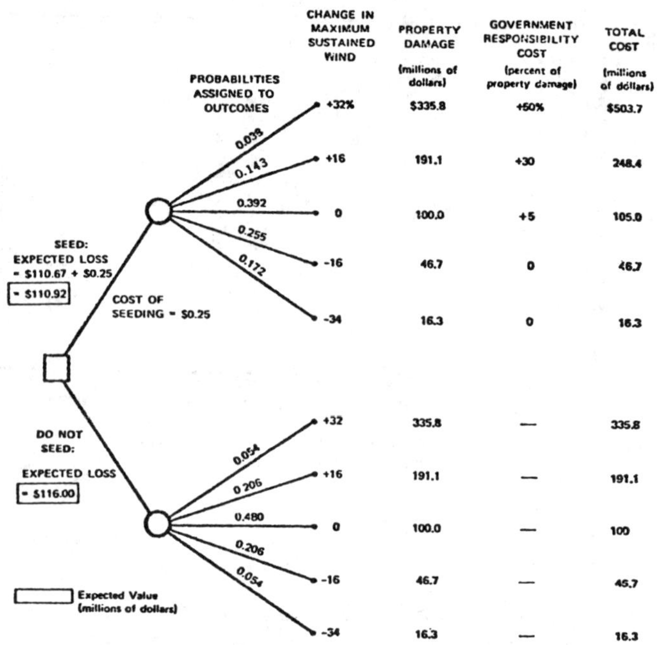

Fig. 10. The seeding decision for the nominal hurricane (government responsibility cost included).

be raised to 0.07 before stochastic dominance of seeding no longer holds. Even if it is raised to 0.20 seeding still has an expected loss 7 percent less than the not-seeding alternative. As stated in [5]: "the results of extensive sensitivity analyses may be summarized as follows: The expected loss in terms of property damage appears to be about 20 percent less if the hurricane is seeded. Varying the assumptions of the analysis causes this reduction to vary between 10 and 30 percent but does not change the preferred alternative."

G. Government Responsibility

If seeding appears to be so economically desirable, why is it not done? The answer must lie in the reluctance of government officials to expose the government and themselves to charges of tampering detrimentally with the weather. It would be surprising if government officials did not have such concern; if the concerns are large in comparison with the objective benefits to society, perhaps it is wise to forgo the benefits using the argument that the gain is not worth the trouble.

To see how large such concerns would have to be before hurricane seeding should be avoided, we introduced the concept of government responsibility cost. This cost is added to the economic cost to represent the trouble that the government would have in explaining apparent unfortunate consequences of seeding. We assigned it by asking, in each such case, what percentage increase in economic losses the official would be willing to incur without seeding before he would prefer to have the original losses and the necessity of explaining them if the storm had intensified by a given amount after seeding. For example, an official might say that he was just indifferent between $150 million in damage from a nonseeded storm and $100 million in damage from a seeded storm that had intensified by 32 percent after seeding. In this case, we would say that the government responsibility cost of seeding a storm that then intensified by 32 percent and produced $150 million of damage was 50 percent.

Fig. 10 shows the seeding decision with government responsibility cost included. A wind increase of 32 percent in a seeded hurricane is associated with a 50-percent increase in property damage in accordance with our discussion. An increase of 16 percent in wind speed for a seeded hurricane is assigned a 30-percent government responsibility cost. Even if seeding is followed by no wind change, a 5-percent government responsibility cost is assigned to account for charges that the hurricane would have diminished if the government had left it alone. Cases where the wind actually decreases after seeding are assigned a zero government responsibility cost, since we would expect few complaints about the decision in this case.

No government responsibility costs are assigned if the storm is not seeded, although there is an argument for doing so. Given that seeding is promising but uncertain, is it not just as irresponsible to refrain from seeding a storm that subsequently intensifies as it is to seed it? Nevertheless, since errors of omission seem to attract less notice than errors of commission, no government responsibility costs are associated with a refusal to seed.

Analysis of the tree in Fig. 10 shows that in spite of the relatively high values for government responsibility costs, the seeding alternative has an expected loss less than that of not seeding. The political costs must be considerably higher than those assumed if the model is to recommend against seeding. In fact, for a hurricane in the $1 billion class, like Betsy (1965) and Camille (1969), government responsibility costs would have to be equivalent to $200 million in property damage before seeding should be avoided.

TABLE VI
SUMMARY OF VALUE OF ADDITIONAL INFORMATION ON EFFECT OF SEEDING[a]

Item	Nominal hurricane used in analysis	%	Single hurricane Season	All future hurricane seasons, discounted
Expected Property Damage without Seeding	116.0	100	220.0	3142
Expected Value of Perfect Information	13.6	11.8	26.0	370
Expected Value of a Field Experiment Consisting of Two Experimental Seedings	5.4	4.7	10.2	146

[a]Only the 50 percent of hurricanes that are assumed to be possible candidates for seeding on the basis of tactical considerations are considered. If all hurricanes are assumed to be candidates for operational seeding, the figures of the last two columns should be doubled. All figures are in millions of dollars.

H. Value of Information and Experimentation

One of the most valuable aspects of decision analysis is its ability to place a monetary value on acquiring information on the uncertain aspects of the decision problem [2]. Ideally, the information would be perfect, producing complete resolution of the uncertainty. Practically, information is provided by tests or experiments that only partially resolve uncertainty.

A value of information analysis was performed on the hurricane seeding decision with the results shown in Table VI. The expected value of perfect information on the effectiveness of seeding is $13.6 million when considering whether to seed one nominal hurricane, $26 million for each hurricane season, and $370 million for all future seasons when discounted at a 7-percent rate. The value of another Debbie-size experiment, which would provide partial information, is $5.4 million for the nominal hurricane, $10.2 million for the season, and $146 million in present value. If the question were whether or not to seed a hurricane in the billion-dollar class, the experiment would be worth about 10 times as much as for the nominal hurricane, or about $50 million for that decision alone. Clearly there is a high value in conducting the hurricane seeding experiments.

I. Conclusion

This analysis served to illuminate the hurricane-seeding decision. The decision cannot be avoided: either the government must assume responsibility for seeding and hence for what appear to be detrimental effects of seeding or it must assume responsibility for not seeding and thus exposing the public to greater losses from hurricanes.

VII. NUCLEAR POWER PLANT SAFETY

The question of nuclear power plant safety offers an important illustration of the use of decision analysis in considering the general issue of social safety.

A. Safety

The question of safety is an aspect of virtually all large human systems. Whether in energy production and distribution, transportation, manufacturing, or space exploration, everyone prefers more safety to less. The issue is not whether any system can be made more safe—the answer is almost always "yes." The question is how to find a proper balance between the benefits and costs of the entire system, including the costs associated with safety.

To carry the discussion one level further, we might try to divide the safety costs into the costs of building and maintaining the safety-ensuring systems and the costs incurred from accidents. However, it is not always easy to identify safety system costs per se as separate from total system costs. For example, while the addition of seatbelts in an automobile could be considered as primarily a safety investment, an improved braking system may enhance both safety and performance: its costs cannot reasonably be considered solely an investment in safety.

A more logical way to proceed is to compare total system benefits less total system costs for each alternative system design. This comparison, while simple in principle, is challenging in practice because many system effects are particularly difficult to detect and evaluate. For example, if we consider the alternative of raising the gasoline tax as an energy conservation measure, we find many potential safety effects. If the increased tax causes less traveling, that would presumably increase safety. There would likely be less transport of dangerous gasoline by tank trucks—another diminished hazard. But would people switch to smaller, more vulnerable cars? As you can see, the net effect on safety of such an alternative poses difficult analysis problems.

Now let us return to the accident cost side of the safety evaluation. When, in spite of safety devices, an accident occurs, how should the social cost be measured? The accident could cause death, injury, sickness, property destruction or loss of use, aesthetic impairment, etc. One approach to evaluation is to assess a monetary value to each unit of these consequences, multiply by the number of units, and sum to obtain the total social costs of the accident. While some might see this procedure as callous, it is important to observe that we are not attempting to value a human life or injury in a moral sense, we are rather assigning values to life and the other loss categories as a means of making social decisions. The values are not measurements, but the results of a decision made by society concerning the balancing of social benefit and social risk. By assessing these values in monetary terms, we are able to determine the monetary amount that should be spent on safety systems, and at a higher level, to choose between entire systems.

To demonstrate, consider two possibilities for our society. In the first we assign a $1 thousand value to the consequence of a human death resulting from an involuntarily assumed risk. In the second we assign a $10 million value. The first assignment would lead to decisions regarding aircraft, roads, fire pro-

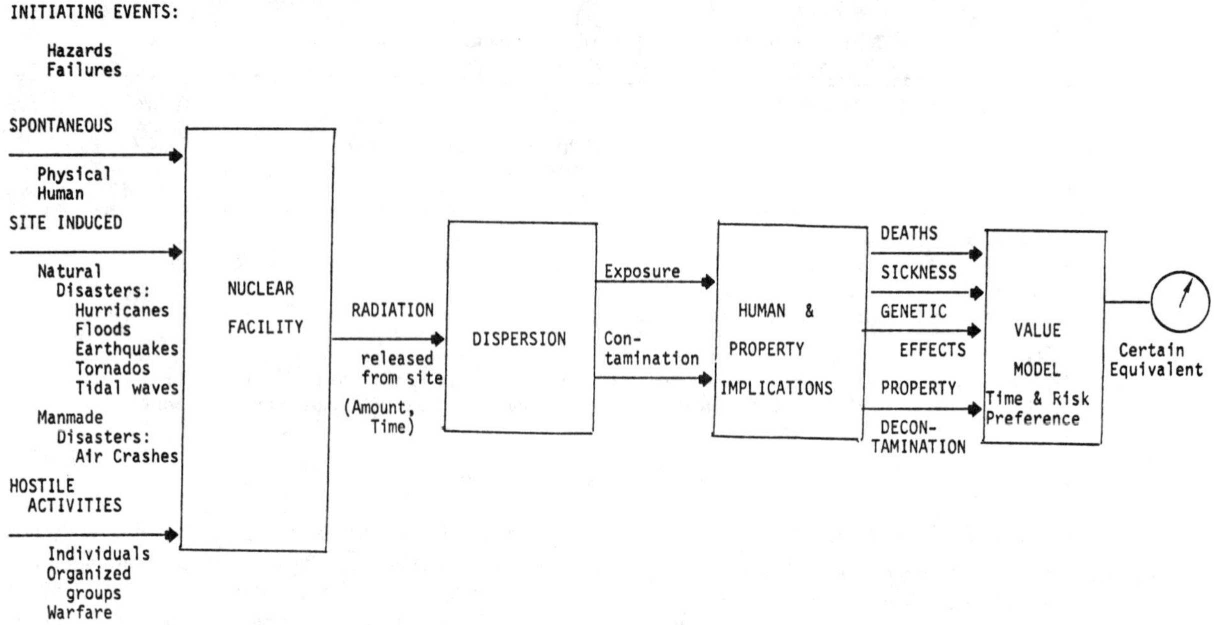

Fig. 11. Nuclear power plant safety—accident model logical structure.

tection, etc., that would soon leave us a nation of the afraid. A popular clamor would arise for "safer" social alternatives. On the other hand, the $10 million assignment would lead to decisions requiring huge sums to be spent on safety equipment. The price of airplane tickets, cars, and electricity would grow until the population again complained about the "waste" of money on "unnecessary" safeguards. Thus the problem is to assign a value to human life (and to injury, etc.) that reflects what we can afford to pay as a society to avoid a death from involuntary risk. Various analyses may be helpful in assessing this value, but in the end, it is a policy decision for society.

Therefore, a logical way to look at the safety problem in context is to select the alternative with the greatest excess of benefits over costs, where the costs due to accidents are evaluated using value assignments made by a social decision process. While the safest system is the one with the lowest accident cost, the most desirable system for society will seldom be the safest.

We shall now present a procedure for evaluating the accident cost of a nuclear power plant. As we have shown, safety is an inseparable part of the process for choosing between system alternatives, such as choosing between fossil and nuclear power generation. Consequently, a safety analysis cannot, by itself, logically determine whether nuclear power generation is a desirable alternative for our society, or even if it is "safe enough" compared to other energy conversion systems.

B. Methodology—Nuclear Power Plant Safety

Assessing nuclear power plant safety may be discussed conveniently with the aid of Fig. 11. Any particular facility is exposed to various hazards. These hazards could be spontaneous failures, such as physical failure of a piece of equipment, or a human error caused by incompetence, inattention, or incapacity. There could also be failure because the reactor is at a particular site. These site-included failures, in turn, could be caused by either natural disasters like windstorms, floods, or earthquakes or by man-made disasters, like airplane crashes. Finally, the reactor might be the target of hostile activities by individuals or organized groups.

It is possible that as a result of those hazards, the nuclear power plant will suffer damage. The question regarding public health and safety is whether a radioactive release from the site occurs, for such a release could be the cause of public loss. The release from the site could be described by the amount of radioactivity of each type emitted in each future time interval.

For a given release, the magnitude of the loss will be governed by the water, by the subsequent spatial and temporal distribution of population in the surrounding area, and by the land use patterns in the area. The dispersion model thus utilizes meteorological, geographic, and demographic information to predict the human exposure and property contamination produced by a given release.

Since humans suffer from differing types and amounts of radiation in different ways, a human implications model is necessary to assess deaths, sickness and genetic damage likely as the consequence of a given exposure. Similarly, the cost of decontaminating or burying property infested with radioactivity is assessed in a property implications model.

Finally, the value model reduces to a single value measure the combined societal evaluation of both human and property losses, in accordance with the discussion of the last section.

C. Model Implementation

The logic of Fig. 11 can be conveniently incorporated in a probability tree. Successive chance nodes of the tree would characterize such uncertain possibilities as the magnitude of an earthquake, the amount of release, the extent of contamination, and the number of deaths. Experts on the various stages could assign probabilities either directly, on the basis of the analysis of data (e.g., records of pipe breakage), or by exercising subsidiary models (e.g., meteorological dispersion simulations). The net result would be a probability distribution on the magnitude of the annual loss that could be produced by a single reactor, or, for certain purposes, by all reactors.

Studies of this kind have been performed by various agencies, but always with serious limitations. They have tended to emphasize the physical causes of accidents and to stop short of considering the value implications of the accident consequences. Studies of the effects of hostile activity ranging from disgruntled employees to vulnerability in time of war have been conspicuously rare. Yet there is no reason why such hazards cannot be analyzed within the same framework used for physical malfunctions. In fact, as improved engineering reduces physical hazards, it becomes increasingly likely that if a serious power plant accident should occur, the cause will be found in human intervention.

D. Summary

Decision analysis models provide a comprehensive and practical framework for investigating safety effects. In choosing between two forms of power generation, safety system costs and uncertainties can be balanced against economic and social benefits. The value of further experiments and the costs of delay can be computed and compared to evaluate a proposed course of action. The controversies that permeate licensing hearings could be reduced to their simplest issues, thus offering the hope that energy now devoted to argument would be applied to resolution.

VIII. Conclusion

We have now seen how the use of extrapersonal models provides a logical and open procedure for making social decisions. Various stakeholders can offer their views and then observe how their contributions would affect the choice of alternatives. Anyone could suggest an alternative, all citizens, either individually or in groups, could influence the values used, and experts would be able to supply information within the realm of their expertise.

The net result would be a more distributive and more effective form of government. We would see more debate on the fundamental issues of the society, rather than on individual laws. For example, the question of whether society has the right to protect mature competent citizens from the consequences of their own actions would have to be faced forthrightly rather than skirted.

A government of law that is not logically based on the information and values of the society is just as arbitrary in its effect on individuals as a government of men.

References

[1] R. A. Howard, "Decision analysis: Applied decision theory," in *The Proceedings of the Fourth International Conference on Operations Research.* New York: Wiley, 1966, pp. 55-71.

[2] —, "The foundations of decision analysis," *IEEE Trans. Syst., Sci., Cybern.*, vol. SSC-4, pp. 211-219, Sept. 1968.

[3] —, "Decision analysis in systems engineering," in *Systems Concepts, Lectures on Contemporary Approaches to Systems.* New York: Wiley, 1973, pp. 51-85.

[4] R. A. Howard and B. R. Judd, "A pilot model for evaluation of economic incentive and direct control policies for limiting automobile air pollution," report for the National Science Foundation under Grant GK-36491, Dec. 1973.

[5] R. A. Howard, J. E. Matheson, and D. W. North, "The decision to seed hurricanes," *Science*, vol. 176, pp. 1191-1202, June 1972.

A Working System for Social Service Delivery

JAMES D. PALMER, SENIOR MEMBER, IEEE

Abstract—The design of a system to provide for delivery of social services to clients and accountability to the state is presented. An introduction to actual systems problems in social services delivery is provided together with the human interaction processes necessary to bring successfully the systems approach to social service delivery. An analysis of case histories has been accomplished to determine common characteristics among actual cases which leads to a classification system for multiproblem families relative to family functioning. A detailed description of the system and the flow diagrams of information, decision, and activity (IDA) points is given. The system is in and working in Lynchburg, Va. and is to be installed and completed in Chesapeake, Va., by the end of 1974. Typical displays of the information system outputs as provided to case workers and supervisors are shown.

Manuscript received March 2, 1974; revised October 2, 1974.

The author is with the Metropolitan State College, Denver, Colo. 80204.

INTRODUCTION

SOCIAL SERVICES comprise a major portion of the total welfare program of this country amounting to more than $11 billion annually and growing at a rate equal to appropriations growth with no limit in sight except as individual states and localities establish arbitrary limits to spending. The delivery of social services; e.g., housing, prenatal care, job training, medical services, counseling, etc., is generally fragmented at the agency level and throughout state government. It is not unusual to find several state-wide agencies providing the same or similar services or for local agencies to be in competition with other local agencies or be unaware of private agencies providing the same services to the same clients. An examination of this complex situation from the systems point of view using the terminal objective of delivering a needed

Reprinted from *Proc. IEEE*, vol. 63, pp. 451–463, Mar. 1975.

service to a client has resulted in a generic system for social service delivery which has been implemented in two agencies. The features of the generic system include immediate emergency service, community and state resource appraisal and evaluation, assessment of client needs and evaluation of the outcomes in behavioral terms. This system is unique in that it retains the human interaction between caseworker and client, while holding up for accountability the use of resources.

The systems approach taken is supported by a management information system which provides complete data on any of the system components to the agency supervisors and directors. Through case-load assignments and queing all clients are assigned to a case worker or are reviewed periodically. This process eliminates case carrying and assures an equitable load distribution for all case workers and most importantly assures each client of service on a regular or periodic basis. Through the assessment of client situation, goals are established for services. Following the development of a service plan prepared by the case worker in consultation with the client, a service delivery package is determined and services made available to the client as appropriate. Evaluation of the effectiveness of the service plan and the services provided is performed at agreed to dates following the service delivery. Actual time may range from a few days for housing, to several months for counseling. The evaluation, as the assessment, provides for establishing client progress or need in behavioral terms so that changes may be determined.

Information concerning existing needs are readily available and a measure of change in clients attributable to the delivery of services becomes a reality. For both state and federal government the ability to assess the efficacy of social services is possible for the first time. The capability to plan for and allocate resources to an agency and measure the outcomes in client movement as well as determine cost is now possible. The prime benefactors are the clients to be served through a planned set of services and the taxpayer who for the first time will have the ability to hold the social service delivery system accountable.

The discussion which follows covers the necessary process aspects of developing a social service delivery system, the development of the generic model of the client/case worker interface activity and the resulting system including samples of the input to and output from the management information system. The system is in full operation in the Lynchburg, Va., and by December 1974 in the Chesapeake, Va., welfare agencies and covers approximately 500 and 7500 welfare cases, respectively.

Social Service Background

Social services, as used in this paper, refers to any service provided to a client of a local social service (or welfare) department, where the services are partially or wholly paid for by federal funds channeled through the Division of General Welfare of Virginia's Department of Welfare and Institutions. Social services range from so-called "hard" services such as the provision of assistance in finding adequate housing to "soft" services such as counseling of clients to assist them in achieving greater self-sufficiency in managing their own lives.

The federal enabling legislation, the Social Security Act of 1933, did not explicitly allow for the provision of social services. It provided funds for financial assistance payments only. However, in addition to their *primary* task of determining eligibility for assistance payments, in a series of gradual steps over the years social workers began providing a wide variety of social services.

The assumption and/or assertion of the professionals in the field was that social services were a necessary adjunct to money payments in order to move clients to a greater degree of self-sufficiency and, on the average, to reduce their dependency on monetary assistance. But, throughout the life of the program, there has been little clear-cut evidence that these claims have validity. There has not even been consensus among professional social workers as to the criteria by which "self-sufficiency" should be measured. In the face of what seems to be overwhelming negative evidence (i.e., the more money spent on social services, the greater the claims for monetary assistance), there has been a growing clamor for the elimination or at least severe reduction in social services offered. Public officials, strongly affected by growing public clamor, have raised serious questions about the value of social services. Also, client groups, viewing some types of social services as attempts to maintain the welfare recipient in a dependent and subservient role, have demanded the elimination of all social services except those defined as "hard" (where "hard" means a service which could be purchased by the client if only he had enough money).

Social Services Delivery Objectives

The original objective in this project was to develop a *set of criteria* for evaluating the effectiveness of social services in moving the client to a greater degree of self-sufficiency. This was expanded to include the development of a *methodology* for evaluating all social services. This latter required the *entire system* of delivering and of managing the delivery of social services to be reoriented, reorganized, and redesigned which would result in improved efficiency and effectiveness in meeting the goal of the Department. For purposes of conceptualization and management, the Social Services System is addressed as three subsystems.

Management subsystem.
Information subsystem.
Delivery subsystem.

The delivery subsystem covers all activities of the service worker as he/she is directly or indirectly engaged in interaction with the client. The delivery subsystem includes the capture of all raw data needed for case management, program management, and "agency" management (at local, state, and federal levels).

The information subsystem processes the raw data captured in the delivery subsystem and stores it for use as management needs for information are established.

The management subsystem has to do with all aspects of local, regional, and state-wide managerial and executive functions associated with the delivery of services and general welfare.

A management study of all major divisions of state government conducted in 1969 clearly showed the need for accountability in all areas of general welfare, particularly in the programs of categorical aid and payments. The study also identified one specific area in which no control, evaluation or accountability was to be found—this area was social services. The response by the Department of Welfare was to develop Virginia uniform welfare reporting system (VUWRS) which was to provide all systems necessary for management and accountability of the welfare programs. VUWRS worked reasonably well in the areas of payments and categorical aid, but did not function in the area of social services and did not recognize that this was a total systems problem.

Problem Statement

In the words of Russell Ackoff the delivery of social services viewed as a system was in fact a *MESS*. There was no way to approach the situation from the traditional point of view and reduce it to a series of manageable parts. At the outset it was clear that to develop a system it would be necessary to make explicit the process involved, the problems and the goals. The major problem lies in the delivery of *needed social services, as available*, to clients in the best possible manner leading the client toward "self-sufficiency." Complicating this was the need for *complete human interaction* between client and case worker. The major problems were defined as follows.

1) The process of delivery of social services was not functioning in a deliberate controlled manner.

2) The welfare client was not well served by the delivery of social services.

3) Accountability was completely missing, due in part to a confusion of funding sources and in part as to what constituted a completed service.

4) Evaluation was not being done.

5) Any system developed to achieve accountability and evaluation *must have* the *support* of the *clients* and *case workers* as well as management at all levels.

6) Costs associated with social services were accelerating at a rate which would consume the entire annual tax base of the Commonwealth of Virginia within ten years.

7) There existed no "case analysis" information from which to judge whether the services provided assisted the client toward self-sufficiency.

8) No resource appraisal or accessibility was available to workers and most resources were not known to clients.

9) Management functions were not being performed.

Systems Approach

From the outset, the needs were recognized for a fast access data base/management information system to support the delivery of social services. In addition, it was necessary to provide for the management of resources and to respond to questions from all levels of management including the governor and legislature.

In practice the concept of service delivery was individualized to the extreme. As a consequence, management was in essence coping with the day-to-day case worker/client problems and control was largely achieved through maintenance of a "full" schedule of clients otherwise termed "case carrying."

In order to bring about changes in the delivery of social services there existed a need for all involved to *participate in the development* of any such system or social service delivery program. An additional complication that had considerable force influencing case workers was the necessity to continue to serve existing case loads and to meet existing reporting requirements both at state and federal levels.

The approach taken was one of total involvement in an open environment leading to the "bottoms-up" or worker/client action interface concept. The "opening" of the Agency caused a great many internal questions, challenges and trauma that moved to the very essence of the social service delivery program, i.e., why are social services to be provided. Once it became possible to discuss service delivery and management style, it became possible to make explicit the goals for service delivery. What evolved was the development of a dynamic adaptive system for service delivery and the explicit statement of short and long term client goals.

Information-Decision-Activity Analysis

The systems analysis was initiated in the Lynchburg Agency with the on-site development of an information-decision-activity (IDA) analysis of all worker/client action interface functions. The process was one of making explicit each function, each point where a decision must be made, each time a resource was called for and each place where information was either generated or required. From this, it became apparent that the workers employed a sophisticated pattern recognition function in establishing service plans for the client.

Pattern recognition was assumed to be a basic underlying principal which, if made explicit, would lead to the fuller definition and understanding of service delivery.

The basic vehicle of involvement, pattern recognition, adaptive/responsive design and bottoms-up approach was the IDA analysis work over time. The initial outcomes of the IDA analysis resulted in the clear understanding of the manner in which the Lynchburg Agency provided social services. The IDA maps provided three basic components·

1) the decisions made by the various staff within a particular service sequence;
2) the information used or needed to make a particular decision;
3) the activity caused by the decision made.

The service sequences examined for the IDA maps were as follows.

1) Child welfare services to unwed mothers.
2) A.D.C. services–medicaid services.
3) Adult services.
4) Adoptions (parent approval process).
5) Adoptions (child selection process and placement).
6) Foster care.
7) Service only.
8) Infant and older child adoptions.

Following this initial identification of the service sequence IDA's, the next step was to assemble these functions into service delivery modules which would identify and separate generic functions from agency style. It is important to note that the entire process was undertaken through *involvement of the case workers*. This has been done in a manner which insures the continuing participation of the case worker in a system of his/her design which in turn has insured the emergency of a design that is implementable at the case worker/client action interface.

Simultaneously with the IDA process, a full inventory of the community resources was established to satisfy the need of the case worker for this information and to outline agency, region and state information requirements in this area.

An example of the outcomes of the IDA process in the area of adult services is shown in Fig. 1.

Case Analysis

The 1963 Federal Guidelines for evaluation and assessment of the effectiveness of the delivery of social services had as a basic requirement, the development of base-line information. The underlying assumption stated succinctly was that unless a

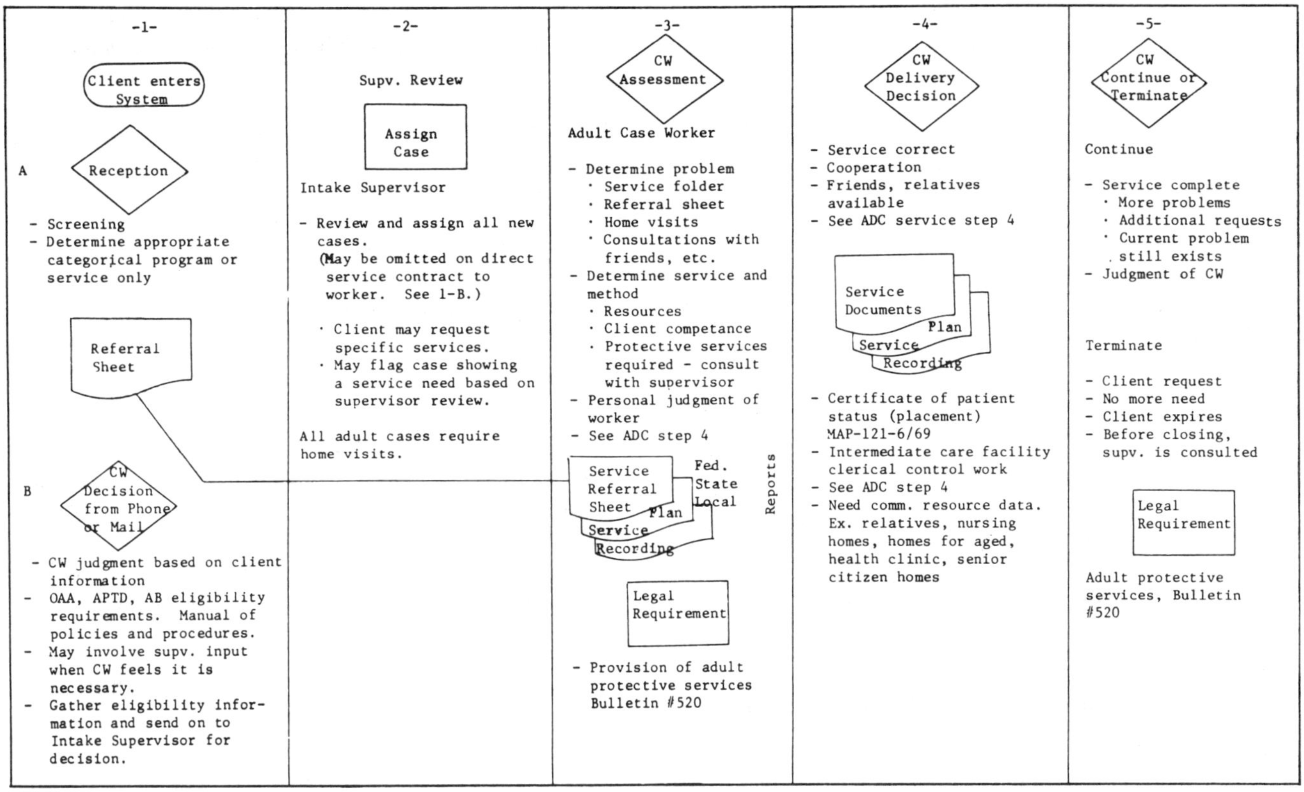

Fig. 1. Adult services–information decision activity.

determination of some index of family functioning was available there would be no ability to measure change as a consequence of receipt of services.

To meet this need and fulfill the basic requirements for evaluation and assessment of the delivery of social services a major effort was undertaken simultaneously with the systems design to develop a set of criteria through which case classification could be established. The program was initiated by requesting of each case worker the analysis of four cases in response to a common questionnaire. Some 3200 cases were analyzed on the basis of the questionnaire.

The major goal of the case analysis project was to isolate and describe discrete characteristics of family functioning from among families receiving social services. A statistical typology of social service recipient families was derived with 16 major groupings identified and characterized by a profile of family functioning difficulties ranging from adult behavior problems to unmet housing needs. The group profiles thus derived provide a statistical mapping of the social service caseload. In order to be useful and used, the typology was reduced to a size and complexity manageable by the case workers on a day-to-day basis. This analysis of family typology showed that nearly all the important problem combinations could be related one to the other in a way which enabled classification of families with virtually no overlap. From this study, four summary measures of family functioning emerged: 1) adult behavior; 2) child behavior; 3) health; and 4) unemployment due to lack of skills. These are the key to the classification scheme of recipient families. The process involves only nine categories and requires no complicated statistical calculation from the social worker, rather only the presence or absence of a particular problem in each area. Table I shows the summary of the major variable combinations and Table II shows the frequency

TABLE I
MAJOR SUMMARY VARIABLE COMBINATIONS

#	Frequency	Adult Behavior Problem	Child Behavior Problem	Unemployment Due To Lack of Skills	Health Problem
1	20.2%	x	x	x	x
2	19.8	x	x		x
3	6.6	x	x		
4	4.0	x		x	x
5	4.6	x			x
6	4.0		x	x	x
7	12.9		x		x
8	10.6		x		
9	4.5				

TABLE II
FREQUENCY OF OCCURRENCE OF SUMMARY VARIABLE COMBINATIONS IN EACH MAJOR GROUPING

COMBINATION	PERCENTAGE OF OCCURRENCE IN GROUP		
	A (DOMINANT GROUP)	B (NEXT GROUP)	C (RESIDUAL)
1	7.3%	40.2%	35.8%
2	10.9	40.2	22.5
3	7.5	1.3	7.5
4	1.6	10.0	7.5
5	3.4	6.7	2.5
6	5.1	-	6.7
7	21.8	-	1.7
8	17.8	-	1.7
9	6.7	-	-

TABLE III
CLASSIFICATION OF GROUPS A7, A8, A9

IF	→ THEN	%
Adult Behavior Problems: absent Child Behavior Problems: absent Skills Shortage : absent Health Problems : absent	A9	3.6
Adult Behavior Problems: absent Child Behavior Problems: present Skills Shortage : absent Health Problems : absent	A8	9.0
Adult Behavior Problems: absent Child Behavior Problems: present Skills Shortage : absent Health Problems : present	A7	11.7
	Total =	24.3%

TABLE IV
PROFILES OF FAMILY FUNCTIONING AMONG GROUPS B4 AND C4

	B4	C4
ADULT BEHAVIOR PROBLEMS	YES	YES
Adult Irresponsibility	80%	89%
Money Management Problems	93	33
Emotional Care of Children	53	44
CHILD BEHAVIOR PROBLEMS	NO	NO
HEALTH PROBLEMS	YES	YES
Adult with Functional Health Problem	93	56
Child with Functional Health Problem	7	22
Unmet Health Needs	67	56
Unemployment Due to Poor Health	40	44
UNEMPLOYMENT DUE TO LACK OF SKILLS	YES	YES
UNMET HOUSING NEEDS	53	44
INCOME BELOW $3000	57	44

TABLE V
CLASSIFICATION OF GROUPS B4 AND C4

STEP 1:

Adult Behavior Problems:	present
Child Behavior Problems:	absent
Skills Shortage :	present
Health Problems :	present

STEP 2:

IF	→ THEN
There are money management problems and there is an adult with a functional health problem.	B4
Otherwise	C4

of occurrence as determined from the analysis of 3200 cases. From Table II it can be seen that variable combinations 7, 8, and 9 allows the classification of three subgroups and will account for nearly one-quarter of all families with children.

What remains are the combinations 1–6. As an example to understand the process, consider two subgroups (B4 and C4) which have the same variable combination. Each of these exhibit problems in the area of adult behavior and health. They both have family members unemployed due to lack of skills, but neither have child behavior problems.

While the two subgroups have the same summary values, it can be shown that there are important individual differences. For example, while almost all the families in B4 have money management problems, only one third of the families in C4 do. Similarly, a family in B4 is almost twice as likely to have an adult with a functional health problem. The question for the case worker is not a proportional value, but the presence or absence of a problem. Certain combinations are much more likely to occur in B4 families than in C4 families. Thus a family with a serious money management difficulties *and* an adult with a functional health problem has a very good chance of being classified as a member of group B4. One way to estimate these chances might be as follows.

> If we assume the presence of money management problems and adults with health problems to be statistically independent, the proportion of B4 families with both will be 87 percent (0.93 X 0.93). Similarly, C4 families will have this combination only 19 percent (0.33 X 0.56) of the time. In other words B4 families will be over *4* times as likely to have both problems.

This process has been applied to each of the various combinations. A partial summary of the results is shown in Tables III–V to give an indication of the power of this methodology in establishing base-line information on family functioning from which to base a measure of change as a consequence of receipt of social service.

SOCIAL SERVICE DELIVERY SYSTEM GENERIC DESIGN CUSTOMIZED STYLE

The design is based upon generic functions common to all service industries and is structured according to the general information, decision, and activities inherent in any service delivery program. It is elastic in scope and flexible in focus and is adaptable to the characteristic style of any human resources project.

OPERATIONAL VISIBILITY-BUDGET ACCOUNTABILITY

The people who work to delivery social services want their working activity to be seen. They want to display what they are doing, what they accomplish, and the resources they have to use. Those who pay for the delivery of social services want to know how the funds they allocate are used and what results are obtained and they want to be able to decide if the results are worth the cost. The system design satisfies both sets of requirements. It takes advantage of the desire for operational visibility and provides a basis for budget accountability. The focus on the need for information of the persons working to deliver services provides evaluative and planning information generated as a byproduct which satisfied the requirements of the management functions which support the delivery process.

STANDARD DATA DESCRIPTION: CLIENT, OPERATIONS, AND RESOURCES

The system design provides for the systematic collection of standard data to describe the characteristics of the clients using the system, the operations performed to deliver services to these clients, and the resources required and used for delivery. This data is processed as needed to gain information to make decisions for the control and support of service delivery activities.

MODULAR DESIGN: EIGHT BASIC SYSTEM FUNCTION (EIGHT DESIGN MODULES)

The design is conceptually organized into eight modules which represent major functions of the service delivery process. Each acts as a convenient focus for organizing the clusters of decisions and activities needed to meet specific intermediate

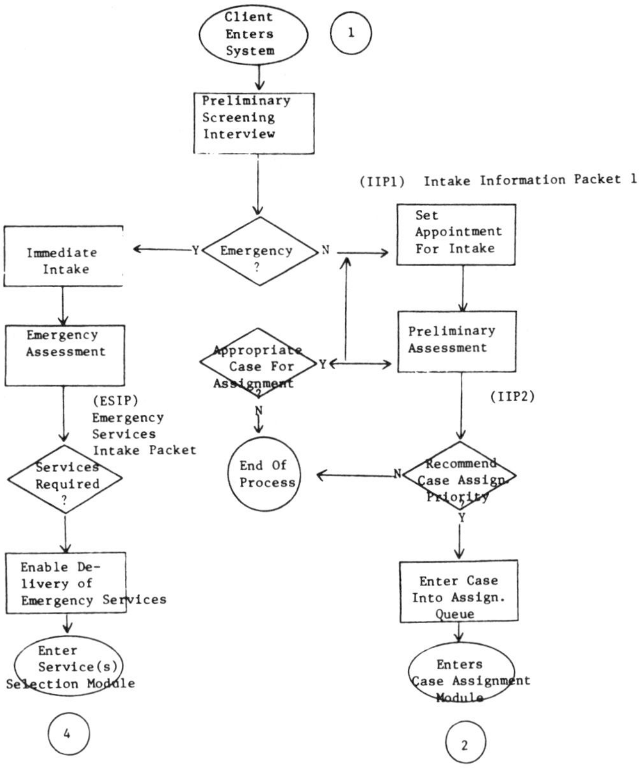

Fig. 2. Intake module.

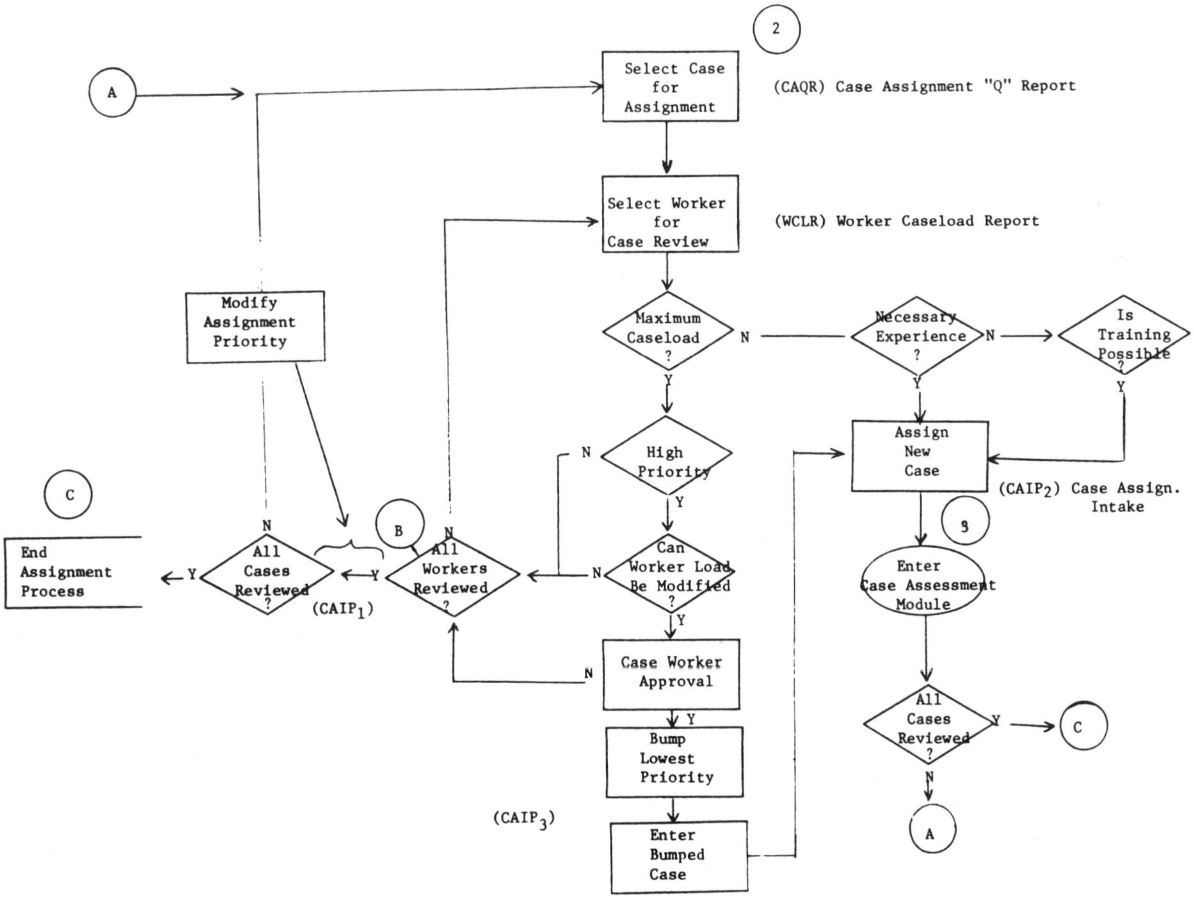

Fig. 3. Case assignment module.

system objectives. The eight modules are as follows.

1) Intake.
2) Case assignment.
3) Assessment.
4) Service selection.
5) Resource appraisal.
6) Service delivery.
7) Service impact evaluation.
8) Case disposition.

The name of each module is intended as a brief description of the major function performed in its operation. No hard and fast sequence of operation is implied other than what is obvious to common sense. Various mixes of function can occur and concurrent operation is natural to several of the modules. For example, assessment might continue during service delivery, etc. The system is "soft shouldered," in that it is designed to adapt to the requirements of individual service delivery situations.

Intake

The purpose of the intake module (Fig. 2) is to facilitate the client's initial use of the service delivery system and establish a priority for that use. One of the unique characteristics of the system design is its explicit concern with priority in the use of resources. This concern starts as the client enters the system. Programmed, concentrated attention is given first to the detection of emergency problems in the client's situation.

Intake Functions

1) Client enters system.
2) Collect case information.
3) Inform client.
4) Detect emergencies.
5) Establish case priority.
6) Refer client.
7) Document client entry.

At intake, the client is introduced to the service delivery system. The information needed to perform this introduction is collected and the client is informed of the services that he may expect to obtain by using the system and how they are delivered. He is given the opportunity to accept or decline these services. If the client elects to receive service, a preliminary assessment of his situation is conducted. The purpose of preliminary assessment is to establish a priority for the client's use of the system and is expressed in terms of the urgency of the client's need for the resources of the system. Preliminary assessment involves the client in identifying his service needs whenever feasible. Documentation of the client's entry into the service delivery system at intake forms the basis for tracking the client's subsequent use of service resources and for managing the activities coincident with that use.

Case Assignment

The purpose of the case assignment module (Fig. 3) is to provide a systematic and visible way to manage the caseload of both individual case managers and the social service agencies which support them. The explicit concern with priorities for resource usage is apparent. Case assignment is a process based upon a designated priority for a client's assess to service resources which serves as a basis for assignment of a social service case manager to each client who wishes to use the service delivery system.

Case Assignment Queue

1) A social service case is either assigned to a case manager or queued for assignment.

2) Highest priority cases are always assigned first.

3) Where the agency caseload exceeds its case management capacity, its lowest priority cases reside in the case assignment queue.

The system design features a unique computer assisted procedure which enhances social service caseload management capability: the case assignment queue. A client whose case is queued for assignment is under the recognizance of the social service agency. While a case is queued, case visibility is maintained through a systematic and periodic scanning process. Client access to the agency is always available and the client is informed that his case is in queue.

Case Assignment Functions

1) Assign case.
2) Transfer case to queue.
3) Maintain caseload balance.
4) Monitor queued caseload.
5) Allocate casework skills.
6) Respond to case status change.

The system design provides the capability to perform these functions. To support this capability, reports are generated which display the information necessary for caseload visibility. The status and activity history of any case in the agency caseload is kept up-to-date and is available for immediate use.

Assessment

The purpose of the assessment module (Fig. 4) is to develop data which describes the patterns of characteristics in a client's situation that depict needs for specific social services. The design of this module promotes the setting of information objectives by the case manager for the direction of efforts to assess the client's situation. It provides for the evaluation of assessment contact based upon the extent to which these objectives are met and allows modification or reformulation of objectives as new information requirements emerge. It enables management and control of assessment contact.

Assessment Functions

1) Set information objectives.
2) Contact information sources.
3) Record contact results.
4) Evaluate contact results.
5) Search for diagnostic pattern.
6) Evaluate emergent pattern.
7) Establish baseline assessment.
8) Provide assessment visibility.

Another of the unique aspects of the system design is the replacement of case dictation as a function of case assessment. The technique that enables this replacement is based upon a battery of standard data elements for use in recording the results of assessment. Each element is in the form of an assessment question with multiple answers. To record a particular assessed characteristic, the appropriate data element is referenced and its value specified using a simple coding technique. It is then entered as input to the computer which supports the system. Display of assessment results is accomplished with computer assistance in the form of an English language statement which describes the assessed characteristic.

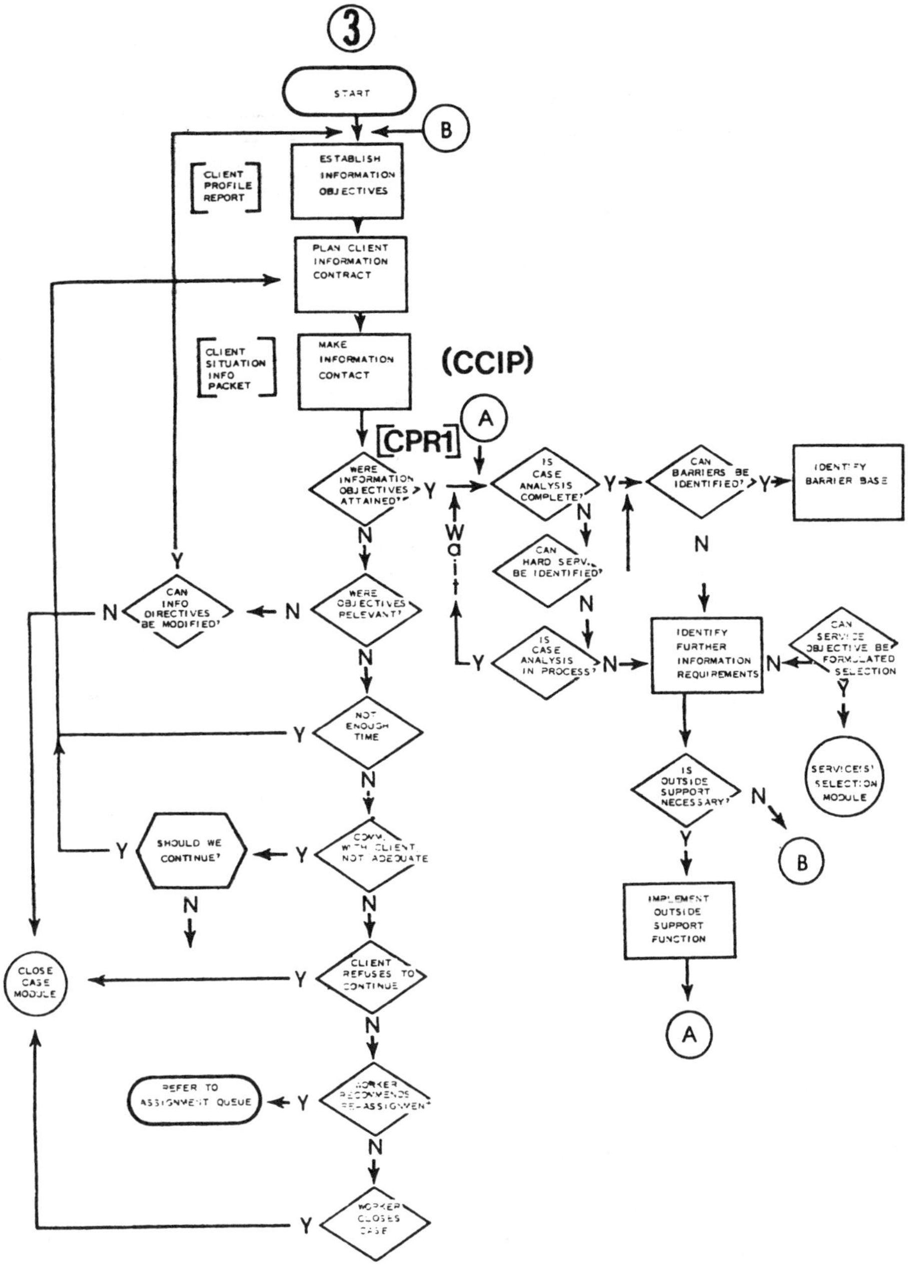

Fig. 4. Assessment module.

Service Selection

The purpose of the service selection module (Fig. 5) is to formulate explicit service delivery objectives based upon the results of assessment and to specify the services which would enable fulfillment of these objectives. The result of service selection is in effect the statement of a *service hypothesis*. It predicts in behavioral terms the change to be expected in particular characteristics of the client's situation as a result of the delivery of specified services, providing effective social service resources can be used.

Service Selection Functions

1) Formulate service delivery objectives.
2) Specify services for delivery.
3) Specify service delivery sequence.
4) Document service hypothesis.

The formulation of service delivery objectives is *directly* based on the results of assessment. Characteristics of the client situation, recognized in the assessment process as making up a problem pattern, are used to define service targets. This definition is expressed as the modification or maintenance of the assessed value of the specific characteristic.

The set of service targets to be reached through the delivery of particular social service constitutes a service objective. All service objectives established for a single service case express a service delivery hypothesis. It remains, then, to join this hypothesis to social service resources that can be used to deliver the specified services, in order to express a service plan.

Resource Appraisal

The purpose of the resource appraisal module (Fig. 6) is to identify the social service resources needed in order to deliver selected services and to appraise those resources that can be

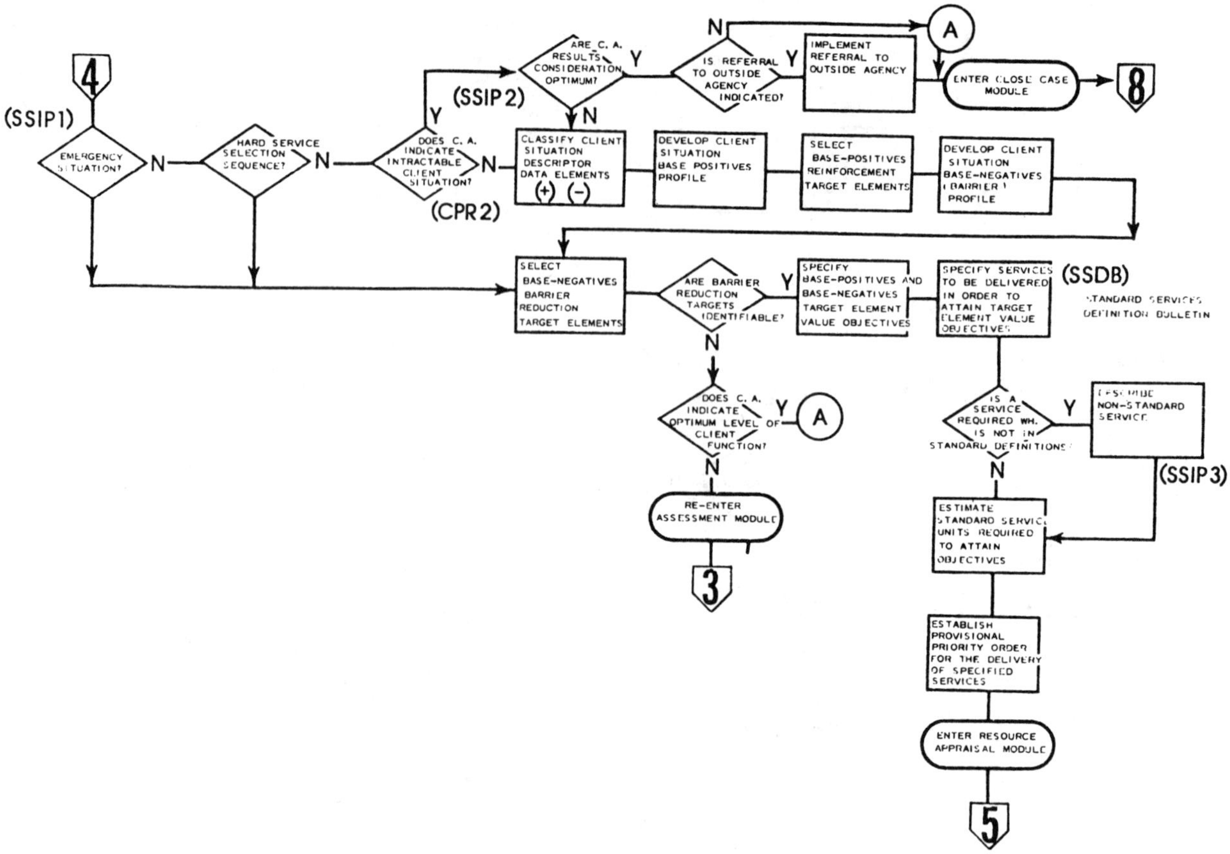

Fig. 5. Service selection module.

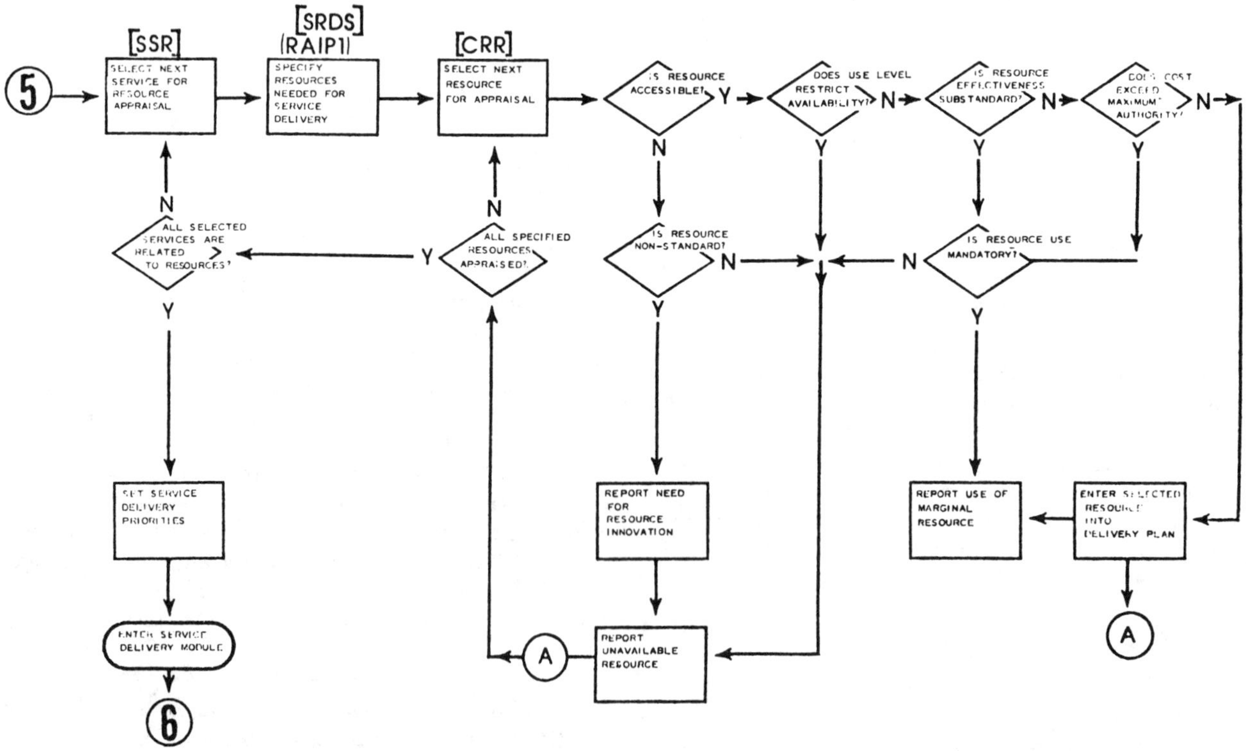

Fig. 6. Resource appraisal module.

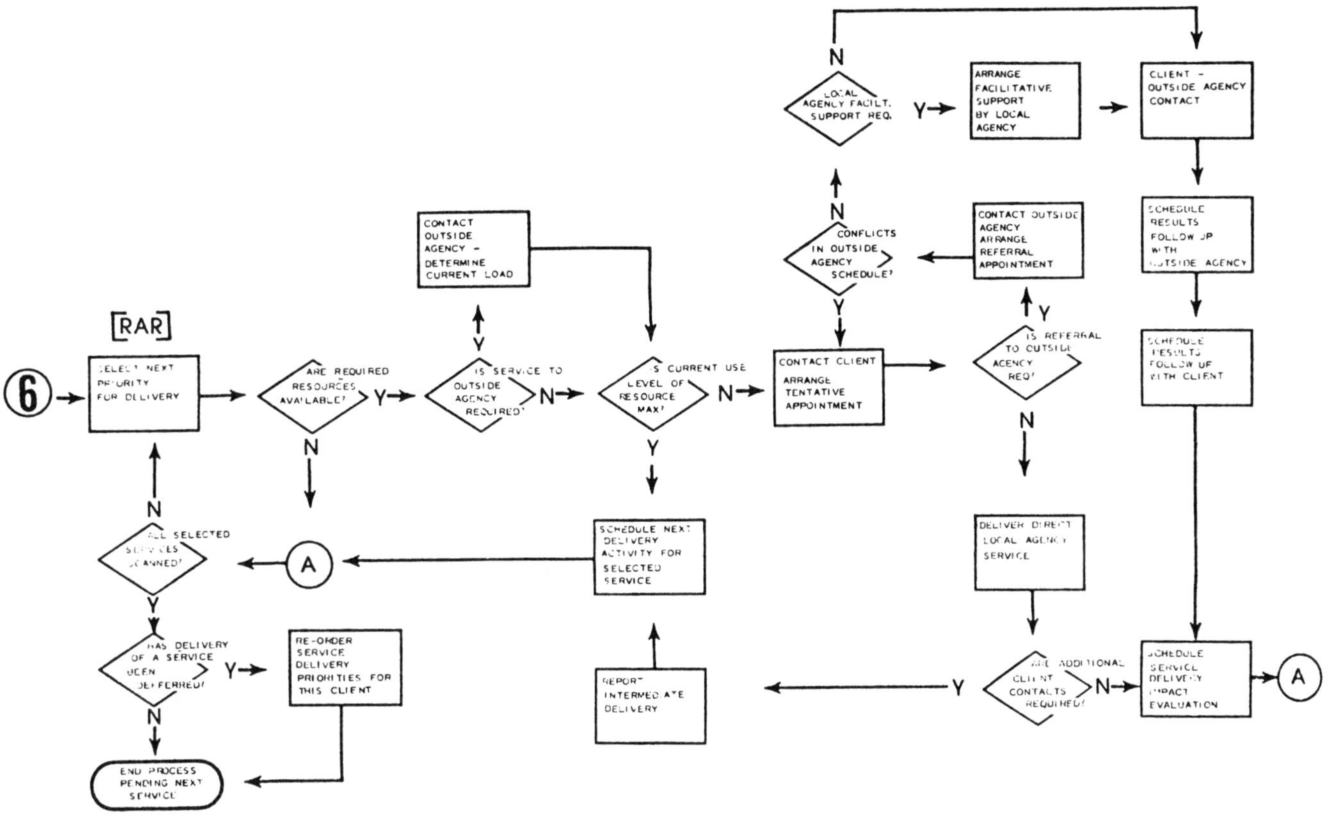

Fig. 7. Service delivery module.

used to meet these needs. The resource identification process consists simply of selecting standard resource categories. There is a two-fold result of the resource appraisal process. First, the gap between service resource needs and effective resources that can be used to meet these needs is explicated. Second, a service plan for the client is completed which specifies the actual social service resources to be used to reach clearly stated service delivery objectives.

Resource Appraisal Functions

1) Identify resource requirements.
2) Analyze resource context.
3) Select actual resources.
4) Specify service activities.
5) Document service plan.

When the actual resources to be used in achieving a specified service delivery objective have been selected, service activities necessary to initiate and maintain the client–resource contact, are explicitly identified, these activities, (for example, referral, transportation, facilitation, etc.) constitute the units of work that must be acquired or supplied by the case manager in order to carry on service delivery. Those service activities which are supplied or purchased represent the direct costs of social service.

After execution of the resource appraisal process, the client's service plan, made up of a service hypothesis plus a statement of resources and service activities is ready for implementation.

Community Resource Component

1) Social service resource inventory.
2) Development and update.
3) Usage and evaluation.

To support the primary service delivery activities of the system, the design includes a specialized community resource component. Its function is centered around a community social services resource inventory which is continuously updated to reflect a current community resource context. Where there are needs for new resources or for expansion or upgrading of existing resources, the necessary development is performed by this function.

Identification of these needs is based upon usage and evaluative data which is posted against the resources which make up the inventory. This data, which reflects service delivery experience with the community resource context, is generated out of the resource appraisal and service impact evaluation activities of the system.

Service Delivery

The purpose of the service delivery module (Fig. 7) is to implement the client's service plan by providing the service activities necessary to enable client contact with selected social service resources and to maintain that contact for the duration of its effective use. Concrete action which joins the client with community social service operations constitutes the service delivery process. It is in this process that the practical exigencies involved in the use of community resources are met. It is here that the myriad and sundry activities take place which facilitate the client's movement toward specified service objectives. The process includes the adaptation of the service plan to accommodate the short-term conditions that block the use of resources, the minor catastrophes that abort carefully developed service sequences, and the changes in the client's situation that make service objectives irrelevant of inappropriate. Service delivery is a dynamic process in continuous inter-

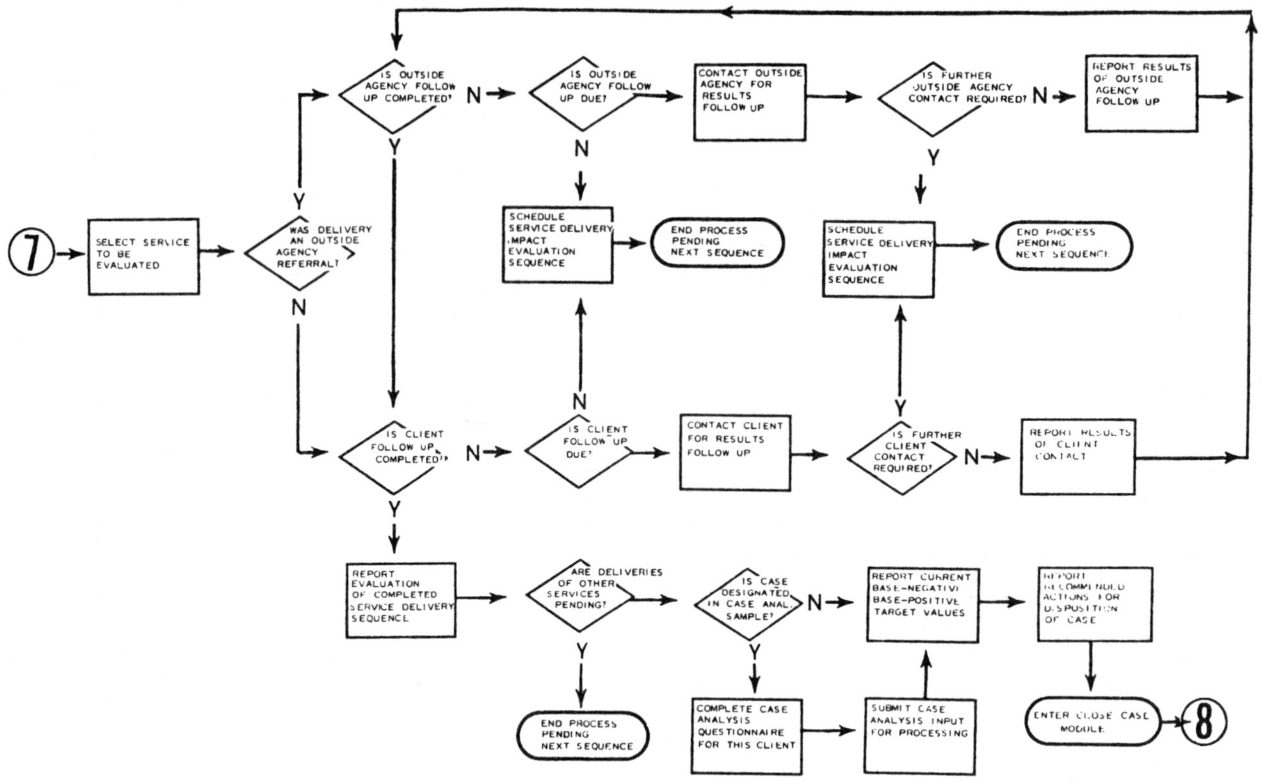

Fig. 8. Service impact evaluation module.

action with the processes of assessment, service selection, and resource appraisal to maintain an optimum response to client needs.

Service Delivery Functions

1) Enable client-resource contact.
2) Monitor service delivery sequences.
3) Document service delivery contacts.

Each significant contact that occurs between the case manager and the client or a community resource agency is recorded. Contact documentation reflects case-member identification and date of contact in conjunction with the service delivered, the resource agency involved, and the service activity expanded. It includes a record of the approximate time used for the contact and an indication of its effectiveness. Contact information is collected throughout the period of service delivery in order to track the client's use of community resources and make the service activities of the case manager visible.

Service Impact Evaluation

The purpose of the service impact evaluation module (Fig. 8) is to follow up the client's use of service resources to determine the results of that use, and to evaluate the impact of the client's situation. The service impact evaluation is carried on in three discrete phases. A critique of the results of resource usage is made by the resource agency performing service, the client receiving the service, and the case manager who arranged the service. This information is fed-back for use by the community resources component of the system.

The characteristics of the client's situation which enter into case classification are reassessed according to established case analysis sampling requirements. This information is used for statistical measurement of case movement. The characteristics of the client situation which were used in formulating service objectives for the client are reassessed to determine the effectiveness of the service plan and the resources used for its implementation.

Service Impact Evaluation Functions

1) Perform "delivery results" follow up.
2) Measure case movement.
3) Evaluate attainment of objectives.
4) Recommend case disposition.
5) Document service impact.

The results of service impact evaluation provide basic intelligence covering the effectiveness of the social service client's interaction with his community resource context. This intelligence includes coverage of case management and other service activity expended to enable the client's use of these resources. Analysis based upon service impact intelligence can be translated into estimates of social benefit. These estimates coupled with cost data derived from statistical analysis of the characteristics of service activity and case management will produce the accountability which is one of the primary objectives of the system.

Case Disposition

The purpose of the case disposition module is to determine how to accommodate a social service case that has been released from caseload of a case manager, transferred from another social service agency, reopened at the request of a client, or is in a transition phase following a completed service delivery sequence.

Conclusion

The generic system described has been installed and is now functioning in Lynchburg, Va., and will be by December 1974 in Chesapeake, Va. The systems are supported by computers

```
PAGE  102                                 CLIENT SERVICE DELIVERY PLAN                                        REPORT NO 05001
                                                                                                              DATE   08/20/74
        PRIORITY 3           WORKER WHC        CASE NAME:   MORRIS              CASE NO.  00000243200
                  SERVICE       HEALTH-PHYSICAL                                                               SVC  HLTP
MEMBER          QUESTION / CHARACTERISTIC       MEMBER   ASSESSED       TARGET        IMPACT    SELECT IMPACT  ID BAT Q  ID A T I
 ID1                                             ID2      VALUE          VALUE         VALUE     DATE   DATE    1  ID NO  2 V V V
CYNTHIA  GENERAL HEALTH STATUS IS                          FAIR         GOOD                     08 74          01  H 001    3 2

         RESOURCE                   SERVICE WORKER ACTIVITIES              APPR   PROJ  ACTUAL  TARGET  PROV     RESRCE    ACT.
                                                                                  EVAL  EVAL
  PRIVATE PSYCHIATRISTS      COUNSELING  REFERRALS   SUPPORTIVE            08 08  02 75         PILLE   DIR      000078  A D I

                  SERVICE       HOUSING                                                                       SVC  HOUS
MEMBER          QUESTION / CHARACTERISTIC       MEMBER   ASSESSED       TARGET        IMPACT    SELECT IMPACT  ID BAT Q  ID A T I
 ID1                                             ID2      VALUE          VALUE         VALUE     DATE   DATE    1  ID NO  2 V V V
CYNTHIA  INTEREST IN RELOCATING                            VERY         XXXXXXXXXXXX             08 74          01  L 058    1 0

         RESOURCE                   SERVICE WORKER ACTIVITIES              APPR   PROJ  ACTUAL  TARGET  PROV     RESRCE    ACT.
                                                                                  EVAL  EVAL
  THE DAILY ADVANCE          COUNSELING  REFERRALS   SUPPORTIVE            08 08  11 74         NDGTE   REF      000054  A D I
  PRIVATE REALTORS           COUNSELING  REFERRALS   SUPPORTIVE            08 08  11 74         NDGTE   REF      000096  A D I
```

Fig. 9. Client service delivery plan.

in the locales and are provided backup through the central ADP agency of the state located in Richmond, Va. To assure full use of the system in its intended operating mode extensive documentation and development of forms was required. A program of case analysis was carried out to determine if it would be feasible to develop base-line information on each client and to establish patterns for aid in developing service programs for clients. A resource bank was developed which would be able to provide for each worker the knowledge relative to the availability of any service, the quality and costs of that service and location to which the client must go to receive the service. The information provided to the case worker and supervisory person is derived from the support system and examples of the type and format of this information is given in Fig. 9.

Procedures manuals have been developed and are being used by case workers and supervisors participating in each phase of the system. A fully developed service resource document has been established which provides access to each service including the cost, availability, quality and location.

The Social Service Delivery System in its present form supports all of the functions of the Lynchburg Agency except financial assistance. This system is being used as the model for state-wide implementation and also as the model for a one-stop broker-advocate human services delivery system. The importance of this system to all levels of social service delivery is several fold. From the point of view of the client immediate access is gained to all services for which that person is eligible. Through the process of intake and assessment the client is in a better position to recognize and try to cope with basic problems. For the case worker, organization comes to what in the past has been a chaotic mess. From assessment through case disposition the worker is in a position to provide for the client from a position of strength gained through knowledge. Emergency situations are determined and met. Service plans are established through cooperative interaction with the client. Evaluation of the effectiveness of the services delivered is provided through the service impact evaluation. The supervisors are provided the opportunity of managing the agency. The concept of case carrying is eliminated.

The major innovation provided by the Social Services Delivery System is in the area of management at the local and state levels. For the first time information is available in a timely fashion and in a form useful to the management function to be performed. Real cost in terms of personnel and services is available. Forecasts of the implications of new policies on both personnel and resource costs can be made. Evaluation of the effectiveness of service as delivered both in terms of the impact on the client and of the effectiveness of service as delivered both in terms of the impact on the client and of the effectiveness of workers is available. It is feasible to discuss cost effectiveness in real terms and provide funding requirements based on assessment of impact of present or new programs.

References

[1] C. Argyris, *Management and Organizational Development.* New York: McGraw-Hill, 1971.

[2] HEW, *Social Worker Manual*, 1963.

[3] W. M. McFeely, "Organization change: Perceptive and realities," Conf. Board, New York, N.Y., Rep. 561, 1972.

[4] A. H. Maslow, "A theory of human motivation," in *Organization Theories.* Columbus, Ohio: Merrill, 1970.

[5] J. D. Palmer, "Developing evaluation criteria for service delivery systems," presented at Massachusetts Conf. Social Welfare, Boston, Mass., Dec. 1971.

[6] J. D. Palmer and J. Deloyht, "A management information system for Virginia," in *Proc. SMC Nat. Conf.*, Nov. 1973.

[7] H. M. Runyan, "Introduction to transition management," Systems Research Co., Boston, Mass., 1973.

[8] The Management Group (TMG), "Computerized service (welfare) delivery systems in Maine," Waltham, Me., 1970.

[9] R. Thoobald, Ed., *The Guaranteed Income, Next Step in Socio-economic Evaluation?* Garden City, N.Y.: Doubleday-Anchor, 1967.

[10] A. Toffler, *Future Shock.* New York: Bantam 1971.

[11] Touche-Ross (T-R), "Use of Computers in Welfare Systems in Washington, Michigan and Virginia," Washington, D.C., 1971.

[12] J. N. Warfield, "An assault on complexity," Battelle Memorial Inst., Columbus, Ohio, Monogr. 3, Apr. 1973.

[13] ——, "Structuring complex systems," Battelle Memorial Inst., Columbus, Ohio, Monogr. 4, Apr. 1974 (many excellent references).

[14] A. K. C. Wong and T. Au, "Regional planning of health care facilities," *IEEE Trans. Syst., Man., Cybern.*, vol. SMC-3, Sept. 1973.

Population in the Club of Rome's World3 Model

WIL THISSEN

Abstract—**The World3 population sector is analyzed, not to criticize the underlying assumptions, but a) to elucidate the inner working and general behavioral properties of the submodel and b) to illustrate the use of simple methods of understanding complicated models. If ample food is available, population tends towards exponential growth owing to positive feedback within the sector itself, whereas negative feedback predominates if food is scarce. The model's carrying capacity for population is primarily determined by food production, equilibrium being mainly achieved by starvation. If income and/or pollution is high, however, equilibrium conditions may be rather different. The effect of introducing age distribution is only slight.**

I. Introduction

SINCE THE publication of Forrester's book, *World Dynamics* [1], and its successors, *The Limits to Growth* [2] and *Dynamics of Growth in a Finite World* [3] by Meadows *et al.*, hundreds of articles have been written about the global issues brought up for discussion by the so-called "Club of Rome." Most of the reactions criticized the philosophy and assumptions underlying the MIT world modeling efforts, or the conclusions drawn and policy recommendations made (see, e.g., [4], [5]). The Sussex team [4] focussed their attention on testing model validity, while other investigators [6] performed mathematical analyses almost without regarding the inner structure of the models and the physical meaning of its variables.

Few have taken the models as they are, and analyzed their *inner* workings to obtain a better understanding of *why* these models behave the way they do. That is what the present series of papers is about.

The analyses presented by Forrester and the Meadows team are, for the greater part, based on the results of computer calculations and proceed mainly along a trial-and-error type of approach. Only few of the various techniques of model analysis known in systems and control engineering circles were utilized. For these and other reasons, a study of the World2 and World3 models was set up, aiming at two main purposes, namely, 1) to provide those interested in the problems with the kind of sober and clarifying information necessary for good comprehension of the inner workings of the world models, and 2) to demonstrate how understanding of the fundamental mechanisms underlying the inner workings of a complicated model can be acquired step by step, using relatively simple methods of systems and control engineering.

Manuscript received July 13, 1976; revised September 20, 1977.

The author was with the Department of Applied Physics, Eindhoven University of Technology, Eindhoven, Netherlands. He will be with the School of Engineering and Applied Science, University of Virginia, Charlottesville, VA 22901.

A part of our findings has already been published in the form of progress reports [7]. In the present issue, the series reporting on our World3 studies is completed. In continuation of the analysis of the capital and resource subsystem [8], this paper discusses the workings of the World3 population sector. The agricultural and persistent pollution sectors are examined in the next paper [9]. Subsequently, the model is studied as a whole, for various conditions, and the conclusions drawn are briefly compared with those formulated by the Meadows team [10]. The completing paper [11] calls attention to the methodological lessons to be drawn from these analyses. The crucial importance of understanding complicated models is emphasized, and a general approach for model analysis is proposed.

The emphasis in this and the two subsequent papers on World3 will be on distinguishing active and dormant parts of the model, and on understanding its general behavioral characteristics. Each sector is examined along different approaches and in increasing detail. It is emphasized that the only purpose of the changes and modifications introduced into the model is to yield new information so as to better understand why the model behaves the way it does. Hence, in this context, it is totally irrelevant whether comparable real-world conditions may exist or not.

The outline of the present paper is as follows. After a concise description of the population submodel, the standard-run behavior of population is examined. Next, a more general investigation into the sector's properties is made by separately studying equilibrium and disequilibrium conditions. Particular emphasis is on discovering the reasons why population grows in World3. Subsequently, the effects of introducing age distribution are investigated, and finally, a summary of the major findings and conclusions is given.

II. Outline of the World3 Population Submodel

A detailed description of the equations, assumptions, and the data on which these were based is given in [3]. Here we will confine ourselves to the information necessary for the proper comprehension of the discussions below. The function of the population sector in World3 is twofold:

1) *it calculates the total population size*, and
2) *it represents the demographic response of the population, through the birth rate and the death rate, to the changing resource supply* [3, p. 30].

As part of the overall World3 model, this sector is coupled to all other subsystems: the capital and resource sectors, the agricultural sector, and the persistent pollution sector. The

Reprinted from *IEEE Trans. Syst., Man, Cybernetics,* vol. SMC-8, pp. 149–159, Mar. 1978.

TABLE I
LIST OF SYMBOLS

AHL	: ASSIMILATION HALF-LIFE (YEARS)
AI	: AGRICULTURAL INPUTS (DOLLARS/YEAR)
AIOPC	: AVERAGE INDUSTRIAL OUTPUT PER CAPITA (DOLLARS/PERSON-YEAR)
AIPH	: AGRICULTURAL INPUTS PER HECTARE (DOLLARS/HECTARE-YEAR)
AL	: ARABLE LAND (HECTARES)
ALI	: ARABLE LAND INITIAL (HECTARES)
ALL	: AVERAGE LIFETIME OF LAND (YEARS)
B	: BIRTHS PER YEAR (PERSONS/YEAR)
CAI	: CURRENT AGRICULTURAL INPUTS (DOLLARS/YEARS)
CBR	: CRUDE BIRTH RATE (BIRTHS/1000 PERSON-YEARS)
CDR	: CRUDE DEATH RATE (DEATHS/1000 PERSON-YEARS)
CMI	: CROWDING MULTIPLIER FROM INDUSTRIALIZATION (DIMENSIONLESS)
CMPLE	: COMPENSATORY MULTIPLIER FROM PERCEIVED LIFE EXPECTANCY (DIMENSIONLESS)
CUF	: CAPITAL UTILIZATION FRACTION
D	: DEATHS PER YEAR (PERSONS/YEAR)
DCFS	: DESIRED COMPLETED FAMILY SIZE (DIMENSIONLESS)
DCFSN	: DESIRED COMPLETED FAMILY SIZE NORMAL (DIMENSIONLESS)
DCPH	: DEVELOPMENT COST PER HECTARE (DOLLARS/HECTARE)
DIOPC	: DELAYED INDUSTRIAL OUTPUT PER CAPITA (DOLLARS/PERSON-YEAR)
DTF	: DESIRED TOTAL FERTILITY (DIMENSIONLESS)
E	: EFFECTIVITY OF INPUTS ALLOCATED TO RESOURCE CONSERVATION (RESOURCE UNITS/DOLLAR)
EHSPC	: EFFECTIVE HEALTH SERVICES PER CAPITA
F	: FOOD (VEGETABLE-EQUIVALENT KILOGRAMS/YEAR)
f_{index}	: TABLE FUNCTION; THE INDEX REFERS TO THE EQUATION NUMBER IN [3]
FALM	: FRACTION OF INPUTS ALLOCATED TO LAND MAINTENANCE (DIMENSIONLESS)
FCAOR	: FRACTION OF CAPITAL ALLOCATED TO OBTAINING RESOURCES (DIMENSIONLESS)
FCAPC	: FERTILITY CONTROL ALLOCATIONS PER CAPITA (DOLLARS/PERSON-YEAR)
FCE	: FERTILITY CONTROL EFFECTIVENESS (DIMENSIONLESS)
FCFPC	: FERTILITY CONTROL FACILITIES PER CAPITA (DOLLARS/PERSON-YEAR)
FIALD	: FRACTION OF INPUTS ALLOCATED TO LAND DEVELOPMENT (DIMENSIONLESS)
FIE	: FAMILY INCOME EXPECTATION (DIMENSIONLESS)
FIOAA	: FRACTION OF INDUSTRIAL OUTPUT ALLOCATED TO AGRICULTURE (DIMENSIONLESS)
FIOAC	: FRACTION OF INDUSTRIAL OUTPUT ALLOCATED TO CONSUMPTION (DIMENSIONLESS)
FIOARC	: FRACTION OF INDUSTRIAL OUTPUT ALLOCATED TO RESOURCE CONSERVATION (DIMENSIONLESS)
FIOAS	: FRACTION OF INDUSTRIAL OUTPUT ALLOCATED TO SERVICES (DIMENSIONLESS)
FPC	: FOOD PER CAPITA (VEGETABLE-EQUIVALENT KILOGRAMS/PERSON-YEAR)
FPU	: FRACTION OF POPULATION URBAN (DIMENSIONLESS)
FR	: FOOD RATIO (DIMENSIONLESS)
HSAPC	: HEALTH SERVICES ALLOCATIONS PER CAPITA (DOLLARS/PERSON-YEAR)
IC	: INDUSTRIAL CAPITAL (DOLLARS)
IFPC	: INDICATED FOOD PER CAPITA (VEGETABLE-EQUIVALENT KILOGRAMS/PERSON-YEAR)
ILF	: INHERENT LAND FERTILITY (VEGETABLE-EQUIVALENT KILOGRAMS/HECTARE-YEAR)
IO	: INDUSTRIAL OUTPUT (DOLLARS/YEAR)
IOPC	: INDUSTRIAL OUTPUT PER CAPITA (DOLLARS/PERSON-YEAR)
λ	: EIGENVALUE
LDR	: LAND DEVELOPMENT RATE (HECTARES/YEAR)
LE	: LIFE EXPECTANCY (YEARS)
LER	: LAND EROSION RATE (HECTARES/YEAR)
LFD	: LAND FERTILITY DEGRADATION (VEGETABLE-EQUIVALENT KILOGRAMS/HECTARE-YEAR-YEAR)
LFDR	: LAND FERTILITY DEGRADATION RATE (1/YEAR)
LFERT	: LAND FERTILITY (VEGETABLE-EQUIVALENT KILOGRAMS/HECTARE-YEAR)
LFRT	: LAND FERTILITY REGENERATION TIME (YEARS)
LMC	: LIFETIME MULTIPLIER FROM CROWDING (DIMENSIONLESS)
LMP	: LIFETIME MULTIPLIER FROM PERSISTENT POLLUTION (DIMENSIONLESS)
LRUI	: LAND REMOVAL FOR URBAN-INDUSTRIAL USE (HECTARE/YEAR)
LUFD	: LABOR UTILIZATION FRACTION DELAYED (DIMENSIONLESS)
LY	: LAND YIELD (VEGETABLE-EQUIVALENT KILOGRAMS/HECTARE-YEAR)
LYMAP	: LAND YIELD MULTIPLIER FROM AIR POLLUTION (DIMENSIONLESS)
LYMC	: LAND YIELD MULTIPLIER FROM CAPITAL (DIMENSIONLESS)
m	: PARAMETER (=F/SO)
MLYMC	: MARGINAL LAND YIELD MULTIPLIER FROM CAPITAL (HECTARES/DOLLAR)
MPAI	: MARGINAL PRODUCTIVITY OF AGRICULTURAL INPUTS (VEGETABLE EQUIVALENT KILOGRAMS/DOLLAR)
MPLD	: MARGINAL PRODUCTIVITY OF LAND DEVELOPMENT (VEGETABLE-EQUIVALENT KILOGRAMS/DOLLAR)
MTF	: MAXIMUM TOTAL FERTILITY (DIMENSIONLESS)
NBR	: NET BIRTH RATE (=CBR-CDR)(1/YEAR)
NBR_4	: NET BIRTH RATE, FOUR-AGE-LEVEL MODEL (1/YEAR)
NFC	: NEED FOR FERTILITY CONTROL (DIMENSIONLESS)
NR	: NONRENEWABLE REOURCES (RESOURCE UNITS)
NRI	: NONRENEWABLE RESOURCES INITIAL (RESOURCE UNITS)
P1	: POPULATION, AGES 0-14 (PERSONS)
P2	: POPULATION, AGES 15-44(PERSONS)
P3	: POPULATION, AGES 45-64(PERSONS)
P4	: POPULATION, AGES 65+ (PERSONS)
PAL	: POTENTIALLY ARABLE LAND (HECTARES)
PALI	: POTENTIALLY ARABLE LAND INITIAL (HECTARES)
PALT	: POTENTIALLY ARABLE LAND TOTAL (HECTARES)
PCRUM	: PER CAPITA RESOURCE USAGE MULTIPLIER (RESOURCE UNITS/PERSON-YEAR)
PFR	: PERCEIVED FOOD RATIO (DIMENSIONLESS)
PJAS	: POTENTIAL JOBS IN AGRICULTURAL SECTOR (PERSONS)
PLE	: PERCEIVED LIFE EXPECTANCY (YEARS)
POP	: POPULATION (PERSONS)
POPE	: POPULATION EQUILIBRIUM VALUE (PERSONS)
POPI	: POPULATION INITIAL (PERSONS)
PPAPR	: PERSISTENT POLLUTION APPEARANCE RATE (POLLUTION UNITS/YEAR)
ppapr	: PERSISTENT POLLUTION APPEARANCE RATE, RELATIVE TO 1970 (DIMENSIONLESS)
PPASR	: PERSISTENT POLLUTION ASSIMILATION RATE (POLLUTION UNITS/YEAR)
ppasr	: PERSISTENT POLLUTION ASSIMILATION RATE, RELATIVE TO 1970 (DIMENSIONLESS)
PPGAO	: PERSISTENT POLLUTION GENERATED BY AGRICULTURAL OUTPUT (POLLUTION UNITS/YEAR)
PPGIO	: PERSISTENT POLLUTION GENERATED BY INDUSTRIAL OUTPUT (POLLUTION UNITS/YEAR)
PPOL	: PERSISTENT POLLUTION (POLLUTION UNITS)
PPOL70	: PERSISTENT POLLUTION IN 1970 (POLLUTION UNITS)
PPOLX	: INDEX OF PERSISTENT POLLUTION (DIMENSIONLESS)
SC	: SERVICE CAPITAL (DOLLARS)
SFPC	: SUBSISTENCE FOOD PER CAPITA (VEGETABLE-EQUIVALENT KILOGRAMS/PERSON-YEAR)
SFSN	: SOCIAL FAMILY SIZE NORM (DIMENSIONLESS)
SO	: SERVICE OUTPUT (DOLLARS/YEAR)
SOPC	: SERVICE OUPUT PER CAPITA (DOLLARS/PERSON-YEAR)
T,TIME,t	: TIME (YEARS)
τ	: TIME CONSTANT (YEARS)
TAI	: TOTAL AGRICULTURAL INVESTMENT (DOLLARS/YEAR)
TF	: TOTAL FERTILITY (DIMENSIONLESS)
UIL	: URBAN-INDUSTRIAL LAND (HECTARES)
UILPC	: URBAN-INDUSTRIAL LAND PER CAPITA (HECTARES/PERSON)
UILR	: URBAN-INDUSTRIAL LAND REQUIRED (HECTARES)
$\underline{V}$	: EIGENVECTOR

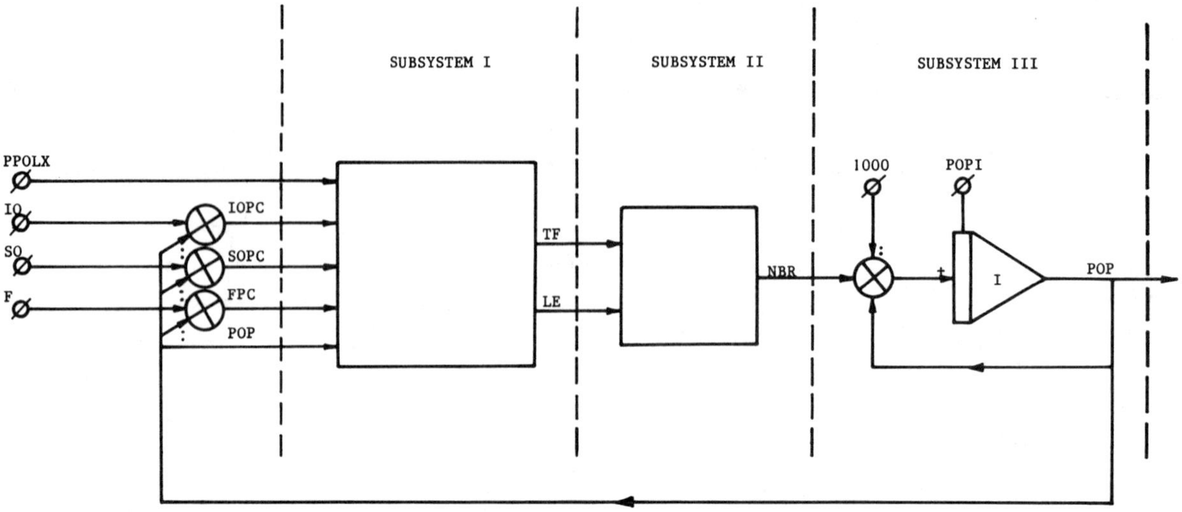

Fig. 1. Block diagram of World3 population model without age distribution; *I* denotes integration. Symbols are explained in Table I.

variable affecting the other subsystems is the total number of people POP.[1] Conversely, the sector is influenced by variables of all other sectors, viz., industrial output IO, service output SO, food production *F* and the persistent pollution ratio PPOLX.

Three basic components may be distinguished, coupled as shown in Fig. 1. Subsystem I calculates total fertility TF and life expectancy LE from PPOLX and POP, and from the per capita values of IO, SO, and *F*. It consists of lags and nonlinear algebraic relationships. The lags, called "delays" by Meadows *et al.*, are linear first- or third-order systems with a static gain equal to 1 [12]. The algebraic relations consist of arithmetic operations (multiplications, divisions, etc.) and table functions.

Subsystem II calculates the net birth rate NBR, that is, the net rate of increase in total population per 1000 people per year, from the values of LE and TF. For this purpose the equation

$$\begin{aligned} \text{NBR} &= \text{CBR} - \text{CDR} \\ &= 1000 * ((0.21 * \text{TF})/30 - 1/\text{LE}) \qquad (1) \end{aligned}$$

is applied in a model without age distribution. The factor 0.21 stands for the fraction of fertile women and 30 is the mean reproductive lifetime.

Finally, subsystem III reflects the calculation of the number of people POP according to

$$\frac{d\,\text{POP}}{dt} = (\text{NBR}/1000) * \text{POP}. \qquad (2)$$

The feedback of POP to subsystem I completes the overall scheme.

Table II gives a full list of the equations. Fig. 2 shows the internal structure of subsystem I. A concise explanation of the relations is given in the Appendix. Since a basic familiarity with the model will be assumed in the discussions below, reading of the Appendix is recommended.

[1] A list of abbreviations and symbols is given in Table I.

TABLE II
EQUATIONS OF WORLD3 POPULATION SUBSYSTEM WITHOUT AGE DISTRIBUTION

STATE EQUATION		
POP	=	B - D
INITIAL CONDITION		
POPI	=	POP(1900) = 1.6×10^9
COUPLING EQUATIONS (CONSTANTS HAVE BEEN SUBSTITUTED)		
CALCULATION OF DEATH RATE		
D	=	POP/LE
LE	=	28 * f_{25}[a] (EHSPC) * f_{20}(FPC/230) * f_{29}(PPOLX) * LMC
LMC	=	1 - f_{27}(IOPC) * f_{26}(POP)
EHSPC	=	SMOOTH (HSAPC,20)[b]
HSAPC	=	f_{21}(SOPC)
SOPC	=	SO/POP
FPC	=	F/POP
IOPC	=	IO/POP
CALCULATION OF BIRTH RATE		
B	=	POP * 0.21 * TF/30
TF	=	MIN(MTF, MTF * (1-FCE) + DTF * FCE)[c]
MTF	=	12 * f_{34}(LE)
FCE	=	f_{45}(FCFPC)
FCFPC	=	DLINF3(FCAPC,20)
FCAPC	=	f_{48}(NFC) * SOPC
NFC	=	MTF/DTF - 1
DTF	=	DCFS * f_{36}(PLE)
PLE	=	DLINF3 (LE,20)
DCFS	=	4 * f_{41}(FIE) * f_{39}(DIOPC)
FIE	=	(IOPC - AIOPC) / AIOPC
AIOPC	=	SMOOTH(IOPC,3)
DIOPC	=	DLINF3(IOPC,20)
INPUT VARIABLES		
IO		(CAPITAL SECTOR)
SO		(CAPITAL SECTOR)
F		(AGRICULTURAL SECTOR)
PPOLX		(PERSISTENT POLLUTION SECTOR)

[a] f_i denotes a table function; the index corresponds to the equation number in [3].
[b] SMOOTH, DELAY3, DLINF3(*A*,*B*) denote lags; *A* is the input variable and *B* is the time constant.
[c] MIN(*A*,*B*) is the minimum of *A* and *B*.

Two further points are, first, in the DYNAMO flow diagram of the complete model [3] the per capita values of IO, SO, and *F* are taken as input variables to the sector. Yet, in consistency with Meadows' analysis of the sector [3, ch. 2],

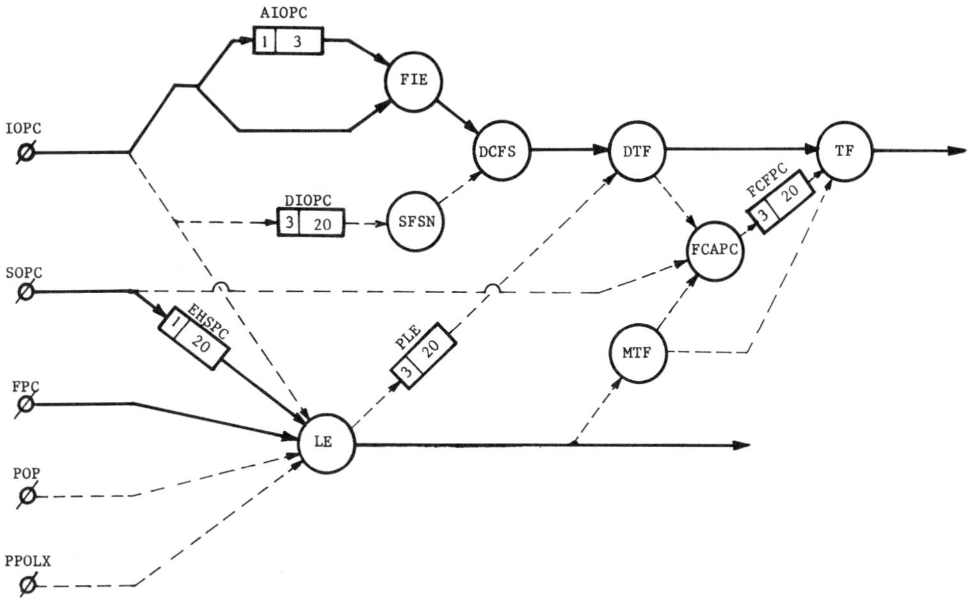

Fig. 2. Internal structure of subsystem I; circles indicate coupling variables and rectangles represent lags, with indication of order (left figure) and time constant. Broken lines indicate weaker influences in standard run.

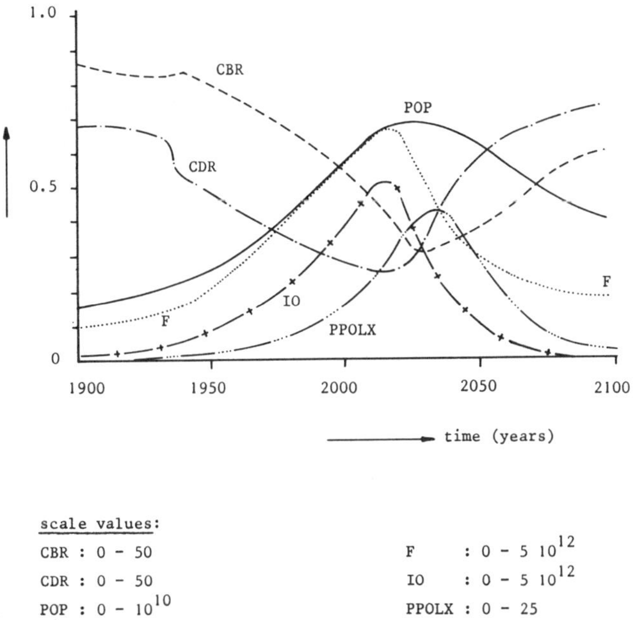

Fig. 3. Standard-run behavior of World3 population sector and input variables.

and with our own discussion of the capital and resource subsystem [8], and to take into account the feedback of POP to its own rate of growth acting via the per capita variables, we will consider IO, SO, and *F* as the inputs to the sector.

Secondly, models with three different degrees of age-specific disaggregation have been presented by Meadows *et al.*, namely, a) a model without age distribution (described above), b) a model distinguishing four age groups, and c) a model with 15 age groups. The differences between these models are in the calculation of the changes in population from the values of TF and LE. Except for the analyses described in Section IV, we used the model with four age groups for our simulations. The effects of introducing four age levels will be considered in detail in Section V.

III. First Explorations of Standard-Run Behavior

The standard- or reference-run behavior of the overall World3 model is obtained by setting a number of parameters to values such that the outcome between 1900 and 1970 corresponds to what is known about reality in that period. Usually, the calculation is continued till the year 2100. The behavior of growth followed by decline, characteristic for the capital and resource subsystem, is also found in the population sector, but the turning point occurs about 15 years later (Fig. 3). The behavior of the first part of the run is essentially the demographic transition (rising income, decreasing crude birth and death rates CBR and CDR). The decline in the second part is incited by a rather sudden, sharp rise in CDR, while the decline of CBR persists for some time before a rather slow increase sets in.

To explore which out of the many relations essentially determine standard-run behavior, various simulations were made. Among the modifications introduced were not only changes in parameters, but also the freezing of variables or functions,[2] changes in initial values, linearization, and analytical treatment. As far as sensitivity tests are concerned, our results were in broad agreement with those obtained by Meadows [3] and by others [6], [13]. Let us briefly review our main findings.

1) Freezing PPOLX from 1970 onwards reveals that, in contrast to what is suggested in [13], pollution has virtually no influence at all on population. The explanation is that, since PPOLX remains lower than 12 in a standard run, LMP (the lifetime multiplier from persistent pollution) deviates less than 2 percent from unity.

2) Freezing IO and SO (close-coupled variables, see [8]) or *F* from 1970 onward reduces the rate of population growth and decline.

[2] Freezing a variable or function means that its value is kept constant from a certain point in time onwards.

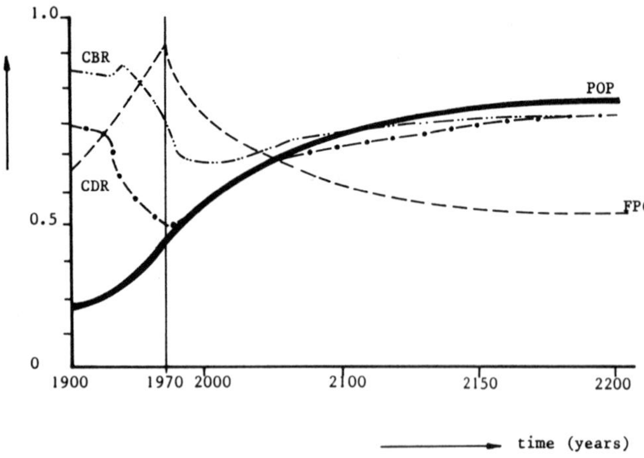

Scale values:
POP : 0 - 10^{10}
CBR : 0 - 50
CDR : 0 - 50
FPC : 0 - 500

Fig. 4. Behavior of World3 population if in standard-run calculation IO, SO, *F*, and PPOLX are frozen from 1970 on. (Note the different time scale.)

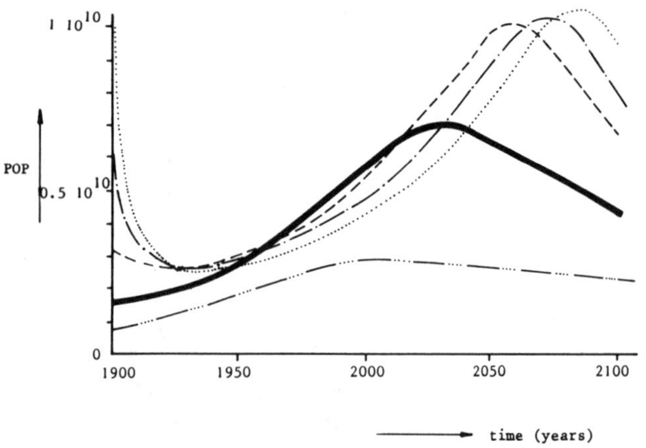

Fig. 5. Behavior of World3 population for various initial values in 1900 (standard-run: heavy line).

3) Freezing all four input variables from 1970 onward fully eliminates population decline, as is shown in Fig. 4. Tentatively, three conclusions were drawn: a) growth and decline are not determined endogenously but are impressed on the sector by the behavior of the inputs IO, SO, and *F*; b) since, in the final situation, food supply is near the subsistence level, the conditions under which population equilibrium is achieved are unattractive; and c) the time constant (larger than 100 years) with which equilibrium is approached, indicates that the sector responds slowly to changes in the inputs. We will return to these subjects below.

4) Quite surprisingly, drastic augmentation of the initial (1900) value of population hardly affects behavior between 1940 and 2000, as is shown in Fig. 5. In contrast, if POP(1900) is reduced with respect to its normal value, POP remains considerably lower than in the standard run during the whole simulation. To obtain the curves shown in Fig. 5, the model without age distribution has been used since the effects it produces are then clearest.

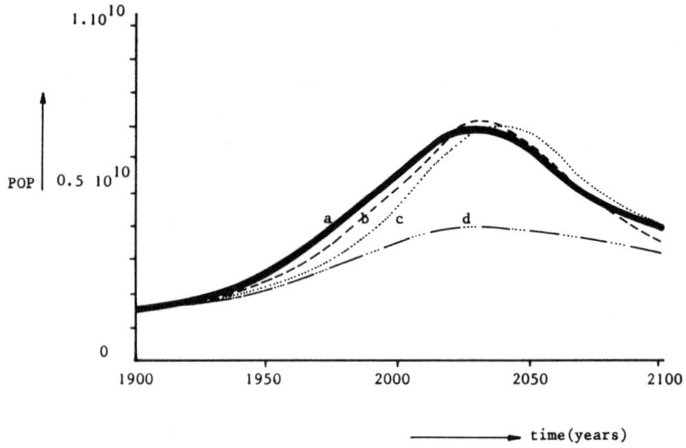

a : standard run
b : 4 simplifications: MTF = 8.5, CMPLE = 1.30, SFSN = 1.06, LMP = 1.
c : simplifications (b), and in addition LMC = 0.96 and FCE = 0.92.
d : all lags omitted

Fig. 6. Standard-run behavior of POP if various simplifications are introduced.

These results point to an asymmetry in the sector's dynamics that will be explained in Section IV. The larger peak values of population if POP(1900) is augmented can be explained from the influence of food scarcity on the capital and resource subsystem. For large values of POP(1900), food is scarce in the first part of the run, explaining the rapid decline in population. Therefore, a large fraction of IO is allocated to agriculture. As a result, the initial growth of capital is moderated, but, since a larger part of resources remains unused, capital growth can persist longer before the decline sets in [8]. Then, population decline also sets in at a later point in time.

5) The influence of MTF, PLE, and SFSN on total fertility TF is insignificant, as is shown in Fig. 6, curve b. If the following constant values, MTF = 8.5, CMPLE = 1.30, and SFSN = 1.06, are substituted in addition to the elimination of the effects of PPOLX, POP is hardly affected.

6) The influences of fertility control effectiveness FCE and of crowding are fairly weak also. Substitution of FCE = 0.90 and LMC = 0.96 in addition to the earlier constant values for MTF, PLE, and SFSN, however, gives rise to somewhat clearer deviations of the standard run, as is shown in curve c of Fig. 6.

7) Modifications in one of the remaining paths cause more serious changes in the model outcome.

8) The effect of omitting the lag AIOPC is pronounced. It causes a weaker growth of population, followed by a more moderate decline. The result of omitting the other lags is similar, but weaker, except for the lag in EHSPC, the omission of which has an opposite effect. Although omission of each lag separately (except for AIOPC) causes only weak modifications, their combined effect is significant, as shown in Fig. 6, curve d.

In Fig. 2, the weaker influences have been indicated by means of broken lines. In summary, it follows that the behavior of fertility in the standard run is mainly determined by the family income expectation FIE, and, to a lesser degree, by the effects of the other influencing variables. Life expectancy is determined by SOPC and FPC, crowding having only a small influence, and pollution none at all. Hence, those factors directly affecting LE, DTF, and TF seem to be the most influential. These conclusions have been confirmed by the results of total linearization of the equations determining the rate of growth of population.

IV. Statics and Dynamics of Population

A. Equilibrium Analysis

We will now discuss a more general investigation into the sector's properties. Equilibrium and disequilibrium conditions are studied separately, first by the application of special test inputs to the overall submodel and subsequently by following a more fundamental approach.

To explore their influence on the final equilibrium size of population, the four sector inputs were frozen at various value combinations. It was tentatively concluded that the size of food production F is the principal determinant of the equilibrium value of POP.

After this, the more general question, "What is the final equilibrium value of POP given any combination of constant IO, SO, F, and PPOLX?" was tackled. POP can only be in equilibrium if NBR $= 0$ (Fig. 1). Hence, the problem is reduced to finding the value(s) of POP that make NBR $= 0$, given the input conditions.

Since the lags do not at all affect equilibrium conditions, they were omitted. Moreover, for reasons of simplicity, a model without age distribution was considered.

The equilibrium value (to be called POPE) was calculated as a function of IO, SO, F, and PPOLX, at first for all the sets of values that these variables assume during the standard run. Fig. 7 shows the curves for POP and POPE in a model without age distribution. Each value of POPE is equal to the equilibrium value that POP would ultimately reach if the values of the inputs were frozen from the corresponding point in time onwards. The POPE curve shows that, if the model's economic variables and food production were stabilized before 2020, the population would grow and ultimately become about twice as large as at the moment of stabilization. In the period of economic decline, the potential for population growth falls rapidly. After 2060 POPE becomes lower than POP. The decline in POP already sets in *before* POPE becomes smaller than POP. This effect is due to the lags incorporated in the sector equations. In a model in which subsystem I contained no lags at all, POP would grow unless POPE $\leq$ POP.

Subsequently, the influences of the four input variables on the standard-run behavior of POPE were compared by sensitivity analysis. The conclusion was that variations in POPE are mainly caused by variations in food production F (an average contribution of about 70 percent), while SO contributes about 20 percent and IO about 10 percent. The influence of PPOLX is virtually nil. For larger variations of the inputs (up to a factor five compared with the standard-run sets of values), the same pattern of sensitivities was found, with only one exception: PPOLX will come into play at values of more than 40 and may then even play a dominant part. This will be explained in Section VI.

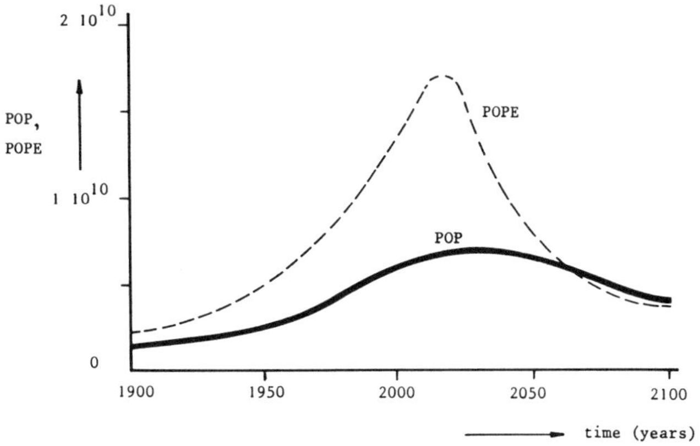

Fig. 7. Standard-run behavior of POP and POPE (no age distribution).

Three further remarks may be made. First, the equilibrium-finding algorithm calculated only one value of POPE. Different equilibria may persist for the same set of input values. Secondly, the calculated values of POPE do not necessarily represent *stable* equilibria. Thirdly, the introduction of age distribution may cause equilibrium conditions to be different. We will return to these questions below.

The overall conclusions are that under normal conditions POPE is mainly determined by F, that the influence of SO is relatively small, that of IO is even less, and that of PPOLX may be neglected for standard-run conditions.

The behavior of POPE tells us in which direction POP will tend, but, as shown in Fig. 7, the difference between POP and POPE may be considerable. Therefore we shall now examine the *dynamics* of the model's population.

B. Dynamics of Population

1) *First Explorations:* Using steps as input, a first impression was obtained of the sector's basic dynamic properties. In general, population responds fairly slowly to changes that cause growth such as an increase in F (time constants of 100 years and more). On the contrary, decline of population is much faster ($\tau \simeq 30$ years). Again, the sector displays asymmetry, which can be explained as follows: all equilibria considered are characterized by a low value of food per capita FPC (about 200). Apparently, equilibrium is mainly achieved by starvation, the death rate rising to an equally high value as the birth rate. Starting from an equilibrium, a reduction of F will enforce starvation, so that the death rate will rise considerably, whereas an increase in F starting from the same equilibrium will only slightly reduce the death rate. The preliminary conclusion is that the nonlinearity in the influence of FPC on the death rate is mainly causing the asymmetry in the sector's dynamics.

2) *Growth Rate of Population under Constant per Capita Conditions:* For constant inputs to subsystem I (Fig. 1), the lags can be omitted from the equations, and in a model without age distribution, NBR can be calculated as an algebraic function of PPOLX, the three per capita variables IOPC, SOPC, and FPC, and POP itself. For a fairly wide range of circumstances, the ratio IO/SO is fairly constant

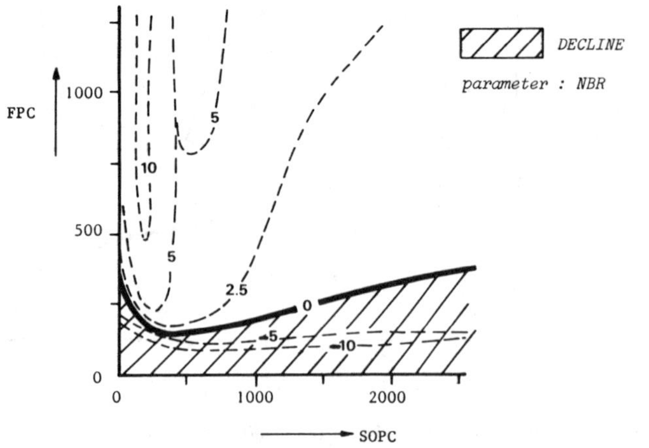

Fig. 8. Relative rate of growth of Population in World3 as function of constant FPC and SOPC.

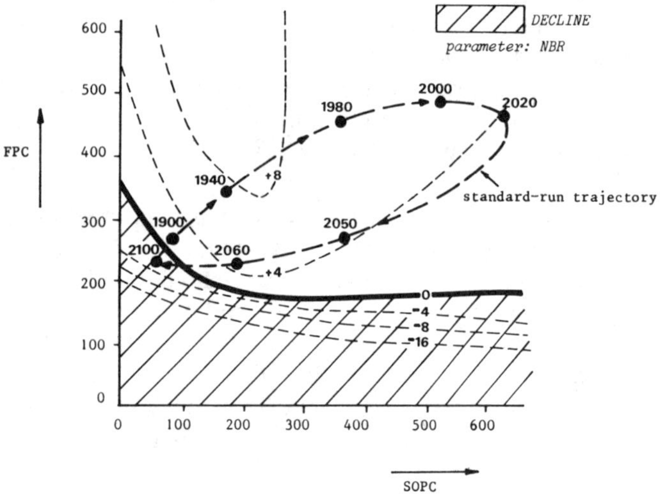

Fig. 9. Relative growth of population for low FPC and SOPC, with sketch of standard-run trajectory.

[8]. Moreover, for more or less normal conditions, the influence of PPOLX may be fully ignored. If, in addition, the direct influence of population on crowding is assumed to be constant, the number of independent variable inputs to subsystem I is reduced to two, namely FPC and SOPC (with IOPC = 2/3 * SOPC).

On the basis of these considerations, NBR was calculated as a function of constant FPC and SOPC, assigning fixed values to PPOLX (zero) and the fraction of population urban FPU (0.4, reflecting a population size of $4 * 10^9$).

The results are shown in Fig. 8, the principal part of which has been redrawn in Fig. 9. A set of curves with constant NBR is given along the FPC and SOPC axes. Clearly, the FPC–SOPC plane is divided into two parts: for all combinations of values of FPC and SOPC *below* the curve for NBR = 0 (solid line), population will decline, while growth occurs for all values *above* this curve. The most surprising feature of the zero-growth curve is that it covers a relatively narrow range of FPC values. If FPC is lower than about 180 kg grain-equivalent per head and per year, population growth is negative, irrespective of SOPC, whereas growth is positive for all FPC values larger than about 400, again irrespective of SOPC. The curve shows also that the nature of the equilibria found in test calculations is representative of all possible equilibria under the conditions considered. Consequently, in the World3 population sector without age distribution, population can, under more or less normal conditions, only be stable for fairly low values of FPC. This explains why the value of POPE is so sensitive to *F*. If *F* is given, POP will inevitably rise or decline until FPC has attained a value around 220 kg grain-equivalent per person per year, and hence, in a first approximation, the value of POPE might be calculated as *F*/220.

At very low levels of material welfare and health care, equilibrium can only be sustained at somewhat higher values of FPC owing to a slight reduction of life expectancy by the effects of crowding. If crowding is eliminated by setting FPU equal to zero, the zero-growth curve remains below FPC = 300 for all values of SOPC considered. Enforcing crowding effects by assuming FPU = 0.8 causes the equilibrium FPC to be as much as 600 for low SOPC.

The results also throw light upon the questions concerning the stability and uniqueness of the equilibrium values POPE: for any combination of values of the sector inputs satisfying PPOLX = 0 and SO = 3/2 * IO, there is only one equilibrium value of POP. All equilibria on the zero-growth curve are stable, since an increase in POP will automatically induce decline, a decrease causing growth. Hence, the POPE values shown before (Fig. 7) are all stable and unique. They represent the model's carrying capacity for population under the exogenous conditions given.

It is important to note that in the calculation of NBR as a function of constant FPC and SOPC values the lags were neglected. In the standard run, NBR is about 1.8 percent per year in 1980 (FPC = 467 and SOPC = 365), while according to Fig. 9 the value would be about 0.7 percent. In addition, the standard-run trajectory sketched in Fig. 9 (heavy broken line) shows that without any lags the population would be growing continuously until 2080, whereas in the actual calculation decline already sets in around 2030. Apparently, since the inputs to subsystem I are not constant at all during the run, the total effect of the lags is important.

Let us, before examining the lags, consider Figs. 8 and 9 more closely. As is clearly shown in Fig. 9, the population decline tends to be dramatic at low values of FPC, whereas growth is weak at high FPC. This once more illustrates and explains the asymmetric dynamics of POP already noted.

In the more general sense, the figures show that at very low income and food per capita, population decline occurs, which turns into growth as income and food per capita become larger. At high levels of income and availability of food, growth is weaker. This is shown more explicitly in Fig. 10, where the relative rate of population growth NBR has been plotted against FPC, going along different lines of constant FPC/SOPC (or *F*/SO). For *constant F* and SO, the system will move along one of these lines until equilibrium has been attained. The curves have been obtained directly from the table underlying Fig. 8. For the standard-like situation, *F*/SO varies between 0.65 and 4. We can now

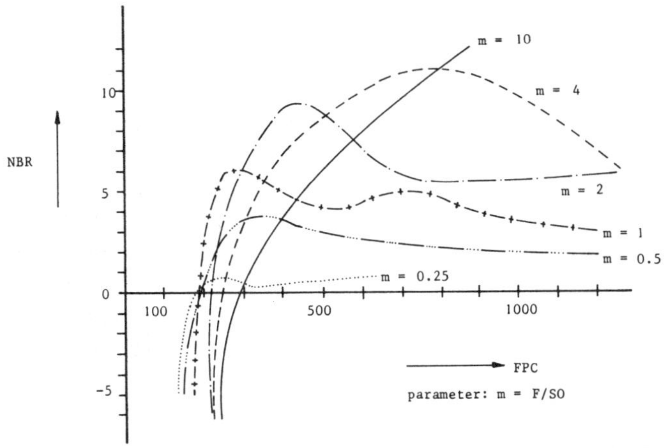

Fig. 10. Relative growth of population as function of food per capita along lines of constant $m = F/SO$.

formulate a more precise conclusion about the causes of population growth in World3. Section-wise linearization of the curves of Fig. 10 according to

$$\text{NBR} = a * \text{FPC} + b \tag{3}$$

(where a and b are constant for each section) leads to the simple equation

$$1000 * \frac{d\ \text{POP}}{dt} = a * F + b * \text{POP}. \tag{4}$$

Hence, the net feedback of POP to its own rate of growth is *positive* if $b > 0$, and *negative* for $b < 0$. The conclusion is that, as long as $0.5 \leq \text{FPC/SOPC} \leq 4$, and, roughly, $\text{FPC} > 300$, the tendency of population to grow exponentially can be explained from the inherent instability of the sector. However, for lower FPC values, the sector acts as a nonlinear lag, and population growth is driven by the growth of the inputs. Since POPE primarily depends on F, the more general conclusion follows that POP is mainly determined by F if FPC is low, but behaves more autonomously if FPC is higher.

However, various simplifying assumptions have been made in the preceding discussion. Notably, the lags in subsystem I (Fig. 1) have been fully ignored. We will examine their influence more closely now.

3) *Influences of the Lags:* As explained in Section III, omission of *all* lags causes considerable deviations from standard-run behavior. This was confirmed by comparison of the figures for NBR under the condition of constant inputs to subsystem I and of the growth rates actually observed during the calculations. The strength and nature of the influences of the five lags were therefore reconsidered starting from the observation that, owing to mechanisms included in the other sectors, *all* the per capita variables (IOPC, SOPC, and FPC) rise or decline most of the time.

Let us first examine the effects of the lag AIOPC, which is used to calculate the family income expectation FIE:

$$\text{FIE} = (\text{IOPC} - \text{AIOPC})/\text{AIOPC}. \tag{5}$$

Because AIOPC is a "delayed" IOPC, FIE represents an average relative rate of change of IOPC. If IOPC is growing, FIE is positive and stimulates the birth rate according to the function relating the desired completed family size DCFS to FIE. Conversely, if IOPC is declining, the birth rate will be reduced.

The effects of the lags DIOPC, PLE, and FCFPC are similar (a decrease in the birth rate owing to rising income is postponed somewhat), but less pronounced, while the lag EHSPC acts in the opposite direction. Quantitatively, the influence of these four lags is about equal, and this explains why total omission of EHSPC, together with either FCFPC, PLE, or DIOPC has hardly any effect. Since the effects of four of the five lags are similar, and the per capita variables usually behave in a similar way, it is obvious why the combined effect of all lags together is considerable.

Another, more detailed analysis by linearization of the influences along the different paths for the 1970 situation showed that the principal effect of the lag AIOPC is to augment the relative population growth by roughly 0.2 percent for each percent growth of IOPC. Because the net contribution of the other four lags is about half as great, we hypothesize that the combined effect of all lags is to change the relative population growth by about 0.4 percent for each percent relative growth of the per capita variables. In the growth phase of the standard run, IOPC rises about 2 percent per year, which would mean an additional population growth of 0.8 percent per year. This result agrees fairly well with the difference observed between the static values (Fig. 9) and the actual growth rates in the standard run.

V. Effects of Age Distribution

For most calculations four age groups have been distinguished by Meadows *et al.*: 0 to 14 years ($P1$, people), 15 to 44 years ($P2$, (re)productive), 45 to 65 ($P3$, productive), and 65 and older ($P4$). The dynamic behavior of $P1$ to $P4$ is calculated according to

$$\frac{dP1}{dt} = a_{11} * P1 + a_{12} * P2, \tag{6}$$

$$\frac{dP2}{dt} = a_{21} * P1 + a_{22} * P2, \tag{7}$$

$$\frac{dP3}{dt} = a_{32} * P2 + a_{33} * P3 \tag{8}$$

$$\frac{dP4}{dt} = a_{43} * P3 + a_{44} * P4, \tag{9}$$

in short,

$$\frac{d\mathbf{P}}{dt} = A * \mathbf{P}. \tag{10}$$

The coefficient a_{12}, accounting for the births entering the first age group $P1$, is a function of total fertility TF, while all other nonzero coefficients, representing mortality and/or transition of people from one age group to the next, are based on age-specific mortality functions depending on LE.

Let us now try to find out what the main differences are between the behavior of the models with and without age distribution. Since subsystems I of both models are identical, we shall focus our attention on the properties of subsystems II and III. Let us define NBR_4 for the four-level model as the net rate of increase of the whole population per 1000 people per year. Then comparison of the behavior of NBR and NBR_4 as a function of TF and LE (subsystem II) must reveal the differences, if any, between the models.

The situation of constant TF and LE will be examined first. NBR is calculated according to (1). Since in (10) all coefficients are constant for constant LE and TF, the solution of these equations follows from the eigenvalues and eigenvectors of matrix A. As all eigenvalues are real, a_{21} and a_{12} being positive, the solution $\boldsymbol{P}(t)$ can be written as

$$\boldsymbol{P}(t) = \sum_{i=1}^{4} \alpha_i * \boldsymbol{V}_i * \exp\,[\lambda_i(t - t_0)], \qquad (11)$$

where λ_i is the ith eigenvalue, $\boldsymbol{V}_i$ the corresponding eigenvector, and α_i are constants. For sufficiently large values of $(t - t_0)$, the term in (11) with the largest λ will be predominant, and all age groups will grow at a relative rate determined by this eigenvalue. Hence, if LE and TF are kept constant, NBR_4 is equal to 1000 times the largest λ of matrix A. The corresponding eigenvector gives the equilibrium ratio $P1 : P2 : P3 : P4$ under the same conditions.

For values of TF between two and eight children per family and LE between 20 and 80 years, NBR_4 and the corresponding age structure were thus computed. In most cases, the static differences between NBR and NBR_4 were found to be less than 0.2 percent per year. NBR_4 tends to be somewhat higher than NBR for TF > 4 and LE > 60, and also for LE < 25, whereas it is somewhat lower for most other values. Only for very low fertility (TF $\simeq$ 2) and LE > 25, is NBR_4 up to 0.5 percent per year lower than NBR. But at low TF and large LE, population growth is only weak in the model without age distribution (see Fig. 8 for large FPC and SOPC values). Hence, in addition to the equilibrium enforced by starvation, another possibility for a stable population may exist in the four-level model. Test calculations have confirmed this hypothesis. The four-level population can also be in equilibrium at high per capita values, crowding acting as the stabilizing factor.

From the eigenvectors it follows that as LE becomes larger, the fraction of POP in $P4$ will become larger and in $P2$ smaller, $P3$/POP and $P1$/POP remaining more or less constant. If TF rises from two to eight, $P1$/POP increases from 0.20 to about 0.45 (irrespective of the value of LE), and the other fractions become smaller.

If LE and/or TF are not constant, subsystem II acts instantaneously in a model with one age level, but, in a more level model the lags in the adaptation of the age structure come into play.

Two strategies for investigating these effects have been followed. First, the solution of (10) for constant LE and TF was studied, and it was found that a new equilibrium age structure is approached exponentially, the time constants being equal to the inverse of the differences of the dominant eigenvalue and the other eigenvalues. This means that subsystem II contains a number of stable dynamic elements, with time constants of 25 years or less.

Secondly, to gain a better insight into the nature of these additional lags, sudden changes were made in the—otherwise constant—values of LE or TF in simulations with the four-level model. It was found that the adaptation of NBR_4 to its new equilibrium value consists of a sudden change, followed by a gradual approach to the new value, which approach may be characterized by a time constant in the order of 15 years. The nature of the response, however, is quite different for the two input variables. The immediate reaction of NBR_4 to a stepwise change in TF is smaller than the ultimate change, whereas the response to a step change in LE displays overshoot followed by a gradual adaptation in the reverse direction. Apparently, the lags inherent in the age structure slow down the effects of changes in TF, whereas the reactions to changes in LE are accelerated. The first effect is well known in demographic literature (see, e.g., [14], [15]), but the second is rather unknown, perhaps because usually only fertility is considered as the variable controlling the size of a population.

A comparison of two age distributions with different LE but equal TF explains the phenomenon. If the value of LE is suddenly changed, mortality and hence CDR will also change at once. Subsequently, the age distribution will gradually adapt itself to the new LE value. In the case of a sudden decrease in LE, $P4$/POP will start to decline, $P2$/POP will start to rise, and, as mortality in $P4$ is considerably higher than in $P2$, the crude death rate will decrease somewhat. The overall result of a downward step in LE is that CDR will jump upwards suddenly, and then decline gradually to its new equilibrium value.

However, since the World3 behavior is hardly affected by the introduction of age distribution, these effects seem to be fairly insignificant. In the standard run TF decreases in the first part of the run, whereas LE rises, so that the initial population growth will tend to be larger in a model *with* disaggregation. Similarly, the fall of POP in the decline phase of the four-level model is somewhat more pronounced, the differences being only slight, however.

VI. Nonstandard Conditions

A number of simplifying assumptions were made to reach most of the conclusions mentioned above. We shall now briefly review the subsystem behavior under different conditions.

A. High Persistent Pollution Ratios

If PPOLX increases above 30, its influence on life expectancy will be considerable. Therefore, high ratios of pollution might radically alter the pattern of growth and decline discussed in Section IV-B. For example, the NBR = 2.5 curve in Fig. 8 shows that if pollution reduces NBR by 2.5 with respect to its value for PPOLX = 0, population equilibrium is possible at almost any value of FPC.

B. Extreme per Capita Values

Fig. 8 shows that NBR is almost zero for large FPC and SOPC values, and this indicates that even minor changes in the structure (such as age distribution) or parameters of the model can lead to alternative equilibria. Because of the properties of the capital sector, however, high welfare cannot be sustained for long unless the model is modified fundamentally [8].

At low SOPC (e.g., less than 50), fertility control effectiveness FCE is low. However, desired total fertility is fairly large (low IOPC), and, because life expectancy is reduced by low SOPC (lack of health services!), the maximum total fertility is relatively low. As a result, the influence of FCE on the birth rate is but small.

C. Alternative Assumptions

The methods and techniques developed in the preceding sections can also be used to study the subsystem properties if changes in model assumptions are introduced. The assumption of a low and constant desired completed family size (DCFS = 2), for instance, radically changes the pattern of growth and decline of population at low IOPC. However, it would be beyond the scope of the present paper to go into further details.

VII. Concluding Remarks

The basic properties of the equations determining population dynamics are in accordance with the well-known demographic transition: high death rates and fast decline if food supply is below subsistence level, lower death rates, but rather high birth rates leading to growth for sufficient food and relatively low income and services per head, and low birth and death rates causing very slow growth only at high welfare levels. The effect of the lags incorporated in the equations is to stimulate the growth of population if welfare and food per head are rising and to counteract growth or reinforce decline if the reverse is the case.

The model's carrying capacity, i.e., the maximum sustainable size of population, is primarily determined by the total food production. Stabilization is mainly caused by a rise in the death rate due to starvation. If the size of POP is far below the model's carrying capacity, population growth tends to be exponential owing to positive feedback within the sector itself, but slow. In contrast, negative feedback is dominant when starvation is approaching, and then changes in POP are mainly determined exogenously, the equilibrium size being primarily a function of food production.

Population growth at high income is only weak. As a result, fairly small modifications, such as the introduction of age distribution, a rise in pollution, or a slight reduction of the desired completed family size may result in rather more attractive population equilibria.

In the standard run, initial population is somewhat lower than its equilibrium value (Fig. 7). Then, owing to the growing food supply, the model's carrying capacity increases rapidly, and until 2000, population grows more or less exponentially, which growth is stimulated by the rise in the output per capita values. After that, growth is moderated somewhat by increasing income per head. About 2020, the turning point is reached in the important capital and agricultural variables, and a fast drop in their per capita values sets in. As a result, population decline sets in since the effect of the lags predominates. Subsequently, food becomes scarce, and decline is enforced. Hence, the growth and decline are not determined endogenously, but impressed on the sector by the capital and resource, and the agricultural sectors.

For standard-run conditions of the various relations, it is only the influence of food and services on the death rate and the influence of family income expectation on the birth rate that mainly determine the behavior of POP. The influence of pollution is virtually nil.

From the methodological point of view it is interesting to call attention to the variety of simple methods that have been used to analyze the sector. The strategy was to first explore a specific type of behavior by means of sensitivity analysis and other computer simulations. After that, a more general approach was followed to study the subsystem's inner workings and behavioral characteristics. Starting from a decomposition into three subsectors, the equilibrium and disequilibrium properties of the sector as a whole and of each of its components were investigated. A more extensive discussion of methods and techniques for understanding complicated models follows in [11].

Appendix
Elucidation of the Sector Equations

Life expectancy LE depends on four factors: health services, food per capita, crowding, and pollution. The effect of health services is calculated from service output per capita SOPC via two table functions and a first-order lag of 20 years. FPC and PPOLX affect LE in a lag-free way, each via a nonlinear table function. A higher value of PPOLX, or a lower value of FPC and/or SOPC will have a diminishing influence on LE. LMC, the lifetime multiplier from crowding, depends on two factors: the fraction of population urban FPU (a function of total population POP) and the crowding multiplier from industrialization CMI, depending on industrial output per capita IOPC.

The total fertility TF represents the average total number of children born per fertile woman over her entire reproductive lifetime. TF is calculated from the desired total fertility DTF, the maximum total fertility MTF (a function of LE), and from the fertility control effectiveness FCE. FCE is determined by the fertility control allocations per capita FCAPC via a third-order lag (20 years). FCAPC is calculated from the need for fertility control NFC, depending on the maximum total fertility MTF and on the desired total fertility DTF, and from SOPC. DTF stands for the average total number of births per woman desired by the individual members of the population, and is the product of the desired completed family size DCFS and of a function of the delayed (third-order, 20 years) life expectancy, compensating for the probability of child mortality. DCFS is the product of the social family size norm SFSN, of a function of the family income expectation FIE, and of a constant normal factor equal to four children per family. SFSN accounts for the tendency to desire a smaller family as income is higher. Via a lag (third-order, 20 years), SFSN is a function of industrial output per capita IOPC. The effect of FIE is to augment DCFS as income perspectives are good, and, conversely, to diminish DCFS if IOPC declines. FIE, which serves as an index for the average growth of income per head, is calculated by comparison of the actual and the delayed IOPC. The lag (AIOPC) is first order, and the time constant is equal to three years.

References

[1] J. W. Forrester, *World Dynamics.* Cambridge, MA: Wright Allen, 1971.

[2] D. H. Meadows, D. L. Meadows, J. Randers, and W. W. Behrens, *The Limits to Growth.* New York: Potomac Associates/Universe Books, 1972.

[3] D. L. Meadows *et al., Dynamics of Growth in a Finite World.* Cambridge, MA: Wright Allen, 1974.

[4] H. S. D. Cole, *et al., Thinking About the Future, A Critique of the Limits to Growth.* Chatto & Windus for Sussex University Press, 1973.

[5] T. W. Oerlemans, M. M. J. Tellings, and H. de Vries, "World dynamics: Social feedback may give hope for the future, *Nature*, vol. 238, pp. 251–255, Aug. 4, 1972.

[6] P. J. Vermeulen and D. C. J. de Jongh, "Growth in a finite world—A comprehensive sensitivity analysis," *Automatica*, vol. 13, pp. 77–84, 1977.

[7] O. Rademaker *et al.*, "Project group 'Global Dynamics'," progress reports numbers 1–7 (1972–1977). Available from O. Rademaker, Eindhoven University of Technology, P.O. Box 513, Eindhoven, Netherlands.

[8] W. Thissen, "Investigations into the World3 model: The capital and resource subsystem," *IEEE Trans. Syst., Man, Cybern.*, vol. SMC-6, no. 7, pp. 455–466, July 1976.

[9] W. Thissen and C. de Mol, "The agricultural and persistent pollution subsystems in World3," this issue, pp. 159–172.

[10] W. Thissen, "Investigations into the World3 model: Overall model behavior and policy conclusions," this issue, pp. 172–182.

[11] ——, "Investigations into the World3 model: Lessons for understanding complicated models," this issue, pp. 183–193.

[12] A. L. Pugh, III, *DYNAMO II User's Manual.* Cambridge, MA: M.I.T. Press, 1970.

[13] P. J. Vermeulen and D. C. J. de Jongh, "Parameter sensitivity of the *Limits to Growth* world model," *Applied Mathematical Modelling*, vol. 1, pp. 29–32, June 1976.

[14] N. Keyfitz, *Introduction to the Mathematics of Population.* Reading, MA: Addison-Wesley, 1968.

[15] J. H. Pollard, *Mathematical Models for the Growth of Human Populations.* New York: Cambridge University, 1973.

System Science and Societal Integrity

JOHN W. SUTHERLAND

Abstract—**Much criticism of the system sciences has been directed at the disparity between the claims we have made about the resolution power of our procedures and instruments, and what we have actually accomplished in the field. The extent to which such a disparity actually exists is discussed, and two strategies are posed for its reduction: a) increasing obedience to the set of procedural dictates under which system philosophy was initially formulated; b) bringing the rates of development of system theory and system methodology into better balance, such that system ontology becomes in part responsive to system epistemology. To the extent that these strategies are employed, the results of system-based research and development exercises should become clearly differentiated from the types of exercises conducted by nonsystem scientists. System science would then be moved further toward fulfilling the unique scientific role to which its philosophical and procedural predicates would appear to assign it.**

Introduction

CRITICS of the system sciences have not been shy about pointing to disparities between what we promise and what we actually accomplish. The most telling criticism is that most system science exercises are virtually indistinguishable from those that might be done by the traditional discipline and procedure-bound scientists we so freely criticize. In short, we tend largely to attack the same problems as do the normal scientists, and we most often attack them in exactly the same way. Particularly, most system science models—like those of traditional science—are either exclusively rhetorical or mathematical in substance; few incorporate statistical or mathematical technology that could not be found in models being built by nonsystem engineers or operations researchers; our constructs are often no less a product of paradigmatic closure than those of the normal scientists, etc. At the conceptual level, many system scientists continue to confuse theory with ontology and epistemology with technology. Thus, if we were to ignore certain semantic eccentricities, there is often nothing of any substantive significance that allows a particular piece of work to be uniquely identified as a proper system science exercise.

It is enough for outside critics of the system sciences to merely mention this and move on. But those of us who rely in some measure on system science as a vehicle for our academic and professional work must consider questions of cause and repair. In this paper, we want to share a set of propositions about the cause of our deficiencies. Particularly, we want to suggest, and in the text elaborate on, the following: While system scientists often pose themselves as uniquely well-qualified to deal with a level of complexity that defies traditional science, we have often neglected the very epistemological dictates that would lend this pose some legitimacy. It is this neglect—perhaps most often inadvertent—that has led us away from the issues of great challenge for which we are responsible in the scheme of science. And because we have so often limited the level of analytical challenge we are willing to accept, we have dampened the rate at which system science technological potential becomes translated into actual instrumental (resolution) power. This is the essential reason why the system sciences have had such sparse influence on policy-level decisions in either government or industry, and consequently such little positive impact on societal integrity. And the consistent incongruence between epistemology and technology is perhaps the decisive reason why system science has been treated with such skepticism by the general academic community.

From another perspective entirely, it is possible to see the deficiencies in the system sciences as the natural result of the extent to which system theory has leapt ahead of—and consequently been uninformed by—system methodology. This puts us, operationally, squarely into the very dilemma from which the progenitors of our field sought to turn us away—into the dilemma of "normal science," where fields of inquiry are rigidly partitioned along both paradigmatic and procedural lines.

If these propositions have any merit as a collective "cause" of system science deficiencies, then two strategies for their repair become immediately evident: a) implement the epistemological platform that belongs uniquely to the system scientist, and is perhaps the only reason for identifying ourselves as such; and b) let the rates of development of system theory and system methodology be adjusted to a better balance, such that ontology becomes at least in part responsive to epistemology.

The extent to which the arguments raised here will be considered depends a great deal on the extent to which the reader shares what must perforce remain an opinion of the author: that what we do in the system sciences really matters, that there are some symptoms of societal debilitation that can only be removed to the extent that system science practice reaches its full potential! Criticism, when the mission is of sufficient weight, is not the same thing as arrogance. And so while I readily apologize for the limits of

Manuscript received November 21, 1978; revised April 11, 1978 and July 24, 1978.

The author was with the Division of Administrative Studies, Rutgers University, New Brunswick, NJ 08903. He is now with the Department of Administrative Sciences, Southern Illinois University, Carbondale, IL 62901.

Reprinted from *IEEE Trans. Syst., Man, Cybernetics,* vol. SMC-8, pp. 837–848, Dec. 1978.

wit and precision that are sure to affect portions of this paper, I cannot really apologize for the ambition that drove it.

The System Science Imperatives

It is possible to find the impetus for the system sciences in the attempt to put back together again a world that generations of analytic science had taken apart.[1] The implication was that the instruments and perspective of empirical, reductionist science were not equal to the task of adequately comprehending or managing the complex systems that have evolved since the 1930's. From its inception, system science had two essential messages. First, the major arms of science—deduction and induction—would have to be cojoined; the result would be a methodology that was free of the biases and strictures of traditional science. Secondly, interfaces would have to be developed between the many disciplines that have a role to play in the management of complex systems: engineering, psychology, sociology, anthropology, ethics, logic, applied mathematics and statistics, economics, etc.[2] The system scientist could come from any discipline and have a wide range of academic and professional ambitions; but he would have to be something more than an engineer, something more than a sociologist or economist. In its most exuberant moments, the early literature asked that the system scientist be part prophet, part poet, and part technician. The prophet in him would see to the moral aspect of his work, and bend his ambitions to the welfare of the broader society and its progeny. As poet, there would be an aesthetic quality to his work. His perspective would have the breadth to see relationships that escaped the discipline-bound scientist. Unlike the instrumentalist, his solutions would be notable for their elegance, not merely their expediency. Unlike the strict empiricist, he would be able to comprehend the virtuosity of every human being, and replace the empty sympathy of the reformer with the empathy that traditional social science often lacks. The technician in him would temper the prophet's fondness for parabolic exercises and the poet's disinterest in operational constructions. Finally, in the person of this remarkable offspring of modern science, qualitative and quantitative analysis would be brought together and put to use to solve problems that neither could attack in isolation. Finally, the long divergent streams of inductive and deductive science would be merged, and science would once again be whole [3, ch. 1], as it was before Greek natural science separated from Greek philosophy.

This was the vision of Popper, Conant, Polanyi, von Bertalanffy, Rapoport, Boulding, *et al.* The common contention was, I believe, that system science would become significant as a fundamentally new epistemology—as a next-generation methodology. It would provide a home for the true interdisciplinarian; it would provide a base where humanists and instrumentalists could intermix and complement each other's emphases; it would be a fertile field for the growth not only of individual scientists, but of the community of science itself [4].

The enthusiasm of these early system scientists usually stopped short of outright arrogance. They criticized many traditional scientists for failing to give their subjects the deserved measure of potential for surprise and confoundation. They worked against the null hypotheses of empiricist epistemology: reductionism, analytic closure, the *ceteris paribus* assumption, decomposition. In their stead they posed a new set of criteria. Rather than reductionism, exercises should be "holistic" in scope. In place of analytic closure there should be an attempt to make as many variables as possible endogenous. The *ceteris paribus* assumption should be abandoned in favor of attempts at the simultaneous manipulation of many determinants, even at the expense of some precision and analytical expedience. Decomposition would be permitted, but only when there was a prior understanding of how the subsequent synthesis could be performed. For them, then, the economies of empiricism were false ones except when the problem at hand was an essentially simple one, and the expediency of traditional analytic methods was suggested to be delusionary unless the subject at hand was inherently deterministic.[3]

There was, then, some merit to the suggestion that system science could begin to rationally resolve many of the issues that had long been condemned to the realm of rhetoric. Indeed, when viewed collectively, the premises of the new epistemology promised to amplify the power of science to resolve significantly complex issues[4]:

a) the interdisciplinary imperative;
b) the demand for constant complementation between quantitative and qualitative instruments;
c) the dictate that we seek scientific efficiency through the exploitation of structural and functional isomorphisms that might exist among otherwise disparate subjects;
d) the alert against evangelism or out-of-hand advocacy of the restrictive paradigms that for so long constrained traditional science (In short, any a priorisms that were allowed to enter system science exercises would take the form of tentative heuristics, not axiomatic constraints.);
e) the commitment to operations in the nexus between inductive and deductive processes, such that reason, insight, and observation could operate as complements rather than competitors;
f) the conviction that the emerging properties of the problem at hand should determine the analytical approach, and not the preferences of any strict methodological platform (with the consequent realization that there may be as many "scientific methods" as there are uniquely defined problem categories);
g) and, again, the caution against attempting to force

[1] See, for example, the remarkable collection of papers in [1].

[2] This was one of the essential points made by von Bertalanffy in [2].

[3] For more on this, see V. E. Frankl, "Reductionism and nihilism," in [1].

[4] For an expanded set of epistemological dictates, see [3, Introduction].

reductionism and analytic procedures on subjects that might better be defined and treated as "organisms" (via, for example, action-research as opposed to algorithmic approaches).

There are other interpretations of system science epistemology that might be less restrictive and more imaginative than given here. But these are the minimal operational dictates for those who would seriously challenge the quandaries of our age. They are, of course, largely gratuitous for those scientists whose interests are confined to essentially simple subjects. To attempt to employ them in the face of well-bounded, well-behaved phenomena would be inappropriate. But if the system scientist is indeed the prime challenger of complexity, they become essential elements of his analytical approach. Yet the thesis we now somewhat reluctantly propose is this: many system scientists continue to pose themselves as uniquely well-equipped to challenge and contain the complexity that traditional science lets pass by, but largely ignore the epistemological predicates that lend this pose some legitimacy. In short, there are anomalies in the system sciences as they are now constituted, and these anomalies are not lost on our critics.

Anomalies in the Development of the System Sciences

There are two generic indictments that can be leveled against the system sciences as they are currently practiced. First, much of the work that tries to pass as *system theory* is often merely ontological in significance. That is, the work develops axiomatic or axiological models that may have either weak or elliptical links to the arena of hypothesis construction and experimentation. Moreover, like most ontological exercises, their components are shielded from validation precisely through the absence of these links. Simply, we often tend to confuse proper theory with prophecy. The second criticism aims at the operational (or applied) system sciences: much work that passes as applied system science is virtually indistinguishable from work that might be done in the normal course of evolution of the social sciences, the engineering sciences, or such fields as operations research and computer science.

The complaint that many system theoretic exercises are actually insular essays in ontology is occasionally well taken. And this complaint applies equally well to some mathematical artifices that are essentially axiomatic in character and to social science or "humanistic" exercises that have a strong axiological underpinning. The axiomatic exercises tend to emerge along a path that parallels the logic and configuration of pure mathematics itself. The other path is that taken by system philosophers. Here, for the most part, an attempt is made to develop a general system theory predicated on a mixture of biology and sociology. The ambition is to make the cell, the individual, the collectivity, etc., all intelligible in the same essential terms of reference, and extends from the work of Laszlo, Ashby, Allport and Miller, *et al.*

While the attempts at a mathematical general system theory are laudable for their ingenuity and discipline, there are some apparent problems. First, many of these constructs are developed not by mathematicians *per se*, but by engineers. As a result, a certain regularity of structure and function is built in at the earliest axiomatic stage. When the constructs are elaborated (as per Forrester, Beer, or Mesarovic) we see a generic system that is hierarchical, fully differentiated (such that each unique function is performed by a uniquely structured subsystem), and one whose relationship to the outside world is regulated via some sort of cybernetic mechanism. Now despite much attention to the matter of contextual determinants, this hierarchical, fully-differentiated, and quasi-cybernetic system pretty much monopolizes the attention of the mathematical system theorists.

But critics such as Trist and Emery have pointed out that contextual (environmental) properties may have the decisive role in determining the properties of an individual organism or system [5]. Their logic, along with other sociotechnical studies,[5] should make it clear that hierarchical structure, full differentiation, and cybernetic behavior are not equally valuable in the face of all possible environments. For example, there are some systems that are not hierarchical but reticular, such that every subsystem or system component is equipotential with respect to every other [7, ch. 2]. There are systems that are not fully differentiated; very simple systems tend, rather, to be segmented, while extremely complex systems may be structurally segmented but functionally differentiated [7, ch. 2] And both von Bertalanffy and this author have tried to show that there are critical theoretical limitations on the cybernetic paradigm. It is gratuitous in the face of essentially simple "random" stimuli, and it does not have the power to resolve behaviors in the face of extremely complex, unprecedented events (or, at the least, it becomes diseconomic as a behavioral modality) [8]. What critics are left with is the thought that while hierarchical structure, differentiation, and cybernetic regulation are mathematically tractable, these attributes may have been introduced at the expense of true "generality."

This complaint is not directed at those mathematical exercises that are methodological rather than definitional in character. For example, the mathematical general system theory developed by Klir (and the work of Zadeh and others in the development of isomorphic referents) cannot be accused of being expedient or overly presumptuous. For the very terms of their inquiries reflect one or more of the epistemological precepts just mentioned. Klir's system referents are developed with an eye toward the different analytical procedures that must be followed as the system at hand turns from determinism to stochasticity, etc. [9]. Thus the significance of Klir's work is mainly in terms of establishing a general theory of problem solving—an essentially epistemological ambition—and not in trying to answer for all instances of system structure and behavior in a single set of equations or set definitions. It may be that significant

[5] The most remarkable of these is perhaps [6].

numbers of real-world mechanisms may be encompassed neatly by such structures, but when they pretend to be *general* system theories they are apt to be indicted for procrusteanism. Excluded from this charge would be the attempts to structure a vocabulary of general system concepts through the elaboration of notational topologies (and the rules for their manipulation).[6] Such exercises, for the most part, seem innocent of operational implications. They are not trying to capture reality of themselves, but rather to provide us with a new nonsubstantive semantics.

Many of the efforts of the rhetoricians to construct a general system theory are perhaps even more open to criticism. Much of this work has been underpinned, often in the most oblique way, by the principles of "open systems" laid down by Katz and Kahn over a decade ago [11]. There is an odd mixture here of biological terms, computer science terminology, and quasi-biology (even a little classical physics gets mixed in). Some of the attributes assigned open systems are merely gratuitous. For example, open systems are defined as net importers of energy from the environment and classed as significant in terms of their "negentropic" properties. The problem here is that while entropy is a concept primarily of teleological significance (except in the face of certain essentially "closed" systems where we can actually measure it), our concern with social systems and organisms, etc., is bounded in time and space. This poses certain logical problems. For example, if negentropy is an essential property of open systems, what do we do with the system that has turned entropic (i.e., is losing order and definition)? If entropy and negentropy refer to processes, then they are largely gratuitous and might better be replaced by more operative concepts. If, on the other hand they are absolute state definitions, then they cannot be employed at all in the domain of the social sciences. As a physicist attending a recent conference on social system theory suggested, "It looks as if they [Katz and Kahn] got their orders of infinity mixed up."

The litany of the open system then moves on to attribute behavior to an essentially first-order feedback-based mechanism that is entirely reminiscent of the cybernetic processes of the engineer. Homeostasis and steady-state behavior is applauded as being a signal of system sophistication. Adaptivity becomes the *sine qua non* of organic survival.[7] But von Bertalanffy had a rather low opinion of homeostatic behavior, seeing instances of operant conditioning and cybernetic operations not as high-order behavior, but as relatively primitive accomplishments. Scholars as diverse as Maslow, Piaget, and Weisskopf would take exception to the assumption that adaptivity is an unmixed blessing.[8] There are even some possibly contradictory properties posed for the open system. For example, at the same time it is postulated as being steady-state, it is also supposed to possess the capability for equifinal behavior. But equifinal behavior is not a logically possible alternative for a system with a stationary configuration; nor would it be possible to restrict an equifinal system to a fixed relationship between inputs and outputs—by definition. It might be useful, therefore to restrict open systems to one or another of these properties, but not both simultaneously. Here again, then, it is possible to suggest that social system science ontology has leapt too far ahead of social system science epistemology.

There are also some faults of exchange between theoretical properties and operational possibilities. For example, Allport's work underlies most of the propositions of Katz and Kahn, and therefore many of the propositions of contemporary social system theorists. But one of Allport's key concepts is that open systems can be defined by "cycles" of behavior—qua events—that tend to close upon themselves and be repeated [14]. Yet do events associated with an open system ever completely close? By definition they would not. And if open systems possess an equifinal capability, even their iterative quality would be merely probable rather than necessary. What it amounts to, then, is that Allport's and Katz and Kahn's insights are extremely useful for the analysis of a subset of the universe of systems, but not for the treatment of systems in general.

I have singled Katz and Kahn out for criticism here not because their work is bad, but rather because I think it is very good when taken as a whole. I use their text in my seminars on social systems, and certainly a majority of social system scientists owe them a debt. But we can see what has happened in the course of their model-building exercises, and by extension what has happened to many others who set about trying to develop an encompassing theory of systems. The substantive aspect of their work was developed largely in the absence of any epistemological considerations. As a result, we have been offered a set of theories of social systems that, in both implication and discipline, are not much advanced beyond the structures developed by Weber and Parsons *et al.* To a certain extent, they are less promising. For most restrict behavior to the adaptive feedback-based modality that, in psychology, passes as operant conditioning; and this has been shown to lead to lower order rather than higher order behaviors (and does not exhaust the behavioral repertoire of any but the simplest organisms) [2, pp. 188, 189]. The potential for organisms or organizations to affect their environment—a potential inherent in the work of K. Lewin [15] done several decades ago—does not really make its appearance in most of our theoretical structures. The emphasis on adaptivity and processual-structural congruence is also reflective of the now largely discarded theory of "functionalism" in anthropology. There is still a tendency to ignore internal sources of system disruption, despite the eloquent warnings given over a decade ago by the eminent social system theorist, W. Buckley [16]. Finally, with the exception of some terminology borrowed from information theory and ecology (and the ubiquitous allusions to the laws of thermodynamics), our social system models—my own included—are not very much different than those developed along the line that traditional sociology has followed since the early nineteenth

[6] See, for example, the Appendix to ch. 5 of [10]: S. E. Seadler, "Ideologics."

[7] For an interesting indication of how thoroughly this assertion has colored work in organization theory, see [12].

[8] See, for example, [13], or Piaget and Inhelder, "The gaps in Empiricism," in [1].

century. And as most of us hopefully know, it is precisely this line of development that has been so eloquently criticized by sociologists themselves, most notably R. Merton [17] and A. Gouldner [18].

My colleagues in the social sciences may thus in large measure be excused their confusion about our system science efforts, for, virtually without exception, the constructs that we have developed suffer from the same type of assumptive closure as those developed by traditional social scientists. Often our constructs even go beyond this, taking recourse in instrumental techniques (like Allport's concept of event-cycles or the borrowing of the servo concept from process control theory) that cannot be used where we suggest, in the face of considerable complexity, and are really not needed in the face of those simple systems where their technical limitations would not be exceeded.

Many of these theoretical shortcomings impose themselves in terms of anomalies or inconsistencies at the operational level. For example, because the qualitative and mathematical system scientists are both so enamoured of the cybernetic paradigm, one would think that considerable integration might take place in actual system analysis exercises. But this has not been the case. Most qualitative system scientists remain uneasy and unschooled in quantitative techniques; others proudly proclaim their innocence of mathematical sophistication as a mark of humanism. Most quantitative scientists appear to have a very difficult time coming to grips with the axiological bases that affect sociobehavioral issues. As a result, the quantitative and qualitative systems scientists tend largely to go their own way, indifferent (and sometimes even inimical) to each others' work. To the mathematical system scientist, the qualitative practitioner often seems more interested in changing the world than in understanding it; the qualitative system scientist is not shy about criticizing the mathematical system scientist for pursuing exercises that are paragons of precision but trivial in implication. It is thus that the old recriminations of the broader scientific community find their way into the new system sciences.

Indeed, the major difference between the traditional and system-based social sciences seems to be merely this: whereas traditional social science pays automatic deference to the assumptive predicates of a Maslow or Marx, the system-based exercises pay homage to the largely nonapodictic constructs of a Miller or Ashby or Laszlo. The work predicated on such constructs is itself nonapodictic (largely incapable of being validated or unvalidated using procedures of normal experimental inquiry), often bereft of any linkage to the quantitative domain and seldom explicitly interdisciplinary in substance. We are thus indeed susceptible to the argument that we are merely sociologists or psychologists or economists trying to pass under a different name.

On the quantitative system science side, our models are seldom significantly different than those that are developed by industrial engineers or operations researchers. Most complex or "fuzzy" determinants are treated exogenously. Few attempts have been made to employ disparate computational forms within a single analytical framework (e.g., simoptimization, a technique used relatively early in the history of the Rand Corporation studies, has still not made any significant appearance in system constructs) [19]. The models of the Forrester and Beer genre use only low-order mathematics and impose considerable constraints on the ability to manipulate critical relational or structural alternatives. We are still sorely restricted in our optimization exercises by the number of state variables that we can accomodate. We have barely begun to exploit the potential of modern algebra or topology, with the exception of a few transportation models based on neural network theory and more modest exercises using catastrophe theory (this latter technique, unfortunately, being largely restricted to operations across a well-behaved manifold, and thus is of little use in the face of significant phenomenal complexity). More pointedly, we, as quantitative system scientists, have not taken a lead in developing procedures by which our instruments might be merged with those of our qualitative counterparts. In fact, there is some legitimacy to the argument that we have done little that would not have been done anyway in the normal course of evolution in traditional engineering. Some contributions have indeed been made in the employment of new search algorithms and in the expansion of the applications of Bayesian statistics and mathematical logic (see, for example, [20]). Yet, at the operational level, most of our work shares the instrumental character of operations research exercises, the preoccupation with ordering means to achieve ends set elsewhere. Or, in those few cases where we develop models of policy significance (with the distinct exception of very forward-looking procedural schemes like interpretive structural modeling),[9] the premises we offer are most often rooted in simple extrapolation or projection. The anomaly between the grand ambitions of our constructs and their relatively unsophisticated technical bases has not been lost on critics of the system sciences.

There is the temptation to discount the criticisms of this section as simple evidence that we are not yet perfect. But some of the criticisms have more serious implications. We must ask, for example, if we are moving toward perfection at anything like the rate we should expect; or is it possible that we are actually retrogressing? Are we lowering our aspirations to draw toward the limitations of our current technology? I shall try to say something about this issue in the final section of this paper. Here, however, I want to explore a more important implication: the possibility that the neglect of our epistemological mission is causing real harm in some areas. In short, are we actually contributing in any way to the debilitation of the societies we pretend to serve?

The Dimensions of Default

I think that a case can be made against us (although, admittedly, many instances of incompetence may be laid at the door of practitioners who call themselves system scientists without any legitimate right to do so). The logic of the

[9] In the interpretive structural modeling process, policy premises are both explicit and rationalized, and not merely as *a priori* constraints. For more on this technique, see [21, ch. 8–16].

indictment is this: there are some societal problems with which traditional science simply cannot cope, but that would be susceptible to attack by a properly constituted system approach, one that adequately reflected our unique epistemological imperatives. But because we have largely ignored this heritage, we have not moved against these problems, and so they remain essentially unchallenged and uncontained. There may, indeed, be other areas in which we have defaulted, but the four to be explored in this section appear to me to be the most serious.

The first of the problems is at once the most important and the most subtle: *the consistent failure of societal systems to produce any significant synthesis of the benefit sets that are potentially available to societal members.* I have dealt with this problem at length elsewhere [22, ch. 6] so shall be rather brief here. The point of the argument is that both historical and contemporary societies tend to emphasize only one of three generic benefit sets—spiritual, social, and material—at the expense of the others. As a result of restricting societal benefits to just one of these three corridors, diminishing marginal returns (satisfactions) affect large proportions of the population, and no societal system has been able to persist over long periods of time in any given benefit configuration.

It is thus that societal change becomes intelligible in terms of an essentially simple replacement process (a pendular function), rather than in terms of real evolution. For history shows us that societal systems change largely by merely displacing one exclusive benefit set with another—with no net gain in satisfaction to system members except in the very short run. Real societal evolution, on the other hand, would appear only with the emergence of systems that sophisticate themselves by calculating the terms of the trade-off among spiritual, social, and material benefits, and thus "manage" their configuration with respect to calculated marginal utilities. In the absence of such sophistication, systems tend to be either emphatically spiritualistic (transcendental), exclusively materialistic (ruled by a calculus of gratification and, ultimately, titilation), or essentially vehicles for the containment of sensual and emotional interchange. Seldom, if ever, has any real *synthesis* appeared, and only recently has the concept of a properly syncretic benefit set been defined [23].

It must be that system managers (irrespective of the particular political modality they adopt) have consistently operated along the path of least analytical resistence, and the passivity of instrumental science has abetted them in this. Specifically, societal management has become characterized mainly by a crude compensatory calculus, where the several benefit sets—and the equilibrators of wealth, status, and power—are viewed as simple substitutes rather than as proper complements. Treating them as substitutes absolves the societal manager from the responsibility of exercising a trade-off logic, which is much more analytically demanding than the calculus of substitution. And considering that normal science is really not prepared to execute the required trade-off logic, the societal managers may perhaps be excused their expediency. But system scientists cannot be let off so easily. For this is precisely the kind of task for which we are uniquely well-equipped. Given the epistemological imperatives of proper system science, the idea of seeking an "optimum at the margin" among "fuzzy" elements does not sound as implausible as it might if we were arguing from the platform of traditional empirical, analytic science. However, from the system science perspective, optimality is a metaphorical rather than measurable attribute.

Some system scientists may be appalled at the idea of a managed society. But societies have always been managed. It is just that for perhaps the first time, we have the opportunity to take societal management out of the hands of those least willing to be informed by either fact, logic, or ethic: the popular ideologue, the agent of *realpolitik*, and the anti-intellectual bureaucrat. Yet in a review of system science literature, this author finds very few studies that even pretend to operate at this level of significance (among the few are [24] and [25]). Indeed, it seems as if we have abandoned the field of societal management and concentrated our efforts on problems that are most likely to occur only within the confines of the corporation or limited government agency. And considering some of the complaints raised earlier, it may be that even there we are doing little that could not be done by traditional administrative scientists.

The second problem area derives from the first: *the tendency for virtually all societal systems to absolve the truly significant issues affecting them—the metaproblems—from any formal problem-solving discipline* (see [10, ch. 4]. A metaproblem entails variables from more than one academic discipline, involves the necessity to impose some sort of balance between two or more apparently competitive ends, and is generally susceptible to moral, axiological, or ethical interpretation. In short, questions of the guns or butter, energy or environment, individualism or collectivism, welfare or Darwinism sort are metaproblems, and they are seldom brought to the scientific arena for treatment. Science thus tends to be occupied with the essentially simpler questions a societal system might pose—with generating means to achieve ends set elsewhere, or with mechanical or deterministic exercises—while the more critical issues go begging for rational attention. There are at at least two reasons for this. First, it is a peculiarity of democratic government to look upon metaproblems as mere value disputes, as instances of the inescapable and largely irresolvable differences of opinion that must be accommodated within any populist state. Compromise—by weighing the relative affection that different constituencies display for the adverse axiological positions—is the essential service that political mechanisms perform. Ideally, the scientific community would challenge the legitimacy of this process by demanding that it be allowed to resolve issues through the manipulation of fact and/or the application of logics (or through the attempt to arrive at quasi-optimal trade-offs through the application of modern mathematical and statistical procedures put in service of qualitative analysis, the first task of which is to raise axiological issues to a new level of abstraction, where they can be treated axiomatically and, ultimately, empirically). But many scientists abet the political process rather than challenge it. Empiricists, for example, perforce remove themselves from value-based issues (e.g., energy versus environment, tuna versus dolphins), largely because the instruments at their disposal cannot operate well with axiological predicates. Rationalists, themselves agents for opinion within the scientific community, have a genuine affection for the process of political compromise, and are themselves often found supporting one or another of the *a priori*, adversarial positions. Many system theorists, captives of the extreme position of Bateson and others about the irreducibility of axiological predicates, express horror at the thought of science ever presuming to disabuse others (or itself) of their axiological anchors, and thus seek to put value disputes beyond the ethical, if not the technological, limits of modern science.

Thus science has traditionally allowed, if not actually reinforced, the tendency to attack societal metaproblems through arrant and uninformed advocacy. If there is any blame to be attached to system science here, it is that we have not honored our promise to serve as translators of abstract instrumental innovations into concrete problem-solving capabilities. Nor have we honored the prophetic role set out for us; rather, we have too often offered ourselves as the servants of power and prevailing opinion rather than their challengers.

This brings us to the second reason why metaproblems remain outside the aegis of science. Instrumental limitations or paradigmatic predilections may legitimately excuse most normal scientists from the responsibility for dealing with axiologically significant issues. And system science, which collectively has the potential to force rational trade-offs, has largely let this potential remain unfulfilled by electing to attack problems that do not really demand full exploitation of the technological state of the art [10, ch. 5]. Our default is, clearly, less innocent than that of normal science, whose paradigmatic preferences exclude its responsibility. Its commitment to reductionistic, analytic, and algorithmic exercises is completely incongruous with the analytical challenges that metaproblems pose. For metaproblems cannot be "solved" unless they are raised to a higher level of abstraction than that at which they appear; it is only in this way that a rational accommodation can be found between the competitive axiological positions. Beyond this, a syncretic analysis is essential, and this goes against the empirical-inductive grain of normal science. Moreover, metaproblems demand a deductive or heuristic closure at the initiation of the solution process, and the normal scientist's discomfort with *a priori* operations marks him as impotent in the domain of the metaproblem. And there is an even more awesome challenge to normal analytical methods. Particularly, no adequate solution to a metaproblem can be generated unless linkages are forged not only between the multiple disciplines involved, but also between the instruments of qualitative and quantitative analysis. While many traditional discipline-bound scientists might be mildly sympathetic to the rationale behind interdisciplinary activities, I have encountered virtually no interest in system science's suggestion that we be prepared to work in the nexus between "methodological" camps. To a significant extent, then, metaproblems escape scientific attention because normal science is not prepared to attend to them. Some scientists may suggest that value-ridden issues are not the proper province of serious science anyway, but I suggest that our instrumental shortcomings make this argument a very transparent conceit indeed. I suspect a much more cogent reason for traditional science's neglect of the metaproblem is that it represents the worst of all possible analytical worlds: it is intractable in the face of positivist epistemology; it argues against quick "completion experiences;" it forbids the scientist to disassociate himself from the effects of his work.

There may, then, be many more or less acceptable reasons for normal science's default in the domain of the metaproblem. But the metaproblem is the essential rationale for the existence of the system sciences, and we ignore it only at the expense of our integrity. For when we elect to work at the same, generally "tractable" problems that monopolize the attention of our nonsystem counterparts, we are passing the metaproblems into the hands of those least qualified to deal with them: the errant evangelist, the politician, and the courts. For the evangelist, the concept of a rational trade-off is anathema; the political procedure is pretty well exhausted by the forcing of a casual compromise between two usually equally ill-considered axiological extremes; and the courts, who are increasingly becoming the *de facto* governors of democratic systems, are generally equipped to deal only with the "semantics" of critical issues, leaving the more critical syntactical aspects entirely unconsidered. Indeed, the judiciary system, with its reliance on exegesis, its restrictive rules of evidence, and its obeissance to procedure and precedent is the institution of society that perhaps *least* reflects the interests and sensitivities of the "scientific method."

The first two consequences of the system sciences' default of its legacy—the failure to evolve properly syncretic, truly sophisticated societal systems and the immunity of metaproblems from rational attack—are operative at a fairly high level of abstraction. But there is a third characteristic of societal systems for which system science practitioners are perhaps more directly and evidently responsible: *the tendency for the essential institutions and organizations of a society to consistently and blatantly exceed optimal scale ofplant*. We system scientists, along with many management scientists whose enthusiasm also occasionally exceeds their judgement, have encouraged centralization and concentration, without really generating a management technology that is sufficient for the disciplining of significantly large enterprises. In so doing, we have contributed in the most direct way to the erosion of the integrity of both the commercial and governmental sectors of the societies we pretend to serve. In short, our propaganda has leapt ahead of our ingenuity, and our constituents suffer for it.

The key index that would be used to judge the contribution of the system sciences in the management sphere would be the schedule of productivity rates across the range of both commercial and public enterprise. Although disaggregation in this area is very difficult, there is no trouble in suggesting that the productivity schedule has dampened over the last fifteen or twenty years. Part of this dampening reflects the deterioration of the American capital base (relative to Asian and European processors), and part is to be accredited to inordinate wages (relative to productivity) distributed to labor. But there are also structural changes, authored by system and management scientists, that are at fault. For example, we have proposed computer-driven management information systems as a mechanism for disciplining larger operations, and have hence abetted the rise of the usually diseconomic conglomerates and integrated complexes (which, only in the rarest case, realize any real synergistic effect, because of the lack of functional interdependence among the components of the conglomerates or vertically integrated firms) [10, ch. 5].

These and other technocratic incursions have a very serious flaw, one which we seldom admit: what most management and decision "systems" do is to impose a sort of quasi-optimal discipline on the lower level operations of an enterprise, without doing much at all for the range of strategic and normative (e.g., goal-setting) functions which are far more critical to an enterprise's effectiveness. In this offense against sound management theory, we have become the willing cohorts of the computer salesmen who have been so successful in seducing executives into the concept that higher management consists of little more than "inducing" strategies from the data base that flows upward through the organization: in "staying on top of things," as it were. It is clearly in the cybernetician's interest to reinforce the computer company's pitch, for the cybernetic logic is basically one that can only benefit from the conceptual equation of management with control [26]. Indeed, it might be suggested that we are the primary force behind the instrumentalization of both the commercial and governmental sectors, for the sum of our technocratic impositions has been to give the impression that judgment, reason, and moral referents can legitimately be stripped from the managerial calculus, without detriment to the larger society. Thus the manipulative risk-averting posture adopted by the modern corporation is a close cousin of the rude *realpolitik* that we find operating at the governmental (and intergovernmental) dimension.

Systems scientists are not, of course, the only contributors to this situation. But we are the very best hope for its reform. What we must guard against, however, is the same type of instrumentalism that we might criticize in traditional management and industrial engineering exercises. We must not oversell our instruments, trying to force essentially simple tools into contexts that exceed their resolution power. Some of us have done this, and have defended ourselves by trying to redefine management in terms that suit our instrumental limitations. For example, the prototypical enterprise for the cybernetician or system dynamics advocate is not the wide-open equifinal system of a von Bertalanffy, but the effectively closed-loop finite-state "machine" seeking adequate adaption to an environment that holds only peril, not opportunity. It thus becomes a creature attempting to oscillate, with minimal deviation, around some referent trajectory or target parameters that just appear from outside, like the magical objectives that we found in our textbook problems as engineering students. The question as to where the goals, objectives, or referent trajectory come from is somehow left unanswered by the system scientist, just as it is left unanswered (but with some legitimacy) by the traditional operations researcher.

But when the system scientist fails to answer for the process by which goals and referent strategies, etc., are set, these normative areas by default become subject to the least promising discipline. In fact, the basic and highest order organizational objectives are usually established by those organization members who have the least comprehension of modern analytical or procedural techniques: the rhetorical executives who still dominate most commercial and governmental enterprise and who, if not actively hostile to scientific management, are at least usually indifferent to it. It is no accident, then that the corporate and governmental giants we helped spawn are generally so ill-managed and so often defiant of the norms of microeconomics. For we have largely just concentrated on advancing those assumptions that are useful to us: that growth is a signature of efficiency; that diminishing returns are not operative in the range of organizational realities; that control is the *sine qua non* of management, and that adaptation is the *sine qua non* of organizational survival. In the process, we have lent legitimacy to the dubious proposition that while operations at the lower level of enterprise should indeed be subjected to scientific discipline, the higher order decision and policy functions can properly remain exercises in reactivity, raw speculation, or casual deduction. But the alternative to the backhanded support of this proposition might be unpalatable to some system scientists, for it would demand that we forward the precept that the most immediate constraints on the size to which differentiated organizations may legitimately grow are the limitations on the ability of existing managerial technology to accommodate complexities. But because we have been so reluctant to broadcast these limitations (and every proper system scientist should be aware of them), we have not challenged nor moved to contain the gigantism that characterizes so many contemporary organizations. As a result, more and more organizations have become diseconomic when viewed against the background of microeconomic normatives and also against the background of aggregate societal welfare (and most also continue to be condemnable from the standpoint of rudimentary humanistic and human relations criteria).

A fourth symptom of societal debilitation for which system scientists must bear some responsibility is *the lack of adequate accountability mechanisms for significantly complex enterprises* (which, as was suggested above, we have in part helped to sponsor). In short, despite the system scientist's preoccupation with the technology of organizational control, and despite our pretensions to considerable sophistication in this area, a larger and larger proportion of societal components gives every evidence of being effectively "out of control."

There are some rather obvious defenses we could offer for this hypothesis, but they would largely be restatements of assertions we have already offered in other areas of this paper. And they would miss what I think is the essential dimension of default: the lack of *a priori* accountability instruments. Simply, the problems and procedures of *a posteriori* accountability and control pretty well exhaust the interests and capabilities of the accountant, the traditional management theorist, and the cybernetician *qua* system scientist. Indeed, the basic postulates that underlie cybernetic logic force the analyst's attention to the after-the-fact event, such that organizational management becomes an effectively reactive, corrective process. Now, again, there is no real problem with this perspective when we confine ourselves to the lower, effectively deterministic levels of enterprise (made such by the "envelope" of certainty passed down in the form of policy constraints from higher level authorities)—or when the system with which we are dealing is an essential mechanism, operating in a relatively placid environment.

But when we try to apply the same accountability logic to more complex enterprises, or to senior levels of the operational hierarchy, then several possible confoundations can occur. First, the results of some decision exercises may not ultimate in the near-time frame, which means that the process which led to the errors may persist over considerable periods of time; secondly, it may be that the "events" that are associated with some strategic or policy-level decision do not close on themselves, or involve ramifications that are too far flung, complicated, or amorphous to be accessible to us (in short, effects become disassociated from their cause); thirdly, as we move towards the highest levels of enterprise, the decisionmakers themselves become the only available accountability agents, and these are unlikely to impose great burdens on themselves, or be too critical of their own misjudgements (the concept of stockholders or trustees, etc., playing a functional role at the apex of the accountability process is largely vapid). In summary, then, as the processes or functions we are attempting to bring to accountability become more complex, the technology of *a posteriori* control becomes more and more tenuous.

However, even if *a posteriori* accountability were feasible at all levels of enterprise, one confoundation would still remain: we could not act to *prevent* errors that occur in the face of unprecedented decision and policy problems—and unprecedented problems make up a significant proportion of the problem repertoire of any complex organization. Now some might suggest that schemes such as capital budgeting and program budgeting act to invoke *a priori* accountability; but we have already suggested, in an earlier paragraph, that technological constraints operate to make such platforms largely gratuitous in the face of considerable complexity. Yet the logic here is good, for such schemes really represent procedural algorithms. And, when a decisionmaker is unlikely to be held accountable for the after-the-fact *results* of his exercises, the only recourse we have is to hold him accountable for the *method* under which he operates—responsible for the process by which he went about trying to solve an operational or policy problem.[10]

[10] For an explanation of *a priori* decision control, see [10, ch. 4].

In more precise terms, a decisionmaker would be responsible for presenting evidence that he acted to *minimize the expected value of decision error*, consistent with the economic dictate that he not spend more on formal analysis than he could expect to gain from the increased expected accuracy. In such a situation, the decisionmaker is absolved from responsibility for unsought consequences, provided that he gave the problem all the analytical discipline it deserved. It is only by asking senior-level personnel to be accountable in terms of a "method audit" that they become at all susceptible to meaningful control—and only through such a technique that we increase the probability of being able to preempt high-level errors. This recourse becomes especially important in the domain of public enterprise, where the lack of a profit impact, or some other relatively neat surrogate for a "bottom line," is often unavailable to us.

The method audit scheme can only become feasible to the extent that a set of *normative* referents are available to the decisionmaker. To a great extent, operations research and industrial engineering have provided such referents only where they are not needed, for the simpler, lower level operational problems. Here, neat transformations of the precepts of microeconomics provide the practitioner with algorithmic instruments such as linear programming and inventory control processes and queuing models, etc. At a somewhat higher level, the instruments of decision theory, rooted in extensions of probability analysis, can discipline relatively pedestrian forecasting, projection, and planning exercises. But operations research and decision theory have little to tell us about the way in which severely stochastic or true "strategic" issues might be disciplined [27]. Indeed, anyone interested in being advised as to the conduct of higher level decision analysis has to break clean from the quantitative-algorithmic domain and enter the world of the rhetorical management theorist who, not always with apology, may be entirely ignorant of mathematical, statistical, or engineering technology.

To a great extent, then, operations research and decision science have defined for themselves the domain of optimization and extrapolation, which serves very well indeed the challenges that might arise at the middle and lower management levels of the complex enterprise. And so far as I can tell, the statisticians and econometricians have provided us with sufficient algorithmic control over well-behaved markets and relatively stable macroeconomic environments. So, partly because nobody else has adopted them, but mostly because we define ourselves to be the primary challengers of complexity, the domain of the higher order decision falls to the system scientist. It is he who must husband the strategic domain, and he who must help provide the normative referents against which the higher order organizational functionaries are to be held accountable. Yet, as anyone who reviews the sum of system literature can quickly see, we seem to spend more time doing jobs that are already being done by decision scientists, operations researchers, and engineers than doing those jobs that are uniquely and solely our responsibilities.

And so it goes, *ad nauseam*. Our propaganda poses us as society's last resort against complexity, confoundation, cataclysm, and quandary. Yet so often we are found rooting in the same essentially simple, tractable fields as our discipline- or methodology-bound counterparts; worse, many of us—susceptible to that same arrogance which for so long has tended to confuse insular science with "pure" science—have chosen to confine our efforts to that airy and largely sterile world of axiomatic architecture. As such, the four dimensions of default that I have sought to define here do not really exhaust the indictments that might be brought against us; other critics of our system science pretensions are both more determined and more ingenious than I. But the message from these critics is that unless we are genuinely willing and able to assume the role we have allocated ourself in the scheme of academic challenge, then we should stop the sham and rejoin the disciplines from which we came.

Restoration and Repair

So far as I can tell, the schedule of rewards available to members of the scientific community is not well-correlated with the degree of difficulty of the problems we elect to attack. Arguing from a materialistic perspective, there is perhaps some legitimate reason why we have, as system scientists, forgone the opportunities for genuine challenge that have been presented to us. I think, however, that the normal sciences have followed the path of least analytical resistance for another reason altogether: they have consistently tended to let ontological apriorities determine both the content and quality of their epistemology. For example, empirical science generated the ontological proposition that all truth lay in the domain of the tangible, the measurable, and the manipulable. Subjects that did not meet these criteria were irrelevant to science. This, of course, would include its own ontology, being primarily aprioristic in substance. In short, empiricist epistemology did not have the ambition or the capability of validating its ontology; it could merely accept it on faith, and set about executing its implications. As a result, early system science was well able to criticize normal science for underestimating the true complexity of many of the subjects it elected to study, and for ignoring a wide range of significant problems and phenomena which could not be accommodated by its methodology. Physics, because of its transition from classical to quantum ontology, is about the only traditional discipline to escape this indictment.

Two subjects which are apparently of great interest to system scientists did not escape: biology and geology. For example, these two fields are still ruled by the ontological precept that *natura non facit saltum* (nature does not proceed by leaps). These fields gave the impression that the "gradualism" of their subjects was induced from observations of nature. Yet the empirical record indicates that both geological and biological evolution have substantially been products of rapid, "catastrophic" discontinuities. Yet the epistemology of these fields never subjected the ontology to examination, and hence developed instruments, models, and assertions that merely responded to the "axiom" of gradualism. The portrait of nature that we derive from such ontological-epistemological constraints is thus very much distorted in favor of a simplicity which doesn't, in fact, exist. As S. J. Gould suggests, "Gradualism is a philosophy of change, not an induction from nature" [28]. The implications of the failure to realize this are only now being recognized in biology and geology.

A similar situation results when we review the history of the social sciences. A. Gouldner explains the origin of sociology's ontology in the desire to advance the cause of stability and conservatism following the French Revolution. As a result, the apriorities of sociology entail such artificial simplifying assumptions as the imperative of shared values among members of social systems, the relegation of potential conflict to the exterior of the system, etc. The result, again, is a body of theory and hypotheses that tends to impose an unwarranted tractability (and insularity) on the subjects treated by sociology [18]. The otiose quality of most economic constructs stems from their roots in 18th century biology (following Smith), where aggregates were viewed as simple sums of their components; this assumption reached its ultimate in Keynesean artifices in this century. In psychology, the Behaviorist ontology has dominated to deny human beings the capability for volitional or autonomous behavior; the result is a consistent understatement of the complexity of human behavior [29].

One explanation for anomalies such as these is that traditional scientists are largely unschooled in matters of ontology and epistemology; most accept the paradigmatic closures of their fields as givens and then forge ahead, content merely to apply or sometimes extend their implications. How guilty are we, as system scientists, of the same behavior? Do the anomalies noted in an earlier section—the anomalies in the development of the system sciences—reflect the dominance of ontology over epistemology in our own field? I think that the answer is yes, in part. But the system scientists (at least those who trouble to concern themselves with the *a priori* predicates of their field) have sometimes gone one step further. Many seem to be enamoured of Bateson's assertion that we are irrevocably condemned to a situation where the *a priori* underpinnings of a field are not separable. Bateson suggests, "In the natural history of the living human being, ontology and epistemology cannot be separated. His (commonly unconscious) beliefs about what sort of world it is will determine how he sees it and acts within it, and his ways of perceiving and acting will determine his beliefs about nature" [30]. While this assertion may adequately reflect the behavioral determinants of the "natural" human being, it cannot be posed as a norm for scientific behavior. For it simply poses ontology and epistemology as potentially co-determinant, and this cannot guarantee any immunity from the anomalies that result when ontology is allowed to rule. Nor does it protect us against those cases where instrumental limitations serve to constrain the ambitions of a discipline. To generalize, it is precisely the failure to properly distinguish between ontological and epistemological predicates that led to the anomalies set out in the second section of this paper. Most efforts to define a general system theory, be they mathematical or rhetorical, simply impose an *a priori* order on the world (an ontology) without troubling to articulate the epistemological predicates that would allow us to validate the normative or axiological premises; in short, our theoretical structures are, as a rule, no more apodictic than are those of traditional science. And, at the operational level, system science *techne* seems largely uninfluenced by any encompassing system science *praxis*. This is perhaps the surest reason why so many system science applications are indistinguishable from work that might be done by traditional practitioners (and, tautologically perhaps, the reason why proper system science has made such little impact on the affairs of everyday business and government).

It is precisely this confusion between ontology and epistemology that opens us to indictments for arrogance. For in the theory-building sector, system science activities become intelligible in essentially the same terms that Kuhn uses to describe the ontologically driven processes of "normal" science [31]. It is also the reason why operational system science cannot escape the strictures that have affected empirical science. For it is at the operational level that the ultimate arrogance appears: that which (following Ashby) suggests that whenever anything is done well, it is system science whether the practitioners know it or not. The result is to turn applied system science into the rough equivalent of the "principles of management" school in the administrative sciences—the school that sees management as little more than the casual distillation of experience, and thus develops practitioners who are largely bereft of both elegance and ethic.

These arguments might help point the way out of the philosophical and methodological wilderness in which the system sciences have been wandering. Particularly, let us recall that Dewey has already raised the final argument against Bateson's denial of differentiation between ontological and epistemological predicates. For Dewey demanded that science realize that apriorities of any order may be subject to the same type of disciplined investigation that we bring to other subjects, that is, apriorities are irrevocably "of this world."[11] Given this possibility, we may now consider the advice given by Bacon as he stood at the threshold of the Age of Reason: Let theory be informed by fact! But in doing this we need not drive ourselves into the same corner as the empiricists, where fact is itself uninformed by either reason, intuition, or inspiration. Nor need we descend into the sterile, existentialistic depths of phenomenology, where every phenomenon is notable primarily for its unadultrated uniqueness (and where strict empiricism is carried to its most excessive form: hypothesis-free analysis). And from these suggestions derives what I think is the essential message that the progenitors of system science tried to tell us: *let the development of a general system theory wait upon the maturation and acceptance of a general system methodology*. In short, let us for once control our ontological enthusiasms so that the *a priori* predicates of our field may be made accountable to an emerging epistemology. This is the essential strategy for removing science from the control of Bateson's "natural" man and Kuhn's "normal" science.

[11] In fact, Dewey tells us, "... knowledge, mind and meaning are part of the same world they have to do with, and are to be studied in the same empirical spirit that animates the natural sciences" [32].

This dictate is implicit in the epistemological predicates reported in the opening section of this paper. The underlying logic is this: there is simply nothing to be gained by entering a field of inquiry constrained by unvalidated ontological (or cosmological) propositions. Consider, for example, that rationalism was discarded by its excessive distrust of any sense data (empiricals); empiricism is now in difficulty because of its unwillingness and inability to deal with conceptual phenomena; and phenomenology is generally neglected because its ontology argues against both the existence and utility of any collective knowledge. The lesson of the failures was not lost on the progenitors of system science. Indeed, it was these very lessons that led them to argue for an ontology (and cosmology) that was essentially *empty* of theoretical or methodological constraints:

a) Search for scientific significance in the domain of empiricals *and* in the realm of concept.
b) Accord both qualitative and quantitative components *a priori* equal access to our emerging analytical constructs.
c) Pursue the "truth" of our system subjects in the nexus between ever more specific and context-dependent deductive inferences and ever more extensive inductive inferences.

This synthetic ontology—unlike, let us say, its rationalist, empirical, or phenomenological predecessors—argues against axiomatic or axiological constraints finding their way into the methodological or procedural domain where they would serve to predetermine election of subjects for study, first premises of hypothetical constructs, etc., and therefore bias results of scientific inquiry. Simply, the "empty" ontology allows the scientist to tailor his analytical approach to the emerging properties of the subject at hand.

These several dictates are primarily of structural significance in that they define (however broadly in this case) the search space over which practitioners may legitimately wander. But every proper epistemology must also be of procedural significance. It is the procedural dictate(s) that provides the dynamic discipline that any active science demands. The procedural dictate must, of course, be in sympathy with the structural dictates. Now we, as modern system scientists, need not invent such a dictate, for a most appropriate prescription was lent us by Aristotle over two millenia ago:

> ... it is the mark of an educated man to look for precision in each class of thing just so far as the nature of the subject admits. It is evidently equally as foolish to accept probabilistic reasoning from a mathematician as to accept rigorous proofs from a rhetorician.

Any aphorism may be subject to different interpretations, but I read the Aristotelian prescription as folding back on the structural dictates we just noted. It, too, demands that the scientific investigator let the emerging properties of the problem at hand determine the analytical strategy and instruments he will employ, and shy away from *a priori* constraints. Turning things a bit to the modern comprehension, Aristotle allows us to consider that the "scientific method" is not a constant, but is actually a continuum of treatments that may be partitioned to reflect the way in which the domain of subject problems and systems may be partitioned.[12] That is, there are as many scientific methods as there are uniquely defined problem or system categories. Thus, when faced with essentially mechanical, deterministic problems, the empirical preferences of the positivist-experimentalist or the algorithmic instruments of the finite-state engineer become the *congruent* analytical response. As the problem at hand or subject system becomes somewhat more complex, taking on distinctly stochastic overtones, then a set of instruments with advanced resolution power is called into play under the criteria of Aristotelian congruence—and the appropriate scientific method becomes that generalized by experimental statistics, and not the hypothesis-free, trial-and-error approach that is at the root of our traditional engineering processes. And as the problem or system becomes still more complex, becoming perhaps severely stochastic or even indeterminate in character, then our criterion of congruence would demand that we give this complexity the humility and caution it deserves; in short, we sacrifice our relatively well-developed optimization and extrapolative-projective instruments in favor of more powerful but less mature tools such as dynamic programming, stochastic-state analysis or game-based modeling. And, just as we shift to the higher order components in our analytical arsenal, so must we shift basic procedural strategies. Particularly, as the problem or system at hand becomes significantly complex, we must depart from the experimentally rooted procedures of normal inductive science and turn to new procedures such as those defined by "action research."[13] At the highest levels of inquiry, then, the system scientist enters the deductive domain, using the immature but important techniques of heuristic analysis and syncretic model-building, perhaps disciplined by instruments such as the Delphi method or consensus statistics [34]. In short, it is at the highest levels of inquiry that the system scientist becomes a creator and manipulator of concepts, developing qualitative "front-end" models to accommodate the complexity that statistical and mathematical instruments cannot.[14]

We are now back at the beginning. This portrait of the idealized system analysis process—from the deterministic to the heuristic—gives eloquent testimony to the wisdom of our progenitors'suggestion that the proper system scientist would have to be part prophet, part poet, and part technician, and that the strict rhetorician and strict mathematician should be excluded from the field. It is this kind of system analysis exercise, free of normative or axiological constraints (except in the form of tentative, initiating heuristics), that gives real substance to the system sciences' claim to be uniquely qualified to attack and contain true complexity. And though much more effort and energy needs to be spent in articulating an epistemology that will be acceptable to the community of system scientists, one thing is compellingly clear: the integrity of the societal systems in which we serve are intimately tied to the integrity of the system sciences themselves.

[12] For an explanation of how this sets up the instrumental arsenal of the system sciences, see [7, ch. 5].

[13] For a recent survey of the concept of action research, see the papers in [33]. An operational interpretation of action research is given in [10, ch. 5].

[14] For an example of the linkages that can be established between qualitative "front-ends" and quantitative constructs, see [22, ch. 5].

References

[1] Koestler and Smithies, *Beyond Reductionism.* New York: Macmillan, 1969.

[2] von Bertalanffy, *General Systems Theory.* New York: Braziller, 1968.

[3] J. W. Sutherland, *A General Systems Philosophy for the Social and Behavioral Sciences.* New York: Braziller, 1973.

[4] E. Laszlo *The Systems View of the World.* New York: Braziller, 1972.

[5] Emery and Trist, "The causal texture of the environment," *Human Relations*, vol. 18, pp. 21–32, 1965.

[6] J. Fried and P. Molnor, "A general model for culture and technology," *Technological Forecasting and Social Change*, vol. 7, no. 1, Mar. 1975.

[7] J. W. Sutherland, *Systems: Analysis, Administration and Architecture.* New York: Van Nostrand Reinhold, 1975, ch. 2.

[8] ——, "Some theoretic limitations on the cybernetic paradigm, *Behavioral Science* vol. 20, no. 3, 1975.

[9] G. Klir, *An Approach to General System Theory.* New York: Van Nostrand Reinhold, 1969.

[10] J. W. Sutherland, *Administrative Decision Making.* New York: Van Nostrand Reinhold, 1977.

[11] Katz and Kahn, *The Social Psychology of Organizations.* New York: Wiley, 1966, particularly pp. 19–26.

[12] S. Terreberry, "The evolution of organizational environments," *Administrative Sciences Quarterly*, vol. 12, Mar. 1968.

[13] W. Weisskopf, *The Psychology of Economics.* Chicago: Univ. Chicago Press, 1955.

[14] F. H. Allport, "A structuronomic conception of behavior," *J. Abnormal and Social Psychology*, vol. 64, pp. 3–30.

[15] K. Lewin, *Field Theory in the Social Sciences.* New York: Harper, 1951.

[16] W. Buckley, *Sociology and Modern System Theory.* Englewood Cliffs, NJ: Prentice-Hall, 1967, especially ch. 1.

[17] R. Merton, *Social Theory and Social Structure.* New York: Free Press, 1968.

[18] A. Gouldner, *The Coming Crisis in Western Sociology.* New York: Basic Books, 1970.

[19] Karr *et al.*, *Simoptimization Research—Phase I* (No. 65-P2.0-1, C.A.C.I.), Rand Corp., Santa Monica, CA, 1965.

[20] K. S. Fu: "Learning system theory," in *System Theory*, Zadeh and Polak, Eds. New York: McGraw-Hill, 1969.

[21] J. Warfield, Societal Systems. New York: Wiley, 1976.

[22] J. W. Sutherland, *Societal Systems: Methodology, Modelling and Management.* New York: Elsevier-North Holland, 1978.

[23] W. Weisskopf, *Alienation and Economics.* New York: E. P. Dutton, 1971.

[24] G. R. Taylor, *Rethink: A Paraprimitive Solution.*

[25] W. I. Thompson, *At the Edge of History: Speculations on the Transformation of Culture.*

[26] Bonini *et al.*, *Management Controls: New Directions in Basic Research.* New York: McGraw-Hill, 1964.

[27] J. W. Sutherland, "Beyond systems engineering," *General Systems*, vol. xviii, 1974.

[28] S. J. Gould, "Evolution: Explosion, not ascent," *New York Times*, Jan. 22, 1978.

[29] J. W. Sutherland, "Beyond behaviorism and determinism," *Fields Within Fields*, vol. 10, Winter 1973–74.

[30] Bateson, "The cybernetics of self: A theory of alchoholism," in *Steps to an Ecology of Mind*, Bateson, Ed. New York: Ballantine, 1972.

[31] T. Kuhn, *The Structure of Scientific Revolutions.* Chicago: Univ. Chicago Press, 1962.

[32] W. V. Quine, *Ontological Relativity and Other Essays.* New York: Columbia Univ. Press, 1969, p. 26.

[33] A. Clark, Ed., *Experimenting with Organizational Life: The Action Research Approach.* New York: Plenum, 1976

[34] J. W. Sutherland, "Attacking indeterminacy: The case for the hypothetico-deductive method and consensus statistics," *Technological Forecasting and Social Change*, vol. 6, 1974.

Some Principles of Knowledge Organization

JOHN N. WARFIELD, FELLOW, IEEE

Abstract—**The substantive complexity of a complex system or issue can be diminished by the introduction of structural complexity to describe that system or issue. To minimize the negative impact of the introduction of structural complexity, graphics should be translatable into prose. To assure that this can be done it is necessary to define a metalanguage that is not only appropriate for use in modeling, but also is adequate to bridge the structural gap to applications. The availability of well-defined structural types helps minimize the impact of structural complexity. Several structural types are defined, and a principle of definition is set forth as a reasonable basis for choosing and distinguishing these types. The set inclusion relation is the most critical one in the metalanguage of structuring in problem definition, because of its key role in the definition of terms, hence of systems or issues. To assure that this relation can be constructively applied, especially with graphics representations, mappings are developed that provide a basis to convert poorly structured information into well-structured information, suitable for a translatable graphic presentation.**

INTRODUCTION

IT IS COMMON practice in systems engineering to use graphics as a means of analyzing, synthesizing, and describing complex systems. While one might suppose that the use of graphics is a way of simplifying the treatment of such systems, another point of view also has merit. From another perspective, the use of graphics in this way amounts to *reducing* substantive complexity by *increasing* structural complexity. The latter point of view has been found to be beneficial in dealing with the organization of knowledge in environmental education [1]. Also this point of view is consistent with the "trade-off" concept that is familiar in systems engineering practice.

The principal benefits of viewing the use of graphics as a trade-off are twofold. First, persons who are unfamiliar with graphics can be advised that the use of graphics does represent a trade-off which allows a reduction of the substantive complexity. This is important in gaining their acceptance of the use of graphics in working with complex issues. Second, a continuous sensitivity to the importance of reducing substantive complexity reminds us that in using graphics no side complexities should be tolerated beyond those that are essential to the effective use of graphics. No new complexities should be introduced as a consequence of the graphics, other than those that have been shown to be inseparable from the graphics.

For purposes of many potential applications, it then becomes essential to devise metalanguages that are keyed to the concept of minimizing the structural complexities introduced by graphics [2].

The organization of knowledge represents an area of application where metalanguages are still not adequate to satisfy the foregoing requirement. Ostensibly designed so that a user can use a computer search to find desired information, one well-known information system that is widely used is said to be effective in the user-search mode only about five percent of the time. In ninety-five percent of the applications, consulting assistance is required if the user is to get any satisfaction from the system. While it may be true that the problem of finding desired information will never yield completely to system design, there is no known evidence to indicate that present systems have incorporated all of the benefits that can be made available through application of design theory.

Many problems in the organization of knowledge go beyond the use of information systems alone. Some of the situations where a good theory of knowledge organization could be applied at a lay level to public issues arise, for example, in technology assessment. Other situations where such a theory could be useful include:

- the development of self-paced learning materials,
- the delegation of prized work assignments in organizations according to the classification of scope and capability of subunits,
- the distinguishing of one federal agency or program from another,
- the preparation of decision packages for use in budgeting,
- the design of libraries,
- the tapping of several disciplines to accomplish interdisciplinary work,
- the definition of complex terms such as "environmental education" and "environment" for purposes of education, management, and administration of the law,
- the clarification of the content and scope of disciplines to entering college students seeking a major field of study.

One approach that has received considerable application is called interpretive structural modeling (ISM) [3]. This approach to structuring combines behavioral concepts, mathematical concepts, group discussion, and computer assistance. It is considered to be a learning process, as well as a means for assisting in the structuring of issues or themes. It has been widely used at federal, regional, state, and local levels in the United States to deal with public issues. It has been used in Australia as an aid in setting public policy for in-service teacher education. Also it has been used in Japan

Manuscript received September 11, 1978; revised February 28, 1979. This work was supported by the National Science Foundation under Grant ENG-75-17609.

The author is with the Department of Electrical Engineering, University of Virginia, Charlottesville, VA 22901.

Reprinted from *IEEE Trans. Syst., Man, Cybernetics,* vol. SMC-9, pp. 317–325, June 1979.

in system design, marketing studies, and planning. The structures produced in these applications typically have been intent structures, priority structures, enhancement structures, process structural models, and intensity structures. Experience with this method has revealed some principles that deserve articulation and wider application, and also has shown some weaknesses in methods of knowledge organization.

It is the purpose of this paper to articulate several principles that have emerged, and to correct some of the weaknesses in the methods of knowledge organization.

The first concept to be elucidated is that of a "translatable graphic." While some readers might find this concept to be quaint, it has been found that this is an essential idea to convey to the majority of practitioners concerned with the organization of knowledge and with public systems analysis.

Then the paper presents a discussion of types of structure, based on the use of digraphs. While most of these structural types have been set forth in [3], it has been found in practice that there is no clear understanding of why these structural types should be considered as basic. A principle of definition of types is given that helps explain why these types are basic. Also the types are needed to help elucidate the later parts of the paper.

Next the relevance of the inclusion relation to the problem definition step of systems engineering [4] is explained, and it is shown that one of the major problems in definition accrues from what we call "structural confusion" that occurs because of the inability to map a digraph back into a unique Venn diagram relationship. This weakens the utility of structural methods, and is the motivation for developing a mapping theory that enables this deficiency to be rectified.

The metalanguage of set theory is shown to be deficient in terms of the requirements of applications, and new types of block representations of sets are introduced. It is shown how these types can be distinguished, and case examples of situations where the newer types are relevant are given.

Finally, new mapping methods are introduced that make use of the block representations, provide for the algebraic elimination of structural confusion, and allow for the effective use of the inclusion relation in structural models.

Translatable Graphics

While we have elected not to reproduce graphics from various works to show how poorly some of them communicate, the reader can readily find examples in most reports that treat public issues. It should be clear that this paper is not recommending that *all* graphics should meet a particular standard of communication. Rather what is recommended here is that whenever graphics are used as a way of diminishing the substantive complexity of a complex issue or system, the graphics should generally meet a communication standard that is set forth here under the heading "translatable graphic."

To introduce the concept of a translatable graphic, Fig. 1 shows four modes of expression. Here it is shown how a prose sentence can be expressed in the conventional way, how it can be replaced with a "graphics sentence," how the same statement can be encoded in set theory notation, and how the idea can be expressed in a Venn diagram.

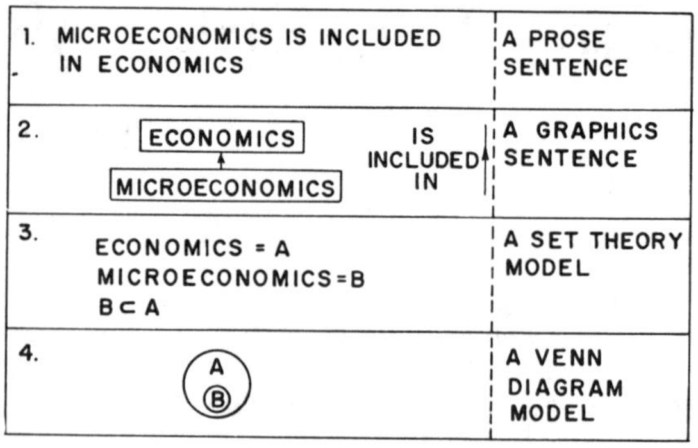

Fig. 1. Four modes of expression.

When large amounts of information are treated in these modes, the structure of the information is masked in a mass of prose, the substance of the information is coded in a mass of set theory notation, the Venn diagrams become unreadable due to excessive overlap, but the structure and substance are both preserved in the digraph-based structural notation, provided the reader can remember the significance of the arrow.

The utility of the digraph-based structures is greatly enhanced when the contextual relation represented by the arrows is transitive [3]. The reason for this is that, when a relation is transitive, any two elements on a digraph structure that are joined by a *path* are also related by the relation, while if the relation is not transitive, then only elements that are joined by an *edge* are so related. This means that the number of graphic sentences represented by a digraph based on a transitive relation may (for large digraph maps) be very much greater than the number of such sentences for a similar structure based on an intransitive relation.

This means that the digraph-based maps enjoy what might be called a "graphical advantage" in that the number of prose sentences represented on one digraph map by graphic sentences can be as great, for example, as would be represented by hundreds of pages of the prose sentences themselves. This is one of the reasons that transitive relations are so useful in structuring complex systems.

It has been illustrated that the underlying structure of digraphs can be developed through the use of set theory [5], particularly with the aid of a form of block representation called a "partition." Because the digraph theory is so readily presented with the use of set theory, it is important to develop a good interface between the two theories.

Unfortunately, there is a problem with the way set theory relates to digraph theory that causes difficulties in developing translatable graphics. This problem is introduced with the aid of Fig. 2, and will be elaborated later in the paper.

Fig. 2 illustrates the transitivity of the set inclusion relation and the intransitivity of the set overlap relation. As will be seen later, there are instances where the same digraph corresponds to quite different Venn diagrams, which means that two-way communication between the two mathematical systems is blocked. This has significance in applications, where translatable digraphs are required.

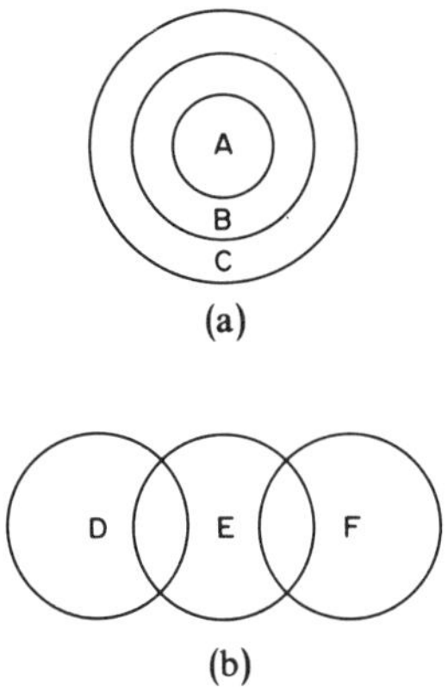

Fig. 2. (a) Illustrating transitivity of set inclusion relation. (b) Illustrating intransitivity of set overlap relation.

With this background, let us now discuss the definition of "translatable graphic." Because this concept is relatively new, it is desired not to freeze the concept too soon. However, one interpretation that can be given to it is the following. A graphic is translatable *if it can be converted unambiguously into prose* (though the prose itself may, of course, be ambiguous). This is the meaning that will be intended throughout the paper.

Examples of such graphics include the products of interpretive structural modeling and sheet music.

Structural Types

It proves to be very beneficial, both in theory and application, to have a clear understanding of structural types of digraph-based structures. While a set of structural types was introduced in [3], experience has shown that there is no widespread concensus on the utility of these types.

In order to help establish the validity of these structural types and to develop types for use in this paper and in later applications, a principle for type designation is set forth, and various structural types are identified in this paper.

The principle that is used in designating types is the following: *If, for any given type of digraph-based structure, a vertex and its adjacent edges are removed, the type of structure should not be changed as a result (except for the trivial cases of each type, which may overlap types).*

With the aid of this principle, one can immediately lend specificity to a class of structures, all digraph-based, that can all represent translatable graphics, and which are useful in applications. All can be regarded as potential knowledge structures.

Fig. 3 shows the simple situation where there is only one "isolated element," and its generalization to a set of isolated elements called an "array." Fig. 4 shows a symbolic representation of a "linear map." Fig. 5 shows a "multilinear map." Fig. 6 shows a "hierarchy." Fig. 7 shows a "cycle." Fig. 8 shows a "multilevel map." Fig. 9 shows that these structural types are related through the contextual relation "is a special case of." This hierarchical representation of structural types should lend further validity to the type designations, since each type not only satisfies the designation principle set forth earlier, but also each type reduces to a

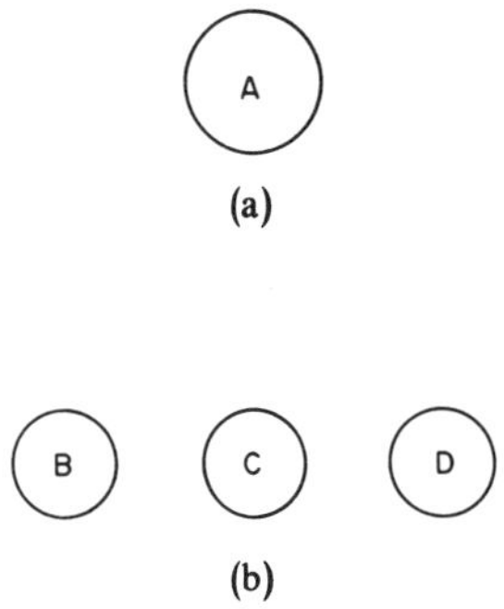

Fig. 3. (a) Isolated element. (b) Array.

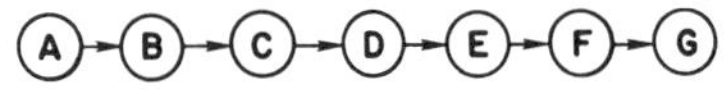

Fig. 4. Linear map.

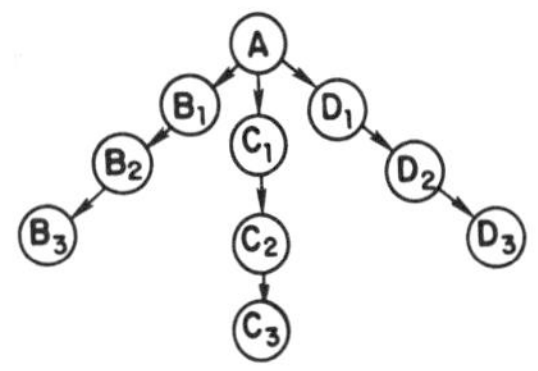

Fig. 5. Multilinear map.

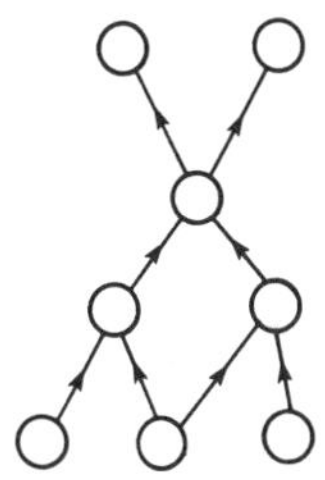

Fig. 6. Hierarchy.

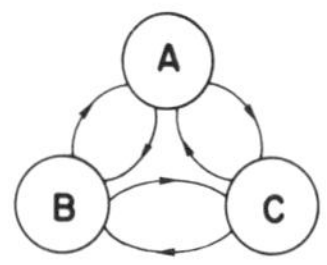

Fig. 7. Cycle.

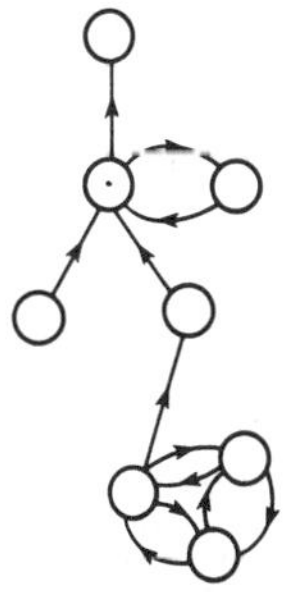

Fig. 8. Multilevel map.

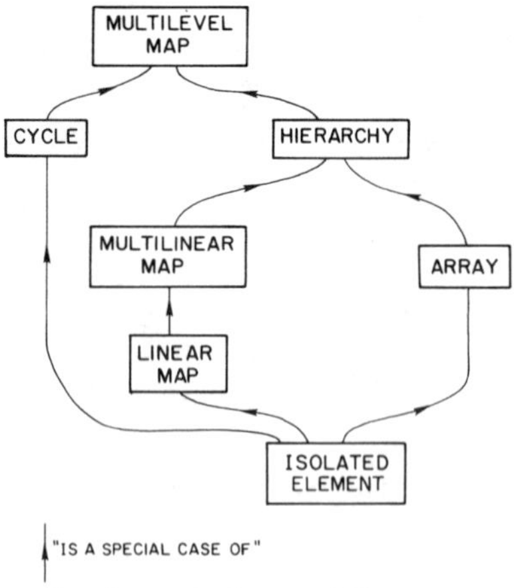

Fig. 9. Hierarchy of structural types of digraph-based structures.

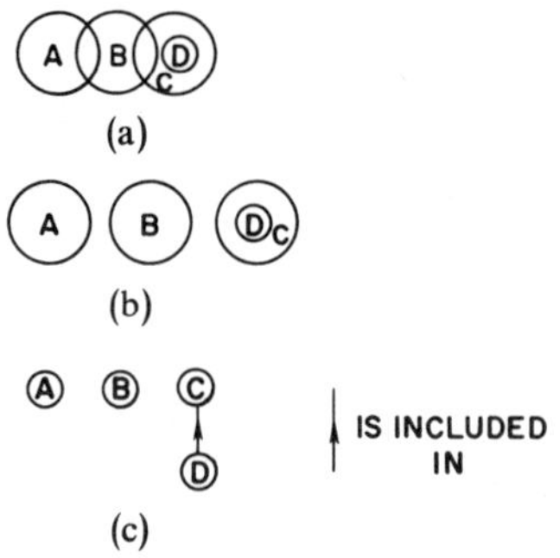

Fig. 10. Illustrating structural confusion.

subordinate type when a sufficient number of vertices and incident edges is removed from it.

Some have argued that a hierarchy should have only one top-level element. There are at least two disadvantages to this narrow interpretation of what constitutes a hierarchy. First, it violates the designation principle which, in itself, seems to be a valuable organizing concept. Second, it makes impossible the conceptualization of a set of disjoint structures as a hierarchy, even when each satisfies the narrow definition, and thus requires an enlargement of the categorization for both theoretical and application purposes. This frustrates discussion and detracts from the utility of theory.

As a final point, one may observe that all nonempty digraph maps fall within the types discussed in this section, and that if a digraph is included in more than one type it is because that digraph represents the limiting case of one of the types. The only exception to this is the multilinear map. This particular type is included only because it appears to fill a need for conversational-mode description in some applications. Were it not for this, the multilinear type would not even be included.

Definitions

The first step in the well-known Hall activity matrix for systems engineering [4] is "Problem Definition." It is asserted that problem definition is a mix of objectivity and subjectivity, and that while subjectivity is involved in making certain choices, objectivity is needed when the word "definition" is construed in its usual meaning. While words do have multiple meanings, one definitely is constrained by the language if communication is sought.

The two most commonly understood approaches to definition, as set forth in philosophy and logic, are definition by extension and definition by intension. In both of these approaches, the *set inclusion relation* is fundamental.

In definition by extension, a class S of entities is defined by offering subsets of the class. Thus the basic statement in this mode of definition is, in set language, $S_i \subset S$.

In definition by intension, an entity E is defined by defining a class S of attributes of the entity. Thus, if S_i is an attribute of E, the statement $S_i \subset S$ is the fundamental statement in this mode of definition, asserting that the attribute S_i is included in the set S of attributes that defines E.

The inclusion relation is readily modeled by a digraph. If there is an entity S having a set $\{S_i\}$ of included subsets, there is a two-level hierarchy with S at the top and the subsets appearing below that top-level element, and with the arrows on the edges representing "is included in."

While two-level hierarchies may be appropriate for many elementary definitions, such hierarchies are totally inadequate for many purposes of definition. For example, in defining "environmental education" for federal programmatic purposes, it was found that a multilevel structure with 45 levels and over 160 elements was needed [1].

In constructing definitions with complex structures, through use of the inclusion relation, the special role of this relation requires that the metalanguages used should give particular attention to this relation in terms of its role in set theory, its role in graphic presentations, and its role in the applications themselves.

Unfortunately it is all too easy to confuse inclusion with partial overlap of sets, and the theory offers no satisfactory way to deal with this. It has already been mentioned that set inclusion is transitive, while set overlap is not, in general. Moreover, Fig. 10 illustrates a situation that we refer to as "structural confusion." Here two Venn diagrams correspond to the same digraph map based on the inclusion relation, but the two Venn diagrams clearly represent situations that are structurally very different. The translation from graphic to graphic is unilateral.

If one extends the concept of "translatable graphic" to mean also that it should be possible to translate the digraph back into a Venn diagram equivalent, it is clear from Fig. 10 that no such translatability exists.

In some applications, people have even construed the partial overlap to mean inclusion, and this clearly is not consistent with transitivity requirements.

Accordingly, it is proposed to carry out operations in the domain of sets that eliminate conceptually the possibility of structural confusion and allow for a unique mapping of a digraph structure into a Venn diagram. This will be the principal aim of the rest of the paper.

Block Representations of Sets

In this section, the standard types of block representations of sets are reviewed and additional types are introduced. The various types are distinguished from one another, and the relationship of the types to graphic representations and to potential applications is described. In the next section, mapping methods are described that use the new types of block representations to simplify structuring of knowledge using the inclusion relation.

Standard Types of Block Representation

A *cover* of a set (also called a "set system") is a finite collection of nonempty subsets (called *blocks*) of a set S, whose set union is S, such that no block in the collection is a subset of any other block in the collection. While no block of a cover is included in any other block of a cover, partial overlap among blocks is permitted in a cover.

A *partition* of a set is a cover, all of whose blocks are mutually disjoint.

Both partitions and covers have been the subject of considerable study in relation to their roles in information handling [6], and special algebras apply to both of them. In order to develop these special algebras, one suspects, restrictions were introduced in defining them. One seldom finds, in applications, that a subdivision of knowledge occurs in the form of a partition. To mention just one example, one imagines that there is considerable overlap among the academic disciplines.

While restrictions in the definitions of these standard types has facilitated the development of rather extensive algebras, these same restrictions can be very inhibiting in applications where one begins by saying, "Construct a collection of subsets that forms a partition on the set."

What seems to be needed is a language that is compatible with set theory, but that also would allow for people to use the terms in applications (or at least have terms that correspond to the set theory language), and where there would be a term for each application concept that seems to be needed.

New Types of Block Representation

The proposed new types of block representation are intended either to serve the need for a term that could be useful in an application setting (not necessarily mathematical), or useful in embedding the application in a useful metalanguage, or both.

The first such term is so loose as to have little potential for algebraic application, but could be very helpful in talking about complex issues. We define a *breakout* of a set S as a finite collection of nonempty blocks of S. This is the most general term that we define, and it is sufficiently general that it can be used to represent what we have been calling a "block representation." Note that a breakout does not even have to include all the subsets or elements of S. It is often true in applications that one is not certain that a complete collection has been attained.

All the other new types that are proposed are *breakouts whose set union is S*. So are the cover and the partition. In addition, other requirements distinguish these other types.

A *field* $\mathscr{F}(S)$ is defined as a breakout of S into n blocks such that

- no two blocks are equal,
- the union of the blocks is equal to S,

Fig. 11. Two types of pair. (a) Nested. (b) Contaminated.

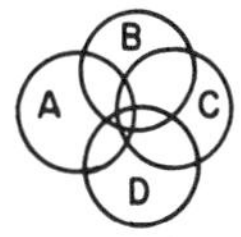

Fig. 12. Contaminated field.

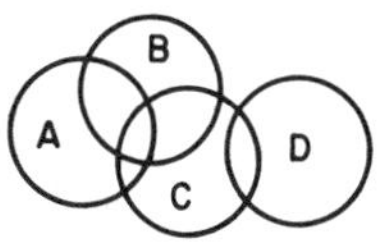

Fig. 13. Partly contaminated field.

- every block intersects at least one other block unless $n = 1$.

Evidently a field is quite different from either a partition or a cover. The only partition that can be a field is a single-block partition. Some covers can be fields, but a field is not necessarily a cover.

It is noted that a field such that no block is a subset of another block is a cover.

It is also noted, for later use, that if a field is structured into a hierarchy using the inclusion relation, then the top-level set of the hierarchy for the field constitutes a cover.

Three types of field appear to have special importance in organizing knowledge. In order to distinguish these types, nomenclature is introduced that relates to the nature of overlap between two sets. Fig. 11 distinguishes the "nested" pair from the "contaminated" pair. A nested pair corresponds to an inclusion relation, while a contaminated pair corresponds to a partial overlap relation.

A field is *contaminated* if and only if *every* pair of blocks in the field is contaminated. Fig. 12 illustrates such a field, using a Venn diagram.

A field is *uncontaminated* if and only if it contains *no* contaminated pairs. Then, necessarily, every block is involved with some other in a nesting situation.

A field is *partly contaminated* if and only if it is neither contaminated nor uncontaminated. An example of a partly contaminated field is shown in Fig. 13.

The mentioned properties of all of these types of block representations, including the standard types, are summarized in Table I.

Graphic Representations

Graphic representation of some of the types of breakout of a set S is easy, and for others is less easy. Table II summarizes some basic concepts. Note that representation of a partition and of an uncontaminated field is straightforward, either with Venn diagrams or with digraphs. For the remaining types, neither type of representation is satisfactory, in general.

TABLE I
PROPERTIES OF BREAKOUTS OF SET S

KEY: Y = must be true; N = must be false; M = may be either true or false

PROPERTY	PARTITION OF S	COVER OF S	UNCONTAMINATED FIELD OF S	PARTLY CONTAMINATED FIELD OF S	CONTAMINATED FIELD OF S	FIELD OF S	BREAKOUT OF S
1. No block is empty, and the number n of blocks is finite.	Y	Y	Y	Y	Y	Y	Y
2. No two blocks are equal.	Y	Y	Y	Y	Y	Y	M
3. The union of the blocks is equal to S.	Y	Y	Y	Y	Y	Y	M
4. No block intersects another block ($n \geq 2$).	Y	M	N	N	N	N	M
5. Every block intersects at least one other block unless n = 1.	N	N	Y	Y	Y	Y	M
6. There is no nesting ($n \geq 2$).	Y	Y	N	N	Y	M	M
7. No block pair ($n \geq 2$) is contaminated, but at least one pair is nested.	N	N	Y	Y	N	M	M
8. There is at least one pair that is either nested or contaminated ($n \geq 2$)	N	M	Y	Y	Y	Y	M
9. Every pair is contaminated ($n \geq 2$).	N	M	N	N	Y	M	M
10. At least one pair is contaminated.	N	M	N	Y	Y	M	M

TABLE II
GRAPHIC REPRESENTATIONS OF BREAKOUTS OF A SET

BREAKOUT OF A SET	GRAPHIC REPRESENTATION
1. Partition	An array, either in a Venn diagram or in a digraph
2. Cover	Venn diagram, with overlap; but no satisfactory digraph due to structural confusion
3. Uncontaminated field	Venn diagram with nesting, digraph in the form of a hierarchy
4. Partly contaminated field	Venn diagram, with overlap and nesting; but no satisfactory digraph, due to structural confusion
5. Contaminated field	Venn diagram, with overlap; but digraph can be misleading due to intransitivity of set overlap relation

Accordingly, what will be presented in the next section of the paper is a method for mapping a field or a cover into a partition by algebraic means. This allows the field or the cover to be represented graphically as an array (without structural confusion), and the latter, in turn, allows for construction of a hierarchy (if desired) through breakout of blocks of the array, or for more general approaches to controlled redundancy (such as the construction of covers).

Before introducing the mapping methods, two examples of potential applications will be discussed to show that the constraints introduced on the mapping methods are aligned with applications.

From Application to Theory

We discuss first the possible application of a knowledge organization theory to reorganize the academic disciplines as a means of eliminating overlap, in order to cut the costs and confusion surrounding the organization of libraries.

Case 1: It is desired that the academic disciplines be reorganized conceptually to eliminate overlap. To avoid additional bureaucracy, the total number of new disciplines shall not exceed the original number, yet all the original ingredients shall be maintained in the new disciplines. The reorganization shall take advantage of the knowledge and the people already available in existing disciplines, and shall therefore satisfy the condition that any new discipline shall either be the same as, or a subset of, one of the original disciplines.

A second case is similar, but refers to the government.

Case 2: Suppose that it is desired to reorganize the government. There is a set of agencies represented by mission statements, and the latter (suitably revised and edited) will make up the original collection. In reorganizing, the total number of agencies shall not be increased. The mission ingredients shall all be retained (to conform to law). The mission of any new agency will be the same as, or a subset of, the mission of one of the old agencies. No partial overlap of missions is wanted in the reorganization (though some may be introduced later, if it is deemed desirable).

These cases illustrate a basic problem that can be described as follows. Given a collection C_1 of n nonempty nonequal subsets of a set S such that the union of the n subsets is equal to S, find a new collection C_2 of r nonempty subsets of S such that the union of the r subsets is also equal to S, and such that

- $r \leq n$ (the *no-proliferation* requirement),
- every block in C_2 is a subset of some block in C_1 (the *continuity* or *block dominance* requirement),
- the subsets in C_2 either
  a) do not overlap at all, or
  b) if overlap occurs, it is in the form of nesting, (the *no-contamination* requirement).

In terms of graphic representations, one can say that the problem is as follows. Beginning with a breakout of a set S that includes all the subsets in S, and which may not be suitable either for representation by Venn diagrams or by digraph-based structures, map the breakout into a form where it can be represented graphically as an array or as a hierarchy, without introducing proliferation, and preserving continuity in the system.

Only the last term is somewhat equivocal from a mathematical perspective, and before proceeding to develop the mapping, this term will be further illuminated.

The term "continuity" as used here appears to have a nice connotation for applications, as can be seen by referring back to the two cases. However, in mathematical terms it means that every block in C_2 is a subset of some block in C_1 and, in the language of block representations, it means that the representation C_1 *dominates* the representation C_2. This expresses, for the whole representation, the condition that each block of C_2 is a subset of some block of C_1.

Mapping Theory

Let $\mathcal{F}(S)$ be a field with n blocks (the field is of size n) or let $\mathcal{C}(S)$ be a cover of size n. Let $\Pi(S)$ be a partition of size m. We will say that a mapping of either $\mathcal{F}$ or $\mathcal{C}$ into Π is *constructive* if and only if

a) $m \leq n$

b) $\mathcal{F} \geq \Pi$, if a field is to be mapped or

$\mathcal{C} \geq \Pi$, if a cover is to be mapped,

where $\geq$ refers to dominance. A constructive mapping meets all the conditions identified in the previous section.

Theorem 1: Any field $\mathcal{F}(S)$ or cover $\mathcal{C}(S)$ can be mapped algebraically into a partition $\Pi(S)$ through a constructive mapping; hence it can be mapped graphically into an array.

Proof: While the theorem is applicable both to fields and to covers, it will be proved using the notation for a field. The same proof will apply to covers. Let $\mathcal{F}(S) = S_1; S_2; \cdots; S_n$, where the $\{S_i\}$ are the blocks of the field. Let $\Pi(S) = T_1; T_2; \cdots; T_n$, where the $\{T_i\}$ are blocks of the partition if and only if they are not empty, and where

$$T_1 = S_1$$
$$T_2 = S_2 - S_2 \cap S_1$$
$$T_3 = S_3 - S_3 \cap [S_1 \cup S_2]$$
$$\vdots$$
$$T_n = S_n - S_n \cap [S_1 \cup S_2 \cup S_3 \cup \cdots \cup S_{n-1}].$$

To prove the theorem, we show

a) the union of the blocks of Π is equal to S,
b) the intersection of any pair of blocks of Π is empty,
c) $\mathcal{F}$ dominates Π.

a) Consider the sequence $Z_1 = T_1$, $Z_2 = T_1 \cup T_2, \cdots, Z_n = T_1 \cup T_2 \cup T_3 \cup \cdots \cup T_n$. By inspection,

$$Z_1 = S_1$$
$$Z_2 = S_1 \cup S_2$$
$$Z_3 = S_1 \cup S_2 \cup S_3$$
$$\vdots$$
$$Z_n = S_1 \cup S_2 \cup S_3 \cup \cdots \cup S_n = S.$$

Thus the union of the blocks is equal to the set S.

b) To see that the intersection of any two blocks is empty it is sufficient to show that in the sequence $T_1, T_2, \cdots, T_n$ no member subset overlaps any of its predecessors in the sequence. We need show this only for some subset T_i $[i \in (2, n)]$.

Now T_i is defined as being equal to S_i, less the intersection of S_i with Z_{i-1}. It has been shown that Z_{i-1} is equal to the union of all the predecessors of T_i in the sequence. Hence, T_i does not overlap the union of its predecessors in the sequence, and therefore, it cannot overlap any of its predecessors.

c) To show that $\mathcal{F}$ dominates Π, it is sufficient to show that S_i dominates T_i $[i \in (1, n)]$. This is readily seen by inspection of the definition of T_i.

Theorem 1 shows that there is a constructive mapping whereby a field or a cover can ultimately be portrayed through an array. In general, the mapping is not unique because the order of the blocks in the field or cover is irrelevant to the proof. Hence, one expects that a variety of arrays could be produced using Theorem 1, and that some could be more useful than others. We shall conclude the present discussion by giving an example of the use of Theorem 1, and then shall move on to ways to take advantage of the nonuniqueness of the result.

Example: Let $\mathcal{F}(S) = A; B; C; D$ where

$$A = \{1, 10, 11, 12, 14\}$$
$$B = \{5, 9, 11, 12, 13\}$$
$$C = \{3, 4, 6, 8, 9, 10\}$$
$$D = \{1, 2, 3, 4, 5, 6, 7, 8\}.$$

In this example, every pair of blocks is contaminated, hence, $\mathcal{F}(S)$ is a contaminated field.

Applying Theorem 1 with $S_1 = A$, $S_2 = B$, $S_3 = C$, and $S_4 = D$, yields

$$T_1 = \{1, 10, 11, 12, 14\}$$
$$T_2 = \{5, 9, 13\}$$
$$T_3 = \{3, 4, 6, 8\}$$
$$T_4 = \{2, 7\}.$$

The contaminated field is mapped into a partition, which can be shown graphically as an array. In this instance, the array has the same size as the field.

Block dominance is observed in that A dominates T_1, B dominates T_2, C dominates T_3, and D dominates T_4.

Roots and Key Blocks of a Cover

We shall define the roots and key blocks of a cover, and subsequently, the remainder of a cover. With these concepts we are able to show an iterative algorithm for mapping a field or cover into a partition whereby the number of blocks of the partition is often smaller than in the field or cover. When the results are applied to a field, advantage is taken of the fact that when a field is represented by a hierarchy in terms of the inclusion relation, the top-level set in the hierarchy is a cover. Hence, a field can be replaced with this top-level set, and a suitable partition can be found to map the field into a partition and, hence, into an array.

Let the blocks of a cover $\mathscr{C}(S)$ be $\{C_i\}$. Let $\overline{C_i}$ represent the union of all of the blocks of $\mathscr{C}$ other than C_i itself. $\overline{C_i}$ is not, in general, the set complement of C_i because of overlap of blocks of $\mathscr{C}$. Then always $C_i \cup \overline{C_i} = S$. We define the *root* R_i of block C_i as

$$R_i = C_i - C_i \cap \overline{C_i}, \qquad i \in (1, n)$$

and we say that the set $\{R_i\}$ makes up the roots of the cover $\mathscr{C}$. Clearly a block of $\mathscr{C}$ always dominates its root.

The intersection of any two roots of a cover is always empty, as is clear from the way in which a root is defined; hence, one can say that a root of a cover is always simple.

The set of nonnull roots may itself be empty; otherwise it is said that the cover has *essential roots*, and every nonnull root is essential. The blocks from which the essential roots are found are called *key blocks* of the cover. The nomenclature reflects the fact that a key block is the *only* block of the cover that includes its essential root. A root is, in general, a subset of S.

If there are no essential roots, then there are no key blocks in the cover, and any block of $\mathscr{C}$ is, in a sense, dispensible, since it is included in the union of all the other blocks.

Example: While the definitions have all been given for a cover, they apply as well to fields. There is some pedagogic value in using a field to illustrate the result, as will be seen. Let a partly contaminated field be given by $\mathscr{F}(S) = A; B; C; D; E; F; G; H; J$, where

$$A = \{1, 2, 3, 4, 5, 6, 7, 8\}$$

$$B = \{1, 10, 11, 12, 14\}$$

$$C = \{3, 4, 6, 8, 9, 10\}$$

$$D = \{5, 9, 11, 12, 13\}$$

$$E = \{3, 4, 5, 6\}$$

$$F = \{11, 12, 13\}$$

$$G = \{1, 10, 11, 12, 13\}$$

$$H = \{6, 7, 8, 9\}$$

$$J = \{7, 8\}.$$

Using the formula $R_i = F_i - F_i \cap \overline{F_i}$, it is found that blocks A and B are key blocks with roots $\{2\}$ and $\{14\}$, respectively, and that all other blocks have empty roots.

Residues and Remainders of a Cover

We define the first residue $\psi_1(S'')$ of a cover $\mathscr{C}(S)$ as that breakout of S whose blocks are given by

$$C_i'' = C_i - C_i \cap U$$

where U is the union of the essential roots of $\mathscr{C}(S)$, and where $S'' = S - S \cap U$. We define the second residue of $\mathscr{C}(S)$ to be the first residue of $\psi_1(S'')$, etc. The empty set is denoted $\varnothing$. Similarly we define the first *remainder* $\theta(S')$ of $\mathscr{C}(S)$ to be that breakout of S whose blocks are given by $C_i' = C_i - C_i \cap V$, where V is the union of the key blocks of $\mathscr{C}(S)$, and where $S' = S - S \cap V$. The second remainder of $\mathscr{C}$ is the first remainder of θ, etc.

Theorem 2: A necessary and sufficient condition that a cover $\mathscr{C}(S)$ shall also be a partition $\Pi(S)$ is that the first residue of $\mathscr{C}(S)$ shall be empty (i.e., all the blocks of the first residue are empty).

Proof:

a) *Necessity:* For all blocks of a partition, $C_i \cap C_j = \varnothing$ if $i \neq j$. Then the root of block C_i is C_i and C_i is its own essential root. Then C_i is a subset of U; hence, $C_i'' = C_i - C_i \cap U = \varnothing$, for all $i \in (1, n)$.

b) *Sufficiency:* Given that $C_i'' = C_i - C_i \cap U = \varnothing$, for all $i \in (1, n)$, then clearly $C_i \subseteq U$, and necessarily $U = S$. Hence, every element $s_i \in S$ satisfies $s_i \in U$. Now let us suppose that $\mathscr{C}$ is not a partition and show a contradiction. If $\mathscr{C}$ is not a partition, then there must be some element $s_i \in S$ that appears in two different blocks of $\mathscr{C}$. However, this same element must appear in some essential root of $\mathscr{C}(S)$. Therefore, it must appear in some key block of $\mathscr{C}(S)$, but then it cannot appear in another block of $\mathscr{C}(S)$, which is a contradiction.

Next we shall use the remainders to construct an algorithm for mapping a field or cover into a partition that is more selective than the mapping achieved using only Theorem 1.

Iterative Array Mapping

We now introduce an algorithm for iterative application of Theorem 1, taking advantage of what has been found about key blocks and remainders. We call this *iterative array mapping*. The process takes a field into a dominated partition which can be mapped as an array, and it takes advantage of means of reducing the number of blocks along the way.

The essential steps are shown in Fig. 14. It is believed that Fig. 14 is self-explanatory and that the following example offers sufficient elaboration.

Example: Let us use the previous example as a base, but proceed slightly differently, according to Fig. 14.

Since blocks E, F, and J are included in other blocks of the field $\mathscr{F}$, they are deleted, leaving only the top-level blocks [a cover $\mathscr{C}(S)$] in the inclusion structure for the blocks of $\mathscr{F}(S)$. The blocks of $\mathscr{C}(S)$ are A, B, C, D, G, and H. We found previously that A and B are the key blocks of this cover. Applying Theorem 1 to just these two blocks, we get $T_1 = S_1 = A = \{1, 2, 3, 4, 5, 6, 7, 8\}$, and $T_2 = B - B \cap A = \{10, 11, 12, 14\}$. These two blocks are stored to be part of the final result.

The first remainder of $\mathscr{C}(S)$ is then formed, and it is defined by the blocks

$$C_1' = \varnothing, \qquad C_2' = \varnothing, \qquad C_3' = \{9\},$$

$$C_4' = \{9, 13\}, \qquad C_5' = \{13\}, \qquad C_6' = \{9\}.$$

Since the first remainder of $\mathscr{C}(S)$ is not empty, the new blocks can be structured using the inclusion relation, and all but the top-level blocks can be deleted. This leaves only the block $C_4' = \{9, 13\}$, which is then a key block, and which is preserved intact when Theorem 1 is applied. The second

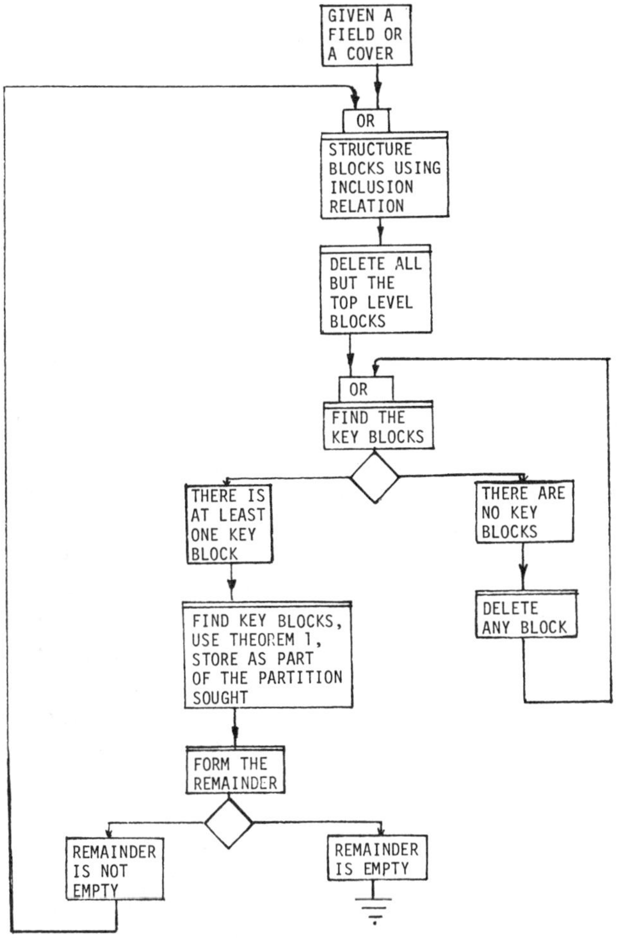

Fig. 14. Structure of algorithm for mapping field or cover onto dominated array.

remainder of $\mathscr{C}$ is empty, and the partition is completed as $\Pi(S) = 1, 2, 3, 4, 5, 6, 7, 8; 10, 11, 12, 14; 9, 13$ giving a three-block partition.

Comments and Conclusions

We have said that there is merit in viewing the use of graphics in working with complex systems as one of diminishing substantive complexity by deliberately increasing structural complexity. In order to help make this trade-off as beneficial as possible, the graphics should often be translatable, so that reading difficulties are not introduced along with the graphics.

We have introduced several structural types, and have given what appears to us to be a convincing principle for designating these types. The types help us describe the structures that appear in applications.

We have shown the critical significance of the set inclusion relation in matters relating to problem definition and to definition in general, and indicated how the use of this relationship can be made somewhat easier by using some new kinds of block representations of sets to simplify the structure of sets before trying to represent sets by digraphs, as in interpretive structural modeling.

Although, in the examples, a set of numerical set elements was used which (by presumption) itself constituted an *a priori* partition on a set S, such a presumption is not necessary in order to apply the results of this paper. On the other hand, to use the results given herein it will often be necessary to do a considerable amount of hard work in order to carry out the algorithms. The difficulties dealt with herein reside in substantial measure in the nature of the problems themselves, and the best that methodology can do is to establish metalanguages and procedures that relieve the user of all but the irreducible minimum of effort that is inherent in the problem.

References

[1] J. N. Warfield, "Portraying the complexities of environmental education and strategies for learning," presented at the Annual Meeting of the American Education Research Association, San Francisco, CA, April 9, 1979 (available from ERIC Clearinghouse on Testing and Measurement, Educational Testing Service, Princeton, NJ, 08540).

[2] ——, "The interface between models and policymakers," *J. Policy Anal. Inform. Syst.*, vol. 3, Jan. 1979.

[3] ——, *Societal Systems: Planning, Policy, and Complexity.* New York: Wiley, 1976.

[4] A. D. Hall, III, "Three-dimensional morphology of systems engineering," *IEEE Trans. Syst. Sci. Cybern.*, vol. SSC-5, no. 2, pp. 156–160, 1969.

[5] J. N. Warfield, "Toward interpretation of complex structural models," *IEEE Trans. Syst., Man, Cybern.*, pp. 405–417, Sept. 1974.

[6] J. Hartmanis and R. E. Stearns, *Algebraic Structure Theory of Sequential Machines.* Englewood Cliffs, NJ: Prentice-Hall, 1966.

A Dynamic Model of Community Growth in a Metropolitan Area

DANIEL A. ENSMINGER, ASSOCIATE MEMBER, IEEE, AND FRANK DiCESARE, MEMBER, IEEE

Abstract—**A two-tiered dynamic model for allocating residential housing growth both among townships in a metropolitan area and among zones within a township is developed. A metropolitan-level model having data aggregated up to the township level is proposed to explain the effects of metropolitan-level controls (such as employment location and accessibility) and township controls (such as the presence of land-use regulations) on residential growth in each township. A township-level model is proposed to accept growth forecasts generated by the metropolitan model and spatially and temporally allocate this growth *within* a given township. A case-study metropolitan area and one township within it are chosen and a data base is developed, using published and raw data sources. Multiple linear regression analysis is used to calibrate and validate these models. Metropolitan-model results indicate good model fit ($R^2 = 0.835$, overall model F value $= 15.9$) with accessibility to employment being the dominant explanatory independent variable with a strong positive correlation (partial $R^2 = 0.71$) to the rate of housing growth. Other independent variables in the model include the property tax rate in the analysis zone (negative correlation), housing value (weak positive correlation), and land-use controls (weak negative correlation). These results support a theory of metropolitan growth that says accessibility to employment weighted by available land for development, availability of existing residential development, and low tax rates promote residential growth. A significant factor is that the presence or absence of land-use controls does not meaningfully affect growth. Model validation is accomplished using an independent set of data. Use of this model is demonstrated by making projections of outcomes of future control actions of interest. This effort has produced a practical calibrated and validated metropolitan-level model whose data requirements are not excessive by metropolitan planning standards. The township model represents the first step toward a methodology of projecting growth at the intratownship level.**

I. Introduction

A. The Contextual Framework

THE PHENOMENON of growth in metropolitan areas of the United States [1], [2] has been with us since the days of the Industrial Revolution. Since World War II much attention has gone to the study of dispersion of this growth out and away from the central cities of metropolitan areas into previously undeveloped areas, areas traditionally rural in nature and devoted to farm use. Initially spurred on by higher wages and the advent of the trolley car [3], suburban land development increased dramatically after World War II chiefly due to the availability of accumulated wartime savings, a high rate of postwar marriage and household formation, and the increased production of automobiles. During the 30 years since the initial postwar expansion other factors have served to spur suburban growth. Among these have been state and federal capital investment in the construction of interstate highways, federal tax legislation making home ownership more attractive, provision of FHA and VA loans for new home purchase, decentralization of industry from the central cities to suburban areas, and the rise of entrepreneurs who have built extensive residential developments on cheap rural land. Such suburban development has had mixed effects on the fiscal and socioeconomic complexion of our metropolitan areas.

For those people able to afford new homes in suburbia and automobiles to get them from home to work, the amenities of suburban living have been apparent. Clean air, ample living space, recreational areas, and low crime and tax rates are some of the advantages initially perceived by suburban dwellers. However, the effects of suburban growth have not been entirely beneficial to the suburbs. The suburban resident faces increased traffic congestion at the end of his daily work trip, which is usually made by automobile to the central city. Suburban growth has generally proceeded according to developers' economic dictates, causing sprawl-type development to occur where the land is cheapest. Such dispersal of substantial amounts of new development has frequently resulted in higher costs of services to suburban residents because of the diseconomies of scale that result from this pattern of development. These higher costs of providing piped water, sanitary sewers, solid-waste pickup, school facilities, and fire and police protection to such a dispersed pattern of housing units, in turn, frequently cause increases in the local property tax rate.

The suburban resident usually moves to his new home because he expects his satisfaction there to be greater than it was before. The desire for open space and the amenities of suburban life may sometimes motivate his choice of residence to a greater degree than such traditional considerations as commuting time to work and price of housing [4]. In the authors' opinion the myriad of problems associated with uncoordinated suburban growth are many times not fully apparent to the resident until he has lived in his new location for several years. The desire to escape the confines of the central city is frequently the paramount reason for his move with little thought given to the long-range consequences [2]. For instance, the initially attractive property tax rates in

Manuscript received April 10, 1979; revised May 11, 1979. This work was supported in part by a National Science Foundation Traineeship and an Urban Mass Transportation Administration Research Grant.

D. A. Ensminger is with The Analytic Sciences Corporation, Six Jacob Way, Reading, MA 01867.

F. DiCesare is with the Electrical and Systems Engineering Department, Rensselaer Polytechnic Institute, Troy, NY 12181.

Reprinted from *IEEE Trans. Syst., Man, Cybernetics,* vol. SMC-9, pp. 587–609, Sept. 1979.

many newly developed suburbs tend to rise rapidly within a few years due to the need for fire and police protection, schools, etc., which cannot be provided by the county government through county revenues alone [2].

According to the Real Estate Research Corporation [5] the very nature of sprawl development forces such increased costs to be incurred. The most common cause of this is the cost of extending water and sewer lines to undeveloped areas. The residents of the new communities to which these utilities are extended must absorb the marginal cost of providing such services, either indirectly through real property taxation or directly (when the developer installs them) through higher housing costs. Such services prove to be very costly because sprawl development increases the amount of piping required to service an entire community. In addition, sufficient school and municipal services must be provided to adequately serve a dispersed residential area, which further boosts costs.

The alternative to the community's providing its own infrastructure and services is to seek these services from the county or state government. From local observation [2] this has resulted in "septic-tank communities" with occasional state police patrols, since the county government is unable to finance all of the community's expanding needs. Once residential densities reach a certain point in these communities, such measures are no longer adequate, as is evidenced by poor sewage drainage and increasing crime rates [2].

Another problem arising from uncoordinated suburban growth is the loss of values and amenities of property resulting from poorly organized or conflicting land uses. This is an especially common occurrence at the urban periphery, since the zoning and other land-use controls prevailing there are usually weak and easily altered to suit parties with sufficient political influence [6]. Carefully drawn zoning plans can enhance a community's tax base and balance of land use, yet the effect of zoning is many times negated by the granting of variances to influential developers without due regard to the larger interest of the community [7]. A balanced and heterogeneous land-use plan can have an important influence on economies of scale by providing a mix of complementary activities such as labor supply and industry, banks and businesses, and schools and open spaces [8].

The commuting pattern of the suburban resident usually involves driving to and from his place of employment each day. This tends to produce great concentrations of vehicles during the rush hour on the routes connecting "bedroom communities" with employment centers, thus slowing traffic, aggravating the air pollution problem, and wasting fuel. All too frequently mass transportation is not economically feasible as a solution to this problem, because the highly dispersed population in the suburban communities makes complete transit coverage prohibitively expensive [3]. In addition, the suburban dweller may develop a certain psychological addiction to using his car, even viewing his arrival at work by auto as a status symbol in extreme cases [2].

In the authors' opinion analyses of metropolitan growth have typically been conducted by higher level governmental agencies who are concerned with evaluating the metropolitan and state-wide growth impacts of large-scale policy decisions. For instance, transportation planners often use growth models to forecast population increases over a metropolitan area resulting from a proposed regional highway master plan [9]. Often such tools fail to accurately project future growth in specific townships within the metropolitan region [10]. Such errors may cause localities that accept the state's projections as accurate to overbuild or underbuild school or municipal service facilities in response to expected future needs, resulting either in unnecessary expenditures or in insufficient availability of services.

The authors feel that many of the existing planning models have failed to produce useful small-scale projections. On the one hand, some large-scale models can project growth with good spatial resolution (at the township level or finer) [11]. However, because these models have large data requirements and long running times on the computer [9], cost considerations dictate that they be run only a limited number of times to generate results for "likeliest-case" scenarios. Because of this they are inflexible. For instance, if one or more highways are added to or deleted from a metropolitan area's original master plan, it is often too expensive or cumbersome to rerun the metropolitan growth allocation model, even though the original projections may well have been rendered invalid due to the alteration of the plan [9]. On the other hand, some simplified planning models ignore some very important causative factors of growth. This is especially true of traditional demographic models [12], which do not include enough causal independent variables to account for population changes due to the combined effects of highway construction, land-use and investment policies, tax rates, and amenity attributes in one community *vis-à-vis* the other communities in the metropolitan area.

Even when local townships are aware of their potential for development, they do not always use those controls available to them to at least insure that their growth is rational; thus sprawl, strip development, and the aforementioned diseconomies of scale have prevailed [2]. It is the authors' belief that this has generally been due to a lack on the part of local officials and planners of a quantitative understanding of the complex socioeconomic, fiscal, and legal forces that govern the suburban land conversion process. However, local officials and citizens are interested in knowing the impacts of proposed actions made at the metropolitan level [2]. For instance, answers are needed to such questions as: what pressure on township growth will occur as a result of changes in accessibility caused by improved highways or transit? The same type of question might be asked about the impact of the expansion or relocation of a major employment center within the metropolitan area. Answers to these questions will allow local policymakers to act in their own interest to take advantage of these pending actions or, at least, to mitigate their adverse impacts.

Given estimates of the local growth impact of metropolitan actions, the local official might then desire estimates of

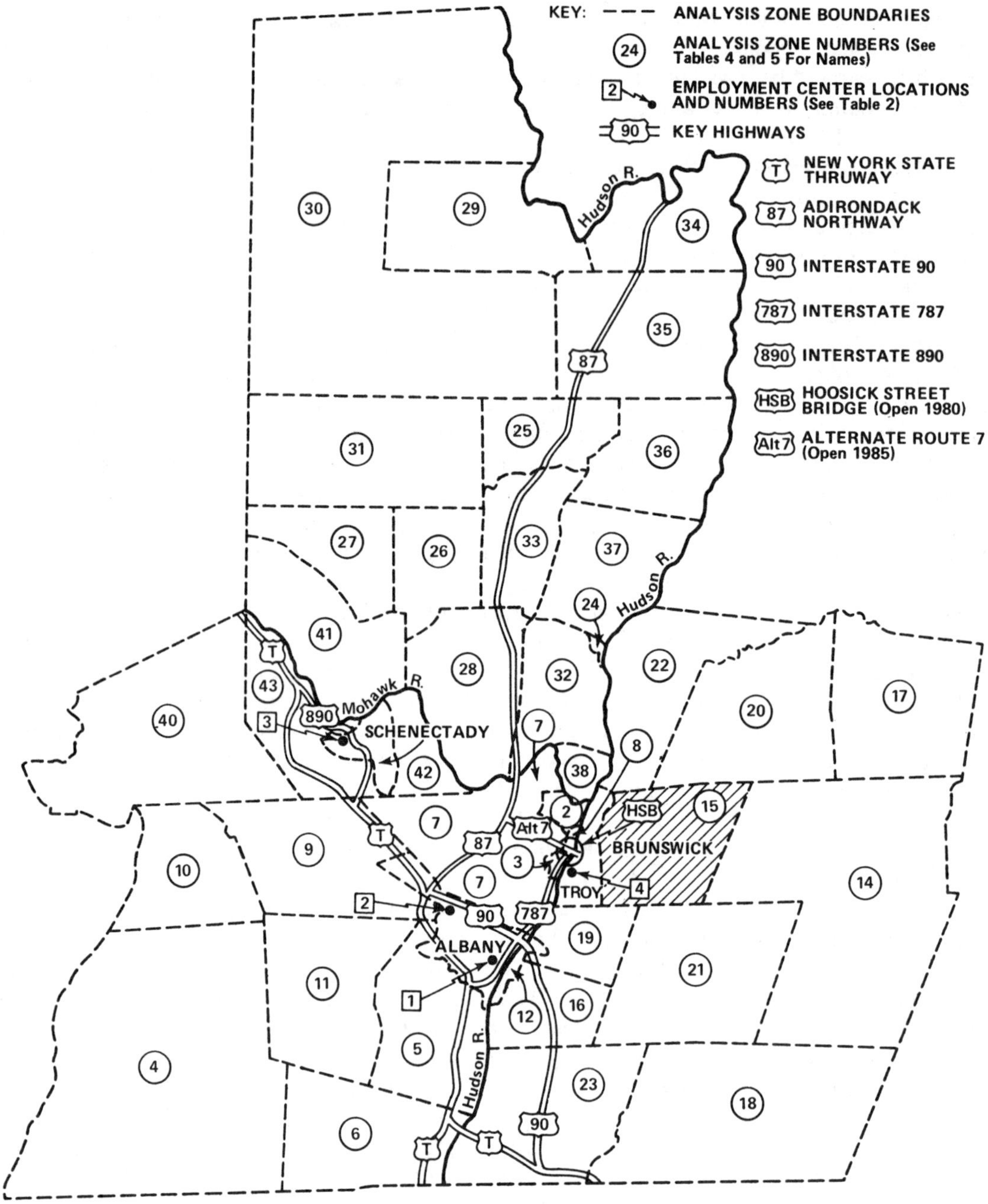

Fig. 1. Case-study metropolitan area.

how local actions can control this expected growth spatially and temporally. The questions he may want answered are as follows. 1) To what extent will public investment in roads and water and sewer facilities control the pattern and staging of this growth? 2) What regulatory provisions, such as zoning or water and sewer requirements, can be made which will cause rational growth? 3) What are the fiscal and socioeconomic impacts of the anticipated growth, i.e., what are the resulting future capital and operating costs required for school, police, fire, and other services and what will the impact be on township, special district, and school tax rates?

It is thus apparent that more accurate and flexible tools for projecting growth and its impacts at the metropolitan and township levels are needed. However, these tools must necessarily be sufficiently inexpensive, flexible, and understandable to local decisionmakers so as to make them competitive with existing planning models.

B. The Case-Study Metropolitan Area

Because of its favorable characteristics and the cooperation of its public officials, the Capital District (Albany–Schenectady–Troy Standard Metropolitan Statistical Area (SMSA) in east central New York State), shown in Fig. 1, was chosen as the case-study metropolitan area for this research. Composed of three central cities, a number of mature suburbs, several communities undergoing rapid growth, and a number of sparsely developed rural townships, the Capital District contained all the elements of a dynamically changing region. The entire region grew from a population of 657 503 in 1960 to 722 094 in 1970 [13], [14].

During this time the three central cities (Albany, Schenectady, and Troy) declined in population, while certain suburban townships that were in close proximity to new interstate highways and/or expanding centers of regional employment grew at a rapid rate.

This metropolitan area was significant as a case study for this work because of previous research in planning models done in one of its cities by the authors [15] and due to some current controversial issues that related to its future patterns of growth. One such issue involved a new bridge being constructed to cross the Hudson River at Troy, which was to link the towns and villages in the eastern portion of the SMSA directly with the metropolitan area's interstate highway grid and main employment centers. A second and related issue concerned the potential redistribution of development as a result of the construction of the Rockefeller Empire State Plaza, a new state capitol facility to house 8000 employees in downtown Albany. Many planners, officials, and interested citizens in the region's semirural townships were worried about the extent of development that might occur locally as a result of these actions [2]. The officials of the Town of Brunswick (shaded area in Fig. 1) were especially concerned and so agreed to cooperate with the authors in this research effort. As a result, the Town of Brunswick was selected as the case-study locality for development of a detailed township model.

C. Objectives

The goal of this research was the development of a simplified model or set of models for use in understanding and assessing spatial and temporal changes in residential growth due to feedback relationships and controlling actions. These models were to be designed to be useful at the suburban township level, where their outputs could be used to derive the fiscal and environmental impacts of such development. These might include measures over time of population within a community, the degree of residential, commercial, and industrial development, availability of public services and open space, and the levels of municipal and school taxes. The worth of these controlling actions could then be subjectively evaluated by local decisionmakers.

The objectives to be met in the model-building process included the development of tools that

1) establish a *causal structure* of residential growth based on the literature and expert opinion using functional forms that can be tested against known data,
2) have sufficient *predictive validity* to make them competitive with existing planning tools in terms of their ability to forecast growth resulting from changes in controlling factors,
3) are *simplified and flexible* enough to permit reasonable ease of use by public officials to test proposed policy changes.

A methodology for addressing these objectives was set forth and each step is discussed as a part of this paper. First, a *survey of the literature* was undertaken to gain insights into the existing theories and causal factors of metropolitan growth, as well as an overview of existing growth models. The section on *model conceptualization* presents the two-tiered hierarchical modeling viewpoint used throughout this work to model both metropolitan and township-level phenomena at a data-efficient level and discusses the simplification process in getting the model structure to interface with the available data. A detailed treatment of the *model-building process* by the use of multiple linear regression analysis is given for the metropolitan model and for the township model in separate sections. Various analytic forms are presented for the metropolitan model to embody existing and proposed theories of growth. The process of data gathering is discussed in detail along with schemes for spatial and temporal partitioning of the data for use in model calibration and validation. Goodness of model fit to the data and predictive validity are assessed. The usefulness of the metropolitan model in projecting growth under a wide variety of action scenarios (including contemplated changes in the region's transportation network) is demonstrated. The final section summarizes the results of model development and the insights gained into understanding the suburban residential development process, along with some suggested future directions for research in this area.

II. Theories of Metropolitan Growth

One can develop from the literature a composite theory of suburban residential growth in a metropolitan area by considering a household's quest for location. In the case of residential growth the land developer acts as a surrogate for the household. His marketability studies assure him (and us) that he is, in general, consistent with the desires of households wishing to locate.

For development to occur there must be suitable land available. The literature has suggested that this factor correlates inversely with land prices [8] and directly with amenity attributes such as spaciousness and lack of crowding [16]. Therefore, land availability can be considered a "surrogate" for these variables and, hence, used to correlate their values with the rate of development. This land should also have access to employment centers. In classical land economics the theory of land rent describes the trade-off between land rent (housing cost) and transportation cost (commuting cost). Among the more prominent recent works in land market theory is Lowdon Wingo's study [17]. This work combined a theoretical analysis of traffic flows and land economics into a mathematical model of the residential land market. In Wingo's theory rents plus transportation costs were considered to be constant, with rents decreasing with distance from the central city as transportation costs increase. A consumption function was postulated which expressed the quantity of land consumed by an individual as a function of the inverse of the price per acre of the land. From these equations price and consumption curves were generated as a function of distance from the central city.

This concept is embodied in the opportunity-accessibility model for land allocation [11]. The Lowry model [18] also quantifies this concept and uses measures of access to employment centers (zones). The model presented in this paper considers the four major employment centers in the case-study metropolitan area and includes factors other than land availability and accessibility as determinants of residential development. These additional factors of potential interest to developers and homebuyers are the value of existing housing [19], property tax rates, and expenditures for services such as education [20]. Further, one might expect that land-use controls such as zoning and subdivision regulations would play a role in residential development [21].

III. Model Conceptualization

A. The Two-Tiered Metropolitan/Township Model

A principal purpose of this work was to develop a model that local officials can use to forecast metropolitan growth and to test policies to control that growth both spatially and temporally. The *simplest* formulation of a model for the dynamic growth of community i is

$$x_i(n+1) = f_i[x_i(n), u_i(n)], \quad (1)$$

where $x_i(n)$ is the state vector of community i (a township, group of townships, or city, referred to generically as a *township*) at iteration n. The component variables may include the number of residential housing units, the number of acres of land available for development, and property tax rates for township i and perhaps for all spatial subsets of township i. $u_i(n)$ is the vector of control policies for township i at iteration n. These policies include zoning and subdivision regulations and water and sewer infrastructure requirements as well as local government capital investment.

One problem with the formulation in (1) is the assumption that the future state of township i depends solely on the present state of township i and the local controls for township i. Common sense and a wealth of literature tell us that a suburban township's growth is affected by regional control policies such as transportation investment and employment-center location and the comparative states of all the townships competing for the regional growth; hence, the decoupled model in (1) is not realistic.

A model formulation including all these factors is

$$x_i(n+1) = f_i[x_1(n), x_2(n), \cdots, x_I(n), v(n), u_1(n), u_2(n), \cdots, u_I(n)] \quad (2)$$

where $v(n)$ is the vector of regional controls at iteration n, and I is the total number of communities considered. While this formulation is relatively complete, it suffers because in order to obtain the resolution desired at the local policy level the number of elements in each state and control vector may be relatively large, and the calibration will require detailed data for I communities. In the case-study metropolitan area inclusion in the model of all spatial subsets of all communities would require data for approximately 20 independent variables in 10 zones (nominally) per township, times 43 townships in the metropolitan area, times at least two points in time for data collection, or over 17000 pieces of data! Thus, we are presented with the following dilemma: any model for a single township cannot incorporate regional controls and the competitive states of other townships that common sense and current knowledge dictate; on the other hand, a regional model which incorporates all the details necessary for local policy analysis will require too much data for calibration.

The compromise adopted was a two-tiered formulation. For projecting the effects of SMSA-wide or aggregate township-level policies, a *metropolitan model* was employed, using the states of each community aggregated to the township level and a relatively long time interval between iterations. For use in community-level policy analysis a *township model* was devised to take both regionally forecast results and local controlling actions into account while projecting intratownship growth patterns at more frequent iterations. The metropolitan model uses the formulation given in (2), where the local state variables and control vectors are aggregated to the township level and where there is no spatial resolution within each township. This yields overall growth of each community as a function of metropolitan policy and the comparative states and controls of all communities in the region. Because of the data available the *temporal* resolution will also be coarse (ten-year intervals between iterations).

At the local level a model similar to (1) can be constructed to give the spatial and temporal resolutions required for policy analysis at the township level. This reformulation of (1) is

$$z_i(m+1) = g_i[z_i(m), q_i(m)] \quad (3)$$

where $z_i(m)$ is the state vector for township i at iteration m (the time interval for this model may be shorter than that of the regional model), and $q_i(m)$ is the control vector which includes local policy controls and the effects of the metropolitan model.

To show the coupling of the metropolitan and local models the local control vector can be written as

$$q_i(m) = \{w_i^1(m), w_i^2(m), \cdots, w_i^J(m), h_i[x_i(n), x_i(n+1)]\} \quad (4)$$

where J is the number of analysis zones within township i; $w_i^1(m), w_i^2(m), \cdots, w_i^J(m)$ are the local control vectors for each of township i's J-analysis zones at iteration m; and h_i is a vector of functions that introduces the effects of metropolitan growth into township i's internal dynamics. The model can be used in either a coupled or decoupled mode, where other regional forecasts or scenarios of the effects of metropolitan growth can be included in the vector of functions h_i.

B. Modeling Methodology

In going from the conceptual model form just presented to the actual model form used, the following methodology was employed:

1) selection of candidate dependent and independent variables (including control and state variables) based on the theory and on data availability;
2) development of a functional model form depicting the

causal relationships shown in the literature and hypothesized by the authors;

3) selection of a case-study area for both the metropolitan and local models;
4) gathering of data in the two case-study jurisdictions;
5) selection of a calibration tool for estimating model parameters and assessing the models' goodness of fit to data and predictive validity;
6) illustration of the models' application by projecting growth resulting from future regional controlling actions of interest.

In choosing the simplest model that would evaluate the effects of metropolitan actions on residential development and its impacts on suburban townships, a first-order difference equation (nonlinear in the state and controls but linear in the parameters) was selected. Then, a single state variable was chosen to best represent the "stock" of development present in each township at each model iteration: the total number of owned and rented housing units (including those both occupied and vacant for sale or rent) in each township. There were several reasons for this choice of state variable. First, data for measuring housing stock are readily available at each calibration/validation iteration. Second, many of the concomitant variables associated with residential development at a given iteration such as population, residentially developed acreage, total developed acreage, developable land remaining, property tax, and housing value are functions of this state variable [28].

The first-order difference equation proposed for this model accounts for system dynamics while keeping the model relatively simple. A modification of the standard state equation format will now be made to introduce the notion of *shift of share* of the state variable:

$$\frac{x_i(n+1)}{\sum_{i=1}^{I} x_i(n+1)} - \frac{x_i(n)}{\sum_{i=1}^{I} x_i(n)} = f[x_i(n), \boldsymbol{v}(n), \boldsymbol{u}_i(n)], \quad (5)$$

where f is unrelated to the vector of functions f_i in (2).

Conceptually, this shift-of-share formulation converts the standard difference equation's prediction of the value of the state of township $i[x_i(n+1)]$ to the prediction of the township's *share of the state* summed over all I townships $[x_i(n+1)/\sum_{i=1}^{I} x_i(n+1)]$. This concept has been used previously in metropolitan models [22], [23] and has several advantages. First, it permits the model to project shares of total metropolitan growth that a given township will possess at future iteration $n+1$, as opposed to projecting the actual magnitude of the township's state at $n+1$. Thus, the model remains valid for allocating any assumed future value of *total* metropolitan development to its townships independent of the *value* of this aggregate amount of metropolitan development. Second, the shift-of-share approach permits the modeler to account for other factors that influence development uniformly in the entire metropolitan area rather than differentially in specific townships by including them in this assumed total. Such factors include inflation, the availability of capital and labor, changes in income tax laws, and changes in federal aid formulas or grant programs [1].

IV. The Metropolitan Model

A. The State Equation

An overall state equation can now be proposed for the metropolitan model which considers all candidate variables of the metropolitan growth process. The state variable of concern will be the number of housing units in each township, as discussed in Section III. The shift of share of this state from one model iteration to the next will be expressed as a function of analytic forms containing both metropolitan-level and aggregated township-level controls, as well as functions of the state variable (such as residential and total developed land present at any iteration). A measure of total land topographically suitable for development is also included, as is a measure of the undeveloped portion of this land at each iteration. Among the primary metropolitan-level controls of interest are large employment center relocation or growth (as measured by employment totals and highway travel times), which follows from the important concept of accessibility to employment established in the literature. Aggregated township-level controls of interest include measures of the presence or absence of zoning and subdivision regulations and fiscal attributes such as property tax rates, housing values, and per-pupil expenditure.

The general equation of the model which projects the shift of share of residential development for each township of the metropolitan area and which is linear in the model parameters is

$$\frac{x_i(n+1)}{\sum_{i=1}^{I} x_i(n+1)} - \frac{x_i(n)}{\sum_{i=1}^{I} x_i(n)} = \gamma_0 + \sum_{k=1}^{K} \gamma_k f_k[x_i(n), \boldsymbol{v}(n), \boldsymbol{u}_i(n)], \quad (6)$$

where

$$x_i(n) = \mathrm{HU}_i(n); \quad (7)$$

$$\boldsymbol{v}(n) = \begin{bmatrix} E(n) \\ \tau_{11}(n) \\ \vdots \\ \tau_{IJ}(n) \end{bmatrix} \quad (8)$$

$\boldsymbol{u}_i(n)$ elements are selected from the set

$$\{D1_i(n), D2_i(n), D3_i(n), \mathrm{PPEXP}_i(n), \mathrm{PBSPC}_i(n)\} \quad (9)$$

and

- i — index of the ith community (township, township group, or city: referred to generically as a *township*) ($i = 1, \cdots, I$);
- n — index of the metropolitan model's iteration;
- $x_i(n)$ — value of the state variable in the ith township at iteration n;
- $\boldsymbol{v}(n)$ — vector of metropolitan control values, averaged over iterations $(n, n+1)$;

$\boldsymbol{u}_i(n)$ vector of values of local controls employed in township i at iteration n;

$\mathrm{HU}_i(n)$ total number of owned and rented housing units (including occupied and vacant available units) in township i at iteration n;

$\boldsymbol{E}(n)$ vector of employment, where element $E_j(n)$ ($j = 1, \cdots, J$) represents the mean number of total metropolitan area residents working in employment center j, averaged over iterations $(n, n+1)$;

$\tau_{ij}(n)$ mean automobile travel time (in minutes) between township i and employment center j averaged over iterations $(n, n+1)$;

$D1_i(n)$, $D2_i(n)$, $D3_i(n)$ values of the three dummy variables denoting presence or absence of land-use controls in township i at iteration n (explained later in this section);

$\mathrm{PPEXP}_i(n)$ mean number of dollars per pupil expended by schools in township i at iteration n, averaged over all grades in all schools within the township;

$\mathrm{PBSPC}_i(n)$ number of acres of publicly held open space present in township i at iteration n;

f_k kth right-hand-side independent variable ($k = 1, \cdots, K$), comprised of a linear or nonlinear function of the state and/or control variables;

γ_k estimated value of the kth model parameter as derived by the regression algorithm (where a term f_k is admitted to the model, $\gamma_k \neq 0$; otherwise, $\gamma_k = 0$).

B. Measures Used for Independent Variables

The following functions describe various attributes of *land use* (here f_k is used as a shorthand representation of $f_k[x_i(n), \boldsymbol{v}(n), \boldsymbol{u}_i(n)]$):

$$f_1 = \mathrm{RA}_i(n); \quad \text{where } \mathrm{RA}_i(n) = g_1[\mathrm{HU}_i(n)] \tag{10}$$

$$f_2 = \mathrm{DA}_i(n); \quad \text{where } \mathrm{DA}_i(n) = g_2[\mathrm{HU}_i(n)] \tag{11}$$

$$f_3 = 1 - \mathrm{DA}_i(n)/\mathrm{LAD}_i \tag{12}$$

$$f_4 = \mathrm{RA}_i(n) \cdot [1 - \mathrm{DA}_i(n)/\mathrm{LAD}_i] \tag{13}$$

$$f_5 = \mathrm{LAD}_i - \mathrm{DA}_i(n) \tag{14}$$

$$f_6 = [\mathrm{LAD}_i - \mathrm{DA}_i(n)]^2 \tag{15}$$

where $\mathrm{RA}_i(n)$ is the total number of acres developed residentially in township i at iteration n; $\mathrm{DA}_i(n)$ is the total number of developed acres (residential, commercial, industrial, and public) in township i at iteration n; LAD_i is the total number of acres of land topographically suitable for development (both developed and undeveloped) in township i (assumed to be a fixed quantity for all time); and g_1, g_2 are functions converting the state variable $\mathrm{HU}_i(n)$ to $\mathrm{RA}_i(n)$ and $\mathrm{DA}_i(n)$, respectively.

Equations (10) and (11) are measures of development already present within a township. They attempt to account for the positive influence of agglomeration on development [24], i.e., the theory that preexisting development in a township makes it attractive for further development. Equations (12), (14), and (15) denote various measures of remaining vacant land available for development within a township. The literature has suggested that this factor correlates inversely with land prices [8] and directly with amenity attributes such as spaciousness and lack of crowding [16]. Therefore, it can be considered a "surrogate" for these variables and, hence, used to correlate their values with the rate of development. Equation (13) is a function that represents the composite effect of the opposing states of agglomeration and sparsity of development on future growth.

The *transportation access-to-employment factor* embodies the observation that most people locate their residence with respect to their place of employment. This factor is cited in the literature on the theory of metropolitan growth [25], [26] and appears in several well-known metropolitan growth models, e.g., the Lowry model [18], the EMPIRIC model [22], and the opportunity-accessibility model [11], and may be expressed as one of the following functions:

$$f_7 = \frac{\sum_{j=1}^{J} E_j(n) h[\tau_{ij}(n)]}{\sum_{i=1}^{I}\sum_{j=1}^{J} E_j(n) h[\tau_{ij}(n)]}, \tag{16}$$

$$f_8 = f_7 f_3, \tag{17}$$

$$f_9 = f_7 f_4, \tag{18}$$

$$f_{10} = f_7 f_5, \tag{19}$$

$$f_{11} = f_7 f_6, \tag{20}$$

$$f_{12} = \frac{\sum_{j=1}^{J} E_j(n+1) h[\tau_{ij}(n+1)]}{\sum_{i=1}^{I}\sum_{j=1}^{J} E_j(n+1) h[\tau_{ij}(n+1)]} - f_7. \tag{21}$$

The function h decreases with increasing τ (showing that the effect of travel time diminishes with increasing distance [11], [18], and [22]) and can be chosen from the following two functions:

$$h[\tau_{ij}(n)] = \begin{cases} \alpha_j e^{-\beta_j \tau_{ij}(n)} \\ \text{or} \\ \alpha_j \tau_{ij}(n)^{-\beta_j}, \qquad \alpha_j, \beta_j > 0 \end{cases} \tag{22}$$

where α_j and β_j are parameters to be estimated.

In (16) township i's *share* of accessibility relative to that of all townships in the metropolitan area is computed. This quantity is expressed as the sum over j of the product of the number employed in center j [$E_j(n)$] times a monotonic decreasing function [$h(\tau_{ij}(n))$] of the mean automobile travel time from township i to employment center j, divided by the sum of these products over all I residential townships. (Transit travel times are not considered because of the small percentage of daily work trips in the case-study metropolitan area made using transit facilities.) This product forms a quantitative measure of the average "field of employment accessibility" [27] at township i over iterations $(n, n+1)$. As employment opportunities in employment center j $[E_j(n)]$ increase, or travel time between areas i and j $[\tau_{ij}(n)]$ decreases, township i accessibility to employment center j is increased, and vice versa. These two quantities are measures of, respectively, the metropolitan-level controls of *employment expansion* (with respect to space and time) and *interstate or arterial highway construction.* It is assumed that changes in travel times from one iteration to another are due only to construction of new and faster roads and that increased congestion due to growth is not a factor. This assumption is justified by the fact that excess capacity exists on the metropolitan arterial network, and the area's total growth in the next several decades is not anticipated to be large enough to consume all this capacity [9]. If this were not the case it would be necessary to include the feedback from growth in the travel times, a nontrivial task.

Other alternative quantitative representations of employment accessibility are given in (17)–(21). Equations (17)–(20) augment the basic share of employment accessibility function f_7 by postmultiplying it by measures of developable land remaining in township i. This makes the accessibility function more realistic in the sense that it diminishes the value of this measure if the remaining developable land is in short supply, regardless of how favorable the township's employment accessibility is. Another function considered in (21) is the *shift of share* of township i's mean accessibility from iterations $(n, n+1)$ to iterations $(n+1, n+2)$. This factor represents the expected future change in township i's share of the metropolitan area's employment accessibility. This embodies the concept that anticipation of future controlling actions can shape land development well in advance of actual highway construction or employment relocation. For instance, land speculators and developers having advance knowledge of highway corridor locations often purchase and subdivide land for future sales in those areas many years prior to construction [2]. Thus, future model control inputs are seen to impact present state variables through this shift-of-share function.

The following *fiscal quantities* have been shown to influence developers and potential homebuyers [20], [28], [29]:

$$f_{13} = \mathrm{HV}_i(n), \quad \text{where}$$

$$\mathrm{HV}_i(n) = g_3[\mathrm{LAD}_i - \mathrm{DA}_i(n)]; \tag{23}$$

$$f_{14} = \mathrm{TAX}_i(n), \quad \text{where}$$

$$\mathrm{TAX}_i(n) = g_4[\mathrm{HU}_i(n)]; \tag{24}$$

$$f_{15} = \mathrm{HV}_i(n) \cdot \mathrm{TAX}_i(n); \tag{25}$$

where $\mathrm{HV}_i(n)$ is the median value of single-family homes in township i at iteration n; $\mathrm{TAX}_i(n)$ is the mean property tax rate in dollars per thousand dollars *full* value (combined county, city or township, special district, and school district tax rates) in township i at iteration n; and g_3, g_4 are assumed functions (possibly nonlinear) determining, respectively, median housing value $[\mathrm{HV}_i(n)]$ as a function of vacant available land, and property tax rate $[\mathrm{TAX}_i(n)]$ as a function of the number of housing units in township i at iteration n. This is an approximate function that may hold in areas that are almost exclusively residential in nature. Commercial and industrial development must also be added to the tax base when they are present.

Equation (23) brings into the model consideration of housing value as a determinant of the rate of development. The literature shows that the price of housing in a township can have either a positive or negative impact on the rate of development there [29]. On the one hand, middle-class housing demand in a given area correlates negatively with housing value, while "snob appeal" can cause upper-class housing development to benefit from high housing values [29]. High property taxes (24) generally slow the rate of development in a township [20]. These taxes can be expressed as a function of $\mathrm{HU}_i(n)$ because the housing units in a township generate projectable per-unit expenditures and return per-unit assessed value to the tax base [28]. Equation (25) provides a third fiscal measure which can be interpreted as the *median annual taxes* paid per housing unit in township i at iteration n.

One of the important factors to consider in this model is the effect of *land-use controls* on the rate of development [1], [7], [8], [21], [30]. Ideally, zoning data would be available for each township and would include a breakdown of the number of acres of land in each zoning classification. For measures of subdivision regulation in those townships having them, data such as the maximum allowed number of lots per subdivision or minimum street width permitted by a township's subdivision regulations could be used. However, data on land-use controls at this fine level could not be easily obtained for all townships in the metropolitan case-study region for the calibration years 1960 and 1970. The only data readily available on a metropolitan-wide basis were the years during which zoning and subdivision regulations had been implemented in each township. Such information could be used to assess the effect of the *presence or absence* of zoning and subdivision regulations on a township's growth rate. This was accomplished by setting up a series of three "dummy variables" $[D1_i(n), D2_i(n)$, and $D3_i(n)]$ which, when taken together, served as a binary representation of the presence or absence of zoning and/or subdivision regulations in township i at iteration n. This dummy variable approach was adopted in order to preserve an interval scale for these variables so that the subsequent regression analysis could include them as ordinal data [31].

With these considerations in mind the following functions were proposed for representing gross measures of land-use controls:

$$f_{16} = D1_i(n) \tag{26}$$

$$f_{17} = D2_i(n) \tag{27}$$

$$f_{18} = D3_i(n) \tag{28}$$

where $D1_i(n)$, $D2_i(n)$, and $D3_i(n)$ denote the presence or absence of land-use controls as follows.

Land-Use Controls in Effect in Township i at Iteration n	$D1_i(n)$	$D2_i(n)$	$D3_i(n)$
None	0	0	0
Zoning	1	0	0
Subdivision	0	1	0
Both	0	0	1

The concept of a "nice place to live" is difficult to embody in any mathematical formulation. However, the *amenity attributes* of one township *vis-a-vis* those of the others in the metropolitan area are important in determining growth rates and should be included in the model [20]. The following measures of amenity, as a result, have been selected from available data for possible inclusion in the model:

$$f_{18} = \text{PPEXP}_i(n) \tag{29}$$

$$f_{19} = \text{PBSPC}_i(n). \tag{30}$$

Per-pupil expenditure (see (29)) is cited by Tiebout [20] as a surrogate measure of perceived township amenity (the greater the per-pupil expenditure, the greater the attractiveness of a township, all other things being equal). Public open space within a township (see (30)) is another quantifiable measure of amenity [16] since a high degree of such space implies availability of ample recreational facilities for township residents.

Some other measures of amenity such as the percentage of housing units within a township served by sewer and water connections might be used [22]. The sewer and water factors had to be dropped from consideration in the metropolitan model because insufficient time-series data existed throughout the entire metropolitan area to permit their use. However, these factors *were* employed in the township-level model.

Other institutional controls such as a township's restricting of development to those areas having preexisting water, sewer, fire, and other services would have been included in the model if data for these types of controls existed. Unfortunately, this kind of land-use control had never been implemented within the case-study metropolitan area, so no data existed for calibrating its effects on residential development rates.

C. Data Sources

In deciding how best to calibrate and validate the general metropolitan state equation (see (6), two opposing criteria applied. On one hand, in order to implement maximum model descriptiveness it was considered desirable to include a large number of analysis zones and employment zones and have annual or biennial model iterations. On the other hand, in order to be able to calibrate and validate this model with available data, the scope of the model had to be limited to a reasonably small number of zones with less frequent model iterations.

Various data sources, summarized in Table I, were employed in the calibration and validation of the metropolitan model. These ranged from published census tract summaries to calculated values of highway travel times. Many times, data obtained were not useful in their original form, necessitating certain aggregations or recomputations based on "best engineering judgment." When this was done checks were made to determine the sensitivity of final model results to changes in these assumptions. All data used in the metropolitan model calibration and validation process are given in [32].

TABLE I
SUMMARY OF DATA SOURCES FOR CALIBRATION AND VALIDATION OF THE METROPOLITAN MODEL

VARIABLE(S)	EQUATION	DATA SOURCE USED	YEARS DATA AVAILABLE
$HU_i(n)$ $E_j(n)$ $HV_i(n)$	7 8 23	U.S. Census [13,14,33]	1960, 1970
$RA_i(n)$ $DA_i(n)$ LAD_i $PBSPC_i(n)$	10 11 12 30	Land Use Surveys [34,35]	1963, 1973
$D1_i(n)$ $D2_i(n)$ $D3_i(n)$	26 27 28	Housing Survey [36]	1960, 1970
$\tau_{ij}(n)$	16	New York State Dept. of Transportation [37,38,39,40]	1965-1990
$TAX_i(n)$	24	New York State Dept. of Audit & Control [41,42]	1958-59, 1972-73
$PPEXP_i(n)$	29	New York State Dept. of Education [43]	1959-60, 1970-71

Data availability played an important role in model specification. The *time interval* between metropolitan model iterations $(n, n + 1)$ was established based on U.S. Census availability. Since a complete census is taken only once every ten years, the interval between model iterations was set at ten years to permit the model to use available census data for calibration. Table I illustrates that the data available from sources other than the census were not always coincident with the census. In these cases interpolation and extrapolation were necessary (see [32] for a full explanation) in order to develop a data base using a common time scale.

For the case-study metropolitan area, the *township* (or city, in urban areas) was used as the basic metropolitan model analysis zone, with occasional aggregation of several townships into one analysis zone. This was done to produce a useful township-level policy analysis tool that would be compatible with *all* the data sources used. The reason for occasional multitownship aggregation was that certain changes in U.S. Census tract boundaries were made between the 1960 and 1970 censuses. Fig. 1 shows the 43 analysis zones used in the metropolitan model development.

After considering several sources of employment data for the case-study metropolitan area [44], [45] it became apparent that the *U.S. Census tract summaries* [13], [14] were the only comprehensive source of time-series employment data at the analysis zone level. However, 1960 values for the

TABLE II
EMPLOYMENT CENTERS CHOSEN FOR THE METROPOLITAN MODEL

j	EMPLOYMENT CENTER*	NUMBER EMPLOYED		
		$E_j(1960)$	$E_j(1970)$	$E_j(n)$ (1960-70)
1	Albany East (Downtown Albany and part of the eastern half of the city)	52 416	55 157	53 787
2	Colonie/State Campus (Western Albany plus Colonie Village)	25 160	33 636	29 398
3	Downtown Schenectady (General Electric Plant)	44 043	45 016	44 530
4	Downtown Troy	22 181	20 831	21 506
	TOTALS	143 800	154 640	

*Refer to Figure 1.

employment levels could be calculated for only a small number of zones, including the number of metropolitan residents employed in each of the three central cities (Albany, Schenectady, and Troy), those employed in the balance of the counties containing these cities, and those employed in all of the remaining county (Saratoga).

The number of $E_j(n)$'s ($j = 1, \cdots, J$) and their location in the metropolitan area were decided by considering trade-offs between the ideal spatial resolution of employment centers versus the availability of employment data. It was decided that *four* employment centers should be used in the case-study metropolitan arca. These are summarized in Table II.

Special census tallies [33] of home-to-work person-trips from tract of origin to city or township of destination were used to compute the number of persons employed in 1970 in each of the four employment centers [$E_j(1970)$]. For 1960, employment totals for downtown Schenectady [$E_3(1960)$] and downtown Troy [$E_4(1960)$] were available directly from the census data [13] but similar totals for downtown Albany [$E_1(1960)$] and the Colonie Village/State Office Campus area [$E_2(1960)$] were not, since employment destination data were available for each central city only and not for each township, city, or city sector, as was the case with the 1970 data. As a result, best engineering judgment was employed to *prorate* the available 1960 employment data for Albany City and the remainder of Albany County between $E_1(1960)$ and $E_2(1960)$. (The sensitivity of final model results to this assumption is discussed in Section IV-E.)

The decade-averaged employment totals in all four centers for the 1960–1970 model calibration/validation period [$E_j(n)$] were computed as

$$E_j(n) = \tfrac{1}{2}[E_j(1960) + E_j(1970)], \qquad j = 1, \cdots, 4. \tag{31}$$

To summarize, the choice of the number and location of employment centers for the metropolitan model was made as a trade-off between the ideal data desired (inclusion of all employment zones) and actual data available. It should be noted that the four central employment zones chosen account for 61.7 percent of the total 1970 metropolitan area's employment (see [14] and Table II). Therefore, model calibration based on these four zones alone is considered to be valid and is not expected to significantly bias values of parameter estimates.

Highway travel times [$\tau_{ij}(n)$ in (8)] were established from data supplied by the New York State Department of Transportation (DOT) [37]. The New York State DOT had data on automobile travel times based on the 1965 Capital District Transportation Study. The 1965 times were based on the actual network of traffic control devices and link speeds existing in 1965, as well as the then current levels of traffic congestion. The following procedure was used to obtain travel times for the 1970–1990 decades.

1) Average the travel times from all DOT traffic analysis zones comprising a metropolitan model analysis zone to each given employment destination in order to arrive at a single travel time from each analysis zone to each employment center.
2) Use the 1965 DOT travel times as a baseline figure. Then, assemble a time-series record of *all* major expressway, bridge, and arterial construction projects (both completed and proposed-but-funded) in the case-study metropolitan area from 1959 to 1990. Compute average link speeds (including the effects of both speed limits and delays due to traffic control devices) for each type of highway. Then add or subtract appropriate travel times from the 1965 figures to generate the *average* $\tau_{ij}(n)$ for the decades 1960–1970, 1970–1980, and 1980–1990, based on this existing and anticipated highway network. Fig. 1 shows the existing expressway network and key planned improvements to it through 1990.

D. Calibration

Calibration of a model involves estimating the values of its parameters so as to minimize the differences between model predictions and observed data [46]. The validity of a model is determined by the degree to which the calibrated model is able to predict values that compare closely to observed data not used in the calibration process [47]. An idealistic approach to socioeconomic model calibration and validation would be to use all available data to calibrate the model, observe actual future outcomes and the controlling actions associated with them, and then compare these outcomes to model predictions resulting from the use of these same controlling actions as inputs to the model. This is generally not a feasible methodology for urban decisionmakers, whose practical considerations require model validation to occur in the present.

To overcome this difficulty one of the most common methods of validation used by modelers is that of *retrospective prediction* [47]. This involves spatial and/or temporal partitioning of historically observed data into two subsets, one of which is used to calibrate the model and the other to validate it. In the case of the metropolitan model, temporal partitioning of the data was not possible because data for the dependent (state) variable and most of the independent variables in the model were available at just two points in time (for 1960 and 1970). This meant that the available data would cover just one model iteration. Hence, *spatial* partitioning of the data was indicated in this case.

In deciding how to divide the 43 analysis zones of the metropolitan model between the calibration and validation subsets, two criteria were established. One was that the calibration subset should contain as many of the zones as possible in order to guarantee as much statistical significance in the regression analyses as possible. Secondly, there should be a minimum of bias introduced into the calibration as a result of this split of data into the two subsets, i.e., a heterogeneous mix of both fast- and slow-growing communities should be included in each subset so that the resulting model would not be biased in favor of either fast- or slow-growing analysis zones. Accordingly, the following procedure was used to divide the 43 analysis zones into calibration and validation subsets.

1) The analysis zones corresponding to the metropolitan area's three central cities [Albany, Schenectady, and Troy] were *excluded* from the calibration and validation subsets (see the discussion later in this section).
2) The remaining 40 analysis zones were sorted in descending order of their observed shifts of share of housing units from 1960–1970. Then every fourth zone (going down this list in the order sorted) was selected for the validation subset (for a total of 10 zones); the remaining 30 zones became the calibration subset.

The accessibility function parameters α_j and β_j (see (22)) were estimated prior to the overall metropolitan model calibration using a separate data set that was available. These parameters could have been estimated simultaneously with the linear parameters in *one* calibration, but this would have necessitated the use of a nonlinear regression technique since α_j and β_j are nonlinear parameters. It was the authors' opinion that the advantages to be gained by the use of linear regression analysis far outweighed any disadvantages (such as biasing of estimated parameter values). This assumption is justified in Section IV-E by testing dependent variable sensitivity with respect to β.

The purpose of function $h[\tau_{ij}(n)]$ (see (22)) was to quantify the assumed *decline* in attractiveness of analysis zone i with respect to employment center j as highway travel time in minutes (τ_{ij}) between these zones *increases*, and to incorporate this factor in the employment accessibility functions (see (16)–(21)). Since the existing work trip travel pattern can be assumed to be dependent on employment accessibility [18] it is reasonable to use this in estimating the values of α_j and β_j and to determine the preferred functional form. Data for T_{ij}, the *number of employment persons-trips* between analysis zone i and employment center j, were available from a special tabulation of the 1970 Census [33] for each of the metropolitan areas's 172 analysis zone–employment center pairings. 1960 and 1970 data for E_j and τ_{ij} (dropping the n index for notational convenience) were obtained from sources summarized in Table I.

The accessibility function for which parameter estimates were obtained was

$$T_{ij} = E_j \cdot \alpha e^{-\beta\tau_{ij}}. \tag{32}$$

This differs from the form of (22) in that the subscripts are dropped from α_j and β_j. The assumption is made that the estimated values of α_j and β_j should not vary substantially from one employment center (i.e., value of j) to another. That is to say, one would not expect the *nature* of a large employment center to influence the rate of decline of its attractiveness with distance. This assumption was also made in the EMPIRIC model [22], [23]. Therefore, the subscripts were dropped from α_j and β_j, data were combined for *all* employment centers, and estimated values of α and β were determined via linear regression after making a logarithmic transformation in (32).

The regression yielded estimates of $\alpha = 0.158$ and $\beta = 0.091$ with a fit of $R^2 = 0.741$ and a normal residual pattern (except for low τ_{ij}) and was significant at the 95-percent confidence level. The inverse power function of τ_{ij} given in (22) was also tested but rejected on the basis of the nonnormal residual pattern obtained.

It should be noted at this point that α can be dropped from the accessibility measure (thus reducing it to $\exp(-\beta\tau_{ij})$). This is because α can be factored outside the summations in both the numerators and denominators of (16)–(21) (since it is no longer j-dependent) and thereby cancelled out of each accessibility equation.

The procedure used for estimating values for the γ_k in the metropolitan model's state equation (see (6)) was *multiple linear regression analysis* [31], [48]. The final model obtained represented the "best" linear combination of the independent variables as determined by the following attributes and the statistics (in parentheses) by which they are measured [48]:

1) goodness of fit to the data (R^2 statistic, residual plots, standard error of the estimate);
2) statistical significance of admitted independent variables and overall model explanatory power (partial and overall F tests);
3) lack of intercorrelation among admitted independent variables (tolerance).

An iterative procedure was used and many intervening models resulted prior to the one presented. One of the interesting results was that models in which the data from the central cities of the metropolitan area, Albany, Schenectady, and Troy, were *excluded* calibrated with a much better fit as measured by both the R^2 and the standard error of the estimate and showed greater statistical significance as measured by the overall F statistic. In view of these results and the intended scope of this research, namely that of developing policy analysis tools for developing towns at the metropolitan "fringe," the three cities were dropped from all subsequent calibration runs.

The assumption concerning partitioning of the data set for calibration and validation was also tested. It was concluded that no significant bias was introduced into the

TABLE III
THE METROPOLITAN MODEL AND RELATED STATISTICS

$$\frac{HU_i(n+1)}{\sum_{i=1}^{I} HU_i(n+1)} - \frac{HU_i(n)}{\sum_{i=1}^{I} HU_i(n)} = 5.73 \times 10^{-3} + \underset{\substack{(69.68)^* \\ [0.515]}}{4.50 \times 10^{-10}} \left[\frac{\sum_{j=1}^{4} E_j(n) e^{-0.091\tau_{ij}(n)} [LAD_i - DA_i(n)]^2}{\sum_{i=1}^{I} \sum_{j=1}^{4} E_j(n)\, e^{-0.091\tau_{ij}(n)}} \right]$$

$$+ \underset{\substack{(5.50)^* \\ [0.402]}}{1.51 \times 10^{-6}} RA_i(n) + \underset{\substack{(0.20) \\ [0.063]}}{8.21 \times 10^{-8}} HV_i(n) - \underset{\substack{(6.87)^* \\ [-0.260]}}{2.21 \times 10^{-4}} TAX_i(n)$$

$$- \underset{\substack{(3.18) \\ [-0.191]}}{1.50 \times 10^{-3}} D1_i(n) - \underset{\substack{(0.07) \\ [-0.027]}}{5.26 \times 10^{-4}} D2_i(n) - \underset{\substack{(0.05) \\ [-0.028]}}{3.21 \times 10^{-4}} D3_i(n)$$

$R^2 = 0.835$

Overall Model F = 15.89, F-Value for 95% Significance = 2.46

Standard Error of the Estimate = 1.64×10^{-3}

Residual Pattern: Normal

NOTE: ( ) - F-Value for Estimated Parameter
( )* - Indicates Significance at the 95% Level of the F-Distribution
[ ] - Beta Weight

model by including in the regressions only the 30 analysis zones chosen for calibration, instead of all 40 noncentral-city zones.

Proceeding with the 30 calibration zones, stepwise regression was employed to select the "best" model from among the 19 f_k functions described in (10)–(30). The minimum F value was set at 1.0 and the tolerance value at 0.4.[1] These values were chosen by trial and error to provide for the admission of statistically valid and relatively uncorrelated independent variables to the model, without being unduly restrictive. (No pair of independent variables was found to be completely uncorrelated; the tolerance level of 0.4 was judged to be a reasonable value for avoiding highly intercorrelated terms.)

This automatic calibration procedure produced a model which was inconsistent with theory. A trial and error process was adopted. This involved substituting combinations of independent variables in the equation until a model was found to "best" fit the data while maintaining statistical significance, low intercorrelations among independent variables, and normal residual patterns. Qualitative judgments become as much a part of this process as did quantitative statistical comparisons. Many alternate models were tried, and a choice (Table III) was finally made from among them.

Referring to Table III, one can verify that the model yields a good fit with an R^2 of 0.835, a significant overall model F statistic of 15.89, and three statistically significant independent variables. The most robust of these is the *share of accessibility weighted by land available, squared*, which has a positive coefficient, an F statistic of 69.98 (only 4.20 is required for statistical significance of this variable), and a significant beta weight[2] of +0.515. This indicates an extremely strong positive correlation between accessibility of a zone to employment (weighted by zonal land available for development, squared) and the shift of share of housing units in that zone—*therefore, good accessibility of available land to employment has been statistically shown to promote residential growth within an analysis zone*, and the rate of development in a zone is sensitive (by reason of the beta weight of 0.515) to accessibility to employment. Property tax rates (TAX) correlate *negatively* with a positive shift of growth, which lends statistical credence to the belief that *zonal growth rates decline as property taxes rise*. Residentially developed land (RA) enters the model with statistical significance and a strongly positive beta weight. Therefore, it can be concluded that there is a desire for agglomeration of homes where homes already exist, i.e., *that existing residential development correlates positively with the shift of share of residential development in a zone and exerts a strong influence on the shift of share.*

In order to test the effects of land-use controls in making model projections, inclusion of $D1$, $D2$, and $D3$ (see (26)–(28)) in the model was desirable. These variables were not brought into the model by statistical criteria and so had to be forced in (as a group, necessarily, since these variables have no meaning singly). Although not all are statistically significant (Table III), the variable $D1$ has a partial-F ratio of 3.18 which places it close to the range of statistical admissibility, and so further justifies its inclusion in the model. These variables correlate negatively with growth,

[1] Tolerance is the proportion of the variance of a variable being considered for inclusion in the model that is not explained by the independent variables already in the regression equation [48].

[2] The beta weight is a standardized regression coefficient, computed by multiplying the estimated coefficient (γ_k in (6)) by the standard deviation of the dependent variable and dividing by the standard deviation of the independent variable f_k (see (6)). It represents the number of changes in standard deviation units of the dependent variable per unit standard deviation change in the independent variable [48].

TABLE IV
METROPOLITAN MODEL CALIBRATION—PREDICTED VERSUS ACTUAL 1970 HOUSING UNITS IN THE 30 CALIBRATION ZONES

i	ANALYSIS ZONE*	OBSERVED 1970 HU_i	PREDICTED 1970 HU_i	ABSOLUTE ERROR	PER CENT ERROR
2	Cohoes	6 540	7 148	608	+9.3
3	Watervliet	4 549	4 635	86	+1.9
5	Bethlehem	7 357	7 572	215	+2.9
6	Coeymans	2 126	1 940	-186	-8.8
7	Colonie	20 798	20 321	-477	-2.2
8	Green Island	1 148	1 334	186	+16.2
10	Knox	555	532	-23	-4.2
11	New Scotland	2 460	2 810	350	+14.2
12	Rensselaer	3 341	2 887	-454	-13.6
14	Berlin-Grafton-Petersbrg	1 246	1 334	88	+7.1
16	East Greenbush	3 289	3 033	-256	-7.8
18	Nassau-Stephentown	1 756	2 062	306	+17.4
19	North Greenbush	2 967	3 181	214	+7.2
21	Poestenkill-Sand Lake	2 781	2 885	104	+3.7
22	Schaghticoke	1 775	2 009	234	+13.2
23	Schodack	3 486	3 832	346	+9.9
24	Mechanicville	2 104	2 116	12	+0.5
26	Ballston	2 031	2 025	-6	-0.3
27	Charlton	1 035	1 035	0	0
29	Corinth	1 711	1 641	-70	-4.1
30	Day-Edbg-Gnfld-Hdly-Prov	1 991	1 929	-62	-3.1
31	Galway-Milton	3 957	3 981	24	+0.6
32	Halfmoon	2 826	1 990	-836	-29.6
33	Malta	1 159	1 217	58	+5.0
35	Northumberland-Wilton	1 367	1 105	-262	-19.1
36	Saratoga	1 261	987	-274	-21.7
40	Duanesburg-Princetown	1 554	1 105	-449	-28.9
41	Glenville	8 760	9 573	813	+9.3
42	Niskayuna	5 663	5 301	-362	-6.4
43	Rotterdam	9 448	9 521	73	+0.8

*Refer to Figure 1.

and the relatively low beta weights given in Table III suggest that *zoning and subdivision regulations are not very important determinants of metropolitan growth.*

Housing value (HV) enters the model as a forced variable. Although it is not statistically significant at the 95-percent level, it was an important factor in rendering the residual plots more nearly normal and so was retained.

Additional credence was lent to the choice of this model by examination of the model's residual versus predicted dependent variable plot and a comparison of predicted versus actual housing units for 1970 in the 30 analysis zones used for calibration (Table IV). Examination of a scatterplot showed the residual pattern to be approximately normally distributed with a mean of zero. Additionally, it was noted that 24 out of the 30 residuals lay within plus or minus one standard error of the estimate (SEE) from the zero line, 28 within 2 SEE, and all 30 within 3 SEE of the line, indicating that the fit to data is good and that there are no extreme outliers. Table IV translates these residuals' values into absolute and percent errors in housing unit counts. The maximum percent error is -29.5 percent for Halfmoon (a rapidly growing township), while most errors fall in the -10 percent to $+10$ percent range. (The township of Charlton in Saratoga Co. had zero error!)

Two other statistical tests, the chi-square goodness-of-fit test and Smirnov's theorem [49], were used to evaluate the fit of the calibrated model to data.

The chi-square test was performed on aggregated cell values of the 1970 observed and predicted number of housing units (Table IV) to test the null hypothesis that each set of values represented a random sample drawn from the same population. (Aggregation of a minimum of *five* housing unit counts to a cell and use of a minimum of *five* cells for the chi-square test was necessary to preserve the assumed approximation of the discrete χ^2 frequency function by the continuous χ^2 curve.) This analysis was performed using data aggregated to six cells, resulting in a χ^2 value of 0.775. The table value of the χ^2 statistic for 1 degree of freedom and $\alpha = 0.95$ is 3.841. Thus, *the null hypothesis of a common distribution holding for both predicted and observed 1970 housing units cannot be rejected at the 0.05 level.*

Smirnov's theorem outlines a procedure for comparing

TABLE V
METROPOLITAN MODEL VALIDATION—PREDICTED VERSUS ACTUAL 1970 HOUSING UNITS IN THE 10 VALIDATION ZONES

i	ANALYSIS ZONE*	OBSERVED 1970 HU_i	PREDICTED 1970 HU_i	ABSOLUTE ERROR	PER CENT ERROR	NO OF STANDARD ERRORS OBSERVED HU_i LIES OFF REGRESSION LINE	{70%, 95%} CONFIDENCE BOUNDS ON PROJECTED 1970 HU_i LOWER	UPPER
4	Berne-Rensslrvl-Westerlo	1 791	1 868	+77	+4.3	-0.199	1 411 975	2 325 2 761
9	Guilderland	6 264	7 046	+782	+12.5	-2.019	6 538 6 052	7 555 8 040
15	Brunswick	3 396	3 486	+90	+2.7	-0.232	2 985 2 506	3 988 4 466
17	Hoosick	2 103	2 470	+367	+17.5	-0.948	2 035 1 619	2 906 3 322
20	Pittstown	1 059	640	-419	-39.6	+1.082	201 (-219)	1 079 1 499
25	Saratoga Springs	6 323	5 528	-795	-12.6	+2.053	4 992 4 481	6 063 6 575
28	Clifton Park	3 876	2 814	-1 062	-27.4	+2.743	2 203 1 620	3 425 4 008
34	Moreau	2 963	2 822	-141	-4.8	+0.364	2 385 1 969	3 258 3 675
37	Stillwater	1 514	1 597	+83	+5.5	-0.214	1 173 769	2 020 2 424
38	Waterford	2 445	3 088	+643	+26.3	-1.660	2 612 2 158	3 564 4 919

*Refer to Figure 1.

two independent samples $(X_1, X_2, \cdots, X_m)$ and $(X_1, X_2, \cdots, X_n)$ to test the null hypothesis that the two samples have the same probability distribution. First, define the empirical distribution function $F_M(x)$ and the random variable $D_{m,n}$ (Smirnov's statistic) as follows [49]:

$$F_M(x) = \frac{1}{M} \cdot (\text{number of values of } X_j \text{ for which } X_j \leq x)$$

$$D_{m,n} = \operatorname*{lub}_{-\infty < x < \infty} |F_m(x) - F_n(x)|.$$

Now, Smirnov's theorem states that if $N = mn/(m + n)$ is large and m/n is neither very close to zero nor very large, then for a given significance level α and for z_α as determined from a table [49], the following equation holds (approximately):

$$P(D_{m,n} < z_\alpha N^{-1/2}) = 1 - \alpha.$$

Thus, if the observed value of $D_{m,n}$ is greater than or equal to $z_\alpha N^{-1/2}$, the null hypothesis can be *rejected* at level α.

For the calibration data in Table IV, $D_{m,n}$ was found to be 0.1, N to be 15, and z_α to be 1.36 for $\alpha = 0.05$. Since $0.1 < 1.36/\sqrt{15}$, Smirnov's theorem says that *the null hypothesis that the two sets of observations have the same distribution cannot be rejected at the 0.05 level.*

E. Validation

This section assesses the *validity* of the metropolitan model, that is, the model's ability to predict values that compare closely to observed data not used in the calibration process. As another aspect of validation, the sensitivity of model outputs to changes in key model parameters (specifically, to various values of β in (32)) was also determined.

Table V compares the predicted number of 1970 housing units to the actual U.S. Census count [14] for each of the ten analysis zones in the validation subset. As was the case with the calibration subset, this group of analysis zones had a wide spectrum of observed shifts of share of housing units over the 1960–1970 decade, thus insuring that a good cross section of the zones had been chosen for validation. The absolute values of the percent errors of prediction range from a low of 2.7 percent (for Brunswick, the case-study township for the local model) to a high of (−) 39.6 percent (for Pittstown, one township away from Brunswick). Among the "rapidly growing" towns of interest are Guilderland (+12.5-percent error) and Clifton Park (−27.4-percent error). Besides Brunswick, other zones having "low" absolute error include Berne–Rensselaerville–Westerlo (+4.3 percent), Moreau (−4.8 percent), and Stillwater (+5.5 percent). During the calibration phase it was noted that many of these errors resulted from the "fine-tuning" that occurred during the trial-and-error design of the final model. Such errors in the validation communities were judged to be an acceptable trade-off for obtaining a model that was highly accurate for a large number of communities (as evidenced by the percentage errors displayed in Table IV and the R^2 and other statistics of Table III).

In addition to percentage error analysis, two other measures can be used: the *standard error of the estimate* (SEE) and the *confidence intervals for the true mean number of housing units, given a single future observation of the independent variable vector*. The standard error is the standard deviation of the residuals of the 30 calibration townships about the shift-of-share regression line. It can be converted from shift-of-share units into housing units by multiplying it by

TABLE VI
1970 POPULATION PROJECTIONS FOR THE TEN METROPOLITAN MODEL VALIDATION ZONES USING CURRENT TECHNIQUES

i	ANALYSIS ZONE†	OBSERVED 1970 POPULATION	PREDICTED 1970 POPULATION	PER CENT ERROR
4	Berne-Rensslrvl-Westerlo	5 810	4 733	-18.5
9	Guilderland	21 208	26 060	+22.9
15	Brunswick	11 193	13 289	+18.7
17	Hoosick	6 651	2 701	-59.4
20	Pittstown	3 905	3 269	-16.3
25	Saratoga Springs	18 845	17 462	-7.3
28	Clifton Park	14 867	9 888	-33.5
34	Moreau	10 411	6 708	-35.6
37	Stillwater	5 023	4 680	-6.8
38	Waterford	7 644	5 528	-27.7

*Source: Reference [10]
†Refer to Figure 1

the total number of housing units to be allocated among the metropolitan area's townships at iteration $n+1$.

Table V shows the number of standard errors between the 1970 observed number of housing units for each validation township and the fitted regression line. Note that five of the township's observed number of housing units lie within 1 SEE of the line, two between 1 and 2 SEE's, and the remaining three between 2 and 3 SEE's. This shows that there are no *extreme* outliers in the data, although there are three townships beyond the 2 SEE limit.

A much more precise measure of the model's predictive validity is the establishment of *prediction confidence intervals* [31]. Confidence bounds were evaluated at both the 70-percent and 95-percent levels for the 1970 projected number of housing units in each of the ten validation communities. These bounds are listed in Table V. Note that *all ten* townships' observed 1970 housing units fall within the 95-percent confidence bounds, while six out of ten fall within the 70-percent bounds. Whether these magnitudes of error are acceptable to planners or elected officials is a matter for them to decide. In the authors' opinion such limits of error are an improvement over existing planning tools. To support this contention Table VI is presented as a representative set of results obtained by current projection techniques.

As a further explanation of the errors of projection noted, there appear to have been forces at work in some of the validation zones that forced some quantum jumps in the shift of share of housing units there. For instance, Clifton Park's rapid development from 1280 housing units in 1960 [13] to 3876 housing units in 1970 [14] was due in large part to the actions of a single developer living there who acquired land and built residential units on a much larger scale than had ever before (or since) been attempted in the Capital District [50]. On the other hand, the model satisfactorily accounted for the 5846 housing unit increase during 1960–1970 in the town of Colonie [13], [14] with an error of −2.2 percent (Table IV). Therefore, the model is seen as being able to reliably model large *absolute* increases in an analysis zone's housing stock. It is only when extreme *percentage* changes take place (such as the 200-percent change in Clifton Park) that the model fails to replicate the data.

Errors in the measurement (or adjustment) of the independent variables could have some bearing on the lack of predictive model accuracy in some of these validation communities. In most cases it is felt that these errors were small compared to the total variance of each of the independent variables about their mean, which retains the validity of the regression approach [51]. In any case, the authors' conclusion is that the strict calibration and validation criteria employed here have created a tool that yields projections of an equal or greater degree of accuracy than existing methods for this metropolitan area.

As a further validation check, the sensitivity of the model to a key parameter was evaluated. The value of β in the accessibility equations (see (16)–(22) and (32)) had been estimated prior to the overall model calibration by a separate linear regression analysis in order to avoid the use of nonlinear estimation methods. To test the sensitivity of the model's outputs to changes in this estimated value ($\beta = 0.09$), the model was rerun for $\beta = 0.08$ and $\beta = 0.10$, values thought [22], [23] to bound the feasible limits of the "true" value of β, for the ten validation zones. In nearly all cases the absolute error introduced by the perturbed value of β was much lower than the previously discussed validation error, leading to the conclusion that *metropolitan model outputs are not sensitive to changes in the value of β about its nominal value of 0.09 and, therefore, this value can be used in the overall model without introducing much bias.* This also supports the decision to use a single estimated value of β for all employment centers in the model calibration process.

Other assumptions that were made in the course of the model-building process were subsequently tested to assess model sensitivity to them. The first assumption concerned the prorating of total Albany City 1960 employment between "Albany East" and "Colonie/State Campus" (see Table II). This had initially been done at a 1/3–2/3 ratio. The most extreme test of this assumption (made using a reversed 2/3–1/3 ratio instead and rerunning the regression using the new values of $E_1(1960)$ and $E_2(1960)$ so obtained) yielded less than 1-percent difference in each of the estimated coefficients. Thus, these assumptions were shown not to bias model outcomes in any significant way.

F. Application

Now that a calibrated and validated metropolitan model has been developed, it can be used to project the impacts of certain metropolitan-level controlling actions that are of current interest to Capital District citizens. These actions are the proposed building of a new bridge across the Hudson River at Troy (Hoosick Street Bridge) and the construction of Alternate Route 7 from the bridge out to the Northway (I87) (refer to Fig. 1). Additionally, employment levels in downtown Albany (employment center $j = 1$) and in downtown Troy ($j = 4$) are projected to increase substantially through 1990 as a result of the continued expansion of the

Rockefeller Empire State Plaza in Albany and the opening of the Uncle Sam Mall in Troy. The case-study Town of Brunswick has a vital interest in this matter [52], since one of its major highways leads directly to the proposed bridge site. The townspeople want to know what the expected impacts on their town's growth will be due to the coupled effect of the bridge (and/or Alternate Route 7) opening and new employment opportunities arising in Albany and Troy. Also, Brunswick might be made more attractive for development if tax rates rise disproportionally in competing communities and sterner land-use controls are instituted elsewhere.

These possibilities formed the basis for six *growth scenarios* that were run using the metropolitan model to assess potential impacts of key controlling actions on Brunswick and other townships in the Capital District. These scenarios ranged from a "*zero-action*" *scenario*, where none of the impending control actions take place, to a "*maximum Brunswick growth*" scenario where all control actions are instituted in such a way as to favor maximum growth in the case-study township. These scenarios were set up by the authors as follows (refer to Fig. 1).

Scenario 1: The Hoosick Street Bridge and Alternate Route 7 are not built until 1990 or later; metropolitan employment increases at a normal rate in all centers; and property tax rates, housing value, and land-use controls stay at their 1970 values (*the "zero-action" scenario*).

Scenario 2: All the conditions present in Scenario 1, except that *Albany Downtown employment* is assumed to rise by 5000 per decade over 1970–1980 and 1980–1990, while Troy employment rises by 2000 per decade (best judgment of authors).

Scenario 3: All the conditions present in Scenario 2, except that the *Hoosick Street Bridge* is built in 1980.

Scenario 4: All the conditions present in Scenario 3, except that *Alternate Route 7* (*Alt 7*) is built in 1985 from the Hoosick Street Bridge to Exit 7 of the Northway (I87).

Scenario 5: All the conditions present in Scenario 4, except that *1980 property tax rates* are assumed to rise 50 percent above their 1970 values in all townships assumed to be competing with Brunswick for growth [32]. Brunswick's taxes are assumed to stay the same.

Scenario 6: All the conditions present in Scenario 5, except that in 1980 *zoning controls* are assumed to be imposed within all townships competing with Brunswick for growth [32] that do not have them in 1970 (the "*maximum-Brunswick" scenario*).

For the model given in Table III, values for total metropolitan housing units $[\sum_{i=1}^{I} HU_i(n)]$ in 1980 and 1990 were derived using data from the New York State Office of Planning Services [53]. The variables $\tau_{ij}(n)$, $TAX_i(n)$, $D1_i(n)$, $D2_i(n)$, and $D3_i(n)$ (Table III) for 1980 are all scenario-dependent, as described above. Parameters LAD_i, β, and land-use densities used to determine $RA_i(n)$ and $DA_i(n)$ are assumed to keep the same values as used in 1960–1970, while $HV_i(n)$ is assumed held at its 1970 value due to its low significance in the model. Note that the terms $RA_i(n)$ and $DA_i(n)$ are functions of the state variable $HU_i(n)$.

A complete set of 1980 and 1990 metropolitan model projections for each of the six scenarios appears in [32]. Table VII summarizes these results for the Towns of Brunswick, Clifton Park, and Guilderland along with 70-percent and 95-percent prediction confidence bounds about the 1980 values. In general, the Town of Brunswick is seen to grow (as expected) as the scenarios become more and more pro-Brunswick, while Clifton Park stabilizes and Guilderland grows at a slow rate. Guilderland's jump in growth in 1990 under Scenario 5 is explained by the fact that its tax rate, though increased, is still very competitive with that of other growing townships. Note that the only effect of the 1980 opening of the Hoosick Street Bridge on Brunswick (Scenario 3) is to increase the 1990 number of housing units by about 40 above the "zero-action" level. This result is contrary to the fears of some Brunswick citizens who, according to the author's observations, have feared that the bridge would cause overdevelopment of the town. Indeed, it is not until the more extreme assumptions of Scenario 5 are made (regarding unfavorable tax rates everywhere else) that Brunswick attains any meaningful growth by 1990. In the authors' view this is due to the fact that the bridge is only the final link in a highway network (I787 and I90 in Fig. 1) that began improving Brunswick's accessibility to the metropolitan area's employment zones during 1969–1974 [39]. Additionally, the bridge increases accessibility to employment for not only Brunswick but also for most neighboring townships. Since this increased accessibility is thus "shared" by a number of townships, no single township receives the full impact of the bridge and suffers a quantum jump in development.

A check was made, additionally, on the sensitivity of model projections to alternate assumptions on river crossing time saved by the Hoosick Street Bridge. The 1990 development impacts on Rensselaer County townships resulting from the halving and doubling of these assumed times were computed (see [32] for greater detail). In the case of Brunswick, a doubling of assumed time saved by the bridge results in a projected mean additional 55 housing units in 1990. This suggests that the model is as sensitive to crossing time saved by the bridge as it is to the bridge construction itself (which introduced 40 housing units into Brunswick). However, the expected incremental growth in Brunswick due to both these controlling actions is seen to be a very small percentage of the current township housing stock.

V. The Township Model

A. Choice of the Case-Study Township

It is of interest to a township to observe exactly which areas within it will grow and which will remain unaffected under both metropolitan-forecast growth scenarios and under other "what-if" situations. This is important in terms of forecasting both development and the values of the concomitant impact variables (e.g., overall composition of land use, public service requirements, and tax rates) within the different areas of the township [28]. The township possesses its own internal dynamics which serve to allocate its projected share of metropolitan growth among these

TABLE VII
METROPOLITAN MODEL PROJECTIONS—1980 AND 1990 PROJECTED NUMBER OF HOUSING UNITS IN SELECTED ANALYSIS ZONES

i	ANALYSIS ZONE*	SCENARIO NUMBER	OBSERVED 1970 HU_i	PROJECTED 1980 HU_i	{70% / 95%} CONFIDENCE BOUND ON PROJECTED 1980 HU_i LOWER	UPPER	PROJECTED 1990 HU_i
9	Guilder-land	1	6 264	8 940	8 282 7 655	9 597 10 225	12 159
		2		8 933	8 277 7 649	9 590 10 217	12 137
		3		8 933	8 277 7 649	9 590 10 217	12 103
		4		8 933	8 277 7 649	9 590 10 217	12 111
		5		8 933	8 277 7 649	9 590 10 217	12 442
		6		8 933	8 277 7 649	9 590 10 217	12 479
15	Brunswick	1	3 396	4 239	3 623 3 035	4 854 5 442	5 293
		2		4 241	3 625 3 038	4 856 5 444	5 301
		3		4 241	3 625 3 038	4 856 5 444	5 333
		4		4 241	3 625 3 038	4 856 5 444	5 333
		5		4 241	3 625 3 038	4 856 5 444	5 941
		6		4 241	3 625 3 038	4 856 5 444	6 139
28	Clifton Park	1	3 876	6 130	5 252 4 414	7 007 7 845	8 957
		2		6 124	5 248 4 411	7 001 7 838	8 938
		3		6 124	5 248 4 411	7 001 7 838	8 914
		4		6 124	5 248 4 411	7 001 7 838	8 921
		5		6 124	5 248 4 411	7 001 7 838	8 949
		6		6 124	5 248 4 411	7 001 7 838	8 870

*Refer to Figure 1

areas in response to certain feedback states, local controls, and those regional actions that have been "internalized" (see (3) and (4)) with respect to the township's internal dynamics.

As was previously mentioned, it was considered useful to model one township within the region as a case study in conjunction with the overall metropolitan allocation model. This allowed the modeling of both intratownship and metropolitan effects at a reasonable level of data collection effort while still maintaining a fairly fine-grained level of study. The town of Brunswick, NY, lying within the Capital District SMSA, was selected as the case-study township.

Brunswick (indexed with the number 15 in Fig. 1) was considered a suitable choice because of the willingness of its elected officials to cooperate in this venture and due to its potential for future growth; hence, the modeling effort could provide tangible benefits to these officials. Brunswick is immediately adjacent to Troy, NY, one of the Capital District's central cities, and faces growth pressures due to a planned bridge that will directly link this township to the metropolitan area's interstate highway grid. Its citizens are concerned with what this means in terms of expected new residential and commercial development. They want to know what, if any, local controls can be implemented in order to shape growth into economically and aesthetically suitable patterns [52].

B. The State Equation

There exist two schools of thought concerning the modeling of residential growth at the township level. One assumes that the characteristics of the actors involved in the land conversion process are more important than the characteristics of the property [54], [55]. This is discussed in the authors' previous work [56], wherein a mathematical representation of this process has been postulated. The other school of thought presumes that the characteristics of the property dominate the process [17], [57], [58]. An ideal representation would be one which included both viewpoints. Since there existed no readily available data on the characteristics of the actors involved in the process, this research was forced to attempt to explain growth at the local level using characteristics tied principally to the physical and socioeconomic attributes of the analysis zone and not to those of the actors.

Therefore, the overall state equation of the intratownship residential growth process will resemble that of the metropolitan model. As discussed in Section III, the state variable is the number of housing units present in each of the analysis zones of the township. The shift of share of this state from one model iteration to the next is expressed as a function of analytic forms containing detailed measures of township-level controls as well as functions of the state variable including, as in the metropolitan model, residential and total developed land present. Total topographically suitable land for development is also included as a measure, as is a measure of the *undeveloped* portion of this land at each iteration. The controls include measures within each township analysis zone of mean travel times to major arteries; sewer, water, and street density; zoning district sizes as a fraction of total analysis zone area; and fiscal measures such as property tax rates and per-pupil expenditure.

The objective of the township model is to take the township's aggregate projected share of metropolitan growth at some future time (output of the metropolitan model) and spatially and temporally distribute it within the township. The metropolitan model accounted for the influence of gross controlling actions that affect growth throughout the metropolitan area. In the local model the local controlling actions can be modeled at a much greater level of precision while employing as exogenous data the outputs of the metropolitan model or those of any assumed scenario of interest.

Recalling the discussion presented in Section III, the general state equation for the J analysis zones in each township i were given (combining (3) and (4)) as

$$z_i(m+1) = \boldsymbol{g}_i\{z_i(m), \boldsymbol{w}_i^1(m), \cdots, \boldsymbol{w}_i^J(m), \boldsymbol{h}_i[\boldsymbol{x}_i(n), \boldsymbol{x}_i(n+1)]\}. \tag{33}$$

To further simplify this equation, we have the following propositions.

1) Let $z_i(m)$ be a $Jx1$ vector for township i (the case-study township) and drop the i subscript for notational convenience. Denote an element of this vector as $z_j(m)$.
2) Let $\boldsymbol{w}_i^1(m), \cdots, \boldsymbol{w}_i^J(m)$, the vectors of local controls for each of the J analysis zones of township i at iteration m, be denoted by $w_j(m)$, again, by dropping the i subscript.
3) Let $\boldsymbol{h}_i[\boldsymbol{x}_i(n), \boldsymbol{x}_i(n+1)] = h_i[x_i(n), x_i(n+1)]$ due to the fact that only *one* state variable per township (number of housing units) was used in the metropolitan model.

Now introduce metropolitan results into the township's dynamics through linear interpolation by letting

$$h_i[x_i(n), x_i(n+1)] = \delta x_i(n) + (1-\delta)x_i(n+1) = \sum_{j=1}^{J} z_j(m+1) \tag{35}$$

for

$$t(n) \le t(m+1) \le t(n+1), \tag{36}$$

where $t(l)$ denotes the year of occurrence of metropolitan or township model iteration l, and

$$\delta = \begin{cases} 0, & t(m+1) = t(n+1) \\ \dfrac{t(n+1) - t(m+1)}{t(n+1) - t(n)}, & t(n) < t(m+1) < t(n+1) \\ 1, & t(m+1) = t(n). \end{cases} \tag{37}$$

Thus, (35)–(37) introduce the metropolitan model's results into the township model (for projection purposes) via $\sum_{j=1}^{J} z_j(m+1)$, which is a linear interpolation between metropolitan model states $x_i(n)$ and $x_i(n+1)$. Alternatively, one could choose his own township growth scenario by exogenously inputting his own choice for $\sum_{j=1}^{J} z_j(m+1)$. (For township model calibration and validation purposes, $\sum_{j=1}^{J} z_j(m+1)$ is determined from the township data.) Equation (33) can be reduced, using the above simplifications and the shift-of-share concept developed earlier, to

$$\frac{z_j(m+1)}{\sum_{j=1}^{J} z_j(m+1)} - \frac{z_j(m)}{\sum_{j=1}^{J} z_j(m)} = g[z_j(m), \boldsymbol{w}_j(m)], \tag{38}$$

where g represents a function unrelated to the vector $\boldsymbol{g}_i$ in (33).

Now, a general equation for projecting the shift of share of residential development in each analysis zone of the case-study township can be presented. This equation, like that of the metropolitan model, is *linear in the model parameters* (though not necessarily in the state and control variables) and can be written as

$$\frac{z_j(m+1)}{\sum_{j=1}^{J} z_j(m+1)} - \frac{z_j(m)}{\sum_{j=1}^{J} z_j(m)} = \alpha_0 + \sum_{k=1}^{K} \alpha_k g_k[z_j(m), \boldsymbol{w}_j(m)]. \tag{39}$$

As was the case in the metropolitan model, functional forms for the g_k can now be proposed and used to estimate the α_0 and the α_k.

C. Measures used for Independent Variables

The following measures were adopted for the independent variables indicated by $g_k[z_j(m), \boldsymbol{w}_j(m)]$ in (39).

1) Measures of Land Use and Accessibility: These were formulated in a manner similar to the metropolitan-model measures presented in (10)–(15) and account for the same influences of available developable land, agglomeration [24], land prices [8], and crowding [16].

2) *Measures of Employment Accessibility:* A measure of accessibility was computed from each analysis zone within the case-study township to the boundary of the city of Troy (see Fig. 1), through which most work trips from the township to employment centers pass. The same metropolitan model functional form used in (16) (with the E_j term removed) was employed and the same theories relating accessibility to development [11], [18], [22], [25], [26] were assumed to apply. The value of $\beta = 0.09$ from the metropolitan model (16) was shown in [32] to be valid for computing township-level accessibility measures.

3) *Fiscal Impact Measures:* The combined property tax rate from all taxing jurisdictions in an analysis zone, divided by the sum of tax rates over all zones, was the primary fiscal impact measure used. This is analogous to the approach adopted in the metropolitan model.

4) *Land-Use Control Measures: Subdivision regulations* were uniform in all parts of the town during the study period. Thus, they would not be expected to differentially affect rates of development in the township's analysis zones. However, *zoning* could be included as a control variable because different percentages of each analysis zone's area were zoned for various land uses. In the interests of normalizing zoning data for both large and small analysis zones, all data were expressed as fractions of total analysis zone area.

5) *Measures of Existing Infrastructure:* At the township level a closer look can be taken at the effects of water, sewer, and street patterns on a zone's rate of development than was possible at the metropolitan level, due to greater data availability. Accordingly, model variables were included for the number of lineal feet of water lines, sewer lines, and roads within each analysis zone, divided by total developable land within the zone. The importance of these variables as determinants of development has been stressed by Clawson [1], the Rockefeller Brothers Fund [7], and Kenney [55].

6) *Amenity Measures:* As in the metropolitan model, measurable attributes of the amenities of an analysis zone were derived. Per-pupil expenditure, normalized for inflation by dividing by the total expenditure in all zones in a given iteration, was discussed in the metropolitan-model section. As a result of the finer grain of study in the township model, the presence of areas zoned for recreational use, school and cemetery use, or both, were available as amenity measures.

D. Data Sources

As in the case of the metropolitan model, certain tradeoffs had to be made between ideal and practical data availability and resolution in the township model. Ideally, in the authors' judgment, time-series data should be available over many analysis zones and years for all states and controls of interest. At the outset of the search for data sources at the township level, however, it became obvious that far less data existed in published form than at the metropolitan level.

The search for township-level data on state and control variables, therefore, centered on the case-study Town of Brunswick's own records. Several valuable data sources were discovered, including town tax rolls, housing deeds, building permit records, and subdivision plat maps. The Town's Comprehensive Plan was also found to be an extremely valuable source of data. It was soon realized that a judicious analysis of this data would yield a valuable time-series data base from which the township's development over a historic period could be modeled. These data sources and the township model variables to which they apply are summarized in the following discussion.

The most valuable single source of township-level data proved to be the *town's tax rolls* [59]. Among the important information contained in them for each taxable parcel were

1) *the tax district code*—determined in which fire, sewer, water, and school districts the property was located;
2) *land and building assessed value*—determined whether property was occupied by a building or was vacant land;
3) *acreage of parcel*—useful in computing total existing developed land;
4) *land-use classification code of parcel*—classified land in residential, commercial, industrial, public, or agricultural categories;
5) *parcel identification number*—helped in tracing parcel development status over time.

On the basis of the fact that conformable Brunswick tax roll records were available only for the years 1968–1974, it was decided that the time frame of the analysis should extend in yearly increments from 1968–1974. Parcels had been assigned unique and sequenced identification numbers in 1969 in alphabetical order by owner. Since parcels usually retained their identification number through 1974 even though sold or built on in the interim, this code was seen as the key to a time-series analysis of development trends in the township. The existence of a tax district code on each record permitted these development trends to be aggregated within specific town school, fire, and water tax district boundaries. Analysis zones within the case-study township were, therefore, delineated largely along these lines.

The 1974 tax records were sorted on the parcel code and a printout of these records was manually compared to the printed 1969 tax rolls. Notations were then made, on a parcel-by-parcel basis, of 1969-to-1974 changes in land-use type, development status, and owner of each parcel in the township's regular tax file (a total of some 4400 records). If a parcel's identification number was found to change during the period 1968–1974, a follow-up search was made using *land deeds* [60], to match each such 1968 parcel with its 1974 development status. Areas for some 80 percent of the township's parcels were also determined at this time by adding acreages found in *subdivision plat maps* [61] and *property deeds* [60] to those areas already appearing on the 1974 tax roll. After this sequence of data comparison and editing procedures was completed, all housing units constructed during the years 1968–1974 could be identified by tax district location. All that remained was to affix the month and date of construction of each. This was done by examination of the town's *building permit* file, which contained this information.

Up to this point data was aggregated to the spatial level of township tax districts. In forming the final 17 analysis zones from these it was considered desirable to aggregate some of the smaller tax zones in the township's urbanized area together to form a single analysis zone and to split some of the larger tax zones in rural areas to form two or three analysis zones. This was done in the interests of keeping the number of analysis zones as large as possible (within the limits of data disaggregation imposed by the data collection process), while not having too few housing units in a zone. Once these analysis zones were designated, the number of housing units and the number of acres of land developed in residential and other categories within them could be determined for each of the years 1968–1974.

The Town of Brunswick's *Comprehensive Plan* [63] proved to be another valuable source of data for the township model. Travel times to the Troy border were determined by measuring distances from the centroids of all analysis zones to this border by the most direct route, and by then applying highway link speeds to generate mean travel times. Land available for development was measured by overlaying the soils and elevations maps from the Comprehensive Plan with analysis zone boundaries on graph paper, and then counting squares. Following Capital District Regional Planning Commission guidelines [35], developable land was taken as all land having less than a 15-degree slope *and* having drainable soil capable of excavation by conventional means.

Values for the footage of water lines, sewer lines, and roads in each analysis zone were derived by direct measurement from the water, sewer, and road maps contained in the Comprehensive Plan [63]. According to the town supervisor [52], no changes to these variables occurred during the six-year analysis period. This was also the case with the travel times, implying that some important township independent variables lacked temporal variation in these controls. However, *spatial* variation still existed and the system dynamics still operated through feedback of the state variable.

Percentages of land zoned by use type in each analysis zone were obtained by overlaying the *township's zoning map* [64] with the analysis zone outlines and, again, counting squares. As before, no temporal variation in the data was observed over the 1968–1974 study period, so these data kept constant values in each analysis zone during this period.

A *tax information sheet* distributed by the Brunswick Receiver of Taxes [65] contained time-series property tax rate data for each school, fire, sewer, and water taxing district in Brunswick. Tax rates in analysis zones having multiple tax districts were prorated in proportion to the number of housing units in each taxing jurisdiction, then normalized by their sum over all analysis zones at each iteration (again, to factor out inflation). Note that this variable did possess *temporal* variation. Data for determining *relative per-pupil expenditure* was determined from the same source [43] used in the metropolitan model. These data, too, had temporal variation.

E. Calibration and Discussion

The data base just described contained a six-year history of dependent and independent variables observable at one-year iterations for each of the 17 analysis zones. Thus, 17 × 6 or 102 individual data records could be used in the calibration-validation process if the interval between iterations were chosen as one year. A serious drawback to this scheme was the fact that housing counts increased by only one or two units from one year to the next in some analysis zones and by significantly more units in other zones having nearly the same characteristics. This would result in poor model fit despite the large number of data points. These small fluctuations could be eliminated in most zones by aggregating the data, hence computing the increase in housing units, over several years. Using only a single model iteration over the full six years spanned by the data, however, would result in just one data record for each of the 17 analysis zones. This would provide insufficient data to accomplish both model calibration *and* validation. In view of these considerations it was decided that the best partitioning scheme for the township data was one in which the housing data were aggregated from 1968–1971 and from 1971–1974. This would permit calibration and validation over two independent data sets each spanning a three-year interval.

Model *calibration* was performed using a combination of stepwise and "forced-variable" linear regression analysis on the 17 records comprising the 1968–1971 subset of the data. Variables admitted to the model included measures relating to *crowding* (housing units per developable acre), *percentage of land zoned for large-lot residential or commercial use*, and *availability of streets* (lineal feet of streets per developable acre). The calibration process was considered a success in that it produced statistically significant parameter estimates.

However, *validation* showed that the model had poor fit in those areas that grew during the calibration interval but not during the validation interval. The case-study township had captured spatial variation in its control variables during its calibration period but lacked sizable temporal variation. In view of this *one must conclude that Brunswick did not serve as a good case-study township for calibrating the local model.* This fact did not become apparent, however, until all data had been gathered. It is entirely possible that a much *simpler* model could fit the Brunswick data as well as the present model. Such a model might include a constant plus a term containing the state variable on the right-hand side to predict the shift of share of housing units in an analysis zone. Alternatively, one could aggregate the shifts of share of housing units into a discrete number of ranges and calibrate a model using the range indices or the mean value of each range as the dependent variable.

VI. Results, Conclusions, and Suggestions for Future Work

A. The Two-Tiered Modeling Approach

A *two-tiered* hierarchical methodology has been stated and explained in mathematical terms. This approach was developed in order to be able to forecast growth within a township in a metropolitan area without requiring detailed data on all other townships in the metropolitan area. Both metropolitan and township-level controlling actions can be included as determinants of township growth via this formulation.

B. The Metropolitan Model

A simplified dynamic model has been developed that explains residential growth in townships in a metropolitan area. The model includes a wide variety of control and feedback variables and serves to support many existing theories of growth while adding some new ones. The overall theory of metropolitan growth supported by this model contains the following positive determinants of residential development.

Accessibility to employment weighted by developable land available—contributes partial R-square of 0.71, is F-significant at 69.7, and has a beta weight of $+0.515$. Reinforces the theories of Alonso [57], Muth [58], Lowry [18], Hill [22], and Chapin and Weiss [25].

Existing residentially developed land—is F-significant at 5.50 and has a beta weight of 0.402. Reinforces the concept of the desire for agglomeration proposed by Hughes [24].

Housing value—included to improve residual patterns, but not a statistically significant term.

The following was found to correlate *negatively* with the rate of residential development.

Property tax rates—F-significant at 6.87 with a beta weight of -0.260. Confirms one of the tenets of the Tiebout hypothesis [20].

The following was found *not* to be a statistically significant factor in shaping development.

Presence of subdivision and zoning controls—have a negative correlation with growth but are not F-significant.

The model's goodness of fit was evaluated by a chi-square test and the Smirnov statistic. The results of both tests held that the null hypothesis of goodness of fit could not be rejected. The model's *predictive validity* was demonstrated by running it against an independent data set and measuring the model's predictions versus observations using percent error, standard error, and confidence interval criteria. Comparison was made between the predictive validity of this model and that of existing planning tools, and in this context the model performance came out looking favorable. *Sensitivity analyses* were also performed on assumptions made in the modeling procedure. The general conclusion reached was that the model calibration was *not* biased by these assumptions.

To demonstrate the model's *flexibility and ease of use* it was employed to project the expected outcomes of certain metropolitan- and local-level policy actions of interest. These included major employment center growth, the addition of superhighways and bridges to the metropolitan highway network, and the imposition of higher tax rates and more stringent land-use controls by localities. The resulting expected number of housing units in each township was projected for the years 1980 and 1990, along with 70-percent and 95-percent confidence bounds for the 1980 figures. These results can be used by interested planners, officials, and citizens in a qualitative context for determining local public policy decisions such as transportation investment and land-use control changes in response to regional growth forces. An example of this usefulness was demonstrated by evaluating the effects on growth of a proposed bridge in the case-study metropolitan area. This bridge would connect several rural and semirural townships on one side of the river more directly with employment centers on the other side. The model results indicated a relatively small effect on these townships due to this bridge because it represented only a minor increment to the accessibility to employment already enjoyed by the various affected townships. For instance, the case-study town of Brunswick, although lying directly on an arterial highway connected to the bridge, was projected to grow less than one percent higher under the bridge scenario than under the no-bridge scenario.

C. The Township Model

This effort also developed an intratownship residential growth allocation model. Although Chapin and Weiss [25] treated development on a small-scale rural analysis zone basis, the authors' model is the first *dynamic* tool in this area to treat a wide variety of input variables. The calibration process was considered a success in that it produced statistically significant parameter estimates. However, validation showed that the model had poor fit in other analysis zones. It was concluded that the lack of temporal data for key independent variables limited the usefulness of the model. The data, in fact, would probably only support a much simpler model. A more general model would be appropriate for a case-study township having greater temporal variation in its dependent and independent variables than that existing in Brunswick's.

D. Suggestions for Future Work

The large data base that has been accumulated for the *metropolitan model* should enable future researchers to

1) extend the flexibility of the model in making projections for any desired scenarios involving the area's transportation network, future employment levels, or other variables of interest;
2) add new data as it becomes available (e.g., 1980 Census) and use it to recalibrate or temporally validate the model;
3) verify the metropolitan model's structure by using data from another metropolitan area to recalibrate it.

Several avenues of research are proposed to follow up on these initial efforts at building a *residential development model at the intratownship level:*

1) recalibrate the model in a case-study township having more temporal variation in the control and state variables;
2) collect data in such a manner that the pertinent *actor* characteristics (especially those of farmers) can be included in a more general model formulation (this will allow a parallel assessment to be made of the relative effects of actor attributes versus site characteristics on the rate of residential development);
3) once a reasonably well-calibrated and validated model has been constructed, add a *fiscal impact model* using the methodology outlined by Schaenman and Muller [28] or DiCesare *et al.* [15].

References

[1] M. Clawson, *Suburban Land Conversion in the United States: An Environmental and Governmental Process.* Baltimore: Johns Hopkins Press for Resources for the Future, Inc., 1971, pp. 33–46, 111–113, 117–127, 184, and 238.

[2] Discussions with Thornton K. Ware, former Director of The Rensselaer County Bureau of Planning (1/74).

[3] H. Blumenfeld, "Transportation in the modern metropolis," in *Internal Structure of the City*, L. S. Bourne, Ed. Toronto: Oxford Univ., 1971, p. 231.

[4] E. W. Butler and E. J. Kaiser, "Prediction of residential movement and spatial allocation," *Urban Affairs Quart.*, vol. 6, no. 4, pp. 477–494, June 1971.

[5] Real Estate Research Corporation, *The Costs of Sprawl* (Detailed Cost Analysis), Apr. 1974.

[6] R. O. Harvey and W. A. V. Clark, "The nature and economics of urban sprawl," *Land Economics*, vol. 41, no. 1, pp. 1–10, Feb. 1965.

[7] The Rockefeller Brothers Fund, *The Use of Land: A Citizens' Policy Guide to Urban Growth*, W. K. Reilly, Ed. New York: Crowell, 1973, pp. 190 and 191.

[8] R. B. Andrews, *Urban Land Economics and Public Policy.* New York: Free Press, 1971, pp. 17, 83, and 84.

[9] Conversations with W. Hebert, Planning and Research Bureau, New York State Department of Transportation.

[10] Capital District Regional Planning Commission, *Population Growth and Development in the Capital District Region.* June 1969, pp. 126–132.

[11] G. T. Lathrop and J. R. Hamburg, "An opportunity-accessibility model for allocating regional growth," *J. Am. Inst. Planners*, vol. 31, no. 2 (Special Issue), May 1965.

[12] N. Keyfitz, "The population projection as a matrix operator," *Demography*, vol. 1, pp. 56–73, 1964.

[13] U.S. Department of Commerce, Bureau of the Census, *1960 Census of Population and Housing—Census Tracts* (Albany–Schenectady–Troy, NY SMSA), Final Rep. PHC(1)-4.

[14] U.S. Department of Commerce, Bureau of the Census, *1970 Census of Population and Housing—Census Tracts* (Albany–Schenectady–Troy, NY SMSA), Final Rep. PHC(1)-4, May 1972.

[15] F. DiCesare, D. A. Ensminger, P. G. VanBuskirk, and K. R. Carlisle, "Policy analysis modeling for community development in small cities," *IEEE Trans. Syst., Man, Cybern.*, vol. SMC-6, no. 8, pp. 515–531, Aug. 1976.

[16] R. N. S. Harris, C. Harrell, and G. S. Tolley, "The residence site choice," *Rev. Econom. Statist.*, vol. L, no. 2, May 1968.

[17] L. Wingo, *Transportation and Urban Land.* Washington, DC: Resources for the Future, Inc., 1961.

[18] I. S. Lowry, "A model of metropolis," Rand Corp., Santa Monica, CA, Memo. RM-4035-RC, Aug. 1964.

[19] B. J. L. Berry and F. E. Horton, *Geographic Perspectives on Urban Systems.* Englewood Cliffs, NJ: Prentice-Hall, 1970, pp. 311–313.

[20] C. M. Tiebout, "A pure theory of local expenditures," in *Readings in Urban Economics*, M. Edel and J. Rothenberg, Eds. New York: Macmillan, pp. 513–523, 1972.

[21] L. B. Sagalyn and G. Sternlieb, *Zoning and Housing Costs.* New Brunswick, NJ: Center for Urban Policy Research, 1973.

[22] D. M. Hill, "A growth allocation model for the Boston region," *J. Am. Inst. Planners*, vol. 31, no. 2 (Special Issue), pp. 111–120, May 1965.

[23] J. W. Dickey *et al.*, *Metropolitan Transportation Planning.* Washington: Scripta Book Co., 1975, pp. 155–173.

[24] J. Hughes, "Social area analysis," in *Readings in Urban Economics and Spatial Patterns*, M. R. Greenberg, Ed. New Brunswick, NJ: Center for Urban Policy Research, pp. 46–47, 1974.

[25] F. S. Chapin, Jr., and S. F. Weiss, Eds., *Urban Growth Dynamics.* New York: Wiley, 1962, pp. 433–437.

[26] J. F. Kain, "The journey to work as a determinant of residential location," in *Models of Employment and Residence Location*, F. J. James, Ed. New Brunswick, NJ: Center for Urban Policy Research, pp. 127–150, 1974.

[27] G. Chadwick, *A Systems View of Planning.* New York: Pergamon, 1971, pp. 204–208.

[28] P. S. Schaenman and T. Muller, *Measuring Impacts of Land Development.* Washington, DC: Urban Institute, 1974, pp. 10–11, 47–50.

[29] W. Z. Hirsch, *Urban Economic Analysis.* New York: McGraw-Hill, 1973.

[30] J. Delafons, *Land-Use Controls in the United States*, 2nd ed. Cambridge: M.I.T., 1969, pp. 27, 42–89, and 172–173.

[31] N. R. Draper and H. Smith, *Applied Regression Analysis.* New York: Wiley, 1966.

[32] D. A. Ensminger, "A two-tiered dynamic policy analysis model of residential development in suburban metropolitan communities," Ph.D. dissertation, Rensselaer Polytechnic Inst., Troy, NY, 1977.

[33] U.S. Department of Commerce, Bureau of the Census, "Special tabulation of employment person-trips from census tract of origin to minor civil division of destination," 1970.

[34] New York State Department of Transportation, Planning and Research Bureau, *Summary of the Findings of the Land Use Inventory*, Capital District Transportation Study, Oct. 1969.

[35] Capital District Regional Planning Commission, *Preliminary Regional Development Plan: A Consultation Document.* Rep. No. CDRPC 100-5, Mar. 1975.

[36] Capital District Regional Planning Commission, *Housing in the Capital District Region 1970–1990.* Dec. 1971, pp. 114–128.

[37] New York State Department of Transportation, Planning and Research Bureau, "Special computer tabulations of automobile travel times from 1965 traffic analysis zone of origin to zone of destination."

[38] New York State Department of Transportation, Planning Division, *1972 Annual Report—Capital District Transportation Study.* July 1973.

[39] Consultation with Mr. Dan Ruddy, New York State Department of Transportation, Contracts Division.

[40] P. Albin, *Arterial Network Capacity Calculation Program*, New York State Department of Transportation, Planning and Research Bureau, Nov. 1970.

[41] New York State Department of Audit and Control, *1959 Tax Atlas of New York State, Comptroller's Studies in Local Finance, Number 2*, 1961, pp. 15–63.

[42] New York State Department of Audit and Control, Division of Municipal Affairs, Bureau of Municipal Research and Statistics, *Overall Real Property Tax Rates, Local Governments in New York State (Fiscal Years Ended in 1973).* 1973, pp. 9–35.

[43] New York State Department of Education, *Annual Education Summary*, 1959–1960, 1970–1971.

[44] U.S. Department of Commerce, Bureau of the Census, *County Business Patterns* (New York) CBP-71-34, 1971.

[45] U.S. Department of Commerce, Bureau of the Census, *Census of Population: 1970. General Social and Economic Characteristics* (New York), Final Rep. PC(1)-C34, 1972.

[46] C. Lee, *Models in Planning.* Oxford: Pergamon, 1973, p. 23.

[47] M. D. Kilbridge, R. P. O'Block, and P. V. Teplitz, *Urban Analysis.* Boston: Graduate School of Business Administration, Harvard Univ., 1970.

[48] N. H. Nie, C. H. Hull, J. G. Jenkins, K. Steinbrenner, and D. H. Bent, *Statistical Package for the Social Sciences*, 2nd Ed. New York: McGraw-Hill, 1975.

[49] Z. W. Birnbaum, *Introduction to Probability and Mathematical Statistics.* New York: Harper & Brothers, 1962, pp. 248–249, 277, and 283–284.

[50] M. Cermak, "Saratoga County: The people explosion," *The Knicker-*

bocker News–Union Star, Feb. 15, 1971.
[51] C. Daniel and F. S. Woods, *Fitting Equations to Data*. New York: Wiley, 1971, p. 32.
[52] Conversations with George Barkley, Brunswick Town Supervisor (6/76).
[53] New York State Office of Planning Services, *Demographic Projections—Households*. Oct. 1974.
[54] E. J. Kaiser and S. F. Weiss, "Public policy and the residential development process," *J. Am. Inst. Planners*, vol. 36, no. 1, pp. 30–37, Jan. 1970.
[55] K. B. Kenney, "The residential land developer and his land purchase decision," Ph.D. dissertation, Univ. North Carolina, 1972, pp. 139–146.
[56] D. A. Ensminger and F. DiCesare, "Analysis tools for evaluating the effects of land-use and fiscal control policies in developing communities," in *Transportation Planning for a Better Environment*, H. Wenzel and P. Stringer, Eds. New York: Plenum, 1976, pp. 103–114.
[57] W. Alonso, *Location and Land Use*. Cambridge: Harvard Univ., 1964.
[58] R. Muth, *Cities and Housing*. Chicago: Univ. Chicago, 1969.
[59] Town of Brunswick, NY, assessment rolls (1968–1974).
[60] Town of Brunswick, NY, land deed records (1968–1974).
[61] Town of Brunswick, NY, subdivision and lot maps (1895–1974).
[62] Town of Brunswick, NY, building permits (1968–1974).
[63] Hans Klunder Associates, *Brunswick, New York—Comprehensive Plan*. 1971.
[64] Town of Brunswick, NY, *Zoning Map*. 1958.
[65] Town of Brunswick, NY, township, school, and special district property tax rates, 1968–1974.

A Multiple Objective Optimization-Based Approach to Choicemaking

CHELSEA C. WHITE, III, MEMBER, IEEE, AND ANDREW P. SAGE, FELLOW, IEEE

Abstract—An approach for choicemaking under certainty, based in part upon multiple objective optimization, is presented in this paper. With this approach clearly inferior alternatives are eliminated and subsets of alternatives are determined which are guaranteed to contain the most preferred alternatives through use of an induced partial order on the set of alternatives. Use of such a relation to order the alternative set partially is reminicent of multiobjective optimization theory (MOOT). Our MOOT based approach is believed behaviorally relevant in that it is interactive and can accommodate a variety of cognitive styles and levels of on-line decisionmaker preference feedback. It does not require elicitation of all criterion or attribute weights as does the multiple attribute utility theory (MAUT) approach. MAUT is compatible with and can be used to complement and potentially enhance the effectiveness of the approach for value elicitation for choicemaking. Application of the MOOT based approach is illustrated by several examples.

I. Introduction

THE FOLLOWING problem statement is central to many different choicemaking problems and may be compatible with several decisionmaker cognitive styles.

P: select, and rank to the extent possible, the N most preferred alternatives from X, the finite set of all alternatives under consideration.

The motivation for problem P is due to the fact that decisionmakers and policymakers, in seeking the most preferred alternative, typically will first request a modest number, perhaps between three and five, of highly preferred alternatives from their analysts. They then will explore further the rationale behind the selection and ranking of these alternatives. The goal of the analysts, therefore, is to provide the N most preferred alternatives and (generally) an explanation of the rationale behind their selection and ranking in such a manner that disbenefits of planning and decisionmaking such as the time, cost, and stress of choicemaking are kept to a tolerably low level. We will assume throughout that the set of alternatives X has previously been identified and perhaps refined with the aid of an elimination by attributes procedure which eliminates clearly undesirable alternatives (Tversky [10]).

A standard approach to solving problem P is multiattribute utility theory (MAUT) (Keeney and Raiffa [4], Edwards [1]), which can be described as follows.

1) Assess a real-valued (utility) function u, defined on the alternative set $X, u: X \to R$, which satisfies the property that the utility of alternative x is greater than the utility of alternative x' if and only if alternative x is preferred to alternative x'; i.e., $u(x) > u(x')$ if and only if $x \succ x'$. Mathematically, the converse is also true. That is $x \succ x'$ if and only if $u(x) > u(x')$. In practice we elicit utility or value functions and from this infer preferences. Behaviorally the concept of preference must preceed that of utility.
2) Compute $u(x)$ for all $x \in X$ and rank order the computed utilities.
3) Identify the alternatives that generate the N largest utility values.

There are conditions,[1] e.g., mutual utility independence, which imply that utility functions having special functional forms, e.g., the multilinear form, adequately model a decisionmaker's preferences (Keeney and Raiffa [4]). Such specialized functional forms significantly reduce the amount of time and effort required to complete the assessment of a multiattribute utility function. However, verification of sufficient conditions to insure existence of these functional forms can be a time and effort consuming task. Also, if such sufficient conditions do not hold, utility functions with functional forms based on these conditions may not represent good quantitative models of decisionmaker preference. Finally, the cognitive style of many decisionmakers may make elicitation of the complete set of attribute weights needed for MAUT very difficult. Thus, although MAUT provides a solution to problem P, accurate execution of step 1) can be a time consuming and stressful activity in practice, especially if there are many attributes under consideration and the decisionmaker is not comfortable with the highly structured elicitation procedure required to determine attribute weights.

Manuscript received December 17, 1979; revised March 11, 1980.

The authors are with the Department of Engineering Science and Systems, School of Engineering and Applied Science, University of Virginia, Charlottesville, VA 22901.

[1]Many authors distinguish between preference (value) functions and preferential independence conditions for the riskless case and utility functions and utility independence conditions for the risky case. We use the word utility to represent both of these.

Reprinted from *IEEE Trans. Syst., Man, Cybernetics*, vol. SMC-10, pp. 315–326, June 1980.

In this paper we propose a behaviorally attractive complementary and alternative approach to solving problem P. This new approach only attempts, a least initially, to order the alternative set partially. This is suggestive of a search for the Pareto-optimal, noninferior, or nondominated set (or frontier) in multiobjective optimization theory (MOOT). We therefore refer to the choicemaking approach to be presented and analyzed in this paper as the *MOOT based* approach. There have been a number of previous studies concerning multiple objective optimization theory (Haimes [3], MacCrimmon [6], MacCrimmon and Siu [5], Starr and Zeleny [9], and Zeleny [12]). However, our approach is fundamentally much more an evaluation, refinement, or ranking approach than an optimization approach.

The primary advantage of at least initially assessing only a partial order on the alternative set is that the often difficult and time consuming effort of trading off noncommensurate objectives, or attributes, may be partially avoided. Avoidance or delay in making trade-offs will eliminate or at least postpone explication of those values and goals that the decisionmaker may wish to leave initially implicit. The disadvantage of a partial order is that the resulting nondominated set of alternatives will contain, in general, more than one alternative, thereby requiring further effort for final choice selection. The reason for determining the partial order, however, is that if the nondominated set is sufficiently small, the decisionmaker may be able to select the most preferred alternative from the nondominated set without any significant further external assessment. Equivalently, the partial order can be thought of as eliminating dominated alternatives from further consideration. We remark that since, for example, the second most preferred alternative may not be in the original nondominated set, determination of only the original nondominated set is not guaranteed to solve problem P.

The MOOT based approach to solving problem P is presented in Section II. The key assumption made is that the decisionmaker can select the most preferred alternative from any given set of nondominated alternatives. This assumption is referred to as the *complete preference feedback* assumption. Three procedures are offered for enhancing the likelihood that the complete preference feedback assumption will hold.

An algorithm for solving problem P for the complete preference feedback case and an example are presented in Section III. Our development of this algorithm, presented in Section IV, is based on use of a directed graph representation of the partial order on the set of alternatives X induced by a vector of objective functions.

In Section IV we modify the statement of problem P and the associated analysis to accommodate the possibility that complete preference feedback may not be available. An algorithm and several examples are presented for this *partial preference feedback* case.

A rather complete hypothetical example is presented in Section V to illustrate several of the notions developed earlier.

II. The MOOT Based Choicemaking Approach—Complete Preference Feedback

In this section we present, discuss, and analyze the MOOT based approach to solving problem P. We make a complete preference feedback assumption: that the decisionmaker can choose the most preferred alternative from any given nondominated set. Procedures are suggested for improving the likelihood that this assumption will hold. This section begins with several preliminary definitions and assumptions.

We assume that M presumably noncommensurate attributes have been identified and that for each attribute m, a real-valued objective function on $X, v_m: X \to R$, has been assessed.[2] For the mth attribute, v_m is equivalent to a total order, $\prec_m$, on X; i.e., x' is preferred to x with respect to the mth attribute ($x \prec_m x'$) if and only if $v_m(x) \leqslant v_m(x')$.

Although the MOOT based approach to choicemaking presented in this paper requires M objective functions to be assessed, compared to only one utility function to be assessed for the MAUT approach, assessment of the complete set of lowest level objectives $\{v_i\}$ is often significantly easier than complete assessment of u. Typically, assessment of a complete scalar utility function will require assessment of M lowest level component utility functions v_i or u_i, $i = 1, 2, \cdots, M$, and trading off all of these utility functions by assessing at least $M-1$ utility or attribute weights. Assessment of $\{v_i\}$ for the MOOT based approach does not require trading off the M lowest level noncommensurate objectives, an activity that utility function assessment cannot avoid. For example, few would disagree that 1) maximizing economic benefit and 2) minimizing environmental pollution are two important objectives in the management of many industrial innovation situations. It is unlikely that determining objective measure functions for these two objectives would be unduly time consuming, stressful, or controversial. Trading off achievement levels for these two objectives, however, may be agonizing.[3] To the extent possible, it may be desirable to avoid, or at least postpone, making such trade-offs, especially if they are not really necessary, as may often be the case.

Let the vector function $v: X \to R^M$ be defined as $v(x) = \{v_1(x), \cdots, v_M(x)\}$. Note that v is equivalent to a partial order on X; i.e., x' is preferred to x ($x \prec x'$) if and only if $v(x) \leqslant v(x')$. Let $v(X) = \{v(x) : x \in X\}$, and assume without significant loss of generality that $v: X \to v(X)$ is one-to-one and onto. The M-vector element $w \in v(X)$ and its corresponding alternative are said to be *nondominated* if there does not exist another element $w' \in v(X)$ such that $w \leqslant w'$ and $w \neq w'$.

The MOOT based choicemaking approach for solving problem P consists of two basic steps.

[2]For notational convenience we will use v_i to represent attributes and objectives for the MOOT based approach and u_i to represent attributes for the MAUT based approach.

[3]Generally, practical determination of objective measure functions requires preference independence among objectives, a requirement common to both MAUT and our MOOT based approach.

1′ Select the most preferred element in $v(X)$ from the nondominated set of $v(X)$, i.e., the set of all nondominated elements in $v(X)$.

2′ If the number of alternatives determined thus far is less than N, say N', then redefine X as X with these N' alternatives removed and go to 1); otherwise stop.

We now discuss several important issues raised by each of these two steps.

A) Step 1′

Two fundamental assumptions must hold in order that Step 1′ can be successfully accomplished:

a) the most preferred element in $v(X)$ can be found in the nondominated set of $v(X)$;

b) the decisionmaker can select the most preferred element in the nondominated set of $v(X)$. This is the complete preference feedback assumption.

We remark that it is particularly desirable for the first assumption to hold since, in general, the fewer the alternatives the decisionmaker has from which to choose, the easier it is to select the most preferred alternatives. We will show that the first assumption holds under a weak and behaviorally relevant hypothesis, which will be presented after a needed definition.

Definition: A scalar *social choice function* is a function $V: v(X) \rightarrow R$ which when composed with the objectives function v produces the utility function u; i.e., V is such that $u(x) = V[v(x)]$ for all $x \in X$.

The scalar social choice function trades off the objectives functions to produce a real-valued measure of preference. In our approach, we wish to avoid a complete explicit assessment of V as much and for as long as possible. To verify that it is sufficient to examine only elements in the nondominated set of $v(X)$ in searching for the most preferred element in $v(X)$, however, we must assume its existence.

Lemma 1: Assume that the social choice function V is isotone (monotonically nondecreasing). Then, there is a vector element w in the nondominated set of $v(X)$ such that $V(w) \geqslant V(w')$ for all $w' \in v(X)$.

Proof: Assume that there is a $\hat{w} \in v(X)$ which is not in the nondominated set of $v(X)$ but which has the property: $V(\hat{w}) > V(\tilde{w})$ for all $\tilde{w}$ in the nondominated set of $v(X)$. This property implies that $\hat{w} \geqslant \tilde{w}$ for all $\tilde{w}$ in the nondominated set of $v(X)$; hence, $\hat{w}$ is a member of the nondominated set of $v(X)$, which is a contradiction.

In light of Lemma 1, we therefore assume throughout the remainder of this paper that the social choice function is isotone in the value function in order to preserve the isotonicity of the utility function u with respect to the partial ordering induced by v on X.

We remark that if mutual utility independence (Keeney and Raiffa [4]) holds, then the multiplicative form of risk preference or utility under risk

$$1 + Ku(x) = \prod_{m=1}^{M} \left[1 + Kk_m u_m(x)\right]$$

represents a special case of an isotone social choice function.

The primary difficulty in successfully accomplishing Step 1′ is that the decisionmaker may not be able to easily select the most preferred element in the nondominated set of $v(X)$, i.e., the complete preference feedback assumption may not be valid, in general. The ease with which this selection can be made depends on the specific decision situation, the efficiency of the attribute or objective structuring procedure, the number of attributes, and the cognitive style of the decisionmaker and decision situation associated task characteristics.

In some cases the nondominated set of $v(X)$ may be sufficiently small and/or its alternative elements may be sufficiently distinguishable, and/or the dimension of the objectives vector v may be sufficiently small for the decisionmaker to select the most preferred alternative without further preference elicitation. For example, a recently developed medical diagnostic decision aid, which uses a specialized and simplified version of the MOOT based approach, typically provides a nondominated diagnostic test set which is sufficiently small to cause a physician little difficulty in selecting a test from the nondominated set without further explicit preference elicitation (White *et al.* [11]). If this is not the case and the decisionmaker finds it difficult to select a most preferred alternative from the nondominated set of $v(X)$, two approaches, not requiring complete preference elicitation, can be considered.

a) Decrease the size of the nondominated set until it is small enough for the decisionmaker to be able to select a most preferred alternative.

b) Modify problem statement P and the desired end result of the analysis in order to accommodate the fact that complete preference feedback is not available.

The second approach will be considered in detail in Section IV and will not be discussed further in this section. We devote the remainder of this section to the first approach.

There appears to be three potentially implementable, general procedures to accomplish the first approach:

1) trading off objectives such as to reduce, in effect, the number of objectives by moving up the objectives tree,

2) determining bounds on the objective trade-off weights,

3) utilizing prior elicited knowledge concerning the functional form of the social choice function.

The first procedure results in trading off some of the objectives. This is equivalent to redefinition of some lower level objectives in terms of higher level objectives and, in this way, "moving up" the objectives tree. Following several preliminary definitions, we will show that, under proper assumptions, trading off some of the objectives is a generally desirable activity, especially where there are many lowest level objectives. Trading off some of the objectives:

a) "strengthens" the induced partial order
b) does not add elements to the set of nondominated elements and often results in a reduction in the number of elements in the nondominated set.

Let $\tilde{v}: X \rightarrow R^N, N \leqslant M$, be defined as $\tilde{v}(x) = f[v(x)]$ for all $x \in X$, where $f: R^M \rightarrow R^N$. Let $\tilde{\prec}$ be the partial order on X induced by $\tilde{v}$. Define $\bar{X}$ and $\tilde{X}$ as the nondominated sets in X generated by the respective partial orders $\prec$ and $\tilde{\prec}$.

We remark that the function f can represent a social choice function (if it is isotone and if $N=1$) or a function that trades off less than all of the objectives functions. For example, let $f(w_1, \cdots, w_M) = \{w_1, \cdots, w_{M-2}, g(w_{M-1}, w_M)\} \in R^{M-1}$, where the function $g: R^2 \rightarrow R^1$ represents a (presumably isotone) "local" social choice function of the two objectives functions w_{M-1} and w_M. The function f can also delete some of the objectives from consideration (let $f(w_1, \cdots, w_M) = \{w_1, \cdots, w_N\}$, where $N < M$). See (Fishburn [2]) for a related discussion which involves lexicographic ordering. Deletion of objectives, however, appears best accomplished as part of the prescreening and prescanning of alternatives, or alternatives refinement step, which would precede the use of the MOOT based procedure.

Lemma 2:

a) If f is isotone, then for any pair $x, x' \in X$, the fact that $x \prec x'$ implies $x \tilde{\prec} x'$.
b) If f is strictly isotone, then $\tilde{X} \subset \bar{X}$.

Proof: We note that $x \prec x'$ if and only if $v(x) \leqslant v(x')$ which implies $\tilde{v}(x) = f[v(x)] \leqslant f[v(x')] = \tilde{v}(x')$ if and only if $x \tilde{\prec} x'$. Thus part a) is shown to be correct.

To verify the correctness of part b) we will prove that $\bar{X}^c \subset \tilde{X}^c$, where the superscript c stands for complementation with respect to X. Let $x \in \bar{X}^c$. Then there exists an $x' \in \bar{X}$ such that $v(x) \leqslant v(x')$ and $v(x) \neq v(x')$. The strict isotonicity of f preserves this property under composition.[4] Hence, $\tilde{v}(x) = f[v(x)] \leqslant f[v(x')] = \tilde{v}(x')$ and $\tilde{v}(x) \neq \tilde{v}(x')$, which implies $x \notin \tilde{X}$.

Part a) of Lemma 2 states that if alternative x' is preferred to alternative x before a trade-off is made, then this preference will also hold after the trade-off is made. Clearly the converse may not be true; for example, let $w = (1,2)$, $w' = (2,1)$, and $f(w) = w_1$. Thus, a trade-off will always "strengthen" (more properly, not weaken) a partial ordering on X (and $\tilde{\prec}$ is said to be stronger than $\prec$) in that there will be at least as many pairs of alternatives having elements that can be related by $\tilde{\prec}$ as could be related by $\prec$.

Part b) of Lemma 2 states that, under the assumption that the trade-off function is strictly isotone, the set of nondominated alternatives resulting from a trade-off will never contain more alternatives than before the trade-off was made. This is desirable since a decisionmaker would not want to spend the time and effort involved in trade-off elicitations if this were to result in an increased number of alternatives to be considered. We remark, however, that strict isotonicity is required. For example, if $f(w) = 1$ for all $w \in v(X)$, then $\tilde{X} = X$.

The procedure of making trade-offs may be accomplished by the use of an attribute or objectives measure template, or tree, the structure of which is generally obtained as part of the impact assessment and analysis phase of a choicemaking effort. As has been demonstrated by Lemma 2, moving up the objectives template by increasing the number of objectives traded off will not increase the size of the nondominated set. It is therefore beneficial to perform as many attribute or objectives trade-offs initially as is comfortably possible (i.e., make the number of objectives M as small as is comfortably possible). Trading off all of the objectives (through use of an assessed social choice function which may be an approximation based on ordinal preference data or a function based on a complete cardinal preference assessment) is mathematically, though not necessarily behaviorally, equivalent to assessing a utility function and would produce a nondominated set consisting of only the most preferred alternative.

The second procedure to decrease the number of elements in the nondominated set of $v(X)$ involves determining bounds on the objective trade-off weights. This procedure would impose at least a partial ordering on the alternatives within a nondominated set. For example, consider the following $M=2$ case. The isotonicity of a social choice function guarantees that if an equipreference curve goes through the point w in Fig. 1, then the remainder of the curve will be in the shaded area. Therefore, the isotonicity of the social choice function does not provide sufficient information to determine how the points w and w' are related by preference. If, however, it could be determined that the equipreference curve would remain in the shaded region depicted in Fig. 2, then we would be assured that w is preferred to w'. For example, if the decisionmaker can state that the weight of objective 2 can never exceed α times the weight of objective 1 and that $\alpha > (w_1 - w_1')/(w_2' - w_2)$, then we can surely assume that w is preferred to w'.

The third procedure for reducing the number of elements in the nondominated set of $v(X)$ requires utilizing what information there exists regarding the functional form of the social choice function V. For example, assume V is linear in the objectives functions; hence the equipreference curves are lines having negative slope as in Fig. 3. Consider three alternatives x_1, x_2, and x_3, each producing objective vectors, $v(x_i)$, of $(0,1)$, $(1,0)$, and $(0.4, 0.4)$, respectively. Clearly, all three alternatives are nondominated; however, it is easy to demonstrate that the linearity of V implies that alternative x_3 can not be preferred to both x_1 and x_2.

Step 2′

Once an alternative has been determined as most preferred in Step 1′, Step 2′ requires that this alternative be

[4] f composed with v is the function $f[v(\cdot)]$.

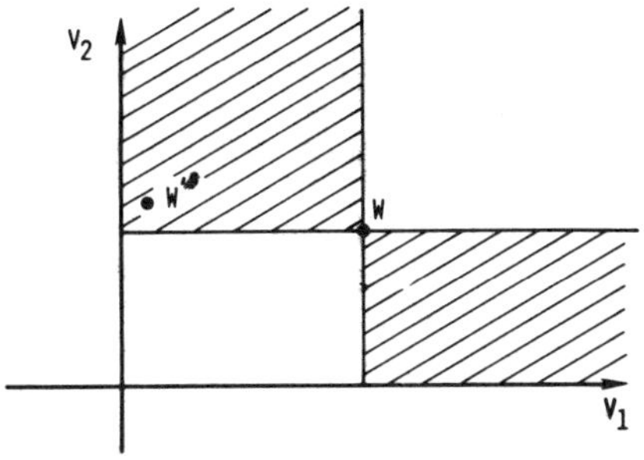

Fig. 1. Bounds on equipreference curves.

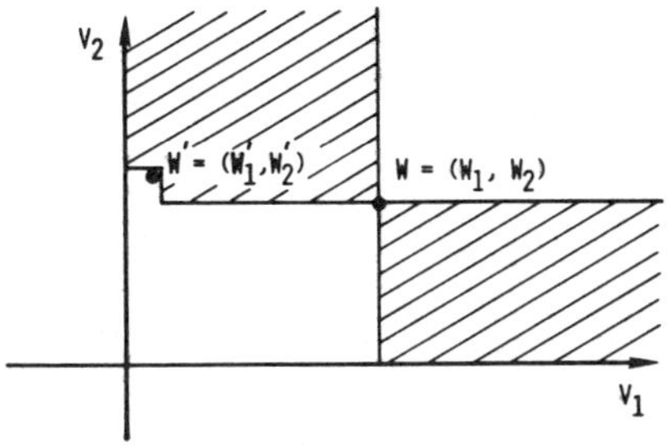

Fig. 2. Reduced region for equipreference curves.

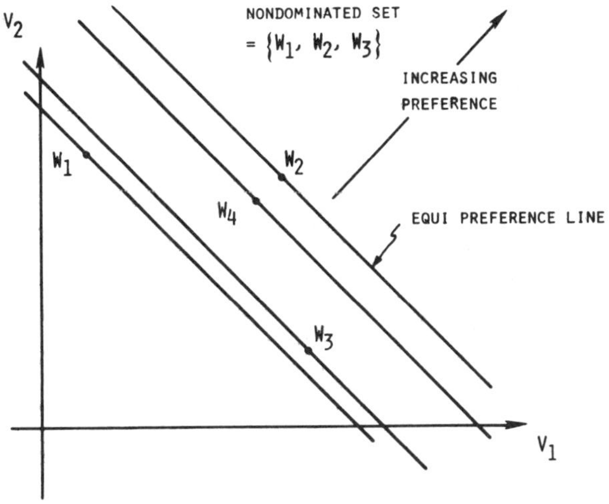

Fig. 3. Dominated and nondominated alternatives.

removed from further consideration. Even though the most preferred alternative must be in the nondominated set, the second most preferred alternative may not be. Fig. 3 illustrates this fact, which is very important in practice for successful use of multiple objective based approaches.

III. Algorithm for the Complete Preference Feedback Case

Following some preliminaries, we present an algorithm in this section for determining the N most preferred alternatives and their ranking from a finite set of alternatives. This algorithm assumes that the decisionmaker can always identify the most preferred element from any given nondominated set.

The algorithm for the complete preference feedback case as well as one other algorithm to be presented in Section IV, will be based on a directed graph, or digraph,

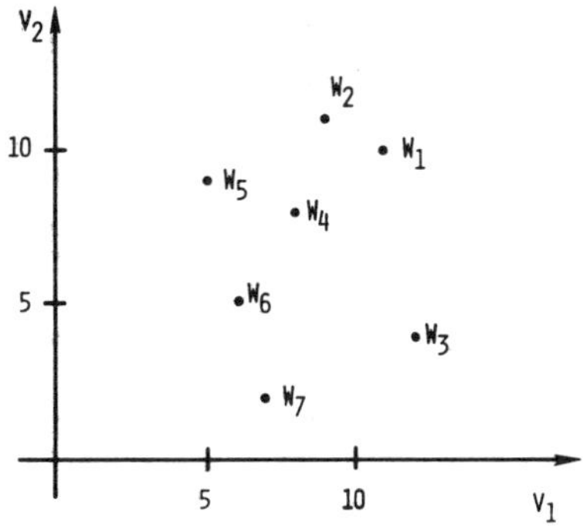

Fig. 4. Elements of $v(X)$ for the example.

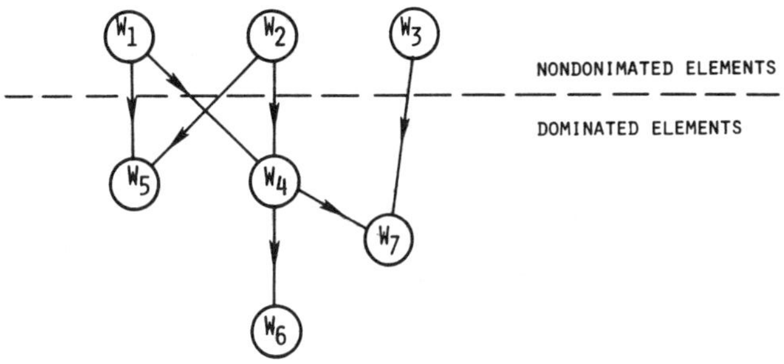

Fig. 5. Initial digraph for the example: contextual relation "dominates."

constructed using the contextual relation "dominates." Details of the construction of such digraphs are given in [8] and elsewhere. We now present an example in order to illustrate the relationship between a set $v(X)$ and its associated digraph.

Example: 1: Assume $M=2$ and that $v(X)$ is composed of the following points, where $w_i=\{w_{i1},w_{i2}\}$:

$$w_1=(11,10)$$
$$w_2=(9,11)$$
$$w_3=(12,4)$$
$$w_4=(8,8)$$
$$w_5=(5,9)$$
$$w_6=(6,5)$$
$$w_7=(7,2).$$

These points are displayed in Fig. 4; the associated digraph is presented in Fig. 5.

Consider the following algorithm.

0) Set $n=1$.
1) Determine the most preferred element in the top level set (or nondominated set) of the digraph. This is equivalent to determining the nth most preferred alternative.
2) Remove the selected element from the digraph. Move up to the top level all second level elements which were predecessors of *only* the removed element.
3) If $n=N$ stop; if not, increase n by one and go to step 1).

We illustrate the above algorithm with the following example.

Example: 2: Assume that $X=\{x_1,\cdots,x_7\}$ and $w_i=v(x_i)$, $i=1,\cdots,7$, as given in Example 1 and depicted in Fig. 4 and 5. We note from Fig. 5 that search for the most preferred alternative can be restricted to those represented by $\{w_1,w_2,w_3\}$ (or equivalently alternatives $\{x_1,x_2,x_3\}$); consequently, the decisionmaker can ignore the elements w_i, $i=4,\cdots,7$, in selecting the most preferred alternative. Assume that the decisionmaker selects x_1 as the most preferred alternative. We then eliminate w_1 to produce Fig. 6. Observe that although elements w_4 and w_5 are no longer dominated by w_1 (which is now removed), they remain dominated by w_2, and hence no new elements move into the nondominated set. A search for the second most preferred element can therefore be restricted to the set $\{w_2,w_3\}$. Assume the decisionmaker selects w_2 as the second most preferred score. Removal of w_2 from Fig. 6 now allows w_4 and w_5 to enter the nondominated set, as shown in Fig. 7. A search for the third most preferred element in $v(X)$ can therefore be restricted to the set of attribute or objective scores $\{w_3,w_4,w_5\}$ and the alternatives x_3, x_4, and x_5 that they represent.

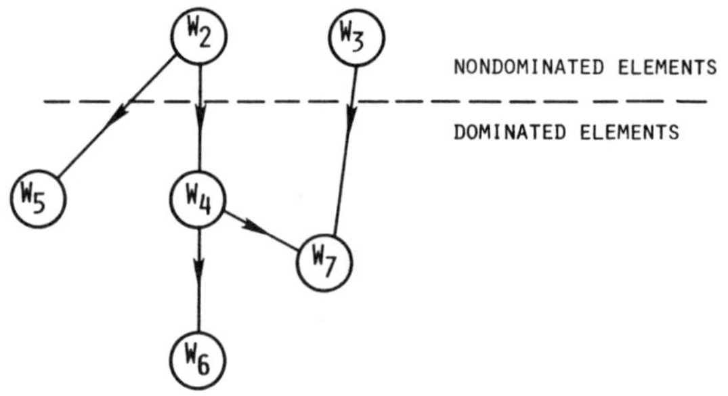

Fig. 6. Digraph with w_1 eliminated.

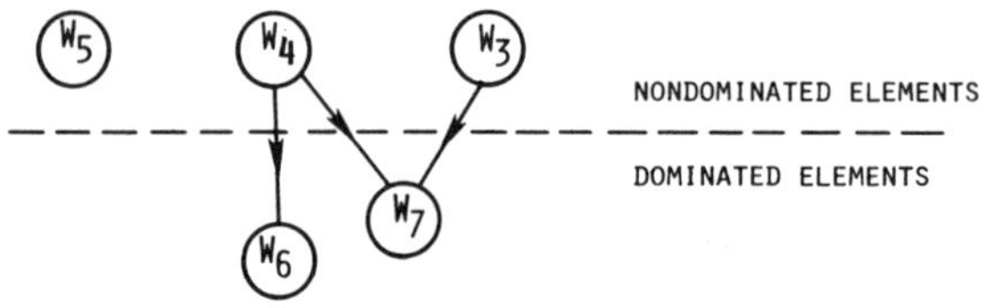

Fig. 7. Digraph with w_1 and w_2 eliminated.

IV. The Partial Preference Feedback Case

As discussed in Section II, the complete preference feedback assumption may not always be satisfied without making difficult trade-offs to insure a sufficiently small dimension for the objectives vector and sufficiently small size of the nondominated set. In order to be able to accommodate the possibility that complete preference feedback may not be available, we modify the statement of problem P as follows.

P': determine $\mathcal{X}_N$, a collection of ordered subsets of X, where

1) each element in $\mathcal{X}_N$ contains N alternatives,
2) each element in $\mathcal{X}_N$ is a candidate for the ordered set of the N most preferred alternatives based on:
   a) the digraph induced by the vector function v
   b) preferences additional to those already expressed in determining the vector-valued function v.

We remark that although the information contained in $\mathcal{X}_N$ is of obvious value for decisionmaking, $\mathcal{X}_N$ may not be the most desirable form in which to present this information. A desirable aggregated form might be $X_N\subset X$, where X_N is the subset that is guaranteed to contain the N most preferred alternatives. The subset X_N is easily obtained by taking the union of subsets in $\mathcal{X}_N$. Observe that in general $\mathcal{X}_N$ cannot be reconstructed from X_N and the digraph because of the additional preference information supplied by the decisionmaker. It is true, however, that $\mathcal{X}_N$ is a subset of the set of all combinations of alternatives in X_N which are compatible with the original digraph.

We observe that any subset of X having cardinality n (i.e., containing n elements) may not be a member of $\mathcal{X}_n$. For example, $\{x_1,x_2,x_7\}$ cannot be a member of $\mathcal{X}_3$ for the set of alternatives considered in Examples 1 and 2 because it is incompatible with the digraph displayed in Fig. 5; i.e., the digraph indicates that x_7 cannot be the third most preferred alternative.

Lemma 2b) demonstrated, for the complete preference feedback case, that trading off objectives would not cause additional alternatives to be included in the nondominated set. We now prove an analogous result for the partial preference feedback case, following several preliminary definitions. Let $\tilde{\mathcal{X}}_N$ be the collection of ordered subsets of X which is defined as $\mathcal{X}_N$ is defined, except $\tilde{\mathcal{X}}_N$ is determined based on a partial order $\tilde{\prec}$ that is stronger than $\prec$.[5] Similarly, we define $\tilde{X}_N\subset X$ as the union of all subsets of X in $\tilde{\mathcal{X}}_N$.

Corollary: Assume that the function f defined in Section II is strictly isotone. Then, $\tilde{\mathcal{X}}_N\subset\mathcal{X}_N$ and $\tilde{X}_N\subset X_N$.

Proof: Clearly, $\tilde{X}_N\subset X_N$ if $\tilde{\mathcal{X}}_N\subset\mathcal{X}_N$. Lemma 2b) implies that $\tilde{\mathcal{X}}_1\subset\mathcal{X}_1$; assume $\tilde{\mathcal{X}}_{n-1}\subset\mathcal{X}_{n-1}$. It is easily shown that Lemma 2b) then implies $\tilde{\mathcal{X}}_n\subset\mathcal{X}_n$, and the result follows by induction.

An algorithm for the determination of $\mathcal{X}_N$ is as follows.

0) Set $\mathcal{X}_0=\varnothing$ and $n=1$.
1) Assume $\mathcal{X}_{n-1}=\{X^1_{n-1},\cdots,X^K_{n-1}\}$ and $\mathcal{X}_n=\varnothing$; set $k=1$.
2) Construct a digraph from the alternatives in $X\sim X^k_{n-1}$.[6]
3) Assume there are L nondominated alternatives in the digraph constructed in step 2). Set $l=1$.
4) Let $Y=\{X^k_{n-1},x\}$, where $x\in X$ is the lth nondominated alternative in the digraph constructed in step 2).
5) Determine whether or not the set $Y\subset X$ should be included in $\mathcal{X}_n$ and include it or exclude it as appropriate. This determination should result from

[5] Refer to Section II for further explication of these sets and subsets.

[6] $A\sim B=A\cap B^c$; that is, $A\sim B$ is the set of all points in A which are not also contained in B.

either direct interaction with the decisionmaker or from preference information, additional to the information contained in the function v, obtained from the decisionmaker.

6) If $l \geq L$, GO TO 7). If $l < L$, let $l = l + 1$ and GO TO 4).
7) If $k \geq K$, GO TO 8). If $k < K$, let $k = k + 1$ and GO TO 2).
8) If $n \geq N$, STOP. If $n < N$, let $n = n + 1$ and GO TO 1).

We can now demonstrate that problem P' generalizes problem P for the case of complete preference feedback. Observe that complete preference feedback would select $\mathcal{X}_1$ to be composed of one subset. This subset would contain only one alternative, the most preferred alternative. We assume $\mathcal{X}_{n-1}$ contains only one ordered subset of X which contains the $n-1$ most preferred alternatives. Then, a quick iteration of the algorithm indicates that step 5) determines the nth most preferred alternative which is included with the subset of $\mathcal{X}_{n-1}$ to result in the single subset contained in $\mathcal{X}_n$ for this complete preference feedback case.

The following four examples illustrate the MOOT based choicemaking algorithm for the partial preference feedback case and several forms of the partial preference information that may be available from the decisionmaker.

Example 3: Consider the set of alternatives and their objective measures which were examined in Examples 1 and 2. We will assume, however, that the decisionmaker can supply no preference feedback beyond the preference information already contained in the function v. This assumption is the opposite extreme of the complete preference feedback case and will be referred to as the *no preference feedback* case. Note that step 5) in the above partial preference feedback algorithm is essentially removed in this case. Thus,

$$\mathcal{X}_1 = \{\{x_1\},\{x_2\},\{x_3\}\},$$

$$\mathcal{X}_2 = \{\{x_1,x_2\},\{x_1,x_3\},\{x_2,x_1\},\{x_2,x_3\}\{x_3,x_1\},\{x_3,x_2\}\},$$

$$\mathcal{X}_3 = \{\{x_1,x_2,x_3\},\{x_1,x_2,x_4\},\{x_1,x_2,x_5\},\{x_1,x_3,x_2\},\{x_2,x_1,x_3\},\{x_2,x_1,x_4\},\{x_2,x_1,x_5\},\{x_2,x_3,x_1\},\{x_3,x_1,x_2\},\{x_3,x_2,x_1\}\}$$

as can easily be seen by inspection of Fig. 5. We note that

$$X_1 = \{x_1,x_2,x_3\}$$

$$X_2 = \{x_1,x_2,x_3\}$$

$$X_3 = \{x_1,x_2,x_3,x_4,x_5\}.$$

It is of interest to observe that if the first three most preferred alternatives are sought, then alternatives x_6 and x_7 can always be excluded from consideration, even if no preference feedback is available from the decisionmaker.

We remark that for the no preference feedback case $\mathcal{X}_N$ can be reconstructed exactly from X_N and the original digraph.

Example 4: Assume that in examining the results of Example 3, the decisionmaker decides that x_4 cannot be a member of X_3. Then, $\mathcal{X}_3$ becomes

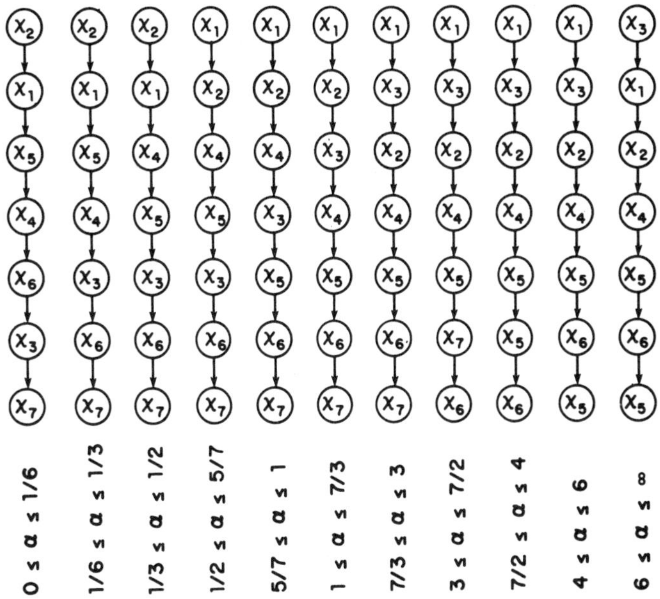

Fig. 8. Preferences and domination relations for linear social choice function.

$$\mathcal{X}_3 = \{\{x_1,x_2,x_3\},\{x_1,x_2,x_5\},\{x_1,x_3,x_2\},\{x_2,x_1,x_3\},\{x_2,x_1,x_5\},\{x_2,x_3,x_1\},\{x_3,x_1,x_2\},\{x_3,x_2,x_1\}\}.$$

Example 5: Assume that the decisionmaker cannot exclude any of the original three nondominated alternatives as the most preferred alternative, given only the information provided in Examples 1 and 2. That is, $\mathcal{X}_1 = \{\{x_1\},\{x_2\},\{x_3\}\}$. For the $n=2$ case, however, we assume that the decisionmaker feels that both x_1 and x_2 cannot be more preferred than x_3, or equivalently that x_3 must either be preferred to x_1 or to x_2. Therefore,

$$\mathcal{X}_2 = \{\{x_1,x_3\},\{x_2,x_3\},\{x_3,x_1\},\{x_3,x_2\}\},$$

which is identical to the $\mathcal{X}_2$ determined in Example 3 with the sets $\{x_1,x_2\}$ and $\{x_2,x_1\}$ excluded. If no further preferences can be expressed, it then follows that

$$\mathcal{X}_3 = \{\{x_1,x_3,x_2\},\{x_2,x_3,x_1\},\{x_3,x_1,x_2\},\{x_3,x_2,x_1\}\}.$$

Thus, $X_1 = X_2 = X_3 = \{x_1,x_2,x_3\}$. We now observe that alternatives x_i, $i = 4,\cdots,7$, can be excluded from consideration in determining the three most preferred alternatives. x_6 and x_7 were already excludable; the information that both alternatives x_1 and x_2 cannot be more preferred than x_3 allows us to exclude x_4 and x_5 from consideration as candidates for the three most preferred alternatives. Note that this exclusion has been accomplished without detailed elicitation of weights.

Example 6: Again consider the data presented in Examples 1 and 2. Assume preference information from the decisionmaker has implied that the three most preferred alternatives are x_1, x_2, and x_4 and that the social choice function is linear. Let $v(x) = \{v_1(x), v_2(x)\}$ and assume that the social choice function is $V[v(x)] = \alpha v_1(x) + v_2(x), \alpha \geq 0$. We can easily obtain the preference or domination digraph of Fig. 8. Since we have assessed that $\mathcal{X}_3 = \{\{x_1,x_2,x_4\}\}$, we see from Fig. 8 that $\frac{1}{2} < \alpha < 1$. It also follows without further preference information that $\mathcal{X}_4 = \{\{x_1, x_2, x_4, x_3\}, \{x_1, x_2, x_4, x_5\}\}$. Note that $\{x_1, x_2, x_4, x_6\} \notin \mathcal{X}_4$ due to the linearity of the social choice function even though x_6 is in the nondominated set of X with alternatives x_1, x_2, and x_4 removed. Thus we see that complementary use of the proposed MOOT based approach with an ordinal linear social choice function has allowed bounding of the linear social choice function and elimination of infeasible alternatives without detailed efforts to elicit all attribute weights. Of course we, indirectly, use many of the attractive and useful features of MAUT in this procedure. However we selectively use portions of the attribute or objectives template and our MOOT based approach to bound and limit criterion or attribute weight elicitation. Here we are able to eliminate x_6 from further consideration and, as a byproduct, determine bounds on a linear social choice function trading off $v_1(x)$ and $v_2(x)$.

V. An Example Application of the MOOT Based Approach

We now present a hypothetical example of an interactive decision support process which utilizes the MOOT based choicemaking approach. It is assumed that the template displayed in Fig. 9 represents the attribute structure of the problem. For example, if the problem under consideration is to purchase a new automobile, the attributes in Fig. 9 might have the following meaning:

A) safety
B) initial cost
C) fuel economy
D) scheduled maintenance expenses
E) unscheduled maintenance expenses
F) resale value
G) attractiveness
H) trunk and passenger compartment capacity
I) desirability
C)–E) operating cost
B)–F) cost.

Note that the template indicates that "cost" is composed of "initial cost," "operating cost," and "resale value."

Assume that six (6) hypothetical alternatives are under consideration, x_1 through x_6. Table I presents alternative scores for each alternative and for each lowest level attribute. These scores would be assessed from experts familiar with the particular attribute and alternative. They correspond to the $w_i = v(x_i)$ function form concepts used in our previous examples.

The objectives measure value for alternative x_1 is at least as great as the objectives measure value for x_5 for every attribute, and therefore x_1 dominates x_5. Similarly, x_2 dominates x_6. Fig. 10 presents the associated digraph. If the intent is to select a single alternative eventually, then both x_5 and x_6 could be deleted from further consideration at this point.[7] If the ultimate intent is to select more than just the best alternative, then all alternatives require further consideration. We also observe that if the two most preferred alternatives are to be selected and $x_1(x_2)$ is considered the most preferred alternative, then $x_6(x_5)$ can be deleted from further consideration.

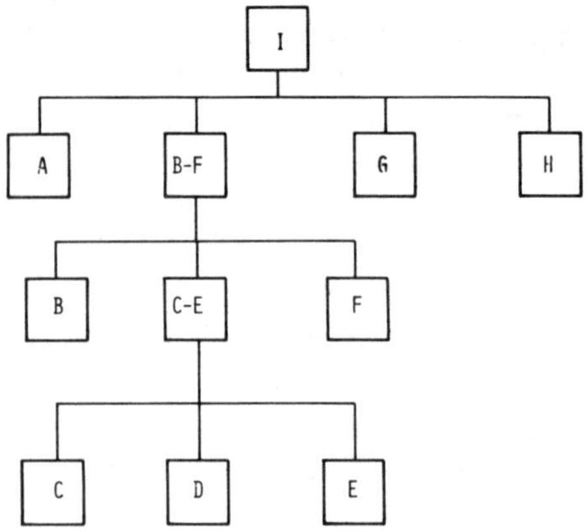

Fig. 9. Attribute structure.

TABLE I
ALTERNATIVE SCORES FOR LOWER LEVEL ATTRIBUTES OR CRITERIA

		ALTERNATIVE SCORES FOR ATTRIBUTES							
		A	B	C	D	E	F	G	H
ALTERNATIVES	x_1	70	100	65	40	80	10	100	60
	x_2	100	40	70	30	100	100	10	100
	x_3	60	35	70	35	10	10	40	50
	x_4	50	0	100	100	0	90	10	100
	x_5	65	40	0	40	75	0	30	55
	x_6	0	35	60	0	90	40	0	0

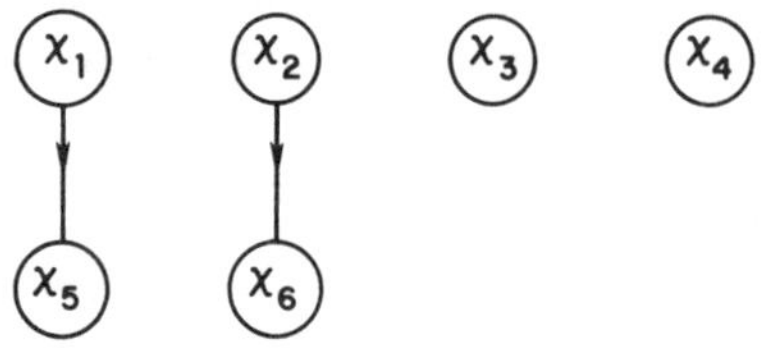

Fig. 10. Domination digraph for Table I.

Fig. 10 may contain a sufficient amount of information for choicemaking; however, if this is not the case, which is very likely since there are eight lowest level attributes, some trade-off assessment becomes necessary. In order to strengthen the partial order on the set of alternatives, we move up the objectives template by performing our first trade-off. The attributes which we assume are first traded

[7]Unless there exists the possibility that x_1 may be "unavailable," in which case x_5 should be retained, or the possibility that x_2 may be "unavailable" in which case x_6 should be retained. Henceforth we will routinely assume potential availability of all alternatives.

TABLE II
TRADE-OFF WEIGHTS FOR ATTRIBUTES C, D, AND E

	ATTRIBUTES C	D	E
BEST ALTERNATIVES	x_4	x_4	x_2
WORST ALTERNATIVES	x_5	x_6	x_4
WEIGHTS	.20	.35	.45

TABLE III
ALTERNATIVE SCORES FOR ALL OBJECTIVE MEASURE VALUES: FIRST TRADE-OFF CASE

ALTERNATIVES	ALTERNATIVE SCORES FOR ATTRIBUTES A	B	C-E	F	G	H
x_1	70	100	63	10	100	60
x_2	100	40	69.5	100	10	100
x_3	60	35	30.75	10	40	50
x_4	50	0	55	90	10	100
x_5	65	40	47.75	0	30	55
x_6	0	35	52.5	40	0	0

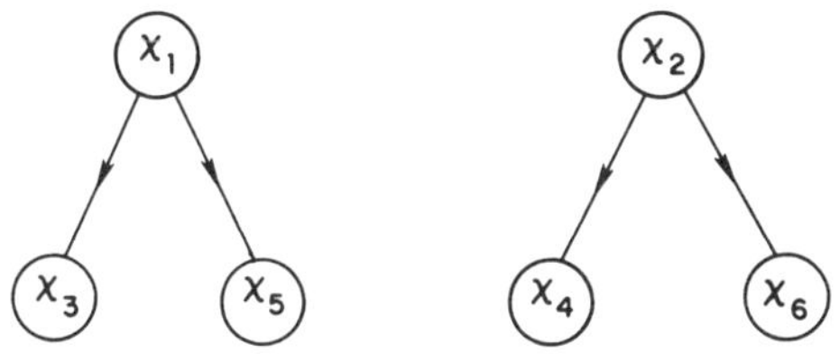

Fig. 11. Domination digraph for Table III.

off are C, D, and E. We note that with respect to each of these three attributes, the best and the worst alternatives are given in Table II. Assume that the decisionmaker provides the following information: the linear weights for comparing the difference between the best alternative and the worst alternative are 0.20, 0.35, and 0.45 for the three attributes C, D, and E, respectively. Table III gives the resulting objectives measure values. We now note that x_1 dominates x_3 and x_5, and x_2 dominates x_4 and x_6. Thus we have the digraph of Fig. 11. Note that if the intent were to determine a single best alternative, it would be sufficient at this point to eliminate alternatives x_3, x_4, x_5, and x_6 from further consideration. A comparison of Figs. 10 and 11 indicates the impact of strengthening the partial order on the set of alternatives by trading off the three attributes C, D, and E. If selecting more than the best alternative were the objectives of the choicemaking effort, once again all alternatives would have to be considered further. We observe also that if the two most preferred alternatives were being sought and $x_1(x_2)$ were chosen as most preferred, then x_4 and x_6 (x_3 and x_4) could be eliminated from further consideration.

TABLE IV
TRADE-OFF WEIGHTS FOR ATTRIBUTES B, C–E, AND F

	ATTRIBUTES B	C-E	F
BEST ALTERNATIVES	x_1	x_2	x_2
WORST ALTERNATIVES	x_4	x_3	x_5
WEIGHTS	.40	.10	.50

TABLE V
ALTERNATIVE SCORES FOR OBJECTIVE MEASURE VALUES: SECOND TRADE-OFF CASE

ALATERNATIVES	ALTERNATIVE SCORES A	B-F	G	H
x_1	70	51.3	100	60
x_2	100	72.95	10	100
x_3	60	22.075	40	50
x_4	50	50.5	10	100
x_5	65	20.775	30	55
x_6	0	39.25	0	0

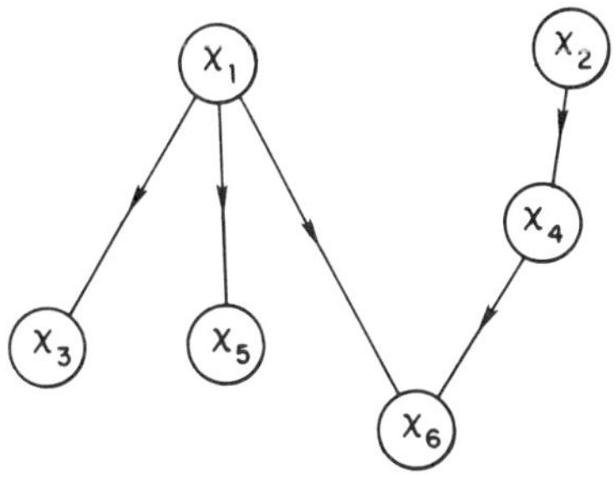

Fig. 12. Domination digraph for Table V.

Fig. 11 may convey enough information to the decisionmaker for alternative selection. If not, as may well be the case since there are now six lowest level objectives, further trade-off assessment is required. In order to further strengthen the partial order on the set of alternatives, we assume that the decisionmaker has decided to move up the template by trading off the attributes B, C–E, and F. The best and worst alternatives for each of these attributes are given in Table IV.

Assume that the decisionmaker provides the following trade-off information: linear weights for comparing the difference between the best alternative and the worst alternative are 0.40, 0.10, and 0.50 for the attributes B, C–E, and F, respectively. Table V provides the resulting objectives measure values. We note that x_1 continues to dominate x_3 and x_5 but now also dominates x_6. Once again, x_2 dominates x_4 and x_6. Also, x_4 dominates x_6. The digraph of these new relationships is presented in Fig. 12.

TABLE VI
Trade-Off Weights for Attributes G and H

	ATTRIBUTES G	ATTRIBUTES H
BEST ALTERNATIVES	x_1	x_2, x_4
WORST ALTERNATIVES	x_6	x_6
WEIGHTS	.10	.90

TABLE VII
Alternative Scores for Objective Measure Values: Third Trade-Off Case

ALTERNATIVES	ALTERNATIVE SCORES A	B-F	G-H
x_1	70	51.3	64
x_2	100	72.95	91
x_3	60	22.075	49
x_4	50	50.5	91
x_5	65	20.775	52.5
x_6	0	39.25	0

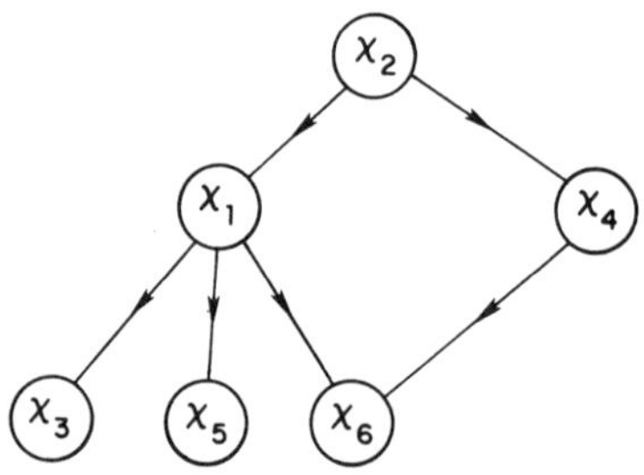

Fig. 13. Domination digraph for Table VI.

From this digraph we observe that x_6 need not be considered further if the objective is to seek the best and second best alternatives. We also note that if the decisionmaker could select x_2 as the most preferred alternative without further assessment, then x_3 and x_5 each would join x_6 as being at best the third most preferred alternative. If x_1 is the most preferred, then x_2, x_3, and x_5 would compete for second best, x_4 could be no better than third best, and x_6 could not be better than the fourth most preferred.

If the second trade-off has not provided the decisionmaker with enough information to make the desired selections, another trade-off assessment is required. We therefore move up the template by trading off at least some of the second level attributes. Assume the decisionmaker elects to trade off the attributes G and H. Table VI lists the best and worst alternatives and their associated incremental weights for each of these two attributes. Table VII lists the resulting objectives measure values, and Fig. 13 presents the associated digraph. It is now clear that x_2 is the most preferred alternative and x_1 and x_4 would com-

TABLE VIII
Trade-Off Weights for Attributes A, G–F, and G–H

	ALTERNATIVES A	B-F	G-H
BEST ALTERNATIVES	x_2	x_2	x_2, x_4
WORST ALTERNATIVES	x_6	x_5	x_6
WEIGHTS	.40	.50	.10

TABLE IX
Alternative Scores for Objective Measures Values: Final Case

ALTERNATIVES	ALTERNATIVE SCORES A-H
x_1	60.05
x_2	85.575
x_3	39.9375
x_4	54.35
x_5	41.6375
x_6	19.625

Fig. 14. Domination digraph for Table IX.

TABLE X
Final Weights for Lowest Level Attributes

A	0.4000
B	0.2000
C	0.0100
D	0.0175
E	0.0225
F	0.2500
G	0.0100
H	0.0900

TABLE XI
ALTERNATIVE SCORES FOR OBJECTIVE MEASURE VALUES: THIRD TRADE-OFF CASE WITH VARIED WEIGHTS

ALTERNATIVE	ALTERNATIVE SCORES A	B-F	GH
x_1	70	51.3	74
x_2	100	72.95	68.5
x_3	60	22.075	46.5
x_4	50	50.5	68.5
x_5	65	20.775	46.25
x_6	0	39.25	0

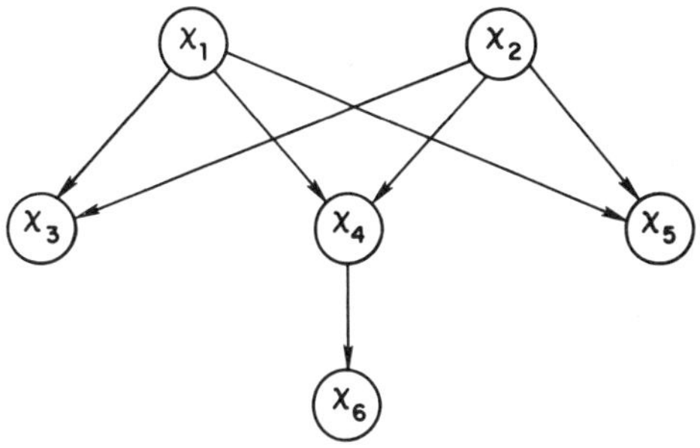

Fig. 15. Domination digraph for Table XI.

TABLE XII
POSSIBILITY SUBSETS FOR N BEST ALTERNATIVES, $N = 1, 2$

TABLE	FIGURE	ORDERED SUBSET CANDIDATES FOR N BEST ALTERNATIVES	SUBSET GUARANTEED TO CONTAIN N MOST PREFERRED ALTERNATIVES
1	9	$X_1 = \{\{x_1\}, \{x_2\}, \{x_3\}, \{x_4\}\}$ $X_2 = \{\{x_1,x_2\}, \{x_1,x_3\}, \{x_1,x_4\}, \{x_1,x_5\}, \{x_2,x_1\}, \{x_2,x_3\}, \{x_2,x_4\}, \{x_2,x_6\}, \{x_3,x_1\}, \{x_3,x_2\}, \{x_3,x_4\}, \{x_4,x_1\}, \{x_4,x_2\}, \{x_4,x_3\}\}$	$X_1 = \{x_1,x_2,x_3,x_4\}$ $X_2 = \{x_1,x_2,x_3,x_4,x_5,x_6\}$
3	10	$X_1 = \{\{x_1\}, \{x_2\}\}$ $X_2 = \{\{x_1,x_2\}, \{x_1,x_3\}, \{x_1,x_5\}, \{x_2,x_1\}, \{x_2,x_4\}, \{x_2,x_6\}\}$	$X_1 = \{x_1,x_2\}$ $X_2 = \{x_1,x_2,x_3,x_4,x_5,x_6\}$
5	11	$X_1 = \{\{x_1\}, \{x_2\}\}$ $X_2 = \{\{x_1,x_2\}, \{x_1,x_3\}, \{x_1,x_5\}, \{x_2,x_1\}, \{x_2,x_4\}\}$	$X_1 = \{x_1,x_2\}$ $X_2 = \{x_1,x_2,x_3,x_4,x_5\}$
6	12	$X_1 = \{\{x_2\}\}$ $X_2 = \{\{x_2,x_1\}, \{x_2,x_4\}\}$	$X_1 = \{x_2\}$ $X_2 = \{x_1,x_2,x_4\}$
9	13	$X_1 = \{\{x_2\}\}$ $X_2 = \{x_2,x_1\}\}$	$X_1 = \{x_2\}$ $X_2 = \{x_1,x_2\}$

pete for second most preferred. If x_4 were chosen to be second best, then x_1 would be the third most preferred alternative, with x_3, x_5, and x_6 competing for fourth best. If, however, x_1 were chosen as second best, x_3, x_4, and x_5 would compete for third best, and x_6 would be at best the fourth most preferred. We now have moved up the attribute tree or template such that there are only three lowest level attributes. They correspond to safety, costs, and attractiveness in terms of beauty and spaciousness. At higher levels in the attribute template, trade-offs become more difficult to make as the "noncommensurateness" of the attributes increases as we move up the tree. Choicemaking at this point, or prior to this point, seems a strong possibility. Interactive computer graphics to present the results of figures and tables such as these to the decisionmaker will be most desirable as aids in choicemaking. It will be very difficult to do this for many objectives which is another reason that appears to support the potential attractiveness of our proposed approach.

In order to order the alternatives totally, if such a total ordering were necessary, trading off all three of the remaining attributes would be required. All hypothetically assessed trade-off information and associated results are provided in Tables VIII and IX and in Fig. 14; Table X lists the final weights for the initial attributes.

We expect that the decisionmaker frequently will not require full trade-off assessment for choicemaking and hence will elect to reach a decision using the decisionmaking support process before the final assessment. Regardless of the stage at which final evaluation is made, however, the decisionmaker is likely to want to know the impact of trade-off weight variation and initial objective measure value variation on the digraph structure. For example, let the weight for attribute G in Table VI be changed from 0.10 to 0.35, thus changing the weight for attribute H to 0.65. The new table of objectives measure values are presented in Table XI and its associated digraph in Fig. 15. This sort of sensitivity analysis should always precede final alternative selection especially where there are any questions concerning subjectivities of either the alternative scores or criterion (attribute) weights.

Table XII presents the $\mathcal{X}_n$ and X_n, $n=1,2$, for this example for each level of trade-off considered. We note that the number of elements in both $\mathcal{X}_n$ and X_n decrease as the number of trade-offs increase, which is in agreement with the corollary presented in Section IV.

VI. Conclusion

We have presented an approach to choicemaking which may substantially reduce the amount of trade-off information required from the decisionmaker relative to the MAUT approach. The approach presented may be used to complement the MAUT approach and as an alternative to it. The most general assumption considered relative to the amount of preference information that should be elicited from the decisionmaker was the partial preference feedback assumption. Two special cases of the partial preference feedback assumption were considered. The complete preference feedback case was considered in detail, and three procedures were presented to enhance the likelihood that this assumption would hold. The no preference feedback case was examined by means of an example. We remark that the partial preference feedback case is, to borrow a phrase from the optimum systems control literature, a closed-loop procedure involving decisionmaker preferences which augment those preferences elicited in construction of the vector of objectives functions. By analogy, the no preference feedback case is an open-loop procedure.

Future research and application of the MOOT based approach to choicemaking will include risk and uncertainty considerations and will indicate behaviorally relevant procedures for implementation of the approach to form a planning and decision support process for aided choicemaking. Contrasts, comparisons, and complementarity with the well-known MAUT approach will, of course, be of interest.

References

[1] W. Edwards, "How to use multiattribute measurements for social decisionmaking," *IEEE Trans. Syst., Man, Cybern.*, vol. SMC-7, pp. 327–340, May 1977.

[2] P. C. Fishburn, "Lexicographic orders, utilities and decision rules," *Management Sci.*, vol. 20, no. 11, pp. 1442–1471, 1974.

[3] Y. Haimes, *Hierarchical Analysis of Water Resources Systems.* New York: McGraw-Hill, 1977.

[4] R. L. Keeney, and H. Raiffa, *Decisions with Multiple Objectives: Preferences and Value Trade Offs.* New York: Wiley, 1976.

[5] K. R. MacCrimmon and J. K. Siu, "Making tradeoffs," *Decision Sci.*, vol. 5, pp. 680–705, 1974.

[6] K. R. MacCrimmon, "An overview of multiple objective decisionmaking," in *Multi Criteria Decisionmaking*, J. L. Cochrane and M. Zeleny Eds. South Carolina: Univ. South Carolina, 1973, pp. 18–44.

[7] H. Raiffa, *Decision Analysis: Introductory Lectures on Choices Under Uncertainty.* Reading, MA: Addison-Wesley, 1968.

[8] A. P. Sage, *Methodology for Large-Scale Systems.* New York: McGraw-Hill, 1977.

[9] M. K. Starr and M. Zeleny, "MCDM-state and future of the arts," *TIMS Studies in the Management Sci.*, vol. 6, pp. 5–29, 1977.

[10] A. Tversky, "Elimination by aspects: A theory of choice," *Phychol. Rev.*, vol. 79, pp. 281–299, 1972.

[11] C. C. White, E. C. Wilson, and A. C. Weaver, "Decision aid development for use in ambulatory health care settings," working paper, Dep. of Engineering Science and Systems, Univ. VA, 1980.

[12] M. Zeleny, *Linear Multiobjective Programming.* New York: Springer, 1974.

Author Index

Subject Index

F

G

H

I

K

L

M

N

O

Editors' Biographies

DR. JAMES D. PALMER, Fellow, Institute of Electrical and Electronics Engineers, received his B.S. and M.S. degrees in electrical engineering from the University of California, Berkeley, CA, in 1955 and 1957, respectively. He received the Ph.D. degree in engineering sciences from the University of Oklahoma in 1963.

He is presently vice president and general manager, Research and Development Division, Mechanical Technology Inc., Latham, NY. He served as professor and director, School of Electrical Engineering, University of Oklahoma, from 1957 to 1966; as professor of electrical engineering and dean of science and engineering, Union College, Schenectady, NY, from 1966 to 1971; as president, Metropolitan State College, Denver, CO, from 1971 to 1978; and as administrator, Research and Special Programs Administration, U.S. Department of Transportation from 1978 to 1979.

Dr. Palmer has served as a consultant to industry, government, and academe. He has served on many boards and national committees, including the U.S. Coast Guard Academy Advisory Committee, Executive Policy Board for the Alaska Natural Gas Pipeline, Volunteers for International Technical Assistance, and the National Commission for Cooperative Education. He has been active in the American Society for Engineering Education, the Accreditation Board for Engineering and Technology, the American Association of State Colleges and Universities, and the IEEE. He has served as President of the Systems, Man, and Cybernetics Society (SMC), and as an officer and AdCom member as well. He is chairman of the IEEE Committee on Large Scale Systems and has served on similar groups for the National Science Foundation.

He received the SMC 1980 award for outstanding contribution in systems science, was named a Case Centennial Scholar, and is a member of several honor societies, including Eta Kappa Nu and Sigma Xi. He has been awarded an honorary Doctor of Public Service degree from Regis College, Denver, CO. He is listed in several biographical reference works, including *Who's Who in America* and *Who's Who in Engineering.* He is widely published in the systems engineering field.

RICHARD SAEKS was born in Chicago, Illinois, in 1941. He received the B.S. in 1964, the M.S. in 1965, and the Ph.D. in 1967 from Northwestern University, Evanston, IL, Colorado State University, Fort Collins, CO, and Cornell University, Ithaca, NY, respectively, all in electrical engineering.

He held the positions of assistant and associate professor of electrical engineering at the University of Notre Dame between 1967 and 1973. Since 1973, he has been on the faculty at Texas Tech University, Lubbock, TX, where he is presently Paul Whitfield Horn Professor of Electrical Engineering, Mathematics and Computer Science. At Texas Tech, he is involved in teaching and research in the areas of fault analysis, large scale systems, and mathematical system theory, and also serves as director of the Institute for Electronic Science and the Joint Services Electronics Program.

Dr. Saeks is the author of four textbooks and some sixty refereed journals. In addition, he has organized numerous conferences, edited four books, and held editorial assignments for several professional journals. He is a Fellow of the Institute of Electrical and Electronics Engineers, and a member of the American Mathematical Society, the Society for Industrial and Applied Mathematics, the American Society for Engineering Education, and Sigma Xi.